Lecture Notes in Computer Science 3743

Commenced Publication in 1973
Founding and Former Series Editors:
Gerhard Goos, Juris Hartmanis, and Jan van Leeuwen

Ivan Lirkov Svetozar Margenov
Jerzy Waśniewski (Eds.)

Large-Scale Scientific Computing

5th International Conference, LSSC 2005
Sozopol, Bulgaria, June 6-10, 2005
Revised Papers

Springer

Volume Editors

Ivan Lirkov
Svetozar Margenov
Bulgarian Academy of Sciences, Institute for Parallel Processing
Acad. G. Bonchev, Bl. 25A, 1113 Sofia, Bulgaria
E-mail: {ivan, margenov}@parallel.bas.bg

Jerzy Waśniewski
Technical University of Denmark
Department of Informatics and Mathematical Modelling
IMM, DTU, Bldg. 321, 2800 Lyngby, Denmark
E-mail: jw@imm.dtu.dk

Library of Congress Control Number: 2006920786

CR Subject Classification (1998): G.1, D.1, D.4, F.2, I.6, J.2, J.6

LNCS Sublibrary: SL 1 – Theoretical Computer Science and General Issues

ISSN 0302-9743
ISBN-10 3-540-31994-8 Springer Berlin Heidelberg New York
ISBN-13 978-3-540-31994-8 Springer Berlin Heidelberg New York

Springer is a part of Springer Science+Business Media

springer.com

Printed in Germany

Typesetting: Camera-ready by author, data conversion by Scientific Publishing Services, Chennai, India
Printed on acid-free paper SPIN: 11666806 06/3142 5 4 3 2 1 0

Preface

The 5th International Conference on Large-Scale Scientific Computations (LSSC 2005) was held in Sozopol, Bulgaria, June 6–10, 2003. The conference was organized and sponsored by the Institute for Parallel Processing at the Bulgarian Academy of Sciences. Partial support was also provided from project BIS-21++ funded by the European Commission in FP6 INCO via grant 016639/2005.

The plenary invited speakers and lectures were:

- O. Axelsson, Eigenvalue Estimates for Preconditioned Saddle Point Matrices
- R. Blaheta, Algebraic Multilevel Methods with Aggregations
- S. Brenner, Additive Multigrid Theory
- C. Carstensen, Review on the Convergence of Adaptive Finite Element Methods
- S. Heinrich, Numerical Analysis on a Quantum Computer
- U. Langer, Inexact Date-Sparse Boundary and Finite Element Domain Decomposition Methods
- R. Lazarov, Discontinuous Galerkin Method as Stabilization Technique for Nonconforming Finite Element Approximations of PDEs
- J. Waśniewski, New Data Storage Formats for Dense Matrices Lead to Variety of High-Performance Algorithms
- Z. Zlatev, Parallel Treatment of General Sparse Matrices

The success of the conference and the present volume in particular are the outcome of the joint efforts of many colleagues from various institutions and organizations. Firstly thanks to all the members of the Scientific Committee for their valuable contribution forming the scientific face of the conference, as well as for their help in reviewing contributed papers. We especially thank the organizers of the special sessions. We are also grateful to the staff involved in the local organization.

Traditionally, the purpose of the conference is to bring together scientists working with large-scale computational models of environmental and industrial problems, and specialists in the field of numerical methods and algorithms for modern high-speed computers. The key lectures reviewed some of the advanced achievements in the field of numerical methods and their efficient applications. The conference lectures were presented by the university researchers and practical industry engineers including applied mathematicians, numerical analysts and computer experts. The general theme for LSSC 2005 was large-scale scientific computing with a particular focus on the organized special sessions.

Special sessions and organizers were the following:

- Multiscale and Multiphysics Computations — P. Bochev, R. Hoppe, R. Lazarov
- Robust Algebraic Multigrid and Hierarchical Preconditioning Methods — R. Blaheta, U. Langer, S. Margenov
- Monte Carlo: Tools, Applications, Distributed Computing — D. Vasileska, M. Nedyalkov, T. Gurov
- Control/Uncertain Systems and Validated Numerics — N. Dimitrova, M. Krastanov, V. Veliov
- Operator Splittings, Their Application and Realization — I. Faragó
- Distributed Numerical Methods and Algorithms for Grid Computing — T. Sakuray, M. Sato, S. G. Petiton
- Environmental Modelling — A. Ebel, K. Georgiev, Z. Zlatev
- Large-Scale Computation of Engineering Problems — P. Minev, O. Iliev
- Numerical Methods for the Schrödinger Equation and Application — H. Kosina
- Advances in Computational Mechanics — S. Brenner, C. Carstensen

Special events comprised:

- "Bulgarian Involvement in European Grid Initiatives" — presentation and discussion
- Grid Help Desk and Demos

More than 120 participants from all over the world attended the conference representing some of the strongest research groups in the field of advanced large-scale scientific computing. This volume contains 80 papers submitted by authors from over 20 countries.

The 6th International Conference, LSSC 2007, will be organized in June 2007.

November 2005

Ivan Lirkov
Svetozar Margenov
Jerzy Waśniewski

Table of Contents

I Plenary and Invited Papers

II Advances in Computational Mechanics and Multi Physics

III Robust Algebraic Multigrid and Hierarchical Preconditioning Methods

IV Monte Carlo: Tools, Applications, Distributed Computing

V Control Systems

VI Uncertain Systems

VII Operator Splitting, Their Application and Realization

VIII Distributed Numerical Methods and Algorithms for Grid Computing

IX Environmental Modelling

X Large Scale Computation of Engineering Problems

XI Numerical Methods for the Schrödinger Equation and Application

XII Contributed Talks

Part I

Plenary and Invited Papers

Algebraic Multilevel Methods with Aggregations: An Overview

Radim Blaheta

Department of Applied Mathematics, Institute of Geonics AS CR,
Studentská 1768, 70800 Ostrava-Poruba, Czech Republic
blaheta@ugn.cas.cz

Abstract. This paper deals with the numerical solution of elliptic boundary value problems by multilevel solvers with coarse levels created by aggregation. Strictly speaking, it deals with the construction of the coarse levels by aggregation, possible improvement of the simple aggregation technique and use of aggregations in multigrid, AMLI preconditioners and two-level Schwarz methods.

1 Introduction

This paper considers *multilevel solvers* for algebraic systems arising from the finite element approximation to selfadjoint elliptic problems. It is well known that nested finite element grids allow to introduce *two-level* and *multilevel* methods for solving the finite element systems. Multigrid methods [13], AMLI preconditioners [1] and Schwarz methods [24] are typical examples of numerical methods exploiting the grid hierarchy.

Algebraic multilevel methods avoid the necessity of nested triangulation of the problem domain and allow algebraic construction of coarser spaces by using mostly only information involved in the matrix of the solved problem. As multilevel iterative methods reduce the error by two complementary tools like relaxation on the fine grid and coarse grid correction for multigrid methods, the algebraic approach also enables to construct the coarse space with approximating properties that are necessary for efficiency of the complementary tool. In this way, the algebraic approach enhances the robustness of the multilevel solution methods.

In this paper, we outline the *aggregation techniques* that constitute one of possible approaches to algebraic multilevel methods. The aggregation technique originally appeared in the context of multigrid methods but can be exploited also in hierarchical algebraic multilevel preconditioners and the two-level Schwarz domain decomposition methods.

The paper provides an overview of applications of the aggregation technique, which undergo an important development during the last decade.

2 AMG with Aggregation

Let us consider the solution of linear systems appearing from the finite element (FE) approximation of elliptic boundary value problems and let $\mathcal{T}_h$ be a FE

I. Lirkov, S. Margenov, and J. Waśniewski (Eds.): LSSC 2005, LNCS 3743, pp. 3–14, 2006.

triangulation which arises as a refinement of a coarser triangulation $\mathcal{T}_H$ of the problem domain Ω. Then the FE system corresponding to the fine triangulation $\mathcal{T}_h$,

$$A_h u_h = b_h, \quad u_h, b_h \in R^{n_h} \tag{1}$$

can be solved by the following iterative *two-grid method*. Its one iteration $u_h^{i+1} = TG(A_h, b_h, u_h^i)$ is described as follows

function $\mathrm{TG}(A_h, b_h, u_h^i = \bar{u})$
$\nu_1 - times: \quad \bar{u} \leftarrow S(A_h, b_h, \bar{u})$ pre-smoothing
$r_H = I_h^H (b_h - A_h \bar{u})$ restriction of the residual
$v_H = A_H^{-1} r_H$ coarse grid correction
$\bar{u} = \bar{u} + I_H^h v_H$ prolongation of the correction
$\nu_2 - times: \quad \bar{u} \leftarrow S(A_h, b_h, \bar{u})$ post-smoothing
return $(u_h^{i+1} = \bar{u})$

Above, the *smoothing* $\bar{u} \leftarrow S(A_h, b_h, \bar{u})$ represents one iteration of an inner iterative (relaxation) procedure like Jacobi, Gauss-Seidel etc. The *coarse grid correction* uses matrix A_H from FE discretization of the solved problem on the coarse grid, restriction I_h^H to the coarse FE space and prolongation I_H^h induced by the natural interpolation between the nested grids. The smoothing procedure should collaborate with the coarse grid correction. Usually S efficiently reduces oscillating error components and produces *smooth error* that can be reduced by the coarse grid correction. Therefore S is called the *smoother*. Note that the introduced two-level method can be naturally extended to the multilevel one.

For a broad class of problems, it can be shown that multigrid methods are highly efficient and even *optimal*, which means that the system (1) is solved in $O(n_h)$ operation. But application of multigrid methods can also meet two drawbacks: it can be difficult or impossible to produce a sequence of auxiliary coarser discretizations of the solved boundary value problems and it can be difficult to produce coarse discretizations collaborating well with the used smoother in the case of problems with certain anisotropy, singularity etc.

These difficulties motivate an interest in *algebraic multigrid methods* (AMG), which construct the prolongation, restriction and coarse matrices by using only the information included in the solved system or very little additional geometric information. In the AMG context, the system at a current level k ($k = 1$ is the finest level) is written as

$$A_k u_k = b_k, \quad u_k, b_k \in R^{n_k} \tag{2}$$

and the coarser level works with $n_{k+1} \times n_{k+1}$ matrix $A_{k+1} = I_k^{k+1} A_k I_{k+1}^k$ defined with the aid of a *prolongation* I_{k+1}^k and a *restriction* I_k^{k+1}. For symmetric positive definite (SPD) problems, we choose $I_k^{k+1} = \left(I_{k+1}^k\right)^T$ ensuring that A_k remains also SPD. Note that in this case $\left\|e_k - I_{k+1}^k v_{k+1}\right\|_{A_k} = \min$ is equivalent to the coarse correction

$$(I_{k+1}^k)^T A_k I_{k+1}^k v_{k+1} = A_{k+1} v_{k+1} = (I_{k+1}^k)^T A_k e_k.$$

One iteration $u_k^{i+1} = MG_\mu(A_k,\, b_k,\, u_k^i)$ of the multilevel AMG method is recursively described as follows

function $MG_\mu(A_k,\, b_k,\, u_k^i = \bar{u})$
  $\nu_1 - times:\ \ \bar{u} \leftarrow S(A_k,\, b_k,\, \bar{u})$ — pre-smoothing
  $r_{k+1} = I_k^{k+1}\,(b_k - A_k\bar{u})$ — residual restriction
  if $\ k+1 = coarsest$ then — coarse grid correction
    $v_{k+1} = A_{k+1}^{-1} r_{k+1}$
  else
    $v_{k+1} = 0$
    $\mu - times:\ \ v_{k+1} = MG_\mu(A_{k+1},\, r_{k+1},\, v_{k+1})$
  end
  $\bar{u} = \bar{u} + I_{k+1}^k v_{k+1}$ — correction prolongation
  $\nu_2 - times:\ \ \bar{u} \leftarrow S(A_k,\, b_k,\, \bar{u})$ — post-smoothing
return$(u_k^{i+1} = \bar{u})$

For discrete PDE problems, AMG has the following advantages:

- there is no need for creating nested grids, it is possible to develop black box solvers,
- instead of seeking of smoothers adapted to the coarse problem, the coarse grid can be adapted to the smoother,
- the size of the coarse problem can be controlled e.g. for balancing the work load on many processors in the case of two-level Schwarz method.

AMG methods can be based on different ideas including the aggregation technique on which we focus our interest. From the other ideas, we can mention AMG based on an C-F decomposition and interpolation developed by A. Brandt, J.W. Ruge, K. Stüben, and others, see e.g. [23].

2.1 Aggregation of Unknowns

We shall restrict our attention to SPD problems, when AMG needs only to define the interpolation I_{k+1}^k. The simplest interpolation and restriction are in the form $I_{k+1}^k = \mathcal{R}^T$, $I_k^{k+1} = \mathcal{R}$ with $n_{k+1} \times n_k$ Boolean matrix $\mathcal{R}$ with just one unity in each column, e.g.

$$\mathcal{R} = \begin{bmatrix} 1 & & & & & & & & 1 & & & \\ & 1 & & & & & & 1 & & & & \\ & & 1 & & & & 1 & & & & & \\ & & & 1 & & & & & & 1 & & \\ & & & & 1 & & & & & & 1 & \\ & & & & & 1 & & & & & & 1 \end{bmatrix}.$$

Definition of $\mathcal{R}$ is equivalent to the division of the set of n_k unknowns into n_{k+1} disjoint groups (*aggregation of unknowns*)

$$\{1, \ldots, n_k\} = \bigcup_{i=1}^{n_{k+1}} G_i, \quad \text{where } G_i = \{j : \mathcal{R}_{ij} = 1\}.$$

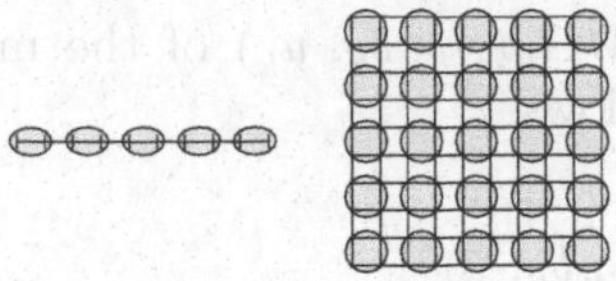

Fig. 1. An aggregation on regular 1D and 2D grids, regular clustering of 2 and 2×2 nodes, respectively

For 1D problems, the aggregation of unknowns can be easily defined by clustering of neighbouring nodes. This clustering can be easily generalized to regular grids in 2D and 3D domains, see e.g. Fig. 1. More general aggregation on irregular grids will be discussed later in Subsection 2.4. For an application of the two-level and multilevel aggregation methods, it is important that the prolongation, restriction and construction of the coarse matrix can be efficiently implemented.

The idea of aggregation in the context of iterative solution methods was already used in [22]. In the context of multigrid methods for solving elliptic boundary value problems, the aggregations were used e.g. in [2, 3, 4] and [9].

2.2 Overcorrection

Let us consider a model 1D or 2D Dirichlet problem for Poisson equation in an interval or a square (see Fig. 1) and the linear finite element discretization of these problems on uniform meshes with mesh size h providing 3 and 5 point stencil, respectively. Then the aggregations can be constructed e.g. by regular clustering of 2 and 2×2 nodes, respectively. In these cases, it is easy to compute the coarser matrices and see that these matrices differ from matrices arising from discretization on the coarser uniform grids with mesh size $2h$ by the factor 2.

We can also consider a 1D Dirichlet problem for the equation $-u'' = f$ in $\langle 0, 1\rangle$, aggregation by regular clustering and approximation of a hat shape error in the energy norm, see Fig. 2. The computed approximation from the coarse space created by aggregation indicates the possible improvement by scaling the correction by the factor 2.

For the 1D model problem, it is also possible to apply the Fourier analysis [2, 4] to show that the smooth error components are only partly re-

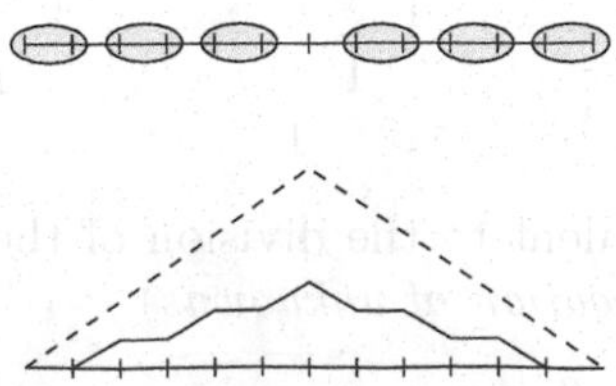

Fig. 2. Correction by aggregation (solid line) computed to a hat shape error (dashed line) for a 1D model problem

duced by the correction from the aggregated space and that the efficiency of multi-level aggregation method can be substantially improved by the *overcorrection*,

$$\bar{x} = \bar{x} + \omega\, I^k_{k+1} v_{k+1} \quad \text{with the scaling factor } \omega > 1.$$

The use of overcorrection was introduced in [3, 4] and [9]. For more general problems, a variational computation of ω was suggested in [3, 4] by using the following algorithm

$$\begin{array}{ll}
v_{k+1} = A^{-1}_{k+1} r_{k+1} & \text{corse grid correction} \\
\bar{v}_k = I^k_{k+1} v_{k+1} & \text{prolongate the correction} \\
\eta - times: \ \bar{v}_k \leftarrow S(A_k, r_k, \bar{v}_k) & \text{smooth the correction} \\
\omega = \langle \bar{v}_k, r_k \rangle \,/\, \langle \bar{v}_k, A_k \bar{v}_k \rangle & \text{compute the scaling factor} \\
\quad = \text{argmin} \left\| A_k^{-1} b_k - (\bar{x} + \omega \bar{v}_k) \right\|_{A_k} & \\
\bar{x} = \bar{x} + \omega\, I^k_{k+1} v_{k+1} \text{ or } \bar{x} = \bar{x} + \omega \bar{v}_k & \text{perform the overcorrection}
\end{array}$$

2.3 Smoothing

Above, we mentioned that the matrix created by aggregation is too stiff. This matrix is a Galerkin type matrix defined with the aid of basis functions in aggregation space, which are sums of basis functions in the original fine FE space or in a previous aggregation space. A difficulty is in a high energy of these aggregation basis functions. The improvement can be find in smoothing the aggregation basis functions, which produces new ones with a lower energy, see e.g. [25, 26, 27].

For a 1D model problem in Fig. 3, the smoothing of the aggregation basis function ϕ^a by $\mathcal{S} = I - \frac{2}{3}A$ produces new basis function $\mathcal{S}\phi^a$ having a larger support but a lower energy. For this model problem, the new smoothed aggregations are even piecewise linear. But it is not true in more general cases.

The process of smoothing can be formalized as follows. Firstly, we define a prolongation $\mathcal{I}^k_{k+1}$ defined by aggregation (*tentative prolongation*). Then, a more efficient prolongation operator is constructed in the form

$$I^k_{k+1} = \mathcal{S}_k \mathcal{I}^k_{k+1} \text{ with } \mathcal{S}_k = I - \omega \Lambda_k^{-1} \bar{A}_k.$$

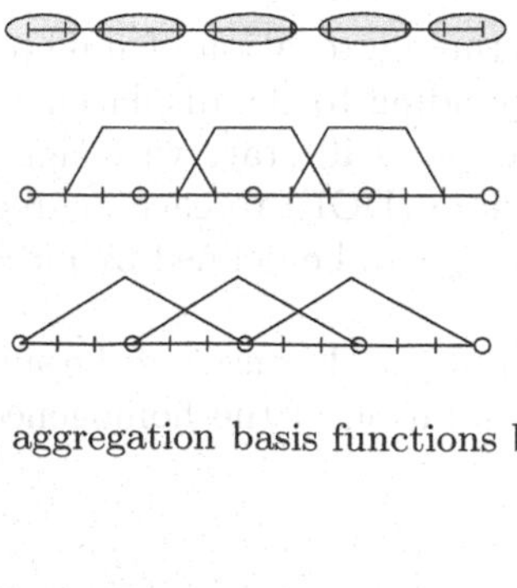

Fig. 3. 1D model problem, aggregation basis functions before and after smoothing

In [26], the *prolongation smoother* S has the components $\omega = \frac{2}{3}$, $\Lambda_k = \mathrm{diag}(A_k)$ and $\bar{A}_k = (\bar{a}_{ij})$ arises from $A_k = (a_{ij})$ by filtering,

$$\bar{a}_{ij} = \left\{ \begin{matrix} a_{ij} & \text{if } |a_{ij}| \geq \varepsilon \sqrt{a_{ii}} \sqrt{a_{jj}} \\ 0 & \text{otherwise} \end{matrix} \right\} \quad \text{for } i \neq j, \qquad \bar{a}_{ii} = a_{ii} - \sum_{i \neq j} (a_{ij} - \bar{a}_{ij}).$$

A heuristic choice of the parameter ε is $\varepsilon = 0.08 \left(\frac{1}{2}\right)^{k-1}$.

In [27], $\omega = \frac{4}{3\lambda_k}$, $\Lambda_k = (P_k^1)^T P_k^1$, where $P_k^1 = I_2^1 \cdots I_k^{k-1}$ and $\lambda_k \geq \rho(\Lambda^{-1} A_k)$, $\bar{A}_k = A_k$. A possible choice is $\lambda_k = 9^{k-1}\rho$, where $\rho \geq \rho(A_1)$. For this choice, the convergence factor q_{MG-L} of L level multigrid can be estimated as follows,

$$q_{MG-L} \leq 1 - \frac{1}{C(L)},$$

where $C(L)$ is a polynomial in L, see [27] for the proof.

2.4 Construction of Aggregations

The construction of aggregations on general meshes with paying the attention to strong couplings between unknowns (smooth error character) can be node or element oriented. A standard node oriented algorithm for creating the aggregates is the following one:

preliminary phase: separate isolated points as individual aggregates,
phase I: repeat until all unaggregated nodes are adjacent to an aggregate:
  a) pick the root node not adjacent to any existing aggregate,
  b) define new aggregate as the root node plus all its neighbours,
phase II: sweep unaggregated nodes into existing aggregates (to which they are connected) or use them to form new aggregates.

Such algorithm can be found e.g. in [26] and has many variants. At first, some measures can be done for not leaving too many nodes for the phase II. Secondly, the connection and neighbourhood can be defined in a *strong* sense to create the coarse problem suitable for handling those error components, which can not be removed by the smoother. A typical strong coupling between the nodes (unknowns) i and j means that

$$|a_{ij}| \geq \varepsilon \sqrt{a_{ii}} \sqrt{a_{jj}}.$$

For elasticity problems, the aggregation of unknowns is restricted to aggregation of unknowns corresponding to the displacements in the same coordinate direction. Alternatively, we can still start with aggregation of the nodes and assign more degrees of freedom (DOF) to each aggregate (see also next subsection). The strength of coupling can be defined by means of blocks corresponding to nodal DOFs, see [16].

A further information about the character of the smooth error can be obtained from an auxiliary iterative solution of the homogeneous variant $A_k = 0$ of the solved problem (2). This information can be used for an improved construction of aggregations, see e.g. [11].

From the other algorithms for the construction of aggregations, we can mention subsequent pairing [9, 19]. The algorithm can be described as follows

step I: repeat until all unaggregated nodes are classified as aggregated pair or singleton:

* pick up a node i and find the node j with the strongest coupling to i. If this coupling is not strength enough classify i as singleton otherwise create a pair $\{i, j\}$.
* aggregate the matrix

step II: apply the previous algorithm to aggregated nodes and aggregated matrix to create generalized quaternion aggregations, etc.

This algorithm creates aggregations similar to aggregations on a regular grid.

The aggregations can be also created by agglomeration of adjacent finite elements. Such approach is described e.g. in [15, 12].

2.5 Enriched Aggregations

For scalar boundary value problems, the aggregation of unknowns is equal to aggregation of nodes, i.e. one DOF is assigned to each aggregate of nodes. For elasticity problems or systems of equations, it is natural to aggregate separately displacements in different directions or unknowns corresponding to different phenomena. In other terminology, more DOFs are assigned to the aggregates. For elasticity, these DOFs can be two or three displacements per aggregate but an additional enhancement is also possible, e.g by adding the rotations [14]. For 2D elasticity, it gives 3 unknowns: the displacements u, v and rotation angle α per aggregate. If (x_T, y_T) is the barycentre of the aggregate, then the prolongation assign the displacement $(u - \alpha(y - y_T), v + \alpha(x - x_T))$ to any node (x, y) of the aggregate.

In the case of aggregation by agglomeration of finite elements, a further enrichment can be done by using low energy eigenvectors corressponding to the agglomeration matrices, see [15, 12].

3 AMLI Preconditioners with Aggregation

Aggregation based AMG methods can be also used as preconditioners, a pioneering work in this respect is [9].

Beside multigrid preconditioners, there is also a class of hierarchical AMLI preconditioners, which use a space decomposition and work separately on the coarse space and its complement. These preconditioners can be also constructed with the aid of aggregation, see [18, 19].

In the case of scalar boundary value problem and the system $Au = b$, $u, b \in R^n$, we start with creating the aggregations $\{G_i : i = 1, \ldots, m\}$ and selecting one node in each aggregation as a C-node. All remaining nodes are considered to be F-nodes. The F-C decomposition induces a decomposition of the matrix A,

$$A = \begin{bmatrix} A_{11} & A_{12} \\ A_{21} & A_{22} \end{bmatrix} = \begin{bmatrix} I & \\ A_{21}A_{11}^{-1} & I \end{bmatrix} \begin{bmatrix} A_{11} & \\ & S_A \end{bmatrix} \begin{bmatrix} I & A_{11}^{-1}A_{12} \\ 0 & I \end{bmatrix}$$

and a preconditioner

$$B = \begin{bmatrix} I & \\ A_{21}P_{11}^{-1} & I \end{bmatrix} \begin{bmatrix} P_{11} & \\ & S \end{bmatrix} \begin{bmatrix} I & P_{11}^{-1}A_{12} \\ 0 & I \end{bmatrix}$$

where $S \sim S_A$ and $P_{11} \sim A_{11}$. In [19], S is given by a scaled aggregation and P_{11} is realized by dynamically constructed MILU factorization. The dynamic feature means that F-nodes, which are problematic for the MILU factorization, are shifted among C-nodes. Multilevel preconditioners then arise by solution of the second pivot block by inner iterations (CG) with the same type of hierarchical preconditioner.

Alternatively, we can create a hierarchical basis (HB) with basis functions

$$\phi_i^{HB} = \sum_{k \in G_i} J_{ik}\, \phi_i^h \quad \text{if } i \in C \text{ and } \phi_i^{HB} = \phi_i^h \text{ if } i \in F,$$

where C and F denote the sets of C-nodes and F-nodes, respectively. The transformation between the standard and hierarchical bases is given by the matrix J, which can be written as follows,

$$J = (J_{ij}) = \begin{bmatrix} I_1 & 0 \\ I_{21} & I_2 \end{bmatrix} \begin{matrix} F \\ C \end{matrix}$$

where I_{21} is a Boolean matrix with one unity per column, I_1 and I_2 are identity matrices of proper dimensions.

The matrix A can be transformed to the hierarchical form A_{HB} and both matrices A and A_{HB} can be written in F-C, F-C ordering as follows

$$A = \begin{bmatrix} A_{11} & A_{12} \\ A_{21} & A_{22} \end{bmatrix}, \quad A_{HB} = J\, A\, J^T = \begin{bmatrix} H_{11} & H_{12} \\ H_{21} & H_{22} \end{bmatrix},$$

where $A_{11} = H_{11}$ and H_{22} is the matrix arising from aggregation of A. This decomposition enables to define both additive and multiplicative preconditioners

$$B_A = J^{-1} \begin{bmatrix} \widetilde{H}_{11} & \\ & \widetilde{H}_{22} \end{bmatrix} J^{-T}, \quad B_A^{-1} = J^T \begin{bmatrix} \widetilde{H}_{11} & \\ & \widetilde{H}_{22} \end{bmatrix}^{-1} J,$$

$$B_M = \begin{bmatrix} I & \\ A_{21}\widetilde{H}_{11}^{-1} & I \end{bmatrix} \begin{bmatrix} \widetilde{H}_{11} & \\ & \widetilde{H}_{22} \end{bmatrix} \begin{bmatrix} I & \widetilde{H}_{11}^{-1}A_{12} \\ & I \end{bmatrix},$$

where $\widetilde{H}_{11} \sim H_{11}$ and $\widetilde{H}_{22} \sim H_{22}$. Some analysis and comparisons with standard AMLI can be done on the basis of the strengthened CBS inequalities, see [7].

4 Schwarz Methods with Aggregation

The algebraic coarse space created by aggregation can be also used in the framework of the *two-level additive and hybrid Schwarz preconditioners.* General form of these preconditioners is as follows

$$B_A = B_0 + B_{1L}, \quad B_{1L} = \sum 1^m B_k \, , \; B_k = R_k^T A_k^{-1} R_k \tag{3}$$

where A_k $(k = 1, \ldots, m)$ are FE matrices of local subproblems and A_0 is a coarse matrix created by aggregation. More details will be provided later. We shall also consider nonsymmetric hybrid preconditioner defined by

$$B_H = B_0 + B_{1L}(I - AB_0) \tag{4}$$

and its symmetrized version

$$B_{SH} = B_0 + (I - B_{1L}A)B_0(I - AB_{1L}), \tag{5}$$

More details about the Schwarz preconditioners can be found e.g. in [24].

Now, let us solve the system (1) arising from a finite element discretization of an elliptic boundary value problem in Ω. Let $\mathcal{T}_h$ be a FE triangulation of the domain Ω and V_h be a corressponding FE space. The triangulation $\mathcal{T}_h$ can be divided into m parts in two steps: firstly $\mathcal{T}_h$ is divided into nonoverlapping sets $\mathcal{T}_k^0$, which are consequently extended to overlapping sets $\mathcal{T}_k^\delta$. We shall denote

$$\Omega_k^0 = \cup\{E : E \in \mathcal{T}_k^0\}, \qquad \Omega_k^\delta = \cup\{E : E \in \mathcal{T}_k^\delta\}.$$

Now, we can define the local FE spaces $V_k \subset V_h$ of admissible functions on Ω which vanish outside Ω_k, matrices A_k and restrictions R_k. Let A_k^δ be the FE matrix arising from assembling the element matrices A_E for $E \in \mathcal{T}_h^\delta$. Then A_k will be the matrix arising from A_k^δ by incorporating homogeneous Dirichlet type boundary conditions on the inner boundary $\partial\Omega_k^\delta \backslash \partial\Omega$. The boundary conditions on the outer boundary $\partial\Omega_k^\delta \cap \partial\Omega$ are given from the solved boundary value problem.

The decomposition $V_h = V_1 + \cdots + V_m$ can be enriched by a coarse space V_0 created algebraically by aggregation, which ensures the numerical scalability with respect to the number of the subdomains. If $G_1, \ldots, G_N$ be the aggregations and $V_h = \mathrm{span}\{\phi_1^h, \ldots, \phi_n^h\}$, where ϕ_i^h are basis functions, then it is possible to define aggregated basis functions ψ_k and the space $V_0 \subset V$ as follows,

$$\psi_k = \sum_{i \in G_k} \phi_i^h \, , \qquad V_0 = \mathrm{span}\{\psi_1, \ldots, \psi_N\} \, .$$

We shall assume that the aggregations are regular, i.e. there is a constant $\bar{\beta}$ such that each $\mathrm{supp}\,\psi_k$ contains a ball with diameter $\bar{\beta}H$, where

$$H \sim \max_k \mathrm{diam}(\mathrm{supp}\,\psi_k).$$

Such construction gives again a stable decomposition $V = V_0 + V_1 + \cdots + V_m$ resulting in numerically scalable preconditioners B_A and B_{SH}. For more details see [10, 17, 6, 21].

We shall conclude this section with some numerical examples. The efficiency of various preconditioners arising from implementation of the described ideas can be compared by solving two boundary value problems in $\Omega = \langle 0,2\rangle \times \langle 0,3\rangle$ with pure homogeneous Dirichlet boundary conditions ($\partial\Omega_D = \partial\Omega$). The first problem is for the Poisson equation, the second one is a model elasticity (plane deformation) problem with the elasticity modulus $E = 1$ and Poisson ratio $\nu = 0.3$. The right hand side is a linear function in both cases.

The problems are discretized by linear triangular FE on a uniform grid with the mesh size $h = 1/30$. The local problems are given on subdomains $\Omega_k = \langle 0,2\rangle \times \langle x_k, x_{k+1}\rangle$ with overlap $\delta = 2h$. The subproblems are solved exactly.

The required numbers of iterations for the accuracy $\varepsilon = 10^{-3}$ and various additive (AP) and hybrid (HP) Schwarz preconditioners can be seen in Tables 1 and 2. The hybrid preconditioners are used in nonsymmetric form in combination with a generalized conjugate gradient method GPCG[1], see [5]. The coarse problem uses either the nested coarse triangular grid with the mesh size $H = 2h$ or the aggregations with clustering 2×2 square macroelements (3×3 nodes). The smoothing was done by $S = I - \frac{2}{3}diag(A)^{-1}\bar{A}$ where $\bar{A}$ was equal to A in both cases. For the elasticity, we test also sparser $\bar{A}$ given by the separate displacement component part of A but both choices give the same results.

Table 1. Poisson equation problem. Numbers of iterations for $\varepsilon = 10^{-3}$. AP=additive preconditioner, HP=hybrid preconditioner + GPCG[1].

Overlap 2h, #subdomains:	4	16	24
c-grid H=3h, AP	7	7	8
c-grid H=3h, HP	6	6	6
aggreg. 2h, AP	13	17	17
aggreg. 2h, HP	10	11	11
smooth. aggreg. 2h, AP	10	11	11
smooth. aggreg. 2h, HP	7	7	8

Table 2. Elasticity problem. Numbers of iterations for $\varepsilon = 10^{-3}$. AP=additive preconditioner, HP=hybrid preconditioner + GPCG[1].

Overlap 2h, #subdomains:	4	16	24
c-grid H=3h, AP	8	8	9
c-grid H=3h, HP	6	7	8
aggreg. 2h, AP	17	20	20
aggreg. 2h, HP	12	13	14
aggreg. 2h-rotat, AP	16	18	19
aggreg. 2h-rotat, HP	11	12	12
smooth. aggreg. 2h, AP	12	14	14
smooth. aggreg. 2h, HP	9	10	10

5 Conclusions

In this paper, we provide an overview of possible applications of the aggregation technique in multilevel methods. Additionally, we can mention application of the aggregation technique in a nonoverlapping Schwarz method with interfaces on the coarse grid, see [6], or a specific aggregations for construction of AMLI preconditioners for nonconforming Crouzeix-Raviart finite elements [8].

Acknowledgement

The support given by the grant 1ET400300415 of the Academy of Sciences of the Czech Republic is greatly acknowledged.

References

1. O. Axelsson, Iterative Solution Methods, Cambridge Univ. Press, Cambridge 1994.
2. R. Blaheta, A multilevel method with correction by aggregation for solving discrete elliptic problems, *Aplikace Matematiky*, **31** (1986), 365–378.
3. R. Blaheta, Iterative methods for solving problems of elasticity, Thesis, Charles University, Prague 1987 (In Czech).
4. R. Blaheta, A multilevel method with overcorrection by aggregation for solving discrete elliptic problems. *J. Comp. Appl. Math.*, **24** (1988), 227–239.
5. R. Blaheta, GPCG — generalized preconditioned CG method and its use with non-linear and non-symmetric displacement decomposition preconditioners. *Numer. Linear Algebra Appl.*, **9** (2002), 527–550.
6. R. Blaheta, Space decomposition preconditioners and parallel solver. In: M. Feistauer et al eds., *Numerical Mathematics and Advanced Applications* (ENUMATH 2003), Springer-Verlag, Berlin 2004, 20–38.
7. R. Blaheta, An AMLI preconditioner with hierarchical decomposition by aggregation, in progress.
8. R. Blaheta, M. Neytcheva, S. Margenov: Uniform estimate of the constant in the strengthened CBS inequality for anisotropic non-conforming FEM systems, *Numerical Linear Algebra with Applications,* **11** (2004), 309–326.
9. D. Braess, Towards Algebraic Multigrid for Elliptic Problems of Second Order. *Computing*, **55** (1995), 379–393.
10. M. Brezina, Robust iterative methods on unstructured meshes, PhD. thesis, University of Colorado at Denver, 1997.
11. M. Brezina, R. Falgout, S. MacLachlan, T. Manteuffel, S. McCormick, J. Ruge, Adaptive Smoothed Aggregation (αSA), *SIAM J. Sci. Comput.,* **25** (2004), 1896–1920.
12. M. Brezina, C. Tong, R. Becker, Parallel Algebraic Multigrids for Structural Mechanics, *SIAM J. Sci. Comput.*, to appear. Also available as LLNL Technical Report UCRL-JRNL-204167, 2004.
13. W. Briggs, V. Henson, and S. McCormick, A Multigrid tutorial, 2nd ed., SIAM, Philadelphia, 2000.
14. V.E. Bulgakov, G. Kuhn, High-Performance Multilevel Iterative Aggregation Solver for Large Finite Element Structural Analysis Problems, *Int. J. Numer. Methods in Engrg,* **38** (1995), 3529–3544.

15. J. Fish and V. Belsky, Generalized Aggregation Multilevel Solver, *Int. J. for Numerical Methods in Engineering*, **41** (1997), 4341–4361.
16. M. Griebel, D. Oeltz, M.A. Schweitzer, An Algebraic Multigrid Method for Linear Elasticity, *SIAM J. Sci. Computing*, **25** (2003), 385–407.
17. E.W. Jenkins, C.T. Kelley, C.T. Miller, C.E. Kees, An aggregation-based domain decomposition preconditioner for groundwater flow. *SIAM J. Sci. Comput.*, **23** (2001), 430–441.
18. Y. Notay, Robust parameter free algebraic multilevel preconditioning, *Numer. Lin. Alg. Appl.*, **9** (2002), 409–428.
19. Y. Notay, Aggregation-based algebraic multilevel preconditioning, to appear in *SIAM J. Matrix Anal. Appl.*, 2005.
20. Y. Notay, Algebraic multigrid and algebraic multilevel methods: a theoretical comparison, *Numer. Lin. Alg. Appl.*, **12** (2005), 419–451.
21. M. Sala, J. Shadid, R. Tuminaro. An Improved Convergence Bound for Aggregation-Based Domain Decomposition Preconditioners. submitted to SIMAX, 2005.
22. R.S. Southwell, Relaxation Methods in Theoretical Physics, Oxford University Press, Oxford, 1946.
23. K. Stüben, Algebraic Multigrid (AMG): An Introduction with Applications. GMD Report 53, March 1999. Also published as An Introduction to Algebraic Multigrid. In Multigrid, Trottenberg et al. (eds), Appendix A. Academic Press: London, (2001), 413–532.
24. A. Toselli, O. Widlund, Domain Decomposition Methods — Algorithms and Theory, *Springer Series in Computational Mathematics*, **34**, Springer, 2005.
25. P. Vanek, Acceleration of Convergence of a Two Level Algorithm by Smooth Transfer Operators, *Appl. Math.*, **37** (1992), 265–274.
26. P. Vanek, J. Mandel, M. Brezina, Algebraic multigrid by smoothed aggregation for second and fourth order elliptic problems. *Computing* **56** (1996), 179–196.
27. P. Vanek, M. Brezina, J. Mandel, Convergence of algebraic multigrid based on smoothed aggregation, *Numerische Mathematik*, **88** (2001), 559–579.

Reduced-Order Modeling of Complex Systems with Multiple System Parameters*

Max Gunzburger and Janet Peterson

School of Computational Sciences, Florida State University,
Tallahassee, FL 32306-4120, USA
{gunzburg, peterson}@csit.fsu.edu

Abstract. The computational approximation of solutions of complex systems such as the Navier-Stokes equations is often a formidable task. For example, in feedback control settings where one often needs solutions of the complex systems in real time, it would be impossible to use large-scale finite element or finite-volume or spectral codes. For this reason, there has been much interest in the development of low-dimensional models that can accurately be used to simulate and control complex systems. Reduced-order modeling approaches based on proper orthogonal decompositions and centroidal Voronoi tessellations are discussed. The important implementation issue of how boundary conditions containing multiple parameters are handled in the reduced-order modeling context is highlighted.

1 Introduction

Computational solutions of (nonlinear) complex systems are expensive to obtain with respect to both storage and CPU costs. As a result, it is difficult if not impossible to deal with a number of situations that require multiple state solutions (e.g., continuation or homotopy methods for computing state solutions, parametric studies of state solutions, and optimization and optimal control problems) and/or real-time state solutions (e.g., feedback control settings). Not surprisingly, a lot of attention has been paid to reducing the costs of the nonlinear state solutions by using reduced-order models for the state; these are (very) *low-dimensional approximations to the state.*

Reduced-order modeling (ROM) has been and remains a very active research direction in many seemingly disparate fields, e.g., to name five: linear algebra (singular value decomposition); statisitics (Karhunen-Loève analysis and clustering); information science (representation, interpolation, and reconstruction); boundary layer theory (in a fluids setting, replacing the Navier-Stokes equations with the simpler Prandtl boundary layer equations); and turbulence modeling (again, in a fluids setting, replacing the Navier-Stokes equations by another complex system, e.g., a k-ϵ or LES model, that is "easier" to approximate).

* Supported in part by Sandia National Laboratories under contracts 233519 and 406670.

I. Lirkov, S. Margenov, and J. Waśniewski (Eds.): LSSC 2005, LNCS 3743, pp. 15–27, 2006.

For a state simulation of a nonlinear partial differential equation, a reduced-order method would proceed as follows. One chooses a reduced basis Ψ_i, $i = 1, \ldots, d$; hopefully, d is very small compared to the usual number of functions used in a finite element approximation or the number of grid points used in a finite difference approximation. Next, one seeks an approximation u_{rom} to the state u of the form

$$u_{rom} = \sum_{j=1}^{d} c_j \Psi_j \in V \equiv \text{span}\{\Psi_1, \ldots, \Psi_d\}.$$

Then, one determines the coefficients c_i, $i = 1, \ldots, d$, by solving the state equations in the set V, e.g., one could find a Galerkin solution of the state equations in a standard way, using V for the space of approximations. The cost of such a computation would be very small if d is small, ignoring the cost of the off-line determination of the reduced basis $\{\Psi_i\}_{i=1}^{d}$.

In control or optimization settings, one is faced with multiple state solves or real-time state solves. If one approximates the state in the reduced, low-dimensional set V, then state solutions will be relatively very cheap to obtain. In an adjoint or sensitivity equation-based optimization method, one would also employ the adjoint or sensitivity equations for the low-dimensional discrete state equations; thus, if d is small, the cost of each iteration of the optimizer would be very small relative to that using full, high-fidelity state solutions. In a feedback control setting, the approximate state equations in the low-dimensional space could possibly be solved in real time.

All reduced bases share some common features. They all require the solution of high-fidelity and therefore very expensive discrete state and/or sensitivity equations; the idea is that these expensive calculations can be done off line before a state simulation or the optimization of the design parameters or feedback control is attempted. Moreover, one hopes that a single reduced basis can be used for several state simulations or in several design or control settings.

All reduced basis functions are global in nature, i.e., the support of the basis functions is global. Therefore, solving the state or sensitivity or adjoint equations with respect to any of the reduced bases requires the solution of dense linear and nonlinear systems; thus, unless the dimension of a reduced basis is "small," it cannot be effectively used.

The question remains about how one determines a reduced basis. One can also ask when and if are reduced-order methods effective. Below, we describe two approaches to ROM: proper orthogonal decomposition (POD) and centroidal Voronoi tessellations (CVT). First, however, we briefly discuss the second question about the effectiveness of ROM's.

It is clear that reduced-order methods should work in an *interpolatory* setting. In a simulation setting, one has the tautology that if the state can be approximated well in the reduced basis V, then one should expect that things will work well. In an optimization setting, if the optimal solution and the path to the optimal solution can be well approximated in the reduced basis V, then one should expect that things will work well; if all the states determined by the feedback process can be well approximated in the reduced basis V, then again one should expect that things will work well. Thus, the reduced basis V should be chosen so that it contains all the features, e.g., the dynamics, of the states encountered during the simulation or the control process. This, of course, requires some intuition about the states to be simulated or about where in parameter space the optimal set of parameters are located.

What happens in an *extrapolatory* setting (for which the reduced basis may not contain sufficient information to accurately approximate the states encountered) is not so clear. Most reduced order simulation and control computations have been done in an interpolatory regime. It is obvious that if the reduced set V does not contain a good approximation to the solution one is trying to obtain, then one cannot hope to successfully determine that solution.

Despite any misgivings about the effectiveness of reduced-order modeling, one realizes that without an inexpensive method for effecting state simulations, it is unlikely that the solution of three-dimensional optimization and control problems involving complex systems, e.g., the Navier-Stokes system, will become routine anytime soon. Thus, it is certainly true that these methods deserve more study from the computational and theoretical points of views.

2 Snapshot Sets

The state of a complex system is determined by parameters that appear in the specification of a mathematical model for the system. Of course, the state of a complex system also depends on the independent variables appearing in the model. Snapshot sets consist of state solutions corresponding to several sets of parameter values and/or the state solution evaluated at several time instants during the evolution process. Sensitivities of the state with respect to parameters appearing in the model may also be included in the snapshot set. Snapshot sets are usually determined by solving the full, very large-dimensional discretized system obtained by, e.g., a finite volume or finite element discretization.

Snapshots sets often contain "redundant" information; therefore, snapshot sets are post-processed to remove as much of the redundancy as possible before they can be used for reduced-order modeling. POD and CVT are simply different ways to post-process snapshot sets.

Since snapshot sets are the underpinning for most usages of POD and CVT reduced-order modeling, we briefly discuss how they are generated in practice. At this time, the generation of snapshot sets is an art and not a science; in fact, it is a rather primitive art. The generation of snapshot sets is an exercise in the design of experiments, e.g., for stationary systems, how does one choose the sets of parameters at which the state (and perhaps the sensitivities) are to be calculated (using expensive, high-fidelity computations) in order to generate the snapshot set? Clearly, some a priori knowledge about the types of states to be simulated or optimized using the reduced-order model is very useful in this regard; the large body of statistics literature on the design of experiments has not been used in a systematic manner for the generation of snapshot sets.

For time-dependent systems, many (ad hoc) measures have been invoked in the hope that they will lead to good snapshot sets; time-dependent parameters (e.g., in boundary conditions) are used to generate states that are "rich" in transients, even if the state of interest depends only on time-independent parameters. In order to generate even "richer" dynamics, impulsive forcing is commonly used, e.g., starting the evolution impulsively and/or introducing impulsive changes in the parameters in the middle of a simulation.

In the future, a great deal of effort needs to be directed towards developing and justifying methodologies for generating good snapshot sets; after all, a POD or CVT basis is only as good as the snapshot set used to generate it.

We next describe how POD and CVT bases are constructed from a snapshot set. We will view snapshots as vectors in Euclidean space. They represent, e.g., the nodal values of a finite element approximation of the solution of the partial differential equation.

3 Proper Orthogonal Decomposition (POD)

We begin with n snapshots $\mathbf{s}_j \in \mathbb{R}^N$, $j = 1, \ldots, n$. Here, N could be the dimension of the finite element approximating space. Let $d \leq n < N$. Then, the POD basis $\{\boldsymbol{\phi}_i\}_{i=1}^d$ of cardinality d is found by successively solving, for $i = 1, \ldots, d$, the problem

$$\lambda_i = \max_{|\boldsymbol{\phi}_i|=1} \frac{1}{n} \sum_{j=1}^{n} |\boldsymbol{\phi}_i^T \mathbf{s}_j|^2 \quad \text{and} \quad \boldsymbol{\phi}_i^T \boldsymbol{\phi}_\ell = 0 \quad \text{for } \ell \leq i-1.$$

Let S denote the $N \times n$ snapshot matrix whose columns are the snapshots $\mathbf{s}_j$, i.e.,

$$S = (\mathbf{s}_1 \,,\, \mathbf{s}_2 \,,\, \ldots \,,\, \mathbf{s}_n).$$

Let K denote the $n \times n$ (normalized) correlation matrix for the snapshots, i.e.,

$$K_{j\ell} = \frac{1}{n} \mathbf{s}_j^T \mathbf{s}_\ell \qquad \text{or} \qquad K = \frac{1}{n} S^T S.$$

Let $\boldsymbol{\chi}_i$ with $|\boldsymbol{\chi}_i| = 1$ denote the eigenvector corresponding to the i-th largest eigenvalue λ_i of K. Then, the i-th POD basis vector is given by $\boldsymbol{\phi}_i = \frac{1}{\sqrt{n\lambda_i}} S \boldsymbol{\chi}_i$. The POD basis is orthonormal, i.e., $\boldsymbol{\phi}_i^T \boldsymbol{\phi}_j = 0$ for $i \neq j$ and $\boldsymbol{\phi}_i^T \boldsymbol{\phi}_i = 1$. POD is closely related to the statistical method known as Karhunen-Loève analysis or the method of empirical orthogonal eigenfunctions or principal component analysis.

POD is also closely related to the singular value decomposition (SVD) of the snapshot matrix S. Let $S = U \Sigma V^T$ denote the SVD of S; then, $\sigma_i^2 = n\lambda_i$ for $i = 1, \ldots, n$, where σ_i denotes the i-th singular value of S. The POD basis vectors are the first d left singular vectors of the snapshot matrix S, i.e., $\boldsymbol{\phi}_i = \mathbf{u}_i$ for $i = 1, \ldots, d$, where $\mathbf{u}_i$ denotes the i-th left singular vector of S.

The POD basis is optimal in the following sense. Let $\{\boldsymbol{\psi}_i\}_{i=1}^n$ denote an arbitrary orthonormal basis for the span of the snapshot set $\{\mathbf{s}_j\}_{j=1}^n$. Let $P_{\psi,d}\mathbf{s}_j$

denote the projection of the snapshot $\mathbf{s}_j$ onto the d-dimensional subspace spanned by $\{\boldsymbol{\psi}_i\}_{i=1}^d$. Clearly we have, for each $j = 1, \ldots, n$, $P_{\psi,d}\mathbf{s}_j = \sum_{i=1}^d c_{ji}\boldsymbol{\psi}_i$, where $c_{ji} = \boldsymbol{\psi}_i^T \mathbf{s}_j$ for $i = 1, \ldots, d$. Let the error be defined by

$$\mathcal{E} = \sum_{j=1}^{n} |\mathbf{s}_j - P_{\psi,d}\mathbf{s}_j|^2.$$

Then, the minimum error is obtained when $\boldsymbol{\psi}_i = \boldsymbol{\phi}_i$ for $i = 1, \ldots, d$, i.e., when the $\boldsymbol{\psi}_i$'s are the POD basis vectors.

The connection between POD and SVD makes it is easy to show that the error of the d-dimensional POD subspace is given by

$$\mathcal{E}_{\text{pod}} = \sum_{j=d+1}^{n} \sigma_j^2 = n \sum_{j=d+1}^{n} \lambda_j, \quad \text{where} \begin{cases} n = \text{number of snapshots} \\ d = \text{dimension of the POD subspace.} \end{cases}$$

If one wishes for the relative error to be less than a prescribed tolerance δ, i.e., if one wants $\mathcal{E}_{\text{pod}} \le \delta \sum_{j=1}^{n} |\mathbf{s}_j|^2$, one should choose d to be the smallest integer such that

$$\frac{\sum_{j=1}^{d} \sigma_j^2}{\sum_{j=1}^{n} \sigma_j^2} = \frac{\sum_{j=1}^{d} \lambda_j}{\sum_{j=1}^{n} \lambda_j} \ge \gamma = 1 - \delta. \tag{1}$$

There have been several variations introduced in attempts to "improve" POD. Weighted POD gives more weight to some members of the snapshot set; this can be accomplished, e.g., by including multiple copies of an "important" snapshot in the snapshot set. POD with derivatives gets more information into the snapshot set in order to get a "better" POD basis; e.g., one adds derivatives or numerical approximations to derivatives (especially time derivatives) of simulated states to the snapshot set. H^1 POD changes the error measure for POD in order to get a "better" POD basis; e.g., one can use H^1 norms and inner products (instead of L^2) in the definition and construction of POD bases. Constrained POD imposes a constraint (e.g., symmetry) on the POD basis. Adaptive POD changes the POD basis when it no longer seems to be working; this requires detection of failure of the POD basis, the determination of new snapshot vectors, and the solution of the eigenvalue or SVD problem for the new correlation or snapshot matrix determined from the new snapshot vectors. Details about POD reduced-order modeling and its variants may be found in, e.g., [1, 2] and the references cited therein.

4 Centroidal Voronoi Tessellations (CVT)

Given a *discrete* set of snapshots $W = \{\mathbf{s}_j\}_{j=1}^n$ belonging to $\mathbb{R}^N$, a set $\{V_i\}_{i=1}^d$ is a tessellation of W if $V_i \subset W$ for $i = 1, \ldots, d$, $V_i \cap V_j = \emptyset$ for $i \ne j$, and $\cup_{i=1}^d V_i = W$, i.e., if $\{V_i\}_{i=1}^d$ is a subdivision of W into d disjoint, covering subsets. Given a set of points $\{\mathbf{z}_i\}_{i=1}^d$ belonging to $\mathbb{R}^N$ (but not necessarily to W), the Voronoi region corresponding to the point $\mathbf{z}_i$ is defined by

$$\widehat{V}_i = \{\, \mathbf{s} \in W \;:\; |\mathbf{s} - \mathbf{z}_i| \le |\mathbf{s} - \mathbf{z}_j| \quad \text{for } j = 1, \ldots, d,\ j \ne i\},$$

where equality holds only for $i < j$. The set $\{\widehat{V}_i\}_{i=1}^d$ is called a Voronoi tessellation or Voronoi diagram of W corresponding to the set of points $\{\mathbf{z}_i\}_{i=1}^d$. The points in the set $\{\mathbf{z}_i\}_{i=1}^d$ are called the generators of the Voronoi diagram $\{\widehat{V}_i\}_{i=1}^d$ of W.

Given a density function $\rho(\mathbf{y}) \geq 0$, defined for $\mathbf{y} \in W$, the mass centroid $\mathbf{z}^*$ of any subset $V \subset W$ is defined by $\sum_{\mathbf{y}\in V} \rho(\mathbf{y})|\mathbf{y} - \mathbf{z}^*|^2 = \inf_{\mathbf{z}\in V^*} \sum_{\mathbf{y}\in V} \rho(\mathbf{y})|\mathbf{y} - \mathbf{z}|^2$, where the sums extend over the points belonging to V; the set V^* can be taken to be V or it can be an even larger set such as all of $\mathbb{R}^N$ in which case $\mathbf{z}^*$ is the ordinary mean

$$\mathbf{z}^* = \frac{\sum_{\mathbf{y}\in V} \rho(\mathbf{y})\mathbf{y}}{\sum_{\mathbf{y}\in V} \rho(\mathbf{y})}.$$

In this case, $\mathbf{z}^* \notin W$ in general.

If $\mathbf{z}_i = \mathbf{z}_i^*$ for $i = 1, \ldots, d$, where $\{\mathbf{z}_i\}_{i=1}^d$ is the set of generating points of the Voronoi tessellation $\{\widehat{V}_i\}_{i=1}^d$ and $\{\mathbf{z}_i^*\}_{i=1}^d$ is the set of mass centroids of the Voronoi regions $\{\widehat{V}_i\}_{i=1}^d$, we refer to the Voronoi tessellation as being a *centroidal Voronoi tessellation* or CVT for short. The concept of CVT's can be extended to more general sets, including regions in Euclidean spaces, and more general metrics.

CVT's are optimal in the following sense. Given the discrete set of points $W = \{\mathbf{s}_j\}_{j=1}^n$ belonging to $\mathbb{R}^N$, we define the error of a tessellation $\{V_i\}_{i=1}^d$ of W and a set of points $\{\mathbf{z}_i\}_{i=1}^d$ belonging to $\mathbb{R}^N$ by

$$\mathcal{F}\big((\mathbf{z}_i, V_i), i = 1, \ldots, d\big) = \sum_{i=1}^{d} \sum_{\mathbf{y}\in V_i} \rho(\mathbf{y})|\mathbf{y} - \mathbf{z}_i|^2.$$

Then, it can be shown that a necessary condition for the error $\mathcal{F}$ to be minimized is that the pair $\{\mathbf{z}_i, V_i\}_{i=1}^d$ form a CVT of W.

CVT's of discrete sets are closely related to optimal K-means clusters so that Voronoi regions and centroids can be referred to as clusters and cluster centers, respectively. The error $\mathcal{F}$ is also often referred to as the variance, cost, distortion error, or mean square error.

There are several algorithms known for constructing centroidal Voronoi tessellations of a given set. Lloyd's method is a deterministic algorithm which is the obvious iteration between computing Voronoi diagrams and mass centroids, i.e., a given set of generators is replaced in an iterative process by the mass centroids of the Voronoi regions corresponding to those generators. MacQueen's method is a very elegant probabilistic algorithm which assigns randomly sampled points into one of d sets or clusters according to the distance to the mean of the clusters. Various other methods based on the minimization properties of CVT's have also been developed. Another probabilistic method which may be viewed as a generalization of both the MacQueen and Lloyd methods and which is amenable to efficient parallelization has also been developed.

CVT's are useful in a variety of applications including optimal quadrature rules; covolume and finite difference methods for PDE's; optimal representation, quantization, and clustering; cell division; data compression; optimal distribution of resources; territorial behavior of animals; optimal placement of sensors and actuators; grid generation in 2D, 3D, and on surfaces; meshfree methods; clustering of gene expression data; and image segmentation and reconstruction. Details about algorithms, applications, and theory of CVT can be found in [1, 2, 3, 5] and the references cited therein.

Since CVT's have been successfully used in several data compression applications, it is natural to examine CVT's in another data compression setting, namely reduced-order modeling.

5 POD, CVT, and Model Reduction

The idea, in both the POD and CVT settings, is to extract, from a given set of snapshots $\{\mathbf{s}_j\}_{j=1}^n$ of vectors in $\mathbb{R}^N$, a smaller set of vectors also belonging to $\mathbb{R}^N$ which "captures" the information contained in the snapshot set. In the POD setting, the reduced set of vectors was the d-dimensional set of POD vectors $\{\phi_i\}_{i=1}^d$. In the CVT setting, the reduced set of vectors is the d-dimensional set of vectors $\{\mathbf{z}_i\}_{i=1}^d$ that are the generators of a centroidal Voronoi tessellation of the set of snapshots. POD produces an optimal reduced basis in the sense that the error $\mathcal{E}$ is minimized; CVT produces an optimal reduced basis in the sense that the error $\mathcal{F}$ is minimized. One can, in principle, determine the dimension d of an effective POD basis, e.g., using the singular values of the correlation matrix. Similarly, one can, in principle, determine the dimension k of an effective CVT basis by examining the (computable) error $\mathcal{F}(\cdot)$.

For the sake of concreteness, suppose that we wish to solve the problem:

$$\begin{array}{l} \text{given } \theta \in \Theta, \text{ find } u(t,x) \in U \text{ such that} \\ N(u;\theta;v) = 0 \;\forall v \in U, \;\text{ a.e. } t \in (0,T), \end{array} \tag{2}$$

where U and Θ are given solution and parameter function spaces, respectively and $(0,T)$ is a specified time interval. The mapping $N : U \times \Theta \times V$ is linear in its third argument and is generally nonlinear in its first two arguments; $N(\cdot;\cdot;\cdot)$ in general involves space and time derivatives and integrals of combinations of its arguments. In (2), θ denotes a set of parameters that serve to specify the problem, u denotes the desired solution, and v denotes a suitable test function. Thus, (2) implicitly defines a solution mapping $u(\theta) : \Theta \to U$. The object of a computational simulation is to find approximations of this mapping.

A finite element method could be used to find approximate solutions of (2). In general, for complex systems of practical interest, the dimension of the approximating finite element space may be in the many thousands or even millions. As a result, it may be very expensive to solve the finite element discrete system for a single set θ of parameter values and it may be prohibitively expensive to solve it for many sets of parameter values. The latter is exactly one of the situations that cause the interest in reduced-order modeling. So, instead of using standard finite element bases, we use POD or CVT reduced bases to find approximations of the solutions of (2).

Let V_{rom} denote the reduced basis space. We view the POD and CVT basis sets $\{\phi_i\}_{i=1}^d$ and $\{\mathbf{z}_i\}_{i=1}^d$, respectively, as coefficient vectors in the expansion of the corresponding basis functions $\{\Phi_i(\mathbf{x})\}_{i=1}^d$ and $\{Z_i(\mathbf{x})\}_{i=1}^d$ in terms of a finite element basis. Then let $V_{rom} = \text{span}\{\Psi_i\}_{i=1}^d$, where, for $i = 1, \ldots, d$, $\Psi_i =$ either Φ_i or Z_i. Then, a reduced-order solution of (2) is found by solving the problem

$$\begin{aligned} &\text{given } \theta \in \Theta, \text{ find } u_{rom}(t,x) \in V_{rom} \text{ such that} \\ &N(u_{rom};\theta;v) = 0 \quad \forall v \in V_{rom}, \text{ a.e. } t \in (0,T). \end{aligned} \tag{3}$$

Since the reduced order solution has the form $u_{rom} = \sum_{j=1}^d c_j(t)\Psi_j(\mathbf{x})$ for some time dependent functions c_j, $j = 1, \ldots, d$, we have that (3) is equivalent to

$$\begin{aligned} &\text{given } \theta \in \Theta, \text{ find } c_j(t), \; j = 1\ldots, d, \text{ such that} \\ &N\left(\sum_{j=1}^d c_j\Psi_j;\theta;\Psi_i\right) = 0 \quad \text{for } i = 1,2,\ldots,d, \; t \in (0,T). \end{aligned} \tag{4}$$

Since, so far, we have only effected spatial discretization, (4) is a system of nonlinear ordinary differential equations for $c_j(t)$, $j = 1\ldots, d$. Thus, further discretization in time is usually needed to solve (4).

POD and CVT bases often (or at least sometimes) do well at capturing the information contained in the snapshot set; in fact, they are designed to do just that in different optimal senses. It is important to note that this does not necessarily imply that these bases do well at approximating solutions of the complex system. We have that $\|u - u_{rom}\| \approx \|u_{snapshot} - u_{rom}\| + \|u - u_{snapshot}\|$. POD and CVT reduced-order modeling often render $\|u_{snapshot} - u_{rom}\|$ small; to make $\|u - u_{snapshot}\|$ small, one has to have "good" snapshots to begin with.

A comparison of the relative performance of POD and CVT-based reduced-order modeling is given in [1, 2].

6 Handling Multiple Parameters in the Boundary Data

Most implementations of POD and CVT-based reduced-order modeling involve a single parameter appearing in the specification of the problem; see, e.g., [1, 2] and the references cited therein. Handling multiple parameters, especially in the data associated with Dirichlet-type boundary conditions, is not a totally straightforward endeavor. We present two methods for handling problems with inhomogeneous Dirichlet boundary data which contain K multiplicative parameters. More details about the methods and more extensive comparisons between the methods can be found in [4].

The first approach is an obvious generalization of the standard approach used in the literature for the one-parameter case. One uses reduced-order basis functions that satisfy homogeneous boundary data. The reduced-order solution is then written as a linear combination of these basis vectors plus another linear combination of particular solutions of the complex system that is specifically chosen to satisfy the desired inhomogeneous boundary conditions. To make this precise, suppose we are using a reduced-order modeling technique to obtain

$u_{rom}(\mathbf{x}, t_n)$ which approximates the solution $u(\mathbf{x}, t)$ at the time t_n to a nonlinear partial differential equation defined in a domain Ω with boundary Γ. We suppose that the boundary conditions

$$u(\mathbf{x}, t) = \begin{cases} \beta_k(t) g_k(\mathbf{x}) & \text{on } \Gamma_k,\ k = 1, \dots, K \\ 0 & \text{on } \Gamma - \cup_{k=1}^{K} \Gamma_k \end{cases} \tag{5}$$

where $\cap_{k=1}^{K} \Gamma_k = 0$ and $\cup_{k=1}^{K} \Gamma_k$ may be a portion of the boundary Γ or the entire boundary. The functions $\{g_k(\mathbf{x})\}_{k=1}^{K}$ are assumed given so that there are K (time-dependent) parameters $\{\beta_k\}_{k=1}^{K}$ that serve to specify the problem. Then, the first method for treating such boundary conditions in the reduced-order modeling context is given as follows.

METHOD I

(i) Generate vectors $\boldsymbol{\alpha}_k \in \mathbb{R}^K$, $k = 1, \dots, K$;
(ii) generate u_k^{p}, $k = 1, \dots, K$, where $u_k^{\mathrm{p}} = (\boldsymbol{\alpha}_k)_i g_i(\mathbf{x})$ on Γ_i, $i = 1, \dots, K$;
(iii) generate snapshots s_j, $j = 1, \dots, n$, which satisfy $s_j = (\boldsymbol{\sigma}_j)_i g_i(\mathbf{x})$ on Γ_i, $i = 1, \dots, K$ for some $\boldsymbol{\sigma}_j \in \mathbb{R}^K$;
(iv) form modified snapshots $\widetilde{s}_j$, $j = 1, \dots, n$, that satisfy homogeneous boundary conditions on all of Γ by subtracting from s_j the linear combination of particular solutions u_k^{p}, $k = 1, \dots, K$, that have the same boundary data as does s_j;
(v) generate a reduced-order basis (POD or CVT) Ψ_j, $j = 1, \dots, d$, from the modified snapshots $\widetilde{s}_j$, $j = 1, \dots, n$; the resulting reduced-basis functions satisfy homegeneous boundary conditions;
(vi) at each time level t_n, determine the appropriate linear combination $\sum_{k=1}^{K} \eta_{k,n} u_k^{\mathrm{p}}$ of the solutions u_k^{p} which satisfy the given boundary conditions $\beta_k(t_n) g_k(\mathbf{x})$ on Γ_k, $k = 1, \dots, K$;
(vii) at each time level t_n, set $u_{rom}(\mathbf{x}, t_n) = \sum_{j=1}^{M} \mu_{j,n} \Psi_j + \sum_{k=1}^{K} \eta_{k,n} u_k^{\mathrm{p}}$ and solve the discrete weak problem tested against each Ψ_i, $i = 1, \dots, d$, to determine $\mu_{j,n}$, $j = 1, \dots, d$.

We leave the discussion of the choices for $\boldsymbol{\alpha}_k$ and u_k^{p} which are used for forcing the snapshots to satisfy homogeneous boundary conditions and for the formation of u_{rom} until we consider some numerical results. These choices can have a significant effect on the quality of the reduced-order solution.

The second method is a more "natural" approach for handling multi-parameter, reduced-order modeling problems. This approach is similar to the standard approach in finite element methods where one does not constrain the trial functions to satisfy the inhomogeneous boundary conditions, but instead one adds additional equations to satisfy those conditions. In this approach we set $u_{rom} = \sum_{j=1}^{m} \mu_{j,n} \Psi_j$, where now Ψ_j, $j = 1, \dots, m$, are reduced-order basis vectors which, in general, do not satisfy homogeneous boundary data. They are determined via either a POD of CVT analysis applied to the unmodified snapshot functions s_j, $j = 1, \dots, n$. Note that the basis vectors Ψ_j do not satisfy the boundary conditions (involving the parameter choice $\boldsymbol{\alpha}_k$) that were used to determine the snapshots. We cannot satisfy the boundary conditions (5) for our problem by simply setting some of the coefficients $\mu_{j,n}$ because the basis functions are global rather than nodal. However, we can add K equations to satisfy the K boundary conditions. The difficulty is that we can no longer test our discrete weak problem against all m basis functions Ψ_j because we would obtain an overdetermined system of equations, *i.e.*, we would have $m + K$ equations in m unknowns.

We recall that in the standard finite element setting we would test against functions which vanish on the boundary. Consequently, in the reduced-order setting, we want to test against $m-K$ linear combinations of the original m reduced basis functions that are linearly independent and that vanish on the boundary. The question then becomes one of finding the appropriate linear combinations of the global basis vectors Ψ_j, $j = 1, \dots, m$. The QR algorithm is useful in this situation, as is evidenced in the following algorithm.

METHOD II

(i) Generate the snapshot functions s_j, $j = 1, \dots, n$, that satisfy $s_j = (\boldsymbol{\sigma}_j)_i g_i(\mathbf{x})$ on Γ_i, $i = 1, \dots, K$, for some $\boldsymbol{\sigma}_j \in \mathbb{R}^K$;
(ii) generate a reduced-order basis (either of POD or CVT type) Ψ_j, $j = 1, \dots, m$ from the snapshots s_j, $j = 1, \dots, n$;
(iii) use the QR algorithm to determine the linear combinations $\widehat{\Psi}_\ell$, $\ell = 1, \dots,$ $d = m - K$ of Ψ_j, $j = 1, \dots, m$, which vanish on the boundary;
(iv) at each time level t_n, set $u_{rom}(\mathbf{x}, t_n) = \sum_{j=1}^m \mu_{j,n}\phi_j$ and solve the m-dimensional system formed by testing the discrete weak problem against each $\widehat{\Psi}_i$, $i = 1, \dots, d = m-K$ plus the additional K equations $u_{rom}(\mathbf{x}_{\gamma_i}, t_n) = \beta_k(t_n) g_k(\mathbf{x}_{\gamma_i})$, where $\mathbf{x}_{\gamma_i}$ is any point on Γ_i.

To clarify step (iii), let $\mathbf{x}_{\gamma_k}$ be a point on Γ_k, $k = 1, \dots, K$, and let B be an $m \times K$ matrix with entries given by

$$B_{ij} = \Psi_i(\mathbf{x}_{\gamma_j}), \quad i = 1, \dots, m, \ j = 1, \dots, K.$$

We now want to determine $d = m-K$ vectors in $\mathbb{R}^m$ that are linearly independent and orthogonal to the span of columns of B; this can, of course, be done by performing a QR decomposition of B. Thus, the last $d = m - K$ columns of the $m \times m$ orthogonal matrix Q determine the coefficients of the linear combination of the basis functions Ψ_j that we use as test functions. Specifically, for $\ell = 1, \dots, d$, we have that $\widehat{\phi}_\ell = \sum_{j=1}^m Q_{j,\ell+K}\Psi_j$.

Note that in step (ii), we determine a reduced basis of dimension $m = d + K$ compared to the d-dimensional basis of Method I. This is because when we compare the two approaches, we want to compare results for which we are using the same number of test functions in each method. For Method II, we "use up" K of the basis functions (to be more precise, K independent linear combinations of those basis function) to satisfy the boundary conditions, so that we need to start with $m = d+K$ basis functions to end up with d test functions. Of course, for Method I, instead of using up K basis functions, we have to use K particular solutions to satisfy the boundary conditions.

6.1 Numerical Results

In this section we present numerical results comparing the two approaches for handling multi-parameters appearing in the boundary conditions. We take as a simple prototype example the nonlinear parabolic problem

$$\frac{\partial u}{\partial t} - \Delta u + f(u) = 0 \quad \mathbf{x} \in \Omega = (0,1) \times (0,1),\ 0 < t \leq 1, \tag{6}$$

where $f(u) = u^2$ or $f(u) = e^u$, along with the initial condition $u(\mathbf{x}, 0) = 0$ and the boundary conditions

$$u(\mathbf{x}, t) = \begin{cases} \beta_1(t) 4x_1(1-x_1) & \text{when } x_2 = 1 \\ \beta_2(t) 4x_1(1-x_1) & \text{when } x_2 = 0 \\ \beta_3(t) 4x_2(1-x_2) & \text{when } x_1 = 0 \\ \beta_4(t) 4x_2(1-x_2) & \text{when } x_1 = 1. \end{cases} \tag{7}$$

Note that $K = 4$ time-dependent parameters $\{\beta_k(t)\}_{k=1}^4$ appear in the boundary conditions.

We use a finite element method to discretize in space; a uniform, triangular grid consisting of 128 triangles along with continuous, piecewise quadratic basis functions is employed. The number of degrees of freedom is 225. In time, we use the backward Euler method. The nonlinear discretized problem is solved using Newton's method. Reduced-order bases are generated using standard POD techniques.

To determine the particular solutions u_i^{p}, $i = 1, \ldots, 4$ (see steps (i) and (ii) of Method I), we choose $\boldsymbol{\alpha}_1 = (3,0,0,0)$, $\boldsymbol{\alpha}_2 = (0,3,0,0)$, $\boldsymbol{\alpha}_3 = (0,0,3,0)$, and $\boldsymbol{\alpha}_4 = (0,0,0,3)$. A total of 300 snapshots are determined for $f(u) = u^2$ and 288 for $f(u) = e^u$. The specific values of the boundary parameter functions $\beta_k(t)$, $k = 1, \ldots, 4$, used to generate the snapshots are described in detail in [4]. Here, we merely mention that the parameter functions are impulsively changed several times during the snapshot generation process; between the changes, the solution is allowed to relax to a steady state.

For Method I, a total of 16 POD vectors were generated from the snapshots while for Method II a total of 20 POD basis vectors were computed. The singular values are plotted in Figure 1. These results can be used in conjunction with (1) to choose an effective value for d, the number of POD basis vectors one uses for ROM. The POD basis functions are used to determine reduced-order solutions of (6)-(7) for

$$\beta_1(t) = \begin{cases} 2t & 0 \leq t \leq 0.5 \\ 2(1-t) & 0.5 \leq t \leq 1, \end{cases} \quad \begin{matrix} \beta_2 = 4(t - t^2) \\ \beta_3 = |\sin 2\pi t|, \quad \beta_4 = |\sin 4\pi t|. \end{matrix} \tag{8}$$

Note that these choices for the parameter functions are quite different than those used to generate snapshots or, for Method I, for generating the particular solutions u_k^{p}, $k = 1, \ldots, 4$. The normalized L^2-errors between the reduced-order solution and the full finite element solution using both methods are given in Figures 2 and 3. Note that the errors are comparable in all cases.

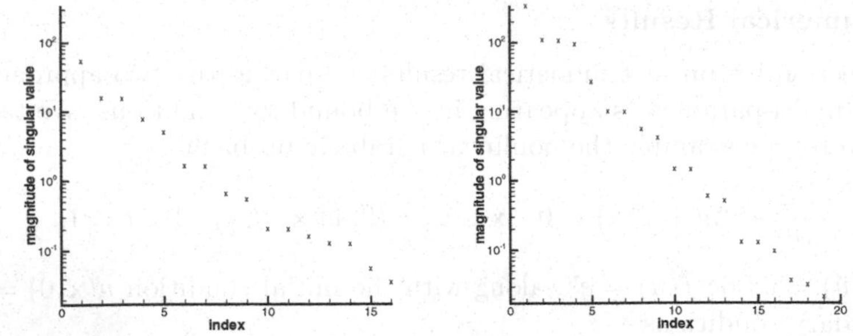

Fig. 1. Comparison of the POD singular values for Method I (left) for which the basis functions satisfy homogeneous boundary data and for Method II (right) where the basis functions do not satisfy homogeneous boundary data

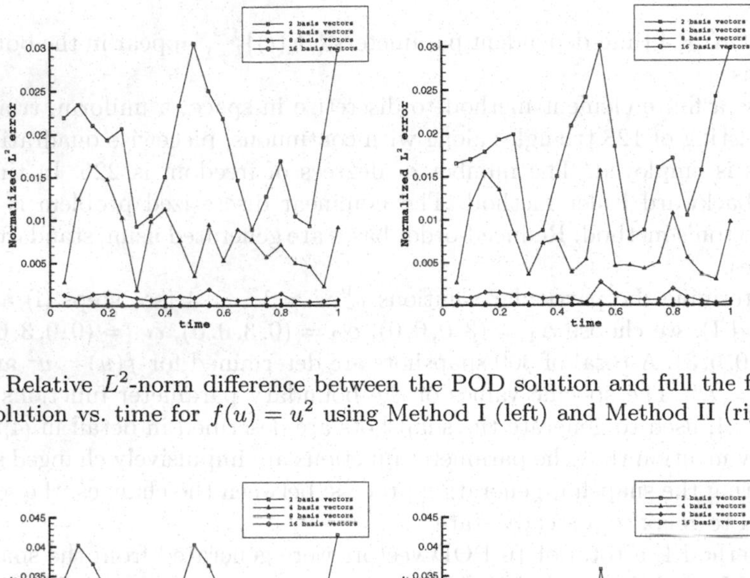

Fig. 2. Relative L^2-norm difference between the POD solution and full the finite element solution vs. time for $f(u) = u^2$ using Method I (left) and Method II (right)

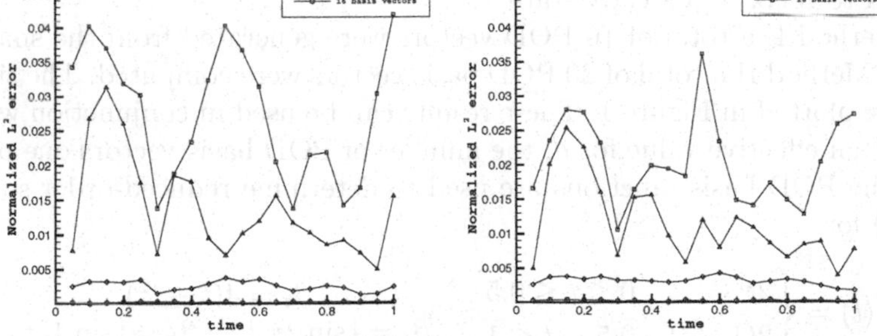

Fig. 3. Relative L^2-norm difference between the POD solution and full the finite element solution vs. time for $f(u) = e^u$ using Method I (left) and Method II (right)

6.2 Discussion

For Method I, the main question is the choice of $\{u^{\mathrm{p}}_{k}\}_{k=1}^{K}$ which are used to make the reduced-order solution satisfy the inhomogeneous boundary conditions. There is clearly a myriad of choices for u^{p}_{k}; a thorough discussion of some of these choices is given in [4].

Another issue is the satisfaction of initial conditions. In our example, we had zero data in the initial conditions so that it was easy to make a reduced-order solution satisfy the initial condition. If the initial data is non-zero, there are several approaches to applying the initial condition, including projecting the initial data onto the reduced-order space. A detailed discussion is given in [4].

Although there seems to be little difference between the results obtained using the two methods, Method II is more straightforward and therefore is easier to implement. Also, Method I is rather sensitive to the choice of $\{u^{\mathrm{p}}_{i}\}_{i=1}^{k}$; Method II avoids having to make this choice. These and some other differences between the two approaches give the edge to Method II. More details are given in [4].

References

1. Burkardt, J., Gunzburger, M., and Lee, H.: Centroidal Voronoi tessellation-based reduced-order modeling of complex systems. SIAM J. Sci. Comp., to appear
2. Burkardt, J., Gunzburger, M., and Lee, H.: POD and CVT-based Reduced-order Modeling of Navier-Stokes flows. Comp. Meth. Appl. Mech. Engrg., to appear
3. Du, Q., Faber, V., and Gunzburger, M.: Centroidal Voronoi tessellations: applications and algorithms. SIAM Review **41** (1999) 637–676
4. Gunzburger, M., Peterson, J., and Shadid, J.: The treatment of multiple parameters appearing in the boundary data for reduced-order modeling applications. International Journal of Numerical Methods in Engineering. To appear.
5. Ju, L., Du, Q., and Gunzburger, M.: Probablistic methods for centroidal Voronoi tessellations and their parallel implementations. J. Parallel Comput. **28** (2002) 1477–1500

Numerical Analysis on a Quantum Computer

Stefan Heinrich

Fachbereich Informatik,
Universität Kaiserslautern,
D-67653 Kaiserslautern, Germany
heinrich@informatik.uni-kl.de
http://www.uni-kl.de/AG-Heinrich

Abstract. We give a short introduction to quantum computing and its relation to numerical analysis. We survey recent research on quantum algorithms and quantum complexity theory for two basic numerical problems — high dimensional integration and approximation. Having matching upper and lower complexity bounds for the quantum setting, we are in a position to compare them with those for the classical deterministic and randomized setting, previously obtained in information-based complexity theory. This enables us to assess the possible speedups quantum computation could provide over classical deterministic or randomized algorithms for these numerical problems.

1 Introduction

A quantum computer is a computing device based on quantum mechanical laws of the (sub)atomic world. The first idea of such a computer is due to Manin [21] in 1980 (see also [22]), and Feynman [7] in 1982. An abstract, theoretical model of quantum computation was developed in 1985 by Deutsch [5]. The breakthrough of quantum computing occurred in 1994, when Shor [31] proved that a quantum computer could factor large integers N in $\mathcal{O}((\log N)^3)$ operations, while no polynomial in $\log N$ classical (deterministic or randomized) algorithm is known. (By "classical" we always mean "non-quantum".) Another important result was obtained by Grover [8] in 1996, who dealt with the following problem: Let $f : \{0, \ldots, N-1\} \to \{0, 1\}$ with the property that there is a unique i_0 with $f(i_0) = 1$. Find this i_0. It is not difficult to show that classically (deterministically or randomized) one needs $\mathcal{O}(N)$ operations. In the quantum setting Grover showed that $\mathcal{O}(\sqrt{N})$ suffice.

This created a challenge to physicists: Find quantum systems suitable for computation, i.e., build a quantum computer. In recent years, various realizations are tested in laboratories. So far only systems with a small number of components (qubits) are possible.

The challenge of quantum computing to mathematicians and computer scientists, on the other hand, is: Find more problems for which quantum algorithms are (provably) better than all classical algorithms. In the sequel, all kinds of discrete problems were investigated. Much less was done for problems of analysis.

I. Lirkov, S. Margenov, and J. Waśniewski (Eds.): LSSC 2005, LNCS 3743, pp. 28–39, 2006.

A natural question is the following: What could a quantum computer bring for the solution of numerical problems? Research in this direction was started in 1998 by Boyer, Brassard, Høyer, Mosca, and Tapp [3, 4], who developed a quantum algorithm for computing the mean of a sequence of numbers. Nayak and Wu [23] showed matching lower bounds. Novak [26] was the first to give a quantum complexity analysis for the integration of functions from Hölder spaces. Computing the mean of p-summable sequences, integration in Sobolev spaces, and the rigorous quantum setting of information-based complexity theory are due to Heinrich [11, 12].

Numerous further recent contributions include Traub, Woźniakowski [33] (path integration), Novak, Sloan, Woźniakowski [27] (integration and approximation in reproducing kernel Hilbert spaces), Heinrich [13, 14] (approximation of Sobolev embeddings), Wiegand [35] (parametric integration), Kacewicz [17, 18] (intitial value problems for systems of ordinary differential equations), Papageorgiou, Woźniakowski [28] (eigenvalue computation for the Sturm-Liouville-problem), Kwas [19] (Feynman-Kac path integrals), and Heinrich [15] (elliptic PDE).

Combining the results above with known results from information-based complexity theory about the classical deterministic and randomized setting [25, 34, 10], one can prove the superiority of quantum algorithms for many of these problems.

In this paper I want to give an introduction to the ideas of quantum computing and survey a few typical recent results concerning basic numerical problems: high dimensional integration and approximation. This will include a comparison of the potential of quantum algorithms with that of deterministic and randomized classical ones.

For further reading on quantum computation we refer to the surveys by Aharonov [1], Ekert, Hayden, and Inamori [6], Shor [32], and the monographs by Pittenger [30], Gruska [9], and Nielsen and Chuang [24]. Basic notions and results in information-based complexity theory can be found in the monographs by Traub, Wasilkowski, and Woźniakowski [34] and Novak [25], and the survey [10] of the randomized setting.

2 Quantum Computing

First we describe the mathematical framework of quantum computing. Let $H_1 := \mathbf{C}^2$ be the two-dimensional complex Hilbert space (the unit sphere of H_1 represents the state space of a **qubit** – quantum bit). Let $\{e_0, e_1\}$ be the unit vector basis. In accordance with quantum mechanical notation we write $|0\rangle$ instead of e_0 and $|1\rangle$ instead of e_1.

The basic quantum computing device is given by an m-qubit-system. Mathematically, it is represented by the tensor product

$$H_m := \underbrace{H_1 \otimes H_1 \otimes \ldots \otimes H_1}_{m},$$

with the basis

$$e_{i_0} \otimes e_{i_1} \otimes \ldots \otimes e_{i_{m-1}} \quad (i_0, i_1, \ldots, i_{m-1}) \in \{0,1\}^m.$$

Thus, H_m is the 2^m-dimensional complex Hilbert space, its unit sphere is the state space of the m-qubit system, and a quantum computation is a trajectory through this state space according to specific rules, which we describe below.

We make the following further notational conventions:

$$e_{i_0} \otimes e_{i_1} \otimes \ldots \otimes e_{i_{m-1}} =: |i_0\rangle\, |i_1\rangle \ldots |i_{m-1}\rangle =: |i\rangle$$

where $i := (i_0 i_1 \ldots i_{m-1})_2 := \sum_{k=0}^{m-1} i_k 2^{m-1-k}$.

The basis states $|i\rangle = |i_0\rangle\, |i_1\rangle \ldots |i_{m-1}\rangle$ represent the classical states of the system, the general quantum states of the m-qubit system are given by superpositions

$$|\xi\rangle = \sum_{i=0}^{2^m-1} \alpha_i\, |i\rangle \qquad \left(\sum_{i=0}^{2^m-1} |\alpha_i|^2 = 1 \right).$$

How to use m-qubit quantum systems for computing? To explain this, let us first consider an example of a classical computation – the addition of two m-bit numbers, which we write as follows (the i's and j's denote the bits of the summands, the k's stand for the bits of the result):

$$\begin{gathered} |i_0\rangle \ldots |i_{m-1}\rangle\, |j_0\rangle \ldots |j_{m-1}\rangle\, |0\rangle\ \ldots |0\rangle \\ \downarrow \\ |i_0\rangle \ldots |i_{m-1}\rangle\, |j_0\rangle \ldots |j_{m-1}\rangle\, |k_0\rangle \ldots |k_m\rangle \end{gathered}$$

This computation is realized using circuits of classical gates (`and`, `or`, `not`, `xor`) in the usual way: add the last bits, then the second last plus the carry bit etc. Let us emphasize here: *Classically, we add two numbers at a time.*

Now, which operations are allowed in a quantum system? Schrödinger's equation implies: all evolutions of a quantum system must be represented by unitary transforms of H_m. **Quantum computing assumes that we are able to perform a number of elementary unitary transforms – (quantum) gates on the system.** A typial set is the following:

One qubit gates: They manipulate only one component of the tensor product $H_1 \otimes H_1 \otimes \ldots \otimes H_1$:

Hadamard gate: $H_1 \to H_1$

$$|0\rangle \to \frac{|0\rangle + |1\rangle}{\sqrt{2}}$$

$$|1\rangle \to \frac{|0\rangle - |1\rangle}{\sqrt{2}}$$

(a unitary transform is uniquely determined by its values on the elements of a basis).

Phase shift: For each parameter $\theta \in [0, 2\pi)$ a quantum gate $H_1 \rightarrow H_1$ is defined by

$$|0\rangle \rightarrow |0\rangle$$
$$|1\rangle \rightarrow e^{\imath\theta}|1\rangle$$

Two qubit gates: They manipulate two components of $H_1 \otimes H_1 \otimes \ldots \otimes H_1$:

Quantum xor gate (also called **controlled-not gate**): $H_1 \otimes H_1 \rightarrow H_1 \otimes H_1$

$$|0\rangle|0\rangle \rightarrow |0\rangle|0\rangle$$
$$|0\rangle|1\rangle \rightarrow |0\rangle|1\rangle$$
$$|1\rangle|0\rangle \rightarrow |1\rangle|1\rangle$$
$$|1\rangle|1\rangle \rightarrow |1\rangle|0\rangle$$

These gates form a universal system: Each unitary transform in H_m can be represented as a finite composition of these gates. If we restrict ourselves to one single phase shift with $\theta = \pi/4$, we obtain an approximately universal system: Each unitary transform can be approximated in the operator norm to arbitrary precision by suitable finite composition of these gates.

So once we can implement these gates we can carry out all unitary transforms (of course, the efficiency of such a representation or approximation is still an issue). Physicists are working on implementations of these gates in various quantum systems such as photons, trapped ions, magnetic resonance systems.

Let us mention two crucial features:

1. These gates can transform classical states into superpositions. Example: the Hadamard gate applied to the first and then to the second qubit

$$|0\rangle|0\rangle \longrightarrow \frac{1}{2}(|0\rangle|0\rangle + |0\rangle|1\rangle + |1\rangle|0\rangle + |1\rangle|1\rangle)$$

2. They act also on superpositions. Examples:

2.1. The quantum xor gate:

$$\alpha_0|0\rangle|0\rangle + \alpha_1|0\rangle|1\rangle + \alpha_2|1\rangle|0\rangle + \alpha_3|1\rangle|1\rangle$$
$$\downarrow$$
$$\alpha_0|0\rangle|0\rangle + \alpha_1|0\rangle|1\rangle + \alpha_2|1\rangle|1\rangle + \alpha_3|1\rangle|0\rangle$$

2.2. Quantum addition of binary numbers (a classical gate implementation can easily be turned into a quantum gate implementation):

$$\sum \alpha_{ij}|i_0\rangle \ldots |i_{m-1}\rangle|j_0\rangle \ldots |j_{m-1}\rangle|0\rangle \ \ldots \ |0\rangle$$
$$\downarrow$$
$$\sum \alpha_{ij}|i_0\rangle \ldots |i_{m-1}\rangle|j_0\rangle \ldots |j_{m-1}\rangle|k_0\rangle \ldots |k_m\rangle$$

Assuming that all $\alpha_{ij} \neq 0$, we see that starting with the superposition of all possible inputs, and carrying out the quantum implementation just once, we obtain (by linearity of the quantum gates) the superposition of the results of all possible inputs.

That is, in the quantum world, we add all possible binary m-digit numbers in parallel.

But does that mean that we have an exponentially powerful parallel computer? No, because we cannot access all components of the superposition! According to quantum mechanics, we have to measure the system, which destroys the superposition. So:

Quantum computing assumes that we are able to access the results of the quantum computation process via measurement (with respect to the canonical basis). Measuring a system in a (superposition) state

$$|\psi\rangle = \sum_{i=0}^{2^m-1} \alpha_i \, |i\rangle \qquad \left(\sum_{i=0}^{2^m-1} |\alpha_i|^2 = 1 \right)$$

results in one of the classical states:

$$|i\rangle \quad \text{with probability} \quad |\alpha_i|^2 \quad (i = 0, \dots, 2^m - 1).$$

Coming back to our example of quantum addition of binary numbers we would get

$$|i_0\rangle \dots |i_{m-1}\rangle \, |j_0\rangle \dots |j_{m-1}\rangle \, |k_0\rangle \dots |k_m\rangle$$

with probability $|\alpha_{ij}|^2$. This is not much of a gain: just one single result, and on top of that a random one!

The reasoning above indeed showed typical features of a quantum computation, but, on the other hand, made it also plausible, that in order to let a quantum computer behave more efficiently than a classical one, more ingenious and sophisticated techniques are required. We will briefly discuss some of them in section 5.

To study numerical problems in the quantum setting, a few more preparations are required. First of all, we shall view a numerical problem as given by an operator $S : F \to G$, the solution operator. Here F is a set (usually a set of functions), G is a normed space (either a space of functions or the scalar field), and $S(f) \in G$ is the (exact) solution of the problem at input $f \in F$.

If we consider the example of numerical integration, we have that F is a set of functions on some domain D, $G = \mathbf{R}$, and

$$S(f) = \int_D f(t)dt.$$

How does the quantum algorithm get information about $f \in F$? It is helpful to look first at the binary case: Let

$$f : \{0, 1, \dots, 2^{m_1} - 1\} \to \{0, 1\}.$$

A classical black box (query, subroutine) produces $f(i)$ at request i, that is, maps $(i, 0) \to (i, f(i))$. Quite similarly, the quantum (binary) query:

$$|i\rangle \, |0\rangle \to |i\rangle \, |f(i)\rangle$$

This mapping has many extensions to a bijection of classical states, and hence, to a unitary operator $Q_f : H_m \to H_m$. The following is customary:

$$Q_f : |i\rangle\,|j\rangle \to |i\rangle\,|j \oplus f(i)\rangle$$

(where $\oplus$ stands for addition modulo 2).

For problems of analysis we have to consider the general case of functions f from a domain D to $\mathbf{R}$ (or, analogously, to $\mathbf{C}$). The appropriate quantum query $Q_f : H_m \to H_m$ is defined as follows:

$$Q_f : |i\rangle\,|j\rangle \to |i\rangle\,|j \oplus \beta(f(\tau(i)))\rangle\,,$$

where $1 \le m_1 < m$, H_m is identified with $H_{m_1} \otimes H_{m-m_1}$,

$$\tau : \{0,\ldots,2^{m_1}-1\} \to D$$

maps indices i to nodes $\tau(i) \in D$, the mapping

$$\beta : \mathbf{R} \to \{0,\ldots,2^{m-m_1}-1\}$$

encodes the real number $f(\tau(i))$ as a binary integer $\beta(f(\tau(i)))$, and $\oplus$ stands for addition modulo 2^{m-m_1} (the choice of m_1, τ and β is part of the algorithm design). Thus we arrived at the **quantum model of computation for numerical problems:**

starting state:
$|i_0\rangle \in H_m$ (a classical state)
computation:
$|i_0\rangle \to U_0\,|i_0\rangle \to Q_f U_0\,|i_0\rangle \to U_1 Q_f U_0\,|i_0\rangle \to \ldots$
$\to U_n Q_f U_{n-1} \ldots Q_f U_1 Q_f U_0\,|i_0\rangle =: |\xi\rangle$
measurement:
$|\xi\rangle = \sum_{i=0}^{2^m-1} \alpha_i\,|i\rangle \to |i\rangle$ with probability $|\alpha_i|^2$
output:
$|i\rangle \to \phi(i) =: A_n(f) \in G$

The U_i represent the composition of the quantum gates applied before, between, and after the queries, and ϕ symbolizes any classical computation performed on the measurement result i to obtain the final output. We call this a quantum algorithm with n queries. The output $A_n(f)$ is a random variable.

We introduce the (probabilistic) error of A_n at input f:

$$e(S, A_n, f) = \inf\{\varepsilon : \mathbf{P}\{\|S(f) - A_n(f)\|_G \le \varepsilon\} \ge 3/4\}\,,$$

and the error of A_n over F by

$$e(S, A_n, F) = \sup_{f\in F} e(S, A_n, f).$$

Note that the choice of the probability threshold 3/4 is inessential: By repeating the algorithm k times and computing the median of the results, the success probability can be increased to $1-2^{-ck}$ for some $c>0$ not depending on k.

The crucial quantity for complexity analysis is the quantum n-th minimal error

$$e_n^{\mathrm{q}}(S,F)=\inf_{A_n} e(S,A_n,F).$$

It gives the minimal possible error among all quantum algorithms which use at most n quantum queries. For the problems we consider here, this is, up to logarithmic factors, also the best possible errors among all algorithms of cost (number of gates, queries, and measurements) at most n. Along with $e_n^{\mathrm{q}}(S,F)$ we shall also consider

$e_n^{\mathrm{det}}(S,F)$, the best possible error among all deterministic classical algorithms with cost (number of arithmetic operations, function values) $\le n$, and
$e_n^{\mathrm{ran}}(S,F)$, the best possible error among all randomized classical algorithms with cost (number of random generator calls, arithmetic operations, function values) $\le n$.

For detailed definitions and references for the respective results in the classical settings we refer to [25, 34, 10].

3 Multivariate Integration

Let $D=[0,1]^d$, $f:[0,1]^d\to\mathbf{R}$,

$$S(f)=I_d f:=\int_{[0,1]^d} f(t)dt$$

and consider the following function classes: Let $r\in\mathbf{N}_0$, $0<s\le 1$ and define the Hölder classes by

$$F=\mathcal{B}(F_d^{r,s})=\{f\in C^r([0,1]^d),\ \|f\|_\infty\le 1,\\ |\partial^\alpha f(x)-\partial^\alpha f(y)|\le |x-y|^s, |\alpha|=r\}.$$

Here $C^r([0,1]^d)$ is the set of r times continuously differentiable functions and ∂^α is the partial derivative corresponding to the multiindex α.

Next define the Sobolev classes for $r\in\mathbf{N}$, $1\le p\le\infty$, satisfying $r/d>1/p$ (Sobolev embedding condition), by

$$F=\mathcal{B}(W_{p,d}^r)=\{f\in L_p([0,1]^d)\ :\ \|\partial^\alpha f\|_{L_p}\le 1,\ |\alpha|\le r\}$$

with ∂^α being the respective weak partial derivative. We will be particularly interested in the behaviour of the complexity in the various settings for large d.

The following result is due to Novak [26].

Theorem 1.

$$e_n^{\mathrm{q}}(I_d,\mathcal{B}(F_d^{r,s}))\asymp n^{-\frac{r+s}{d}-1}$$

Let us compare this with the classical deterministic and randomized setting:

$$e_n^{\text{det}}(I_d, \mathcal{B}(F_d^{r,s})) \asymp n^{-\frac{r+s}{d}}$$
$$e_n^{\text{ran}}(I_d, \mathcal{B}(F_d^{r,s})) \asymp n^{-\frac{r+s}{d}-1/2}$$

We write $a_n \asymp b_n$ for sequences of nonnegative reals (a_n) and (b_n) if there are constants $c_1, c_2 > 0$, $n_0 \in \mathbf{N}$, such that $c_1 a_n \le b_n \le c_2 a_n$ for all $n \in \mathbf{N}$ with $n \ge n_0$.

It is interesting to look at these rates for small $(r+s)/d$ (that is, for large d). We see that in the classical deterministic setting, the exponent is negligible, meaning there is no chance to solve this problem deterministically. This is the well-known curse of dimension. In the classical randomized settting the situation is different: The exponent is always smaller than $-1/2$ even for very small $(r+s)/d$, corresponding to the fact that randomization (Monte Carlo integration) allows to overcome the curse of dimension. Comparing, finally, the quantum setting with the classical randomized setting, we have essentially a quadratic speedup (the exponent close to $-1/2$ is replaced by an exponent close to -1). This is the same sort of speedup as in the Grover search algorithm mentioned in the beginning.

Novak's result settled the Hölder case, that is, function classes related to the maximum norm, but what about function spaces involving the L_2, or, more generally, the L_p norm for $1 \le p < \infty$. Since L_2 is the natural space for Monte Carlo algorithms, it was an open, interesting question whether Monte Carlo algorithms could possibly be as good as quantum algorithms, say for $p = 2$? Furthermore, it was known that for $p = 1$, Monte Carlo algorithms do not yield a speedup over (classical) deterministic algorithms. Will quantum algorithms do so? These questions are answered in the following theorem from [12]. To suppress inessential for the purpose of our survey logarithmic factors, we write $a_n \asymp_{\log} b_n$ if there are constants $c_1, c_2 > 0$, $n_0 \in \mathbf{N}$, $\alpha_1, \alpha_2 \in \mathbf{R}$ such that

$$c_1(\log(n+1))^{\alpha_1} a_n \le b_n \le c_2(\log(n+1))^{\alpha_2} a_n$$

for all $n \in \mathbf{N}$ with $n \ge n_0$.

Theorem 2. *Let* $1 \le p < \infty$, $r, d \in \mathbf{N}$, $r/d > 1/p$. *Then*

$$e_n^{\text{q}}(I_d, \mathcal{B}(W_{p,d}^r)) \asymp_{\log} n^{-r/d-1}.$$

In the classical deterministic setting we have

$$e_n^{\text{det}}(I_d, \mathcal{B}(W_{p,d}^r)) \asymp n^{-r/d},$$

while in the classical randomized setting the following holds:

$$e_n^{\text{ran}}(I_d, \mathcal{B}(W_{p,d}^r)) \asymp n^{-r/d-1/2} \quad \text{if} \quad 2 \le p < \infty$$
$$e_n^{\text{ran}}(I_d, \mathcal{B}(W_{p,d}^r)) \asymp n^{-r/d-1+1/p} \quad \text{if} \quad 1 \le p < 2.$$

We see that the same spedup of the quantum over the classical randomized setting (a gain of $-1/2$ in the exponent) holds through for all $p \geq 2$. For $1 \leq p < 2$ the gain is even greater, reaching -1 for $p = 1$, the case where classical randomization does not yield any gain over classical deterministic algorithms.

4 Approximation of Sobolev Embeddings

It was well-known that Monte Carlo methods are especially suited for problems whose output is a scalar (integration, computation of functionals of solutions of integral equations). The integration results presented above are of this kind, leaving the question how quantum algorithms would behave if the output were not a scalar, but a function. A particularly typical situation is function approximation — we are asked to compute an approximation to a function using (a limited number of) values of that function.

Let $1 \leq p, q \leq \infty$,

$$S = J_{pq} : W_p^r([0,1]^d) \to L_q([0,1]^d), \quad J_{pq}(f) = f,$$

and put $F = \mathcal{B}(W_{p,d}^r)$. Thus, given $f \in \mathcal{B}(W_{p,d}^r)$, we seek to approximate f in the norm of $L_q([0,1]^d)$. The following was shown in [13, 14].

Theorem 3. *Let $r, d \in \mathbb{N}$, $1 \leq p, q \leq \infty$ and assume $r/d > \max(1/p, 2/p - 2/q)$. Then*

$$e_n^q(J_{pq}, \mathcal{B}(W_p^r(D))) \asymp_{\log} n^{-r/d}.$$

Again it is instructive to compare with the classical deterministic and randomized setting:

$$e_n^{\text{det}}(J_{pq}, \mathcal{B}(W_{p,d}^r)) \asymp e_n^{\text{ran}}(J_{pq}, \mathcal{B}(W_{p,d}^r)) \asymp \begin{cases} n^{-r/d} & \text{if} \quad p \geq q \\ n^{-r/d+1/p-1/q} & \text{if} \quad p < q. \end{cases}$$

We observe a possible improvement of n^{-1} (for $p = 1$, $q = \infty$) of quantum algorithms over the classical deterministic and randomized case. This is the same speedup as in Theorem 2 for $p = 1$. We also see that there are regions of the parameter domain where the speedup is smaller, and others, where there is no speedup at all.

5 Further Comments

Let us summarize the results in a table (suppressing again constants and logarithmic factors).

	deterministic	random	quantum
Integration			
$\mathcal{B}(F_d^{r,s})$	$n^{-(r+s)/d}$	$n^{-(r+s)/d-1/2}$	$n^{-(r+s)/d-1}$
$\mathcal{B}(W_{p,d}^r)$, $2 \le p \le \infty$	$n^{-r/d}$	$n^{-r/d-1/2}$	$n^{-r/d-1}$
$\mathcal{B}(W_{p,d}^r)$, $1 \le p < 2$	$n^{-r/d}$	$n^{-r/d-1+1/p}$	$n^{-r/d-1}$
Approximation			
$\mathcal{B}(W_{p,d}^r) \to L_q$, $p \ge q$	$n^{-r/d}$	$n^{-r/d}$	$n^{-r/d}$
$\mathcal{B}(W_{p,d}^r) \to L_q$, $p < q$	$n^{-r/d+1/p-1/q}$	$n^{-r/d+1/p-1/q}$	$n^{-r/d}$

We mentioned in section 2 that a naive view on quantum computation does not bring us very far. So what *are* the algorithmic methods that make quantum computers superior to classical ones? There is, first of all, the quantum Fourier transform, a highly efficient implementation of the discrete Fourier transform on a quantum system. Based on this, Shor [31] developed a technique of estimating eigenvalues of unitary operators, which eventually lead to his seminal results on factoring. The crucial idea of the Grover search is an iterative amplification of the amplitude of the state we are interested in (the state $|i_0\rangle$ with i_0 such that $f(i_0) = 1$). Finally, Boyer, Brassard, Høyer, Mosca, and Tapp [3, 4], combined this by estimating the eigenvalues of the Grover transform using the Shor approach and this way produced an efficient counting algorithm (estimating the number of 1's in a huge sequence of bits, or, equivalently, estimating the mean of a sequence of bits). For a good, self-contained exposition of these basic techniques see [24]. The lower bound results by Nayak and Wu [23] are derived by the polynomial method [2]: the succcess probability is a polynomial (in the bits the mean of which is to be computed) of degree at most the number of queries. Interesting from the point of view of approximation theory: Nayak and Wu use the Bernstein and Markov inequalities for polynomials to get their result.

Novak's Theorem 1 is built on these results, combining them with techniques from information-based complexity [25, 34], in particular for the lower bound proof. The upper bound is shown by adopting a technique from Monte Carlo methods: separation of the main part (control variate).

These were L_∞-results exclusively. The step from $p = \infty$ to arbitrary p needed for Theorem 2 is based on respective results for mean computation in finite dimensional l_p^N spaces [11, 16]. Those are achieved by splitting the function into dyadic levels, distributing queries over levels, combining decay of means and precise error estimates for counting. In the case of $1 \le p < 2$ a combination with the Grover search is used. A new discretization technique (inspired by Maiorov's technique [20] from approximation theory), reducing integration to a sequence of mean computation problems [12], leads to Theorem 2.

Related techniques (Grover search, multilevel splittings, discretization) are also used in [13, 14] to prove the upper bounds in Theorem 3. The lower bound technique is a new one: multiplicativity of minimal quantum errors. This was inspired by functional analysis — the multiplicativity of s-numbers [29].

References

1. D. Aharonov. Quantum computation — a review, in: D. Stauffer (Ed.) *Annual Review of Computational Physics*, vol. VI, World Scientific, Singapore, 1998, see also http://arXiv.org/abs/quant-ph/9812037.
2. R. Beals, H. Buhrman, R. Cleve, M. Mosca, R. de Wolf. Quantum lower bounds by polynomials, *Proceedings of 39th IEEE FOCS*, 352–361, 1998, see also http://arXiv.org/abs/quant-ph/9802049.
3. M. Boyer, P. Brassard, P. Høyer, A. Tapp. Tight bounds on quantum searching, *Fortschritte der Physik*, **46**, 493–505, 1998, see also http://arXiv.org/abs/quant-ph/9605034.
4. G. Brassard, P. Høyer, M. Mosca, A. Tapp. Quantum amplitude amplification and estimation, In: *Quantum Computation and Quantum Information: A Millennium Volume*, AMS Contemporary Mathematics Series, **305**, 2002, see also http://arXiv.org/abs/quant-ph/0005055.
5. D. Deutsch. Quantum theory, the Church-Turing principle and the universal quantum computer, *Proc. R. Soc. Lond.*, Ser. A 400, 97–117, 1985.
6. A. Ekert, P. Hayden, H. Inamori. Basic concepts in quantum computation, 2000, see http://arXiv.org/abs/quant-ph/0011013.
7. R. Feynman. Simulating physics with computers, *Int. J. Theor. Phys.*, **21**, 467–488, 1982.
8. L. Grover. A fast quantum mechanical algorithm for database search, *Proc. 28 Annual ACM Symp. on the Theory of Computing*, 212–219, ACM Press New York, 1996, see also http://arXiv.org/abs/quant-ph/9605043.
9. J. Gruska. *Quantum Computing*, McGraw-Hill, London, 1999.
10. S. Heinrich. Random approximation in numerical analysis, in: K. D. Bierstedt, A. Pietsch, W. M. Ruess, D. Vogt (Eds.), *Functional Analysis*, Marcel Dekker, New York, 1993, 123–171.
11. S. Heinrich. Quantum summation with an application to integration, *J. Complexity*, **18**, 1–50, 2002, see also http://arXiv.org/abs/quant-ph/0105116.
12. S. Heinrich. Quantum integration in Sobolev classes, *J. Complexity*, **19**, 19–42, 2003, see also http://arXiv.org/abs/quant-ph/0112153.
13. S. Heinrich. Quantum Approximation I. Embeddings of Finite Dimensional L_p Spaces,*J. Complexity*, **20**, 5–26, 2004, see also http://arXiv.org/abs/quant-ph/0305030.
14. S. Heinrich. Quantum Approximation II. Sobolev Embeddings, *J. Complexity*, **20**, 27–45, 2004, see also http://arXiv.org/abs/quant-ph/0305031.
15. S. Heinrich. The quantum query complexity of elliptic PDE, in preparation.
16. S. Heinrich, E. Novak. On a problem in quantum summation, *J. Complexity*, **19**, 1–18, 2003, see also http://arXiv.org/abs/quant-ph/0109038.
17. B. Kacewicz. Randomized and quantum algorithms yield a speed-up for initial-value problems, *J. Complexity*, **20**, 821-834, 2004, see also http://arXiv.org/abs/quant-ph/0311148.

18. B. Kacewicz. Improved bounds on the randomized and quantum complexity of initial-value problems, see http://arXiv.org/abs/quant-ph/0405018.
19. M. Kwas. Complexity of multivariate Feynman-Kac path integration in randomized and quantum settings, see http://arXiv.org/abs/quant-ph/0410134.
20. V. E. Maiorov. Discretization of the problem of diameters, *Usp. Mat. Nauk*, **30**, No. 6 (186), 179–180, 1975 (in Russian).
21. Yu. I. Manin. Computable and uncomputable, *Sovetskoye Radio*, Moscow, 1980 (in Russian).
22. Yu. I. Manin. Classical computing, quantum computing, and Shor's factoring algorithm, 1999, see http://arXiv.org/abs/quant-ph/9903008.
23. A. Nayak, F. Wu. The quantum query complexity of approximating the median and related statistics, *STOC, May 1999*, 384–393, see also http://arXiv.org/abs/quant-ph/9804066.
24. M. A. Nielsen, I. L. Chuang. *Quantum Computation and Quantum Information*, Cambridge University Press, Cambridge, 2000.
25. E. Novak. *Deterministic and Stochastic Error Bounds in Numerical Analysis*, Lecture Notes in Mathematics, **1349**, Springer-Verlag, Berlin, 1988.
26. E. Novak. Quantum complexity of integration, *J. Complexity*, **17**, 2–16, 2001, see also http://arXiv.org/abs/quant-ph/0008124.
27. E. Novak, I. H. Sloan, H. Woźniakowski. Tractability of approximation for weighted Korobov spaces on classical and quantum computers, *Found. Comput. Math.*, **4**, 121–156, 2004, see also http://arXiv.org/abs/quant-ph/0206023.
28. A. Papageorgiou, H. Woźniakowski. Classical and quantum complexity of the Sturm-Liouville eigenvalue problem, see http://arXiv.org/abs/ quant-ph/0502054.
29. A. Pietsch. *Eigenvalues and s-Numbers*, Cambridge University Press, 1987.
30. A. O. Pittenger. *Introduction to Quantum Computing Algorithms*, Birkhäuser, Boston, 1999.
31. P. W. Shor. Algorithms for quantum computation: Discrete logarithms and factoring. *Proceedings of the 35th Annual Symposium on Foundations of Computer Science*, IEEE Computer Society Press, Los Alamitos, CA, 124–134, 1994, see also http://arXiv.org/abs/quant-ph/9508027.
32. P. W. Shor. Introduction to quantum algorithms, 2000, see http://arXiv.org/abs/quant-ph/0005003.
33. J. F. Traub, H. Woźniakowski. Path integration on a quantum computer, *Quantum Information Processing*, **1** 5, 365–388, 2002, see also http://arXiv.org/abs/quant-ph/0109113.
34. J. F. Traub, G. W. Wasilkowski, H. Woźniakowski. *Information-Based Complexity*, Academic Press, New York, 1988.
35. C. Wiegand. Quantum complexity of parametric integration, *J. Complexity*, **20**, 75–96, 2004, see also http://arXiv.org/abs/quant-ph/0305103.

Quantum Monte Carlo Simulations of Solid ^{4}He

P.A. Whitlock[1] and S.A. Vitiello[2]

[1] Computer and Information Science Department,
Brooklyn College, CUNY Brooklyn, NY 11210, USA
whitlock@brooklyn.cuny.edu

[2] Instituto de Física, Universidade Estadual de Campinas,
13083 Campinas — SP, Brazil
vitiello@ifi.unicamp.br

Abstract. Recent experimental investigations [20] of solid ^{4}He have been interpreted as showing possible superfluidity in the solid at low temperatures, below 0.2 K. A solid behaving this way, exhibiting both long range translational order and superfluidity, has been called a supersolid phase. The existence of a supersolid phase was proposed many years ago [1], and has been discussed theoretically. In this paper we review simulations of the solid state of bulk ^{4}He at or near absolute zero temperature by quantum Monte Carlo techniques. The techniques considered are variational calculations at zero temperature which use traditional Bijl-Dingle-Jastrow wavefunctions or more recently, shadow wavefunctions; Green's function Monte Carlo calculations at zero temperature; diffusion Monte Carlo, and finally, the finite temperature path integral Monte Carlo method. A brief introduction to the technique will be given followed by a discussion of the simulation results with respect to solid helium.

1 Introduction

After many years of investigation, the properties of the solid phases of bulk, ^{4}He, were felt to be well understood [3]. At temperatures near absolute zero, ^{4}He exists in both the solid and fluid states. The crystalline structure is known to exhibit hexagonal closest packing (hcp) rather than the face centered cubic packing expected for a three-dimensional hard sphere system. Even at absolute zero, the atoms exhibit zero point motion around their lattice positions which leads to a very "loose" solid at densities near melting. In the 1960's, Andreev and Lifshitz [1] proposed that such quantum solids could sustain superfluidity. One indicator of the onset of superfluidity is the Bose-Einstein condensate, the fraction of the atoms condensed into the zero momentum state. The condensed atoms acquire quantum mechanical coherence over macroscopic length scales. Quantum Monte Carlo simulations observed a Bose-Einstein condensate of several percent in liquid ^{4}He systems [34]; but only detected a condensate in a quantum solid when the atoms were interacting with a Yukawa potential [6].

Interest has been recently renewed by the torsional oscillator experiments of Kim and Chan [20]. Ultrahigh-purity ^{4}He was confined in a torsion cell and subjected to pressures between 26 and 66 bars to reach the solid phase. A non-classical rotational inertia fraction that can be associated with superflow was

I. Lirkov, S. Margenov, and J. Waśniewski (Eds.): LSSC 2005, LNCS 3743, pp. 40–52, 2006.

observed at temperatures below 230 milliKelvin. These observations have lead to renewed interest in the measurements of the Bose-Einstein condensate and other measures of superfluidity in simulations of the properties of solid helium.

^{4}He systems may be studied theoretically by solving the appropriate Schrödinger or Bloch equation. At absolute zero, the behavior of an N body helium system is described by the eigenfunction of the Schrödinger equation in 3N dimensional space:

$$[-\nabla^2 + V(R)]\Psi_0(R) = E_0\Psi_0(R) \tag{1}$$

where $R \equiv \{\mathbf{r}_i \mid i = 1, \ldots, N\}$ and the $\mathbf{r}_i$ are the positions of the individual atoms. V(R) represents the interaction potential between the atoms in the system. The term $[-\nabla^2 + V(R)]$ is the Hamiltonian, H, for the system and is written here in dimensionless form. Knowledge of the physical relationships between the atoms can be built into a parametrized mathematical form for a trial wavefunction, $\Psi_T(R)$, an approximation to $\Psi_0(R)$. The variational energy can be minimized with respect to the parameters through the Monte Carlo evaluation of the expectation value of the ground energy, E_0. This technique is referred to as variational Monte Carlo (VMC). Approaches where the simulation results are subject only to statistical uncertainties are referred to as Quantum Monte Carlo (QMC) methods. In the Green's function Monte Carlo (GFMC) method the integral transform of the Schrödinger equation, (1), is iterated by performing a random walk in the configuration space of the N atoms to yield an asymptotically exact solution. Such a solution can also be obtained by sampling a short time Green's function followed by an extrapolation of the results to account for the time step errors introduced by the approximation. This technique is known as diffusion Monte Carlo (DMC). Finally, finite temperature systems may be studied by considering the Bloch Equation:

$$[-\nabla^2 + V(R) + \partial/\partial t]\varrho_B(R,t) = 0, \tag{2}$$

where $\varrho_B(R,t)$ is the many-body density matrix. Path integrals [12, 16] Monte Carlo (PIMC) simulations can be performed and for small enough temperature intervals, the density matrix can be compared to the ground state eigenfunction from (1).

In the following sections, the results of applying the techniques outlined above to the simulation of solid ^{4}He will be described.

2 Variational Monte Carlo Methods

Given a trial wavefunction, $\Psi_T(R)$, an estimator for the variational energy,

$$E_T = \frac{\langle\Psi_T|H|\Psi_T\rangle}{\langle\Psi_T|\Psi_T\rangle} \geq E_0, \tag{3}$$

is an upper bound to the true ground state energy E_0 of the system. The lowest variational energy is obtained through a minimization process with respect to the parameters in $\Psi_T(R)$. In the coordinate representation (3) becomes:

$$E_T = \frac{\int dR \Psi_T^*(R) H \Psi_T(R)}{\int dR |\Psi_T(R)|^2} = \int dR \pi(R) E_L(R), \tag{4}$$

where $dR \equiv d^3\mathbf{r}_1 d^3 \ldots \mathbf{r}_N$. The last term of the above equation, $E_L(R)$, the local energy, is obtained by multiplying and dividing the numerator of (3) by $\Psi_T(R)$,

$$E_L(R) = \frac{H\Psi_T(R)}{\Psi_T(R)}, \tag{5}$$

and $\pi(R)$ is a normalized probability distribution function,

$$\pi(R) = \frac{|\Psi_T(R)|^2}{\int dR |\Psi_T(R)|^2}. \tag{6}$$

2.1 Trial Wavefunctions

As point out by Feenberg [11] a plausible general form for the exact ground-state wavefunction of a system of N interacting bosons is

$$\Psi(R) = \prod_{i<j} f_2(r_{ij}) \prod_{i<j<k} f_3(i,j,k) \prod_{i<j<k<l} f_4(i,j,k,l) \cdots \tag{7}$$

$$= \exp\frac{1}{2}\left[\sum_{i<j} u(r_{ij}) + \sum_{i<j<k} u_3(i,j,k) + \sum_{i<j<k<l} u_4(i,j,k,l)\right] \cdots \tag{8}$$

In the liquid phase, the simplest variational function, the so called Bijl-Dingle-Jastrow or Jastrow trial function, considers only a single term of the above expression: $u_2(r_{ij})$. The first computer simulation for a system of helium atoms was performed by McMillan [22] using $u_2(r) = b/r^5$ and reasonable results were obtained. A better approximation to the variational wavefunction which included three-body correlations [28], $u_3(i,j,k)$, led to an improvement of about 10% in the estimated values of the energies.

For the solid phase, the usual approach was to write the trial wavefunction as

$$\psi_{T sol}(R) = \prod_{i<j} f_2(r_{ij}) \Phi(R), \tag{9}$$

where $\Phi(R)$ is a model function, ideally, a permanent of localized single particle orbitals. However, since the effect of the quantum statistics on the energy is minor, $\Phi(R)$ is left unsymmetrized. Gaussian orbitals were used in simulations performed by Hansen and Levesque [14] with reasonable results. The inclusion of triplet correlations, $f_3(i,j,k)$, in the trial wavefunction lead to an improvement of about 15% in the simulations results [32]. The introduction of higher-order correlations in a trial wavefunction has become feasible by introducing the shadow wavefunction [30], discussed more fully below.

The functional form of correlation factors in a trial wavefunction can be fully optimized. The idea is to write the correlation factors as a sum of the elements of a basis set [31]. For the two-body correlation factor,

$$f(r) = \sum_{n=1}^{M} c_n f_n(r), \tag{10}$$

where the f_n are functions of the basis set and the c_n are variational parameters. If the basis is well chosen, a small value of M is sufficient to recover all the energy associated with the correlation under consideration. For (10), a very suitable basis is the one obtained by solving for the lowest M energy states of the Schrödinger-like equation involving a pair of helium atoms,

$$\left(-\frac{\hbar}{2m}\nabla^2 + V(r)\right) f_n(r) = \lambda_n f_n(r). \tag{11}$$

The boundary conditions are such that at a distance d, chosen as a cutoff or eventually as a variational parameter, the functions f_n go smoothly to 1 or to a function that gives the correct long range behavior of the system. One of the advantages of this method is to automatically obtain an optimal correlation at small values of r. Since, the wavefunction is small when $r \to 0$, it is difficult to sample this very important region of configuration space. Thus, the usual Monte Carlo optimization of the trial function, does not perform well at small r.

2.2 Monte Carlo Techniques

The simulation starts by sampling the normalized probability distribution $\pi(R)$ of (6), *i. e.*, by constructing a sequence of points $\{R_i | i = 1, \ldots, M\}$ in the configuration space. More formally we require [19] that R_i belong to the sequence with probability given by

$$Pr\{R_i\} = \int_{\Omega} dR \pi(R) \tag{12}$$

for any $\Omega \subset \Omega_0$ of the sample space Ω_0. The sampling in most cases is performed using the Metropolis [23] algorithm.

If we consider M independent samples, the variational energy is estimated as

$$E_M = \frac{1}{M} \sum_{i=1}^{M} E_L(R_i). \tag{13}$$

In the limit of large M we have $E_M \to E_T$. This energy is obtained without any uncontrolled approximations or nonconvergence for any form of the wavefunction and is subject only to statistical uncertainties of $\mathcal{O}(M^{-1/2})$. The statistical error is easily estimated. Variance reduction techniques, *e. g.* importance sampling, can be used to reduce the multiplicative factor that appears in the calculation of the error. Other properties of the system can also be readily estimated.

2.3 Variational Results on Solid ^{4}He

The earliest variational calculations could not differentiate between a crystal with fcc packing and one with hcp order. Chester [10] showed that a Jastrow wavefuntion, $\Psi_T(R) = \prod_{i<j} f(r_{ij})$, supports a Bose-Einstein condensate. However, this pair product form without the localization provided by $\phi(r)$, as in (9), only crystallizes at a very high density [15]. Therefore the use of gaussian one-body orbitals, $\phi(r)$, was introduced. This, however, precluded the observation of a Bose-Einstein condensate.

Recently, Vitiello considered in great detail the question of the ground-state structure of solid helium using the most recent he-he potential of interaction. Performing careful variational calculations and employing reweighting, he was able to show that the hcp order is favored in the ^{4}He system [29].

2.4 Shadow Wavefunction Calculations

The construction of trial functions based on the inclusion of auxiliary variables, "shadow particles", is a very successful approach within the variational methods. These trial functions are particular representations of the Feenberg form [11], where one is able to introduce tractable correlations up to the number of particles

$$\Psi_{Sh}(R) = \prod_{i<j} f(r_{ij}) \prod_{i<j<k} f^{(3)}_{ijk} \cdots \prod_{i<j\ldots<w} f^{(N)}_{ij\ldots w}. \tag{14}$$

The variational shadow wavefunction is defined in terms of an integral over auxiliary variables $S \equiv \{\mathbf{s}_i | i = 1, \ldots, N\}$ in the whole space

$$\Psi_{Sh}(R) = \int \Xi(R, S) dS, \tag{15}$$

where Ξ is a function that includes a factor of the Jastrow form dependent solely on the configuration space coordinates R, a Gaussian coupling between the space variables $\mathbf{r}_i$ and the auxiliary variables $\mathbf{s}_i$, and a term of the Jastrow form that correlates the $\mathbf{s}_i$ among themselves:

$$\Xi(R, S) = \exp\left(-\sum_{i<j} \frac{1}{2}\left(\frac{b}{r_{ij}}\right)^5 - \sum_i C|\mathbf{r}_i - \mathbf{s}_i|^2 - \sum_{i<j} \gamma V(\delta s_{ij}) \right). \tag{16}$$

In this formulation $\Psi_{Sh}(R)$ depend on the He-He interacting potential and four variational parameters: b, C, γ and δ. Since the auxiliary variables, due to the last term of Eq. (16), are isomorphic to the coordinates of a system of particles interacting through V, we call the auxiliary variables shadow particles.

A trial wavefunction that can correlate all the atoms in the system is important by itself. In addition, there are strong physical motivations to deal with such a class of variational wavefunctions: Feynman's path integrals in imaginary time and justifications from projection methods.

Shadow wavefunctions have enabled the investigation of disorder phenomena in solid ^{4}He such as vacancies [24] or the interfacial region between a solid and a liquid at coexistence [25] by variational calculations. This is possible because with the shadow wavefunction approach both the fluid phase and solid phases can be described, without the introduction of single particle orbitals. Moreover the Bose symmetry is manifestly maintained and so relaxation around non-localized defects are allowed.

3 Green's Function Monte Carlo

If the potential energy in (1) is bounded from below $V(R) \geq -V_0$, (1) can be rewritten as:

$$[-\nabla^2 + V(R) + V_0]\Psi_0(R) = (E_0 + V_0)\Psi_0(R). \quad (17)$$

A Green's function,

$$[-\nabla^2 + V(R) + V_0]G(R, R_0) = \delta(R - R_0) \quad (18)$$

can be derived with the same boundary conditions as $\Psi_0(R)$ and used to transform (17) into an integral equation:

$$\Psi_0(R) = (E_0 + V_0)\int G(R, R')\Psi_0(R')dR'. \quad (19)$$

Since the ground-state wavefunction and Green's function for a Bose system are non-negative; the ground state wavefunction and approximations to it may be treated as probability distribution functions. The Green's function may also be used as a distribution function for R conditional on the previous position R'. The integral version of the Schrödinger equation, (19), is solved by a Neumann iteration starting with a zeroth order approximation, such as a trial wavefunction optimized in a variational calculation. A population of points $\{R'\}$ is sampled from $\Psi_T(R')$ and a new set of points $\{R\}$ is sampled from $(E_t + V_0)G(R, R')$ where E_t is an approximation to the ground-state energy. As this process is repeated, at the nth iteration the set of points $\{R'\}$ has been sampled from $\psi^{(n)}$ and the next generation of points, n+1, is sampled from:

$$\psi^{(n+1)}(R) = (E_t + V_0)\int G(R, R')\psi^{(n)}(R')dR'. \quad (20)$$

Equation (20) defines one step of a random walk whose asymptotic distribution is $\Psi_0(R)$. Since the simulation system is composed of N atoms with periodic boundary conditions, the Schrödinger equation has a discrete spectrum and the iterations are guaranteed to converge.

The procedure may be made computationally more efficient and the variance reduced by employing an importance sampling transformation [18]. Let $\Psi_T(R)$ be a trial wavefunction which may be the same as $\psi^{(0)}(R)$ and $\bar{\Psi}(R) = \Psi_T(R)\Psi(R)$, then (19) becomes

$$\bar{\Psi}(R) = (E + V_0)\int [\Psi_T(R)G(R, R')/\Psi_T(R)]\bar{\Psi}(R')dR'. \quad (21)$$

The sequence of functions obtained by iteration of the integral equation converges to $\Psi_T(R)\Psi_0(R)$ and E_t is chosen such that the random walk is stable.

Unfortunately, the Green's function, (18), is not known analytically owing to the complexity of the boundary conditions. However, to implement the algorithm represented by (20) or (21), it is not necessary to know the full Green's function;

it is sufficient to develop a method to sample configurations from $G(R,R')$. The Green's function may be written as,

$$G(R,R') = \int_0^\infty G(R,R',\tau)d\tau \tag{22}$$

$G(R,R',\tau)$ is the Green's function for a Bloch equation, (2),

$$(H + \partial/\partial\tau)G(R,R',\tau) = \delta(R - R')\delta(\tau). \tag{23}$$

For a given configuration, R_0, a finite domain, $D(R_0)$, may be chosen such that the potential of interaction, V(R), is bounded from above within the domain by the constant, $U(R_0)$. A domain Green's function may then be defined:

$$[-\nabla^2 + U(R_1) + \partial/\partial\tau]G_U(R_1,R_0,\tau) = \delta(R_1 - R_0)\delta(\tau) \tag{24}$$

subject to the boundary condition that $G_U(R_1,R_0,\tau) = 0$ whenever R_1 is on the boundary or outside of $D(R_0)$. Physically, (24) represents a diffusion process of a particle in a domain with a constant absorption rate and a perfectly absorbing boundary. Multiplying (18) by $G_U(R_1,R_0,\tau)$ and (24) by $G(R,R_1,\tau)$, integrating both equations over R_1 and subtracting the two resulting equations yields:

$$\begin{aligned} G(R,R_0,\tau) &= G_U(R,R_0,\tau) \\ &+ \int_{\partial D(R_0)} \left(\frac{\partial G_U(R_1,R_0,\tau-\tau_0)}{\partial n}\right) G(R,R_1,\tau_0)dR_1 \\ &+ \int_{D(R_0)} (U(R_0) - V(R_1))G_U(R_1,R_0,\tau-\tau_0)G(R,R_1,\tau_0)dR_1 \end{aligned} \tag{25}$$

Equation (26) is a linear integral equation for $G(R,R_0,\tau)$ in terms of $G_U(R,R_0,\tau)$. In the second term of the right hand side of (26), the boundary condition for $G_U(R_1,R_0,\tau)$ has been used to convert a volume integral into a surface integral over $\partial D(R_0)$ and the derivative normal to the domain's surface, $\partial/\partial n$ has been introduced. The domain, $D(R_0)$ may be chosen in any convenient way; in particular, as the Cartesian product of three-dimensional spheres or cubes centered at R_0. $G_U(R,R_0,\tau)$ is known analytically and may be interpreted as a conditional probability distribution function. Thus, points $\{R\}$ may be sampled by a random walk governed by (26) for any given set $\{R_0\}$.

An asymptotically unbiased estimator for the energy is given by

$$E_m = \frac{\int \Psi^n(R)H\Psi_T(R)dR}{\int \Psi^n(R)\Psi_T(R)dR}. \tag{26}$$

Except for statistical sampling and convergence errors, (26) is an exact estimator for the ground state energy. For other properties of the physical system, F(R), that do not commute with the Hamiltonian, a "mixed" estimator may be defined as

$$\langle F\rangle_m = \frac{\int \Psi^n(R)F(R)\Psi_T(R)dR}{\int \Psi^n(R)\Psi_T(R)dR}. \tag{27}$$

If the trial wavefunction $\Psi_T(R)$ is "close" to the actual ground state wavefunction, $\Psi_T(R) = \Psi_0(R) + \delta\psi(R)$, then a linear extrapolation may be used to estimate the exact value to within an order δ^2:

$$\langle F \rangle_x = 2\langle F \rangle_m - \langle F \rangle_T \tag{28}$$

where $\langle F \rangle_T$ is the variational value calculated with $\Psi_T(R)$. It was shown [34] that this extrapolation process gave the same expected value as the random walk based on the "forward walking" algorithm but with much smaller statistical errors. The extrapolated value was also shown to be independent of the trial wavefunction used.

As in variational calculations, the result of the GFMC simulations is a wavefunction represented as an ensemble of configurations of atomic positions. Through (27) and (28), the Bose-Einstein condensate fraction can be obtained for the helium system. The fraction of particles in the zero-momentum state is given by the asymptotic limit of $n(\mathbf{r})$,

$$n_0 = \lim_{r \to \infty} n(\mathbf{r}). \tag{29}$$

The one-body density matrix, $n(\mathbf{r})$, is a measure of the change in the wavefunction for given displacement r and is the fourier transform of the momentum distribution, $n(\mathbf{k})$:

$$\begin{aligned} n(\mathbf{r}) &= \int e^{\mathbf{i}\mathbf{k}\cdot\mathbf{r}} n(\mathbf{k}) d\mathbf{k} \\ &= \left\langle \frac{\Psi(r_1, r_2, \cdots, r_i + r, \cdots, r_n)}{\Psi(r_1, r_2, \cdots, r_i, \cdots, r_n)} \right\rangle . \end{aligned} \tag{30}$$

The first calculation of the Bose-Einstein condensate in solid ^{4}He using the GFMC method [7] involved the trial wavefunction of Eq. (9) and the Lennard-Jones potential of interaction [4]. It was of course recognized that the form of the trial wavefunction that was used as an importance function might bias the results and, not surprisingly, the condensate fraction was less than 1%. Additional calculations [34], showed that the Lennard Jones potential itself was inadequate to describe the helium system. A further investigation of the Bose-Einstein condensate concluded that Ceperley, *et. al.* [6] observed a condensate because they used a translationally invariant wavefunction for the importance function. The GFMC simulations were repeated using a more realistic form for the potential of interaction, the HFDHE2 potential [2] and improved wave-functions [17]. However, no condensate fraction within statistical error was observed in the solid phase [35, 36].

In a variational calculation using a shadow wavefunction, when the density of a system of helium atoms reaches the appropriate value, a state with translationally broken symmetry is spontaneously produced. Thus, it was natural to introduce shadow wavefunctions as importance functions in GFMC calculations. In order to perform these calculations, Whitlock and Vitiello [33] made an extension to the GFMC method such that the shadow degrees of freedom where updated using the Metropolis algorithm according to the probability distribution of (16). Despite good results for some of the properties of the system, the variance of the calculation did not encourage further attempts to compute the condensate fraction in the solid phase. However the idea of using the shadow wavefunction ideas in a QMC method seems promising.

3.1 Diffusion Monte Carlo

The time dependent Schrödinger equation in imaginary time $t \to it/\hbar$,

$$\frac{\partial\psi(R,t)}{\partial t} = -(-\nabla^2 + V(R) - E_t)\psi(R,t), \tag{31}$$

is equivalent to the classical diffusion equation with sources represented by $V(R)$. In (31), the Hamiltonian H has been written as the sum of the kinetic energy T, $-\nabla^2$, plus the potential energy $V(R)$ displaced by a trial energy E_t, which does not change the description of the state of the system.

In a short time approximation, the Green's function for (31) can be written to $\mathcal{O}(t)$ as,

$$G(R,R',\delta t) \approx \langle R|\mathrm{e}^{-V(R)\delta t/2}\mathrm{e}^{-T\delta t}\mathrm{e}^{-V(R')\delta t/2}\mathrm{e}^{E_t\delta t}|R'\rangle. \tag{32}$$

It is possible to rewrite the above expression as the product of a rate term,

$$w(R,R',\delta t) = \exp(-(V(R)+V(R'))\frac{\delta t}{2} + E_t\delta t), \tag{33}$$

times a propagator, identified as the Green's function for ordinary diffusion,

$$G_d(R,R',\delta t) = \langle R|\mathrm{e}^{-T\delta t}|R'\rangle = (4\pi\delta t)^{-3N/2}\exp\left(-\frac{(R-R')^2}{4\delta t}\right). \tag{34}$$

In a simulation, for each R' in a given set of configurations, a new R is easily sampled from G_d and weighted by $w(R,R',\delta t)$. By repeating these steps and performing a suitable extrapolation to $t \to 0$, the results will yield an estimate of the ground-state energy if $E_t \approx E_0$. This is shown by writing the formal solution of the time dependent Schrödinger equation as

$$\psi(R,t) = \sum_i \varphi_i(R)\mathrm{e}^{-i(E_i-E_t)t/\hbar}, \tag{35}$$

where the $\varphi_i(R)$ are an orthogonal basis set.

The method presented so far is very inefficient due to the branching process and because the random walk may explore unimportant regions of the configuration space. Here again an importance sampling transformation as in (21) allows the simulations to converge faster and more efficiently. If (31) is multiplied by a trial wavefunction ψ_T, it can be written in the coordinate representation as

$$\frac{\partial\bar{\psi}(R,t)}{\partial t} = -\left(-\nabla^2 + \nabla\cdot\mathbf{F}(\mathbf{R}) + \mathbf{F}(\mathbf{R})\cdot\nabla - (E_t - E_L(R)\right)\bar{\psi}(R,t), \tag{36}$$

where $\bar{\psi}(R,t) = \psi_T(R)\psi(R,t)$, $E_L(R)$ is the local energy and $\mathbf{F}(R) = 2\nabla\ln\psi_T(R)$. If we compare the above expression with equation (31), it still includes a branching process, given by $\mathcal{V} = E_L(R) - E_t$. The diffusion process has a superimposed drift velocity given by the two last terms of the expression, $-\nabla^2 + \nabla \cdot \mathbf{F}(R) + \mathbf{F}(R) \cdot \nabla$.

By taking the short time approximation, as before we can write:

$$\bar{G}(R, R', \delta t) = \bar{W}(R, R', \delta t)\bar{G}_d(R, R', \delta t), \tag{37}$$

where

$$\bar{W}(R, R', \delta t) = \exp\left(-(E_L(R) + E_L(R'))\frac{\delta t}{2} + E_t \delta t \right), \tag{38}$$

and

$$\bar{G}_d(R, R', \delta t) = (4\pi\delta t)^{-3N/2} \exp\left(-\frac{(R - R' - \delta t \mathbf{F}(R))^2}{4\delta t} \right). \tag{39}$$

Simulations that include importance sampling converge to $\psi_T \psi$. Instead of computing $V(R)$, now we calculate E_L which approaches a constant as $\psi_T(R)$ goes to the true eigenfunction of the system. This is important since the simulations become much more stable. Moreover, the drift guides the random walk to the important regions of the configuration space.

4 Path Integral Monte Carlo

All static and, in principle, dynamic properties of a many-body quantum system, such as ^{4}He, at thermal equilibrium, may be obtained from the density matrix, $\varrho(R, R', \beta)$, the solution to the Bloch equation, (2). β represents an inverse temperature or "imaginary time", $\beta = 1/kT$. The solution to the Bloch equation can be written in the coordinate representation as:

$$\varrho(R, R', \beta) =< R|e^{-\beta H}|R' > \tag{40}$$

For distinguishable particles, the density matrix is non-negative for all values of its arguments and can be interpreted as a probability distribution function. If two density matrices are convoluted together, a density matrix at a lower temperature results:

$$< R|e^{-(\beta_1+\beta_2)H}|R' >= \int < R|e^{-\beta_1 H}|R'' >< R''|e^{-\beta_2 H}|R' > dR''. \tag{41}$$

The integral over R'' may be evaluated using a generalization of the Metropolis sampling algorithm [26, 8]. Starting at a sufficiently high temperature, the density matrix may be accurately written as an expansion in one and two-body density matrices. Then, multiple convolutions can be performed to reduce the temperature to near absolute zero.

The density matrix for a boson system such as ^{4}He is obtained from the distinguishable particle density matrix by using the permutation operator to project out the symmetric component,

$$< R|e^{-(\beta H)}|R' >_B = \frac{\sum_{\wp} < \wp R|e^{-(\beta H)}|R' >}{N!} \tag{42}$$

The sum over permutations is performed by a Monte Carlo technique.

To calculate the momentum distribution requires obtaining the off-diagonal parts of the density matrix. In one method, an atom is displaced off the diagonal by a distance $\mathbf{r}$ while the other atoms and permutation are held fixed while n($\mathbf{r}$), (30), is computed. This method is very accurate at small r. In a second method, one atom is again off the diagonal, but the distance between the two ends of the path for that atom is allowed to vary. This allows the calculation of $n(\mathbf{r})$ at large $\mathbf{r}$ [9].

Ceperley and Bernu [5] have found that the superfluid density observed in PIMC simulations of solid ^{4}He are strongly affected by the size of the system simulated. A 48 atom system exhibits a 1.2% superfluid density at 55 bars pressure while a 180 atom systems has zero superfluid density. They conclude that the phenomenon observed by Kim and Chan can not be explained by vacancies or interstitials in the equilibrium bulk ^{4}He system.

4.1 Path Integral Ground State Calculations

Ground state expectations values at finite temperatures can be efficiently calculated by using a path integral ground state Monte Carlo method [27]. The integral equation in imaginary time equivalent to the Schrödinger equation is

$$\psi(R,t) = \int G(R,R',t-t_0)\psi(R',t_0)dR'. \tag{43}$$

In the above equation, $G(R,R',t)$ is the propagator of (23). As was seen in the previous section, this propagator is viewed as density matrix operator corresponding to an inverse temperature β and simulated by path integrals.

The difference in the present method compared to PIMC is that a truncated path on a trial wavefunction is considered instead of periodic boundary conditions in imaginary time as the trace of $G(R,R',t)$ requires. Since the ground state eigenfunction can be obtained by filtering a suitable trial function ψ_T

$$\psi_0(R) = \lim_{t\to\infty}\psi(R,t) = \lim_{t\to\infty}\int G(R,R',t)\psi_T(R')dR', \tag{44}$$

the ground sate expectation value of any operator can be written as

$$\langle\mathcal{O}\rangle = \frac{\langle\psi_t|G(t)\mathcal{O}G(t')|\psi_T\rangle}{\langle\psi_t|G(t)G(t')\psi_T\rangle}. \tag{45}$$

If the convolution of the density matrix of (41) is divided into N time steps, $\beta/N = t/N = \delta t$,

$$G(R,R',t) = \int dR_1 dR_2\cdots dR_{N-1}\rho(R,R_1,\delta t)\rho(R_1,R_2,\delta t)\rho(R_{N-1},R',\delta t) \tag{46}$$

and substituted in (45), we obtain

$$\langle\mathcal{O}\rangle = \frac{\int\prod_{i=0}^{N} dR_i\mathcal{O}(R_M)\psi_T(R_0)\left(\prod_{i=0}^{N-1}\rho(R_i,R_{i+1},\delta t)\right)\psi_T(R_N)}{\int\prod_{i=0}^{N} dR_i\psi_T(R_0)\left(\prod_{i=0}^{N-1}\rho(R_i,R_{i+1},\delta t)\right)\psi_T(R_N)}, \tag{47}$$

where $R_0 = R$, $R_M = R'$ and R_M is an internal time slice. For a converged calculation, if the operator is placed on the extreme edges of the path one gets a mixed estimator. If R_M is in the middle, the exact expectation value of the ground state is obtained. The paths are sampled using the Metropolis algorithm. Samples that do not include coordinates of the trial wavefunction are performed as in PIMC.

Galli and Reatto [13] have employed this formalism using the shadow wavefunction as $\Psi_T(R)$ to study confined solid ^{4}He. Their model system contains a large static spherical object that uses a purely repulsive potential to prevent the helium atoms from reaching the center of the simulation cell. As a consequence of the periodic boundary conditions this correspond to a static lattice of hard core spheres. They observe that the freezing pressure increases to about 38 atm. This behavior is comparable to that found in experiments of ^{4}He confined in vycor [21]. In addition they observe that the disorder induced by the mismatch between the ^{4}He crystalline structure and the static hard spheres induces delocalization. This is a necessary condition to have off-diagonal long range order in the system. Also, the presence of a Bose-Einstein condensate requires delocalization of the atoms. These results could be relevant in explaining the observation of a supersolid phase for ^{4}He in vycor [21].

However, to date, no quantum Monte Carlo studies have observed delocalization or Bose-Einstein condensation in the pure bulk solid ^{4}He.

References

1. A.F. Andreev and L.M. Lifshitz: *Soviet Phys. JETP* **29** (1969) 1107
2. R.A. Aziz, V.P.S. Nain, J.S. Carley, W.L. Taylor, and G.T. McConville: An accurate intermolecular potential for helium. *J. Chem. Phys.* **70** (1979) 4330–4342
3. K.H. Bennemann and J.B. Ketterson, eds.: The Physics of Liquid and Solid Helium. Wiley, New York (1976)
4. J. de Boer and A. Michels: Contribution to the Quantum Mechanical Theory of the Equation of State and the Law of Corresponding States. Determination of the Law of Force of Helium. *Physica (Utrecht)* **5** (1938) 945–957
5. D.M. Ceperley and B. Bernu: Ring Exchanges and the Supersolid Phase of ^{4}He. *Phys. Rev. Lett.* **93** (2004) 155303-1–4
6. D.M. Ceperley, G.V. Chester, and M.H. Kalos: Monte Carlo study of the ground state of bosons interacting with Yukawa potentials. *Phys. Rev. B* **17** (1978) 1070–1081
7. D.M. Ceperley, G.V. Chester, M.H. Kalos, and P.A. Whitlock: Monte Carlo Studies of Crystalline Helium. *Journal de Physique* **39 Colloque C6** (1978) 1298–1304
8. D.M. Ceperley and E.L. Pollock: Path-integral computation of the low-temperature properties of liquid ^{4}He. *Phys. Rev. Lett.* **56** (1986) 351–354
9. D.M. Ceperley and E.L. Pollock: The momentum distribution of normal and superfluid liquid ^{4}He. *Can. J. Phys.* **65** (1987) 1416
10. G.V. Chester: Speculations on Bose-Einstein Condensation and Quantum Crystals. *Phys. Rev. A* **2** (1970) 256–258
11. E.Feenberg: Ground state of an interacting boson system. *Ann. Phys. (N.Y.)* **84** (1974) 128
12. R.P. Feynman: The lambda-Transistion in Liquid Helium. *Phys. Rev.* **90** (1953) 1116–1117

13. D.E. Galli and L.Reatto: The shadow path integral ground state method: study of confined solid ^{4}He. *J.Low Temp. Phys.* **136** (2004) 343–359
14. J.P. Hansen and D. Levesque: Ground state of solid helium–4 and –3. *Phys. Rev.* **165** (1968) 293–299
15. J.P. Hansen and E.L. Pollock: Ground-State Properties of Solid Helium-4 and -3. *Phys. Rev. A* **5** (1972) 2651–2665
16. M. Kac: Probability and Related Topics in Physical Science. Interscience, New York (1959)
17. M.H. Kalos, M.A. Lee, P.A. Whitlock, and G.V. Chester: Modern potentials and the properties of condensed ^{4}He. *Phys. Rev. B* **24** (1981) 115–130
18. M.H. Kalos, D.Levesque, and L.Verlet: Helium at zero temperature with hard-sphere and other forces. *Phys. Rev.A* **9** (1974) 2178–2195
19. M.H. Kalos and P.A. Whitlock: *Monte Carlo Methods Volume I: Basics.* John Wiley, New York (1986).
20. E. Kim and M.H.W. Chan: Observations of Superflow in Solid Helium. *Science* **305** (2004) 1941–1944
21. E.Kim and M.H.W. Chan: Probable observation of a supersolid helium phase. *Nature* **427** (2004) 225–227
22. W.L. McMillan: Ground state of liquid ^{4}He. *Phys. Rev.* **138** (1965) A442–A451
23. N.Metropolis, A.W. Rosenbluth, M.N. Rosenbluth, A.H. Teller, and E.Teller: Equation of state calculations by fast computing machines. *J.Chem. Phys.* **21** (1953) 1087–1092
24. F.Pederiva, G.V. Chester, S.Fantoni, and L.Reatto: Variational study of vacancies in solid ^{4}He with shadow wave functions. *Phys. Rev. B* **56** (1997) 5909-5917
25. F.Pederiva, A.Ferrante, S.Fantoni, and L.Reatto: Homogeneous nucleation of crystalline order in superdense liquid ^{4}He. *Phys. Rev. B* **52** (1995) 7564–7571
26. E.L. Pollock and D.M. Ceperley: Simulation of quantum many-body systems by path-integral methods. *Phys. Rev. B* **30** (1984) 2555–1568.
27. A.Sarsa, K.E. Schmidt, and W.R. Magro: A path integral ground state method. *J.Chem. Phys.* **113** (2000) 1366–1371
28. K.E. Schmidt, M.H. Kalos, M.A. Lee, and G.V. Chester: Variational Monte Carlo calculations of liquid ^{4}He with triplet correlations. *Phys. Rev. Lett.* **45** (1980) 573–576
29. S.A. Vitiello: Relative stability of hcp and fcc crystalline structures of ^{4}He, *Phys. Rev. B* **65** (2002) 214516–214520
30. S.A. Vitiello, K. Runge, and M.H. Kalos: Variational calculations for solid and liquid ^{4}He with a "shadow" wavefunction. *Phys. Rev. Lett.* **60** (1988) 1970–1972
31. S.A. Vitiello and K.E. Schmidt: Optimization of ^{4}He wave functions for the liquid and solid phases. *Phys. Rev. B* **46** (1992) 5442–5447
32. S.A. Vitiello and K.E. Schmidt: Variational Methods for ^{4}He using a modern he-he potential. *Phys. Rev. B* **60** (1999) 12342–12348
33. S.A. Vitiello and P.A. Whitlock: Green's function Monte Carlo algorithm for the solution of the Schrödinger equation with the shadow wave function. *Phys. Rev. B* **44** (1991) 7373–7377
34. P.A. Whitlock, D.M. Ceperley, G.V. Chester, and M.H. Kalos: Properties of liquid and solid ^{4}He. *Phys. Rev. B* **19** (1979) 5598–5633
35. P.A. Whitlock and R.M. Panoff: One-body density matrix and the momentum density in ^{4}He and ^{3}He. Proc. of the 1984 Workshop on High-Energy Excitations in Condensed Matter, ed. R.N. Silver. LA-10227-C Vol II (1984)
36. P.A. Whitlock and R.M. Panoff: Accurate momentum distributions from computations on ^{3}He and ^{4}He. *Can. J. Phys.* **65** (1987) 1409–1415

Parallel Treatment of General Sparse Matrices

Zahari Zlatev

National Environmental Research Institute,
Frederiksborgvej 399, P.O. Box 358, DK-4000 Roskilde, Denmark

Abstract. The discretization of large mathematical models, which arise in many fields of science and engineering, leads to the solution of long sequences of systems of linear algebraic equations. These systems are often very large (up to many millions of equations). Therefore, it is desirable to achieve high performance when such systems (with coefficient matrices the order of which is greater than or equal to one million) are treated on modern high-speed computers. In order to achieve high performance, it is absolutely necessary to exploit efficiently:

– the sparsity of the coefficient matrices of these systems,
– the caches in the multi-hierarchical memory of the modern high-speed computers,

and

– the power of the modern parallel architectures.

An algorithm, in which those three tasks are successfully resolved, has been developed and tested. This algorithm is described and many results obtained by using this algorithm are presented and discussed. Some comparisons with the well-known code *SuperLU* for the treatment of sparse matrices are presented. The results of these comparisons show clearly that the option of the new code, in which small non-zero elements are dropped, is much faster than *SuperLU*. Some plans for further improvements are discussed in the end of the paper.

1 Treatment of General Sparse Matrices

The discretization of large-scale mathematical models arising in different fields of science and engineering leads to the solution of very large systems of linear algebraic equations. It will be assumed here that the Gaussian elimination is used in the solution of the systems of linear algebraic equations (but many of the ideas can also be applied when other methods are used, as, for example the Householder method or the Givens method). By the use of the Gaussian elimination the original matrix A is decomposed into two triangular factors L and U (some decompositions of A are to be obtained also when other methods are used).

Sparse matrix techniques have to be used during the Gaussian elimination, because the coefficient matrices of this systems are, as mentioned above, very large. The use of a sparse matrix algorithm means that one stores only the non-zero elements of the matrix and works only with the non-zero elements. The number NZ of the non-zero elements is very small in the beginning of the

I. Lirkov, S. Margenov, and J. Waśniewski (Eds.): LSSC 2005, LNCS 3743, pp. 53–64, 2006.

computations ($NZ \leq kN$, where N is the order of the matrix, which is assumed to be greater than 10^6 in this paper, while k is a small integer; as a rule $k \leq 20$). However, non-zero elements are created during the computations (in positions where originally zero elements were located). The number of these new non-zero elements, fill-ins, can be very large. Therefore, special devices are to be used in order to keep the number of fill-ins small, i.e. in the efforts to preserve better the sparsity of the treated matrix during the computations. The pivotal strategy and different dropping criteria are the major devices which can successfully be used in order to reduce the number of fill-ins.

It should be pointed out here that in the pivotal strategies for sparse matrices not only an attempt is made to keep the number of fill-ins small, but an attempt to preserve the stability of the computations is also carried out. These two requirements, the preservation of the stability and keeping the sparsity are working in opposite direction. Therefore, the pivotal strategy for general sparse matrices is always a compromise between these two requirements.

A special parameter $RELTOL$, the relative drop-tolerance, is used to decide which elements are to be dropped. If $RELTOL = 0$, then no non-zero element is dropped. Positive values of $RELTOL$, $0 < RELTOL < 1$, will result in dropping, during the computation of the factors L and U by the Gaussian elimination, all elements a_{ij} that are in absolute value smaller than the product of $RELTOL$ and the largest in absolute value element in row i. The use of positive values of $RELTOL$ may result in an efficient preservation of the sparsity, but the factors L and U calculated in this way are normally inaccurate, which leads to an inaccurate solution of the system of linear algebraic equations. Therefore, the obtained by using dropping factorization of matrix A should be used as a preconditioner in some iterative method.

The pivotal strategy and the dropping devices will not be further discussed in this paper. It will be assumed that

- some good pivotal strategy,
- some efficient device for dropping small elements,

and

- some fast and accurate preconditioned iterative method

have been selected. More details about the choice of a pivotal strategy, a dropping device and a preconditioned iterative method can be found in Zlatev [9, 10] and [12].

2 Cache Problems for Very Large Sparse Matrices

It will be assumed in the remaining part of this paper that the matrices that are to be handled are *very large*, which mean that the order N is greater than 10^6. Cache problems arise when the matrices are very large, because their structure is as a rule very irregular. Therefore, it is worthwhile to develop some methods by which the problems related to the use of the cache memory can, at least, be reduced. One of the methods used in Zlatev [12], implemented in the code *y12m3*, can easily be modified for efficient runs of modern computers with multi-hierarchical caches. The main ideas on which these modification are based will be discussed in this section.

In *y12m3* the matrix is reordered (before the start of the computations) by row-column permutations, which are made to move the non-zero elements in the lower part of the matrix as close as possible to the main diagonal. This reordering allows us to divide the matrix into block-rows with approximately the same number of rows in each of them. Moreover, it is possible to specify the number q of the desired block-rows. When the reordering is prepared and the matrix is divided into the desired number q of block-rows, the Gaussian elimination is performed in three steps (which are fully described in [12]):

(a) factorizing the block-rows,

(b) producing zeros in the appropriate locations in the unfinished during the previous step rows,

and

(c) reordering again the matrix and factorizing the unfinished matrix in the lower right-hand corner or the reordered matrix.

If the reordering made before the start of the computation is a good one, then the numbers of unfinished rows that are to be treated in (b) is small and the unfinished matrix obtained after the reordering in (c) is also small. Therefore the major part of the computational work in *y12m3* consists in the treatment of the block-rows in step (a) and improvements in this step are most desirable. Such improvements were achieved in the following way. Assume that the non-zero elements are stored in a one-dimensional array ALU of length NN, where $NN \leq NZ$ (in fact, NZ must be considerably larger than NN in order to ensure some *elbow room* for fill-ins; see again Zlatev [12]). Assume also that the number of block-rows that will be used in the run is q (the block-rows contain approximately equal number of rows). Then array ALU is divided into q equal parts and the non-zero elements of each block-row are stored in one of these parts. Arrays, in which information about the non-zero elements (about row numbers, column numbers, etc.) is stored, are also to be divided, in the same way, into q parts.

When block-row i, $i = 1, 2, \ldots, q$, is executed only the data stored in a small part of array ALU, the part where the non-zero elements of block-row i are stored, participate in the computations (together with the data in the corresponding parts of the arrays in which information about the non-zero elements is stored). This means that if the number q of block-rows selected is sufficiently large, then there is a good chance that the amount of the data participating in computations (when block-row i, $i = 1, 2, \ldots, q$, is executed) is not very large and, thus, these data will stay longer in cache. However, the number of blocks should not be too large, because in such a case the work that has to be done in steps (b) and (c) see above, is also becoming larger. The numerical experiments show clearly that if the matrix is very large, then the use of a large number of block-rows leads to considerable savings in computing time.

3 Parallel Computations

It is obvious that the block-rows that are treated in step (a) of the algorithm sketched in the previous section can be run in parallel. There are also parallel tasks in steps (b) and (c).

An attempt to parallelize some parts of the preliminary reordering of the non-zero elements of the matrix as well as some parts of the preconditioned iterative method by only inserting compiler options in the code was not very successful. Numerical results will be given in the next section in order to demonstrate this fact. However, the numerical results show also that the computational work spent to reorder the matrix and to solve the system by a preconditioned iterative method is much smaller than the computational work needed in the factorization part, i.e. than the computational work in steps (a), (b), and (c). This shows that the successful parallelization of the factorization of matrix A will ensure efficiency when the number of processors used is not very large.

4 Numerical Experiments with Very Large Matrices

The numerical algorithm discussed in previous sections has been tested by using different matrices produced by one of the matrix generators described in Zlatev [12]. There are several advantages when such generators are used:

(1) one can produce arbitrarily many matrices (while the number of test-matrices in the available data-bases of general sparse matrices is limited; see, for example, the matrices given in Davis [1]),
(2) one can vary the size of the matrices,
(3) one can vary the sparsity pattern,
(4) one can vary the number of non-zero elements,
and
(5) one can vary the condition number of the matrices tested.

All experiments were carried out on SUN computers at the Danish Center for Scientific Computing (DCSC). Some experiments on IBM computers at the University of Århus gave quite similar results.

4.1 Checking the Utilization of the Cache Memory

The ability of the code *y12m3* to exploit efficiently the cache memory has been tested by performing a series of runs in which the number of block-rows is varied. Results obtained when both small and very large matrices were tested will be presented in this sub-section.

Varying the Number of Block-Rows for Small Matrices. Systems containing 512 000 linear algebraic equations are used. The number of the non-zero elements is 5 120 110 (i.e. the average number of non-zero elements per row is about 10). The number q of block-rows is varied in the range from 4 to 512.

Table 1. Computing times (measured in seconds) obtained in the solution of a system of 512000 linear algebraic equations with different values of parameter q (different numbers of block-rows). The number of non-zero elements in the coefficient matrix is 5120110. $RELTOL = 0.1$ is used.

Block-rows	ORD time	FACT time	SOLV time	TOTAL time
4	7.34	647.52	4.68	659.54
32	7.79	73.16	4.72	85.65
128	7.87	17.58	4.85	30.30
512	7.80	53.70	5.63	67.03

Some results are shown in Table 1 (the computing times are given in seconds). Computing times measured in seconds are also used in all remaining tables. If small non-zero elements are to be dropped, then this is always achieved (in the remaining part of the paper) by setting $RELTOL = 0.1$. The notation that is applied in all the tables can be explained as follows:

(A) ORD time is the sum of the time needed for the reordering of the matrix and the time needed to divide the matrix into block-rows.

(B) FACT time is the time need to obtain the approximate LU-factorization, which is used to solve the system directly if $RELTOL = 0$ or as a preconditioner if $0 < RELTOL < 1$.

(C) SOLV time is the time needed
- if $RELTOL = 0$, then to obtain a the solution directly by performing a back substitution step

or
- if $0 < RELTOL < 1$, then to calculate a starting approximation to the solution and to improve it by performing iterations until a prescribed accuracy (an accuracy requirement of 10^{-7} was actually used in all experiments) is achieved by the preconditioned iterative method chosen (the Modified Orthomin Method from Zlatev [12] was actually used in all runs reported in this paper).

(D) TOTAL time is the total computing time spent in the run.

Several conclusions can be drawn by studying the results in Table 1 (similar conclusions can be drawn by using the results of many other runs which were performed).

1. The ORD times and the SOLV times do not depend too much on the parameter q (on the number of block-rows).
2. The FACT times (and, therefore, also the TOTAL times) depend on the choice of q. For this matrix, the best result is obtained with the choice $q = 128$. Using a small number of blocks (see the results in Table 1 for $q = 4$) is very expensive. It should be mentioned here that the use of a very large number q of blocks is also inefficient.
3. When the FACT time is the best one (i.e. when $q = 128$), the sum of the ORD time and the SOLV time is comparable with the FACT time.

Table 2. Computing times (measured in seconds) obtained in the solution of a system of 16384000 linear algebraic equations with different values of parameter q (different numbers of block-rows). The number of non-zero elements in the coefficient matrix is 163840110. $RELTOL = 0.1$ is used.

Block-rows	ORD time	FACT time	SOLV time	TOTAL time
4	259	679563	176	679999
32	270	86244	178	86691
256	256	10845	183	11301
512	268	5390	179	5837
1024	275	2887	185	3348

Varying the Number of Block-Rows for Large Matrices. The number of equations in the system of linear algebraic equations is increased 32 times, i.e. from 512000 to 16384000. The number of non-zero elements in the matrix is increased from 5120110 to 163840110, i.e. the average number of non-zero elements per row is again about 10. This very large system was also solved by using different values of q. Results are shown in Table 2.

1. It is clearly seen that both the ORD times and the SOLV times practically do not depend on the parameter q (on the number of block-rows).
2. The FACT times (and, therefore, also the TOTAL times) depend on the choice of q. For this matrix, the best result is obtained with the choice $q = 1024$. Using a small number of blocks is very expensive.
3. The FACT time is the largest part of the TOTAL time also with the best choice of q (i.e. when $q = 1024$).

The results obtained in many other runs show the same trends. We shall give an example with even bigger matrix in order to illustrate this statement. The number of equations in the system of linear algebraic equations is increased 4 times, i.e. from 16384000 to 65536000. The number of non-zero elements in the matrix is increased from 163840110 to 655360110, i.e. the average number of non-zero elements per row is again about 10. Results are shown in Table 3. It is a computational disaster to run this matrix with $q = 4$. Even the run with 32 block-rows is very difficult when the matrix is so large.

Table 3. Computing times (measured in seconds) obtained in the solution of a system of 65536000 linear algebraic equations with different values of parameter q (different numbers of block-rows). The number of non-zero elements in the coefficient matrix is 655360110. $RELTOL = 0.1$ is used.

Block-rows	ORD time	FACT time	SOLV time	TOTAL time
32	1497	1401211	936	1403646
512	1491	87338	891	89721
1024	1550	44664	1002	47216

Table 4. Computing times (measured in seconds) obtained in the solution of a system of 65536000 linear algebraic equations when different numbers of processors are used. The number of non-zero elements in the coefficient matrix is 655360110. These runs were performed by using $RELTOL = 0.1$. The number of blocks used in these runs is $q = 512$.

Processors	ORD time	FACT time	SOLV time	TOTAL time
1	1085	70964	676	72724
2	1008	35532	540	37081
4	979	17871	473	19323
8	959	9050	441	10452

Table 5. Speed-ups obtained in the solution of a system of 65536000 linear algebraic equations when different numbers of processors are used. The number of non-zero elements in the coefficient matrix is 655360110. These runs were performed by using $RELTOL = 0.1$ The number of blocks used in these runs is $q = 512$. The computing times for the same runs are given in Table 4.

Processors	ORD time	FACT time	SOLV time	TOTAL time
2	1.08 (54%)	1.99 (99%)	1.25 (63%)	1.96 (98%)
4	1.11 (28%)	3.97 (99%)	1.43 (36%)	3.76 (94%)
8	1.13 (14%)	7.84 (98%)	1.53 (19%)	6.96 (87%)

Running Very Large Matrices in Parallel. The largest example, i.e. the matrix of order 65536000 and with 655360110 non-zero elements has been run by using different number of processors. The results from these runs are given in Table 4 (computing times) and in Table 5 (speed-ups). It should be mentioned here that the results given in 4 and Table 5 were run on an upgraded and slightly faster version of the computer used to produced the results given in Table 3.

It is seen, as mentioned above, that good speed-ups are achieved in the factorization part. The speed-ups for the total times are also good (because the factorization times are much larger that the times spent for the remaining parts of the computational work).

The speed-ups for the ORD times and the SOLV times are rather poor. We have not developed special techniques for the parallelization of these two parts (excepting the fact that some directives were inserted before several loops in these parts; however, the effects of this attempt to obtain parallelization are, as mentioned above, minimal). It is not very clear at present how to improve the parallel performance in these two parts. Fortunately, the computing time spent in these two parts is much smaller than the computing time spent in the factorization part when the matrix is very large. Nevertheless, some efforts have to be carried out in order to improve the parallelization in the ordering part and in the solution part when large systems of linear algebraic equations the coefficient matrices of which are general sparse matrices.

5 Comparisons with Another Sparse Matrix Code

The numerical results obtained with the sparse code *y12m3*, in which an attempt to utilize better the cache memory, indicate that very considerable reductions of the computing time can be achieved the following two conditions are satisfied:

(a) when the matrices are very large

and

(b) when the number of block-rows is carefully chosen.

The natural question, which has to be answered, is: *could we conclude (by using the numerical results presented in the previous section) that algorithm based on using block-rows is efficient?* Unfortunately, the results that were presented in the previous section do not allow us to draw such a conclusion, because the algorithm studied by us might be slower (even when used with the best division of the matrix into block-rows) than other codes for sparse matrices. Therefore, it is necessary to compare the performance of our algorithm with the performance of a good code for general sparse matrices.

We have selected the code *SuperLU*. This code is well-known, has been used by many scientists and can be down-loaded from the Internet (this can be done by searching for "SuperLU" using the search-machine Google). Documentation of the code is also available on the Internet, but the code is also well-described in several papers:

- The version of the *SuperLU*, which is best designed for sequential computers, is described in Demmel et al. [4].
- The version of *SuperLU*, which is best designed for shared memory computers, is described in Demmel et al. [5].
- The version of *SuperLU* which is best designed for distributed memory computers, is described in Li and Demmel [8].

Dropping of small non-zero elements is not used in *SuperLU*. Therefore, we had to use much smaller matrices in this section.

5.1 Comparing the Performance of the Two Codes on One Processor

Comparisons of the codes *y12m3* and *SuperLU* by performing runs on one processor were carried carried out in the beginning. Some of the results obtained in the experiments are presented in Table 6. It is clear that the following conclusions can be drawn:

- If the code *y12m3* is run without dropping, then it is at least comparable with the performance of *SuperLU*.
- The code *y12m3* is performing much better when it is used with the option for dropping small non-zero elements (dropping is carried out by using $RELTIOL = 0.1$ in Table 6).

Table 6. Comparison of computing times obtained by *SuperLU* with results obtained by the code in which the cache memory is exploited by using block-rows (*y12m3*). The latter code was run both with dropping and without dropping of small non-zero elements. $RELTOL = 0.1$ is used to drop small elements. Dropping is switched off with $RELTOL = 0$. The numbers of blocks used are given in brackets.

Order	SuperLU	Without dropping	With dropping
128000	33	21 (16)	8 (32)
256000	67	45 (32)	14 (64)
512000	139	112 (32)	30 (128)
1024000	277	237 (64)	68 (256)
2048000	560	551 (96)	179 (512)

It should be mentioned here that the utilization of the cache memory in the code *y12m3* will require some search for the number of blocks that gives best results. However, it has been shown in the previous section that this option is performing best when the matrices treated are very large. If this is the case, then one has as a rule to run many different scenarios. Therefore, it is worthwhile to try in the beginning to find a good value of the number of blocks and then to carry out the whole computational process by using this number of blocks.

The choice of a good value of the drop-tolerance $RELTOL$, when the code *y12m3* is used, may also require some search in the beginning of the computations. The results in the previous section and in this section show clearly that a good value of $RELTOL$ can give great savings in both computing time and storage. In fact, some of the problems solved in the previous section can be treated successfully only by using a proper positive value of $RELTOL$.

5.2 Parallel Runs with the Two Codes

As stated in the previous sub-section, dropping of small non-zero elements is not used in *SuperLU*. Therefore, we had to use small matrices in this sub-section (as in the comparisons performed in the previous sub-section). In fact, the largest matrix which could be run by *SuperLU* when up to eight processors are used is of order 1024000 (it was possible to solve a system with 2096000 when one processor only is used, i.e. the storage used by *SuperLU* seems to be increased when the number of processors is increased). It should be mentioned here that larger matrices (of order up to 4096000) can be handled when *y12m3* is used.

Results obtained when a system of 1024000 linear algebraic equations is solved are given in Table 7 and Table 8. The number of non-zero elements is 10240110; i.e. again the average number of non-zero elements per row is 10.

Computing times obtained in the parallel runs are given in Table 7. Speed-ups are given in Table 8.

It is seen (see the third column in Table 8) that the direct solution version of the code *y12m3* is not performing very well when the matrices are not very large. The reason for this is perhaps the fact that the blocks are not very well balanced in this case. The situation becomes slightly better when the precondi-

Table 7. Comparison of computing times obtained by *SuperLU* with results obtained by the code in which the cache memory is exploited by using block-rows (*y12m3*). The latter code was run both with dropping and without dropping of small non-zero elements. $RELTOL = 0.1$ is used to drop small elements (the number of blocks is $q = 512$ in this case). Dropping is switched off by setting $RELTOL = 0$ (the number of blocks is $q = 80$ in this case; this choice is probably not optimal).

Processors	SuperLU	Without dropping	With dropping
1	277	217	92
2	162	158	62
4	115	143	45
8	101	140	41

Table 8. Speed-ups obtained by *SuperLU* with results obtained by the code in which the cache memory is exploited by using block-rows (*y12m3*). The latter code was run both with dropping and without dropping of small non-zero elements. $RELTOL = 0.1$ is used to drop small elements (the number of blocks is $q = 512$ in this case). Dropping is switched off with $RELTOL = 0$ (the number of blocks is $q = 80$ in this case; this choice is probably not optimal).

Processors	SuperLU	Without dropping	With dropping
2	1.71 (86%)	1.37 (69%)	1.48 (74%)
4	2.41 (60%)	1.52 (38%)	2.04 (51%)
8	2.74 (34%)	1.55 (19%)	2.24 (28%)

tioned version is used (see the fourth column of Table 8). However, the speed-ups obtained by *SuperLU* are also in this case better (see the second column of Table 8).

On the other hand, the computing times for the direct version of the code *y12m3* are comparable with the computing times for **SuperLU** (compare the results given in the second and third columns in Table 7), while the preconditioned version is performing clearly better for this matrix (compare the results given in the second and fourth columns in Table 7).

6 Conclusions and Remarks

The discretization of modern large-scale mathematical models leads to huge computational tasks. The treatment of general sparse matrices is crucial when such tasks are to be solved on modern parallel computers. It was shown in this paper that at least some of the problems can be resolved when

– the cache memories of the available computer are efficiently utilized

and

– the potential power of the parallel architecture is efficiently utilized.

It was shown that the proper utilization of the cache memories and of the parallel abilities of the computer available allowed us to resolve huge computational tasks which cannot be handled when this is not done.

The big question is: *will it be possible to improve further the algorithm used in the code y12m3?* Several improvements can possibly be achieved by performing the following actions:

- improving the reordering procedure which is carried out before the beginning of the computations,
- performing in a better way the division into block-rows,
- parallelizing the preconditioned iterative method,

and

- developing an MPI version of the code *y12m3*.

There are some plans to prepare an improved version of the code *y12m3* in which the above requirements are taken into account.

It has been assumed that the Gaussian elimination is used in the solution of the systems of linear algebraic equation. It must be emphasized here, however, that the ideas are general and most of them can easily be applied also when some other methods for solving very large systems of linear algebraic equations (as, for example, the methods that are discussed in Zlatev, [11] and [12]).

Furthermore, some of the ideas are applicable also when other methods for the treatment of general sparse matrices are used (as the methods for treating general sparse matrices that are discussed in Davis and Davidson [2], Davis and Yew [3], George and Ng [7], and Duff et al. [6]).

Acknowledgements

Two grants (CPU-1002-27 and CPU-1101-17) from the Danish Centre for Scientific Computing (DCSC) gave us access to the IBM computers at the Århus University and the SUN computers at the Technical University of Denmark. The members of the staff of DCSC helped us to resolve some difficult problems related to the efficient exploitation of the grids on IBM and SUN computers.

The work on this project was partly supported by the NATO Scientific Programme (Grant No. CLG 980505).

References

1. Davis, T. A.: Sparse matrix collection, University of Florida, Department of Computer and Information Science and Engineering, Gainesville, FL, USA, 32611–6120, http://www.cise.ufl.edu, 2005.
2. Davis, T. A. and Davidson, E. S.: Pairwise reduction for the direct parallel solution of sparse unsymmetric sets of linear equations, IEEE Trans. Comput., **37** (1988), 1648-1654.
3. Davis, T. A. and Yew, P.-C.: A nondeterministic parallel algorithm for general unsymmetric sparse LU factorization, SIAM J. Matrix Anal. Appl., **3** (1990), 383–402.

4. Demmel, J. W., Eisenstat, S. C., Gilbert, J. R., Li, X. S., and Liu, J. W. H.: A supernodal approach to sparse partial pivoting, SIAM Journal of Matrix Analysis and Applications, **20** (1999), 720–755.
5. Demmel, J. W., Gilbert, J. R., and Li, X. S.: An asynchronous parallel supernodal algorithm for sparse Gaussian elimination, SIAM Journal of Matrix Analysis and Applications, **20** (1999), 915–952.
6. Duff, I. S., Erisman, A. M., and Reid, J. K.: *Direct methods for sparse matrices.* Oxford University Press, Oxford-London, 1986.
7. George, J. A. and Ng, E.: An implementation of Gaussian elimination with partial pivoting for sparse systems, SIAM J. Sci. Statist. Comput., **6** (1985), 390–405.
8. Li, X. S. and Demmel, J. W.: Super_LU_DIST: A scalable distributed-memory sparse direct solver for unsymmetric linear system, ACM Transactions on Mathematical Software, **29** (1999), 110–140.
9. Zlatev, Z.: On some pivotal strategies in Gaussian elimination by sparse technique, SIAM Journal on Numerical Analysis, **17** (1980), 18–30.
10. Zlatev, Z.: Use of iterative refinement in the solution of sparse linear systems, SIAM Journal on Numerical Analysis, **19** (1982), 381–399.
11. Zlatev, Z.: General scheme for solving linear algebraic problems by direct methods, Applied Numerical Mathematics, **1** (1985), 176–186.
12. Zlatev, Z.: Computational methods for general sparse matrices, Kluwer Academic Publishers, Dordrecht-Boston-London, 1991.

Part II

Advances in Computational Mechanics and Multi Physics

Numerical Methods for Transport Problems in Microdevices

Assyr Abdulle[1] and Sabine Attinger[2]

[1] University of Basel, Department of Mathematics, Rheinsprung 21, CH-4051 Basel, Switzerland
assyr.abdulle@unibas.ch

[2] Computational Environmental Systems, UFZ Leipzig-Halle, Germany and Institute for Geoscience, University of Jena, Germany
sabine.attinger@ufz.de

Abstract. Transport processes of large biomolecules in microdevices are of high interest for biological research and biomedical applications. Such devices have been proposed recently to address the problem of separating different components of macromolecules. From the mathematical point of view, such problems can be modeled by multiscale partial differential equations, which can be analyzed and numerically simulated. In this paper we propose a framework to model a class of transport problems which arise in microdevices and discuss their numerical simulation.

1 Introduction

Transport of biomolecules in microdevices provides molecular biologists novel opportunities for sample analysis. Without attempting to be exhaustive, we mention [6, 8, 9, 10], where separation procedures of large biomolecules based on transport in microarray have been proposed. In these experiments, microfabricated silicon arrays replace the agarose gel used in electrophoresis [16]. We also mention [13, 15], where electrophoretic transport of DNA in microarrays have been studied.

In this paper we discuss the simulation of transport processes of macromolecules as proposed in [6, 8, 9, 10], in the multiscale modeling approach proposed recently in [3]. The numerical techniques follow a general methodology for the numerical solution of multiscale transport problems proposed in [5].

2 Multiscale Transport in Microdevices

In this section we introduce the multiscale advection-diffusion equation, coupled with a multiscale elliptic equation, modeling the transport of the macromolecules in microdevices.

Let Ω be a domain in $\mathbb{R}^2$ representing the microdevice or a sample of it. In the following, we will consider microdevices with periodic asymmetric obstacles as sketched in Figure 1. We consider the evolution of a scalar physical field, a mass density, $c(t, x) : I \times \Omega \longrightarrow \mathbb{R}$ advected by a velocity field $v : \Omega \longrightarrow \mathbb{R}^2$

I. Lirkov, S. Margenov, and J. Waśniewski (Eds.): LSSC 2005, LNCS 3743, pp. 67–75, 2006.

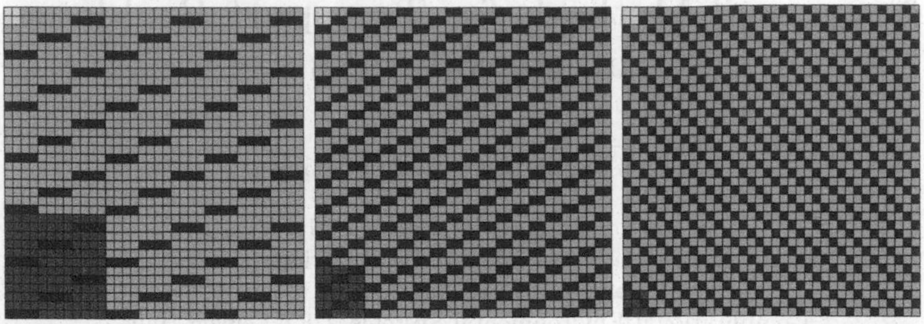

Fig. 1. Sample of microdevices with various geometries. The initial particles location is at the left upper corner. The shadowed square at the lower left corner represents a typical periodic structure of the array.

and submitted to molecular diffusion. Standard conservation laws yield the advection-diffusion equation $\frac{\partial c}{\partial t} + \nabla \cdot (vc) = D\Delta c$, together with suitable initial and boundary conditions. In the sequel we will assume that the diffusion tensor D is a constant, but we emphasize that our numerical method also apply for non-scalar or non-constant diffusion tensors. Following [9, 10], we suppose that the velocity field is induced by a difference of (electrical) potential u and given by $v = -\rho k \nabla u$, where k is the electrical conductivity and ρ the charge density of the electrical array which we assume for simplicity to be constant and set to one.

In our transport problems in microdevices, we have two typical length scale involved. A small length scale $y = x/\varepsilon$, which represents a self-similar structure of the device, and a large length scale x at which the transport behavior is observed. Assuming that the concentration and the velocity field depend on both the macro and micro length scale, we will denote them as $c^\varepsilon(t,x)$ and $v^\varepsilon(x)$ where we add a superscript ε to emphasize the dependence of these quantities on ε. The obstacles of the microdevice introduce a micro-scale variability of the conductivity and we write it as $k^\varepsilon = k(x/\varepsilon) = k(y)$. In Figure 1 the shadowed square (at the left lower corner) represents a typical self-similar structure of the microdevice (which will also be called a "periodic cell"), where ε represents the length of this cell. Thus, the conductivity coefficients k^ε is periodic, more precisely, $k(y)$ is 1-periodic in $Y = (0,1)^2$ (see [3] for details).

The velocity field is given by an elliptic equation for the potential u

$$-\nabla \cdot (k^\varepsilon \nabla u^\varepsilon) = f,$$

where f is a source term and where we also add a superscript ε to u, to emphasize its dependence on ε.

Summarizing, the evolution of an initial concentration of particles in a microdevice (see Figure 1) is given by the following coupled equations

$$\frac{\partial c^\varepsilon}{\partial t} + \nabla \cdot (v^\varepsilon c^\varepsilon) = D\Delta c^\varepsilon, \tag{1}$$

$$v^\varepsilon = -k^\varepsilon \nabla u^\varepsilon, \tag{2}$$

$$-\nabla \cdot (k^\varepsilon \nabla u^\varepsilon) = f, \tag{3}$$

together with suitable initial and boundary conditions for equations (1) and (3). In order to have a unique solution for equation (3), we will assume the tensor k^{ε} to be coercive and bounded.

The direct numerical solution of (1-3) is difficult, often impossible due to the work needed to resolve the small scale of the problems. The typical size of the obstacles is a few micrometers, whereas the size of the microdevices is of a few centimeters, so that even in $2D$ the degrees of freedom with a mesh resolving the small scale can be of order $10^8 - 10^9$ for realistic simulations.

Remark 1. To analyze the behavior of particles following equations (1-3), a widely used framework is the homogenization theory [7]. In the present context of microdevices, the homogenization procedure has been described in [3]. With suitable assumption it can be shown for incompressible velocity, $\nabla \cdot v^{\varepsilon} = 0$, that c^{ε} converges in the L^2 norm towards $\bar{c}$ when $\varepsilon \to 0$, where $\bar{c}$ is given by (see [14] and the reference therein)

$$\frac{\partial \bar{c}}{\partial t} + \bar{v} \cdot \nabla \bar{c} = \nabla \cdot \bar{D} \nabla \bar{c}, \tag{4}$$

and where $\bar{v}$ is a mean velocity and $\bar{D}$ is an effective diffusivity tensor involving the solution of so-called cell problems (see also [3]).

3 Numerical Method

We describe in this section the numerical procedures proposed in [5] which will be applied to our transport problems in microdevices in order to overcome the problem of resolving the fully detailed equations (1-3). This method is based on the combination of the following ideas.

(A) An approximation of the velocity field $v^{\varepsilon,h}$ of the small scale velocity v^{ε}, which captures the fine scale information at a much lower cost than the one needed for the full resolution of the elliptic problem (3).

(B) A high order Chebyshev method (ROCK) for the time integration of the equation (1) for the concentration (where the velocity field is replaced by its numerical counterpart obtained in (A)). Discretizing the spatial variables of (1) by finite differences, leads to a large system of ordinary differential equations (ODEs). Such systems of ODEs are known to be stiff since standard explicit solvers, as for example the forward Euler method, will have a restriction of the step-size due to stability issues. Using implicit solvers avoid such problems but at the cost of solving linear systems of dimension $n \times n$, where n is the dimension of the system of ODEs (corresponding to the number of grid points in Ω chosen for the spatial discretization). The ROCK2 and ROCK4 methods, belonging to the class of Chebyshev methods, are explicit, avoiding linear algebra problems, possess extended stability properties for diffusion dominated advection-diffusion problems, avoiding the severe step-size restriction for standard explicit solvers and have variable step-size. These ROCK methods recently proposed [1, 2] are of order 2 and 4.

We will briefly describe (A) and we refer the reader to [1, 2, 5] for more details and for a description of (B). We consider the elliptic problem (3) together with zero Dirichlet boundary conditions, for simplicity of the presentation. Applying a standard finite element method to the variational form of (3) requires usually a meshsize $h < \varepsilon$ for convergence, i.e., to resolve the small scale of the problem. The FE-HMM (Finite Element Heterogeneous Multiscale Method) for elliptic homogenization problems introduced in [11] and analyzed in [4] and [12] is based on the following ideas. Define a quasi-uniform macro triangulation $\mathcal{T}_H$ of Ω, assumed to be a convex polygon. By macro triangulation we mean that the size H of a triangle $K \in \mathcal{T}_H$ can be much larger than ε. In each macro triangle K, we define a micro sampling domain K_ε of size ε as sketched in the Figure below.

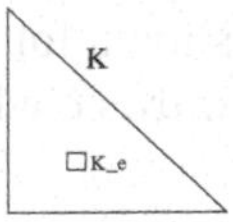

Associated to the macro triangulation, we define a Macro Finite Element space $S_0^1(\Omega, \mathcal{T}_H)$, the subset of functions of $H_0^1(\Omega)$ which are piecewise linear on each triangle K. We also define a quasi-uniform triangulation $\mathcal{T}_h$ of the sampling domain K_ε, which will be referred as micro triangulation, since $T \in \mathcal{T}_h$ is of size $h < \varepsilon$. Associated to the micro triangulation, we define a Micro Finite Element space $S_{per}^1(K_\varepsilon, \mathcal{T}_h)$, the subset of functions in $H^1(K_\varepsilon)$ which are piecewise linear on each triangle T, periodic in K_ε and have vanishing mean $\int_{K_\varepsilon} v^h dx = 0$ (see [4] for a precise definition).

We define a modified macro bilinear form for $u^H, v^H \in S_0^1(\Omega, \mathcal{T}_H)$

$$B(u^H, w^H) := \sum_{K \in \mathcal{T}_H} \frac{|K|}{|K_\varepsilon|} \int_{K_\varepsilon} \nabla u^h \ k(x/\varepsilon)(\nabla w^h)^T dx, \tag{5}$$

where $|K|$, $|K_\varepsilon|$ denote the measure of K and K_ε, respectively. Note that the factor $|K|/|K_\varepsilon|$ is a scaling factor. The macro FE-HM solution is defined by the following variational problem: find $u^H \in S_0^1(\Omega, \mathcal{T}_H)$ such that

$$B(u^H, w^H) = \langle f, w^H \rangle, \quad \forall w^H \in S_0^1(\Omega, \mathcal{T}_H). \tag{6}$$

To compute the bilinear form, one needs to know u^h (or w^h), which is the solution of the following micro problem: find u^h such that $(u^h - u^H) \in S_{per}^1(K_\varepsilon, \mathcal{T}_h)$ and

$$b_{K_\varepsilon}(u^h, z^h) := \int_{K_\varepsilon} \nabla u^h a(x_k, x/\varepsilon)(\nabla z^h)^T dx = 0 \quad \forall z^h \in S_{per}^1(K_\varepsilon, \mathcal{T}_h). \tag{7}$$

It can be shown that the problem (6) is well posed and has a unique solution [4, 12]. The macro solution u^H converges in the H^1 norm to the homogenized solution of equation (3) without the need of computing explicitly the homogenized equations [4, 12]. The saving of the method, compared to the full resolution, comes from the fact that the small scale problems are solved only on sampling domains within each macro triangle, usually much smaller than the overall domain. The extension for the non uniformly periodic case, $k^\varepsilon = k(x, x/\varepsilon)$, has also been studied [4, 12]. For periodic problems the convergence is independent of ε. Note that since the cell problems within each macro element are independent, these problems can be solved in parallel. The algorithm does not rely on periodic problems, although the error analysis in the nonperiodic case is mostly open. We refer to [4] and [12] for the analysis of the FE-HMM.

So far, we have obtained a numerical approximation of the homogenized solution of (3), that is, the solution of the limit equation (3), when $\varepsilon \to 0$. However, the known micro solution u^h on the sampling domain obtained from (7), and the macro solution u^H obtained from (6), allow to reconstruct an approximation of the small scale solution u^ε in the following way

$$u^{\varepsilon,h}|_K := u^H + (u^h - u^H)|_K^{\#}, \tag{8}$$

where $|_K^{\#}$ denotes the periodic extension of the fine scale solution $(u^h - u^H)$, available in K_ε, on each macro element K. To give a convergence estimate, we define a broken H^1 norm $\|u\|_{\bar{H}^1(\Omega)} := (\sum_{K \in \mathcal{T}_H} \|\nabla u\|^2_{L^2(K)})^{1/2}$, since $u^{\varepsilon,h}$ can be discontinuous across the macro elements K. The following convergence result has been obtained in [4].

Theorem 1. *Let u^ε be the solution of problem (3), let u^H be the solution of problem (6) and consider $u^{\varepsilon,h}$ defined in (8). Then*

$$\|u^\varepsilon - u^{\varepsilon,h}\|_{\bar{H}^1(\Omega)} \leq C(\sqrt{\varepsilon} + H + (h/\varepsilon)), \tag{9}$$

where H is the size of the triangulation of the macro finite element space $S_0^1(\Omega, \mathcal{T}_H)$ and h is the size of the triangulation of the micro finite element space $S^1_{per}(K_\varepsilon, \mathcal{T}_h)$.

Notice that if we denote by $M = \dim S^1_{per}(K_\varepsilon)$, then the mesh size of the micro finite element space on K_ε (of measure $|K_\varepsilon| = \varepsilon^2$) is given by $h \simeq \varepsilon M^{-\frac{1}{2}}$. Therefore, the quantity h/ε does not depend on ε but only on the dimension of $S^1_{per}(K_\varepsilon)$, i.e., the degrees of freedom of the micro finite element space.

We define a numerical approximation of the velocity field $v^\varepsilon := -k^\varepsilon \nabla u^\varepsilon$ by

$$v^{\varepsilon,h}|_K := -k(x/\varepsilon)\nabla u^{\varepsilon,h} \text{ for } K \in \mathcal{T}_H, \tag{10}$$

where $u^{\varepsilon,h}$ is defined in (8). Since k^ε is assumed to be bounded (see Section 1), with help of Theorem 1 we obtain

$$\|v^\varepsilon - v^{\varepsilon,h}\|_{\bar{L}^2(\Omega)} \leq C(\sqrt{\varepsilon} + H + h/\varepsilon), \tag{11}$$

where we also used a mesh-dependent norm.

Remark 2. A numerical procedure to simulate the effective or homogenized equation (4) can also be obtained in the present framework. The approximation of the mean velocity $\bar{v}$ involves the homogenized tensor and the homogenized solution of equation (3). The macro solution u^H of the FE-HMM approximates the homogenized solution of (3) and the homogenized tensor can be recovered from the known micro solution in the sampling domain, during the assembly procedure of the FE-HMM. We refer to [5] for more details. An approximation of the effective diffusion tensor $\bar{D}$ can be obtained via perturbation theory [3].

4 Numerical Experiments

We present in this section numerical experiments illustrating the use of the described method for a transport problem in a microdevice. Let $\Omega = (0,1)^2$ represent the microdevice. This square domain is surrounded by points electrodes sketched in Figure 2. The corners of the device are insulated. The molecules, supposed to be negatively charged (as for example the DNA molecules), are injected into the top corner of the device and are propelled due to the electrical field as explained in Section 1.

We follow for our example [9, 10] and we suppose that the velocity field $v^\varepsilon = -k^\varepsilon \nabla u^\varepsilon$ is divergence free. The equation for the potential u^ε is given by

$$-\nabla \cdot (k^\varepsilon \nabla u^\varepsilon) = 0 \text{ in } \Omega \tag{12}$$

$$u^\varepsilon|_{\Gamma_{D_0}} = u^-, \quad u^\varepsilon|_{\Gamma_{D_1}} = u^+ \tag{13}$$

$$n \cdot \left(k(\frac{x}{\varepsilon})\nabla u^\varepsilon\right)|_{\Gamma_N} = 0 \quad \text{on } \Gamma_N := \partial\Omega \backslash \Gamma_N, \tag{14}$$

where $\Gamma_{D_0} := \{(x_1,x_2); x_1 = 0, x_2 \in [\alpha,\beta]\} \cup \{(x_1,x_2); x_1 \in [\alpha,\beta], x_2 = 1\}$, $\Gamma_{D_1} := \{(x_1,x_2); x_1 = 1, x_2 \in [\alpha,\beta]\} \cup \{(x_1,x_2); x_1 \in [\alpha,\beta], x_2 = 0\}$. The Neumann boundary condition (14) represents the insulated regions, whereas the Dirichlet conditions Γ_{D_i} the charged sites of the boundary, chosen as $u^- = 0$, $u^+ = 1$.

We discretize Ω with a meshsize $h = 1/400$ in each spatial direction. We chose $\alpha = 0.25, \beta = 0.75$ and k^ε to be given by the geometry of the left sample

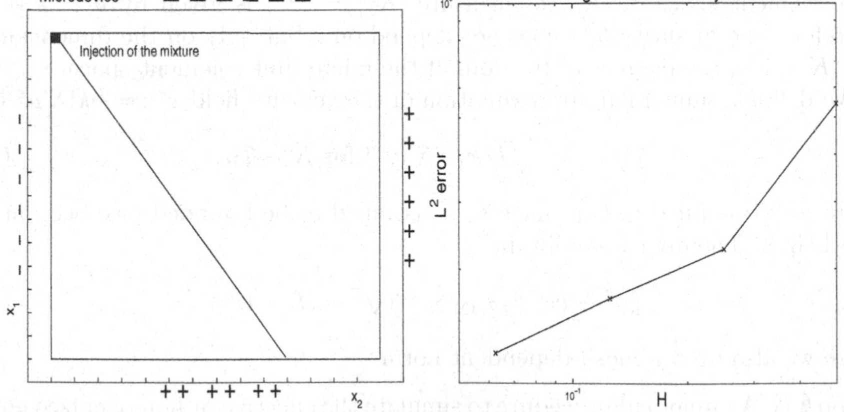

Fig. 2. Illustration of the trajectory of particles injected in a microdevice surrounded with electrodes (left picture). Convergence rate of the reconstructed velocity $v^{\varepsilon,h}$ towards v^ε for macro mesh refinement $H = 1/2, 1/4, 1/8, 1/16$ (right picture).

in Figure 1. We set $k^\varepsilon(x) = 0.01$ in the obstacles and $k^\varepsilon(x) = 1$ outside the obstacles. We chose $\varepsilon = 0.0625$, which represents the size of the periodic cell of the sample. Notice that with the method described in Section 3 we could have chosen a much smaller value for ε compared to the overall domain Ω. We chose this value in order to compute a reference solution via scale resolution with a standard FEM. Recall that the small scale approximation $v^{\varepsilon,h}$ is obtained as explained in Section 3, by computing a macro solution u^H and performing the reconstruction procedures (8) and (10). The convergence of $v^{\varepsilon,h}$ towards v^ε depends on both macro and micro meshes. For the micro mesh within each sampling domain, we take $h = 1/400$ and for the macro mesh we take successively $H = 1/2, 1/4, 1/8, 1/16$. The amount of work needed to obtain a velocity approximation is to solve $2(1/H)^2$ micro problems and one macro problem of size $\sim (1/H)^2$, while the full resolution of the fine scale requires, with the same h as used for the sampling domain, 160801 degrees of freedom. Since the tensor k^ε is uniformly periodic (i.e. it does not depend on the macro variable) it is sufficient to solve one micro problem. This is however no longer possible in the more general situation where $k^\varepsilon = k(x, x/\varepsilon)$.

With the numerical reconstructed velocity field, we then discretize the spatial variable of the transport problem

$$\frac{\partial c^\varepsilon}{\partial t} + \nabla \cdot (v^\varepsilon c^\varepsilon) = D \Delta c^\varepsilon, \tag{15}$$

and integrate the semi-discrete system obtained with the ROCK4 method as explained in Section 3. We take zero Dirichlet boundary conditions and we perform the transport simulation on a sub-domain $\Omega_s = [0.1125, 0.8875] \times [0.1125, 0.8875]$ in order to minimize the influence of the artificial boundary conditions. The initial concentration is set to be $c_0(x) = 1$ for $x \in B$ and 0 otherwise, where $B = [0.1125, 0.1375] \times [0.1125, 0.1375]$. For the diffusion coefficient we chose $D = 0.001$.

In Figure 2 we study the convergence in the L^2 norm of $v^{\varepsilon,h}$ towards a reference solution for v^ε (left picture) for macro mesh refinement $H = 1/2, 1/4, 1/8, 1/16$. The errors are computed in Ω_s, the same subdomain as used for the transport simulation. We see in Figure 2 that $v^{\varepsilon,h}$ converges nicely, first linearly than with a slightly slower rate. The cell problems have been solved with sufficient precision as to avoid the influence of the h/ε term in (11). Notice that since ε is not very small, the influence of the term $\sqrt{\varepsilon}$ may appear in further macro mesh refinement. However, this influence becomes smaller when $\varepsilon \to 0$, while in this case the full resolution becomes harder.

Finally, we simulate the transport problem (15) with the initial and boundary conditions described above. We compare in Figure 3 the evolution of the particles advected by a reference velocity field (obtained via scale resolution) and a reconstructed velocity field obtained with a very coarse macro mesh chosen as $H = 1/4$. We perform the time integration for $t \in [0, 1.2]$ and record the solution at discrete time $t = 0, 0.3, 0.6, 0.9, 1.2$, to compare the evolution of the two transport problems. We see in Figure 3 that the solution with the reconstructed velocity field agrees very well with the reference solution, illustrating the

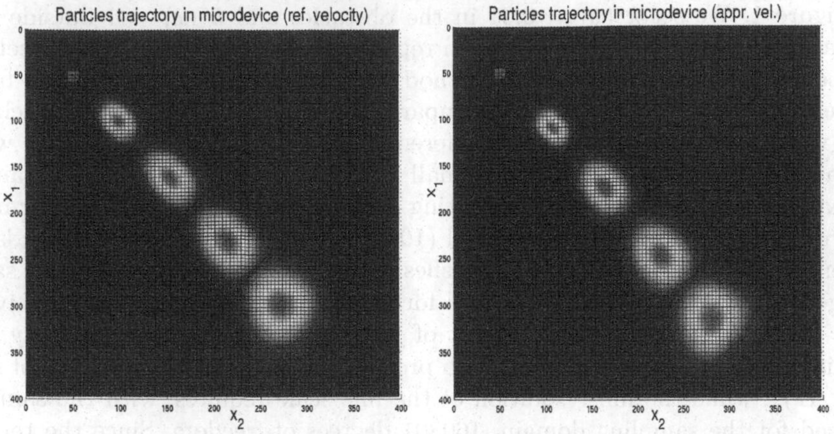

Fig. 3. Transport simulation in the microdevice with a reference velocity field (left) and a reconstructed velocity field (right), $H = 1/4$. The cloud of particles represent the concentration evolution at time $t = 0, 0.3, 0.6, 0.9, 1.2$ starting from the top left corner.

efficiency of the proposed method. We also note that the particles are distracted from the 45^0 diagonal, i.e., the effective conductivity tensor becomes anisotropic due to the micro arrays. We refer to [3] where this observation has been discussed and explained.

Acknowledgments

The work of A. Abdulle is partially supported by the Swiss National Foundation under grant 200021-103863/1.

References

1. A. Abdulle and A.A. Medovikov, Second order Chebyshev methods based on orthogonal polynomials, *Numer. Math.* **90**, 1–18, 2001.
2. A. Abdulle, Fourth order Chebyshev methods with recurrence relation, *SIAM SISC*, Vol.23, No. 6, pp. 2041–2054, 2002.
3. A. Abdulle and S. Attinger, Homogenization methods for Transport of DNA particles in Heterogeneous Arrays, *Lect. Notes in Comp. Sci. and Eng.*, **39**, 23–34, 2004.
4. A. Abdulle, On a-priori error analysis of Fully Discrete Heterogeneous Multiscale FEM, *SIAM Multiscale Model. Simul.*, **4** 2, 447–459, 2005.
5. A. Abdulle, Multiscale methods for advection-diffusion problems, to appear *AIMS, Discrete and Continuous Dynamical Systems*, 2005.
6. R.H. Austin, N.Darnton, R. Huang, J. Sturm, O. Bakajin, T. Duke, Ratchets: the problems with boundary conditions in insulating fluids, *Appl. Phys. A*, **75**, 279–284, 2002.

7. A. Bensoussan, J.-L. Lions and G. Papanicolaou, *Asymptotic Analysis for Periodic Structures*, North Holland, Amsterdam, 1978.
8. Chou et al., Sorting biomolecules with microdevices, *Electrophoresis*, **21**, 81–90, 2000.
9. T.A. Duke and R.H. Austin, Microfabricated Sieve for the Continuous Sorting of Macromolecules, *Phys. Rev. Lett.* **89** 7, 1552–1555, 1998.
10. D. Ertas, Lateral Separation of Macromolecules and Polyelectrolytes in Microlithographic Arrays, *Phys. Rev. Lett.* **80** 7, 1548–1551, 1998.
11. W. E and B. Engquist, The Heterogeneous Multi-Scale Methods, *Comm. Math. Sci.*, **1** 1, 87–132, 2003.
12. W. E, P. Ming and P. Zhang, Analysis of the heterogeneous multi-scale method for elliptic homogenization problems, *J. Amer. Math. Soc.* **18**, 121–156, 2004.
13. S. K. Kassegne et. all, Numerical modeling of transport and accumulation of DNA on electronically active biochips, *Sensors and Actuators B*, **94**, 81–98, 2003.
14. A. J. Majda and P. R. Kramer, Simplified models for turbulent diffusion: Theory, numerical modelling and physical phenomena, *Physics Reports*, **314**, 237–574, 1999.
15. R. Radtkey et. all, Rapid, high fidelity analysis of simple sequence repeats on an electronically active DNA microchip, *Nucleic Acids Research*, **28** 7, 2000.
16. D. Rickwood and B.D. Hames, *Gel electrophoresis of the nucleid Acids: A Practical Approach*, Oxford University Press, Oxford, 1990.

Discretization of Integro-Differential Equations Modeling Dynamic Fractional Order Viscoelasticity

K. Adolfsson[1], M. Enelund[1], S. Larsson[2], and M. Racheva[3]

[1] Dept. of Appl. Mech., Chalmers Univ. of Technology, SE–412 96 Göteborg, Sweden
[2] Dept. of Mathematics, Chalmers Univ. of Technology, SE–412 96 Göteborg, Sweden
[3] Dept. of Mathematics, Technical University of Gabrovo, 5300 Gabrovo, Bulgaria

Abstract. We study a dynamic model for viscoelastic materials based on a constitutive equation of fractional order. This results in an integro-differential equation with a weakly singular convolution kernel. We discretize in the spatial variable by a standard Galerkin finite element method. We prove stability and regularity estimates which show how the convolution term introduces dissipation into the equation of motion. These are then used to prove a priori error estimates. A numerical experiment is included.

1 Introduction

Fractional order operators (integrals and derivatives) have proved to be very suitable for modeling memory effects of various materials and systems of technical interest. In particular, they are very useful when modeling viscoelastic materials, see, e.g., [3].

Numerical methods for quasistatic viscoelasticity problems have been studied, e.g., in [2] and [8]. The drawback of the fractional order viscoelastic models is that the whole strain history must be saved and included in each time step. The most commonly used algorithms for this integration are based on the Lubich convolution quadrature [5] for fractional order operators. In [1,2], we develop an efficient numerical algorithm based on sparse numerical quadrature earlier studied in [6].

While our earlier work focused on discretization in time for the quasistatic case, we now study space discretization for the fully dynamic equations of motion, which take the form of an integro-differential equation with a weakly singular convolution kernel. A similar equation but with smooth kernel was studied in [7]. The singular kernel requires a different approach. Inspired by [4] we introduce appropriate function spaces and prove stability estimates for both the continuous and discrete problems. These are used to prove a priori error estimates. Finally, we present a numerical example for a two-dimensional viscoelastic body. Time-discretization, sparse quadrature, and a posteriori error estimates are subject to future investigations.

I. Lirkov, S. Margenov, and J. Waśniewski (Eds.): LSSC 2005, LNCS 3743, pp. 76–83, 2006.

2 Viscoelastic Equations of Motion

Assuming isothermal and isotropic conditions, the fractional order linear viscoelastic constitutive equation for the stress $\boldsymbol{\sigma}$ can be written in convolution integral form as

$$\begin{aligned}\boldsymbol{\sigma}(t) &= \boldsymbol{\sigma}_0(t) - \int_0^t \beta(t-s)\boldsymbol{\sigma}_1(s)\,\mathrm{d}s, \quad \text{with} \\ \boldsymbol{\sigma}_0(t) &= 2\mu_0\boldsymbol{\epsilon}(t) + \lambda_0 \mathrm{tr}(\boldsymbol{\epsilon}(t))\mathrm{I}, \quad \boldsymbol{\sigma}_1(t) = 2\mu_1\boldsymbol{\epsilon}(t) + \lambda_1 \mathrm{tr}(\boldsymbol{\epsilon}(t))\mathrm{I},\end{aligned} \tag{1}$$

where $\lambda_0 > \lambda_1 > 0$ and $\mu_0 > \mu_1 > 0$ are the elastic constants of Lamé type, $\boldsymbol{\epsilon}$ is the strain and β is the convolution kernel

$$\beta(t) = -\frac{\mathrm{d}}{\mathrm{d}t}\Big(\mathrm{E}_\alpha(-(t/\tau)^\alpha)\Big) = \frac{\alpha}{\tau}\Big(\frac{t}{\tau}\Big)^{\alpha-1}\mathrm{E}'_\alpha(-(t/\tau)^\alpha). \tag{2}$$

Here $\mathrm{E}_\alpha(x) = \sum_{k=0}^{\infty} \frac{x^k}{\Gamma(1+\alpha k)}$ is the Mittag-Leffler function of order α.

In the convolution kernel (2), τ is a relaxation constant and $\alpha \in (0,1)$ is the order of the fractional derivative. The convolution kernel is weakly singular and $\beta \in L_1(0,\infty)$ with $\int_0^\infty \beta(t)\,\mathrm{d}t = 1$. The fractional order model represents a fading memory because the convolution kernel in (2) is a strictly decreasing function (i.e., $\mathrm{d}\beta/\mathrm{d}t < 0$). The Lamé constants in (1) can be expressed as

$$\mu_0 = \frac{E_0}{2(1+\nu)}, \quad \mu_1 = \frac{E_1}{2(1+\nu)}, \quad \lambda_0 = \frac{E_0\nu}{(1+\nu)(1-2\nu)}, \quad \lambda_1 = \frac{E_1\nu}{(1+\nu)(1-2\nu)},$$

where ν is the Poisson ratio, E_0 is the instantaneous uniaxial elastic modulus, while $E_0 - E_1 > 0$ can be identified as the relaxed uniaxial modulus. For convenience we introduce $\gamma = \frac{\mu_1}{\mu_0} = \frac{\lambda_1}{\lambda_0} = \frac{E_1}{E_0} < 1$, and note that $\boldsymbol{\sigma}_1 = \gamma\boldsymbol{\sigma}_0$.

We are now in the position to formulate the viscoelastic dynamic problem. The basic equations in strong form are

$$\begin{aligned}
&\rho\ddot{\mathbf{u}}(\mathbf{x},t) - \nabla\cdot\boldsymbol{\sigma}_0(\mathbf{u};\mathbf{x},t) \\
&\qquad + \int_0^t \beta(t-s)\nabla\cdot\boldsymbol{\sigma}_1(\mathbf{u};\mathbf{x},s)\,\mathrm{d}s = \mathbf{f}(\mathbf{x},t) && \text{in } \Omega\times I, \\
&\mathbf{u}(\mathbf{x},0) = \mathbf{u}_0(\mathbf{x}) && \text{in } \Omega, \\
&\dot{\mathbf{u}}(\mathbf{x},0) = \mathbf{v}_0(\mathbf{x}) && \text{in } \Omega, \\
&\mathbf{u}(\mathbf{x},t) = \mathbf{0} && \text{on } \Gamma_\mathrm{D}\times I, \\
&\boldsymbol{\sigma}(\mathbf{u};\mathbf{x},t)\cdot\mathbf{n}(\mathbf{x}) = \mathbf{g}(\mathbf{x},t) && \text{on } \Gamma_\mathrm{N}\times I,
\end{aligned} \tag{3}$$

where ρ is the (constant) mass density, $\mathbf{f}, \mathbf{g}$ represent the volume and surface loads, respectively, $\mathbf{u}$ is the displacement vector, $\boldsymbol{\sigma}_0$ and $\boldsymbol{\sigma}_1$ are the stresses according to (1), and the strain is defined through the usual linear kinematic relation $\boldsymbol{\epsilon} = \frac{1}{2}\Big(\nabla\mathbf{u} + (\nabla\mathbf{u})^\mathrm{T}\Big)$.

In order to give the equations (3) a convenient mathematical formulation, we let $\Omega \subset \mathbf{R}^d$, $d = 2,3$, be a bounded domain with $\partial\Omega = \bar{\Gamma}_\mathrm{D} \cup \bar{\Gamma}_\mathrm{N}$, $\Gamma_\mathrm{D} \cap \Gamma_\mathrm{N} = \emptyset$, meas $(\Gamma_{\rm D}) > 0$, and we define $H = L_2(\Omega)^d$ with its usual inner product and norm denoted by $(\cdot,\cdot)$ and $\|\cdot\|$. We also define $V = \{\mathbf{v} \in H^1(\Omega)^d : \mathbf{v} = \mathbf{0} \text{ on } \Gamma_{\rm D}\}$ and the bilinear form

$$a(\mathbf{v},\mathbf{w}) = \int_\Omega \big(2\mu_0 \epsilon_{ij}(\mathbf{v})\epsilon_{ij}(\mathbf{w}) + \lambda_0 \epsilon_{ii}(\mathbf{v})\epsilon_{jj}(\mathbf{w})\big)\,\mathrm{d}\mathbf{x}, \quad \mathbf{v},\mathbf{w} \in V.$$

It is well known that a is coercive on V. The corresponding operator $A\mathbf{u} = -\nabla\cdot\boldsymbol{\sigma}_0(\mathbf{u})$, defined together with the homogeneous boundary conditions in (3) $(\mathbf{g} = \mathbf{0})$, so that $a(\mathbf{u},\mathbf{v}) = (A\mathbf{u},\mathbf{v})$ for sufficiently smooth $\mathbf{u},\mathbf{v} \in V$, can be extended to a self-adjoint, positive definite, unbounded linear operator on H.

The equation of motion (3) can then be written in weak form: Find $\mathbf{u}(t) \in V$ such that $\mathbf{u}(0) = \mathbf{u}_0$, $\dot{\mathbf{u}}(0) = \mathbf{v}_0$ and, with $\langle \mathbf{g},\mathbf{v}\rangle_{\Gamma_{\rm N}} = \int_{\Gamma_{\rm N}} \mathbf{g}\cdot\mathbf{v}\,\mathrm{d}S$,

$$\begin{aligned} \rho(\ddot{\mathbf{u}}(t),\mathbf{v}) + a(\mathbf{u}(t),\mathbf{v}) - \gamma \int_0^t \beta(t-s) a(\mathbf{u}(s),\mathbf{v})\,\mathrm{d}s \\ = (\mathbf{f}(t),\mathbf{v}) + \langle \mathbf{g}(t),\mathbf{v}\rangle_{\Gamma_{\rm N}}, \quad \forall \mathbf{v} \in V. \end{aligned} \tag{4}$$

We next introduce the spatially semidiscrete finite element method. Let $V_h \subset V$ be a standard piecewise linear finite element space based on a triangulation of Ω. The finite element problem is to find $\mathbf{u}_h(t) \in V_h$ such that $\mathbf{u}_h(0) = \mathbf{u}_{h,0}$, $\dot{\mathbf{u}}_h(0) = \mathbf{v}_{h,0}$ and

$$\begin{aligned} \rho(\ddot{\mathbf{u}}_h(t),\mathbf{v}_h) + a(\mathbf{u}_h(t),\mathbf{v}_h) - \gamma \int_0^t \beta(t-s) a(\mathbf{u}_h(s),\mathbf{v}_h)\,\mathrm{d}s \\ = (\mathbf{f}(t),\mathbf{v}_h) + \langle \mathbf{g}(t),\mathbf{v}_h\rangle_{\Gamma_{\rm N}}, \quad \forall \mathbf{v}_h \in V_h. \end{aligned} \tag{5}$$

3 Stability Estimates

By adapting the analysis in [4] we can show existence and uniqueness of solutions of (3) by means of the theory of strongly continuous semigroups. We leave the details to a forthcoming paper and show only the main ingredient, namely, that the convolution term introduces dissipation into the equation. We introduce the function

$$\xi(t) = 1 - \gamma \int_0^t \beta(s)\,ds,$$

which is decreasing with $\xi(0) = 1$, $\lim_{t\to\infty} \xi(t) = 1-\gamma$, so that $\xi(t) \ge 1-\gamma > 0$. We also use the norms

$$\|\mathbf{v}\|_l = \|A^{l/2}\mathbf{v}\| = \sqrt{(\mathbf{v}, A^l\mathbf{v})}, \quad l \in \mathbf{R}.$$

Theorem 1. *Let* $\mathbf{u}$ *be the solution of* (4) *with sufficiently smooth data* $\mathbf{u}_0, \mathbf{v}_0, \mathbf{f}$, $\mathbf{g}$, *and denote* $\mathbf{v} = \dot{\mathbf{u}}$ *and* $\mathbf{w}(t,s) = \mathbf{u}(t) - \mathbf{u}(t-s)$. *Then, for any* $l \in \mathbf{R}$, $T > 0$, *we have the identity*

$$\begin{aligned}
&\rho\|\mathbf{v}(T)\|_l^2 + \xi(T)\|\mathbf{u}(T)\|_{l+1}^2 \\
&\quad + \gamma\int_0^T \beta(t)\|\mathbf{u}(t)\|_{l+1}^2\,\mathrm{d}t + \gamma\int_0^T \beta(s)\|\mathbf{w}(T,s)\|_{l+1}^2\,\mathrm{d}s \\
&\quad + \gamma\int_0^T\int_0^t [\beta(s)-\beta(t)]\,\mathrm{D}_s\|\mathbf{w}(t,s)\|_{l+1}^2\,\mathrm{d}s\,\mathrm{d}t \\
&= \rho\|\mathbf{v}_0\|_l^2 + \|\mathbf{u}_0\|_{l+1}^2 + 2\int_0^T (\mathbf{f}, A^l\mathbf{v})\,\mathrm{d}t + 2\int_0^T \langle \mathbf{g}, A^l\mathbf{v}\rangle_{\Gamma_\mathrm{N}}\,\mathrm{d}t,
\end{aligned} \tag{6}$$

where all terms on the left-hand side are non-negative. Moreover, for $l=0$,

$$\begin{aligned}
&\rho^{\frac12}\|\dot{\mathbf{u}}(T)\| + (1-\gamma)^{\frac12}\|\mathbf{u}(T)\|_1 \le C\Big[\rho^{\frac12}\|\mathbf{v}_0\| + \|\mathbf{u}_0\|_1 \\
&\quad + \rho^{-\frac12}\int_0^T \|\mathbf{f}\|\,\mathrm{d}t + (1-\gamma)^{-\frac12}\Big(\max_{[0,T]}\|\mathbf{g}\|_{L_2(\Gamma_\mathrm{N})} + \int_0^T \|\dot{\mathbf{g}}\|_{L_2(\Gamma_\mathrm{N})}\,\mathrm{d}t\Big)\Big].
\end{aligned} \tag{7}$$

We remark that if $\mathbf{g}=\mathbf{0}$, then we can kick back for any l and estimate $\|\dot{\mathbf{u}}(T)\|_l + \|\mathbf{u}(T)\|_{l+1}$, see Theorem 2 below.

Proof. Equation (4) can be written in the form

$$\begin{aligned}
\rho(\dot{\mathbf{v}}(t),\boldsymbol{\psi}) + \xi(t)a(\mathbf{u}(t),\boldsymbol{\psi}) + \gamma\int_0^t \beta(s)a(\mathbf{w}(t,s),\boldsymbol{\psi})\,\mathrm{d}s& \\
= (\mathbf{f}(t),\boldsymbol{\psi}) + \langle \mathbf{g}(t),\boldsymbol{\psi}\rangle_{\Gamma_\mathrm{N}},\quad \forall\boldsymbol{\psi}\in V.&
\end{aligned} \tag{8}$$

Taking $\boldsymbol{\psi} = A^l\mathbf{v}(t)$ in (8) and integrating in t we get:

$$\begin{aligned}
&\rho\int_0^T (\mathrm{D}_t\mathbf{v}, A^l\mathbf{v})\,\mathrm{d}t + \int_0^T \xi(t)(A\mathbf{u}, A^l\mathbf{v})\,\mathrm{d}t \\
&+\gamma\int_0^T\int_0^t \beta(s)(A\mathbf{w}(t,s), A^l\mathbf{v}(t))\,\mathrm{d}s\,\mathrm{d}t = \int_0^T (\mathbf{f}, A^l\mathbf{v})\,\mathrm{d}t + \int_0^T \langle\mathbf{g}, A^l\mathbf{v}\rangle_{\Gamma_\mathrm{N}}\,\mathrm{d}t.
\end{aligned}$$

We consider each term on the left-hand side. For the first term we have:

$$\rho\int_0^T (\mathrm{D}_t\mathbf{v}, A^l\mathbf{v})\,\mathrm{d}t = \frac{\rho}{2}\int_0^T \mathrm{D}_t\|\mathbf{v}\|_l^2\,\mathrm{d}t = \frac{\rho}{2}\|\mathbf{v}(T)\|_l^2 - \frac{\rho}{2}\|\mathbf{v}_0\|_l^2. \tag{9}$$

For the second one we have:

$$\begin{aligned}
&\int_0^T \xi(t)(A\mathbf{u}, A^l\mathbf{v})\,\mathrm{d}t = \frac12\int_0^T \xi(t)\mathrm{D}_t\|\mathbf{u}\|_{l+1}^2\,\mathrm{d}t \\
&\quad = \frac12\xi(T)\|\mathbf{u}(T)\|_{l+1}^2 - \frac12\|\mathbf{u}_0\|_{l+1}^2 + \frac{\gamma}{2}\int_0^T \beta(t)\|\mathbf{u}(t)\|_{l+1}^2\,\mathrm{d}t.
\end{aligned} \tag{10}$$

For the third term, using $\mathbf{v}(t) = \dot{\mathbf{u}}(t) = \mathrm{D}_t\mathbf{w}(t,s) + \mathrm{D}_s\mathbf{w}(t,s)$, we get:

$$\begin{aligned}\int_0^T \int_0^t \beta(s) \left(A\mathbf{w}(t,s), A^l\mathbf{v}(t)\right) \mathrm{d}s\,\mathrm{d}t \\ = \frac{1}{2}\int_0^T \int_0^t \beta(s) \left(\mathrm{D}_t\|\mathbf{w}(t,s)\|_{l+1}^2 + \mathrm{D}_s\|\mathbf{w}(t,s)\|_{l+1}^2\right) \mathrm{d}s\,\mathrm{d}t.\end{aligned} \tag{11}$$

In the first term we change the order of integration:

$$\begin{aligned}\frac{1}{2}\int_0^T \int_s^T \beta(s)\mathrm{D}_t\|\mathbf{w}(t,s)\|_{l+1}^2 \,\mathrm{d}t\,\mathrm{d}s \\ = \frac{1}{2}\int_0^T \beta(s)\|\mathbf{w}(T,s)\|_{l+1}^2 \,\mathrm{d}s - \frac{1}{2}\int_0^T \beta(s)\|\mathbf{w}(s,s)\|_{l+1}^2 \,\mathrm{d}s.\end{aligned}$$

Using that

$$\frac{1}{2}\int_0^T \beta(s)\|\mathbf{w}(s,s)\|_{l+1}^2 \,\mathrm{d}s = \frac{1}{2}\int_0^T \int_0^t \beta(t)\mathrm{D}_s\|\mathbf{w}(t,s)\|_{l+1}^2 \,\mathrm{d}s\,\mathrm{d}t,$$

we can write

$$\begin{aligned}\int_0^T \int_0^t \beta(s) \left(A\mathbf{w}(t,s), A^l\mathbf{v}(t)\right) \mathrm{d}s\,\mathrm{d}t = \frac{1}{2}\int_0^T \beta(s)\|\mathbf{w}(T,s)\|_{l+1}^2 \,\mathrm{d}s \\ -\frac{1}{2}\int_0^T \int_0^t \left[\beta(t) - \beta(s)\right] \mathrm{D}_s\|\mathbf{w}(t,s)\|_{l+1}^2 \,\mathrm{d}s\,\mathrm{d}t.\end{aligned} \tag{12}$$

To show the positivity of the last term we consider, for $0 < \varepsilon < t$, the integral

$$\begin{aligned}\int_\varepsilon^t \left[\beta(t) - \beta(s)\right] \mathrm{D}_s\|\mathbf{w}(t,s)\|_{l+1}^2 \,\mathrm{d}s = -\left[\beta(t) - \beta(\varepsilon)\right] \|\mathbf{w}(t,\varepsilon)\|_{l+1}^2 \\ + \int_\varepsilon^t \beta'(s)\|\mathbf{w}(t,s)\|_{l+1}^2 \,\mathrm{d}s \le \beta(\varepsilon)\|\mathbf{w}(t,\varepsilon)\|_{l+1}^2,\end{aligned}$$

where we have taken into account that $\beta'(s) \le 0$ and $\beta(t) \ge 0$. Using

$$\mathbf{w}(t,\varepsilon) = \mathbf{w}(t,0) + \int_0^\varepsilon \mathrm{D}_s\mathbf{w}(t,s)\,\mathrm{d}s = \int_0^\varepsilon \mathrm{D}_s\mathbf{w}(t,s)\,\mathrm{d}s,$$

and the Cauchy-Schwarz inequality we get

$$\|\mathbf{w}(t,\varepsilon)\|_{l+1}^2 \le \left(\int_0^\varepsilon \|\mathrm{D}_s\mathbf{w}(t,s)\|_{l+1}\,\mathrm{d}s\right)^2 \le \int_0^\varepsilon \frac{\mathrm{d}s}{\beta(s)} \int_0^\varepsilon \beta(s)\|\mathrm{D}_s\mathbf{w}(t,s)\|_{l+1}^2 \,\mathrm{d}s,$$

and consequently

$$\int_\varepsilon^t \left[\beta(t) - \beta(s)\right] \mathrm{D}_s\|\mathbf{w}(t,s)\|_{l+1}^2 \,\mathrm{d}s \le \int_0^\varepsilon \frac{\beta(\varepsilon)}{\beta(s)}\,\mathrm{d}s \int_0^\varepsilon \beta(s)\|\mathrm{D}_s\mathbf{w}(t,s)\|_{l+1}^2 \,\mathrm{d}s.$$

But $\dfrac{\beta(\varepsilon)}{\beta(s)} \leq 1$, which yields $\displaystyle\int_0^\varepsilon \frac{\beta(\varepsilon)}{\beta(s)}\,\mathrm{d}s \leq \int_0^\varepsilon \mathrm{d}s = \varepsilon$, so that

$$\int_\varepsilon^t [\beta(t)-\beta(s)]\,\mathrm{D}_s\|\mathbf{w}(t,s)\|_{l+1}^2\,\mathrm{d}s \leq \varepsilon \int_0^\varepsilon \beta(s)\|\mathrm{D}_s\mathbf{w}(t,s)\|_{l+1}^2\,\mathrm{d}s.$$

According to [4] we have $\int_0^T \beta(s)\|\mathrm{D}_s\mathbf{w}(t,s)\|_{l+1}^2\,\mathrm{d}s < \infty$ provided that the data are sufficiently smooth ($A^{l+2}\mathbf{u}_0 \in H$, $A^{l+1}\mathbf{v}_0 \in H$, and certain conditions on $\mathbf{f}$, g). Letting $\boldsymbol{\varepsilon} \longrightarrow \mathbf{0}$ we get

$$\int_0^t [\beta(t)-\beta(s)]\,\mathbf{D}_s\|\mathbf{w}(t,s)\|_{l+1}^2\,\mathbf{d}s \leq \mathbf{0}. \tag{13}$$

From (9), (10), (12), and (13) we conclude (6), from which (7) follows easily, in view of the trace inequality and the fact that the energy norm $\|\cdot\|_1$ is equivalent to the $\boldsymbol{H}^1$ norm. □

The previous proof applies also to the finite element problem (5) if we use the orthogonal projection $P_h : H \to V_h$ and the operator $A_h : V_h \to V_h$ defined by

$$a(\mathbf{v}_h, \mathbf{w}_h) = (A_h\mathbf{v}_h, \mathbf{w}_h) \quad \forall \mathbf{v}_h, \mathbf{w}_h \in V_h,$$

and use the discrete norms

$$\|\mathbf{v}_h\|_{h,l} = \|A_h^{l/2}\mathbf{v}_h\| = \sqrt{(\mathbf{v}_h, A_h^l\mathbf{v}_h)}, \quad \mathbf{v}_h \in V_h,\ l \in \mathbf{R}.$$

It is sufficient to prove the discrete stability with boundary data $\mathbf{g} = \mathbf{0}$.

Theorem 2. *Let* $\mathbf{u}_h$ *solve* (5) *with* $\mathbf{g} = \mathbf{0}$. *Then, for any* $l \in \mathbf{R}$, $T > 0$, *we have*

$$\begin{aligned}\rho^{\frac{1}{2}}\|\dot{\mathbf{u}}_h(T)\|_{h,l} &+ (1-\gamma)^{\frac{1}{2}}\|\mathbf{u}_h(T)\|_{h,l+1} \\ &\leq C\Big[\rho^{\frac{1}{2}}\|\mathbf{v}_{h,0}\|_{h,l} + \|\mathbf{u}_{h,0}\|_{h,l+1} + \rho^{-\frac{1}{2}}\int_0^T \|P_h\mathbf{f}\|_{h,l}\,\mathrm{d}t\Big].\end{aligned} \tag{14}$$

4 A Priori Error Estimates

Let $R_h : V \to V_h$ be the Ritz projection defined by

$$a(R_h\mathbf{v} - \mathbf{v}, \mathbf{v}_h) = 0 \quad \forall \mathbf{v}_h \in V_h. \tag{15}$$

In this section we assume the elliptic regularity estimate

$$\|\mathbf{v}\|_{H^2} \leq C\|A\mathbf{v}\| \quad \forall \mathbf{v} \in D(A), \tag{16}$$

so that the following error estimates can be proved (by duality)

$$\|R_h\mathbf{v} - \mathbf{v}\|_{H^l} \leq Ch^{m-l}\|\mathbf{v}\|_{H^m}, \tag{17}$$

for all integers $0 \leq l < m \leq 2$. The elliptic regularity (16) holds, for example, for the pure Dirichlet problem ($\Gamma_{\mathrm{D}} = \partial\Omega$) when Ω is a convex polygonal domain. For more general boundary conditions and domains the situation is more complicated and we refrain from a discussion of this.

Theorem 3. *Let* $\mathbf{u}$ *and* $\mathbf{u}_h$ *be the solutions of* (4) *and* (5), *respectively, and denote* $\mathbf{e} = \mathbf{u}_h - \mathbf{u}$. *Then, with* C *depending on* ρ, γ,

$$\begin{aligned}
\|\dot{\mathbf{e}}(T)\| &\le C\Big(\|\mathbf{v}_{h,0} - \mathbf{v}_0\| + \|\mathbf{u}_{h,0} - R_h\mathbf{u}_0\|_{H^1}\Big) \\
&\quad + Ch^2\Big(\|\mathbf{v}_0\|_{H^2} + \|\dot{\mathbf{u}}(T)\|_{H^2} + \int_0^T \|\ddot{\mathbf{u}}\|_{H^2}\,\mathrm{d}t\Big), \\
\|\mathbf{e}(T)\|_{H^1} &\le C\Big(\|\mathbf{v}_{h,0} - \mathbf{v}_0\| + \|\mathbf{u}_{h,0} - \mathbf{u}_0\|_{H^1}\Big) \\
&\quad + Ch\Big(\|\mathbf{v}_0\|_{H^1} + \|\mathbf{u}_0\|_{H^2} + \|\mathbf{u}(T)\|_{H^2} + \int_0^T \|\ddot{\mathbf{u}}\|_{H^1}\,\mathrm{d}t\Big), \\
\|\mathbf{e}(T)\| &\le C\Big(\|\mathbf{v}_{h,0} - \mathbf{v}_0\| + \|\mathbf{u}_{h,0} - \mathbf{u}_0\|\Big) \\
&\quad + Ch^2\Big(\|\mathbf{v}_0\|_{H^2} + \|\mathbf{u}_0\|_{H^2} + \|\mathbf{u}(T)\|_{H^2} + \int_0^T \|\ddot{\mathbf{u}}\|_{H^2}\,\mathrm{d}t\Big).
\end{aligned}$$

Proof. In the usual way we split the error $\mathbf{u}_h - \mathbf{u} = \boldsymbol{\theta} + \boldsymbol{\rho}$, where $\boldsymbol{\theta} = \mathbf{u}_h - R_h\mathbf{u}$, $\boldsymbol{\rho} = R_h\mathbf{u} - \mathbf{u}$. In view of (17) it is sufficient to estimate $\boldsymbol{\theta}$. From (4), (5), and (15) we have

$$\rho(\ddot{\boldsymbol{\theta}}, \mathbf{v}_h) + a(\boldsymbol{\theta}, \mathbf{v}_h) - \gamma \int_0^t \beta(s) a(\boldsymbol{\theta}(t-s), \mathbf{v}_h)\,\mathrm{d}s = -\rho(\ddot{\boldsymbol{\rho}}, \mathbf{v}_h) \quad \forall \mathbf{v}_h \in V_h.$$

Applying the stability estimate (14) in Theorem 2 with $l = 0$, and using the fact that $\|\cdot\|_{h,0} = \|\cdot\|$ and that $\|\cdot\|_{h,1}$ is equivalent to $\|\cdot\|_{H^1}$, we get

$$\|\dot{\boldsymbol{\theta}}(T)\| + \|\boldsymbol{\theta}(T)\|_{H^1} \le C\Big(\|\dot{\boldsymbol{\theta}}(0)\| + \|\boldsymbol{\theta}(0)\|_{H^1} + \int_0^T \|P_h\ddot{\boldsymbol{\rho}}\|\,\mathrm{d}t\Big),$$

where C depends on ρ, γ. Similarly, with $l = -1$, we have

$$\|\dot{\boldsymbol{\theta}}(T)\|_{h,-1} + \|\boldsymbol{\theta}(T)\| \le C\Big(\|\dot{\boldsymbol{\theta}}(0)\|_{h,-1} + \|\boldsymbol{\theta}(0)\| + \int_0^T \|P_h\ddot{\boldsymbol{\rho}}\|_{h,-1}\,\mathrm{d}t\Big).$$

Using that $\|\cdot\|_{h,-1} \le C\|\cdot\|$, $\mathbf{e} = \boldsymbol{\theta} + \boldsymbol{\rho}$, $\|\boldsymbol{\theta}(0)\| \le \|\mathbf{e}(0)\| + \|\boldsymbol{\rho}(0)\|$, we obtain the desired estimates. □

5 Numerical Example

The purpose of the present numerical method is demonstrated by solving the dynamic viscoelastic equations in (3) for a two-dimensional structure under plane strain condition. Initial conditions, boundary conditions and model parameters read:

$$\begin{aligned}
&\mathbf{u}(\mathbf{x}, 0) = \mathbf{0}\,\mathrm{m}, \quad \dot{\mathbf{u}}(\mathbf{x}, 0) = \mathbf{0}\,\mathrm{m/s}, \quad \mathbf{f}(\mathbf{x}, t) = \mathbf{0}\,\mathrm{N/m^3}, \\
&\mathbf{u}(\mathbf{x}, t) = \mathbf{0}\,\mathrm{m} \text{ at } x_1 = 0\,\mathrm{m}, \quad \mathbf{g}(\mathbf{x}, t) = (0, -1)\Theta(t)\,\mathrm{Pa} \text{ at } x_1 = 1.5\,\mathrm{m}, \\
&\gamma = 0.5, \quad E_0 = 10\,\mathrm{MPa}, \quad \alpha = 0.5, \quad \nu = 0.3, \quad \rho = 40\,\mathrm{kg/m^3},
\end{aligned}$$

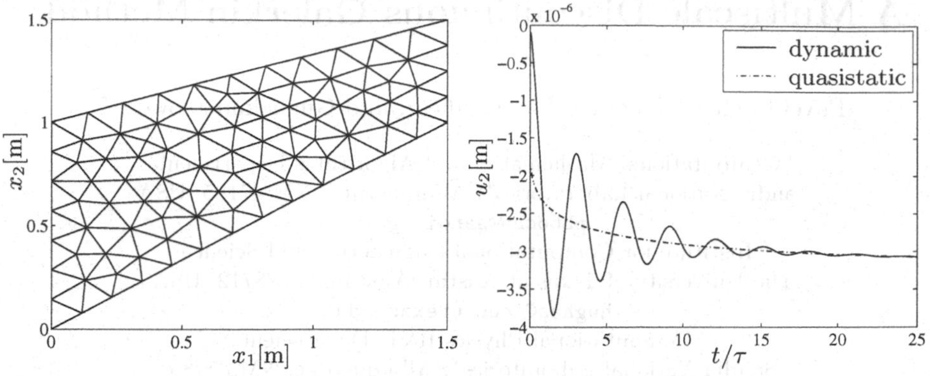

Fig. 1. The left figure shows the spatial discretization. The right figure shows the computed vertical displacement at the point (1.5,1.5) m.

where Θ is the Heaviside function. The geometry and the spatial discretization are shown in Figure 1. Figure 1 also shows the computed vertical displacement versus nondimensional time at the point $(1.5, 1.5)$ m. For comparison the quasistatic (i.e., neglecting inertia, $\rho\ddot{\mathbf{u}} \approx 0$) solution is included. As expected, the two solutions coincide for large times.

References

1. K. Adolfsson, M. Enelund, and S. Larsson, *Adaptive discretization of an integro-differential equation with a weakly singular convolution kernel*, Comput. Methods Appl. Mech. Engrg. **192** (2003), 5285–5304.
2. K. Adolfsson, M. Enelund, and S. Larsson, *Adaptive discretization of fractional order viscoelasticity using sparse time history*, Comput. Methods Appl. Mech. Engrg. **193** (2004), 4567–4590.
3. R. L. Bagley and P. J. Torvik, *Fractional calculus–a different approach to the analysis of viscoelastically damped structures*, AIAA Journal **21** (1983), 741–748.
4. R. H. Fabiano and K. Ito, *Semigroup theory and numerical approximation for equations in linear viscoelasticity*, SIAM J. Math. Anal. **21** (1990), 374–393.
5. C. Lubich, *Convolution quadrature and discretized operational calculus. I.*, Numerische Mathematik **52** (1988), 129–145.
6. W. McLean, V. Thomée, and L. B. Wahlbin, *Discretization with variable time steps of an evolution equation with a positive-type memory term*, Journal of Computational and Applied Mathematics **69** (1996), 49–69.
7. A. K. Pani, V. Thomée, and L. B. Wahlbin, *Numerical methods for hyperbolic and parabolic integro-differential equations*, J. Integral Equations Appl. **4** (1992), 533–584.
8. S. Shaw and J. R. Whiteman, *A posteriori error estimates for space-time finite element approximation of quasistatic hereditary linear viscoelasticity problems*, Comput. Methods Appl. Mech. Engrg. **193** (2004), 5551–5572.

A Multiscale Discontinuous Galerkin Method

Pavel Bochev[1], Thomas J.R. Hughes[2], and Guglielmo Scovazzi[3]

[1] Computational Mathematics and Algorithms Department,
Sandia National Laboratories*, Albuquerque, NM 87185, USA
pbboche@sandia.gov

[2] Institute for Computational Engineering and Science,
The University of Texas at Austin, Austin, TX 78712, USA
hughes@ices.utexas.edu

[3] Computational Physics R&D Department,
Sandia National Laboratories*, Albuquerque, NM 87185
gscovaz@sandia.gov

Abstract. We propose a new class of Discontinuous Galerkin (DG) methods based on variational multiscale ideas. Our approach begins with an additive decomposition of the discontinuous finite element space into continuous (coarse) and discontinuous (fine) components. Variational multiscale analysis is used to define an interscale transfer operator that associates coarse and fine scale functions. Composition of this operator with a donor DG method yields a new formulation that combines the advantages of DG methods with the attractive and more efficient computational structure of a continuous Galerkin method. The new class of DG methods is illustrated for a scalar advection-diffusion problem.

1 Introduction

Discontinuous Galerkin (DG) methods offer several important computational advantages over their continuous Galerkin counterparts. For instance, DG methods are particularly well-suited for application of h and p-adaptivity strategies. DG methods are also felt to have advantages of robustness over conventional Galerkin methods for problems of hyperbolic type [13, 11, 12]. There has also been recent interest in applying DG to elliptic problems so that advective-diffusive phenomena can be modeled; see Brezzi *et al.* [3], Dawson [7], and Hughes, Masud and Wan [10]. For a summary of the current state-of-the-art and introduction to the literature we refer to [1] and [6].

Despite the increased interest in DG methods, there are shortcomings that limit their practical utility. Foremost among these is the size of the DG linear system. Storage and solution cost are, obviously, adversely affected, which seems the main reason for the small industrial impact the DG method has had so far.

In [8] we proposed a new multiscale DG method that has the computational structure of a standard continuous Galerkin method. In this paper we extend

* Sandia is a multiprogram laboratory operated by Sandia Corporation, a Lockheed Martin Company, for the United States Department of Energy under contract DE-AC04-94-AL85000.

I. Lirkov, S. Margenov, and J. Waśniewski (Eds.): LSSC 2005, LNCS 3743, pp. 84–93, 2006.

this idea to a general multiscale framework for DG methods. Our approach starts with an additive decomposition of a given discontinuous finite element space into continuous (coarse) and discontinuous (fine) components. Then, variational multiscale analysis is used to define an interscale transfer operator that associates coarse and fine scale functions. Composition of this operator with a donor DG method yields a new formulation that combines the advantages of DG methods with the attractive and more efficient computational structure of a continuous Galerkin method. Variational multiscale analysis leads to a natural definition of local, elementwise problems that allow for an efficient computation of the interscale operator.

2 Notation

Throughout this paper Ω will denote an open bounded region in $\mathbb{R}^n$, $n = 2, 3$ with a polyhedral boundary $\partial\Omega$. We recall the standard Sobolev spaces $L^2(\Omega)$ and $H^1(\Omega)$. Let $\mathcal{T}_h$ be a regular partition of Ω into finite elements K that contains only regular nodes [4]. For simplicity, we limit our discussion to two space dimensions. Extension to three dimensions is straightforward.

Every element $K \in \mathcal{T}_h$ is an image of a reference element $\hat{K}$ that can be a triangle $\hat{T}$ or a square $\hat{Q}$. The vertices $\mathbf{v}$ and the edges $\mathbf{e}$ of K form the sets $V(K)$ and $E(K)$, respectively; $V(\mathcal{T}_h) = \cup_{K\in\mathcal{T}_h} V(K)$, $E(\mathcal{T}_h) = \cup_{K\in\mathcal{T}_h} E(K)$, Γ_h^0 is the set of all internal edges and Γ_h is the set of all edges on $\partial\Omega$.

The local space. The reference space $S^{p(\hat{K})}(\hat{K})$ on $\hat{K}$ is defined as follows:

$$S^{p(\hat{K})}(\hat{K}) = \begin{cases} \varphi = \sum\limits_{i,j} a_{ij}\xi_1^i\xi_2^j\,, \quad 0 \le i,j \le p(\hat{K});\ i+j \le p(\hat{K}) & \text{if } \hat{K} = \hat{T} \\ \varphi = \sum\limits_{i,j} a_{ij}\xi_1^i\xi_2^j\,, \quad 0 \le i,j \le p(\hat{K}) & \text{if } \hat{K} = \hat{Q} \end{cases} \tag{1}$$

The local element spaces $S^{p(K)}(K)$ are defined by a mapping of the reference space (1) to the physical space.

The discontinuous finite element space. Given two integers $0 \le p_{\min} < p_{\max}$ we consider the following finite element subspace of $L^2(\Omega)$

$$\Phi_h(\Omega) = \left\{ \varphi_h \in L^2(\Omega) \,|\, \varphi_{h|_K} \in S^{p(K)}(K),\ \ p_{\min} \le p(K) \le p_{\max};\ \forall K \in \mathcal{T}_h \right\}. \tag{2}$$

We will assume that $p_{\min} \ge 1$. Note that $\Phi_h(\Omega)$ is a formal union of the local spaces $S^{p(K)}(K)$.

The continuous finite element space. The additive decomposition of $\Phi_h(\Omega)$ is induced by a finite element subspace $\overline{\Phi}_h(\Omega)$ of $H^1(\Omega)$, defined with respect to the same partition $\mathcal{T}_h$ of Ω into finite elements. The space $\overline{\Phi}_h(\Omega)$ can be defined in many possible ways. However, to ensure H^1 conformity, functions in this space are constrained to be continuous across element interfaces; see [5]. Here, for simplicity we consider a minimal choice of $\overline{\Phi}_h(\Omega)$ given by (see Fig. 2)

$$\overline{\Phi}_h(\Omega) = \{\overline{\varphi}_h \in H^1(\Omega) \,|\, \overline{\varphi}_h|_K \in S^1(K)\}\,. \tag{3}$$

In $\overline{\Phi}_h(\Omega)$ we consider a nodal basis $\{\overline{V}_{\overline{\mathbf{v}}}\}$; $\overline{\mathbf{v}} \in V(\mathcal{T}_h)$ such that $\overline{V}_{\overline{\mathbf{v}}_i}(\overline{\mathbf{v}}_j) = \delta_{ij}$. The basis functions have local supports given by $\mathrm{supp}(\overline{V}_{\overline{\mathbf{v}}}) = \cup_{\overline{\mathbf{v}} \in V(K)} K$. For $K \in \mathrm{supp}(\overline{V}_{\overline{\mathbf{v}}})$, $\overline{V}_{\overline{\mathbf{v}}}|_K = V_{\mathbf{v}}$ where $\mathbf{v} \in V(K)$ is the local vertex that corresponds to the global vertex $\overline{\mathbf{v}} \in V(\mathcal{T}_h)$. Owing to the assumption $p_{\min} \geq 1$ the space $\overline{\Phi}_h(\Omega)$ is contained in $\Phi_h(\Omega)$. While the actual choice of $\overline{\Phi}_h(\Omega)$ and the resulting decomposition will have an impact on the accuracy of the multiscale DG, it will not affect formulation of the overall framework.

Orientations, jumps and averages. We briefly review the relevant notation following the Brezzi conventions. We assume that all edges in $E(\mathcal{T}_h)$ are endowed by orientation. A convenient way to orient an edge is to pick a normal direction to that edge; see Fig. 2. An element can be oriented by selecting one of the two possible normal directions to its boundary ∂K. Without loss of generality, all elements are oriented by using the outward normal.

An internal edge $\mathbf{e} \in \Gamma_h^0$ is shared by exactly two elements. The outward normal on one of these elements will coincide with the normal used to orient $\mathbf{e}$; we call this element K^-. The outward normal on the other element will have the opposite direction to the normal on $\mathbf{e}$; we call this element K^+; see Fig. 2. Edge orientation also induces partition of the boundary of an internal element into $\partial^+ K$, consisting of all edges whose normal direction coincides with the outer

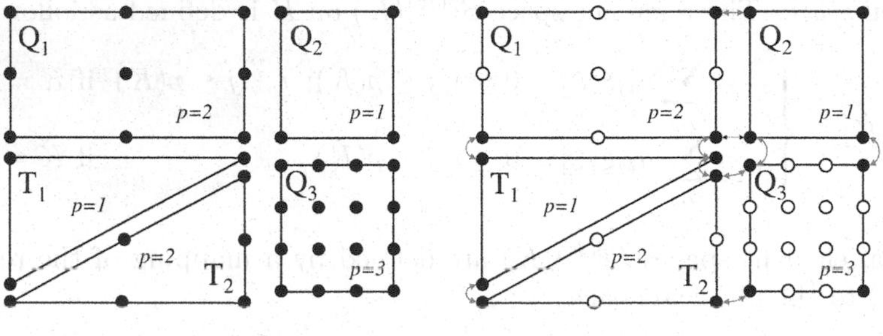

Fig. 1. The space $\Phi_h(\Omega)$ (left) and the corresponding minimal C^0 space $\overline{\Phi}_h(\Omega)$ (right)

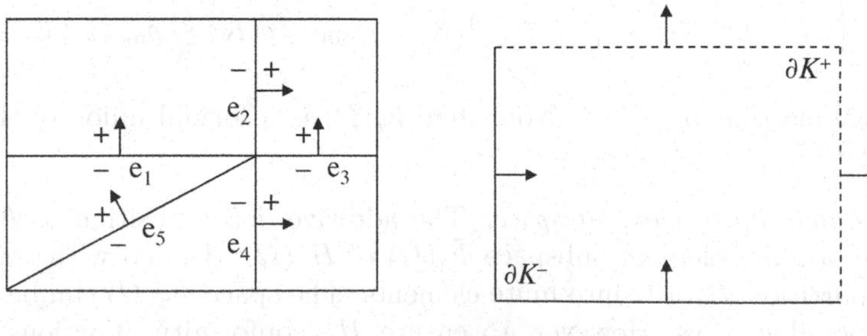

Fig. 2. Orientation of internal edges in $\mathcal{T}_h$ and $+/-$ elements with respect to an edge (left). Partition of element boundary into $\partial^+ K$ and $\partial^- K$ (right).

normal on ∂K and $\partial^- K$, consisting of all edges $\mathbf{e}$ whose normal direction is opposite to the outer normal on ∂K.

Let φ be a scalar field, and $\varphi^\pm := \varphi|_{K^\pm}$. For $\mathbf{e} \in \Gamma_h^0$ we define the *average* and the *jump* as $\langle\varphi\rangle := \frac{1}{2}(\varphi^+ + \varphi^-)$ and $[\varphi] := \varphi^+ \boldsymbol{n}^+ + \varphi^- \boldsymbol{n}^-$, respectively. Analogously, if $\mathbf{u}$ is a vector field, $\langle\mathbf{u}\rangle := \frac{1}{2}(\mathbf{u}^+ + \mathbf{u}^-)$ and $[\mathbf{u}] := \mathbf{u}^+ \cdot \boldsymbol{n}^+ + \mathbf{u}^- \cdot \boldsymbol{n}^-$. Note that, by definition of "$[\cdot]$", the jump of a scalar quantity is a vector and the jump of a vector quantity is a scalar. For edges belonging to Γ_h, $[\varphi] = \varphi\, \boldsymbol{n}$ and $\langle\mathbf{u}\rangle = \mathbf{u}$. It will not be necessary to define $\langle\varphi\rangle$ and $[\mathbf{u}]$ on the boundary Γ, because they are never utilized.

3 Multiscale Discontinuous Galerkin Method

We consider an abstract linear boundary value problem

$$\mathcal{L}(\mathbf{x}, D)\varphi = f \text{ in } \Omega \quad \text{and} \quad \mathcal{R}(\mathbf{x}, D)\varphi = g \text{ on } \Gamma . \tag{4}$$

The multiscale DG framework for problem (4) has two basic components. The first is a *donor* DG formulation for (4): *find* $\varphi_h \in \Phi_h(\Omega)$ *such that*

$$B_{DG}(\varphi_h; \psi_h) = F_{DG}(\psi_h) \quad \forall \psi_h \in \Phi_h(\Omega) . \tag{5}$$

In (5), $B_{DG}(\cdot;\cdot)$ is a continuous bilinear form $\Phi_h(\Omega) \times \Phi_h(\Omega) \mapsto \mathbb{R}$ and $F_{DG}(\cdot)$ is a bounded linear functional $\Phi_h(\Omega) \mapsto \mathbb{R}$. We assume that (5) has a unique solution φ_h that depends continuously on the data and converges (in a suitable norm) to all sufficiently smooth solutions φ of (4). The second component is an interscale transfer (or expansion) operator

$$T : \overline{\Phi}_h(\Omega) \mapsto \Phi_h(\Omega) . \tag{6}$$

We assume that T is a bounded linear operator, however, it is not required to be surjective, or invertible. Thus, in general $T(\overline{\Phi}_h(\Omega))$ will be a proper subspace of the discontinuous space $\Phi_h(\Omega)$.

We define the Multiscale DG (MDG) method by a composition of the donor DG scheme with the interscale transfer operator T: *find* $\overline{\varphi}_h \in \overline{\Phi}_h(\Omega)$ *such that*

$$B_{DG}(T\overline{\varphi}_h; T\overline{\psi}_h) = F_{DG}(T\overline{\psi}_h) \quad \forall \overline{\psi}_h \in \overline{\Phi}_h(\Omega) . \tag{7}$$

Substitution of discontinuous test and trial functions in the donor DG method by $T\overline{\psi}_h$ and $T\overline{\varphi}_h$ reduces the number of degrees-of-freedom in the MDG formulation to that of a standard Galerkin method posed on $\overline{\Phi}_h(\Omega)$. Since $T(\overline{\Phi}_h(\Omega)) \subset \Phi_h(\Omega)$, (7) occupies a middle ground between a DG and a CG method for (4).

3.1 Definition of the Interscale Operator

The definition of the interscale operator T is key to a robust, efficient and accurate MDG method. For instance, it is desirable to compute T locally on each element. To discuss definition of this operator assume that

$$B_{DG}(\varphi_h; \psi_h) = \sum_{K \in \mathcal{T}_h} B_K(\varphi_h; \psi_h) + \sum_{\mathbf{e} \in \Gamma_h} B_\Gamma(\varphi_h; \psi_h) + \sum_{\mathbf{e} \in \Gamma_h^0} B_{\mathbf{e}}\left(\{\varphi_h^-, \varphi_h^+\}; \{\psi_h^-, \psi_h^+\}\right) \tag{8}$$

where $B_K(\cdot;\cdot)$ is a bilinear *local* element form defined for every $K \in \mathcal{T}_h$, $B_\Gamma(\cdot;\cdot)$ is a bilinear form defined on $\mathbf{e} \in \Gamma_h$, and $B_\mathbf{e}(\{\cdot\};\{\cdot\})$ is an *edge* bilinear form defined for $\mathbf{e} \in \Gamma_h^0$.

To define T we proceed to formally split functions $\varphi_h \in \Phi_h(\Omega)$ into a continuous ("coarse" scale) part $\overline{\varphi}_h \in \overline{\Phi}_h(\Omega)$ and a discontinuous ("fine" scale) component $\varphi'_h \in \Phi_h(\Omega)$, viz. $\varphi_h = \overline{\varphi}_h + \varphi'_h$. Then, (5) takes the following form:

$$\begin{aligned} B_{DG}(\overline{\varphi}_h;\overline{\psi}_h) + B_{DG}(\varphi'_h;\overline{\psi}_h) &= F_{DG}(\overline{\psi}_h) \quad \forall \overline{\varphi}_h \in \overline{\Phi}_h(\Omega) \\ B_{DG}(\varphi'_h;\psi'_h) + B_{DG}(\overline{\varphi}_h;\psi'_h) &= F_{DG}(\psi'_h) \quad \forall \psi'_h \in \Phi_h(\Omega) \end{aligned} \tag{9}$$

The first line in (9) is the coarse scale equation. The second line is the fine scale equation that will be used to define T. Treating the coarse scale function as data we write this equation as: *find* $\varphi'_h \in \Phi_h(\Omega)$ *such that*

$$B_{DG}(\varphi'_h;\psi'_h) = F_{DG}(\psi'_h) - B_{DG}(\overline{\varphi}_h;\psi'_h) \quad \forall \psi'_h \in \Phi_h(\Omega). \tag{10}$$

We restrict (10) to an element K by choosing test functions $\psi'_h \in S^{p(K)}(K)$ that vanish outside of this element. With the above selection of a test function, $(\psi'_h)^+ = \chi(\partial^- K)\psi'_h$ and $(\psi'_h)^- = \chi(\partial^+ K)\psi'_h$ where $\chi(\cdot)$ is the characteristic function. Using these identities and that $(\overline{\varphi}_h)^+ = (\overline{\varphi}_h)^- = \overline{\varphi}_h$, for a C^0 function, the restricted fine scale problem can be expressed as follows: *find* $\varphi'_h \in S^{p(K)}(K)$ *such that*

$$\begin{aligned} &B_K(\varphi'_h;\psi'_h) + B_\Gamma(\varphi'_h;\psi'_h) + \sum_{\mathbf{e}\in E(K)} B_\mathbf{e}\left(\{(\varphi'_h)^-,(\varphi'_h)^+\};\{\chi(\partial^+K)\psi'_h,\chi(\partial^-K)\psi'_h\}\right) \\ &= F_{DG}(\psi'_h) - B_K(\overline{\varphi}_h;\psi'_h) - B_\Gamma(\overline{\varphi}_h;\psi'_h) \\ &\quad - \sum_{\mathbf{e}\in E(K)} B_\mathbf{e}\left(\{\overline{\varphi}_h,\overline{\varphi}_h\};\{\chi(\partial^+K)\psi'_h,\chi(\partial^-K)\psi'_h\}\right) \quad \forall \psi'_h \in S^{p(K)}(K). \end{aligned} \tag{11}$$

Problem (11) relates fine scales to the coarse scales, but remains coupled to the contiguous elements through the numerical flux terms in (11). Therefore, it does not meet our criteria for localized computation of the interscale transfer operator T. However, we make the important observation that our goal is not to solve the DG problem (9) but rather use it to define a local computation procedure for T that maps $\overline{\varphi}_h$ into the local space $S^{p(K)}(K)$. We note that this objective is reminiscent of other applications of variational multiscale analysis in which the fine scale problem is used for *estimation* rather than *approximation* of the unresolved solution component. This process can be accomplished by a modification of the numerical flux inherited from the donor DG formulation, or by using a new flux defined only in terms of the local function $\varphi'_h \in S^{p(K)}(K)$. Let $B'_\mathbf{e}(\{\cdot\};\{\cdot\})$ be the new numerical flux. The local fine scale problem obtained from (11) is: *find* $\varphi'_h \in S^{p(K)}(K)$ *such that*

$$\begin{aligned} &B_K(\varphi'_h;\psi'_h) + B_\Gamma(\varphi'_h;\psi'_h) + \sum_{\mathbf{e}\in E(K)} B'_\mathbf{e}\left(\{\varphi'_h\};\{\psi'_h\}\right) \\ &= F_{DG}(\psi'_h) - B_K(\overline{\varphi}_h;\psi'_h) - B_\Gamma(\overline{\varphi}_h;\psi'_h) \\ &\quad - \sum_{\mathbf{e}\in E(K)} B_\mathbf{e}\left(\{\overline{\varphi}_h,\overline{\varphi}_h\};\{\chi(\partial^+K)\psi'_h,\chi(\partial^-K)\psi'_h\}\right) \quad \forall \psi'_h \in S^{p(K)}(K). \end{aligned} \tag{12}$$

Problem (12) is a local equation that can be solved on an element by element basis. This problem defines an operator $T_K : \overline{\Phi}_h(\Omega) \mapsto S^{p(K)}(K)$ that maps any given C^0 finite element function $\overline{\varphi}_h$ to a function in the local element space $S^{p(K)}(K)$. Therefore,

$$T : \overline{\Phi}_h(\Omega) \mapsto \Phi_h(\Omega); \quad T|_K = T_K \quad \forall K \in \mathcal{T}_h \tag{13}$$

defines an interscale transfer operator T for the MDG method. The abstract variational equation (7) and the local problem (12) complete the definition of the MDG framework.

4 Multiscale DG for a Scalar Advection-Diffusion Problem

We consider a model advection diffusion problem written in conservative form:

$$\begin{cases} \nabla \cdot (F_a + F_d) = f \text{ in } \Omega; & -(F_a + F_d) \cdot \boldsymbol{n} = h^- \text{ on } \Gamma_n^- \\ \varphi = g \text{ on } \Gamma_g; & -(F_d) \cdot \boldsymbol{n} = h^+ \text{ on } \Gamma_n^+ \end{cases} \tag{14}$$

where $F_d = -\kappa \nabla \varphi$ and $F_a = \mathbf{a}\varphi$ denote diffusive and advective flux, respectively. The total flux is $F = F_a + F_d$. The Neumann boundary condition can be written compactly as $-(\chi(\Gamma_n^-) F_a + F_d) \cdot \boldsymbol{n} = h$; where $h = \chi(\Gamma_n^-) h^- + \chi(\Gamma_n^+) h^+$.

4.1 A Donor DG Method for the Model Problem

When dealing with advection-diffusion problems it is profitable to coordinate edge orientations with the advective direction. Given an edge $\mathbf{e}$ we choose the normal $\boldsymbol{n}_\mathbf{e}$ for which $\boldsymbol{n}_\mathbf{e} \cdot \mathbf{a} \geq 0$. A general weighted residual form of a Discontinuous Galerkin method for (14) is given by: *find* $\varphi \in \Phi_h(\Omega)$ *such that*

$$\begin{aligned} 0 = \sum_{i=1}^{n_{el}} -\int_{K_i} (F_i \cdot \nabla \psi + f\psi)\, d\Omega + \int_{\Gamma_n} (\chi(\Gamma_n^+) F_a \cdot \boldsymbol{n} - h)\psi dl + \int_{\Gamma_g} (F \cdot \boldsymbol{n})\psi dl + \\ \int_{\Gamma_g} \epsilon(\varphi - g) W(\psi) dl + \sum_{\mathbf{e} \in \Gamma_h^0} \int_{\mathbf{e}} \Big(F_b^h(\varphi^+; \varphi^-) \cdot [\psi] + F_c^h(\psi^+; \psi^-) \cdot [\varphi] + \alpha [\varphi][\psi] \Big)\, dl \end{aligned} \tag{15}$$

for all $\psi \in \Phi_h(\Omega)$. Above, $W(\psi)$ is a weight function that enforces the Dirichlet boundary condition weakly,

$$F_b^h \stackrel{\text{def}}{=} s_{11} F_a^h + s_{12} F_d^h \quad \text{and} \quad F_c^h \stackrel{\text{def}}{=} s_{21} F_a^h + s_{22} F_d^h \tag{16}$$

are numerical models of the total flux across $\mathbf{e} \in \Gamma_h^0$ and

$$F_a^h \stackrel{\text{def}}{=} F_a^h(\varphi^+, \varphi^-) \quad \text{and} \quad F_d^h \stackrel{\text{def}}{=} F_d^h(\varphi^+, \varphi^-) \tag{17}$$

are constitutive relations for the advective and the diffusive fluxes across $\mathbf{e}$ in terms of the solution states φ^+ and φ^- from the two elements that share $\mathbf{e}$. The component bilinear forms in (8) can be easily identified from (15):

$$B_K(\varphi;\psi) = \int_K -F_K \cdot \nabla\psi \, d\Omega \tag{18}$$

$$B_\Gamma(\varphi;\psi) = \int_{\Gamma_n} (\chi(\Gamma_n^+)F_a \cdot \boldsymbol{n})\psi \, dl + \int_{\Gamma_g} (F \cdot \boldsymbol{n})\psi \, dl + \epsilon \int_{\Gamma_g} \varphi W(\psi) \, dl \tag{19}$$

$$B_{\mathbf{e}}\left(\{\varphi^+;\varphi^-\};\{\psi^+;\psi^-\}\right) = \int_{\mathbf{e}} \left(F_b^h(\varphi^+;\varphi^-)\cdot[\psi] + F_c^h(\psi^+;\psi^-)\cdot[\varphi] + \alpha[\varphi][\psi]\right) dl \,. \tag{20}$$

particular donor DG method is obtained from (15) by specification of ϵ, α, the numerical fluxes in (16)–(17) for the internal edges Γ_h^0, and the weight function $W(\psi)$.

We set $\epsilon = \alpha = \delta\kappa/h_\perp$, where $\delta > 0$ is non-dimensional parameter and $h_\perp = (\text{meas}(K^+) + \text{meas}(K^-))/(2\ \text{meas}(\mathbf{e}))$. Roughly speaking, $h_\perp$ is a length scale in the direction perpendicular to the edge e, close to the length of the segment joining the barycenters of K^- and K^+.

A standard choice for F_a^h is the upwinded advective flux $F_a^h(\psi^+;\psi^-) = F_a(\psi^-) = \mathbf{a}\varphi^-$. Possible choices for the numerical diffusive flux are the averaged flux $F_d^h(\psi^+;\psi^-) = \langle F_d(\psi)\rangle = -\frac{1}{2}(\kappa\nabla\psi^+ + \kappa\nabla\psi^-)$ or the upwinded flux $F_d^h(\psi^+;\psi^-) = F_d(\psi^-) = -\kappa\nabla\psi^-$. To define F_b^h and F_c^h we set $s_{11} = s_{12} = 1$, $s_{21} = 0$ and $s_{22} = s \in \{-1, 0, +1\}$ in (16). This leads to two different donor DG methods: DG-A which uses averaged diffusive flux, and DG-B which uses the upwinded version of that flux; see [8]. Flux and weight function definitions for the two methods are summarized in Table 1.

The effect of the parameter s has been extensively studied in the discontinuous Galerkin literature (see Arnold *et al.* [1], Baumann and Oden [2], and Hughes *et al.* [9]). The symmetric formulation ($s = -1$) is adjoint-consistent, guaranteeing optimal L_2-convergence rates in the diffusive limit. Ostensibly, the skew formulation ($s = +1$) has superior stability properties but the ϵ and α-terms can be used to improve the stability behavior of the neutral (i.e., $s = 0$) and symmetric formulations. For more details about the implementation of the donor DG and numerical results we refer to [8].

For DG-B the numerical flux F_b^h is simply the upwinded total flux $F(\varphi^-)$. DG-A and DG-B have the same element form $B_K(\cdot;\cdot)$ (given by (18)) and the same boundary form:

$$B_\Gamma(\varphi;\psi) = \int_{\Gamma_n} (\chi(\Gamma_n^+)F_a \cdot \boldsymbol{n})\psi \, dl + \int_{\Gamma_g} (F \cdot \boldsymbol{n})\psi \, dl + \epsilon \int_{\Gamma_g} \varphi \underbrace{(\psi - s\kappa\nabla\psi \cdot \boldsymbol{n})}_{W(\psi)} \, dl \tag{21}$$

The internal edge form for DG-A is

$$B_{\mathbf{e}}\left(\{\varphi^+;\varphi^-\};\{\psi^+;\psi^-\}\right) = \int_{\mathbf{e}} \alpha[\varphi][\psi] \, dl + \int_{\mathbf{e}} \Big(\underbrace{(\mathbf{a}\varphi^- - (\kappa\nabla\varphi^+ + \kappa\nabla\varphi^-)/2)}_{F_b^h}\cdot[\psi] - \underbrace{s(\kappa\nabla\psi^+ + \kappa\nabla\psi^-)/2}_{F_c^h}\cdot[\varphi]\Big) \, dl \tag{22}$$

while for DG-B this form is given by

$$B_{\mathbf{e}}\left(\{\varphi^+;\varphi^-\};\{\psi^+;\psi^-\}\right) = \int_{\mathbf{e}} \alpha[\varphi][\psi] \, dl + \int_{\mathbf{e}} \Big(\underbrace{(\mathbf{a}\varphi^- - \kappa\nabla\varphi^-)}_{F_b^h}\cdot[\psi] - \underbrace{s\kappa\nabla\psi^-}_{F_c^h}\cdot[\varphi]\Big) \, dl \,. \tag{23}$$

Table 1. Specialization of fluxes and weight function for the donor DG methods

Function	DG-A	DG-B
$F_b^h(\varphi^+;\varphi^-)$	$F_a(\varphi^-)+\langle F_d(\varphi)\rangle$	$F_a(\varphi^-)+F_d(\varphi^-)$
$F_c^h(\psi^+;\psi^-)$	$s\langle F_d(\psi)\rangle$	$sF_d(\psi^-)$
$W(\psi)$	$\psi+sF_d(\psi)\cdot\boldsymbol{n}$	

4.2 The Interscale Operator

We develop a consistent approach that reduces the edge form $B_{\mathbf{e}}(\{\cdot\};\{\cdot\})$ in the donor DG method to a form defined in terms of the local (fine scale) variable φ' and test function ψ'. In doing so we aim to preserve as much as possible from the structure of the donor DG method in the local problem.

For this purpose we redefine the calculation of the jump, the average and the states $\varphi^{\pm}$, $\psi^{\pm}$ as follows: given $\psi\in S^{p(K)}(K)$ its states are defined by

$$\psi^+=\chi(\partial^-K)\psi \quad\text{and}\quad \psi^-=\chi(\partial^+K)\psi \tag{24}$$

its jump is the vector

$$[\psi]=\boldsymbol{n}_K\psi\,, \tag{25}$$

and its average is the function itself:

$$\langle\psi\rangle=\psi\,. \tag{26}$$

The rules in (24)-(26) have the following interpretation. To compute the states and the jump of ψ, extend by zero to a function $\psi_0\in L^2(\Omega)$. Then $[\psi_0]=\boldsymbol{n}^+\chi(\partial^-K)\psi_0+\boldsymbol{n}^-\chi(\partial^+K)\psi_0=\boldsymbol{n}_K\psi_0$. Definition (26) can be motivated by noting that for affine elements ψ can be trivially extended to a function$\psi_\infty\in C^\infty(\Omega)$ for which $\langle\psi_\infty\rangle=\frac{1}{2}(\psi_\infty+\psi_\infty)=\psi_\infty$ giving (26). The local definitions of the numerical fluxes obtained through (24)-(26) are summarized in Table 2.

Local Problem for DG-A. The localized edge form for DG-A method is

$$B'_{\mathbf{e}}(\{\varphi\};\{\psi\})=\int_{\mathbf{e}}\Big(\underbrace{(\mathbf{a}\chi(\partial^+K)\varphi-\kappa\nabla\varphi)}_{F_b^h}\cdot\boldsymbol{n}_K\psi-\underbrace{s\kappa\nabla\psi}_{F_c^h}\cdot\boldsymbol{n}_K\varphi+\alpha\varphi\psi\Big)\,d\mathbf{l}\,. \tag{27}$$

The last two terms can be combined into a single weight function $W_\alpha(\psi)=\alpha\psi-s\kappa\nabla\psi\cdot\boldsymbol{n}_K$. Thus, the local problem obtained from DG-A is: given a $\overline{\varphi}\in\overline{\Phi}_h(\Omega)$ *find* $\varphi'\in S^{p(K)}(K)$ *such that*

Table 2. Specialization of fluxes for the local problem

Function	DG-A	DG-B
$F_b^h(\varphi)$	$F_a(\chi(\partial^+K)\varphi)+F_d(\varphi)$	$F_a(\chi(\partial^+K)\varphi)+F_d(\chi(\partial^+K)\varphi)$
$F_c^h(\psi)$	$sF_d(\psi)$	$sF_d(\chi(\partial^+K)\psi)$

$$B_K(\varphi';\psi') + B_\Gamma(\varphi';\psi') + \sum_{\mathsf{e}\in\partial K}\int_{\mathsf{e}}\Big((\mathbf{a}\chi(\partial^+K)\varphi' - \kappa\nabla\varphi')\cdot\boldsymbol{n}_K\psi' + \varphi' W_\alpha(\psi')\Big)\,dl$$

$$= F_{DG}(\psi') - B_K(\overline{\varphi};\psi') - B_\Gamma(\overline{\varphi};\psi')$$

$$-\sum_{\mathsf{e}\in\partial K} B_{\mathsf{e}}\left(\{\overline{\varphi},\overline{\varphi}\};\{\chi(\partial^+K)\psi',\chi(\partial^-K)\psi'\}\right) \quad \forall\psi'\in S^{p(K)}(K). \qquad (28)$$

Remark 1. This local problem is identical to the one used in [8].

Local Problem for DG-B. For DG-B we have the localized edge form:

$$B'_{\mathsf{e}}(\{\varphi\};\{\psi\}) = \int_{\mathsf{e}}\Big(\underbrace{\chi(\partial^+K)(\mathbf{a}\varphi - \kappa\nabla\varphi)}_{F_b^h}\cdot\boldsymbol{n}_k\psi - s\underbrace{\chi(\partial^+K)\kappa\nabla\psi}_{F_c^h}\cdot\boldsymbol{n}_K\varphi + \alpha\varphi\psi\Big)\,dl. \qquad (29)$$

The last two terms can be combined into the weight function $W_\alpha^-(\psi) = \alpha\psi - s\chi(\partial^+K)\nabla\psi\cdot\boldsymbol{n}_K$, which is an "upwinded" version of $W_\alpha(\psi)$. The local problem is: given a $\overline{\varphi}\in\overline{\Phi}_h(\Omega)$ *find* $\varphi'\in S^{p(K)}(K)$ *such that*

$$B_K(\varphi';\psi') + B_\Gamma(\varphi';\psi') + \sum_{\mathsf{e}\in\partial K}\int_{\mathsf{e}}\Big(\chi(\partial^+K)(\mathbf{a}\varphi' - \kappa\nabla\varphi')\cdot\boldsymbol{n}_K\psi' + \varphi' W_\alpha(\psi')\Big)\,dl$$

$$= F_{DG}(\psi') - B_K(\overline{\varphi};\psi') - B_\Gamma(\overline{\varphi};\psi')$$

$$-\sum_{\mathsf{e}\in\partial K} B_{\mathsf{e}}\left(\{\overline{\varphi},\overline{\varphi}\};\{\chi(\partial^+K)\psi',\chi(\partial^-K)\psi'\}\right) \quad \forall\psi'\in S^{p(K)}(K). \qquad (30)$$

5 Conclusions

In this work we extended the DG method developed in [8] to a general framework for multiscale DG methods that have the computational structure of continuous Galerkin methods. This represents a solution to a fundamental and long-standing problem in discontinuous-Galerkin technology, namely, restraining the proliferation of degrees-of-freedom. Numerical results reported in [8] indicate that for a scalar advection-diffusion equation the new method at least attains, and even somewhat improves upon, the performance of the associated continuous Galerkin method. Within the framework of the multiscale discontinuous Galerkin method, the local problem provides a vehicle for incorporating the necessary stabilization features such as discontinuity capturing and upwinding. There seems to be a potential connection here with ideas from wave propagation methods based on solutions of the Riemann problem, which is worth exploring in more detail.

The MDG formulation can be also viewed as an approach that enables uncoupling of *storage* locations of the data from the *computational* locations where this data is used. For example, one can envision a situation where information is stored at the nodes and then mapped to flux and circulation degrees-of-freedom by the operator T. Such an extension of MDG appears to be a fruitful direction for further research.

Acknowledgements

The authors would like to thank Giancarlo Sangalli and the anonymous referees for helpful remarks. The work of T. J. R. Hughes was supported by Sandia Contract No. A0340.0 with the University of Texas at Austin. This support is gratefully acknowledged.

References

1. D. N. Arnold, F. Brezzi, B. Cockburn and L. D. Marini, Unified analysis of discontinuous Galerkin methods for elliptic problems, SIAM Journal on Numerical Analysis, **39** (5), (2002), 1749–1779
2. C.E. Baumann and J.T. Oden, A discontinuous *hp* finite element method for convection-diffusion problems, Comp. Meth. Appl. Mech. Engrg., **175**, (1999) 311–341
3. F. Brezzi, G. Manzini, D. Marini, P. Pietra, and A. Russo, Discontinuous Galerkin approximations for elliptic problems, Numerical methods in PDEs, **16** (4), (2000), 365–378
4. C. Schwab, $p-$ and $hp-$ finite element methods. Theory and applications in solid and fluid mechanics. Clarendon Press, Oxford, 1998.
5. P. Ciarlet, The finite element method for elliptic problems. Classics in applied mathematics, SIAM, Philadelphia, 2002.
6. B. Cockburn, G. E. Karniadakis, and C.-W. Shu (eds.), Discontinuous Galerkin Methods: Theory, Computation and Applications, Springer-Verlag, Lecture Notes in Computational Science and Engineering, **11**, (2000)
7. C. Dawson, The P^{k+1}-S^k local discontinuous Galerkin method for elliptic equations, SIAM J. Num. Anal., **40** (6), (2003), 2151–2170
8. T.J.R. Hughes, G. Scovazzi, P. Bochev and A. Buffa, A multiscale discontinuous Galerkin method with the computational structure of a continuous Galerkin method. Comp. Meth. Appl. Mech. Engrg., submitted
9. T.J.R. Hughes, G. Engel, L. Mazzei and M.G. Larson, A comparison of discontinuous and continuous Galerkin methods based on error estimates, conservation, robustness and efficiency, In: Discontinuous Galerkin Methods: Theory, Computation and Applications, Edited by Cockburn, B. and Karniadakis, G.E. and Shu, C.-W., Lecture Notes in Computational Science and Engineering, **11**, Springer-Verlag, (2000).
10. T. J. R. Hughes, A. Masud, and J. Wan, A stabilized mixed discontinuous Galerkin method for Darcy flow, Comp. Meth. Appl. Mech. Enrgr. To appear.
11. C. Johnson, U. Nävert and J. Pitkäranta, Finite element methods for linear hyperbolic problems, Comp. Meth. Appl. Mech. Engrg., **45**, (1984), 285–312
12. C. Johnson, and J. Pitkäranta, An analysis of the discontinuous Galerkin method for a scalar hyperbolic equation, Mathematics of Computation, **46** (173), (1986), 1–26
13. W. H. Reed and T. R. Hill, Triangular mesh methods for the neutron transport equation, LA-UR-73-479, Los Alamos Scientific Laboratory, (1973)

On the Discontinuous Galerkin Method for Friedrichs Systems in Graph Spaces

Max Jensen

Oxford University Computing Laboratory,
Parks Road, Oxford OX1 3QD, UK
Max.Jensen@comlab.ox.ac.uk

Abstract. Solutions of Friedrichs systems are in general not of Sobolev regularity and may possess discontinuities along the characteristics of the differential operator. We state a setting in which the well-posedness of Friedrichs systems on polyhedral domains is ensured, while still allowing changes in the inertial type of the boundary. In this framework the discontinuous Galerkin method converges in the energy norm under *h*- and *p*-refinement to the exact solution.

1 Introduction

Friedrichs systems are first-order linear boundary value problems which allow the study of a wide range of hyperbolic, parabolic and elliptic differential equations in a unified framework [1]. Because of this unifying approach, Friedrichs systems provide tools for the study of mixed-type problems, i.e. boundary value problems, which change their type depending on the position in the domain. For instance, equations in compressible gas dynamics can be transformed into Friedrichs systems, where regions of supersonic flow correspond to a locally hyperbolic differential operator, while subsonic regions correspond to a local model of elliptic type [2].

In general, the solution of a Friedrichs systems is not contained in a Sobolev space. Instead it belongs to the associated graph space, i.e. it is weakly differentiable along the characteristics of the differential operator. Functions in the graph space may be discontinuous. In addition, poles in the solution, due to type-changes of the differential operator or the boundary conditions, may lead to a loss of the integration-by-parts rule in its classical sense [3]. This is in close connection to the question of well-posedness of Friedrichs systems [4, 5, 6, 7, 8, 9, 10].

In 1973 Reed and Hill [11, 12] introduced the discontinuous Galerkin method (DGFEM) to solve the neutron transport equation. Already in this paper, numerical experiments make the good approximation and stability properties of the DGFEM for boundary value problems with discontinuous solution apparent.

Assuming shape-regularity, LeSaint and Raviart prove in [13] for meshes with triangular and quadrilateral elements the suboptimal $L^2(\Omega)$-error bound

$$\|u - u_{\mathrm{DG}}\|_{L^2(\Omega)} \leq C\, h^p\, \|u\|_{W^{p+1,2}(\Omega)}, \qquad C > 0,$$

I. Lirkov, S. Margenov, and J. Waśniewski (Eds.): LSSC 2005, LNCS 3743, pp. 94–101, 2006.

for solutions u in $W^{p+1,2}(\Omega)$ and DG solutions u_{DG}. Johnson and his coworkers [14, 15] show for equations with non-constant coefficients and certain Friedrichs systems an improved $\mathcal{O}(h^p)$ bound in the DG energy norm. Bey and Oden [16] extend the analysis to non-uniform p. In the framework by Houston, Schwab and Süli [17, 18] the exact solution u is only required to be elementwise of Sobolev regularity. Thus u may be discontinuous along element edges.

The afore-mentioned publications have in common that their a priori analysis is restricted to solutions which are of elementwise or global Sobolev regularity. Solutions with discontinuities across elements are not covered, for which already Reed and Hill and also others [16, 18] highlighted the competitive performance of the DG method with numerical experiments.

In this publication we address the convergence of the discontinuous Galerkin method in graph spaces. We base our analysis on Friedrichs systems which allow typical changes in the inertial type of the boundary conditions such as between in- and outflow components, but which at the same time satisfy basic requirements such as the integration-by-parts formula.

2 Friedrichs Systems

Let Ω be a bounded Lipschitz domain in $\mathbb{R}^n$. Let $m \in \mathbb{N}$. Given a tensor $B \in W^{1,\infty}(\Omega)^{m\times m\times n}$ and a matrix $C \in L^\infty(\Omega)^{m\times m}$, we consider the differential operator

$$\mathcal{L} : v \mapsto \partial_k(B_{ijk}\, v_j) + C_{ij} v_j,$$

making use of the Einstein summation convention and assuming that v and the coefficients B and C are real-valued. We denote by ν the unit outward normal of Ω and by $B(\nu)$ the matrix $B(\nu)_{ij} = B_{ijk}\,\nu_k$. With $D_{ij} := C_{ij} + \frac{1}{2}\,\partial_k B_{ijk}$, the symmetry condition

$$B_{ijk} = B_{jik}, \qquad i, j \in \{1, \ldots, m\},\ k \in \{1, \ldots, n\}, \tag{1}$$

implies for $v, w \in H^1(\Omega)^m$

$$\langle \mathcal{L}v, w\rangle_\Omega + \langle v, \mathcal{L}w\rangle_\Omega = 2\langle Dv, w\rangle_\Omega + \langle B(\nu)\, v, w\rangle_{\partial\Omega}, \tag{2}$$

where $\langle\cdot,\cdot\rangle_\Omega$ and $\langle\cdot,\cdot\rangle_{\partial\Omega}$ are the L^2-scalar products on Ω and $\partial\Omega$. If (1) is satisfied and there is a constant $\gamma > 0$ such that $v \cdot D\,v - \gamma v \cdot v$ is positive on Ω, we call $\mathcal{L}$ accretive. Notice that $\langle \mathcal{L}v, v\rangle_\Omega \geq \gamma\,\|v\|_{L^2(\Omega)^m}$ for $v \in H^1_0(\Omega)^m$.

Boundary operators $J : \partial\Omega \to \mathbb{R}^{m\times m}$ are semi-admissible if $R := J + \frac{1}{2}\,B(\nu)$ is positive semi-definite [1]. Then, due to (2),

$$\langle \mathcal{L}v, v\rangle_\Omega + \langle Jv, v\rangle_{\partial\Omega} = \langle Dv, v\rangle_\Omega + \langle Rv, v\rangle_{\partial\Omega} \geq \gamma\,\|v\|_{L^2(\Omega)^m} \tag{3}$$

for $v \in H^1(\Omega)^m$. Let J^T be the transpose of J. Given the formal adjoint operator $\mathcal{L}' : v \mapsto -B_{jik}\partial_k v_j + C_{ji} v_j$, the adjoint boundary operator $J' = J^\mathsf{T} + B(\nu)$ satisfies

$$\langle \mathcal{L}v, w\rangle_\Omega + \langle Jv, w\rangle_{\partial\Omega} = \langle v, \mathcal{L}'w\rangle_\Omega + \langle v, J'w\rangle_{\partial\Omega}, \qquad v, w \in H^1(\Omega)^m.$$

Let $f \in L^2(\Omega)^m$. One says that a function $u \in L^2(\Omega)^m$ solves the boundary value problem $\mathcal{L}u = f$, $Ju = 0$ weakly if for all $v \in C^1(\overline{\Omega})^m$ with $J'v = 0$ the identity

$$\langle f, v\rangle_\Omega = \langle u, \mathcal{L}'v\rangle_\Omega \tag{4}$$

holds.

Friedrichs [1] proves that a weak solution always exits if $\mathcal{L}$ is accretive and J is semi-admissible. Clearly, $\mathcal{L}u$ is equal to f in the sense of distributions. Therefore the solution u belongs to the graph space of $\mathcal{L}$. That is the set

$$H(\mathcal{L}, \Omega) := \{v \in L^2(\Omega)^m : \mathcal{L}v \in L^2(\Omega)^m\},$$

which is normed by

$$\|v\|_{\mathcal{L}}^2 := \|v\|_{L^2(\Omega)^m}^2 + \|\mathcal{L}v\|_{L^2(\Omega)^m}^2.$$

In what sense u satisfies the boundary conditions is more intricate. We assume initially that u belongs to $C^0(\overline{\Omega})^m$ to delay the definition of a trace operator. Because the test functions in (4) are contained in $\ker J'$, the test space may be too small to ensure that $Ju = 0$. We call $P_J, P_{J'} \in \mathbb{R}^{m\times m}$ a pair of projections if

$$P_J + P_{J'} = I \qquad \text{and} \qquad P_J P_{J'} = P_{J'} P_J = 0.$$

Semi-admissible boundary operators J are called admissible if for each $x \in \partial\Omega$ there is a pair of projections $P_J, P_{J'} \in \mathbb{R}^{m\times m}$ such that

$$J(x) = -P_J(x)^\top B(\nu, x) \quad \text{and} \quad J'(x) = B(\nu, x) P_{J'}(x).$$

Under sufficient regularity, e.g. for each $v \in C^1(\overline{\Omega})^m$ there is a $\dot{v} \in C^1(\overline{\Omega})^m$ such that $P_J v = \dot{v}$ on $\partial\Omega$, admissibility of J guarantees $Ju = 0$. Boundary value problems consisting of an accretive differential operator and admissible boundary operators are called Friedrichs systems.

The following example, which is an adaptation of [3], shows that for weak solutions the integration-by-parts formula is in general not valid.

Example 1. Let $\Omega = (0, 1)^2$ and

$$\mathcal{L}v := \mathcal{L}_{\mathrm{CR}}v + v, \qquad \mathcal{L}_{\mathrm{CR}}v := \begin{pmatrix} -\partial_x & \partial_y \\ \partial_y & \partial_x \end{pmatrix} \begin{pmatrix} v_1 \\ v_2 \end{pmatrix}.$$

The boundary conditions

$$J|_{x=1} := \begin{pmatrix} 1 & 1 \\ 0 & 0 \end{pmatrix}, J|_{x=0} := \begin{pmatrix} 0 & 0 \\ 1 & 1 \end{pmatrix}, J|_{y=1} := \begin{pmatrix} 0 & -1 \\ 0 & 0 \end{pmatrix}, J|_{y=0} := \begin{pmatrix} 0 & 1 \\ 0 & 0 \end{pmatrix}$$

are admissible. Since $\mathcal{L}_{\mathrm{CR}}$ is the Cauchy-Riemann operator and the function $v(\phi, r) := r^{-1/2}(\cos\phi/2, -\sin\phi/2)$ in polar coordinates (ϕ, r) represents the holomorphic function $z^{-1/2}$, it follows that $\mathcal{L}_{\mathrm{CR}}v = 0$ and $\mathcal{L}v = v$. Let $\psi \in$

$C^1(\overline{\Omega})$ be a radially symmetric function with support in the unit ball and which is equal to 1 in a neighbourhood of the origin. Then $u := \psi v \in L^2(\Omega)^m$ satisfies pointwise and weakly the homogeneous boundary conditions $Ju = 0$ on $\partial\Omega$ and $\mathcal{L}u = f$ with $f := (\psi(r) - \psi'(r))v \in L^2(\Omega)^m$.

For bounded smooth functions w which satisfy the homogeneous boundary conditions, the operator $i\mathcal{L}_{\mathrm{CR}}$ is self-adjoint and therefore $\int_\Omega \mathcal{L}_{\mathrm{CR}} w \cdot w\, dx = 0$. In contrast, $\int_\Omega \mathcal{L}_{\mathrm{CR}} u \cdot u\, dx = \pi/4$. Consequently, formula (2) is not valid for $u \in H(\mathcal{L}, \Omega)$.

The loss of the integration-by-parts formula has far-reaching implications on the analysis of the discontinuous Galerkin method. It is, for instance, used for the definition of the energy norm.

Insight why the formula fails is given by the trace operator of $H(\mathcal{L}, \Omega)$. We report relevant properties of the operator, but refer for details to [19]. The trace operator

$$\mathcal{T} : H(\mathcal{L}, \Omega) \to H^{-1/2}(\partial\Omega)^m, v \mapsto \langle B(\nu)v, \cdot\rangle$$

is bounded, but in general not surjective. We equip the trace space $H(\mathcal{T}, \partial\Omega) := \operatorname{im} \mathcal{T}$ with the norm

$$\|v\|_{\mathcal{T}} := \inf\{\|w\|_{\mathcal{L}} : w \in H(\mathcal{L}, \Omega) \text{ and } \mathcal{T}w = v\}.$$

The terminology trace operator and trace space for $\mathcal{T}$ and $H(\mathcal{T}, \partial\Omega)$ is justified by the observation that all mappings $\mathcal{J} : H(\mathcal{L}, \Omega) \to V$ which vanish on $H_0^1(\Omega)^m$ can be factorised in the form $\mathcal{J} = \dot{\mathcal{J}} \circ \mathcal{T}$ where $\dot{\mathcal{J}} : H(\mathcal{T}, \partial\Omega) \to V$ is continuous and V is any abstract normed vector space. Up to homeomorphy only $H(\mathcal{T}, \partial\Omega)$ has this property.

Example 2. Let $\Omega := \{(x, y) \in \mathbb{R}^2 : y > |x| \text{ and } y < 1\}$ and $\mathcal{L}v = \partial_x v$. Then $y^{-1/2} \in H(\mathcal{L}, \Omega)$. The only admissible boundary conditions with respect to $\mathcal{L}$ are inflow conditions. Yet for them, $\langle Jv, v\rangle_{\partial\Omega}$ diverges. Thus $(v, w) \mapsto \langle Jv, w\rangle_{\partial\Omega}$ is not continuous on $H(\mathcal{L}, \Omega) \times H(\mathcal{L}, \Omega)$ and (3) cannot be continuously extended from $H^1(\Omega)^m$ to $H(\mathcal{L}, \Omega)$.

The properties of the trace space are connected to the eigenvalues of $B(\nu)$. Due to (1), for each $x \in \partial\Omega$ there is an orthogonal transformation X and a diagonal matrix Λ such that $B(\nu) = X^\mathsf{T}\Lambda X$. Substituting in Λ negative entries by 0 gives Λ_+. The positive and negative semi-definite components of $B(\nu)$ are $B_+(\nu) := X^\mathsf{T}\Lambda_+X$ and $B_-(\nu) := B(\nu) - B_+(\nu)$, respectively. The absolute part is $|B|(\nu) := B_+(\nu) - B_-(\nu)$. A change in the rank of $B_+(\nu)$ or $B_-(\nu)$ is termed a change in the inertial type of $B(\nu)$.

The space $L_B^2(\partial\Omega)$ is the set of all integrable functions $v : \partial\Omega \to \mathbb{R}^m$ for which the norm

$$\|v\|_B^2 := \int_{\partial\Omega} v \cdot |B|(\nu)v\, dx$$

is finite. The space $L^\infty_B(\partial\Omega)^{m\times m}$ consists of the matrices in $L^\infty(\partial\Omega)^{m\times m}$ which define an endomorphism on $L^2_B(\partial\Omega)$.

Traces $\langle B(\nu)v,\cdot\rangle$ contained in $H^{-1/2}(\partial\Omega)^m \setminus L^2(\partial\Omega)^m$ can arise through a coupling of in- and outflow components. Pointwise we understand under in- and outflow components the eigenspaces of $B(\nu)$ associated to negative and positive eigenvalues, respectively. For instance, in Example 2 in- and outflow boundary are coupled through the sign change of $B(\nu)$ at the origin. Traces in $H^{-1/2}(\partial\Omega)^m \setminus L^2(\partial\Omega)^m$ are not limited to domains with corners. Coupling with tangential components of $\mathcal{L}$ has comparable effects.

Example 3. Let $\Omega = \{(x,y) \in \mathbb{R}^2 : x > 0\}$ and

$$\mathcal{L}v = \partial_x \begin{pmatrix} -1 & 0 & 0 \\ 0 & 1 & 0 \\ 0 & 0 & 0 \end{pmatrix} v + \frac{\partial_y}{\sqrt{2}} \begin{pmatrix} 0 & 0 & -1 \\ 0 & 0 & 1 \\ -1 & 1 & 0 \end{pmatrix} v.$$

The trace space of $u \in H(\mathcal{L},\Omega)$ is equal to

$$\{(v_1 - v_2, v_1 + v_2, 0) : v_1 \in H^{1/2}(\partial\Omega)^m \text{ and } v_2 \in H^{-1/2}(\partial\Omega)^m\}$$

with the intrinsic norm $(\|v_1\|^2_{H^{1/2}(\partial\Omega)^m} + \|v_2\|^2_{H^{-1/2}(\partial\Omega)^m})^{1/2}$.

To provide a basis for definition of the discontinuous Galerkin method for Friedrichs systems, we introduce the additional condition that there is a factorisation of $B(\nu)$ of the form $B(\nu) = (R + R^\top)F$. Then one can ensure that the solution of the Friedrichs system is unique and contained in the closure $H(\mathcal{L}, B, \Omega)$ of $H^1(\Omega)^m$ in the norm

$$(\|v\|^2_{\mathcal{L}} + \|v\|^2_B)^{1/2}, \qquad v \in H^1(\Omega)^m.$$

Functions in $H(\mathcal{L}, B, \Omega)$ have a trace in $L^2_B(\partial\Omega)$ and satisfy formula (2).

Theorem 1. *Let $\mathcal{L}$ be an accretive operator and J be semi-admissible. Suppose that there are two projections $P_1, P_2 \in L^\infty_B(\partial\Omega)^{m\times m}$ such that $J = -B(\nu)P_1$ and $J' = B(\nu)P_2$. We also adopt the hypothesis that there is an $F \in L^\infty_B(\partial\Omega)^{m\times m}$ such that $B(\nu) = (R + R^\top)F$. Then for each $f \in L^2(\Omega)^m$ and $g \in L^2_B(\partial\Omega)$ there is a unique function $u \in H(\mathcal{L}, B, \Omega)$ which solves $\mathcal{L}u = f$ and $Ju = Jg$. Furthermore, u depends continuously on f and g.*

For details we refer to [19]. We remark that P_1 and P_2 are not necessarily a pair of projections. Also note that we assume $g \in L^2_B(\partial\Omega)$ and not $g \in H(T, \partial\Omega)$.

Example 4. Let H be the Heaviside function. Selecting

$$(P_1)_{ij} := X_{ki} H(\Lambda_{kk}) X_{kj}, \quad (P_2)_{ij} := X_{ki} H(-\Lambda_{kk}) X_{kj}, \quad F := P_2 - P_1$$

shows that inflow boundary conditions satisfy the requirements set in Theorem 1.

3 The Discontinuous Galerkin Method

Let $T = \{\kappa_1, \kappa_2, \ldots, \kappa_N\}$ be a decomposition of Ω into polyhedral elements κ_i. Suppose that all $\kappa \in T$ are an affine image of a fixed master element $\hat{\kappa}$, i.e. $\kappa = F_\kappa(\hat{\kappa})$ for all $\kappa \in T$, where F_κ is an injective affine mapping and where $\hat{\kappa}$ is either the open unit simplex or the open unit hypercube in $\mathbb{R}^n$. We denote by $\mathcal{P}_k$ the space of polynomials on $\hat{\kappa}$ with total degree less or equal k. If $\hat{\kappa}$ is the hypercube then we also consider the space $\mathcal{Q}_k$ of tensor-polynomials on $\hat{\kappa}$ with degree less or equal k in each coordinate direction. Let $p = (p_1, p_2, \ldots, p_N)$ be a vector which associates to each element κ_i the polynomial degree p_i. We consider the finite element spaces

$$S(T,p) = \{v \in L^2(\Omega) : v|_\kappa \circ F_\kappa \in \mathcal{R}_{p_\kappa}\},$$

where $\mathcal{R}_k$ is either $\mathcal{P}_k$ or $\mathcal{Q}_k$.

The finite element spaces $S(T,p)$ are contained in the broken graph space

$$H(\mathcal{L}, B, T) := \bigoplus_{\kappa \in T} H(\mathcal{L}, B, \kappa).$$

At the boundary $\partial\kappa_i \cap \partial\kappa_j$ between the element κ_i and a neighbour κ_j, a member v of $H(\mathcal{L}, B, T)$ has, in general, two distinct traces: one from the restriction $v|_{\kappa_i}$ and one from $v|_{\kappa_j}$. We denote the internal trace$(v|_{\kappa_i})|_{\partial\kappa_i}$ of κ_i by $v^+$and the external trace$(v|_{\kappa_j})|_{\partial\kappa_i \cap \partial\kappa_j}$ of κ_i by v^-. Altogether the external trace v^- is composed from the traces of all elements neighbouring κ_i. The difference $v^+ - v^-$ is denoted by $[v]$.

For $v, w \in H(\mathcal{L}, B, T)$ let

$$B_{\mathrm{DG}}(v,w) = \langle \mathcal{L}v, w\rangle_\Omega + \langle Jv, w\rangle_{\partial\Omega} + \sum_{\kappa \in T} \langle B_-(\nu)[v], w^+\rangle_{\partial\kappa \setminus \partial\Omega},$$

$$\ell_{\mathrm{DG}}(w) = \langle f, w\rangle_\Omega + \langle Jg, w\rangle_{\partial\Omega}.$$

The integration-by-parts formula

$$B_{\mathrm{DG}}(v,v) = \langle Dv, v\rangle_\Omega + \langle Rv, v\rangle_{\partial\Omega} + \tfrac{1}{2}\textstyle\sum_{\kappa\in T} \langle |B|(\nu)[v], [v]\rangle_{\partial\kappa\setminus\partial\Omega}$$

induces the energy norm $\|v\|_{\mathrm{DG}} := \sqrt{B_{\mathrm{DG}}(v,v)}$ on $H(\mathcal{L}, B, T)$. We remark that $H(\mathcal{L}, B, T)$ is not complete in this norm.

Let T' be another finite element mesh on Ω and suppose that $v \in H(\mathcal{L}, B, T) \cap H(\mathcal{L}, B, T')$. The energy norm is mesh-independent in the sense that

$$\begin{aligned} &\langle Dv, v\rangle_\Omega + \langle Rv, v\rangle_{\partial\Omega} + \tfrac{1}{2}\textstyle\sum_{\kappa\in T} \langle |B|(\nu)[v], [v]\rangle_{\partial\kappa\setminus\partial\Omega} \\ &= \langle Dv, v\rangle_\Omega + \langle Rv, v\rangle_{\partial\Omega} + \tfrac{1}{2}\textstyle\sum_{\kappa\in T'} \langle |B|(\nu)[v], [v]\rangle_{\partial\kappa\setminus\partial\Omega}. \end{aligned}$$

The positive definiteness of B_{DG} implies that there is a unique discontinuous Galerkin solution $u_{\mathrm{DG}} \in S(T,p)$ to

$$B_{\mathrm{DG}}(u_{\mathrm{DG}}, w) = \ell_{\mathrm{DG}}(w) \qquad \forall w \in S(T,p).$$

The solution satisfies the stability estimate

$$\|u_{\mathrm{DG}}\|_{\mathrm{DG}} \leq \gamma^{-1}\|f\|_{L^2(\Omega)^m} + C\|g\|_B.$$

The constant $C > 0$ depends on the boundary operator J but not on the approximation space [19].

The next theorem shows that the discontinuous Galerkin method converges in the energy norm under h- and p-refinement.

Theorem 2. *The discontinuous Galerkin solution satisfies the bound*

$$\|u - u_{\mathrm{DG}}\|_{\mathrm{DG}} \leq C \inf\{\|u - v\|_{\mathcal{L}} : v \in S(T,p) \cap H(\mathcal{L}, \Omega)\}.$$

The constant $C > 0$ is independent of p and T.

The proof relies on Galerkin orthogonality and on the factorisation of the boundary operator with F, cf. [19].

References

1. Friedrichs, K.O.: Symmetric positive linear differential equations. Comm. Pure Appl. Math. **11** (1958) 333–418
2. Dautray, R., Lions, J.L.: Mathematical analysis and numerical methods for science and technology. Vol. 3. Springer-Verlag, Berlin (1990)
3. Moyer, R.D.: On the nonidentity of weak and strong extensions of differential operators. Proc. Amer. Math. Soc. **19** (1968) 487–488
4. Lax, P.D., Phillips, R.S.: Local boundary conditions for dissipative symmetric linear differential operators. Comm. Pure Appl. Math. **13** (1960) 427–455
5. Sarason, L.: On weak and strong solutions of boundary value problems. Comm. Pure Appl. Math. **15** (1962) 237–288
6. Phillips, R.S., Sarason, L.: Singular symmetric positive first order differential operators. J. Math. Mech. **15** (1966) 235–271
7. Rauch, J.: Symmetric positive systems with boundary characteristic of constant multiplicity. Trans. Amer. Math. Soc. **291** (1985) 167–187
8. Rauch, J.: Boundary value problems with nonuniformly characteristic boundary. J. Math. Pures Appl. (9) **73** (1994) 347–353
9. Secchi, P.: A symmetric positive system with nonuniformly characteristic boundary. Differential Integral Equations **11** (1998) 605–621
10. Secchi, P.: Full regularity of solutions to a nonuniformly characteristic boundary value problem for symmetric positive systems. Adv. Math. Sci. Appl. **10** (2000) 39–55
11. Reed, W.H., Hill, T.R.: Triangular mesh methods for the neutron transport equation. Proceedings of the 1973 Conference on Mathematical Models and Computational Techniques of Nuclear Systems, Fifth biennial topical meeting of the mathematical and computational division of the American Nuclear Society, vol. 1, session 1 (1973) 10–31
12. Reed, W.H., Hill, T.R.: Triangular mesh methods for the neutron transport equation. Technical Report LA-UR-73-479 (1973)

13. LeSaint, P., Raviart, P.A.: On a finite element method for solving the neutron transport equation. In: Mathematical aspects of finite elements in partial differential equations (Proc. Sympos., Math. Res. Center, Univ. Wisconsin, Madison, Wis., 1974). Academic Press, New York (1974) 89–123. Publication No. 33
14. Johnson, C., Pitkäranta, J.: Convergence of a fully discrete scheme for two-dimensional neutron transport. SIAM J. Numer. Anal. **20** (1983) 951–966
15. Johnson, C., Nävert, U., Pitkäranta, J.: Finite element methods for linear hyperbolic problems. Comput. Methods Appl. Mech. Engrg. **45** (1984) 285–312
16. Bey, K.S., Oden, J.T.: *hp*-version discontinuous Galerkin methods for hyperbolic conservation laws. Comput. Methods Appl. Mech. Engrg. **133** (1996) 259–286
17. Houston, P., Schwab, C., Süli, E.: Stabilized *hp*- finite element methods for first-order hyperbolic problems. SIAM J. Numer. Anal. **37** (2000) 1618–1643
18. Houston, P., Süli, E.: *hp*-adaptive discontinuous Galerkin finite element methods for first-order hyperbolic problems. SIAM J. Sci. Comput. **23** (2001) 1226–1252
19. Jensen, M.: Discontinuous Galerkin methods for Friedrichs systems with irregular solutions. PhD thesis, Oxford University (2004) `http://web.comlab.ox.ac.uk/oucl/research/na/theses.html`

$\mathcal{H}$-Matrix Techniques for Stray-Field Computations in Computational Micromagnetics

Nikola Popović[1] and Dirk Praetorius[2]

[1] Department of Mathematics and Statistics,
Boston University,
111 Cummington Street,
Boston, MA 02215, USA
Popovic@math.bu.edu

[2] Institute for Analysis and Scientific Computing,
Vienna University of Technology,
Wiedner Hauptstraße 8-10,
A-1040 Vienna, Austria
Dirk.Praetorius@tuwien.ac.at

Abstract. A major task in the simulation of micromagnetic phenomena is the effective computation of the stray-field $\mathbf{H}$ and/or of the corresponding energy, where $\mathbf{H}$ solves the magnetostatic Maxwell equations in the entire space. For a given FE magnetization $\mathbf{m}_h$, the naive computation of $\mathbf{H}$ via a closed formula typically leads to dense matrices and quadratic complexity with respect to the number N of elements. To reduce the computational cost, it is proposed to apply $\mathcal{H}$-matrix techniques instead. This approach allows for the computation (and evaluation) of $\mathbf{H}$ in linear complexity even on adaptively generated (or unstructured) meshes.

1 Basic Micromagnetics

Let $\Omega \subset \mathbb{R}^d$ be the bounded spatial domain of a ferromagnet. Then, the magnetization $\mathbf{m} : \Omega \to \mathbb{R}^d$ induces the so-called stray-field [9] (or demagnetization field) $\mathbf{H} : \mathbb{R}^d \to \mathbb{R}^d$, which is the solution of the magnetostatic Maxwell equations

$$\operatorname{curl} \mathbf{H} = 0 \quad \text{and} \quad \operatorname{div} \mathbf{B} = 0 \quad \text{on } \mathbb{R}^d. \tag{1}$$

Here, $\mathbf{B} = \mathbf{H} + \mathbf{m}$ denotes the magnetic induction, with $\mathbf{m}$ extended by zero to $\mathbb{R}^d \backslash \Omega$. Stokes' Theorem implies $\mathbf{H} = -\nabla u$, with a potential u that solves

$$\operatorname{div} (-\nabla u + \mathbf{m}) = 0 \quad \text{in } \mathcal{D}'(\mathbb{R}^d). \tag{2}$$

Thus, there holds

$$u = \sum_{j=1}^{d} \frac{\partial G}{\partial x_j} * \mathbf{m}_j \quad \text{for any } \mathbf{m} \in L^2(\Omega)^d, \tag{3}$$

I. Lirkov, S. Margenov, and J. Waśniewski (Eds.): LSSC 2005, LNCS 3743, pp. 102–110, 2006.

where G is the Newtonian kernel defined by $G(x) = (2\pi)^{-1} \log |x|$ for $d = 2$ and by $G(x) = -(4\pi)^{-1}|x|^{-1}$ for $d = 3$, respectively. Therefore, the components of the stray-field can be written as convolutions in the sense of Calderón-Zygmund. The operator $\mathcal{P} : L^2(\Omega)^d \to L^2(\Omega)^d$ mapping $\mathbf{m}$ onto the corresponding stray-field $\mathbf{H}$ is an orthogonal projection. Details and the precise mathematical setting can be found in [12].

The remainder of this paper is organized as follows: Section 2 introduces the stiffness matrix $\mathbf{A}$ arising from the FE discretization of (2). Section 3 recalls the definition of $\mathcal{H}^2$-matrices and indicates how $\mathbf{A}$ can be approximated by a $d \times d$ block matrix $\mathbf{A}_{\mathcal{H}}$ consisting of $\mathcal{H}^2$-matrix type blocks. Sections 4 and 5 contain our main results: In Section 4, we prove that $\mathbf{A}_{\mathcal{H}}$ can in fact be interpreted as a global $\mathcal{H}^2$-matrix approximation for $\mathbf{A}$. Section 5 provides an a priori analysis of the corresponding approximation error for a quite general class of FE discretizations. Some numerical experiments in Section 6 conclude the work.

2 Stray-Field Discretization

In FE simulations of micromagnetic phenomena, one usually restricts oneself to a finite dimensional subspace $\mathcal{S}_h$ of $L^2(\Omega)$. Fix a basis $\{\phi_j\}_{j=1}^N$ of $\mathcal{S}_h$. Then, the functions $\Phi_{[j,\alpha]} = \phi_j \mathbf{e}_\alpha$, with $\mathbf{e}_\alpha \in \mathbb{R}^d$ the α-th standard unit vector, define a basis of $\mathcal{S}_h^d \subset L^2(\Omega)^d$. To fix a numbering of these basis functions, we set $[j, \alpha] = j + (\alpha - 1)N$ for $1 \le \alpha \le d$. Now, for an FE discretization of $\mathcal{P}$, one has to compute the corresponding stiffness matrix $\mathbf{A} \in \mathbb{R}^{dN \times dN}$ defined by

$$\mathbf{A}_{jk} = \int_\Omega (\mathcal{P}\Phi_j)(x)\, \Phi_k(x)\, dx. \tag{4}$$

Actually, we consider the individual blocks $\mathbf{A}^{\alpha\beta} \in \mathbb{R}^{N \times N}$ of $\mathbf{A}$ separately, where

$$\mathbf{A}^{\alpha\beta}_{jk} = \int_\Omega (\mathcal{P}\Phi_{[j,\alpha]})(x)\, \Phi_{[k,\beta]}(x)\, dx \quad \text{for } 1 \le \alpha, \beta \le d. \tag{5}$$

Lemma 1. *The matrix $\mathbf{A}$ is a symmetric $d \times d$ block matrix. Furthermore, each of the $N \times N$ blocks $\mathbf{A}^{\alpha\beta}$ of $\mathbf{A}$ also is symmetric.*

Proof. The symmetry of $\mathbf{A}$ is a consequence of the L^2 orthogonality of $\mathcal{P}$. The symmetry of the blocks $\mathbf{A}^{\alpha\beta}$ follows from Calderón-Zygmund theory [12, Proposition 6.1]. □

3 Blockwise $\mathcal{H}^2$-Approximation of A

When applying $\mathcal{H}$-matrix techniques to approximate $\mathbf{A}$, and hence to reduce the cost of computing $\mathbf{H}$, one possibility is to treat each block $\mathbf{A}^{\alpha\beta}$ of $\mathbf{A}$ individually, as is done in [10]. To that end, one requires a classical integral representation of the associated far field.

Lemma 2. *Given a basis function ϕ_j of $\mathcal{S}_h$, let* supp (ϕ_j) *denote its support. If* supp $(\phi_j) \cap$ supp $(\phi_k) = \emptyset$ *for* $1 \leq j, k \leq N$, *then*

$$\mathbf{A}_{jk}^{\alpha\beta} = \int_\Omega \int_\Omega \partial_{\alpha\beta} G(x-y)\phi_j(y)\phi_k(x)\,dy\,dx, \tag{6}$$

with $\partial_{\alpha\beta}G$ *the second derivative of* G.

Proof. On supp (ϕ_k), there holds $\mathcal{P}(\phi_j \mathbf{e}_\alpha)\cdot(\phi_k \mathbf{e}_\beta) = \partial_\beta(\partial_\alpha G * \phi_j)\phi_k$. By classical convolution results, we have $\partial_\alpha G * \phi_j \in \mathcal{C}^1(\mathbb{R}^d \backslash \mathrm{supp}\,(\phi_j))$ and $\partial_\beta(\partial_\alpha G * \phi_j) = \partial_{\alpha\beta} G * \phi_j$. This concludes the proof. □

The kernel functions $\kappa^{\alpha\beta}(x,y) := \partial_{\alpha\beta}G(x-y)$ appearing in (6) are asymptotically smooth. Therefore, each $\mathbf{A}^{\alpha\beta}$ can be approximated by an $\mathcal{H}^2$-matrix obtained from tensorial interpolation of $\kappa^{\alpha\beta}$ [3].

Let $\mathcal{I} = \{1, \dots, N\}$ denote the index set corresponding to the basis $\{\phi_j\}_{j=1}^N$. Given a cluster $\sigma \subseteq \mathcal{I}$, let $\cup\sigma := \bigcup\{\mathrm{supp}\,(\phi_j) \,:\, j \in \sigma\}$, and let $B_\sigma = \prod_{\ell=1}^d [a_\ell, b_\ell] \subset \mathbb{R}^d$ denote the box of minimal size containing $\cup\sigma$. For $j \in \sigma$, fix an element $x_j \in$ supp (ϕ_j) (e.g., the center of mass).

From $\mathcal{I}$, we build a so-called cluster tree $\mathbb{T}$ by binary space partitioning [1], calling the function `clustertree` via `clustertree`$(\mathcal{I}, \emptyset)$:

```
function clustertree(σ, var T)
if |σ| ≤ 1
  return
else
  split B_σ along longest edge into boxes B_σ^1, B_σ^2 of equal volume
  if σ_1 := {j ∈ I : x_j ∈ B_σ^1} ∉ {∅, σ}
    add σ_1, σ_2 := σ\σ_1 to T
    call clustertree(σ_1, var T)
    call clustertree(σ_2, var T)
  end
end
```

Having constructed $\mathbb{T}$, we generate a block partitioning $\mathbb{P}$ for $\mathcal{I}\times\mathcal{I}$ as follows: For a fixed parameter $\eta > 0$, we call $(\sigma,\tau) \in \mathbb{T}\times\mathbb{T}$ an *admissible (far field) block* if

$$\mathrm{diam}\,(B_\sigma \times B_\tau) \leq \eta\, \mathrm{dist}\,(B_\sigma, B_\tau). \tag{7}$$

Otherwise, (σ,τ) is an *inadmissible (near field) block*. Now, the following recursive function, called by `partition`$(\mathcal{I},\mathcal{I},\mathbb{T},\mathrm{var}\,\mathbb{P})$, partitions $\mathcal{I}\times\mathcal{I}$ into admissible blocks $(\sigma,\tau) \in \mathbb{P}_{\mathrm{far}}$ and inadmissible blocks $(\sigma,\tau) \in \mathbb{P}_{\mathrm{near}}$; clearly, $\mathbb{P} = \mathbb{P}_{\mathrm{far}} \cup \mathbb{P}_{\mathrm{near}}$. Here, sons$(\sigma)$ denotes the set of all sons of $\sigma \in \mathbb{T}$ with respect to $\mathbb{T}$.

```
function partition(σ, τ, T, var P)
if (σ, τ) admissible
  add (σ, τ) to P_far
elseif sons(σ) ≠ ∅
  if sons(τ) ≠ ∅
    for all (σ', τ') ∈ sons(σ) × sons(τ) call partition(σ', τ', T, var P)
```

```
    else
      for all σ' ∈ sons(σ) call partition(σ', τ, T, var P)
    end
elseif sons(τ) ≠ ∅
    for all τ' ∈ sons(τ) call partition(σ, τ', T, var P)
else
    add (σ, τ) to P_near
end
```

Note that for $(\sigma, \tau) \in \mathbb{P}_{\text{far}}$, Lemma 2 applies. Since $\kappa^{\alpha\beta}$ is smooth on $B_\sigma \times B_\tau$, we may replace it by its tensorial Čebyšev interpolation,

$$\kappa^{\alpha\beta}(x,y) \approx \kappa^{\alpha\beta}_{\sigma\tau}(x,y) := \sum_{m,n=1}^{p^d} \kappa^{\alpha\beta}(x_m^\sigma, x_n^\tau) L_m^\sigma(x) L_n^\tau(y). \tag{8}$$

Here, L_m^σ and L_n^τ are tensorial Lagrange polynomials of overall degree p^d (i.e. of degree p in each of the coordinate directions), with corresponding interpolation nodes $x_m^\sigma \in B_\sigma$ and $x_n^\tau \in B_\tau$, respectively. This leads to an approximation of $\mathbf{A}^{\alpha\beta}$ via

$$\mathbf{A}^{\alpha\beta}|_{\sigma\times\tau} \approx \mathbf{A}^{\alpha\beta}_{\mathcal{H}}|_{\sigma\times\tau} := V_\sigma M^{\alpha\beta}_{\sigma\tau} V_\tau^T \quad \text{for } (\sigma,\tau) \in \mathbb{P}_{\text{far}}, \tag{9}$$

where $(V_\sigma)_{jm} = \int_\Omega \phi_j L_m^\sigma \, dx$ and $(M^{\alpha\beta}_{\sigma\tau})_{mn} = \kappa^{\alpha\beta}(x_m^\sigma, x_n^\tau)$. Moreover, there holds the additional hierarchy

$$V_\sigma|_{\sigma'} = V_{\sigma'} T_{\sigma'\sigma} \quad \text{for } \sigma' \in \text{sons}(\sigma),$$

with a transfer matrix T given by $(T_{\sigma'\sigma})_{mn} = L_n^\sigma(x_m^{\sigma'})$. The following complexity estimate is a standard result from $\mathcal{H}^2$-matrix theory [1].

Theorem 1. *Given $\mathbb{T}$ and $\mathbb{P}$ as constructed above, define the sparsity constant $\mathrm{C_{sp}} = \max_{\sigma\in\mathbb{T}} \#\{\tau \in \mathbb{T} : (\sigma,\tau) \in \mathbb{P}\}$. Assume that for $(\sigma,\tau) \in \mathbb{P}_{\text{near}}$ and $(j,k) \in \sigma \times \tau$, each entry $\mathbf{A}^{\alpha\beta}_{jk}$ can be computed with complexity $\mathcal{O}(1)$. Then, for $\mathbf{A}^{\alpha\beta}_{\mathcal{H}}$, the assembly, storage, and matrix-vector multiplication can be performed with complexity $\mathcal{O}(\mathrm{C_{sp}} p^{2d} N)$.* □

4 Global $\mathcal{H}^2$-Approximation of A

We now define an approximation $\mathbf{A}_{\mathcal{H}}$ for the stiffness matrix $\mathbf{A}$ by replacing all blocks $\mathbf{A}^{\alpha\beta}$ by their $\mathcal{H}^2$-matrix approximants $\mathbf{A}^{\alpha\beta}_{\mathcal{H}}$. In fact, we show that $\mathbf{A}_{\mathcal{H}}$ can be interpreted as a global $\mathcal{H}^2$-approximation for $\mathbf{A}$ if one takes into account the following considerations:

- In contrast to the previous section, we now consider the index set $\widehat{\mathcal{I}} = \{1, \ldots, dN\}$. Given the cluster tree $\mathbb{T}$ built from $\mathcal{I} = \{1, \ldots, N\}$, we make d copies $\mathbb{T}_\alpha$ of $\mathbb{T}$ which correspond to the indices $\mathcal{I}_\alpha = \{[1,\alpha], \ldots, [N,\alpha]\}$. This gives us a cluster tree $\widehat{\mathbb{T}}$ for $\widehat{\mathcal{I}}$. The root of $\widehat{\mathbb{T}}$ has precisely the d son branches $\mathbb{T}_\alpha$.

- We use the same admissibility condition (7) as before, but replace ϕ_j in the definition of $\cup\sigma$ by Φ_j.
- Finally, when $\sigma = \widehat{\mathcal{I}}$ and $\sigma' = \mathcal{I}_\alpha$, the transfer matrices $T_{\sigma'\sigma}$ are just the identities.

Theorem 2. *The partitioning $\widehat{\mathbb{P}}$ induced by $\widehat{\mathbb{T}}$ and the modified admissibility condition coincides blockwise with the partitioning $\mathbb{P}$ for $\mathbf{A}_{\mathcal{H}}^{\alpha\beta}$. Therefore, and with the above definition of the additional transfer matrices needed, $\mathbf{A}_{\mathcal{H}}$ is an $\mathcal{H}^2$-matrix.* □

This important result allows us to apply any algorithm from $\mathcal{H}^2$-matrix theory to $\mathbf{A}$ as a whole. In particular, this concerns algorithms for the preconditioning or recompression of $\mathcal{H}^2$-matrices or the assembly of $\mathbf{A}$ by use of adaptive cross approximation [11]. So far, we were only able to apply the respective algorithms blockwise, i.e. to each $\mathbf{A}^{\alpha\beta}$ individually [10].

5 Approximation Error Estimate

It remains to study the approximation error which results from replacing $\mathbf{A}^{\alpha\beta}$ by $\mathbf{A}_{\mathcal{H}}^{\alpha\beta}$. For $\kappa^{\alpha\beta}(x,y) = \partial_{\alpha\beta}G(x-y)$ and $\kappa_{\sigma\tau}^{\alpha\beta}$ as defined in (8), there holds

$$\|\kappa^{\alpha\beta} - \kappa_{\sigma\tau}^{\alpha\beta}\|_{\infty,B_\sigma\times B_\tau} \le C\,\Lambda(p)\,(1+2/\eta)^{-p},$$

with $\eta > 0$ as in (7). Here, $\Lambda(p)$ grows logarithmically with p, and C is a numerical constant which depends only linearly on $\operatorname{dist}(B_\sigma, B_\tau)^{-d}$, cf. [2, 10]. In particular, the error decreases exponentially with the approximation order p. Now, given $C_{\mathcal{H}} > 0$, choose p large enough such that

$$\|\kappa^{\alpha\beta} - \kappa_{\sigma\tau}^{\alpha\beta}\|_{\infty,B_\sigma\times B_\tau} \le C_{\mathcal{H}} \quad \text{for all } (\sigma,\tau) \in \mathbb{P}_{\text{far}}. \tag{10}$$

As a first direct consequence, we obtain

Theorem 3. *The matrix error in the Frobenius norm satisfies*

$$\|\mathbf{A}^{\alpha\beta} - \mathbf{A}_{\mathcal{H}}^{\alpha\beta}\|_F \le C_{\mathcal{H}} N \max_{j=1,\dots,N} \|\phi_j\|_{L^1}^2.$$ □

More interesting than the matrix error, however, is the error for the corresponding bilinear forms: For a discrete magnetization $\mathbf{m}_h \in \mathcal{S}_h^d$, let $\widehat{\mathbf{m}}_h \in \mathbb{R}^{dN}$ denote the coefficient vector with respect to the basis functions $\Phi_{[j,\alpha]}$. Then, replacing $\mathbf{A}$ by $\mathbf{A}_{\mathcal{H}}$ corresponds to replacing the bilinear form $a(\mathbf{m}_h, \mathbf{n}_h) = \widehat{\mathbf{m}}_h \cdot \mathbf{A}\widehat{\mathbf{n}}_h$ by

$$a_{\mathcal{H}}(\mathbf{m}_h, \mathbf{n}_h) := \widehat{\mathbf{m}}_h \cdot \mathbf{A}_{\mathcal{H}}\widehat{\mathbf{n}}_h \quad \text{for } \mathbf{m}_h, \mathbf{n}_h \in \mathcal{S}_h^d.$$

The error analysis for these bilinear forms requires some additional assumptions on the basis functions ϕ_j: First, assume that

$$\sum_{j\in\sigma} |\operatorname{supp}(\phi_j)| \le C_{\text{loc}} \Big| \bigcup_{j\in\sigma} \operatorname{supp}(\phi_j) \Big| \tag{11}$$

for any $\sigma \subseteq \{1, \dots, N\}$, with some $C_{\text{loc}} > 0$. Moreover, let $C_{\text{stab}} > 0$ be a constant such that for any coefficient vector $x \in \mathbb{R}^N$,

$$C_{\text{stab}}^{-1} \Big\| \sum_{j=1}^{N} x_j \phi_j \Big\|_{L^2}^2 \leq \sum_{j=1}^{N} \|x_j \phi_j\|_{L^2}^2 \leq C_{\text{stab}} \Big\| \sum_{j=1}^{N} x_j \phi_j \Big\|_{L^2}^2. \tag{12}$$

Note that assumptions (11) and (12) are quite natural. For a triangular mesh and the corresponding P^1 hat functions, (11) is essentially an assumption on the angles in the triangulation. Moreover, the usual FE bases satisfy (12) even for a quite general class of meshes [7]. In particular, both (11) and (12) are clearly satisfied for piecewise constant basis functions, with $C_{\text{loc}} = 1 = C_{\text{stab}}$.

The following theorem is our second main result:

Theorem 4. *Under the above assumptions, there holds*

$$|a(\mathbf{m}_h, \mathbf{n}_h) - a_{\mathcal{H}}(\mathbf{m}_h, \mathbf{n}_h)| \leq C_{\mathcal{H}} C_{\text{loc}} C_{\text{stab}}^2 |\Omega| d^2 \, \|\mathbf{m}_h\|_{L^2} \|\mathbf{n}_h\|_{L^2} \tag{13}$$

for all $\mathbf{m}_h, \mathbf{n}_h \in \mathcal{S}_h^d$.

Proof. For $\mathbf{m} \in \mathcal{S}_h^d$ and $\sigma \subseteq \mathcal{I}$, we write $\mathbf{m}_\sigma := \sum_{\ell \in \sigma} \widehat{\mathbf{m}}_\ell \Phi_\ell$. With this notation, the error $e = a - a_{\mathcal{H}}$ for the bilinear forms reads

$$e(\mathbf{m}, \mathbf{n}) = \sum_{(\sigma,\tau) \in \mathbb{P}_{\text{far}}} \int_{\cup\sigma} \int_{\cup\tau} \mathbf{m}_\sigma(x) \big(\kappa^{\alpha\beta} - \kappa_{\sigma\tau}^{\alpha\beta}\big)(x, y) \, \mathbf{n}_\tau(y) \, dy \, dx.$$

From (10) and the Hölder and Cauchy Inequalities, one obtains

$$\begin{aligned} |e(\mathbf{m}, \mathbf{n})| &\leq C_{\mathcal{H}} \sum_{(\sigma,\tau) \in \mathbb{P}_{\text{far}}} |\cup\sigma|^{1/2} |\cup\tau|^{1/2} \|\mathbf{m}_\sigma\|_{L^2} \|\mathbf{n}_\tau\|_{L^2} \\ &\leq C_{\mathcal{H}} \Big(\sum_{(\sigma,\tau) \in \mathbb{P}_{\text{far}}} |\cup\sigma| |\cup\tau| \Big)^{1/2} \Big(\sum_{(\sigma,\tau) \in \mathbb{P}_{\text{far}}} \|\mathbf{m}_\sigma\|_{L^2}^2 \|\mathbf{n}_\tau\|_{L^2}^2 \Big)^{1/2}. \end{aligned}$$

Finally, a direct calculation shows that these sums can be dominated by $d^2 C_{\text{loc}}^2 |\Omega|^2$ and $d^2 C_{\text{stab}}^4 \|\mathbf{m}\|_{L^2}^2 \|\mathbf{n}\|_{L^2}^2$, respectively. □

6 Numerical Experiments

To underline our theoretical results, we performed numerical experiments for the Landau-Lifshitz minimization problem in the large-body limit [8]. We discretized the corresponding Euler-Lagrange equations by a Galerkin method with $\mathcal{T}$-piecewise constant ansatz and test functions, where $\mathcal{T}$ is a triangulation of $\Omega \subset \mathbb{R}^2$ by rectangular elements which admits hanging nodes. More specifically, we considered a ferromagnetic rod $\Omega = (-0.5, 0.5) \times (-2.5, 2.5)$, with the uniform initial mesh consisting of $N = 20$ squares. For the corresponding numerical analysis and an effective implementation, the reader is referred to [5, 6].

The experiments were conducted using the HLib software package by S. Börm and L. Grasedyck of the Max-Planck-Institute for Mathematics in the Sciences (Leipzig, Germany). We utilized a Compaq/HP AlphaServer ES45 under Unix, with 32 GB of RAM and four Alpha EV68 CPUs running at 1 GHz each. Implementational details can be found in [10, 11]; in particular, the $\mathcal{H}^2$-factorization of a block $\mathbf{A}^{\alpha\beta}|_{\sigma\times\tau}$ was stored only if this was cheaper than storing the exact matrix block.

Tables 1–3 contain experimental results (assembly times, storage requirements, and the error $\|\mathbf{A} - \mathbf{A}_\mathcal{H}\|_2$ computed by a power iteration) for uniform mesh-refinement. Moreover, we compared $\mathbf{A}_\mathcal{H}$ to the matrix $\mathbf{A}_\mathcal{H}^{\rm rcp}$ obtained by adaptive $\mathcal{H}^2$-recompression [4] of $\mathbf{A}_\mathcal{H}$. Note that $\mathbf{A}_\mathcal{H}^{\rm rcp}$ provides almost the same accuracy as $\mathbf{A}_\mathcal{H}$, but typically requires only 70% of the storage, and that the recompression times are negligible in comparison to the respective assembly times.

Finally, in Tables 4–6, we give the corresponding results for a sequence of adaptively generated meshes. Here, $\mathcal{T}^{(j+1)}$ is obtained from $\mathcal{T}^{(j)}$ as follows: First,

Table 1. Assembly times (left) and recompression times (right) for $\mathbf{A}_\mathcal{H}$ ([ms/N], uniform mesh-refinement)

N / p	**2**	**3**	**4**	**5**	**6**
320	2.1	2.1	2.1	2.7	2.7
1280	4.3	4.3	4.3	6.0	6.1
5120	5.8	5.8	6.1	11.2	11.2
20480	6.7	6.7	6.8	13.8	13.9
81920	7.0	7.0	7.0	15.2	15.3

N / p	**2**	**3**	**4**	**5**	**6**
320	0.1	0.1	0.1	0.1	0.2
1280	0.1	0.1	0.2	0.2	0.3
5120	0.1	0.2	0.3	0.4	0.6
20480	0.2	0.2	0.4	0.6	1.0
81920	0.2	0.2	0.5	0.7	1.2

Table 2. Storage requirements of $\mathbf{A}_\mathcal{H}$ (left) and of $\mathbf{A}_\mathcal{H}^{\rm rcp}$ (right) ([kB/N], uniform mesh-refinement). For comparison, we give the values for the full matrix $\mathbf{A}$.

N / p	**2**	**3**	**4**	**5**	**6**	**full**
320	5.9	6.4	7.7	9.5	11.7	10.0
1280	12.3	13.5	16.8	20.9	25.4	40.0
5120	17.8	21.1	30.0	40.5	50.9	160.0
20480	21.0	26.0	39.5	56.9	76.4	640.0
81920	22.8	28.9	45.2	67.0	92.9	2560.0

N / p	**2**	**3**	**4**	**5**	**6**
320	5.8	5.9	6.0	7.4	7.5
1280	12.1	12.3	12.8	17.0	17.3
5120	17.4	18.4	20.7	32.4	33.1
20480	20.4	21.7	24.3	41.9	43.4
81920	22.1	23.6	27.4	47.3	49.3

Table 3. Errors $\|\mathbf{A} - \mathbf{A}_\mathcal{H}\|_2$ (left) and $\|\mathbf{A} - \mathbf{A}_\mathcal{H}^{\rm rcp}\|_2$ (right) (uniform mesh-refinement)

N / p	**2**	**3**	**4**	**5**	**6**
320	2.7_{-5}	2.9_{-6}	3.3_{-7}	3.5_{-8}	4.0_{-9}
1280	1.5_{-5}	1.6_{-6}	1.7_{-7}	1.8_{-8}	2.1_{-9}
5120	5.5_{-6}	5.8_{-7}	6.0_{-8}	6.1_{-9}	6.7_{-10}
20480	1.7_{-6}	1.8_{-7}	1.8_{-8}	1.8_{-9}	2.1_{-10}

N / p	**2**	**3**	**4**	**5**	**6**
320	3.7_{-5}	3.6_{-6}	2.9_{-7}	2.3_{-8}	2.0_{-9}
1280	2.5_{-5}	2.0_{-6}	2.6_{-7}	2.0_{-8}	1.9_{-9}
5120	8.4_{-6}	9.2_{-7}	1.3_{-7}	8.0_{-9}	6.7_{-10}
20480	2.4_{-6}	3.0_{-7}	4.2_{-8}	2.2_{-9}	2.7_{-10}

Table 4. Assembly times (left) and recompression times (right) for $\mathbf{A}_{\mathcal{H}}$ ([ms/N], adaptive mesh-refinement)

N / p	2	3	4	5	6
308	2.0	2.0	2.1	2.3	2.4
1244	4.4	4.4	4.4	6.2	6.2
6548	5.0	5.1	8.5	9.2	13.5
26204	6.1	6.1	10.9	12.6	18.4
117524	7.0	7.2	13.3	13.4	22.5

N / p	2	3	4	5	6
308	0.1	0.1	0.1	0.1	0.2
1244	0.1	0.1	0.1	0.2	0.3
6548	0.2	0.2	0.3	0.5	0.6
26204	0.2	0.3	0.3	0.8	0.9
117524	0.2	0.3	0.5	0.9	1.3

Table 5. Storage requirements of $\mathbf{A}_{\mathcal{H}}$ (left) and of $\mathbf{A}_{\mathcal{H}}^{\text{rcp}}$ (right) ([kB/N], adaptive mesh-refinement). For comparison, we give the values for the full matrix **A**.

N / p	2	3	4	5	6	full
308	5.7	6.1	7.3	9.0	11.5	9.6
1244	12.7	13.9	17.3	21.0	25.0	38.9
6548	18.0	23.8	34.2	46.5	54.9	204.6
26204	21.8	30.7	46.6	66.0	80.4	818.9
117524	24.9	33.9	54.0	76.8	101.3	3672.6

N / p	2	3	4	5	6
308	5.5	5.6	5.6	6.5	6.6
1244	12.3	12.6	12.8	17.3	17.5
6548	16.8	18.0	26.5	29.1	37.9
26204	20.8	30.7	35.8	40.0	53.1
117524	23.6	25.9	41.2	43.8	65.6

Table 6. Errors $\|\mathbf{A}-\mathbf{A}_{\mathcal{H}}\|_2$ (left) and $\|\mathbf{A}-\mathbf{A}_{\mathcal{H}}^{\text{rcp}}\|_2$ (right) (adaptive mesh-refinement)

N / p	2	3	4	5	6
308	3.9_{-5}	3.9_{-6}	6.1_{-7}	3.9_{-8}	1.9_{-9}
1244	2.5_{-5}	2.8_{-6}	2.9_{-7}	2.7_{-8}	4.1_{-9}
6548	1.4_{-5}	1.6_{-6}	1.5_{-7}	1.6_{-8}	2.5_{-9}
26204	5.1_{-6}	5.8_{-7}	5.5_{-8}	6.8_{-9}	9.7_{-10}

N / p	2	3	4	5	6
308	1.6_{-4}	1.3_{-5}	2.4_{-6}	1.0_{-7}	4.6_{-9}
1244	1.6_{-4}	7.5_{-6}	1.1_{-6}	4.7_{-8}	8.6_{-9}
6548	3.1_{-5}	2.6_{-6}	2.1_{-7}	2.6_{-8}	3.2_{-9}
26204	6.8_{-6}	6.4_{-7}	5.7_{-8}	6.8_{-9}	9.7_{-10}

$\mathcal{T}^{(j)}$ is refined uniformly. Then, in a second step, we additionally refine either the four corner elements of Ω (for j even) or all the elements along the edges of Ω (for j odd). Meshes of this type are observed in [6] for an adaptive mesh-refining strategy based on a residual a posteriori error estimate.

References

1. Börm, S., Grasedyck, L., Hackbusch, W.: Introduction to Hierarchical Matrices with Applications, Eng. Anal. with B. Elem. **27** (2003) 405–422.
2. Börm, S., Löhndorf, M., Melenk, M.: Approximation of Integral Operators by Variable-Order Interpolation, Numer. Math. **99** (2005), 605 – 643.
3. Börm, S., Hackbusch, W.: $\mathcal{H}^2$-Matrix Approximation of Integral Operators by Interpolation, Appl. Numer. Math. **43** (2002) 129–143.
4. Börm, S., Hackbusch, W.: Data-Sparse Approximation by Adaptive $\mathcal{H}^2$-Matrices, Computing **69** (2002) 1–35.

5. Carstensen, C., Praetorius, D.: Effective Simulation of a Macroscopic Model for Stationary Micromagnetics, Comput. Methods Appl. Mech. Engrg. **194** (2005) 531–548.
6. Carstensen, C., Praetorius, D.: Numerical Analysis for a Macroscopic Model in Micromagnetics, SIAM J. Numer. Anal. **42** (2005), 2633 – 2651.
7. Dahmen, W., Faermann, B., Graham, I., Hackbusch, W., Sauter, S.: Inverse Inequalities on Non-Quasi-Uniform Meshes and Application to the Mortar Finite Element Method, Math. Comp. **73** (2004) 1107–1138.
8. DeSimone, A.: Energy Minimizers for Large Ferromagnetic Bodies, Arch. Rational Mech. Anal. **125** (1993) 99–143.
9. Hubert, A., Schäfer, R.: Magnetic Domains, Springer-Verlag, Berlin (1998).
10. Popović, N., Praetorius, D.: Applications of $\mathcal{H}$-Matrix Techniques in Micromagnetics, Computing **74** (2005), 177 – 204.
11. Popović, N., Praetorius, D.: The $\mathcal{H}^2$-ACA Algorithm for the Efficient Treatment of the Magnetostatic Potential, in preparation (2005).
12. Praetorius, D.: Analysis of the Operator Δ^{-1}div Arising in Magnetic Models, Z. Anal. Anwendungen **23** (2004) 589–605.

Part III

Robust Algebraic Multigrid and Hierarchical Preconditioning Methods

An Agglomerate Multilevel Preconditioner for Linear Isostasy Saddle Point Problems

Erik Bängtsson and Maya Neytcheva

Uppsala University, Department of Information Technology,
Box 337, SE-751 05 Uppsala, Sweden
{erikba, maya}@it.uu.se

Abstract. This paper discusses preconditioners for the iterative solution of nonsymmetric indefinite linear algebraic systems of equations as arising in modeling of the purely elastic part of glacial rebound processes. The iteration scheme is of inner-outer type using a multilevel preconditioner for the inner solver. Numerical experiments are provided showing a robust behavior.

1 Introduction

The need to approximate the Schur complement of a matrix often arises in the context of constructing preconditioners for iterative solution methods for linear systems of equations. One example is the framework of block factorized preconditioners. These are based on some block 2×2 form of the original matrix A and its exact factorization

$$A = \begin{bmatrix} A_{11} & A_{12} \\ A_{21} & A_{22} \end{bmatrix} = \begin{bmatrix} I & 0 \\ A_{21}A_{11}^{-1} & I \end{bmatrix} \begin{bmatrix} A_{11} & A_{12} \\ 0 & S \end{bmatrix}, \qquad (1)$$

where $S = A_{22} - A_{21}A_{11}^{-1}A_{12}$ is the Schur complement. A preconditioner to A is then sought in the form

$$R = \begin{bmatrix} I & 0 \\ A_{21}P^{-1} & I \end{bmatrix} \begin{bmatrix} P & A_{12} \\ 0 & Q \end{bmatrix}, \qquad (2)$$

where P is an approximation of A_{11} and Q is an approximation of S.

The block 2×2 structure of the matrix can be obtained in various ways.

(i) It can correspond to a splitting of the mesh nodes and the corresponding degrees of freedom into *fine* and *coarse*, related to a refinement of some given (coarse) mesh.
(ii) It can be due to a splitting of the nodes into fine and coarse, where the coarse mesh may be obtained via some agglomeration technique.
(iii) The splitting can be based also on splitting the matrix graph into independent sets.

I. Lirkov, S. Margenov, and J. Waśniewski (Eds.): LSSC 2005, LNCS 3743, pp. 113–120, 2006.

All these techniques can be applied recursively, leading to a multilevel preconditioner, in which the block Q in Equation (2) is again written in the same block-factorized form. Setting (i) is the framework of the Algebraic Multilevel Iteration (AMLI) methods, originally developed in the hierarchical basis functions (HBF) framework ([3] and followup work). An example of a method from class (ii) is the AMGe method [8]. As representatives of class (iii) we mention [14] and [7].

Another context where approximations of a Schur complement matrix are required is that to precondition saddle point matrices

$$\mathcal{A} = \begin{bmatrix} M & B^T \\ B & -C \end{bmatrix}, \tag{3}$$

which arise when solving Stokes problem, Oseen's problem, or linear systems from constraint optimization. For these matrices, one of the frequently used preconditioner is of block lower- or upper-triangular form, for example

$$\mathcal{D} = \begin{bmatrix} D_1 & 0 \\ B & -D_2 \end{bmatrix}, \tag{4}$$

where $D_1 \approx M$ and D_2 is an approximation of the negative Schur complement $S = C + BD_1^{-1}B^T$. For more details on the spectral properties of $\mathcal{D}^{-1}\mathcal{A}$ we refer to [2], and to [6] for a recent survey on preconditioners for saddle point matrices.

When constructing an approximation Q of the true Schur complement S, where the latter is in general a full matrix, we aim at achieving

1 a good approximation of S, for example, in the symmetric case, satisfying a spectral equivalence relation of the form $\underline{\beta} S \leq Q \leq \overline{\beta} S$, where $\underline{\beta}$, $\overline{\beta}$ do not depend on problem or discretization parameters;
2 a sparse matrix Q, to save computational cost;
3 an approximation, which can easily be handled in a parallel environment.

In some cases it is known how to obtain a good quality approximation for the Schur complement. For example, for red-black orderings on regular meshes, A_{11} becomes diagonal and even the exact Schur complement is computed at low cost. In some applications it is enough to approximate A_{11} by its diagonal or by some sparse approximate inverse of A_{11}. For other problems S can be approximated on a differential operator level, as done for the Oseen's problem in [9]. For the Stokes problem it is known that a good approximation of $BM^{-1}B^T$ is the pressure mass matrix. For the AMLI-type of methods, the usual approximation of S is the coarse mesh stiffness matrix.

In Section 2 we describe a method to construct Q, possessing all properties 1–3. In Section 3 we describe a problem, which leads to the solution of systems with saddle point matrices. To precondition those, we use a preconditioner of the form (4), where in addition, systems with D_1 and D_2 are solved using a preconditioner of the form (2).

Although the analysis of the above mentioned Schur complement matrices is done so far for symmetric positive definite (spd) matrices only, we include some tests with non-symmetric matrices, which indicate that the validity of the theoretical results can be extended beyond the class of spd problems.

2 Schur Complement Approximation Obtained Via Assembly of Element Schur Complements

In this paper we stay in the framework (i), i.e., we assume that two (or more) nested meshes are obtained from a given coarse mesh by regular refinements. The two-by-two structure of A corresponds to a splitting of the mesh nodes into coarse (belonging to the coarse mesh) and fine (the rest). The system matrix A arises from a finite element (FE) discretization of a (system of) partial differential equation(s), and it is assembled from local element matrices A^e.

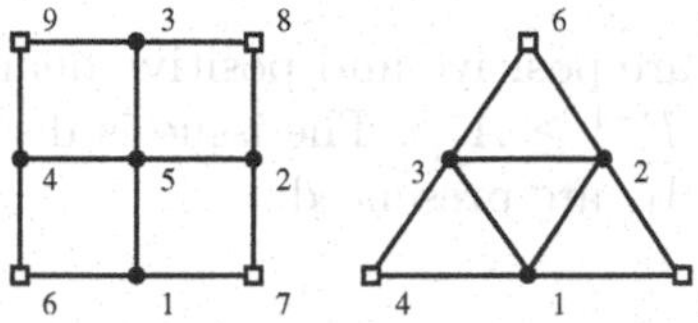

Fig. 1. A macroelement on a quadrilateral and a triangular mesh

We assume that e represents a macroelement, depicted in Figure 1 for triangular and quadrilateral meshes. The local macroelement stiffness matrix admits the same 2×2 block structure as the global matrix, i.e.

$$A^e = \begin{bmatrix} A^e_{11} & A^e_{12} \\ A^e_{21} & A^e_{22} \end{bmatrix}.$$

We then assemble an approximate Schur complement S_a from local Schur complements, $S_a = \sum_e S^e$, $S^e = A^e_{22} - A^e_{21} {A^e_{11}}^{-1} A^e_{12}$, computed exactly on macroelement level (originally proposed in [10]). The so-constructed approximation S_a possesses some attractive properties.

(a) Clearly, S_a is sparse by construction.
(b) In [10] it is shown that for symmetric positive definite matrices, S_a is spectrally equivalent to the true Schur complement S, namely the relation $\underline{\beta} S \leq Q \leq S$ holds for second order elliptic problems. The estimated $\underline{\beta}$ in [10] is not optimal with respect to some problem parameters and can be improved. In what follows, we assume that $\underline{\beta}$ is independent of discretization and problem parameters, such as anisotropy, for instance.
(c) From the construction of S_a, which is the same as for the global stiffness matrix, it is clear that the parallelization techniques applied to handling FE matrices are applicable for S_a as well.
(d) S_a inherits the properties of A and automatically generates symmetric or nonsymmetric approximations of S, which is not true for some of the other above mentioned techniques. For instance, $C + B\text{diag}(M)^{-1}B^T$ will be symmetric even if M is nonsymmetric.

In addition, the framework can be used recursively and permits multilevel extensions. As we see from the numerical experiments, it works well even for some nonsymmetric problems. The theory, however, is not yet completed. When constructing a multilevel preconditioner of type (2), not using the hierarchical basis, and in order to assure that R is positive definite and $\kappa(R^{-1}A)$ is independent of the mesh size, several issues have to be addressed. The approximation P cannot be chosen as an arbitrary p.d. approximation of A_{11}. To observe this, we let $\beta S \leq S_a \leq S$, and then we have

$$\begin{aligned}\mathbf{v}^T R\mathbf{v} &= \mathbf{v}^T A\mathbf{v} - \mathbf{v}_1^T(A_{11} - P)\mathbf{v}_1 + \mathbf{v}_2^T(S_a - S)\mathbf{v}_2 + \mathbf{v}_2^T A_{21}(P^{-1} - A_{11}^{-1})A_{12}\mathbf{v}_2 \\ &= (\mathbf{v}^T A\mathbf{v} - \mathbf{v}_2^T S\mathbf{v}_2) + \mathbf{v}_2^T S_a\mathbf{v}_2 + \mathbf{v}_2^T A_{21}(P^{-1} - A_{11}^{-1})A_{12}\mathbf{v}_2 \\ &+ \mathbf{v}_1^T(A_{11} - P)\mathbf{v}_1. \end{aligned} \tag{5}$$

Hence, the last two terms are positive and positive definiteness of R is guaranteed, only if $P \leq A_{11}$ and $P^{-1} \geq A_{11}^{-1}$. The issue is discussed in [11] and some safe ways to approximate A_{11} are presented.

3 Numerical Experiments

We test the above preconditioning methods on a problem from modeling of glacial rebound, including only the purely elastic response of the viscoelastic earth to glaciation and deglaciation.

A 2D flat Earth model is subjected to a Heaviside load of a 1000 km wide and 2 km thick ice sheet. The size of the domain is 10000 km width and 4000 km depth and the boundary conditions (b.c.) are homogenous Dirichlet b.c. on the boundary $y = -4000$ km, symmetry b.c. on the boundary $x = 0$, and homogeneous Neumann b.c. on the boundary $x = 10000$ km and on the boundary $y = 0, x > 1000$ km. The domain is discretized with a quasi-regular quadrilateral finite element mesh.

The model is described by the following system of PDEs

$$\begin{aligned} -2\mu\Delta\mathbf{u} - \mu\nabla \times \nabla\mathbf{u} - \nabla(\mathbf{b}\cdot\mathbf{u}) + \mathbf{c}(\nabla\cdot\mathbf{u}) - \mu\nabla p &= \mathbf{0} \\ \mu\nabla\cdot\mathbf{u} - \tfrac{\mu}{\lambda}p &= 0, \end{aligned} \tag{6}$$

where $\mathbf{u} = [u\ v]^T$ are the displacements, and p is the kinematic pressure.

The first order terms in Equation (6), describe the *advection of pre-stress* $(\nabla(\mathbf{b}\cdot\mathbf{u}))$, where $\mathbf{b}$ is the pre-stress gradient, and the *buoyancy* $(\mathbf{c}(\nabla\cdot\mathbf{u}))$, arising from density changes. The material parameters $\mu = E/2/(1+\nu)$ and $\lambda = 2\mu\nu/(1-2\nu)$ are the Lamé coefficients, where $\nu \in [0, 0.5]$ is the Poisson number and $E = 400$ GPa is the Youngs modulus. For details on the model, and how Equation (6) is related to the linearized Navier-Stokes equation, see [5], and the references therein.

For the discretization we use standard bilinear basis functions both for $\mathbf{u}$ and p, and a consistently stabilized finite element formulation of Equation (6) (see [5]). The problem reads as follows.

Find $\mathbf{u}_h \in \mathbf{V}^h \subset H_0^1(\Omega_h)$ and $p_h \in P^h \subset P \equiv \{p \in L^2(\Omega_h) : \int_{\Omega_h} \mu p = 0\}$ such that

$$\begin{aligned} a(\mathbf{u}_h, \mathbf{v}_h) + b(\mathbf{v}_h, p_h) &= \langle \mathbf{v}_h, l\rangle \quad \forall \mathbf{v}_h \in \mathbf{V}^h \\ b(\mathbf{u}_h, q_h) - c(p_h, q_h) - \sigma d(p_h, q_h) &= -\sigma e(q_h, \mathbf{u}_h, p_h, \mathbf{b}, \mathbf{c}) \quad \forall q_h \in P^h, \end{aligned} \tag{7}$$

where $e(q, \mathbf{u}, p, \mathbf{b}, \mathbf{c}) = \int_\Omega \left[q\Delta(\mathbf{u}\cdot\mathbf{b}) - \frac{\mu}{\lambda} pq\nabla\cdot\mathbf{c}\right]$ and $d(p,q) = \int_\Omega \left(\mu + \frac{\mu^2}{\lambda}\right)\nabla p\cdot\nabla q$. The bilinear form $\langle\cdot,\cdot\rangle$ is a traction term, arising from the loading boundary condition.

For this problem, $\mathbf{b} = [0, \rho g]$ and $\mathbf{c} = [0, \rho g]$, where g is the gravitational acceleration and $\rho = 3000\ kg\ m^{-3}$. For these $\mathbf{b}$ and $\mathbf{c}$, the bilinear form $a(\mathbf{u}, \mathbf{v})$ in Equation (7) is coercive, and thus, the discrete solution exists and is unique.

For the finite element discretization we use the C++ package `deal.II` [4], and for the numerical linear algebra the package PETSc [12]. The tests are performed on a Sun Ultra-Sparc IV 1050 MHz processor running under Sun Solaris 9.

The matrix in the so-arising linear system is of the form (3) and it is solved using the generalized conjugate gradient-minimized residual (GCG-MR) method. The iterative method is preconditioned with $\mathcal{D}$ and solved until a relative stopping criterion 10^{-6} is achieved.

To define a preconditioner D_1 for M, we order the displacements $\mathbf{u}$ using the so-called *separate displacement ordering* (sdo), i.e., we order all displacements in the x-direction first. This introduces a 2×2 block structure in M, $M = \begin{bmatrix} M_{11} & M_{12} \\ M_{21} & M_{22}\end{bmatrix}$. We recall that M is non-symmetric, and that it is a sum of $a(\mathbf{u}, \mathbf{v}) = \widehat{a}(\mathbf{u}, \mathbf{v}) + \bar{a}(\mathbf{u}, \mathbf{v})$, where $\widehat{a}(\mathbf{u}, \mathbf{v}) = \int_\Omega [2\mu \sum_{k=1}^2 (\nabla u_k)\cdot(\nabla v_k) - \mu(\nabla\times\mathbf{u})\cdot(\nabla\times\mathbf{v})]$ and $\bar{a}(\mathbf{u}, \mathbf{v}) = \int_\Omega [-\nabla(\mathbf{u}\cdot\mathbf{b})\mathbf{v} + (\nabla\cdot\mathbf{u})(\mathbf{c}\cdot\mathbf{v})]$.

We test three preconditioners when solving with M, D_1, $\widehat{D_1}$ and $\widetilde{D_1}$. The bilinear form $a(\mathbf{u}, \mathbf{v})$ is dominated by the elastic part $\widehat{a}(\mathbf{u}, \mathbf{v})$. This motivates the choice of the preconditioner $\widehat{\mathcal{D}}$, where M is solved by an inner iteration method, preconditioned by a symmetric block-diagonal matrix $\widehat{D}_1$ ($\widehat{d}_{1ij} = \widehat{a}(\mathbf{u}_i, \mathbf{v}_j)$), and the preconditioner (2) is applied to the latter matrix.

Due to Korn's inequality, $\widehat{a}(\mathbf{u}, \mathbf{v})$ is spectrally equivalent to the scaled vector Laplacian $\widetilde{a}(\mathbf{u}, \mathbf{v}) = \int_\Omega 2\mu\sum_{k=1}^2 (\nabla u_k)\cdot(\nabla v_k)$. This motivates the choice of the preconditioner $\widetilde{\mathcal{D}}$, in which the inner solver is preconditioned by a block-diagonal matrix $\widetilde{D}_1 = \begin{bmatrix} \widetilde{D}_1^{(1)} & \\ & \widetilde{D}_1^{(2)}\end{bmatrix}$. The preconditioner (2) is applied to each of the diagonal blocks.

In the preconditioner $\mathcal{D}$, the inner solver for M is preconditioned with (2). M is nonsymmetric, so we construct the preconditioner without a theoretical justification. Nevertheless, the numerical results show that it is stable and scalable, and this is also the case for the inner solver for the nonsymmetric matrix D_2. Whenever there is no risk of confusion, the ˜ and ˆ are omitted and $\mathcal{D}$ and D_1 indicate generic preconditioner for $\mathcal{A}$ and M. The diagonal blocks of $\mathcal{D}$, D_1 and D_2, are solved with GCG-MR, to relative stopping criteria τ and 10^{-6}, correspondingly.

The block-factorized multilevel preconditioners for the inner solvers are stabilized with two inner iterations on every second level, and the block P in (2) is approximated with an incomplete LU-factorization (ILUT), (see [13]). The parameters for ILUT are; drop tolerance – 10^{-5}, column pivot – 10^{-2}, fill-in — equal to the maximum number of nonzero elements per row in the underlying matrix.

Table 1. Dependence of the preconditioner on τ, the relative tolerance criterion in the inner solver of D_1. The problem size $N = 31395$ and $\nu = 0.5$.

	$\mathcal{D}$		$\widehat{\mathcal{D}}$		$\widetilde{\mathcal{D}}$	
τ	# iter	t (s)	# iter	t (s)	# iter	t (s)
5e-1	18(3,1)	19.94	16(2,1)	16.25	20(2,1)	16.56
1e-1	13(4,1)	20.27	13(3,1)	17.65	15(4,1)	20.79
1e-2	12(6,1)	26.11	12(5,1)	22.19	13(8,1)	28.66
1e-3	12(8,1)	32.34	12(7,1)	26.77	12(11,1)	36.48
1e-4	12(10,1)	38.81	12(8,1)	32.12	12(14,1)	45.92
1e-5	12(12,1)	45.25	12(10,1)	37.85	12(17,1)	54.52
1e-6	12(14,1)	51.56	12(12,1)	43.75	12(19,1)	62.76

The numerical experiments illustrate the performance of the proposed preconditioner $\mathcal{D}$, depending on the accuracy of the inner solver for D_1, τ, the Poisson number ν, the problem size N and the number of levels l in (2). The notations in the tables are the following. The first iteration count in every column is the number of outer iterations, the figures in the parentheses are the average number of inner iteration for D_1 and D_2, and the last figure is the time to set up the preconditioner and the solve with $\mathcal{A}$ once.

Table 1 shows how the performance of $\mathcal{D}$ is affected by the accuracy of the inner solver for D_1, and one finds that the smallest τ gives the fastest solution time. The performance of D_2 is extraordinary good as only one iteration is needed to reach the required convergence. This convergence rate is, as shown in Tables 2–4, independent of problem size, Poisson number and the number of mesh levels.

Table 2 depicts how the number of outer iterations rises as the material becomes incompressible. This follows well the result observed in [5]. The increase in solution time agrees with the number of outer iterations since the inner iteration count is unaffected by the growth in ν.

Table 3 shows the dependence on the problem size. For large enough problems $N > 120000$, a jump in the inner iteration count, similar to what is observed in [10] for nearly incompressible material, is seen for $\mathcal{D}$ and $\widehat{\mathcal{D}}$, destroying the scalability of those preconditioners. For $\widetilde{\mathcal{D}}$ the solver nearly scales with N both in iteration count and solution time.

The results in Table 4 show that there exist a balance between the number of levels and the solution time. The cost to solve a larger coarse mesh matrix

Table 2. Dependence of the preconditioner on the Poisson number ν. The problem size $N = 31395$ and $\tau = 5 \cdot 10^{-1}$.

	$\mathcal{D}$		$\widehat{\mathcal{D}}$		$\widetilde{\mathcal{D}}$	
ν	# iter	t (s)	# iter	t (s)	# iter	t (s)
0.2	11(2,1)	12.97	10(2,1)	12.67	13(2,1)	12.49
0.3	12(2,1)	15.13	13(2,1)	13.77	14(2,1)	12.94
0.4	14(3,1)	16.74	14(2,1)	14.85	17(2,1)	14.49
0.5	18(3,1)	19.94	16(2,1)	16.25	20(2,1)	16.56

Table 3. Performance of the preconditioner, depending on the problem size N. The Poisson number $\nu = 0.5$ and $\tau = 5 \cdot 10^{-1}$.

	$\mathcal{D}$		$\widehat{\mathcal{D}}$		$\widetilde{\mathcal{D}}$	
N	# iter	t (s)	# iter	t (s)	# iter	t (s)
8019	18(2,1)	3.71	16(2,1)	3.29	19(2,1)	3.2
31395	18(3,1)	19.94	16(2,1)	16.25	20(2,1)	16.56
124227	22(6,1)	181.9	17(3,1)	98.9	18(2,1)	74.94
494211	24(12,1)	1754.97	23(10,1)	1355.83	18(3,1)	392.38

Table 4. Performance of the preconditioner, depending on the number of levels l. The Poisson number $\nu = 0.5$ and $\tau = 5 \cdot 10^{-1}$.

	$\mathcal{D}$		$\widehat{\mathcal{D}}$		$\widetilde{\mathcal{D}}$	
	# iter	t (s)	# iter	t (s)	# iter	t (s)
	$N = 124224$					
$l = 2$	16(3,1)	95.45	16(3,1)	95.69	18(2,1)	79.64
$l = 3$	15(3,1)	86.66	15(3,1)	84.3	18(2,1)	76.11
$l = 4$	19(5,1)	132.83	17(3,1)	91.29	18(2,1)	73.26
$l = 5$	19(5,1)	143.63	17(3,1)	94.49	18(2,1)	75.08
	$N = 494211$					
$l = 2$	18(5,2)	778.85	19(5,2)	808.47	18(2,2)	406.16
$l = 3$	19(5,2)	810.08	20(5,2)	818.88	17(2,2)	385.98
$l = 4$	20(11,2)	1392.75	22(10,2)	1305.02	18(3,2)	414.64
$l = 5$	21(10,2)	1350.98	21(11,2)	1399.93	18(3,2)	421.53

is balanced by less overhead in the short recurrence, which will be even more pronounced for parallel implementations.

4 Conclusions

We present a preconditioner for a nonsymmetric saddle point problem, which exhibits a robust behavior provided that its diagonal blocks well approximate the pivot block M, and the (negative) Schur-complement of the system matrix.

The block D_2 in the outer preconditioner approximates this Schur complement, and it is assembled from element Schur matrices. Numerically, this approximation shows to work well.

The blocks D_1 and D_2 are solved with an inner iterative solver, preconditioned with an algebraic multilevel preconditioner. For both blocks, approximated (coarse mesh) Schur complements are assembled from element Schur complements.

The numerical experiments show also that a low accuracy for the inner solver for M, combined with a short recursion in the multilevel preconditioner, leads to the shortest solution time. For not strongly nonsymmetric blocks M, the choice of a block-diagonal, spectrally equivalent and symmetric preconditioner for M gives an outer preconditioner that is robust and scalable.

References

1. Axelsson, O., Barker, V.A., Neytcheva, M., Polman, B.: Solving the Stokes problem on a massively parallel computer. Math. Model. Anal. **4** (2000) 1–22
2. Axelsson, O., Neytcheva, M.: Preconditioning methods for constrained optimization problems. Num. Lin. Alg. Appl. **10** (2003) 3–31
3. Axelsson, O., Vassilevski, P.S.: Algebraic multilevel preconditioning methods. I. Numer. Math. **56**:2–3 (1989) 157–177
4. Bangerth, W., Hartmann, R., Kanschat, G.: `deal.II` Differential Equations Analysis Library, Technical Reference, IWR, `http://www.dealii.org`
5. Bängtsson, E., Neytcheva, M.: Numerical simulations of glacial rebound using preconditioned iterative solution methods. Appl. Math. **50**:3 (2005) 183–201
6. Benzi, M., Golub, G.H., Liesen, J.: Numerical solution of saddle point problems. Acta Mathematica (2005) 1–137
7. Botta, E.F.F., Wubs, F.W.: Matrix renumbering ILU: an effective algebraic multilevel ILU preconditioner for sparse matrices. SIAM J. Matrix anal. Appl. **20**:4 (1999) 1007–1026.
8. Jones, J.E., Vassilevski, P.S.: AMGe based on element agglomeration. SIAM J. Sci. Comput. **23**:1 (2001) 100–133
9. Kay, D., Loghin, D., Wathen, A.: A preconditioner for the steady-state Navier-Stokes equations. SIAM J. Sci. Comput. **24**:1 (2002) 237–256
10. Kraus, J.K.: Algebraic multilevel preconditioning of finite element matrices using local Schur complements. Num. Lin. Alg. Appl. **12** (2005) 1–19
11. Notay, Y.: Using approximate inverses in algebraic multilevel methods. Num. Math. **80**:3 (1998) 397–417
12. Portable, Extensible Toolkit for Scientific computation (PETSc) suite, Mathematics and Computer Science Division, Argonne National Laboratory, `www-unix.mcs.anl.gov/petsc/`
13. Saad, Y.: ILUT: a Dual Threshold Incomplete LU Factorization. Num. Lin. Alg. Appl. **1** (1994) 387–402
14. Saad, Y., Suchomel, B.: ARMS: an algebraic recursive multilevel solver for general sparse linear systems. Num. Lin. Alg. Appl. **9** (2002) 359–378

On the Utilization of Edge Matrices in Algebraic Multigrid

J.K. Kraus

Johann Radon Institute for Computational and Applied Mathematics,
Altenbergerstraße 69, A-4040 Linz, Austria

Abstract. We are interested in the design of efficient algebraic multigrid (AMG) methods for the solution of large sparse systems of linear equations arising from finite element (FE) discretization of second-order elliptic partial differential equations (PDEs). In particular, we introduce the concept of so-called "edge matrices", which–in the present context–are extracted from the individual element matrices. This allows for the construction of spectrally equivalent approximations of the original stiffness matrix that can be utilized in the framework of AMG.

The edge matrices give rise to modify the definition of "strong" and "weak" connections (edges), which provides a basis for selecting the coarse-grid nodes in algebraic multigrid methods. Moreover, a reproduction of edge matrices on coarse levels offers the opportunity to combine classical coarsening algorithms with effective (energy minimizing) interpolation principles involving small-sized "computational molecules" (small collections of edge matrices). This yields a flexible and robust new variant of AMG, which we refer to as AMGm.

1 Introduction

We are concerned with the solution of large-scale systems of linear equations

$$A\mathbf{u} = \mathbf{f} \tag{1}$$

where A is symmetric and positive definite (SPD). Moreover, we want to assume that A stems from assembling small-sized symmetric positive semidefinite (SPSD) matrices A_T, e.g., element matrices, which typically arise out of finite element modeling of self-adjoint elliptic boundary-value problems (BVPs).

In many instances (of this huge class of problems) algebraic multigrid (AMG) methods [2, 11] can be used to build highly efficient and robust linear solvers [5, 11, 12]. AMG using element interpolation (AMGe) [3, 7, 8], so-called spectral AMGe [4], and AMG based on smoothed aggregation [13, 14] have even broadened the range of applicability of the classical AMG algorithm [11]. These more recent developments are based on techniques of energy-minimizing interpolation (or prolongation). Our approach can be viewed as a further development (generalization) of the so-called element preconditioning technique introduced in [6, 10].

I. Lirkov, S. Margenov, and J. Waśniewski (Eds.): LSSC 2005, LNCS 3743, pp. 121–129, 2006.

The computation of edge matrices, we are proposing in the present paper, is motivated by the fact that they provide a good starting point for building efficient AMG components, while keeping their set-up costs low. We suggest a modification of the concept of "strong" and "weak" connections, as it is used in the process of coarse-grid selection (and interpolation) with classical AMG. The interpolation component in our approach is very similar to the element interpolation known from AMGe methods. However, the *computational molecules* involved in the arising local min-max problem are assembled from edge matrices in our case.

First tests indicate the robustness of the considered method to which we refer as AMGm (Algebraic MultiGrid based on computational molecules).

2 The Concept of Edge Matrices in AMG

2.1 Edge Matrices

Let A_T be an SPSD $(nd) \times (nd)$ element matrix of a Lagrangian finite element. Here, n denotes the number of nodes of the element T, and d denotes the number of degrees of freedom (dofs) associated with each node, i.e., $d = 1$ corresponds to the case of scalar problems. Further, let $\mathcal{E}_T := \{e_{ij} : 1 \leq i < j \leq n\}$ denote the set of (topological) edges of the element T.

Definition 1. *An $(nd) \times (nd)$ matrix E_{ij} whose entries are zero except for a $(2d) \times (2d)$ submatrix corresponding to the dofs associated with nodes $\{i, j\}$ defining an edge $e_{ij} \in \mathcal{E}_T$, $1 \leq i < j \leq n$, is called an edge matrix if it preserves the kernel of A_T, i.e., $E_{ij}\mathbf{v} = \mathbf{0}$ for all vectors $\mathbf{v} \in \ker(A_T)$.*

One can ask for the class of SPSD element matrices A_T that can be split exactly (disassembled) into SPSD edge matrices, i.e., for a characterization of SPSD matrices that allow for a *semipositive splitting.* For the scalar case $d = 1$ one can prove the following result, see Reference [9].

Theorem 1. *A symmetric (element) matrix A_T has a representation*

$$A_T = \sum_{e_{ij} \in \mathcal{E}_T} E_{ij} \tag{2}$$

with E_{ij} being SPSD ($E_{ij} \geq 0$) if and only if the L-ation of A_T is SPSD. If, in addition, A_T is singular, the splitting (2) is unique.

Remark 1. Note that if $A_T = (a_{ij})_{i,j}$ is not an L-matrix, then there is a unique L-matrix $C_T = (c_{ij})_{i,j}$ such that $|a_{ij}| = |c_{ij}|$ for all i, j. We say that C_T is the L-ation of A_T.[1]

---

[1] A real $N \times N$ matrix that has only nonnegative diagonal entries and nonpositive off-diagonal entries is called an L-matrix.

An immediate consequence of Theorem (1) is:

Corollary 1. *For $d = 1$, a singular SPSD matrix $A_T = (a_{ij})_{i,j}$ has an exact semipositive splitting if and only if it is a singular M-matrix.*[2]

Remark 2. For $d > 1$, the matrices that allow for an exact semipositive splitting give rise to a generalization of symmetric (singular) M-matrices.

However, in case an exact semipositive splitting of A_T into edge matrices does not exist (the case of non-M matrices), one can relax the problem and admit general edge matrices. Alternatively, instead of requiring the exact splitting (2), one can switch to the following Problem (P1).

(P1): *Find a set of edge matrices $\{E_{ij}\}$ that provides an approximate splitting*

$$A_T = B_T + R_T = \sum_{e_{ij} \in \mathcal{E}_T} E_{ij} + R_T, \quad E_{ij} \geq 0 \quad \forall e_{ij} \in \mathcal{E}_T \tag{3}$$

which minimizes the general spectral condition number $\kappa(A_T, B_T) := \lambda_{\max}/\lambda_{\min}$:

$$\lambda_{\max} := \inf\{\lambda : \ \mathbf{x}^T A_T \mathbf{x} \leq \lambda \mathbf{x}^T B_T \mathbf{x} \quad \forall \mathbf{x}\} \tag{4}$$

$$\lambda_{\min} := \sup\{\lambda : \ \mathbf{x}^T A_T \mathbf{x} \geq \lambda \mathbf{x}^T B_T \mathbf{x} \quad \forall \mathbf{x}\}. \tag{5}$$

Dealing with scalar second-order elliptic PDEs the individual element (stiffness) matrices throughout have a one-dimensional kernel (spanned by the constant vector $(1, 1, \ldots, 1)^T$). Thence, Corollary (1) implies that the solution of the constrained minimization problem (P1)-((3)–(5)) results in the *best approximation* of A_T by a singular M-matrix B_T in the sense of minimizing its (general) spectral condition number. From this viewpoint the above problem has already been considered in References [6, 10]. Symbolic methods turned out to speed up the (numerical) solution of this kind of (local) low-dimensional optimization problems considerably [10] (for the case $n = 3, 4$ and $d = 1$). The resulting (global) M-matrix B, assembled from the B_T's, was used to build a preconditioner for A, or, more precisely, (classical) AMG was applied to B instead of A.

Regarding (coupled) systems of second-order elliptic PDEs, e.g., in structural mechanics, the solution of Problem (P1), but now for $d > 1$, is the key to generalize this approach in a proper way. As one easily finds, even the element matrices arising in the plane-stress elasticity problem (approximating the x- and y-displacements in the finite element space of piecewise linear functions) cannot exactly be split into edge matrices (and thus an exact semipositive splitting of A_T does not exist a fortiori). We therefore propose to solve Problem (P1) as a basis for constructing efficient AMG methods for SPD non-M matrices (in this more general setting of edge matrices).

[2] According to Reference [1] a symmetric matrix C that has nonpositive offdiagonal entries is an M-matrix if and only if it is nonnegative definite.

2.2 Reference to AMGm

In a recent work [9], a new variant of algebraic multigrid (AMGm) has been considered (for scalar problems). The major new aspects of this approach are:

Definition of strong connections (strong edges). Based on the knowledge of edge matrices the concept of "strong" edges can be established:

Definition 2. *(Direct connections) Any two nodes i and j are said to be directly connected iff there is an edge $\{i,j\}$ connecting nodes i and j; let E_{ij} denote the corresponding edge matrix.*

Now for every loop of length 3 (triangle) in the *algebraic grid* with direct connections (edges) $\{i,j\}$, $\{j,k\}$, and $\{k,i\}$ we consider the molecule

$$M^{(i,j,k)}{:=}E_{ij} + E_{jk} + E_{ki}, \tag{6}$$

which in general is a $(3d) \times (3d)$ matrix. Furthermore, for $d \geq 1$, let

$$\mathcal{M}^{\triangle}{:=}\{M^{(i,j,k)} = (C_{pq})_{p,q} : C_{pp} \neq 0 \quad \forall p = 1,2,3\} \tag{7}$$

be the set of all such local matrices given as the sum of three edge contributions (for edges that form a triangle) for which the three $d{\times}d$ diagonal blocks (associated with the nodes i, j, k) are SPD. Then the following definition provides a (symmetric!) strong connectivity relation ("strong" edges).

Definition 3. *The strength of a (direct) connection $\{i,j\}$ is measured by*

$$s_{ij} := \min\{1, \min_{M^{(i,j,k)} \in \mathcal{M}^{\triangle}} \left\{ \frac{\|E_{ij}\|}{2 \cdot \sqrt{\|C_{p_i p_i}\| \cdot \|C_{p_j p_j}\|}} \right\}\} \tag{8}$$

where connections with $s_{ij} \geq \theta$ are said to be strong, $0 < \theta < 1$ (e.g., $\theta = 1/4$). Here p_i and p_j denote the local numbers associated with nodes i and j, i.e., $1 \leq p_i \equiv p(i), p_j \equiv p(j) \leq 3$, and $C_{p_i p_i}$ and $C_{p_j p_j}$ are the corresponding $d{\times}d$ blocks in the diagonal of $M^{(i,j,k)}$.

Remark 3. For the scalar case $d = 1$, (regarding a triangular mesh) formula (8) reduces to

$$s_{ij} := \min\{1, \min_{M^{(i,j,k)} \in \mathcal{M}^{\triangle}} \{|c_{p_i p_j}| / \sqrt{|c_{p_i p_i} c_{p_j p_j}|}\}\} \tag{9}$$

and essentially yields the energy cosine of the abstract angle between the i-th and j-th (nodal) basis function.

Locally energy-minimizing interpolation (based on molecules). The task is to define suitable computational molecules, assembled from edge matrices, for building interpolation. Assume that "weak" and "strong" edges have been identified, the coarse grid has been selected, and a set of edge matrices is

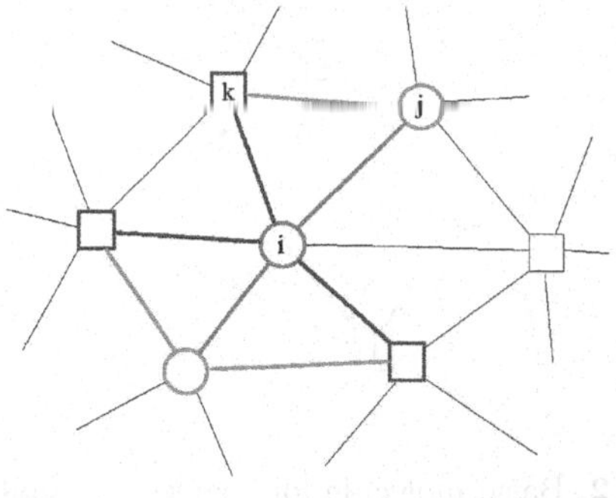

Fig. 1. Formation of interpolation molecule

available. Then for any f-node i (to which interpolation is desired) we define a so-called interpolation molecule

$$M(i) := \sum_{k \in \mathcal{S}_i^c} E_{ik} + \sum_{j \in \mathcal{N}_i^f : \exists k \in \mathcal{S}_i^c \cap \mathcal{N}_j} E_{ij} + \sum_{k \in \mathcal{S}_i^c \cap \mathcal{N}_j : j \in \mathcal{N}_i^f} E_{jk}. \tag{10}$$

This molecule arises from assembling all edge matrices associated with three types of edges: The first sum corresponds to the strong edges connecting node i to some coarse direct neighbor k (*interpolatory edges*), i.e., $k \in \mathcal{S}_i^c$. The second sum represents edges connecting the considered f-node i to any of its fine direct neighbors j being directly connected to at least one c-node $k \in \mathcal{S}_i^c$, i.e., $j \in \mathcal{N}_i^f : \exists k \in \mathcal{S}_i^c \cap \mathcal{N}_j$. Finally, the last sum in (10) corresponds to these latter mentioned connections (edges) between fine direct neighbors j and strongly connected coarse direct neighbors k of node i, i.e., $k \in \mathcal{S}_i^c \cap \mathcal{N}_j : j \in \mathcal{N}_i^f$. The formation of an interpolation molecule is illustrated in Figure 1.

Now, let

$$M(i) = M = \begin{pmatrix} M_{ff} & M_{fc} \\ M_{cf} & M_{cc} \end{pmatrix} \tag{11}$$

be the interpolation molecule where the 2×2 block structure in (11) corresponds to the n_M^f f-nodes and the n_M^c c-nodes the molecule is based on. Assuming that $M(i)$ is SPSD and M_{ff} is SPD the interpolation

$$P_{fc} := -M_{ff}^{-1} M_{fc} \tag{12}$$

then provides the minimum-energy extension (harmonic extension) with respect to $M(i)$. For a more detailed discussion, see [9].

Reproduction of edge matrices (for coarse-edges). We suggest the following practicable procedure for an inexpensive computation of coarse-edge matrices:

- Firstly, one generates a coarse edge connecting any pair of c-nodes $\{i, j\}$ if either or both of the statements below are true:

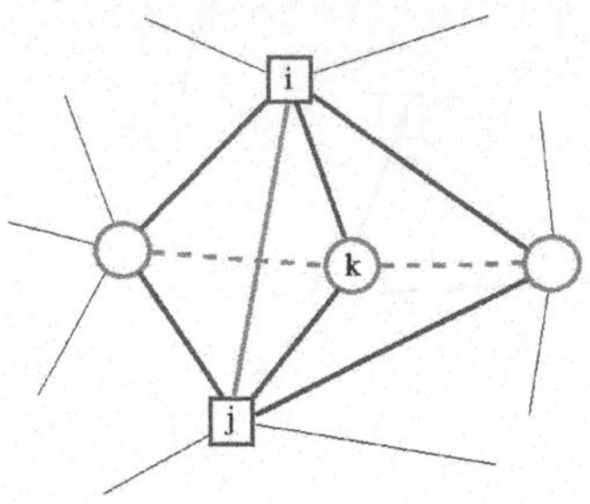

Fig. 2. Basic molecule for coarse-edge matrix

1. Nodes i and j are directly connected on the fine grid $\mathcal{D}$.
2. There exists an f-node k such that nodes i and j both are strongly connected to k, i.e., $\{i, j\} \subset \mathcal{S}_k^c$.

– Secondly, for any coarse edge $\{i, j\}$ one forms a specific computational molecule $M^{(i,j)}$, which is a *pre-stage* for the computation of the corresponding edge matrix; here, the molecule $M^{(i,j)}$ accumulates the contributions (edge-matrices) of all edges that yield paths of length two, starting in node i, passing some f-node k, and ending in node j, including also the contribution of a *direct* edge $\{i, j\}$ if there is one, see Figure 2.
– Finally, one generates the coarse-edge matrix by computing the Schur complement of $M^{(i,j)}$ with respect to its two c-nodes (eliminating all dofs associated with all of its f-nodes), i.e.,

$$E_{ij}^c := M_{cc}^{(i,j)} - M_{cf}^{(i,j)} \left(M_{ff}^{(i,j)}\right)^{-1} M_{fc}^{(i,j)}. \tag{13}$$

Thus, another particular class of molecules can be used for computing coarse-edge matrices. As can be seen from Figure 2, the important pairs of edges in these molecules are those connecting f-nodes k to both of the c-nodes i and j, which are going to be connected via a coarse edge. Additionally, one could take into account the (fine) edges connecting the corresponding f-nodes among each other, as indicated by the dotted lines in Figure 2.

AMGm. Regarding the multilevel algorithm, we notice that the AMGm method agrees with classical AMG, except for the coarse-grid selection and the interpolation component, which are controlled by edge matrices in case of AMGm. One can also view this as involving an auxiliary problem–the one determined by the edge matrices–in the coarsening process. Similar to AMGe, we use an interpolation rule that is based on local energy minimization (12) but now with respect to the molecules (10), which replace the local neighborhood matrices (assembled from certain element matrices) as used in AMGe. The coarse-grid matrices, however, are still computed via the usual Galerkin tripple matrix product, i.e., $A_{k+1} = P_k^T A_k P_k$ at all levels $k = 0, 1, \ldots, l-1$.

3 Numerical Results

The numerical results presented in this section are for the model problem:

Problem 1.

$$-\nabla \cdot [C\nabla u] = f \quad \text{in} \quad \Omega \subset \mathrm{R}^3 \tag{14}$$

$$u = g \quad \text{on} \quad \Gamma_D \subset \partial\Omega \tag{15}$$

$$\frac{\partial u}{\partial n} = 0 \quad \text{on} \quad \Gamma_N \subset \partial\Omega \setminus \Gamma_D, \tag{16}$$

where

$$C = \begin{pmatrix} 1 & 0 & 0 \\ 0 & \epsilon & 0 \\ 0 & 0 & \epsilon \end{pmatrix}, \quad 0 < \epsilon \leq 1, \quad f = 0, \quad \Omega = (-3,3)^3 \setminus (\Omega_1 \cup \Omega_2),$$

$$\Omega_1 = (0.2, 0.3) \times (-0.5, 0.5)^2, \qquad \Omega_2 = (-0.3, -0.2) \times (-0.5, 0.5)^2,$$

$$\Gamma_N = \partial[(-3,3)^3], \quad \Gamma_D = \partial\Omega_1 \cup \partial\Omega_2,$$

and

$$g = \begin{cases} 1 & \text{on} \quad \partial\Omega_1 \\ -1 & \text{on} \quad \partial\Omega_2 \end{cases}$$

We obtained similar convergence rates for rotated diffusion equations, i.e., the case in which the matrix C is no longer diagonal, see Reference [9].

Problem 1 was discretized using piecewise linear nodal-basis functions to build finite element spaces of increasing dimension. The underlying (unstructured) triangular meshes were generated using the NETGEN[3] mesh generator, which is integrated in the software package NGSolve[4].

Note that the considered global stiffness matrices are not contained in the class of M-matrices. Especially for small values of the parameter ϵ, or, equivalently, strong mesh anisotropy, the positive off-diagonal entries gain weight and the variation from an M-matrix increases.

All linear systems were solved using the conjugate gradient method applying a single AMGm iteration for preconditioning (PCG). The threshold parameter θ was 1/6 in all computations.

We compared V- and W-cycles, performing one respectively two symmetric Gauß-Seidel pre- and post-smoothing step(s). In order to reduce computational complexity, the considered W-cycle was truncated to a V-cycle on every other level; it is therefore denoted as W(1,1)* hereafter. Table 1 contains the number of PCG iterations that was required to reduce the residual norm by a factor 10^{-8}, the average convergence factor ρ, as well as the *grid complexity* σ^{Ω} and the *operator complexity* σ^{A}. [5]

---

[3] http://www.hpfem.jku.at/netgen/index.html

[4] http://www.hpfem.jku.at/ngsolve/index.html

[5] σ^{Ω} is the ratio of the total number of points on all grids to that on the fine grid, whereas σ^{A} is the ratio of the total number of nonzero entries in all matrices to that in the fine-grid matrix.

Table 1. AMGm based on semipositive splittings

#elements		3269		26152		209216		1673728	
#levels		2		4		6		8	
		#it.	ρ	#it.	ρ	#it.	ρ	#it.	ρ
$\epsilon = 1$:	V(1,1)	7	0.05	9	0.12	11	0.17	12	0.21
	V(2,2)	5	0.02	6	0.04	7	0.07	9	0.11
	W(1,1)*	7	0.05	8	0.10	10	0.14	10	0.16
	σ^{Ω}		1.28		1.40		1.42		1.42
	σ^{A}		1.80		2.65		2.93		3.04
$\epsilon = 0.5$:	V(1,1)	7	0.06	10	0.14	12	0.19	14	0.25
	V(2,2)	5	0.02	7	0.05	8	0.09	9	0.12
	W(1,1)*	7	0.06	9	0.11	10	0.15	11	0.19
	σ^{Ω}		1.29		1.40		1.42		1.43
	σ^{A}		1.84		2.64		2.94		3.06
$\epsilon = 0.1$:	V(1,1)	9	0.11	12	0.20	14	0.25	17	0.33
	V(2,2)	6	0.04	7	0.07	9	0.13	12	0.21
	W(1,1)*	9	0.11	11	0.18	11	0.19	12	0.21
	σ^{Ω}		1.34		1.47		1.49		1.48
	σ^{A}		1.99		2.97		3.31		3.32
$\epsilon = 0.05$:	V(1,1)	10	0.16	13	0.24	16	0.30	20	0.39
	V(2,2)	7	0.07	10	0.14	10	0.15	14	0.26
	W(1,1)*	10	0.16	12	0.19	12	0.21	13	0.23
	σ^{Ω}		1.38		1.54		1.54		1.52
	σ^{A}		2.10		3.32		3.59		3.59
$\epsilon = 0.01$:	V(1,1)	12	0.20	21	0.41	27	0.50	32	0.56
	V(2,2)	8	0.10	14	0.26	21	0.41	24	0.46
	W(1,1)*	12	0.20	15	0.29	15	0.29	15	0.29
	σ^{Ω}		1.45		1.64		1.64		1.61
	σ^{A}		2.24		3.68		4.09		4.09

References

1. Berman, A., Plemmons, R.J.: Nonnegative Matrices in Mathematical Sciences. Academic Press, New York, 1979.
2. Brandt, A.: Algebraic multigrid theory: the symmetric case. Appl. Math. Comput. **19** (1986) 23–56.
3. Brezina, M., Cleary, A.J., Falgout, R.D., Henson, V.E., Jones, J.E., Manteuffel, T.A., McCormick, S.F., Ruge, J.W.: Algebraic multigrid based on element interpolation (AMGe). SIAM J. Sci. Comput. **22** (2000) 1570–1592.
4. Chartier, T., Falgout, R.D., Henson, V.E., Jones, J.E., Manteuffel, T., McCormick, S.F., Ruge, J.W., Vassilevski, P.S.: Spectral AMGe (ρAMGe). SIAM J. Sci. Comput. **25** (2004) 1–26.
5. Cleary, A.J., Falgout, R.D., Henson, V.E., Jones, J.E., Manteuffel, T.A., McCormick, S.F., Miranda, G.N., Ruge, J.W.: Robustness and scalability of algebraic multigrid. SIAM J. Sci. Stat. Comput. **21** (2000) 1886–1908.
6. Haase, G., Langer, U., Reitzinger, S., Schöberl, J.: Algebraic multigrid methods based on element preconditioning. International Journal of Computer Mathematics **78** (2004) 575–598.

7. Henson, V.E., Vassilevski, P.: Element-free AMGe: General algorithms for computing the interpolation weights in AMG. SIAM J. Sci. Comput. **23** (2001) 629–650.
8. Jones, J.E., Vassilevski, P.: AMGe based on element agglomeration. SIAM J. Sci. Comput. **23** (2001) 109–133.
9. Kraus, J.K., Schicho, J.: Algebraic multigrid based on computational molecules, 1: Scalar elliptic problems. RICAM-Report No. 2005–05, Linz, 2005. Submitted to Computing.
10. Langer, U., Reitzinger, S., Schicho, J.: Symbolic methods for the element preconditioning technique. In: Proc. SNSC Hagenberg, U. Langer and F. Winkler, eds., Springer, 2002.
11. Ruge, J.W., Stüben, K.: Efficient solution of finite difference and finite element equations by algebraic multigrid (AMG). In: Multigrid Methods for Integral and Differential Equations, D.J. Paddon and H. Holstein, eds., The Institute of Mathematics and Its Applications Conference Series, Clarendon Press, Oxford, 1985, 169–212.
12. Stüben, K.: Algebraic multigrid (AMG): experiences and comparisons. Appl. Math. Comput. **13** (1983) 419–452.
13. Vaněk, P., Brezina, M., Mandel, J.: Convergence of algebraic multigrid based on smoothed aggregation. Numer. Math. **88** (2001) 559–579.
14. Vaněk, P., Mandel, J., Brezina, M.: Algebraic multigrid based on smoothed aggregation for second and fourth order problems. Computing **56** (1996) 179–196.

Comparison of Geometrical and Algebraic Multigrid Preconditioners for Data-Sparse Boundary Element Matrices

U. Langer and D. Pusch

Institute of Computational Mathematics, Johannes Kepler University Linz, Austria

Abstract. We present geometric (GMG) and algebraic multigrid (AMG) preconditioners for data-sparse boundary element matrices. Data-sparse approximation schemes such as adaptive cross approximation (ACA) yield an almost linear behavior in N_h, where N_h is the number of (boundary) unknowns. The treated system matrix represents the discretized single layer potential operator (SLP) resulting from the interior Dirichlet boundary value problem for the Laplace equation. It is well known, that the SLP has converse spectral properties compared to usual finite element matrices. Therefore, multigrid components have to be adapted properly. In the case of GMG we present convergence rate estimates for the data-sparse ACA version. Again, uniform convergence can be shown for the V-cycle.

Iterative solvers dramatically suffer from the ill-conditioness of the underlying system matrix for growing N_h. Our multigrid-preconditioners avoid the increase of the iteration numbers and result in almost optimal solvers with respect to the total complexity. The corresponding numerical 3D experiments are confirming the superior preconditioning properties for the GMG as well as for the AMG approach.

Keywords: integral equations of the first kind, single layer potential operator, boundary element method, adaptive cross approximation, algebraic multigrid, geometrical multigrid, preconditioners, iterative solvers.

1 Introduction

In this paper we are concerned with the fast solution of data-sparse boundary element equations by geometrical and algebraic multigrid methods.

The application of iterative solvers only will be reasonable, if the drawback of dense matrices can be overcome. In the last years different sparse approximation techniques for boundary element matrices have been developed. The multipole method [14], the panel clustering method [7], the $\mathcal{H}$-matrix approach [6] and wavelet techniques [9] are certainly now the most popular ones. In our paper we will consider the adaptive cross approximation (ACA) method suggested by M. Bebendorf and S. Rjasanow [1,2]. The basic idea is to decompose the system matrix into its near-field and far-field contributions. Finding an appropriate low-rank approximation for the far-field matrix yields a data-sparse BEM matrix approximating the original dense matrix in such a way that the discretization

I. Lirkov, S. Margenov, and J. Waśniewski (Eds.): LSSC 2005, LNCS 3743, pp. 130–137, 2006.

error is not affected. In conclusion, the application of a sparse representation algorithm allows us to realize the matrix-by-vector multiplication in almost $O(N_h)$ operations.

Boundary element matrices originating from the discretization of the single layer potential lead to ill-conditioned system matrices with a condition number of order $O(h^{-1})$. Thus, it is obvious that we need appropriate preconditioning techniques in order to avoid the steady rise of the number of iterations for finer and finer discretizations. In [11,10] we introduced algebraic multigrid preconditioners for dense BEM matrices as well as for large-scaled data-sparse BEM matrices. In this paper we focus on the comparison between the GMG and AMG approach. Moreover, we give a convergence result for the geometric version of our multigrid approach.

The paper is organized as follows: Section 2 gives a brief overview on the considered single layer potential operator and its properties. In addition, the ACA-method is briefly described. In Section 3, we introduce the multigrid components designed for ACA-matrices and give convergence results for the geometrical variant. Some results of our numerical studies are presented in Section 4. Finally, we end up with some conclusions and discuss further investigations in Section 5.

2 Problem Formulation and the ACA-Method

Let $\Omega \subset \mathbb{R}^d$ (d=2,3) be a bounded, simply connected domain with one closed boundary piece $\Gamma = \partial\Omega$ that is supposed to be sufficiently smooth. We consider the boundary element technique by means of the interior Dirichlet problem for Laplace's equation:

$$\begin{aligned} -\Delta u(x) &= 0 \qquad x \in \Omega \\ u(x) &= g(x) \quad x \in \Gamma \end{aligned} \tag{1}$$

Once the Neumann and Dirichlet data are available, it is possible to formulate the solution of the interior Dirichlet equation by the representation formula

$$\sigma(y)u(y) = \int\limits_{\Gamma} \frac{\partial u}{\partial n_x}(x)E(x,y)ds_x - \int\limits_{\Gamma} u(x)\frac{\partial E}{\partial n_x}(x,y)ds_x \tag{2}$$

where n_x denotes the unit outward normal vector and $E(x,y)$ is the fundamental solution for the Laplace equation, i.e. in $\mathbb{R}^3$ we have $E(x,y) = \frac{1}{4\pi}\frac{1}{|x-y|}$. For $y \in \Omega$ we have $\sigma(y) = 1$, for $y \notin \bar{\Omega}$ it changes to $\sigma(y) = 0$. In the case of $y \in \Gamma$ and Γ is sufficiently smooth we will obtain $\sigma = 1/2$, that is still valid for applying Galerkin discretization on $C^{0,1}$ domains. In that case the first integral defines the single layer potential operator $V : H^{-1/2}(\Gamma) \mapsto H^{1/2}(\Gamma)$. In addition the second integral gives the double layer potential operator $K : H^{1/2}(\Gamma) \mapsto H^{1/2}(\Gamma)$. It can be shown that the single layer potential operator is symmetric and positive definite. These and other properties can be found in e.g. [15].

Applying Galerkin discretization with the use of piecewise constant trial functions leads to the matrix equation

$$V_h \underline{v}_h = \underline{f}_h = (\frac{1}{2} I_h + K_h) \underline{g}_h \tag{3}$$

where $\underline{g}_h$ is the discrete Dirichlet data obtained by linear interpolation, $(V_h)_{ij} = \int_{\Gamma_j} \int_{\Gamma_i} E(x,y) ds_x ds_y$ and $(K_h)_{ij} = \int_{\Gamma_i} \int_{\Gamma} \frac{\partial E}{\partial n_x}(x,y) \psi_j(x) ds_x ds_y$ with the linear trial function ψ_j. At this point we have to notice that V_h is still fully populated and the condition number is of order $O(h^{-1})$. To overcome the drawback of dense matrices we replace the system matrix with some approximation matrix provided by the ACA-algorithm. On the contrary to other matrix approximation techniques, an explicit description of the integral kernel is not necessary. More precisely, only a procedure for evaluating selected matrix entries has to be available. The rest are simple algebraic operations.

The basic idea is to decompose the computational domain into smaller clusters $\mathcal{D}_i$ and classify the interaction of two clusters into a near-field part and a far-field part of the generated matrix, respectively. Based on geometrical information we split the index set $I = \{1, ..., N_h\}$ into index clusters $t_i \subset I$ which corresponds to the partitioning of the domain $\Omega = \bigcup_i \mathcal{D}_i$. In order to select the blocks which can be approximated by low-rank matrices, we give an admissibility condition that classifies clusters-pairs into a near-field part and a far-field part.

Definition 1. *Let* $(\mathcal{D}_1, \mathcal{D}_2)$ *be a cluster pair with* $\mathcal{D}_1, \mathcal{D}_2 \subset \mathbb{R}^d$, *then* $(\mathcal{D}_1, \mathcal{D}_2)$ *is called* η *- admissible if*

$$\operatorname{diam} \mathcal{D}_2 \leq \eta \operatorname{dist}(\mathcal{D}_1, \mathcal{D}_2). \tag{4}$$

As usual $\operatorname{dist}(X, Y) = \inf\{|x - y|, x \in X, y \in Y\}$.

Both, the clustering procedure and the approximation algorithm will cause a overall complexity of $O(\epsilon^{-\alpha} N_h^{1+\alpha})$ with an arbitrarily small positive α. In [2] one can find the appropriate algorithms and more detailed information. Since the proposed adaptive cross approximation technique provides a low-rank approximation of V_h^{far} consisting of submatrices which are η-admissible we obtain the result

$$\widetilde{V}_h = V_h^{near} + \widetilde{V}_h^{far}. \tag{5}$$

Starting from this representation we are able to present an appropriate construction of multigrid methods in the next section. Finally, we refer to [1,2] for more detailed proofs and further remarks concerning the ACA-technique.

3 Multigrid Methods

In the previous section we showed, that our system matrix coincides with the approximated discretized single layer potential operator $\widetilde{V}_h$, which is the most interesting case concerning our multigrid approach. Hence, we have to solve $\widetilde{V}_h \underline{v}_h = \underline{f}_h$ in $\mathbb{R}^{N_h}$ with $\underline{v}_h$ are the unknown Neumann data and $\underline{f}_h$ the corresponding right-hand side. In order to make multigrid methods really efficient, it is necessary to adapt the multigrid components properly according to the underlying physical problem and variational formulation. In the following we are discussing the multigrid components by means of a twogrid algorithm. The indices h and H denote the fine grid and coarse grid quantities, respectively.

In fact, the efficiency of multigrid methods depends on a clever interaction of smoothing sweeps on the fine level and coarse grid correction on the coarse level. Once a grid hierarchy (GMG) or a matrix hierarchy (AMG) is available we can apply multigrid methods like the well-known V-cycle presented in Algorithm 1. The coarsest level is denoted by the variable COARSELEVEL therein.

Algorithm 1 *Multigrid V-Cycle*
$MG(\underline{u}_\ell, \underline{f}_\ell, \ell)$
if $\ell = COARSELEVEL$
  calculate $\underline{u}_\ell = (V_\ell)^{-1}\underline{f}_\ell$ *by some coarse grid solver*
else
  smooth ν_F *times on* $V_\ell \underline{u}_\ell = \underline{f}_\ell$
  calculate the defect $\underline{d}_\ell = \underline{f}_\ell - V_\ell \underline{u}_\ell$
  restrict the defect to the next coarser level $\ell + 1 : \underline{d}_{\ell+1} = P_\ell^\top \underline{d}_\ell$
  set $\underline{u}_{\ell+1} \equiv 0$
  call $MG(\underline{u}_{\ell+1}, \underline{d}_{\ell+1}, \ell + 1)$
  prolongate the correction $\underline{s}_\ell = P_\ell \underline{u}_{\ell+1}$
  update the solution $\underline{u}_\ell = \underline{u}_\ell + \underline{s}_\ell$
  smooth ν_B *times on* $V_\ell \underline{u}_\ell = \underline{f}_\ell$
end if

Since the single layer operator represents a pseudo-differential operator of order minus one, the eigenvalues and eigenvectors act conversely compared to those of finite element matrices. Therefore, standard smoothing procedures like damped Jacobi or Gauß-Seidel does not provide a satisfying smoothing sweep. Bramble, Leyk and Pasciak [3] present an appropriate approach to this problem class of operators. In order to reduce the highly oscillating components of the error we introduce a matrix $A_h \in \mathbb{R}^{N_h \times N_h}$ being some discretization of the Laplace-Beltrami operator on the boundary Γ. Consequently, we obtain a smoothing iteration of the form

$$\underline{u}_h \leftarrow \underline{u}_h + \tau_h \cdot A_h(\underline{f}_h - \widetilde{V}_h \underline{u}_h) \tag{6}$$

with a well chosen damping parameter τ_h, see e.g. [10].

In the case of algebraic multigrid we need a matrix hierarchy which represents a 'virtual' grid on each level. Therefore, we first construct prolongation operators $P_h : \mathbb{R}^{N_H} \mapsto \mathbb{R}^{N_h}$ by exploiting a sparse auxiliary matrix B_h which includes geometrical information [11]. Then, we are applying Galerkin's method to obtain the system matrix $V_H = P_h^\top V_h P_h$ on the coarse level. In addition, the restriction of a fine ACA matrix $\widetilde{V}_h$ immediately leads to matrices on the coarse level

$$V_H^{near} = P_h^\top V_h^{near} P_h, \qquad \widetilde{V}_H^{far} = \sum_{i=1}^{N_B} \sum_{j=1}^{r_i} P_h^\top u_j^i \; (P_h^\top v_j^i)^\top \tag{7}$$

where N_B denotes the number of admissible blocks and r_i the rank of the ith block. Due to the exact preserving of representation (5) on the coarse grid, we are able to use the same ACA-datastructures in our numerical realization.

On the other hand in the geometrical version of our multigrid approach a nested mesh-hierarchy is available. In this case we are calculating the discretized single layer potential on each grid separately. Strictly speaking, we apply the ACA-algorithm level by level to obtain the approximated single layer potential operators. Again we provide a set of data-sparse system matrices, which are used within the V-cycle.

In order to obtain results for convergence, we verify conditions on the approximated single layer potential operator $\widetilde{V}_h$. Based on theoretical results in [4] which are weaker than the regularity and approximation conditions, we have to show the spectral equivalence inequalities

$$c_1(V_h\underline{v},\underline{v}) \leq (\widetilde{V}_h\underline{v},\underline{v}) \leq c_2(V_h\underline{v},\underline{v}) \quad \underline{v} \in \mathbb{R}^{N_h} \tag{8}$$

and an approximation result in the sense

$$|(V_h\underline{v},\underline{w}) - (\widetilde{V}_h\underline{v},\underline{w})| \leq c_0\lambda_J^{-\beta/2}||\underline{v}_h||_{V_h}\,||\underline{w}_h||_{V_h} \quad \underline{v}_h,\underline{w}_h \in \mathbb{R}^{N_h}. \tag{9}$$

In the last inequality λ_J denotes the largest eigenvalue of the induced operator $\mathcal{V}$ defined by $(\mathcal{V}v_h, w_h)_{-1} = (V_h\underline{v}_h,\underline{w}_h)$ with the functions v_h, w_h described by the basis coefficients $\underline{v}_h,\underline{w}_h$. Moreover, β is a arbitrary small positive parameter. It can be proofed, that (8) holds with the spectral constants

$$\begin{aligned} c_1 &= (1 - \epsilon\sqrt{N_h}\kappa(V_h)) \\ c_2 &= (1 + \epsilon\sqrt{N_h}\kappa(V_h)). \end{aligned} \tag{10}$$

Furthermore, we can show that the estimate

$$|(V_h\underline{v}_h,\underline{w}_h) - (\widetilde{V}_h\underline{v}_h,\underline{w}_h)| \leq c_0\epsilon h^\gamma\lambda_J^{-\beta/2}||\underline{v}_h||_{V_h}||\underline{u}_h||_{V_h} \quad \underline{v}_h,\underline{w}_h \in \mathbb{R}^{N_h} \tag{11}$$

is valid, where ϵ is the accuracy from the ACA-approximation. Nevertheless, the upper bound still depends on the typical mesh size h and whose exponent $\gamma = -(d+\beta+4)/2$ additionally includes the dimension d of the boundary parameterization of $\partial\Omega$. With an appropriate choose of ϵ one can cancel out the h-dependency. However, in our numerical experiments we kept ϵ fix and cannot observe a negative influence anyway. From these estimates and the general convergence theory given in [4] we can immediately proof uniform convergence of the V-cycle.

4 Numerical Studies

In order to show the efficiency of the suggested multigrid approach we present some results in 3D for the interior Dirichlet boundary value problem for the Laplace equation. The Galerkin boundary element matrices are generated by

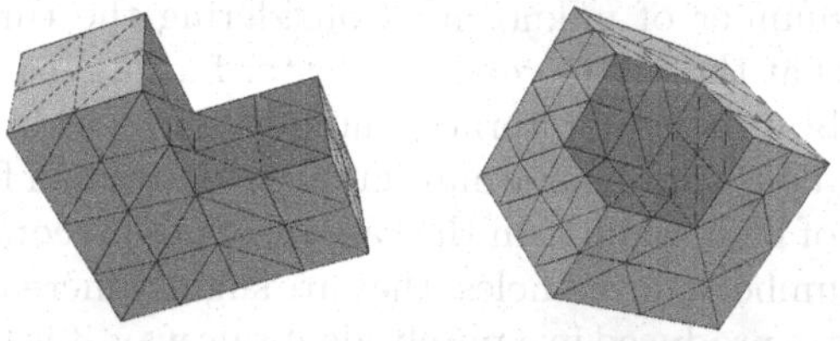

(a) L-Shape (b) Fichera Corner

Fig. 1. 3D Geometries

Table 1. Assembling $\widetilde{V}_h$ and Setup Times for L-Shape

Number of Unknowns	AMG (sec) Assembling $\widetilde{V}_h$	AMG (sec) Galerkin Projection	GMG (sec) Assembling $\widetilde{V}_h$	GMG (sec) Matrix-hierarchy
7168	77	15	32.5	6.9
28672	158	30	158	40

the software package OSTBEM developed by O. Steinbach, cf. [16], the AMG-preconditioner is realized within the software package PEBBLES [8].

For our numerical comparison of the geometrical multigrid preconditioner and the algebraic multigrid preconditioner, we choose a few rather simple 3D geometries, see Figure 1. Nevertheless, these domains include a wide spectrum of problem classes, for e.g. edges, corners and non-convex domains. First of all, we compare the times for constructing the AMG matrix hierarchy by Galerkin projection and building up the ACA matrices for GMG on the coarser grids. These CPU-times are almost of the same order, see Table 1. It is obvious, that most of the assembling time is needed for construction the system matrix $\widetilde{V}_h$. Secondly, we compare the numbers of iterations, that are needed within the preconditioned conjugate gradient (PCG) method. Moreover, the CPU-time of one single PCG-iteration for different numbers of unknowns are listed in Table 2. One can clearly observe the expected almost linear increase of the CPU-times for one iteration

Table 2. Key data for AMG/GMG Preconditioner

Number of Unknowns	AMG PCG-Cycle (sec)	AMG Iterations	GMG PCG-Cycle (sec)	GMG Iterations
L-Shape				
1792	0.1	6	0.1	7
7168	0.8	6	0.6	7
28672	4.2	9	2.9	7
Fichera-Corner				
1920	0.1	14	0.1	15
7680	0.8	15	0.6	15
30720	5.0	17	3.2	15

with respect to the number of unknowns. Considering the time for one PCG-iteration, we notice, that the GMG version is faster for larger problems.

Furthermore, we obtain constant iteration numbers for a wide range of problem sizes. That implies that our data-sparse multigrid preconditioner for the single layer potential operator is of high quality. In the case of AMG preconditioning we also have small iteration numbers, nevertheless they are slightly increasing. Because the coarser matrix levels are produced in a purely algebraic way, it is hardly possible to preserve corresponding 'virtual' coarse grids of the original geometry.

5 Conclusions and Further Remarks

In this paper we presented a geometrical multigrid and algebraic multigrid approach for the solution of large-scale boundary element equations. For that purpose an approximation of the boundary element matrices is absolutely essential. Our numerical experiments have been realized by the adaptive cross approximation technique which guarantees that the effort for storing the matrices and for a single matrix-by-vector multiplication can be reduced to almost $O(N_h)$. The discretized single layer potential operator yields symmetric positive definite matrices in the original dense version as well as in the ACA representation. Therefore, the system of boundary element equations can be solved by means of multigrid preconditioned CG-algorithms. Due to the sparse representation of our matrices, we had to adapt each component of our AMG-algorithm properly. In order to set up the matrix hierarchy and the corresponding transfer operators an auxiliary matrix was constructed for the AMG method. On the other hand the matrices were built accordingly to the grid hierarchy in the GMG method. The smoothing procedure was realized by the proposed BLP-smoother for pseudo-differential operators of order minus one.

The overall algorithm provides interesting numerical results. One can notice small constant iteration numbers for the GMG method and also small (but slightly increasing) iteration numbers for the AMG approach. That confirms the high quality of our multigrid preconditioners. In addition, the CPU time for a single iterative step almost grows like $O(N_h)$. As expected, the GMG variant is faster than the AMG version. We mention that efficient multigrid preconditioner for the discrete single layer potential operator are very important as building blocks in primal and dual domain decomposition preconditioners [5,12,13].

Last but not least we would like to acknowledge the Austrian Science Fund 'Fond zur Förderung der wissenschaftlichen Forschung (FWF)' for supporting this work under grant P14953 'Robust Algebraic Multigrid Methods and their Parallelization'.

References

1. M. Bebendorf. Approximation of Boundary Element Matrices. *Numerische Mathematik*, 86:565–589, 2000.
2. M. Bebendorf and S. Rjasanov. Adaptive Low-Rank Approximation of Collocation Matrices. *Computing*, 70(1):1–24, 2003.

3. J. H. Bramble, Z. Leyk, and J. E. Pasciak. The Analysis of Multigrid Algorithms for Pseudo-Differential Operators of Order Minus One. *Math. Comp.*, 63(208):461–478, 1994.
4. J. H. Bramble and X. Zhang. *The analysis of multigrid methods*, volume VII of *Handbook for Numerical Analysis*, 173–415. P. Ciarlet and J.L.Lions, North-Holland, Amsterdam, 2000.
5. G. Haase, B. Heise, M. Kuhn, and U. Langer. Adaptive domain decomposition methods for finite and boundary element equations. In *Boundary Element Topics* (ed. by W. Wendland), 121–147, Berlin, 1998. Springer-Verlag.
6. W. Hackbusch. A Sparse Matrix Arithmetic based on $\mathcal{H}$-Matrices. *Computing*, 62(2):89–108, 1999.
7. W. Hackbusch and Z. P. Nowak. On the fast matrix multiplication in the boundary element method by panel clustering. *Numer. Math.*, 54(4):463–491, 1989.
8. Johannes Kepler University Linz, Institute of Computational Mathematics. *PEBBLES — User's Guide*, 1999. http://www.numa.uni-linz.ac.at/Research/Projects/pebbles.html.
9. C. Lage and C. Schwab. Wavelet Galerkin Algorithms for Boundary Integral Equations. *SIAM J. Sci. Comput.*, 20:2195–2222, 1999.
10. U. Langer and D. Pusch. Data-Sparse Algebraic Multigrid Methods for Large Scale Boundary Element Equations. *Applied Numerical Mathematics*, 54:406–424, 2005.
11. U. Langer, D. Pusch, and S. Reitzinger. Efficient Preconditioners for Boundary Element Matrices Based on Grey-Box Algebraic Multigrid Methods. *International Journal for Numerical Methods in Engineering*, 58(13):1937–1953, 2003.
12. U. Langer and O. Steinbach. Boundary element tearing and interconnecting methods. *Computing*, 71(3):205–228, 2003.
13. U. Langer and O. Steinbach. Coupled boundary and finite element tearing and interconnecting methods. In *Proceedings of the 15th Int. Conference on Domain Decomposition* (ed. by R. Kornhuber, R. Hoppe, J. Periaux, O. Pironneau, O. Widlund and J. Xu, *Lecture Notes in Computational Sciences and Engineering*, **40**, 83–97, Heidelberg, 2004. Springer.
14. V. Rokhlin. Rapid solution of integral equations of classical potential theory. *J. Comput. Phys.*, 60(2):187–207, 1985.
15. O. Steinbach. *Numerische Näherungsverfahren für elliptische Randwert-probleme: Finite Elemente und Randelemente.* Teubner, Stuttgart, Leipzig, Wiesbaden, 2003.
16. O. Steinbach. *Stability Estimates for Hybrid Coupled Domain Decomposition Methods, Lectures Notes in Mathematics*, **1809** Springer-Verlag, Berlin, 2003.

An Adaptive Multigrid Strategy for Convection-Diffusion Problems

Daniela Vasileva[1], Anton Kuut[2], and Pieter W. Hemker[2]

[1] Institute of Mathematics and Informatics, Bulgarian Academy of Sciences,
Acad. G. Bonchev str., bl. 8, BG-1113 Sofia, Bulgaria
vasileva@math.bas.bg

[2] Centrum voor Wiskunde en Informatica (CWI),
P.O. Box 94079, NL-1090 GB Amsterdam, The Netherlands
{P.W.Hemker, A.Kuut}@cwi.nl

Abstract. For the solution of convection-diffusion problems we present a multilevel self-adaptive mesh-refinement algorithm to resolve locally strong varying behavior, like boundary and interior layers. The method is based on discontinuous Galerkin (Baumann-Oden DG) discretization. The recursive mesh-adaptation is interwoven with the multigrid solver. The solver is based on multigrid V-cycles with damped block-Jacobi relaxation as a smoother. Grid transfer operators are chosen in agreement with the Galerkin structure of the discretization, and local grid-refinement is taken care of by the transfer of local truncation errors between overlapping parts of the grid.

We propose an error indicator based on the comparison of the discrete solution on the finest grid and its restriction to the next coarser grid. It refines in regions, where this difference is too large. Several results of numerical experiments are presented which illustrate the performance of the method.

Keywords: convection-dominated problems, adaptive refinement, multigrid, discontinuous Galerkin method.

1 Introduction

Recently new interest arose in application of discontinuous Galerkin (DG) methods for the solution of partial differential equations of convection-diffusion type. An important reason is their ability to conveniently handle difficulties related to grid- and order-adaptation. This motivates our present work on self-adaptive DG discretisation which is combined with a multigrid (MG) method so that optimal efficiency can be expected.

A detailed description of the multigrid approach and the corresponding smoothing analysis in the case of discontinuous Galerkin methods with constant coefficients can be found in [5, 6, 7].

The paper is organized as follows. Section 2 concerns the governing equation and its discretisation. The third section describes the multigrid h-adaptive refinement algorithm. The adaptive criterion is presented in the fourth section and the last section contains results from numerical experiments.

I. Lirkov, S. Margenov, and J. Waśniewski (Eds.): LSSC 2005, LNCS 3743, pp. 138–145, 2006.

2 Discontinuous Galerkin Discretisation

We consider the linear boundary value problem:

$$-\varepsilon\Delta u + \nabla\cdot(\mathbf{b}u) + cu = f \ \text{ in } \ \Omega\subset\mathbb{R}^d, \ \ d=1,2,3, \qquad (1)$$
$$u(\mathbf{x}) = u_0(\mathbf{x}) \ \text{ on } \ \Gamma=\partial\Omega,$$

where $\mathbf{x}=(x_1,\ldots,x_d)$, $\varepsilon>0$ is a small parameter, the coefficients $\mathbf{b}(\mathbf{x}) = (b_1(\mathbf{x}),\ldots,b_d(\mathbf{x}))\in (C^1(\Omega))^d$, $c(\mathbf{x})\geq 0$, $c(\mathbf{x})\in L_\infty(\Omega)$ and the right-hand side $f(\mathbf{x})\in L_2(\Omega)$. We assume that Ω allows a regular partitioning $\Omega_h=\{\Omega_e \mid \cup_e \Omega_e=\Omega,\ \Omega_i\cap\Omega_j=\emptyset,\ i\neq j\}$, into equally sized square cells Ω_e of size h.

As weak form for (1) we use Baumann-Oden's [2, 1] discontinuous Galerkin formulation: find $u\in H^1(\Omega_h)$, such that

$$L(u,v)=F(v) \ \text{for all} \ v\in H^1(\Omega_h)\,, \qquad (2)$$

where $H^1(\Omega_h)$ is the broken Sobolev space

$$H^1(\Omega_h)=\left\{u\in L_2(\Omega)\mid u|_{\Omega_e}\in H^1(\Omega_e),\ \forall\Omega_e\in\Omega_h\right\},$$

$$L(u,v) = \sum_{\Omega_e\in\Omega_h}\left(\int_{\Omega_e}(\varepsilon\nabla u\cdot\nabla v - \nabla v\cdot\mathbf{b}u + cuv)\,d\mathbf{x} + \int_{\partial\Omega_e\setminus\Gamma_-} vu^-\mathbf{b}\cdot\mathbf{n}\,ds\right)$$
$$+\int_{\Gamma_{\text{int}}\cup\Gamma}\Big(\varepsilon\langle\nabla v\rangle\cdot[u]-\varepsilon\langle\nabla u\rangle\cdot[v]\Big)\,ds\,,$$
$$F(v) = \sum_{\Omega_e\in\Omega_h}\int_{\Omega_e} vf\,d\mathbf{x} + \int_\Gamma \varepsilon\nabla v\cdot\mathbf{n}u_0\,ds - \int_{\Gamma_-} vu_0\mathbf{b}\cdot\mathbf{n}\,ds\,,$$

$\Gamma_-=\{\mathbf{x}\in\partial\Omega\mid(\mathbf{b}\cdot\mathbf{n})(\mathbf{x})<0)$ denotes the instream boundary of the domain Ω and $\mathbf{n}$ is the unit outward pointing normal on the boundary. With u^-, the 'upwind' value of u is denoted, defined by $u^-=\lim_{\epsilon\downarrow 0}u(\mathbf{x}-\epsilon\mathbf{b}(\mathbf{x}))$. The interior cell boundaries are denoted by $\Gamma_{\text{int}}=\cup_e\partial\Omega_e\setminus\partial\Omega$.

The jump operator $[\cdot]$ for a scalar valued function $w(\mathbf{x})$ and the average operator $\langle\cdot\rangle$ for a vector valued function $\boldsymbol{\tau}(\mathbf{x})$ are defined at the common interface $\Gamma_{i,j}=\overline{\Omega}_i\cap\overline{\Omega}_j$ between two adjacent cells Ω_i and Ω_j by

$$[w(\mathbf{x})]=w(\mathbf{x})|_{\partial\Omega_i}\mathbf{n}_i+w(\mathbf{x})|_{\partial\Omega_j}\mathbf{n}_j,\quad \langle\boldsymbol{\tau}(\mathbf{x})\rangle=\frac{1}{2}(\boldsymbol{\tau}(\mathbf{x})|_{\partial\Omega_i}+\boldsymbol{\tau}(\mathbf{x})|_{\partial\Omega_j})\,.$$

For the DG discretisation of (2) we take: find $u_h\in S_h$ such that

$$L(u_h,v_h)=F(v_h)\ \text{ for all }\ v_h\in S_h, \qquad (3)$$

where $S_h=\{\sum_e\phi_e,\ \phi_e\in\mathcal{P}_3(\Omega_e),\ \Omega_e\in\Omega_h\}$ denotes the space of piecewise cubic polynomials on the partitioning Ω_h. In order to introduce a basis of S_h we first take the polynomials basis on the one-dimensional unit interval,

$$\phi_{2n+m}(t)=t^{n+m}(1-t)^{n+1-m},\quad n=0,1,\quad m=0,1. \qquad (4)$$

Then, on the unit cube in $\mathbb{R}^d$ we use a basis of tensor-product polynomials based on (4) and a basis for $\mathcal{P}_3(\Omega_e)$ is obtained by the usual affine mapping.

The coefficients $\mathbf{b}$, c and the right-hand side f are approximated using the same set of basic functions (or a set of lower order). The discretisation (3) yields a linear system $L_h u_h = f_h$, where the matrix L_h has a diagonal block structure with blocks of size $4^d \times 4^d$. We order the basic functions *cell-wise* or *point-wise* (for details see [5, 6, 7]), depending on the equation coefficients and h.

3 MG Realization on the Adaptive Grid

For the discontinuous Galerkin method we describe the application of an h-self-adaptive multi-level algorithm [3, 4].

The algorithm to determine the mesh is closely connected with the discretization and consist of several stages. In the first stage the equation is discretized and solved on the global coarsest grid Ω_0 *with step size* h_0. In later stages, $k = 1, 2, \cdots$, some cells of Ω_{k-1} are selected for refinement. These cells are all divided into 2^d smaller cells of equal size, which together form the grid Ω_k on a subset of $\overline{\Omega_{k-1}}$. The solution on Ω_k is interpolated from Ω_{k-1} and several relaxation sweeps are made on the interior of Ω_k, followed by a coarse grid correction on the whole of Ω_{k-1}. Thus, by recursive application, multigrid V-cycles are made. More details are given below.

By the recursive construction of the meshes Ω_k we see that the meshes cover nested areas $\overline{\Omega_k} \subset \overline{\Omega_{k-1}}$, and that all cells in these meshes form a tree-structure. In this tree, for $k > 0$ each cell has one *father* and possibly 2^d *children.*

Restrictions and prolongations. Given the nested partitioning $\{\Omega_k\}$ for the domain $\Omega = \overline{\Omega_0}$, on each mesh Ω_k we have a space of piecewise cubic polynomials S_k and the restriction of S_k to $\overline{\Omega_{k+1}}$ is a subset of S_{k+1}. This induces a natural prolongation $P_{k+1,k} : S_k \to S_{k+1}$ on $\overline{\Omega_{k+1}}$. The restriction operator *for the residues*, $\bar{R}_{k,k+1} : S_{k+1} \to S_k$ is defined as the adjoint of $P_{k+1,k}$. Because of the consistency of these operators with the DG discretisation, the Galerkin relation exists between the discretisation on the coarse and the fine grid $L_k = \bar{R}_{k,k+1} L_{k+1} P_{k+1,k}$. We use another restriction $R_{k,k+1}$ *for the solution*, which preserves the function values and the derivatives at the coarse cell vertices. It is a left-inverse of the prolongation, $R_{k,k+1} P_{k+1,k} = I_k$, where I_k is the identity operator on S_k.

The internal boundaries. In the case of local refinement the finer grid Ω_k usually covers only a part of the domain, covered by the coarser grid Ω_{k-1}. So some cells $\Omega_e \in \Omega_k$ have no neighbours on the same grid at some of their faces, but these faces will be not on the boundary $\partial\Omega$. We call this the *internal boundary.* We take care that no internal boundaries coincide for different levels k, i.e., if $\Omega_e \in \Omega_k$, then all the neighbours of its father $F(\Omega_e)$ exist on Ω_{k-1}. This can always be ensured performing some additional refinements of neighbouring cells.

At the internal boundaries for the discretization on Ω_k we take Dirichlet boundary conditions, derived by interpolation from Ω_{k-1}.

The relative truncation error. If u_k is the solution of the fine grid system, i.e., $L_k u_k = f_k$, then its restriction to the coarser grid Ω_{k-1} satisfies

$$L_{k-1} R_{k-1,k} u_k = \bar{R}_{k-1,k} f_k + \tau_{k-1,k}(u_k)\,,$$

where the relative local truncation error $\tau_{k-1,k}(y_k)$ is defined by

$$\tau_{k-1,k}(y_k) = L_{k-1}R_{k-1,k}y_k - \bar{R}_{k-1,k}L_k y_k \,.$$

During the computation, however, we do not know the fine grid solution u_k, but only an approximation $\tilde{u}_k$. So to obtain an accurate solution on the coarse grid, that corresponds to the solution on the fine grid (where this exists), we solve the coarse grid system $L_{k-1}\tilde{u}_{k-1} = \tilde{f}_{k-1}$, where

$$\tilde{f}_{k-1} = \begin{cases} f_{k-1} & \text{on } \overline{\Omega_{k-1}} \setminus \overline{\Omega_k}\,, \text{ where no finer grid exists,} \\ \bar{R}_{k-1,k}\tilde{f}_k + \tau_{k-1,k}(\tilde{u}_k) & \text{on } \overline{\Omega_k}\,, \text{ where } \tilde{u}_k \text{ is the current approximation on the finer grid.} \end{cases}$$

MG iteration. Each multigrid V-cycle on level k, denoted by $MS(k, \nu_1, \nu_2)$, consists of the following steps:

1. Perform ν_1 (pre-) relaxation steps (damped block-Jacobi relaxation) on the discrete system $L_k\tilde{u}_k = \tilde{f}_k$, taking as initial approximation
$$\tilde{u}_k = \begin{cases} R_{k,k+1}\tilde{u}_{k+1} & \text{on } \overline{\Omega_{k+1}}\,, \\ \tilde{u}_k & \text{on } \overline{\Omega_k} \setminus \overline{\Omega_{k+1}}\,; \end{cases}$$
2. If $k > 0$
   - perform $MS(k-1, \nu_1, \nu_2)$;
   - Compute the correction $\tilde{u}_k = \tilde{u}_k + P_{k,k-1}\left(\tilde{u}_{k-1} - R_{k-1,k}\tilde{u}_k\right)$;
3. With the current approximation for $\tilde{u}_k$ perform ν_2 (post-) relaxation steps on the discrete system $L_k\tilde{u}_k = \tilde{f}_k$.

On the coarsest level $MS(0, \nu_1, \nu_2)$ consists of $\nu_1 + \nu_2$ relaxation sweeps on Ω_0. Because this is a very coarse grid, an alternative way to solve $L_0\tilde{u}_0 = \tilde{f}_0$ is by Gaussian elimination.

4 Adaptation Criterion

On level Ω_0 all cells are refined at least once. Let K be the current number of grid levels. In order to decide which unrefined cells in Ω_k, $k = 1, 2, \ldots, K$, in the structure will be further refined, we compare the current solution $u_k(\mathbf{x})$ and its restriction to the previous level, $R_{k-1,k}u_k(\mathbf{x})$. The reason for this choice is the following. Let $u(\mathbf{x})$ be the exact solution of our problem (1) and let $R_l u(\mathbf{x})$ be its restriction to Ω_l, where (similar to $R_{l,l+1}$) the restriction $R_l : H^1(\Omega) \to S_l$ preserves the function values and the derivatives at the vertices of each $\Omega_e \in \Omega_l$. Then, with $u \in C^M(\Omega_e)$,

$$\|R_{l+1}u - R_l u\|_{C(\Omega_e)} < C\|u\|_{C^M(\Omega_e)} 2^{-lM}\,, \quad \text{for } M = 1, 2, 3, 4.$$

From this we derive that for piecewise C^M-functions, $M \le 4$, and *asymptotically for large k and l*

$$\|R_{l+1}u - R_l u\|_{C(\Omega_e^k)} \lessapprox q^{l+1-k}\|R_k u - R_{k-1}u\|_{C(\Omega_e^{k-1})} \quad \text{for } l \ge k, \tag{5}$$

with $q = 2^{-M}$. Here $\Omega_e^k \in \Omega_k$ and $\Omega_e^{k-1} := F(\Omega_e^k)$ is its father. For finite $l \geq k$ estimate (5) may be not true: then always a smooth function $u(\mathbf{x})$ can be constructed such that $\|R_k u - R_{k-1} u\|_{C(\Omega_e^{k-1})} = 0$, but $\|R_{l+1}u - R_l u\|_{C(\Omega_e^k)} > 0$ for $l \geq k$.

Let $\tilde{R}u$ be the restriction of u to the unrefined cells of $\cup_{k=1}^K \Omega_k$, where for each unrefined cell $\Omega_e^k \in \Omega_k$ we define $\tilde{R}u := R_k u$, then under the assumption of (5) we estimate $\|u - \tilde{R}u\|_{C(\Omega_e^k)}$ only by using $\|R_k u - R_{k-1}u\|_{C(\Omega_e^{k-1})}$ and q. As during the computation we do not know $R_k u$ and $R_{k-1}u$, we use the best available approximant u_k. With $r_k := \|u_k - R_{k-1,k} u_k\|_{C(\Omega_e^{k-1})}$,

$$\|u - R_k u\|_{C(\Omega_e^k)} \leq \sum_{l=k}^{\infty} \|R_{l+1}u - R_l u\|_{C(\Omega_e^k)} \lessapprox \sum_{l=k}^{\infty} q^{l+1-k} r_k = \frac{q}{1-q} r_k \,.$$

With T a desired tolerance and $r_k \frac{q}{(1-q)} \leq \mathsf{T}$ for all unrefined $\Omega_e^k \in \cup_{k=1}^K \Omega_k$, we have

$$\|u - \tilde{R}u\|_{C(\Omega)} = \max_{\Omega_e^k} \|u - R_k u\|_{C(\Omega_e^k)} \lessapprox \max_{\Omega_e^k} r_k \frac{q}{(1-q)} \leq \mathsf{T}.$$

Notice that we can estimate the local smoothness of the solution by estimating q from computable equivalents of (5). Let

$$r_l := \|R_l u_k - R_{l-1} u_k\|_{C(\Omega_e^{l-1})}, \quad l \leq k\,,$$

where $\Omega_e^{l-1} := F(\Omega_e^l)$, $l \leq k$. Then asymptotically we expect $q = \lim_{l\to\infty} r_l / r_{l-1}$. Based on this relation and additional heuristics we arrive at an estimated local value q_{est}. Then we introduce the local error estimate $\eta(\Omega_e^k) := r_k \frac{q_{\text{est}}}{(1-q_{\text{est}})}$ and we refine cell Ω_e^k if

$$\eta(\Omega_e^k) > \mathsf{T} \quad \text{or} \quad q_{\text{est}} \geq 1.$$

As the amount of work, needed to compute r_l in the 2D and 3D case, can not be neglected, in the examples below we only estimate r_l even in the 1D case.

5 Examples

In all examples in this section, the gridrefinements started on an initial grid with mesh size 1. Ten multigrid sweeps with 1 pre- and 1 post-smoothing iteration ($\nu_1 = \nu_2 = 1$) are performed in each stage.

Example 1. We consider the one dimensional equation

$$\varepsilon u'' + (x-1)u' = -\varepsilon\pi^2 \cos(\pi(x-1)) - \pi(x-1)\sin(\pi(x-1)), \quad x \in (0,2),$$

with Dirichlet boundary conditions, corresponding to the following exact solution

$$u(x) = \operatorname{erf}\left((x-1)/\sqrt{2\varepsilon}\right) + \cos(\pi(x-1)).$$

In Fig. 1 the exact solution and the difference between the exact and the approximate solution are plotted for $\varepsilon = 0.0001$ and $\mathsf{T} = 0.01$. The corresponding

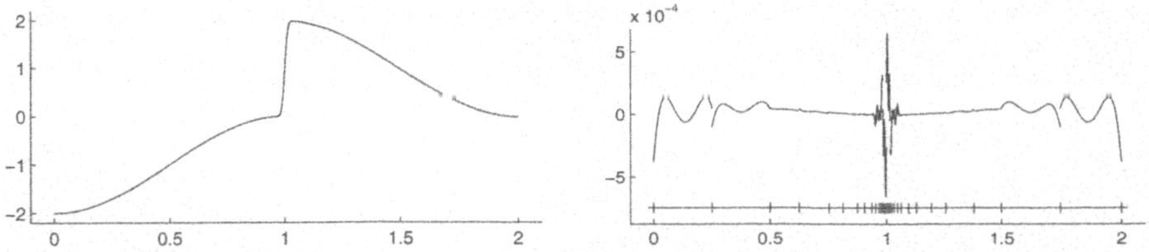

Fig. 1. The exact solution for Example 1 and the corresponding error for $\mathsf{T} = 0.01$

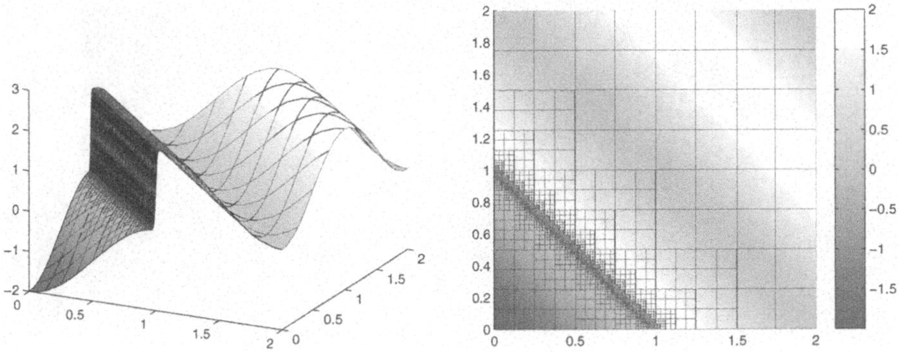

Fig. 2. The approximate solution for Example 2

grid is shown at the bottom of the right picture: 8 levels are used in order to achieve the prescribed tolerance. Similar results are obtained for $\mathsf{T} = 0.001$, then 9 levels are used. In both cases the grid is properly refined in the interior layer area and the C-norm of the error is less than T. The total number of cells in the final grid is respectively $N = 28$ and $N = 32$.

Example 2. In two dimensions a similar problem is considered, with an interior layer skew to the mesh:

$$\varepsilon\left(\frac{\partial^2 u}{\partial x^2}+\frac{\partial^2 u}{\partial y^2}\right)+(x+y-1)\left(\frac{\partial u}{\partial x}+\frac{\partial u}{\partial y}\right)=f(x,y), \quad x\in(0,2),\ y\in(0,2),$$

$$f(x,y)=-2\varepsilon\pi^2\cos(\pi(x+y-1))-2\pi(x+y-1)\sin(\pi(x+y-1)),$$

with Dirichlet boundary conditions, corresponding to the following exact solution

$$u(x,y)=\operatorname{erf}\left((x+y-1)/\sqrt{2\varepsilon}\right)+\cos(\pi(x+y-1)).$$

In Fig. 2 the approximate solution is plotted for $\varepsilon = 0.0001$ and $\mathsf{T} = 0.01$. The grid is refined around the interior layer and 8 levels are used to achieve the prescribed tolerance. The C-norm of the difference between the exact and the approximate solution is less than the prescribed tolerance.

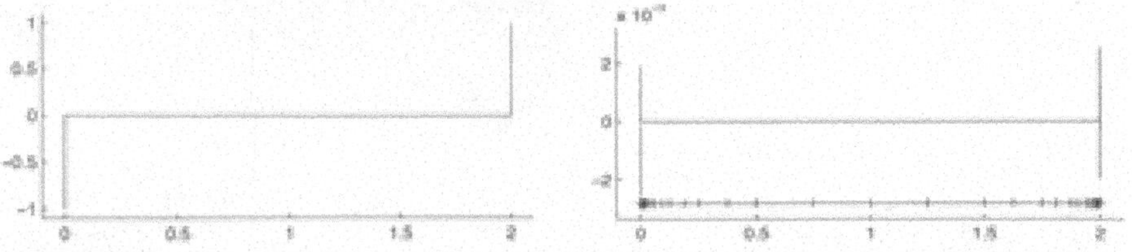

Fig. 3. The exact solution for Example 3 and the corresponding error for T = 0.01

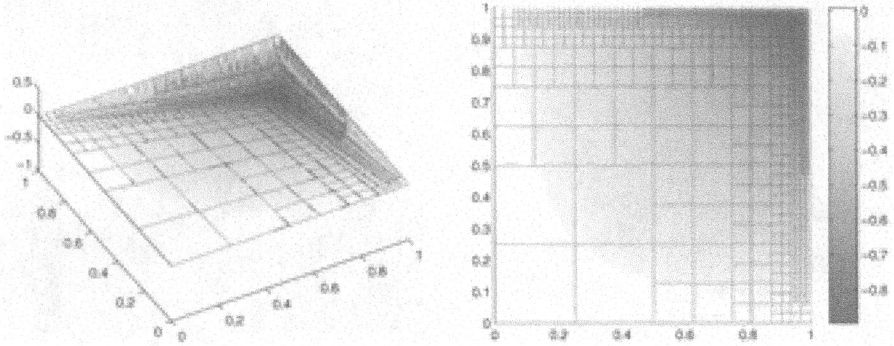

Fig. 4. The approximate solution for Example 4

Example 3. The third example is a one-dimensional problem with a turning point and two boundary layers:

$$\varepsilon u'' - (x-1)u' - u = 0, \quad x \in (0,2),$$

with Dirichlet boundary conditions, corresponding to the following exact solution

$$u(x) = \frac{\operatorname{erf}\left((x-1)/\sqrt{2\varepsilon}\right)\exp((x-1)^2/2\varepsilon)}{\operatorname{erf}\left(1/\sqrt{2\varepsilon}\right)\exp(1/2\varepsilon)}.$$

In Fig. 3 the exact solution and the difference between the exact and the approximate solution are plotted, for $\varepsilon = 0.0001$ and $\mathsf{T} = 0.01$. To achieve the result the algorithm uses 14 levels and the resulting C-norm of the error is less than 0.01. If we require $\mathsf{T} = 0.001$, 15 levels are used and the maximal error is less than 0.001.

Example 4. The following two dimensional problem is considered

$$\varepsilon\left(\frac{\partial^2 u}{\partial x^2} + \frac{\partial^2 u}{\partial y^2}\right) - x\frac{\partial u}{\partial x} - y\frac{\partial u}{\partial y} - 2u = f(x,y), \quad x \in (0,1),\ y \in (0,1),$$

the right-hand side $f(x,y)$ and Dirichlet boundary conditions correspond to the exact solution

$$u(x,y) = -xy(1-\exp((x-1)/\varepsilon)(1-\exp((y-1)/\varepsilon).$$

The approximate solution for $\varepsilon = 0.01$ and $\mathsf{T} = 0.01$ is plotted in Fig. 4. The grid is properly refined around the boundary layers and the C-norm of the error is less than 0.01. The algorithm uses 8 levels. Note, in this case the solution is sufficiently smooth and if we take $q = 2^{-4}$, almost the same grid refinement is achieved.

6 Conclusion

One- and two-dimensional numerical experiments demonstrate that the proposed self-adaptive mesh-generation, embedded in a multigrid strategy and applied with the Baumann-Oden discontinuous Galerkin method, can be successfully used for the automatic resolution of boundary and interior layers in the solution of convection-dominated problems. The strategy is based on the comparison of the numerical approximation on the finest and the one-but-finest grid. Thus, it makes use of the local regularity of the solution. The method used to estimate the local regularity, q_{est}, is still based on heuristic arguments. Although the numerical experiments are quite satisfying, the mathematical motivation needs a more solid theoretical base. This is a subject for further research.

Acknowledgment. D. Vasileva acknowledges the fellowship from the European Consortium for Informatics and Mathematics (ERCIM) for her nine months work at CWI.

References

1. Baumann, C.E.: An *hp*-adaptive discontinuous finite element method for computational fluid dynamics. PhD thesis, the University of Texas at Austin (1997)
2. Baumann, C.E., Oden, J.T.: A discontinuous *hp* finite element method for convection-diffusion problems. Comput. Methods Appl. Mech. Engrg., **175** (1999) 311–341
3. Brandt, A.: Multi-level adaptive solutions to boundary value problems. Math. Comp., **31** (1977) 333–390
4. Hemker, P.W.: On the structure of an adaptive multi-level algorithm. BIT, **20** (1980) 289–301
5. Hemker, P.W., Hoffmann, W., van Raalte, M.H.: Two-level Fourier analysis of a multigrid approach for discontinuous Galerkin discretisation. SIAM J. Sci. Comp., **25** (2004) 1018–1041
6. Hemker, P.W., van Raalte, M.H.: Fourier two-level analysis for higher dimensional discontinuous Galerkin discretisation. Computing and Visualization in Science, **7** (2004) 159–172.
7. van Raalte, M.H.: Multigrid Analysis and Embedded Boundary conditions for discontinuous Galerkin Discretization. PhD thesis, the University of Amsterdam (2004)

Part IV

Monte Carlo: Tools, Applications, Distributed Computing

Femtosecond Evolution of Spatially Inhomogeneous Carrier Excitations Part I: Kinetic Approach

M. Nedjalkov[1], T. Gurov[2], H. Kosina[3], D. Vasileska[4], and V. Palankovski[1]

[1] AMADEA Group, Institute for Microelectronics, Technical University of Vienna, Gusshausstrasse 27-29/E360, A-1040 Vienna, Austria
[2] Institute for Parallel Processing, Bulgarian Academy of Sciences, G. Bontchev str. Bl.25A 1113 Sofia, Bulgaria
[3] Institute for Microelectronics, Technical University of Vienna, Gusshausstrasse 27-29/E360, A-1040 Vienna, Austria
[4] Department of Electrical Engineering, Arizona State University, Tempe, AZ 85287-5706, USA

Abstract. The ultrafast evolution of optically excited carriers which propagate in a quantum wire and interact with three dimensional phonons is investigated. The equation, relevant to this physical problem, is derived by a first principle approach. The electron-phonon interaction is described on a quantum-kinetic level by the Levinson equation, but the evolution problem becomes inhomogeneous due to the spatial dependence of the initial condition. The initial carrier distribution is assumed Gaussian both in energy and space coordinates, an electric field can be applied along the wire. A stochastic method, described in Part II of the work, is used for solving the equation. The obtained simulation results characterize the space and energy dependence of the evolution in the zero field case. Quantum effects introduced by the early time electron-phonon interaction are analyzed.

1 The Coupled Electron-Phonon System

We consider a system of electrons which interact with the lattice vibrations. The electric forces which accelerate the electrons are due to the structure potential and the applied bias, Coulomb interaction between the electrons is neglected. The description of the system is provided by both the electron and the phonon degrees of freedom. We derive the Wigner equation for the coupled electron-phonon system. The corresponding Hamiltonian is given by the free electron part H_0, the structure potential $V(\mathbf{r})$, the free-phonon Hamiltonian H_p, and the electron-phonon interaction H_{e-p}:

$$H = H_0 + V + H_p + H_{e-p} =$$
$$-\frac{\hbar^2}{2m}\nabla_{\mathbf{r}} + V(\mathbf{r}) + \sum_{\mathbf{q}} b^\dagger_{\mathbf{q}} b_{\mathbf{q}} \hbar\omega_{\mathbf{q}} + i\hbar \sum_{\mathbf{q}} \mathcal{F}(\mathbf{q})(b_{\mathbf{q}} e^{i\mathbf{qr}} - b^\dagger_{\mathbf{q}} e^{-i\mathbf{qr}}) \quad (1)$$

I. Lirkov, S. Margenov, and J. Waśniewski (Eds.): LSSC 2005, LNCS 3743, pp. 149–156, 2006.

Here $b^{\dagger}_{\mathbf{q}}$ and $b_{\mathbf{q}}$ are the creation and annihilation operators for the phonon mode $\mathbf{q}$, $\omega_{\mathbf{q}}$ is the energy of that mode, and $\mathcal{F}(\mathbf{q})$ is the electron-phonon coupling element, which depends on the type of phonon scattering analyzed. The state of the phonon subsystem is presented by the set $\{n_{\mathbf{q}}\}$ where $n_{\mathbf{q}}$ is the occupation number of the phonons in mode $\mathbf{q}$. The representation of the basis set is provided by the vectors $|\{n_{\mathbf{q}}\},\mathbf{r}\rangle = |\{n_{\mathbf{q}}\}\rangle|\mathbf{r}\rangle$.

The considered structure is a quantum wire, formed by potential barriers which confine the electron system in the plane normal to the wire. In this plane, at low temperatures, the system occupies the ground state Ψ. A homogeneous electric field E can be applied along the direction of the wire z. It holds:

$$H_0 + V(\mathbf{r}) = H_{\perp} + H_z = H_{0\perp} + V_{\perp} + H_{0z} + V(z); \qquad H_{\perp}\Psi = E_{\perp}\Psi,$$

$V(z) = -eEz$ and $|\mathbf{r}\rangle = |\mathbf{r}_{\perp}\rangle|z\rangle$. The generalized electron-phonon Wigner function is defined by the Fourier transform of the density operator $\hat{\rho}$:

$$f_w(z,p_z,\{n_{\mathbf{q}}\},\{n'_{\mathbf{q}}\},t) = \int \frac{d\mathbf{r}'}{2\pi\hbar} e^{-ip_z z'/\hbar} \langle z + \frac{z'}{2}, \{n_{\mathbf{q}}\}|\langle \mathbf{r}'_{\perp}|\hat{\rho}_t|\mathbf{r}'_{\perp}\rangle|\{n'_{\mathbf{q}}\}, z - \frac{z'}{2}\rangle$$

The evolution problem is separated with respect to the normal and z coordinates as follows:

$$\hat{\rho} = |\Psi\rangle\langle\Psi|\hat{\rho}_{tz}; \qquad \langle \mathbf{r},\{n_{\mathbf{q}}\}|\hat{\rho}_t|\{n'_{\mathbf{q}}\},\mathbf{r}'\rangle = \Psi^*(\mathbf{r}'_{\perp})\Psi(\mathbf{r}_{\perp})\rho(z,z',\{n_{\mathbf{q}}\},\{n'_{\mathbf{q}}\},t)$$

Assuming that Ψ is normalized to unity it is obtained:

$$f_w(z,p_z,\{n_{\mathbf{q}}\},\{n'_{\mathbf{q}}\},t) = \int \frac{dz'}{2\pi\hbar} e^{-ip_z z'/\hbar} \rho(z + \frac{z'}{2}, z - \frac{z'}{2}, \{n_{\mathbf{q}}\},\{n'_{\mathbf{q}}\},t)$$

The equation of motion of f_w is obtained from the von-Neumann equation for the density matrix:

$$\frac{\partial f_w(z,p_z,\{n_{\mathbf{q}}\},\{n'_{\mathbf{q}}\},t)}{\partial t} =$$
$$\frac{1}{i2\pi\hbar^2} \int dz' \int d\mathbf{r}_{\perp} e^{-ip_z z'/\hbar} \langle z + \frac{z'}{2}, \{n_{\mathbf{q}}\}|\langle \mathbf{r}_{\perp}|\,[H,\hat{\rho}_t]_-\,|\mathbf{r}_{\perp}\rangle|\{n'_{\mathbf{q}}\}, z - \frac{z'}{2}\rangle$$

The right hand side, evaluated for each term of the Hamiltonian (1) gives rise to the generalized Wigner equation for the confined electron system. The equation couples an element $f_w(...,\{n\},\{m\},t)$ to four neighborhood elements for any phonon mode $\mathbf{q}$. For any such mode $n_{\mathbf{q}}$ can be any integer between 0 and infinity and a sum over $\mathbf{q}'$ couples all modes. Of interest is the reduced or electron Wigner function defined as the trace over the phonon coordinates:

$$f_w(z,p_z,t) = \sum_{\{n_{\mathbf{q}}\}} f_w(z,p_z,\{n_{\mathbf{q}}\},\{n_{\mathbf{q}}\},t)$$

A closed equation for the reduced Wigner function is obtained by a set of assumptions and approximations: A diagonal element is linked to elements, called first off-diagonal elements, which are diagonal in all modes but the current mode $\mathbf{q}'$ of the summation. In this mode the four neighbors of $n_{\mathbf{q}'}, n_{\mathbf{q}'}$ namely $n_{\mathbf{q}'}\pm 1, n_{\mathbf{q}'}$ and $n_{\mathbf{q}'}, n_{\mathbf{q}'}\pm 1$ are concerned. For convenience we denote the phonon state, obtained from $\{n_{\mathbf{q}}\}$ by increasing or decreasing the phonons in mode $\mathbf{q}'$ by unity as $\{n_{\mathbf{q}}\}^{\pm}_{\mathbf{q}'}$. The first off-diagonal elements are linked to elements which in general are placed further away from the diagonal ones by increasing or decreasing the phonon number in a second mode, $\mathbf{q}''$, by unity. These are the second off-diagonal elements. The only exception is provided by two contributions which recover diagonal elements. They are obtained from $(\{\{n_{\mathbf{q}}\}^{+}_{\mathbf{q}'}\}^{-}_{\mathbf{q}''}, \{n_{\mathbf{q}}\})$ and $(\{n_{\mathbf{q}}\}^{+}_{\mathbf{q}'}, \{n_{\mathbf{q}}\}^{+}_{\mathbf{q}''})$ in the case when the two phonon modes coincide: $\mathbf{q}' = \mathbf{q}''$. The first approximation is to keep only such terms in the equations for the first off-diagonal terms. Then the equations for the diagonal and the four first off-diagonal terms form a closed system. Furthermore, the four first off-diagonal equations can be solved and substituted in the diagonal one. The obtained equation contains only diagonal terms such as $f_w(z, p_z, \{n_{\mathbf{q}}\}, \{n_{\mathbf{q}}\}, t)$ and $f_w(z, p_z, \{n_{\mathbf{q}}\}^{\pm}_{\mathbf{q}'}, \{n_{\mathbf{q}}\}^{\pm}_{\mathbf{q}'}, t)$. The next major assumption is that the phonon system remains in equilibrium during the evolution. This allows to take the trace on the phonon coordinates and to obtain an equation for the reduced Wigner function $f_w(z, p_z, t)$. After few steps of transformations, which include the settings

$$\sum_{\mathbf{q}'} = \frac{V}{(2\pi)^3}\int d\mathbf{q}'; \qquad k_z = p_z/\hbar; \qquad k'_z = k_z - q_z$$

and a conversion to an integral form, the equation reads:

$$f(z,k_z,t) = f(z(0),k_z(0),0) + \int_0^t dt' \int_0^{t'} dt'' \int d\mathbf{q}'_\perp \int dk'_z \times \tag{2}$$
$$\Big[S(k'_z,k_z,t',t'',\mathbf{q}'_\perp) f(z(t'') + \frac{\hbar q'_z}{2m}(t'-t''), k'_z(t''), t'') -$$
$$S(k_z,k'_z,t',t'',\mathbf{q}'_\perp) f(z(t'') + \frac{\hbar q'_z}{2m}(t'-t''), k_z(t''), t'') \Big]$$

$$S(k'_z,k_z,t',t'',\mathbf{q}'_\perp) = \frac{2V}{(2\pi)^3}|G(\mathbf{q}'_\perp)\mathcal{F}(\mathbf{q}'_\perp, k_z - k'_z)|^2 \times$$
$$\Bigg[(n(\mathbf{q}')+1) cos\left(\int_{t''}^{t'} d\tau \frac{1}{\hbar}\Big(\epsilon(k_z(\tau)) - \epsilon(k'_z(\tau)) + \hbar\omega_{\mathbf{q}'} \Big) \right)$$
$$+ \; n(\mathbf{q}') cos\left(\int_{t''}^{t'} d\tau \frac{1}{\hbar}\Big(\epsilon(k_z(\tau)) - \epsilon(k'_z(\tau)) - \hbar\omega_{\mathbf{q}'} \Big) \right) \Bigg]$$

Equation 2 generalizes the Levinson equation [1] for the case of a space dependent initial condition. The Newton trajectories, initialized at z, k_z, t, are governed by the electric force F:

$$z(t'') = z - \frac{1}{m}\int_{t''}^{t} p_z(\tau)d\tau; \qquad k_z(t'') = k_z - F(t - t''); \qquad F = \frac{eE}{\hbar} \tag{3}$$

The shape of the wire affects the electron-phonon coupling through the factor G:

$$G(\mathbf{q}'_\perp) = \int d\mathbf{r}_\perp e^{i\mathbf{q}'_\perp \mathbf{r}_\perp} |\Psi(\mathbf{r}_\perp)|^2$$

2 Phase Space Transform

The equation reveals a very inconvenient from a numerical point of view property, namely that a solution for a phase space point (z, k_z) at t is linked with the solutions on the trajectories (3). Thus the simulation domain grows with the force F and the evolution time t in both position and wave vector subspaces. The following transform ($k_{z1} = k_z, k'_z$) is suggested to cope with this problem:

$$k^t_{z1} = k_{z1} - Ft; \quad k_{z1}(\tau) = k^t_{z1} + F\tau; \quad f(z, k_z, t) = f(z, k^t_z + Ft, t) \stackrel{\text{def}}{=} f^t(z, k^t_z, t)$$

The following equalities can be easily shown:

$$f(z, k_z(t''), t'') = f(z, k^t_z + Ft'', t'') = f^t(z, k^t_z, t'')$$

$$z(t'') = z - \left(\frac{\hbar k^t_z}{m} + \frac{\hbar F}{2m}(t + t'')\right)(t - t'')$$

and $q'_z = k_z - k'_z = k^t_z - k'^t_z$.

$$\epsilon(k'_z(\tau)) - \epsilon(k_z(\tau)) = \epsilon(k'^t_z) - \epsilon(k^t_z) - \frac{2\hbar^2}{2m}F.q'_z\tau$$

All terms in the equation are now expressed as functions of k^t_z, k'^t_z. By omitting the superscripts of the arguments it is obtained:

$$f(z, k_z, t) = f(z - \frac{\hbar k_z}{m}t + \frac{\hbar F}{2m}t^2, k_z, 0) + \int_0^t dt'' \int_{t''}^t dt' \int d\mathbf{q}'_\perp \int dk'_z \times \tag{4}$$
$$\Big[S(k'_z, k_z, t', t'', \mathbf{q}'_\perp) f\left(z - \frac{\hbar k_z}{m}(t - t'') + \frac{\hbar F}{2m}(t^2 - t''^2) + \frac{\hbar q'_z}{2m}(t' - t''), k'_z, t''\right)$$
$$-S(k_z, k'_z, t', t''\mathbf{q}'_\perp) f\left(z - \frac{\hbar k_z}{m}(t - t'') + \frac{\hbar F}{2m}(t^2 - t''^2) - \frac{\hbar q'_z}{2m}(t' - t''), k_z, t''\right)\Big]$$

$$S(k'_z, k_z, t', t'', \mathbf{q}'_\perp) = \frac{2V}{(2\pi)^3}|G(\mathbf{q}'_\perp)\mathcal{F}(\mathbf{q}'_\perp, k_z - k'_z)|^2 \times$$
$$\Big[(n(\mathbf{q}') + 1)cos\left(\frac{\epsilon(k_z) - \epsilon(k'_z) + \hbar\omega_{\mathbf{q}'}}{\hbar}(t' - t'') + \frac{\hbar}{2m}F.q'_z(t'^2 - t''^2)\right)$$
$$+n(\mathbf{q}')cos\left(\frac{\epsilon(k_z) - \epsilon(k'_z) - \hbar\omega_{\mathbf{q}'}}{\hbar}(t' - t'') + \frac{\hbar}{2m}F.q'_z(t'^2 - t''^2)\right)\Big]$$

3 Simulation Results and Discussions

We first consider the numerical properties of (4). The integration is in the wave vector space, while the real space variable enters as a parameter modified by the two time integrals. The advantage of (4) as compared to (2) is that the wave vector variable is decoupled from the time variable and thus the integration domain can be maintained independent of the force and the evolution time. Despite that the numerical challenges posed by (4) are heavy already in the homogeneous problem [2]. In the latter case one of the time integrals can be assigned to S. Furthermore another integration in the wave vector space can be spared due to symmetry considerations. In the inhomogeneous problem it is no more possible to assign the time integral to S due to the t' dependence of f in the right hand side of (4). The physical origin of this dependence is associated with the finite duration of the electron-phonon interaction: the real space trajectory is modified by the half of the phonon wave vector q'_z times the duration $t' - t''$ of the interaction. Thus in the general case each iteration step increases the dimensionality by five more integrals and thus the computational burden.

The equation accounts for interesting quantum effects demonstrated by the presented simulation results. Considered is a $GaAs$ material with a single polar optical phonon having a constant energy $\hbar\omega$. The electric field is zero. The initial condition is a product of two Gaussian distributions of the energy and space. The k_z^2 distribution corresponds to a generating laser pulse with an excess energy of about $150meV$. The z distribution is centered around zero. A quantum wire with a rectangular cross section is assumed. At very low temperature the physical system has a transparent semiclassical behavior. We recall the major results of the homogeneous case [2]. Semiclassical electrons can only emit phonons and loose energy equal to a multiple of the phonon energy $\hbar\omega$. They evolve according to an energy distribution, patterned by replicas of the initial condition shifted towards low energies. Such electrons cannot appear in the region above the initial distribution. The quantum solutions demonstrate two effects of deviation from the semiclassical behavior. The replicas are broadened and the broadening reduces with the time. A finite density of electrons appears in the semi-classically forbidden region above the initial condition. These effects are due to the lack of the energy conserving delta function, which is build up by the cosine function in S for long evolution times.

In the inhomogeneous case the wave vector (and respectively the energy) and the density distributions are given by the integrals

$$f(k_z, t) = \int \frac{dz}{2\pi} f_w(z, k_z, t); \qquad n(z, t) = \int \frac{dk_z}{2\pi} f_w(z, k_z, t)$$

Figure 1 shows the redistribution of the initial electrons after 50 femtoseconds evolution time as a function of the proportional to the energy quantity k_z^2. A window of values for k_z^2 and f is chosen, where the broadening of first replica and the finite density of electrons with energies above the initial condition is well visible. Figure 2 shows the distribution in the whole simulation domain after 150 femtoseconds evolution. The first replica becomes sharper, but still broadened

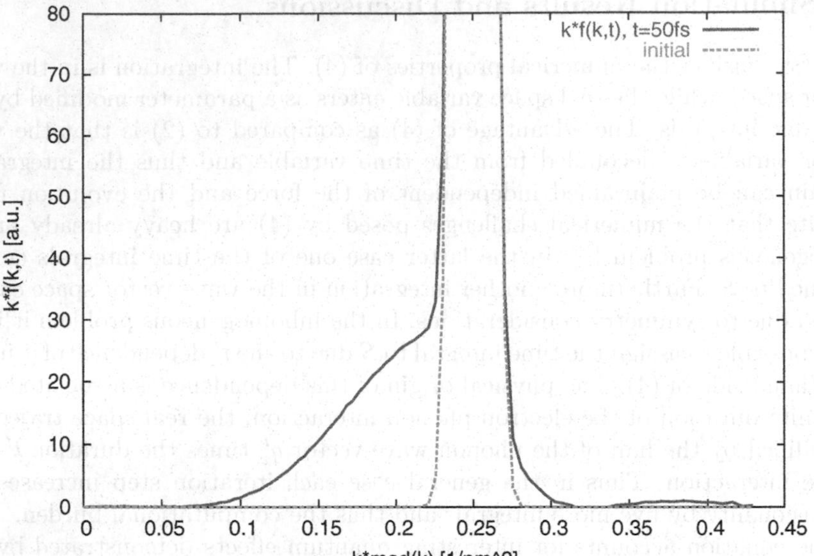

Fig. 1. Initial condition and energy distribution at $50fs$ evolution time

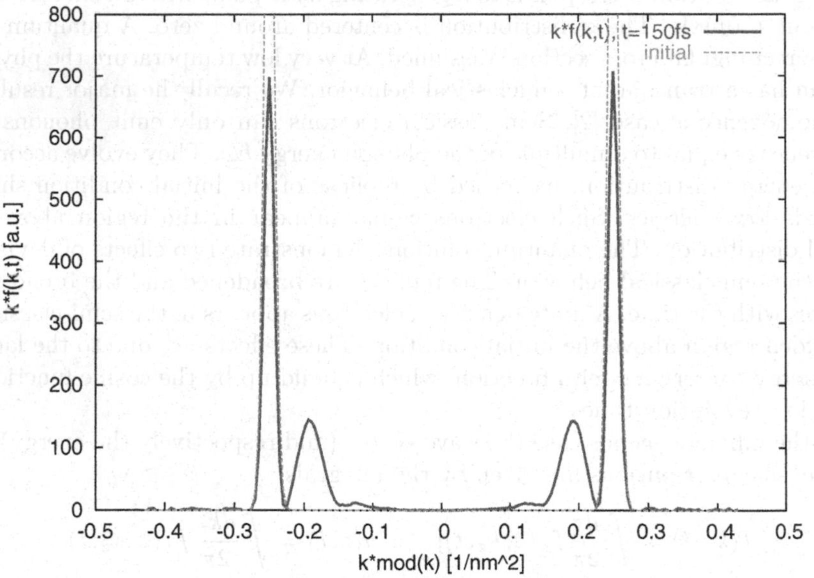

Fig. 2. Initial condition and energy distribution at $150fs$ evolution time

with respect to the initial condition. Also the place of the second replica can be recognized. In the absence of electric field there is a symmetry in the k_z directions. The behavior is analogous to the homogeneous case despite that the

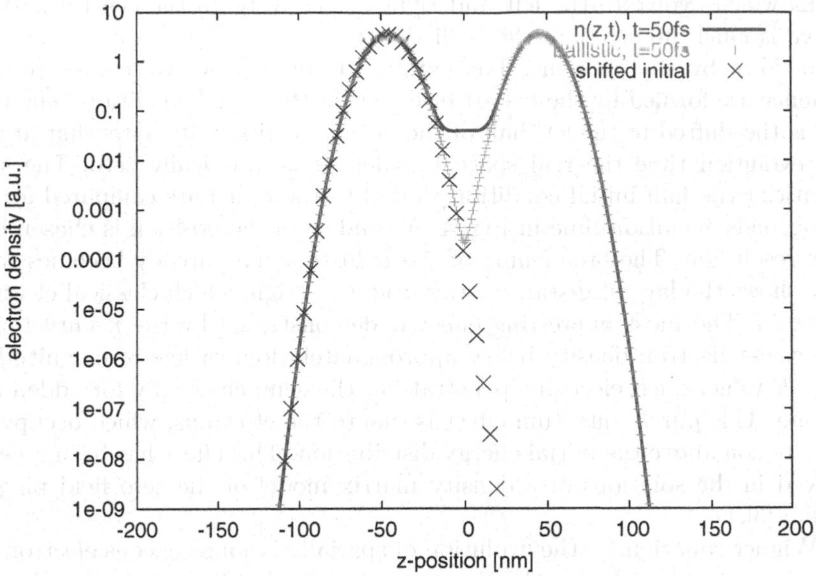

Fig. 3. Electron density after 50fs evolution time

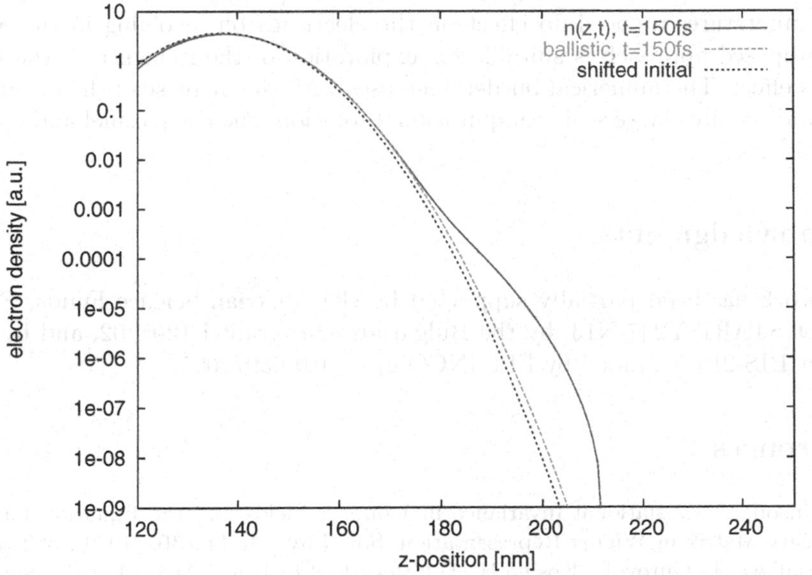

Fig. 4. Electron density after 150fs evolution time

coupling constant is now modified by G. Figure 3 compares the electron density with (n) and without (ballistic) electron-phonon interaction for 50 femtoseconds evolution time. The initial peak at the origin splits into two symmetric distributions which evolve to the left and right respectively. In the central part the n curve is much higher than the ballistic curve due to the electrons which are slowed down by the phonons. The external fronts of the two curves coincide and hence are formed by the fastest electrons in the initial condition. The third curve is the shifted to the left half of the initial condition. It shows that at such small evolution time the real space broadening is practically zero. The same quantities (the half initial condition shifted to the right) are compared for 150 femtoseconds evolution time in Fig. 4. A window in the position is chosen for a better resolution. The broadening of the ballistic curve already becomes sensible. It shows the largest distance away from the origin which classical electrons can attain. The most interesting effect is demonstrated by the n curve: there is an excess electron density below approximately four orders of magnitude of the peak value. Such electrons penetrate in the semi-classically forbidden spatial zone. This purely quantum effect is due to the electrons, which occupy the energy region above the initial energy distribution. This effect has been recently observed in the solutions of a density matrix model of the zero field physical problem [3].

A Wigner equation for the evolution of spatially inhomogeneous electron distribution excited by a laser pulse in a quantum wire has been derived and solved by a Monte Carlo approach. A transformation is proposed which fixes the problem with the spreading integration domain in presence of electric field. It is shown that the quantum character of the electron-phonon interaction causes at low temperatures a speed-up effect on the electron front evolving in the wire. The proposed approach is suitable for exploration of the influence of the field on this effect. The numerical burden increases with the increase of the evolution time and requires large scale computational solutions such as parallel and GRID technologies.

Acknowledgments

This work has been partially supported by the Austrian Science Funds, FWF Project START Y247-N13, by the Bulgarian NSF grant I-1201/02, and by the project BIS-21++ funded by FP6 INCO grant 016639/2005.

References

1. Levinson I, Translational Invariance in Uniform Fields and the Equation for the Density Matrix in Wigner Representation, Sov. Phys. JETP, **30**, (1970), 362–367
2. Nedjalkov M., Gurov T., Kosina H., Whitlock P., Statistical Algorithms for Simulation of Quantum Kinetics in Semiconductors, Large-Scale Scientific Computing,ed. by S. Margenov, J. Wasniewski, P. Yalamov, Springer, (2001), 183–190
3. Herbst M., Glanemann M., Axt V., and Kuhn T., "Electron-Phonon Quantum Kinetics for Spatially Inhomogenenous Excitations," Physical Review B, **67**, (2003) 195305:1–18

Femtosecond Evolution of Spatially Inhomogeneous Carrier Excitations Part II: Stochastic Approach and Grid Implementation

T. Gurov[1], E. Atanassov[1], I. Dimov[1], and V. Palankovski[2]

[1] Institute for Parallel Processing, Bulgarian Academy of Sciences, Acad. G. Bontchev, Bl.25A, 1113 Sofia, Bulgaria
[2] AMADEA Group, Institute for Microelectronics, Technical University of Vienna, Gusshausstrasse 27-29/E360, A-1040 Vienna, Austria

Abstract. We present a stochastic approach for solving the quantum-kinetic equation introduced in Part I. A Monte Carlo method based on backward time evolution of the numerical trajectories is developed. The computational complexity and the stochastic error are investigated numerically. Variance reduction techniques are applied, which demonstrate a clear advantage with respect to the approaches based on symmetry transformation. Parallel implementation is realized on a GRID infrastructure.

1 Introduction

A Wigner equation for nanometer and femtosecond transport regime has been previously derived from a three equations set model based on the generalized Wigner function [7]. The complete equation poses serious numerical challenges so that we consider limiting versions of the equation corresponding to simplified physical. Two limiting cases, namely the Wigner-Boltzmann equation [8] and the homogeneous Levinson (or Barker-Ferry) equation [5, 4] have been analyzed with various Monte Carlo (MC) approaches. In particular, spherical and cylindrical transformations have been used to reduce the dimensions in the momentum space. The equation derived in Part I presents a third limiting case, where the electron-phonon interaction is described on the quantum-kinetic level by the Levinson equation, but the evolution problem becomes inhomogeneous due to the spatial dependence of the initial condition. The problem is relevant e.g. for description of the ultrafast dynamics of optically generated carriers. Particularly we consider a quantum wire, where the carriers are confined in the plane normal to the wire by infinite potentials. At low temperatures the carriers remain in the ground state in the normal plane. Thus, the evolution is described in the two dimensional phase space of the carrier wave vector and the position, while the phonons are three dimensional. The initial carrier distribution is assumed Gaussian both in energy and space coordinates, and an electric field can be

I. Lirkov, S. Margenov, and J. Waśniewski (Eds.): LSSC 2005, LNCS 3743, pp. 157–163, 2006.

applied along the wire. We recall the integral form of the derived in Part I quantum-kinetic equation:

$$f_w(z,k_z,t) = f_w(z - \frac{\hbar k_z}{m}t + \frac{\hbar \mathbf{F}}{2m}t^2, k_z, 0) + \int_0^t \mathrm{d}t'' \int_{t''}^t \mathrm{d}t' \int d\mathbf{q}'_\perp \int dk'_z \times \quad (1)$$

$$\left[S(k'_z, k_z, t', t'', \mathbf{q}'_\perp) f_w\left(z - \frac{\hbar k_z}{m}(t-t'') + \frac{\hbar \mathbf{F}}{2m}(t^2 - t''^2) + \frac{\hbar q'_z}{2m}(t'-t''), k'_z, t''\right)\right.$$

$$\left. - S(k_z, k'_z, t', t'', \mathbf{q}'_\perp) f_w\left(z - \frac{\hbar k_z}{m}(t-t'') + \frac{\hbar \mathbf{F}}{2m}(t^2 - t''^2) - \frac{\hbar q'_z}{2m}(t'-t''), k_z, t''\right)\right]$$

$$S(k'_z, k_z, t', t'', \mathbf{q}'_\perp) = \frac{2V}{(2\pi)^3} |G(\mathbf{q}'_\perp)\mathcal{F}(\mathbf{q}'_\perp, k_z - k'_z)|^2 \times$$

$$\left[(n(\mathbf{q}')+1)\cos\left(\frac{\epsilon(k_z) - \epsilon(k'_z) + \hbar\omega_{\mathbf{q}'}}{\hbar}(t'-t'') + \frac{\hbar}{2m}\mathbf{F}.q'_z(t'^2 - t''^2)\right)\right.$$

$$\left. + n(\mathbf{q}')\cos\left(\frac{\epsilon(k_z) - \epsilon(k'_z) - \hbar\omega'_{\mathbf{q}}}{\hbar}(t'-t'') + \frac{\hbar}{2m}\mathbf{F}.q'_z(t'^2 - t''^2)\right)\right]$$

Here, $f(z, k_z, t)$ is the Wigner function described in the $2D$ phase space of the carrier wave vector k_z and the position z, and t is the evolution time.

$F = e\mathbf{E}/\hbar$, where $\mathbf{E}$ is a homogeneous electric field along the direction of the wire z, e being the electron charge and $\hbar$ - the Plank's constant.

$n'_{\mathbf{q}} = 1/(\exp(\hbar\omega'_{\mathbf{q}}/\mathcal{K}T) - 1)$ is the Bose function, where $\mathcal{K}$ is the Boltzmann constant and T is the temperature of the crystal, corresponds to an equilibrium distributed phonon bath.

$\hbar\omega'_{\mathbf{q}}$ is the phonon energy which generally depends on $\mathbf{q}' = \mathbf{q}'_\perp + q'_z = \mathbf{q}'_\perp + (k_z - k'_z)$, and $\varepsilon(k_z) = (\hbar^2 k_z^2)/2m$ is the electron energy.

$\mathcal{F}$ is obtained from the Fröhlich electron-phonon coupling by recalling the factor $i\hbar$ in the interaction Hamiltonian, Part I:

$$\mathcal{F}(\mathbf{q}'_\perp, k_z - k'_z) = -\left[\frac{2\pi e^2 \omega'_{\mathbf{q}}}{\hbar V}\left(\frac{1}{\varepsilon_\infty} - \frac{1}{\varepsilon_s}\right)\frac{1}{(\mathbf{q}')^2}\right]^{\frac{1}{2}},$$

where (ε_∞) and (ε_s) are the optical and static dielectric constants. The shape of the wire affects the electron-phonon coupling through the factor

$$G(\mathbf{q}'_\perp) = \int d\mathbf{r}_\perp e^{i\mathbf{q}'_\perp \mathbf{r}_\perp} |\Psi(\mathbf{r}_\perp)|^2 ,$$

where Ψ is the ground state of the electron system in the plane normal to the wire.

In the inhomogeneous case the wave vector (and respectively the energy) and the density distributions are given by the integrals

$$f(k_z, t) = \int \frac{dz}{2\pi} f_w(z, k_z, t); \quad n(z,t) = \int \frac{dk_z}{2\pi} f_w(z, k_z, t). \quad (2)$$

Our aim is to estimate these quantites, as well as the Wigner function (1) by MC approach.

2 Monte Carlo Approach

Consider the problem of evaluating the following functional of the solution of the integral equation (1)

$$J_g(f) \equiv (g,f) = \int_0^T \iint_D g(z,k_z,t) f_w(z,k_z,t) dz dk_z dt, \tag{3}$$

by a MC method. Here we specify that the phase space point (z, k_z) belongs to a rectangular domain $D = (-Q_1, Q_1) \times (-Q_2, Q_2)$, and $t \in (0, T)$. The arbitrary function $g(z, k_z, t)$ belongs to the space L_∞.

Now (1) can be written in the following form:

$$\begin{aligned} f_w(z,k_z,t) &= f_w(z - z(k_z,t), k_z, 0) + \\ &+ \int_0^t dt'' \int_{t''}^t dt' \int_G d^3\mathbf{k}' \{K_1(k_z, \mathbf{k}', t', t'') f_w(z + h(k_z, q'_z, t, t', t'', \mathbf{F}), k'_z, t'')\} \\ &+ \int_0^t dt'' \int_{t''}^t dt' \int_G d^3\mathbf{k}' \{K_2(k_z, \mathbf{k}', t', t'') f_w(z + h(k_z, -q'_z, t, t', t'', \mathbf{F}), k_z, t'')\}, \end{aligned} \tag{4}$$

where

$$z(k_z, t) = \frac{\hbar k_z}{m} t - \frac{\hbar \mathbf{F}}{2m} t^2,$$

$$h(k_z, q'_z, t, t', t'', \mathbf{F}) = -\frac{\hbar k_z}{m}(t - t'') + \frac{\hbar \mathbf{F}}{2m}(t^2 - t''^2) + \frac{\hbar q'_z}{2m}(t' - t''),$$

$$K_1(k_z, \mathbf{k}', t', t'') = S(k'_z, k_z, t', t'', \mathbf{q}'_\perp) = -K_2(\mathbf{k}', k_z, t', t''),$$

and $\int_G d^3\mathbf{k}' = \int d\mathbf{q}'_\perp \int_{-Q_2}^{Q_2} dk_z$. We note that the Neumann series of such integral equations as (4) converges [5]. Thus, the functional (3) can be evaluated by a MC estimator [6].

Let us construct a biased MC estimator for evaluating the functional (3) using backward time evolution of the numerical trajectories in the following way:

$$\xi_s[J_g(f)] = \frac{g(z,k_z,t)}{p_{in}(z,k_z,t)} W_0 f_w(., k_z, 0) + \frac{g(z,k_z,t)}{p_{in}(z,k_z,t)} \sum_{j=1}^{s} W_j^\alpha f_w\left(., k_{z,j}^\alpha, t_j\right), \tag{5}$$

where

$$f_w\left(., k_{z,j}^\alpha, t_j\right) =$$

$$\begin{cases} f_w\left(z + h(k_{z,j-1}, k_{z,j-1} - k_{z,j}, t_{j-1}, t'_j, t_j, \mathbf{F}), k_{z,j}, t_j\right), & \text{if } \alpha = 1, \\ f_w\left(z + h(k_{z,j-1}, k_{z,j} - k_{z,j-1}, t_{j-1}, t'_j, t_j, \mathbf{F}), k_{z,j-1}, t_j\right), & \text{if } \alpha = 2, \end{cases}$$

$$W_j^\alpha = W_{j-1}^\alpha \frac{K_\alpha(k_{z,j-1}, \mathbf{k}_j, t'_j, t_j)}{p_\alpha p_{tr}(\mathbf{k}_{j-1}, \mathbf{k}_j, t'_j, t_j)}, \quad W_0^\alpha = W_0 = 1, \quad \alpha = 1, 2, \quad j = 1, \ldots, s.$$

The probabilities $p_\alpha, (\alpha = 1, 2)$ are chosen to be proportional to the absolute value of the kernels in (4). The initial density $p_{in}(z, k_z, t)$ and the transition density $p_{tr}(\mathbf{k}, \mathbf{k}', t', t'')$ are chosen to be tolerant[1] to the function $g(z, k_z, t)$ and

[1] $r(x)$ is tolerant to $g(x)$ if $r(x) > 0$ when $g(x) \neq 0$ and $r(x) \geq 0$ when $g(x) = 0$.

the kernels, respectively. The first point $(z, k_{z,0}, t_0)$ in the Markov chain is chosen using the initial density, where $k_{z,0}$ is the third coordinate of the wave vector $\mathbf{k}_0$. Next points $(k_{z,j}, t'_j, t_j) \in (-Q_2, Q_2) \times (t_j, t_{j-1}) \times (0, t_{j-1})$ of the Markov chain:

$$(k_{z,0}, t_0) \to (k_{z,1}, t'_1, t_1) \to \cdots \to (k_{z,j}, t'_j, t_j) \to \cdots \to (k_{z,s}, , t'_s, t_s)$$

do not depend on the position z of the electrons. They are sampled using the transition density $p_{tr}(\mathbf{k}, \mathbf{k}', t', t'')$ as we take only the z-coordinate of the wave vector $\mathbf{k}$. Note the time t'_j conditionally depends on the selected time t_j. The Markov chain terminates in time $t_s < \varepsilon_1$, where ε_1 is a fixed small positive number called a truncation parameter.

In order to evaluate the functional (3) by N independent samples of the estimator (5), we define a Monte Carlo method

$$\frac{1}{N}\sum_{i=1}^{N}(\xi_s[J_g(f)])_i \xrightarrow{P} J_g(f_s) \approx J_g(f), \tag{6}$$

where $\xrightarrow{P}$ means stochastic convergence as $N \to \infty$; f_s is the iterative solution obtained by the Neumann series of (4), and s is the number of iterations. The relation (6) still does not determine the computational algorithm. To define a MC algorithm we have to specify the initial and transition densities, as well the modeling function (or sampling rule). The modeling function describes the rule needed to calculate the states of the Markov chain by using uniformly distributed random numbers in the interval $(0, 1)$.

Here, the transition density is chosen:

$$p_{tr}(\mathbf{k}, \mathbf{k}', t', t'') = p(\mathbf{k}'/\mathbf{k})p(t, t', t'')$$

$$p(t, t', t'') = p(t, t'')p(t'/t'') = \frac{1}{t}\frac{1}{(t - t'')}$$

$$p(\mathbf{k}'/\mathbf{k}) = c_1/(\mathbf{k}' - \mathbf{k})^2$$

(c_1 is the normalized constant). Thus, if we know t, the next times t'' and t' are computed by using the inverse-transformation rule. The wave vector $\mathbf{k}'$ are sampled in the same way as described in [3]. The difference here is that we have to compute all three coordinates of the wave vector although we need only the third coordinate.

The choice of $p_{in}(z, k_z, t)$ depends on the function $g(z, k_z, t)$. The cases when

$$(i)\quad g(z, k_z, t) = \delta(z - z_0)\delta(k_z - k_{z,0})\delta(t - t_0),$$

$$(ii)\quad g(z, k_z, t) = \frac{1}{2\pi}\delta(k_z - k_{z,0})\delta(t - t_0),$$

$$(iii)\quad g(z, k_z, t) = \frac{1}{2\pi}\delta(z - z_0)\delta(t - t_0),$$

are of special interest, because they estimate the values of the Wigner function (1), the wave vector and the density distribution (2) in fixed points.

3 Grid Implementation and Numerical Results

The computational complexity of an MC algorithm can be measured by the quantity $CC = N \times \tau \times M(s_{\varepsilon_1})$. The number of the random walks, N, and the average number of transitions in the Markov chain, $M(s_{\varepsilon_1})$, are related to the stochastic and systematic errors [5]. The mean time for modeling one transition, τ, depends on the complexity of the transition density functions and on the sampling rule, as well as on the choice of the random number generator (rng).

It is proved [5, 4] that the stochastic error has order $O(N^{-1/2} \exp(c_2 t))$, where t is the evolution time and c_2 is a constant depending on the kernels of the quantum kinetic equation under consideration. This estimate shows that when t is fixed and $N \to \infty$ the error decreases, but for large t the factor $\exp(c_2 t)$ looks ominous. Therefore, the algorithm solves an NP-hard problem concerning the evolution time. To solve this problem for long evolution times with small stochastic error we have to combine both MC variance reduction techniques and distributed or parallel computations.

It is well known that the MC algorithms are very convenient for implementations on parallel computer systems [1], because every realization of the MC estimator can be done independently and simultaneously. Although MC algorithms are well suited to parallel computation, there are a number of potential problems. The available computers can run at different speeds; they can have different user loads on them; one or more of them can be down; the rng's that they use can run at different speeds; etc. On the other hand, these rng's must produce independent and non-overlapping random sequences. Thus, the parallel realization of the MC algorithms is not a trivial process on different parallel computer systems.

By using the grid environment provided by the EGEE project middleware[2] [2] we were able to reduce the computing time of the MC algorithm under consideration. The simulations are parallelized on the existing Grid infrastructure by splitting the underlying random number sequences. The numerical results discussed in Fig. 1 are obtained for zero temperature and GaAs material parameters: the electron effective mass is 0.063, the optimal phonon energy is $36meV$, the static and optical dielectric constants are $\varepsilon_s = 10.92$ and $\varepsilon_\infty = 12.9$. The initial condition is a product of two Gaussian distributions of the energy and space. The k_z^2 distribution corresponds to a generating laser pulse with an excess energy of about $150meV$. The z distribution is centered around zero. A quantum wire with orthogonal cross section is assumed.

[2] The Enabling Grids for E-sciencE (EGEE) project is funded by the European Commission and aims to build on recent advances in grid technology and develop a service grid infrastructure which is available to scientists 24 hours-a-day. The project aims to provide researchers in both academia and industry with access to major computing resources, independent of their geographic location. The EGEE project identifies a wide-range of scientific disciplines and their applications and supports a number of them for deployment. To date there are five different scientific applications running on the EGEE Grid infrastructure. For more information see http://public.eu-egee.org/.

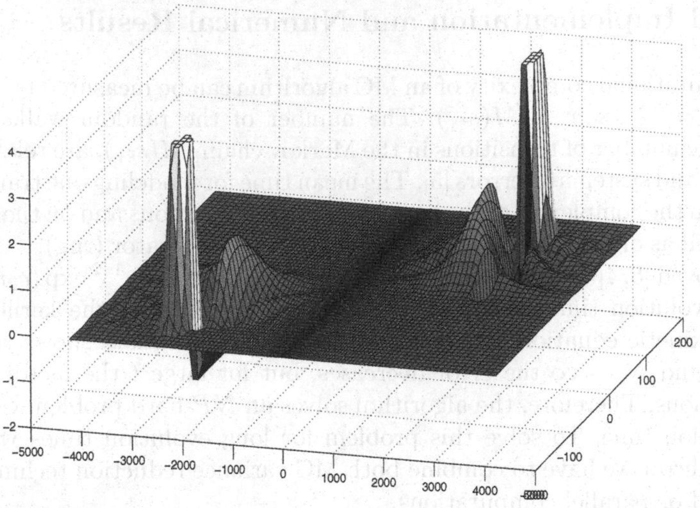

Fig. 1. Evolution of optically generated distribution of electrons in a quantum wire. The electric field E is 0, the number of the random walks per point is $N = 10$ millions, t is 150 femtoseconds.

Table 1. The CPU time (seconds) for all 800×260 points, the speed-up, and the parallel efficiency. The number of random walks is $N = 100000$. The evolution time is $100\,fs$.

Number of CPUs	CPU Time (s)	Speed-up	Parallel Efficiency
2	9790	-	-
4	4896	1.9996	0.9998
6	3265	2.9985	0.9995

The solution $f(z, k_z, t)$ is estimated in 800×260 points (see Figure 1) that are symmetrically located in the rectangular domain $(-Q_1.Q_1) \times (-Q_2, Q_2)$, where $Q_1 = 400 \times 10^9 m^{-1}$ and $Q_2 = 66 \times 10^7 m^{-1}$.

In our research, the MC algorithm has been implemented in C. The SPRNG library has been used to produce independent and non-overlapping random sequences [9]. Successful tests of the algorithm were performed at the Bulgarian SEE-GRID[3] sites using the Resource Broker at ULAKBIM — Turkey. The MPI

[3] South Eastern European GRid-enabled eInfrastructure Development (SEE-GRID) project is funded by the European Commission and aims to provide specific support actions to pave the way towards the participation of the SE European countries to the Pan-European and worldwide Grid initiatives. The SEE-GRID consortium consists of eleven contractors: ten representatives or incubators of National Grid initiatives (NGIs) from SE European countries and CERN. The consortium contractors that represent NGIs are: GRNET (Greece), SZTAKI (Hungary), ICI (Romania), IPP (former CLPP, Bulgaria), TUBITAK (Turkey), INIMA (Albania), BIHARNET (Bosnia-Herzegovina), Ss Cyril and Methodius University (FYRoM), University of Belgrade (Serbia-Montenegro), RBI (Croatia). For more information see http://www.see-grid.org/.

implementation was MPICH 1.2.6, and the execution is controlled from the Computing Element via the Torque batch system.

The timing results for evolution time t=100 femtoseconds are shown in Table 1. The parallel efficiency is close to 100%.

4 Conclusion

A Wigner equation for the evolution of spatially inhomogeneous electron distribution excited by a laser pulse in a quantum wire has been solved by a Monte Carlo approach. Numerical results are obtained for the Wigner function without applied electric field, but the proposed approach is suitable for exploration of the influence of the field. The numerical problem is CPU intensive and thus suitable for implementation on Grid infrastructures, especially since high parallel efficiency is achieved.

Acknowledgments. This work has been supported by the Bulgarian NSF grant I-1201/02, by the project BIS-21++ funded by FP6 INCO grant 016639/2005, and by the Austrian Science Funds, FWF Project START Y247-N13.

References

1. I. Dimov, O. Tonev, Monte Carlo Algorithms: Performance Analysis for Some Computer Architectures, *J. of Comp. and Appl. Mathematics.* **48**, 253–277, 1993.
2. *EGEE GRID Middleware*, http://lcg.web.cern.ch/LCG/Sites/releases.html
3. T.V. Gurov, I.T. Dimov, A Parallel Monte Carlo Method For Electron Quantum Kinetic Equation, *Large scale scientific computing*, I. Lirkov, S. Margenov, J. Waśniewski, P. Yalamov eds., *Lecture notes in computer sciences*, **2907**, Springer-Verlag, 151–161, 2004.
4. T.V. Gurov, M. Nedjalkov, P.A. Whitlock, H. Kosina, S. Selberherr, Femtosecond Relaxation of Hot Electrons by Phonon Emission in Presence of Electric Field, *Physica B*, **314**, 301–304, 2002.
5. T.V. Gurov, P.A. Whitlock, An Efficient Backward Monte Carlo Estimator for Solving of a Quantum Kinetic Equation with Memory Kernel, *Mathematics and Computers in Simulation*, **60**, 85–105, 2002.
6. Mikhailov G.A.: New Monte Carlo Methods with Estimating Derivatives. Utrecht, The Netherlands, 1995.
7. M. Nedjalkov, R. Kosik, H. Kosina, S. Selberherr, A Wigner Equation for Nanometer and Femtosecond Transport Regime, Proceedings of the 2001 First IEEE Conference on Nanotechnology, IEEE, Maui, Hawaii, 277–281, 2001.
8. M. Nedjalkov, H. Kosina, S. Selberherr, C. Ringhofer, D.K. Ferry, Unified particle approach to Wigner-Boltzmann transport in small semiconductor devices, *Physical Review B*, **70**, 115319–115335, 2004.
9. *Scalable Parallel Random Number Generators Library for Parallel Monte Carlo Computations*, SPRNG 1.0 and SPRNG 2.0 – http://sprng.cs.fsu.edu.

Noise Calculation in the Semiclassical Framework: A Critical Analysis of the Monte Carlo Method and a Numerical Alternative

Christoph Jungemann and Bernd Meinerzhagen

Institut für Elektronische Bauelemente und Schaltungstechnik,
TU Braunschweig, Postfach 33 29, 38023 Braunschweig, Germany
C.Jungemann@tu-bs.de
http://www.nst.ing.tu-bs.de

Abstract. Noise modeling in the semiclassical framework of the Boltzmann transport equation (BTE) is analyzed. The usual approach to solve the BTE, the Monte Carlo method, is found to be prohibitively CPU intensive for technically relevant frequencies below 100GHz. A numerical alternative based on a spherical harmonics expansion of the BTE is presented, of which the CPU time does not depend on the frequency. In addition, this approach allows to solve the Langevin-type BTE, which gives more physical insight into noise. This is demonstrated for some relevant device applications.

1 Introduction

Electronic noise is one of the limiting factors for the performance of analog circuits [1] and the most advanced models for devices are based on the semiclassical Boltzmann transport equation (BTE) [2]. The method of choice for solving the BTE is the Monte Carlo (MC) approach [3, 4] which solves the BTE in the time domain and inherently contains noise. By Fourier transformation of the correlation functions into the frequency domain the important power spectral densities (PSD) are obtained. But recently it has been shown that it is difficult to calculate electronic noise in the technically important frequency range below 100GHz with this time-domain based approach [5, 6]. At such frequencies the CPU time is at least inversely proportional to the minimum frequency investigated and thus prohibitively long.

An alternative approach is based on the spherical harmonics expansion (SHE) of the BTE [7, 8, 9]. In this case the BTE can be solved directly in the frequency domain avoiding the problems of MC at low frequencies [10]. In addition, it is possible to solve the Langevin Boltzmann equation (LBE) with SHE [11]. The LBE, which is equivalent to the BTE w.r.t. to the solutions obtained [2], has the advantage that the Langevin approach is well-known and the problems encountered in [10] can be easily avoided.

First, noise calculation in the framework of the BTE is discussed. Second, the Langevin approach is presented and its advantages outlined. Third, the problems

I. Lirkov, S. Margenov, and J. Waśniewski (Eds.): LSSC 2005, LNCS 3743, pp. 164–171, 2006.

of the MC approach for noise are investigated, and finally results of the SHE method are presented. For the sake of brevity only the case of stationary and spatially homogeneous systems is discussed in Sec. 2 for which the real space coordinates can be neglected and the BTE is linear.

2 Theory

Electronic noise in a stationary system is characterized by the one-sided PSD S_{xy} which is the Fourier transform of the corresponding correlation function of the two microscopic quantities x and y [12, 2]

$$S_{xy}(\omega) = S^*_{yx}(\omega) = 2\int_{-\infty}^{\infty} \langle [x(t) - \langle x \rangle][y(0) - \langle y \rangle] \rangle \exp(-\mathrm{i}\omega t)\mathrm{d}t \ , \tag{1}$$

where $\omega = 2\pi f \geq 0$ is the angular frequency, the asterisk denotes the complex conjugate, and $\langle\rangle$ an expectation. The correlation function of the fluctuations is given by

$$\langle x(t)y(0) \rangle - \langle x \rangle \langle y \rangle = \iint x(\boldsymbol{k}) \left(p(\boldsymbol{k}, t; \boldsymbol{k}', 0) - p(\boldsymbol{k})p(\boldsymbol{k}') \right) y(\boldsymbol{k}') \mathrm{d}^3k \mathrm{d}^3k' \tag{2}$$

with the joint probability density

$$p(\boldsymbol{k}, t; \boldsymbol{k}', 0) = \begin{cases} p^>(\boldsymbol{k}, t|\boldsymbol{k}', 0)p(\boldsymbol{k}') & \text{for} \quad t \geq 0 \\ p^>(\boldsymbol{k}', -t|\boldsymbol{k}, 0)p(\boldsymbol{k}) & \text{for} \quad t < 0 \end{cases} . \tag{3}$$

The stationary single-particle probability density $p(\boldsymbol{k})$ and the conditional probability density (CPD) $p^>(\boldsymbol{k}, t|\boldsymbol{k}', 0)$, which gives the probability that a particle, which was at time zero in the state $\boldsymbol{k}'$, is found at t in the state $\boldsymbol{k}$ for $t \geq 0$, are both normalized

$$\int p(\boldsymbol{k})\mathrm{d}^3k = 1 \quad , \quad \int p^>(\boldsymbol{k}, t|\boldsymbol{k}', 0)\mathrm{d}^3k = 1 \tag{4}$$

and must both satisfy the linear BTE (the CPD w.r.t. to the first set of arguments)

$$\left\{ \frac{\partial}{\partial t} - \frac{q}{\hbar}\boldsymbol{E}\nabla \right\} p = \hat{W}\{p\} \tag{5}$$

with the scattering integral

$$\hat{W}\{p\} = \frac{\Omega}{(2\pi)^3} \int W(\boldsymbol{k}|\boldsymbol{k}'')p(\boldsymbol{k}'') - W(\boldsymbol{k}''|\boldsymbol{k})p(\boldsymbol{k})\mathrm{d}^3k'' \ , \tag{6}$$

where $\boldsymbol{E}$ is the electric field, $W(\boldsymbol{k}|\boldsymbol{k}'')$ the transition rate, and Ω the system volume [13]. The initial condition of the CPD is given at $t = 0$ by

$$p^>(\boldsymbol{k}, 0|\boldsymbol{k}', 0) = \delta(\boldsymbol{k} - \boldsymbol{k}') \ . \tag{7}$$

The Fourier transform of the joint probability density can be rearranged with (3)

$$
\begin{aligned}
P(\boldsymbol{k},\boldsymbol{k}',\omega) &= \int_{-\infty}^{\infty} [p(\boldsymbol{k},t;\boldsymbol{k}',0) - p(\boldsymbol{k})p(\boldsymbol{k}')] \exp(-\mathrm{i}\omega t)\mathrm{d}t \\
&= \int_{0}^{\infty} [p(\boldsymbol{k}',t;\boldsymbol{k},0) - p(\boldsymbol{k}')p(\boldsymbol{k})] \exp(\mathrm{i}\omega t)\mathrm{d}t \\
&+ \int_{0}^{\infty} [p(\boldsymbol{k},t;\boldsymbol{k}',0) - p(\boldsymbol{k})p(\boldsymbol{k}')] \exp(-\mathrm{i}\omega t)\mathrm{d}t \\
&= \int_{0}^{\infty} [p^{>}(\boldsymbol{k}',t|\boldsymbol{k},0) - p(\boldsymbol{k}')] \exp(\mathrm{i}\omega t)\mathrm{d}t\, p(\boldsymbol{k}) \\
&+ \int_{0}^{\infty} [p^{>}(\boldsymbol{k},t|\boldsymbol{k}',0) - p(\boldsymbol{k})] \exp(-\mathrm{i}\omega t)\mathrm{d}t\, p(\boldsymbol{k}') \\
&= P^{>*}(\boldsymbol{k}',\boldsymbol{k},\omega)p(\boldsymbol{k}) + P^{>}(\boldsymbol{k},\boldsymbol{k}',\omega)p(\boldsymbol{k}') \qquad (8)
\end{aligned}
$$

with

$$P^{>}(\boldsymbol{k},\boldsymbol{k}',\omega) = \int_{0}^{\infty} [p^{>}(\boldsymbol{k},t|\boldsymbol{k}',0) - p(\boldsymbol{k})] \exp(-\mathrm{i}\omega t)\mathrm{d}t \; . \qquad (9)$$

The corresponding PSD is given by

$$S_{xy}(\omega) = S_{xy}^{>}(\omega) + S_{yx}^{>*}(\omega) \qquad (10)$$

with

$$S_{xy}^{>}(\omega) = \iint x(\boldsymbol{k})P^{>}(\boldsymbol{k},\boldsymbol{k}',\omega)p(\boldsymbol{k}')y(\boldsymbol{k}')\mathrm{d}^3k\mathrm{d}^3k' \; . \qquad (11)$$

Only in the case of an autocorrelation function ($x = y$) does this expression reduce to the result given in [10, (22)]

$$S_{xx}(\omega) = S_{xx}^{>}(\omega) + S_{xx}^{>*}(\omega) = 2\Re\left\{S_{xx}^{>}(\omega)\right\} \; . \qquad (12)$$

With the MC method the CPD can be directly evaluated. In the case of a numerical solver (e.g. SHE) this is very CPU intensive, because the CPD has two sets of arguments. In order to reduce the complexity, conditional microscopic quantities are introduced [14, 10]

$$P_y^{>}(\boldsymbol{k},\omega) = \int P^{>}(\boldsymbol{k},\boldsymbol{k}',\omega)p(\boldsymbol{k}')y(\boldsymbol{k}')\mathrm{d}^3k' \; , \qquad (13)$$

which can be evaluated by solving the BTE in the frequency domain

$$\left\{\mathrm{i}\omega - \frac{q}{\hbar}\boldsymbol{E}\nabla\right\} P_y^{>} = \hat{W}\{P_y^{>}\} + p(\boldsymbol{k})[y(\boldsymbol{k}) - \langle y\rangle] \; , \qquad (14)$$

where the additional term on the RHS appears due to the half-sided Fourier transform (cf. (9)) of the derivative with respect to time [15]. The PSD reads now

$$S_{xy}(\omega) = \int x(\boldsymbol{k})P_y^{>}(\boldsymbol{k},\omega) + y(\boldsymbol{k})P_x^{>*}(\boldsymbol{k},\omega)\mathrm{d}^3k \; . \qquad (15)$$

Thus to calculate the PSD, (5) has to be solved for $p(\boldsymbol{k})$ and (14) for all conditional microscopic quantities of interest.

An alternative approach is based on the LBE [2]

$$\left\{\frac{\partial}{\partial t} - \frac{q}{\hbar}\boldsymbol{E}\nabla\right\} p = \hat{W}\{p\} + \xi(\boldsymbol{k}, t) , \tag{16}$$

where the probability density itself is now a fluctuating quantity. ξ is a Langevin force with zero mean and delta-correlation in time

$$\mathrm{E}\left\{\xi(\boldsymbol{k}, \mathrm{t})\xi(\boldsymbol{k}', \mathrm{t}')\right\} = \mathrm{S}_{\xi\xi}(\boldsymbol{k}, \boldsymbol{k}')\delta(\mathrm{t} - \mathrm{t}') . \tag{17}$$

The white and symmetric PSD of the Langevin force is given by

$$\begin{aligned} S_{\xi\xi}(\boldsymbol{k}, \boldsymbol{k}') = \frac{2\Omega}{(2\pi)^3} \Bigg[\int W(\boldsymbol{k}''|\boldsymbol{k})p(\boldsymbol{k}) + W(\boldsymbol{k}|\boldsymbol{k}'')p(\boldsymbol{k}'')\mathrm{d}^3k''\, \delta(\boldsymbol{k} - \boldsymbol{k}') \\ - W(\boldsymbol{k}|\boldsymbol{k}')p(\boldsymbol{k}') - W(\boldsymbol{k}'|\boldsymbol{k})p(\boldsymbol{k}) \Bigg] \\ = S_{\xi\xi}(\boldsymbol{k}', \boldsymbol{k}) \end{aligned} \tag{18}$$

The corresponding Green's functions $G(\boldsymbol{k}, \boldsymbol{k}', \omega)$ are the solutions of a modified BTE

$$\left\{\mathrm{i}\omega - \frac{q}{\hbar}\boldsymbol{E}\nabla\right\} G = \hat{W}\{G\} + \delta(\boldsymbol{k} - \boldsymbol{k}') . \tag{19}$$

With the Wiener-Lee theorem [12] the PSD of the single-particle probability density reads

$$S_{pp}(\boldsymbol{k}, \boldsymbol{k}', \omega) = S^*_{pp}(\boldsymbol{k}', \boldsymbol{k}, \omega) = \iint G(\boldsymbol{k}, \boldsymbol{k}_1, \omega) S_{\xi\xi}(\boldsymbol{k}_1, \boldsymbol{k}'_1) G^*(\boldsymbol{k}', \boldsymbol{k}'_1, \omega)\mathrm{d}^3k_1\mathrm{d}^3k'_1 \tag{20}$$

and the PSD for x, y

$$S_{xy}(\omega) = \iint x(\boldsymbol{k}) S_{pp}(\boldsymbol{k}, \boldsymbol{k}', \omega) y(\boldsymbol{k}')\mathrm{d}^3k\mathrm{d}^3k' . \tag{21}$$

Similar to the case of the BTE, Green's functions are defined for the microscopic quantities to reduce the CPU time

$$G_x(\boldsymbol{k}', \omega) = \int x(\boldsymbol{k}) G(\boldsymbol{k}, \boldsymbol{k}', \omega)\mathrm{d}^3k , \tag{22}$$

which can be calculated directly with the adjoint technique [16]. S_{xy} reads now

$$S_{xy}(\omega) = \iint G_x(\boldsymbol{k}, \omega) S_{\xi\xi}(\boldsymbol{k}, \boldsymbol{k}') G^*_y(\boldsymbol{k}', \omega)\mathrm{d}^3k\mathrm{d}^3k' . \tag{23}$$

Thus in contrast to (15), it is possible to separate the source of the fluctuation, which is due to particle scattering (cf. (18)), and the propagation of the fluctuations, which are described by the Green's functions. In this sense the LBE approach gives more physical insight into noise than the BTE. The numerical effort is in both cases the same and much smaller than for a MC simulation. Furthermore, the CPU time does not depend on the frequency and even zero frequency is accessible. This is important for the investigation of low-frequency noise, which is almost impossible to simulate by MC.

The above presented approaches are not limited to the calculation of electronic noise. They can be also used to investigate the efficiency of different MC algorithms [17, 18]. Under stationary conditions MC results are often averaged over time. The variance of such an average is given with Macdonald's function by

$$\sigma_x^2(T) = \int_0^\infty S_{xx}(\omega)\frac{1-\cos(\omega T)}{\pi(\omega T)^2}\mathrm{d}\omega \approx \frac{S_{xx}(0)}{2T}\,, \tag{24}$$

where T is the total simulation time and the approximation holds for a sufficiently large T. Therefore, the efficiency of an MC algorithm is proportional to the PSD of the estimated variable.

3 Results

In Fig. 1 the PSD of the velocity is shown for undoped bulk Si. Both approaches, SHE and MC yield the same results, where SHE is at least two orders of magnitude faster than MC, although (19) was solved for 41 different frequencies and a large set of microscopic quantities. At lower frequencies the advantage of the SHE approach becomes increasingly larger. Results for a simulation including a generation/recombination process with a life time of 5ns are shown in Fig. 2. Due to the life time of 5ns the PSD depends even below 1GHz on the frequency. No

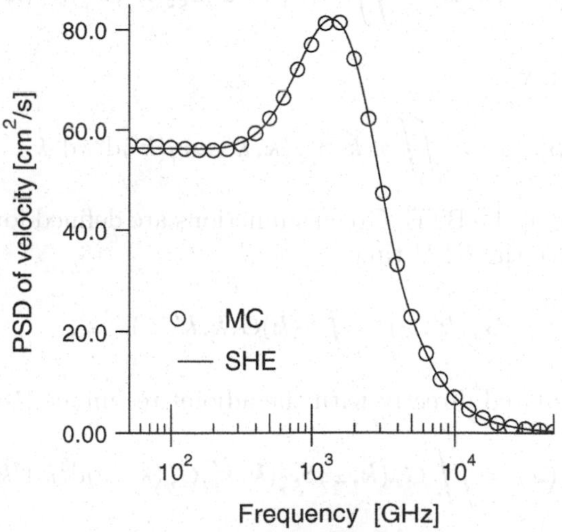

Fig. 1. Power spectral density of the longitudinal velocity fluctuations for an electric field of 30kV/cm in undoped Si

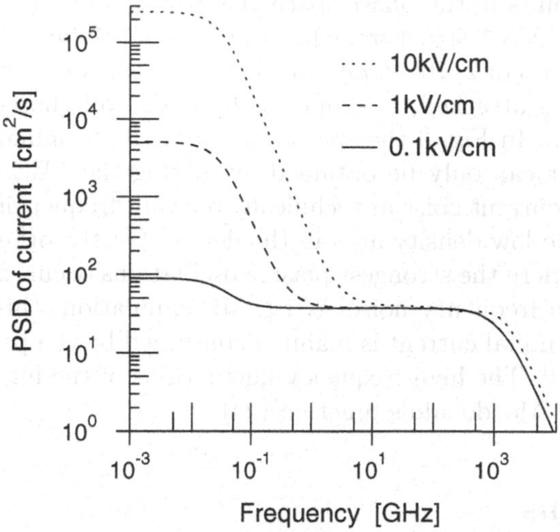

Fig. 2. Power spectral density of the normalized current fluctuations for three electric fields for bulk Si doped with $10^{17}/\text{cm}^3$ and a generation/recombination process with a life time of 5ns at 300K as obtained by SHE

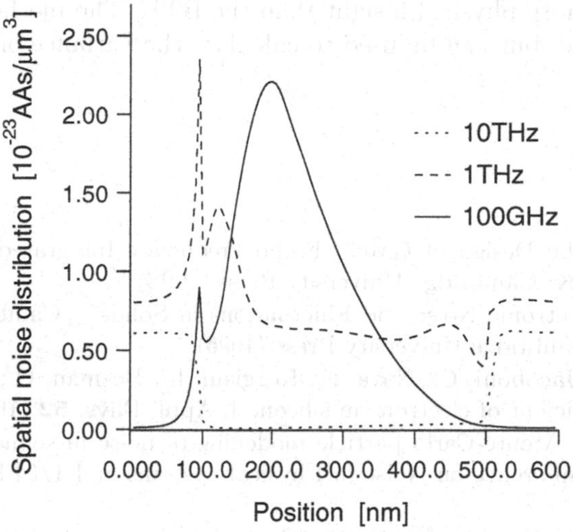

Fig. 3. Spatial distribution of the terminal current noise for the N^+NN^+ structure at 300K and 1.0V bias

MC simulations were performed, because these would entail CPU times many orders of magnitude larger than the one for Fig. 1, which was already four days.

In the case of device simulations the problems of the MC method for frequencies below 100GHz become worse, but also the CPU time of SHE increases, because the additional real space dimensions lead to a huge increase in the number of grid points in the phase space. Therefore, only 1D results are shown for SHE. A 1D N^+NN^+ structure, which consists of a 400nm long region with a doping concentration of $2 \cdot 10^{15}/cm^3$, to which on both sides 100nm long regions of $5 \cdot 10^{17}/cm^3$ are attached, is simulated by SHE and the current flows into the $\langle 100 \rangle$ direction. In Fig. 3 the spatial origin of the terminal current noise is shown. This result can only be obtained by solving the LBE. It clearly shows that the terminal current noise at technically relevant frequencies below 100GHz originates from the low density area in the device. On the other hand, the high density regions, where the strongest plasma oscillations occur, dominate the less relevant ultra-high frequency noise. W.r.t. MC simulations this means that the variance of the terminal current is mainly determined by the particles in the low density regions [13]. The high frequency fluctuations of the high density regions are filtered out by Macdonald's function (24).

4 Conclusions

It has been demonstrated that the MC method is not necessarily the best method for noise calculation. Especially in the case of technically relevant frequencies below 100GHz MC becomes prohibitively CPU intensive. In this case SHE is a viable alternative with the additional benefit that the LBE can be solved, which provides more physical insight than the BTE. The method is not limited to electronic noise, but can be used to calculate the variance of all kinds of MC variables.

References

1. Lee, T. H.: The Design of CMOS Radio Frequency Integrated Circuits, 1st ed. Cambridge, UK: Cambridge University Press (1998)
2. Kogan, S.: Electronic Noise and Fluctuations in Solids. Cambridge, New York, Melbourne: Cambridge University Press (1996)
3. Brunetti, R., Jacoboni, C., Nava, F., Reggiani, L., Bosman, G., Zijlstra, R. J. J.: Diffusion coefficient of electrons in silicon. J. Appl. Phys. **52** (1981) 6713–6722
4. Moglestue, C.: Monte-Carlo particle modelling of noise in semiconductors. In International Conference on Noise in Physical Systems and 1/f Fluctuations (1983) 23–25
5. Jungemann, C., Neinhüs, B., Decker, S., Meinerzhagen, B.: Hierarchical 2–D DD and HD noise simulations of Si and SiGe devices: Part II—Results. IEEE Trans. Electron Devices **49** (2002) 1258–1264
6. Gonzalez, T., Mateos, J., Martin-Martinez, M. J., Perez, S., Rengel, R, Vasallo, B. G., Pardo, D.: Monte Carlo simulation of noise in electronic devices: limitations and perspectives. Proceedings of the 3rd International Conference on Unsolved Problems of Noise (2003) 496–503
7. Goldsman, N., Lin, C., Han, Z., Huang, C.: Advances in the spherical Harmonic-Boltzmann-Wigner approach to device simulation. Superlattices and Microstructures **27** (2000) 159–175

8. Gnudi, A., Ventura, D., Baccarani, G., Odeh, F.: Two-Dimensional MOSFET Simulation by Means of a Multidimensional Spherical Harmonics Expansion of the Boltzmann Transport Equation. Solid–State Electron. **36** (1993) 575–581
9. Jungemann, C., Graf, P., Zylka, G., Thoma, R., Engl, W. L.: New highly efficient method for the analysis of correlation functions based on a spherical harmonics expansion of the BTE's Green's function. In Proc. IWCE, Portland, Oregon (1994) 45–48
10. Korman, C. E., Mayergoyz, I. D.: Semiconductor noise in the framework of semiclassical transport. Phys. Rev. B **54** (1996) 17 620–17 627
11. Jungemann, C., Meinerzhagen, B.: A Legendre polynomial solver for the Langevin Boltzmann equation. In Proc. IWCE (2004) 22–23
12. Papoulis, A: Probability, Random Variables, and Stochastic Processes, 4th ed. McGraw–Hill (2001)
13. Jungemann, C., Meinerzhagen, B.: Hierarchical Device Simulation, ser. Computational Microelectronics, S. Selberherr, Ed. Wien, New York: Springer (2003)
14. K. M. van Vliet, K. M.: Markov approach to density fluctuations due to transport and scattering. I. Mathematical Formalism*. J. Math. Phys. **12** (1971) 1981–1998
15. Bronstein, I. N., Semendjajew, K. A.: Taschenbuch der Mathematik. Stuttgart: B. G. Teubner (1991)
16. Branin, F. H.: Network sensitivity and noise analysis simplified. IEEE Transactions on circuit theory **20** (1973) 285–288
17. Jungemann, C., Meinerzhagen, B.: Analysis of the stochastic error of stationary Monte Carlo device simulations. IEEE Trans. Electron Devices **48** (2001) 985–992
18. Jungemann, C., Meinerzhagen, B.: In-advance CPU time analysis for stationary Monte Carlo device simulations. IEICE Trans. on Electronics **E86-C** (2003) 314–319

Quasi-monte Carlo Methods for Investigating Electrostatic Properties of Organic Pollutant Molecules in Solvent

Aneta Karaivanova[1] and Nikolai A. Simonov[2,3]

[1] IPP - Bulgarian Academy of Sciences, Acad. G. Bonchev, Bl. 25A, 1113 Sofia, Bulgaria
anet@parallel.bas.bg
http://parallel.bas.bg/~anet

[2] School of Computational Science, Florida State University, Tallahassee, FL 32306, USA
simonov@csit.fsu.edu
http://osmf.sscc.ru/LabSab/nas

[3] Institute of Computational Mathematics and Mathematical Geophysics SB RAS, Novosibirsk, Russia

Abstract. The problem of computing electrostatic properties of a pollutant molecule in solvent is considered. To solve it, the 'walk-on-the-boundary' algorithm is applied. For the problems considered in the article, this Monte Carlo method has advantages when compared to random walks on spheres, balls or a grid, and is competitive to conventional computational methods. To accelerate the convergence, we study the properties of the algorithm when quasirandom sequences instead of pseudorandom numbers are used to construct the walks on the boundary. The results of numerical experiments confirm theoretical estimates of the convergence rate improvement.

1 Introduction

Electrostatics is one of the fundamental interactions that determines the structure, stability, binding, chemical and physical properties of pollutants' molecules, especially of organic origin. It is essential to take electrostatics into account when investigating numerically the mechanisms and strength of interaction of pollutant with the environment and biological structures. One of the most important mathematical problems that arises in the process of investigating reactive properties of large organic molecules in solvent is the problem of calculating their internal electrostatic energy, point values of the potential and the electrostatic forces that a molecule is subjected to in the dielectric medium.

There are different approaches to describing the electrostatic field on the molecular level. One of the possible and widely used models is a continuum model [1]. For a given charge distribution $\rho(x)$, the electrostatic potential is determined as a solution to Poisson's equation

$$-\nabla\epsilon\nabla u(x) = \rho(x) \ , \ x \in \mathbb{R}^3 \ , \tag{1}$$

I. Lirkov, S. Margenov, and J. Waśniewski (Eds.): LSSC 2005, LNCS 3743, pp. 172–180, 2006.

where ϵ is a position-dependent permittivity. In biophysics applications, the geometry of a problem is taken into account by thinking of a molecule as a cavity, G, with point charges and constant ϵ inside. The exterior is considered as a dielectric medium with different permittivity and some charge distribution. Clearly, certain boundary conditions on the surface of a molecule have to be added.

The common approach to solving these kinds of problems is the finite-difference technique (see e.g. [1, 2]); boundary element and finite element methods are successfully used as well. It is essential to note, however, that in the molecular electrostatics computations there is often no need in finding the entire solution. Usually, only point values of the electrostatic potential and the electric field have to be computed. This clears the way to application of competitive Monte Carlo algorithms as alternative methods of treating computationally boundary value problems. For elliptic equations, the most efficient and commonly used are the walk on spheres (WOS) method [3] and random walk on the boundary algorithm [4, 5].

It is worth noting, however, that Monte Carlo methods were well established for the equations with constant coefficients and classical boundary conditions. In molecular electrostatics problems, we have to deal with coupling different equations through continuity boundary conditions. Recently, we proposed some new Monte carlo techniques for solving such boundary-value problems [6, 7]. Our approach was based on the combination of WOS, walk-in-subdomains methods [8] and finite-difference approximation of the boundary condition. Later, we implemented more subtle and efficient treatment of boundary conditions. The results will be published soon. We found out that the efficiency of WOS-based Monte Carlo algorithm essentially depends on the charge density in the exterior of the molecule. With no charges present and convex G, the random walk on the boundary algorithm becomes more efficient [9, 10]. In these papers, we also investigated the effect of using quasirandom sequences instead of pseudorandom numbers. We found out that walk-on-the-boundary algorithm is very very suited for such transformation and that the estimate's convergence improved as well as we expected.

In this paper, we continue our investigation of quasirandom walks on the boundary for solving boundary-value problems of molecular electrostatics. We consider the non-convex case and different versions of the algorithm. For the geometries with low probability of multiple boundary intersections, we found out that the random walk-on-the-boundary algorithm and its quasirandom counterpart work well and efficiently.

The paper is organized as follows. The formulation of the problems are given in §1. The random walk on the boundary method is described in §2. In §3 the use of quasirandom sequences is discussed, and some numerical results are presented.

2 Formulation of the Problem

In calculating the internal electrostatic energy of a molecule, we accept the model which is commonly used in bio-molecular electrostatics computations [1]. The molecule in question is considered as a union of a large number of intersecting balls (atoms) that constitutes a compact set $G \in \mathbb{R}^3$. Every spherical atom has its electrical charge, q_m, which is positioned at its center, x_m, and r_m is the radius of this atom (ball). Hence, the electrostatic potential, $u(x)$, satisfies Poisson's equation (1) inside G for the particular charge density $\rho(x) = \sum_{m=1}^{M} q_m \delta(x - x_m)$. Here, the dielectric permittivity, $\epsilon = \epsilon_i$, is constant.

Usually, the molecule is surrounded by some dielectric (e.g., water). The classical approach is to treat the exterior medium as continuous with some constant permittivity, ϵ_e. Assume that there are no dissolved ions outside the molecule. This means that the electrostatic potential, u, satisfies the Laplace equation ((1) with zero charge density ρ) in $G_1 \equiv \mathbb{R}^3 \setminus \bar{G}$. The continuity conditions on the boundary that couple the potential values inside and outside the molecule are:

$$u_i(y) = u_e(y) \ , \ \epsilon_i \frac{\partial u_i}{\partial n(y)} = \epsilon_e \frac{\partial u_e}{\partial n(y)} \ , \ y \in \partial G \ . \tag{2}$$

Here, for convenience, we denote u_i as the solution to (1) in the interior of G, and u_e as the solution in the exterior, G_1. We also assume that $u_e(x) \to 0$ as $|x|$ goes to infinity.

In the linear case, the electrostatic free energy of the molecule is given by [1]

$$E = \frac{1}{2} \sum_{m=1}^{M} u_m q_m \ ,$$

where u_m is the regular part of the electrostatic potential at the center of the m-th atom. This means that in calculation of E we take $u_m = u^{(0)}(x_m)$. Here, $u^{(0)}$ comes from the available explicit decomposition of the potential inside G:

$$u(x) = u^{(0)}(x) + g(x) \ , \tag{3}$$

where $g(x) = \sum_{m=1}^{M} \frac{q_m}{4\pi\epsilon_i} \frac{1}{|x - x_m|}$.

With the assumption that ∂G is smooth enough, it is possible to represent the regular part of the solution in the form of the single-layer potential [11, 12]

$$u^{(0)}(x) = \int_{\partial G} \frac{1}{2\pi} \frac{1}{|x - y|} \mu(y) d\sigma(y) \ . \tag{4}$$

Taking into account boundary conditions (2) and discontinuity properties of the single-layer potential's normal derivative [11], we arrive at the integral equation for the unknown density, μ:

$$\mu = -\lambda_0 \mathcal{K} \mu + f \ , \tag{5}$$

which is valid almost everywhere on ∂G. Here, $\lambda_0 = \frac{\epsilon_e - \epsilon_i}{\epsilon_e + \epsilon_i}$, and the kernel of the integral operator $\mathcal{K}$ is $\frac{1}{2\pi} \frac{\cos \phi_{yy'}}{|y - y'|^2}$, where $\phi_{yy'}$ is the angle between the external normal vector $n(y)$ and $y - y'$. The free term of this equation equals $\lambda_0 \frac{\partial g}{\partial n(y)}$, and it can be computed analytically. Since $\lambda_0 < 1$, and the spectral radius of $\mathcal{K}$ is equal to 1, the Neumann series for (5) converges (see, e.g., [11, 12]), and it is possible to calculate the solution as

$$u^{(0)}(x) = \sum_{i=0}^{\infty} \left(h_x, (-\lambda_0 \mathcal{K})^i f\right) , \tag{6}$$

where $h_x(y) = \frac{1}{2\pi} \frac{1}{|x-y|}$. Usually, however, $\epsilon_e \gg \epsilon_i$ and, hence, $1 - \lambda_1 = \frac{2\epsilon_i}{\epsilon_e + \epsilon_i} \ll 1$. Here, $\lambda_1 = -\lambda_0$ is the eigenvalue of the integral operator in (5) with the largest modulus. This means that convergence in (6) is rather slow.

To speed up the convergence in (6), we apply the method of spectral parameter substitution (see, e.g., [13], and [5] for Monte Carlo algorithms based on this method). This means that we consider the parameterized equation $\mu_\lambda = \lambda(-\lambda_0 \mathcal{K})\mu_\lambda + f$ and analytically continue its solution given by the Neumann series for $|\lambda| < 1/\lambda_0$. This goal can be achieved by substituting in λ its analytical expression in terms of another complex parameter, η, and representing μ_λ as a series in powers of η.

In this particular case, it is possible to use the substitution $\lambda = \frac{2\eta}{\lambda_0(1-\eta)} \equiv \chi(\eta)$, and hence

$$u_0(x) = \sum_{i=0}^{n} l_i^{(n)} (-\lambda_0)^i \left(h_x, \mathcal{K}^i f\right) + O(q^{n+1}) , \tag{7}$$

where $q = \frac{\lambda_0}{2+\lambda_0} = \frac{\epsilon_e - \epsilon_i}{3\epsilon_e - \epsilon_i} < \frac{1}{3}$, and $l_i^{(n)} = \sum_{j=i}^{n} \binom{i-1}{j-1} (2|\lambda_0|)^i q^j$. The rate, q, of geometric convergence of the transformed series in powers of η at the point $\eta_0 = \chi^{-1}(1)$ is determined by the ratio of $|\eta_0|$ and $L = \min_i |\chi^{-1}(-\lambda_i/\lambda_0)|$. Here, $1/\lambda_i$ are eigenvalues of $\mathcal{K}$ and $L = 1$ [11,13].

Given a desired computational accuracy, we calculate the number of terms needed in (7). Thus, the problem reduces to computing a finite number of multidimensional integrals.

3 Random Walks on the Boundary

To construct Monte Carlo estimates for $u_0(x)$, it is sufficient to calculate the integral functionals $I_i(x) = \left(h_x, \mathcal{K}^i f\right)$ of iterations of the integral operator. Here, the domain of integration is $[\partial G]^{i+1}$.

Let G be a convex domain. In this case, the kernel, $k(y, y')$, of $\mathcal{K}$ corresponds to the uniform in a solid angle distribution of the point y as viewed from the point y'. This means that the most convenient way to construct a random estimate for double integral $(h_x, \mathcal{K}f)$ is to use for choosing y' some density p_0 that conforms with f, and to sample y in accordance with the conditional density $p(y' \to y) = k(y, y')$. This construction leads to using so-called 'direct' estimates for $I_i(x)$ [5]. Therefore

$$I_i(x) = \mathbb{E} Q_i h_x(y_i) \,, \tag{8}$$

where $Y = \{y_0, y_1, \ldots\}$ is the Markov chain of random points on the boundary ∂G, with the initial density p_0 and transition density $p(y_i \to y_{i+1}) = \frac{1}{2\pi} \frac{\cos \phi_{y_{i+1} y_i}}{|y_{i+1} - y_i|^2}$. Here, random weights are $Q_i = Q_0 = \frac{f(y_0)}{p_0(y_0)}$, $i = 1, 2, \ldots, n$. Therefore, $O(q^{n+1})$-biased estimate for $u^{(0)}$ is

$$\theta = \sum_{i=0}^{n} l_i^{(n)} (-\lambda_0)^i Q_i h_x(y_i) \,. \tag{9}$$

Consider now a non-convex molecule G. In this case, first, there is a possibility of multiple intersections of a straight line with the boundary, and, second, $\cos \phi_{y_{i+1} y_i}$ can be negative. There are two different approaches to overcome this difficulty. Let $n_i(y_i)$ be the number of these crossings, y_i excluded. The first solution is to choose the next point of Markov chain y_{i+1} randomly from n_i intersections. This corresponds to the transition density $p_1 = \frac{1}{n_i}|p|$ and therefore $Q_{i+1} = Q_i \; n_i \; \mathrm{sign}(\cos \phi_{y_{i+1} y_i})$. The second approach is to use a branching Markov chain, thus taking into account all intersections. In this case the transition density is $p_2 = |p|$, weights $Q_{i+1} = Q_i \; \mathrm{sign}(\cos \phi_{y_{i+1} y_i})$ can only change their signs but not their moduli, and every point y_i gives birth to $n_i(y_i)$ points in the next generation.

There is no exact criterion that could determine which of these two approaches is more efficient. On the one hand, the random choice of the next point leads to a fixed length of the Markov chain, which results in smaller computational time needed to construct one sample of the solution's estimate. On the other hand, the weight factor in this approach can be as large as $\prod_1^{n+1} n_i$, and this could result in large variance. The estimate based on construction of a branching Markov chain has the opposite properties. Branching could result in small variance, but it also requires longer computations needed to obtain an estimate's sample.

In both approaches, the construction of the Markov chain, Y, is based on its geometrical interpretation. Given a point, y_i, we simulate a random isotropic direction ω_i and find the next point y_{i+1} as the intersection of this direction with the boundary surface ∂G. It is well known that different procedures can be used to choose $\omega_i = (\omega_{i,1}, \omega_{i,2}, \omega_{i,3})$. We consider the procedure based on the direct simulation of the longitudinal angle. Normally, an acceptance-rejection method would be used. But since we plan to use quasirandom numbers, this is inadvisable (see, e.g. [14]). So we use the following algorithm: $\omega_{i,3} = 1 - 2\alpha_{i,1}$, $\varphi_i = 2\pi\alpha_{i,2}$, $d = \sqrt{1 - \omega_{i,3}^2}$, $\omega_{i,1} = \sin \varphi_i / d$, $\omega_{i,2} = \cos \varphi_i / d$, where the α_i are standard uniform pseudorandom numbers in the unit interval.

4 Quasirandom Walks on the Boundary

In this section we discuss how to use quasirandom numbers for solving the boundary-value problem (1), (2). To construct Monte Carlo estimates, in §1, we reformulated the original problem into the problem of solving integral equation (5). So, in order to make use of these representations when constructing quasirandom estimates, we have to refer to a Koksma-Hlawka type inequality for integral equations, [15]:

$$\left| u[Y^*] - \frac{1}{N} \sum_{1}^{N} \theta^*[Y^*] \right| \leq V(\theta^*)\, D_N^*(Q) \,, \tag{10}$$

where Q is a sequence of quasirandom vectors in $[0,1)^s$, $s = d \times T$, and d is the number of QRNs in one step of a random walk, T is the maximal number of steps in a single random walk, and θ^* corresponds to an estimate $\theta[Y]$ based on the random walk Y^* generated from Q by a one-to-one map.

This inequality ensures convergence when θ^* is of bounded variation in the Hardy-Krause sense, which is a very serious limitation. But even when this condition is satisfied, the predicted rate of convergence is very pessimistic due to the high (and, strictly speaking, possibly unbounded) dimension of the quasirandom sequence in the general Monte Carlo method for solving these integral equations, (e.g. (6)). To avoid this limitation, we consider variants of the method with each random walk having fixed length. Clearly, the smaller the dimension of Q, the better the rate estimate in (10).

Guided by this reasoning, we used the representation (7) and the correspondent random estimate θ, to construct quasirandom solutions to the original boundary-value problems.

It is essential to note that despite the improved rate of convergence of our quasirandom-based calculations, constants in the error estimates are hard to calculate. On the contrary, the statistical nature of Monte Carlo solutions makes it possible to determine confidence intervals almost exactly.

5 Numerical Tests

We performed numerical tests with the simple model problem of finding the potential and its derivatives (electric field components) for a molecule in the shape of the dumbbell made of two spheres. To solve this problem, we fix the length of series to be $n = 6$ that provides a 0.1% bias. For a convex domain, the number of random (or quasirandom) points needed to construct the estimate will also be fixed: $2(n + 1)$. In non-convex domains the length (dimension) of random (or quasirandom) sequence varies for different Markov chain trajectories. We can only give an upper estimate for this number. In the algorithm with random choice of the next point in the Markov chain the dimension of a sequence is less than or equal to $3(n + 1)$. In the branching random walk on boundary, the number of points depends on the geometry of the domain. Let

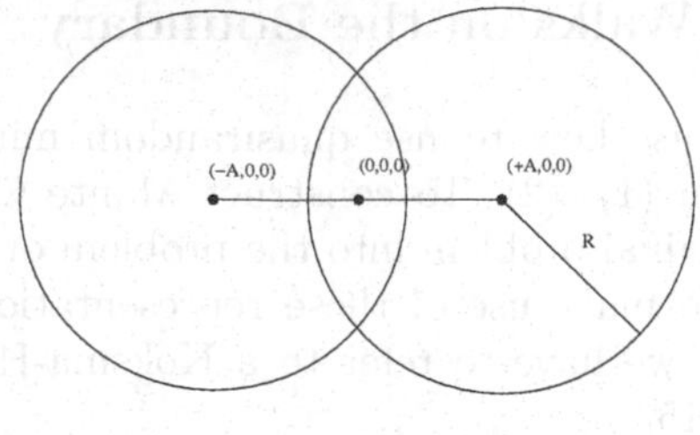

Fig. 1. Test problem

Table 1. Approximate values of energy, potential and X derivative for different number of random trajectories

N	Energy	Potential at (A,0,0)	X derivative at (A,0,0)
10^2	-0.3861	-0.2880	-0.5601
10^3	-0.3915	-0.4007	-0.4508
10^4	-0.3421	-0.3424	-0.3107
10^5	-0.3436	-0.3435	-0.3309
10^6	-0.3441	-0.3419	-0.3340
10^7	-0.3432	-0.3426	-0.3346
statistical error ($N = 10^7$)	0.0003	0.0008	0.0012
Exact values	-0.3434	-0.3434	-0.3343

$n_{max}+1$ be the maximal possible number of intersections of a straight line with the boundary. Then the length of a sequence is bounded from above by n_{max}^{n+1}.

To use quasirandom sequences efficiently we chose the different approach. Every direction in one generation was sampled using the same pair of numbers. This introduced correlation between the points but did not compromised the convergence.

6 Conclusions

In this paper we presented a successful application of quasirandom sequences in constructing the walk on boundary Markov chain that was used for solving non-standard elliptic boundary-value problem in the whole space with the boundary conditions given on the non-convex surface. The problem was reduced to calculating a small number of multidimensional integrals. This is the key point for the successive use of quasirandom sequences. The computational results with Sobol sequences show the better rate of convergence than the statistical one that was obtained by using pseudorandom numbers. We found out that the accuracy of the quasirandom walk-on-the-boundary method is better and the advantage of this method is significant.

Table 2. Approximate values of energy, potential and X derivative for different number of quasirandom trajectories

N	Energy	Potential at (A,0,0)	X derivative at (A,0,0)
10^2	-0.3545	-0.3747	-0.3163
10^3	-0.3348	-0.3359	-0.3408
10^4	-0.3426	-0.3392	-0.3415
10^5	-0.3443	-0.3434	-0.3362
10^6	-0.3436	-0.3434	-0.3352
10^7	-0.3433	-0.3433	-0.3343
Exact values	-0.3434	-0.3434	-0.3343

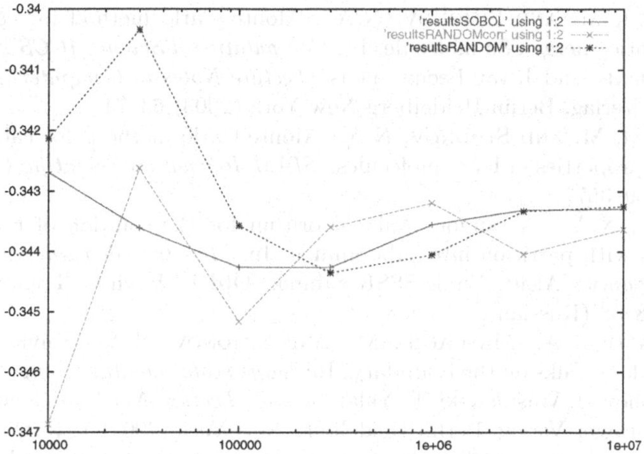

Fig. 2. Dependence of energy on the number of simulated samples for different choices of sequences

Results of our computational experiments with a model molecule confirm that Monte Carlo and quasi-Monte Carlo methods can be successfully applied to calculating electrostatic properties of pollutant molecules in solvent. The choice between WOS and walk-on-the-boundary algorithms depends on the form of the molecule and the presence of dissolved ions outside the molecule. It is also worth noting that WOS-based algorithm is, in general, more simple to implement numerically. This, however, depends on the geometry of the boundary.

Acknowledgement

The work of A. Karaivanova is supported by the Ministery of Education and Science of Bulgaria under Grant No. I1405/04.

References

1. DAVIS M.E. AND MCCAMMON J.A.: Electrostatics in biomolecular structure and dynamics. *Chem. Rev.* **90** (1990) 509–521.
2. NICHOLS A. AND HONIG B.: A rapid finite-difference algorithm, utilizing successive over-relaxation to solve the Poisson-Boltzmann equation. *J. Comput. Chem.* **12** (1991) 435–445.
3. MULLER M.E.: Some continuous Monte Carlo methods for the Dirichlet problem. *Ann. Math. Statist.* **27** (1956) 569–589.
4. SABELFELD, K.K.: Vector algorithms of the Monte Carlo method for solving systems of 2-nd order elliptic equations and Lame equation. *Doklady USSR Acad.Sci.* **262** (1982) 1076–1080 (Russian).
5. SABELFELD, K.K. AND SIMONOV, N.A.: *Random walks on boundary for solving PDEs.* VSP, Utrecht, The Netherlands (1994).
6. MASCAGNI, M. AND SIMONOV, N.A.: Monte Carlo method for calculating the electrostatic energy of a molecule. In: *Computational Science ICCS 2003*, G. Goos, J. Hartmanis, and J. van Leeuwen eds., *Lecture Notes in Computer Science*, **2657**. Springer-Verlag, Berlin Heidelberg New York (2003) 63–74.
7. MASCAGNI, M. AND SIMONOV, N.A.: Monte Carlo methods for calculating some physical properties of large molecules. *SIAM Journal on Scientific Computing.* **26** (2004) 339–357.
8. SIMONOV, N.A.: A random walk algorithm for the solution of boundary value problems with partition into subdomains. In: *Metody i algoritmy statisticheskogo modelirovanija*, Akad. Nauk SSSR Sibirsk. Otdel., Vychisl. Tsentr, Novosibirsk (1983) 48–58 (Russian).
9. KARAIVANOVA, A., MASCAGNI, M., AND SIMONOV, N.A.: Solving BVPs using quasirandom walks on the boundary. In: *Large scale scicntific computing*, I. Lirkov, S. Margenov, J. Waśniewski, P. Yalamov eds., *Lecture Notes in Computer Science*, **2907**. Springer-Verlag, Berlin Heidelberg New York (2004) 162–169.
10. KARAIVANOVA, A., MASCAGNI, M. AND SIMONOV, N.A.: Parallel Quasirandom Walks on the Boundary. *Monte Carlo Methods and Applications.* **10**, 3–4 (2004) 311–320.
11. GÜNTER, N.M.: *Potential theory, and its applications to basic problems of mathematical physics.* New York, Ungar (1967).
12. GOURSAT, E.: *A course in mathematical analysis. Volume III, part 2.* New York, Dover Publications (1964).
13. KANTOROVICH, L.W., KRYLOV, V.I.: *Approximate Methods of Higher Analysis.* Interscience, New York (1964).
14. JÄCKEL, P.: *Monte Carlo Methods in Finance.* John Wiley & Sons (2002).
15. CHELSON, P.: *Quasi-Random Techniques for Monte Carlo Methods.* Ph.D. dissertation, The Claremont Graduate School (1976).

Parallel Realization of Grid-Free Monte Carlo Algorithm for Boundary Value Problems

R.Y. Papancheva*

Institute for Parallel Processing, Bulgarian Academy of Sciences,
Acad. G. Bonchev, bl. 25A, 1113 Sofia, Bulgaria
rumi@parallel.bas.bg

Abstract. In many areas of the science there is a need to evaluate a functional of the solution of a given problem directly without computing the solution itself. The problem here is a linear functional of the solution of an elliptic boundary value problem to be estimated. Such kind of problems are similar to the air pollution problems in environmental sciences, where a rough estimate of the solution is acceptable. For practical computations it means that the relative error is about 5% – 10%. To solve this problem a grid–free Monte Carlo (MC) algorithm is used. The algorithm makes use of a Monte Carlo procedure called "Walk on the balls". Here we consider parallel realizations of the considered grid–free MC algorithm. Various numerical results are obtained by the implementation of the proposed parallel algorithms on several high performance machines: IBM +p690 Regata system and Sun Fire 15K server. One can see that the efficiency of the proposed parallel algorithm is close to 100%.

1 Formulation of the Problem

Consider the functional $J(u)$:

$$J(u) \equiv (g, u) = \int_{\Omega} g(x)u(x)dx, \tag{1}$$

where $\Omega \subset R^3$ and $x = (x_1, x_2, x_3) \in \Omega$ is a point in the Euclidean space R^3. The functions $u(x)$ and $g(x)$ belong to the Banach space X and to the adjoint space X^*, respectively. Let $X = L_1(\Omega)$. Then $X^* = L_\infty(\Omega)$ [4].

The problem under consideration consists of the calculation of the functional $J(u)$, where $u(x)$ is the solution of the following boundary value problem:

$$Mu(x) = -\Phi(x), \qquad x \in \Omega, \quad \Omega \subset R^3 \tag{2}$$

$$u(x) = \psi(x), \qquad x \in \partial\Omega, \tag{3}$$

* The author would like to acknowledge the support of the European Commission's Research Infrastructures activity of the Structuring the European Research Area program, contract number RII3-CT-2003-506079 (HPC-Europa). This work is supported also by grant I-1402/2004 from the Bulgarian NSF.

I. Lirkov, S. Margenov, and J. Waśniewski (Eds.): LSSC 2005, LNCS 3743, pp. 181–188, 2006.

and the operator M is defined by

$$M = \sum_{i=1}^{3} \left(\frac{\partial^2}{\partial x_i^2} + b_i(x) \frac{\partial}{\partial x_i} \right) + c(x). \qquad (4)$$

If in the closed domain $\overline{\Omega} \in A^{(1,\lambda)}$ the coefficients of the operator M satisfy the conditions: $b_i, i = 1,2,3,\ c \in C^{(0,\lambda)}(\overline{\Omega})$, $c \leq 0$ and $\Phi \in C^{(0,\lambda)}(\Omega) \bigcap C(\overline{\Omega})$, $\psi \in C(\partial\Omega)$, the problem (2) - (3) has an unique solution $u(x) \in C^2(\Omega) \bigcap C(\overline{\Omega})$ [1]. Definitions for the classes $A^{(k,\lambda)}$ and $C^{(k,\lambda)}$ can be found in [8].

The MC algorithm used here is based on a local integral representation of the solution in the form $u(x) = Ku(x) + f(x)$. From Green's formula when the Levy's function is used the local integral representation of the solution of our problem can be obtained in the following form [5]:

$$u(x) = \int_{B(x)} M_y^* L_p(y,x) u(y) dy + \int_{B(x)} L_p(y,x) \Phi(y) dy. \qquad (5)$$

$B(x)$ is the maximal ball with center at the point x, lying in the domain Ω. $L_p(y,x)$ is the Levy's function [8] of the problem considered. Its explicit form is:

$$L_p(y,x) = \mu_p(R) \int_r^R \left(\frac{1}{r} - \frac{1}{\rho} \right) p(\rho) d\rho,$$

where $r =\mid x - y \mid$, $\mu_p(R) = \frac{1}{4\pi q_p(R)}$, $q_p(R) = \int\limits_0^R p(\rho) d\rho$, R is the radius of $B(x)$, and $p(r)$ is a density function.

In the integral equation (5) M^* is the adjoint operator to the operator M. The components of the vector-function $\mathbf{b}(x)$ are assumed to satisfy the conditions given above and $div\mathbf{b}(x) = 0$. The Levy's function and its derivatives vanish on the boundary. With these conditions satisfied it is proved that the adjoint operator has the following form [2]: $M_y^* L_p(y,x) = \mu_p(R) \frac{p(r)}{r^2} - \mu_p(R) c(y) \int\limits_r^R \frac{p(\rho)}{\rho} d\rho + \frac{\mu_p(R)}{r^2} \left[c(y) r + \sum\limits_{i=1}^{3} b_i(y) \frac{y_i - x_i}{r} \right] \int\limits_r^R p(\rho) d\rho.$

2 Monte Carlo Algorithm

The MC estimator with mathematical expectation equal to $J(u)$ is

$$\Theta[g] = \frac{g(\xi_0)}{\pi(\xi_0)} \sum_{j=0}^{\infty} Q_j f(\xi_i), \qquad (6)$$

where $Q_0 = 1, Q_j = Q_{j-1} k(\xi_{j-1}, \xi_j) / p(\xi_{j-1}, \xi_j)$,

$$f(x) = \begin{cases} \int\limits_{B(x)} L_p(y,x) \Phi(y) dy, & x \in \Omega \backslash \partial\Omega, \\ \psi(x), & x \in \partial\Omega, \end{cases}$$

and $\xi_0, \xi_1, ...$ is a Markov chain in Ω with initial density function $\pi(x)$ and transition densities $p(x,y)$, which are tolerant[1] to $g(x)$ and $k(x,y)$, respectively. We have to calculate the value: $\zeta_N = 1/N \sum_{s=1}^{N} \Theta[g]_s$, where $\Theta[g]_s$ is the value of the random variable $\Theta[g]$ obtained over the s-th trajectory. The Monte Carlo procedure called "Walk on the balls" is used for the construction of these Markov's chains. The process starts at point $\xi_0 = x \in \Omega$, which is chosen correspondingly with some initial density function $\pi(x)$. The next random point is determined with transition density function $p(x,y)$. In [2] it was proved that the kernel, $k(x,y) = M_y^* L_p(y,x)$, of the integral equation (5) inside the ball $B(x)$ can be used as a transition density in the Markov chain when it is non-negative. This condition is satisfied in case when the density function is $p(r) = e^{-kr}$, $k = b^* + Rc^*$, where $b^* = \max_{x \in \Omega} \mid \mathbf{b}(x) \mid$, and R is the radius of the maximal ball, lying inside Ω [6]. The function $p(x,y)$ in spherical coordinates can be expressed as:

$$p(r, \mathbf{w}) = \frac{\sin\theta}{4\pi} \frac{p(r)}{q_p(R)} \overline{p}_r(\mathbf{w}),$$

where $\overline{p}_r(\mathbf{w}) = 1 + \left[\frac{|\mathbf{b}(x+r\mathbf{w})|\cos(\mathbf{b},\mathbf{w}) + c(x+r\mathbf{w})r}{p(r)}\right] \int_r^R p(\rho)d\rho - \frac{c(x+r\mathbf{w})r^2}{p(r)} \int_r^R \frac{p(\rho)}{\rho} d\rho$.
For simulating random variable with density function $\overline{p}_r(\mathbf{w})$ we use the selection algorithm [2]. Since $\overline{p}_r(\mathbf{w}) \leq 1 + \frac{b^*}{p(r)} \int_r^R p(\rho)d\rho = h(r)$, the function $h(r)$ can be used as a majorant.

To ensure the convergence of the ball process, an ε-strip of the boundary is introduced and the process terminates when the point falls into it. Here follows the algorithm for one random walk [2]. This algorithm is presented first by Sipin [5]. The ball process is similar to the well known spherical process [3, 7, 9].

Algorithm 1

1. **Calculate** *the radius* $R(x)$.

2. **Sample** *the jump* r *with density* $p(r)/q_p(R)$.

3. **Calculate** *the function* $h(r)$.

4. **Compute** *the independent realizations* $\mathbf{w_j}$ *of a unit isotropic vector in* $\mathbf{R}^3$.

5. **Compute** *the independent realizations* γ_j *of a uniformly distributed random variable in the interval* $[0,1]$.

6. **Repeat** *the steps 4 and 5 until the parameter* j_0 *is defined from the condition:* $j_0 = \min\{j : h(r)\gamma_j \leq \overline{p}_r(\mathbf{w_j})\}$. *The random vector* $\mathbf{w_{j_0}}$ *has the density* $\overline{p}_r(\mathbf{w})$.

7. **Calculate** *the next random point* y *by the formula* $y = x + r\mathbf{w_{j_0}}$.

8. **Stop** *the algorithm when the random process reaches the* ε*-strip of the boundary. If* $y \notin \partial\Omega_\varepsilon$ *then the algorithm has to be repeated for* $x = y$.

---

[1] The density function $p(x)$ is tolerant to the function $f(x)$, if $p(x) > 0$ in all points $x \in \Omega$, where $f(x) \neq 0$.

3 Parallel Realizations

The progress of contemporary parallel architectures coupled with new technologies leads to the development and the improvement of efficient parallel algorithms solving various problems. If A is one parallel algorithm, then let $t_p(A)$ be the CPU time of this algorithm on p processors. The productivity of the algorithm A is characterized by the parameters speed–up $S_p(A)$ and efficiency $E_p(A)$, defined as follows: $S_p(A) = t_1(A)/t_p(A), S_1(A) = 1; \quad E_p(A) = S_p(A)/p, E_1(A) = 1$. In the case when A is a MC parallel algorithm an exact value for $t_p(A)$, in general, is impossible to be obtained. Practically the mathematical expectation of the time is estimated and the speed–up of the algorithm is given by the formula: $S_p(A) = Mt_1(A)/Mt_p(A)$.

A portable parallel code is written in Fortran and C. It is based on the Message Passing Interface (MPI) - standard that solves the problem under consideration. The Scalable Parallel Random Number Generator (SPRNG) library for Monte Carlo simulation is used [13].

Concerning the MC algorithms all the particular realizations of the random variables used can be obtained on different processors without any communications between them during the computations. Communications occur at the beginning of the algorithm when the initialization of the processes have to be done, and at the end when the results obtained on every processor have to be collected to finalize the work. For this data transfer, one can use point-to-point connections or collective communications among the processors. In our case, one process needs to send the same data to all processes in the group. MPI provides the broadcast primitive to accomplish this task. Its use is more natural than the usage of the point-to-point scheme. What distinguishes the collective communication from the point-to-point communication is that it always involves every process in the specified communicator. Collective communication routines have been built by using point-to-point communication routines. A great deal of hidden communication takes place with collective communication. The performance of the two approaches depends greatly on the particular implementation and optimizations made by MPI over the concrete hardware architectures.

4 Numerical Results

The parallel grid–free MC algorithm studied and analyzed here solves the elliptic boundary value problem (2)–(4) with

$$\Phi(x) = 0, \quad x \in \Omega = [0,1]^3, \quad \psi(x) = e^{a_1x_1+a_2x_2+a_3x_3}, \quad (x_1, x_2, x_3) \in \partial\Omega,$$

and $b_1(x) = a_2a_3(x_2 - x_3)$, $b_2(x) = a_3a_1(x_3 - x_1)$, $b_3(x) = a_1a_2(x_1 - x_2)$, $c(x) = -(a_1^2 + a_2^2 + a_3^2)$, a_1, a_2, a_3 are parameters of the problem. Different values for the parameters a_i and ε-strip are considered. The functional $J(u)$ estimates the value of the solution of the considered elliptic problem at the point $x_0 = (0.5, 0.5, 0.5)$, i.e. $g(x) = \pi(x) = \delta(x - x_0)$. This problem is similar

Table 1. HPCx, $N = 12800000$, LCG

NP	Time	Speed–up	Efficiency
1	567.771107	1	1
2	284.153646	1.998113	0.999056
4	142.230000	3.991919	0.997979
8	71.082815	7.987459	0.998432
16	35.618116	15.940514	0.996282
32	17.808856	31.881391	0.996293
64	8.946657	63.461816	0.991590
128	4.692989	120.982833	0.945178
256	2.286821	248.279599	0.969842
512	1.162261	488.505475	0.954112
1024	0.600173	946.012411	0.923840

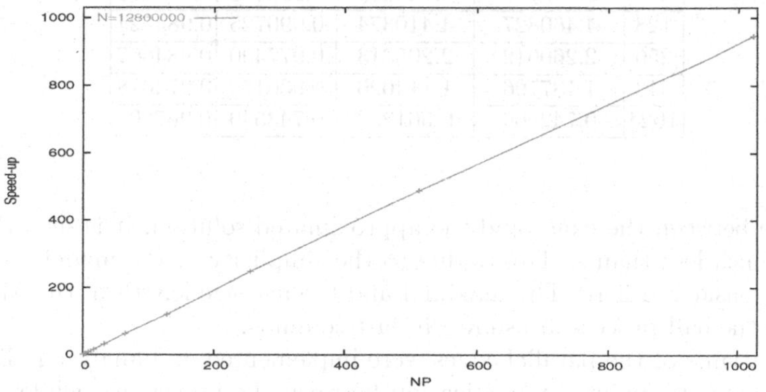

Fig. 1. Speed–up, HPCx, LCG

to problems that arise in environmental mathematics and describe the pollution transport due to advection and diffusion and take into account the deposition of pollution in areas free of emission sources.

Using the access to the high performance facilities at EPCC (Edinburgh Parallel Computer Center), various numerical results have been obtained on IBM +p690 Regata system (HPCx), each Regatta system frame consisting of 32 1.7 GHz POWER4 processors. Table 1 shows the dependence between the number of processors (NP) and the CPU time (Time). N is the number of the Markov's chains, or the number of the realizations of the random variable whose mathematical expectation coincides with the value of the functional $J(u)$. We can see the very high efficiency of the parallel algorithm. Its speed–up is shown on Figure 1 when the LCG (64 Bit Linear congruential Generator with Prime Addent) is used.

Table 2 gives some information about the accuracy of the considered algorithm. $\widetilde{u}$ is the calculated value of the solution of our problem, and "error" is the

Table 2. $u(x_0) = 1.454991$, HPCx, $N = 12800000$, $\varepsilon = 0.01$, $a = 0.25$, LCG

$\widetilde{u}$	**Error**	**maximal length**	**average length**
1.455326	0.000335	401	36.313202
1.455310	0.000318	381	36.304678

Table 3. Comparison between collective and point–to–point communication, HPCx, $N = 12800000$, LFG

NP	**Time PtoP**	**Time Coll**	**Eff PtoP**	**Eff Coll**
1	565.690130	555.904211	1	1
2	282.885402	278.569411	0.999857	0.997784
4	141.571208	139.448884	0.998949	0.996609
8	70.933412	69.536864	0.996868	0.999297
16	35.379827	34.820882	0.999316	0.997792
32	17.775869	17.748940	0.994483	0.978762
64	8.896796	8.725489	0.993493	0.995474
128	4.460827	4.410374	0.990725	0.984724
256	2.260612	2.205313	0.977490	0.984667
512	1.187796	1.114026	0.930179	0.974618
1024	0.742966	0.561282	0.743549	0.967205

difference between the exact and the approximated solution. It is seen that the error is much less than ε. This is due to the simplicity of the model numerical problem considered here. The maximal and the average length of the Markov's chains in the ball process are shown in last columns.

Two variants of the parallel codes were implemented and analyzed. The first one uses point–to–point communication between the processors, while the second one is developed on the base of the collective communications among the processors. The CPU time and the efficiency obtained are presented in Table 3 and are shown on Figure 2. The LFG is the Modified Lagged Fibonacci Generator. We can see that the productivity of the two algorithms is practically the same. When 1024 processors are used, the runtime of the algorithm decreases and slight decrease in the efficiency is seen. The explanation is that when the CPU time is small and number of the processors is large the efficiency is not so good as the most of CPU time is used for communications among the processors.

Analogous results are obtained on Sun Fire server (Lomond) with 52 UltraSPARC III Cu 1.2 GHz processors. Table 4 and Figure 3 illustrate the same tendency. Figure 4 shows the relation between runtimes of the parallel grid–free MC algorithm on HPCx and Lomond. One can see that HPCx is almost twice faster.

The proposed parallel realizations of the considered MC algorithm have very high efficiency and almost linear speed-up. Depending on the architecture and the optimizations available the appropriate communications between the processors at the initial and at the final step of the algorithm can be used.

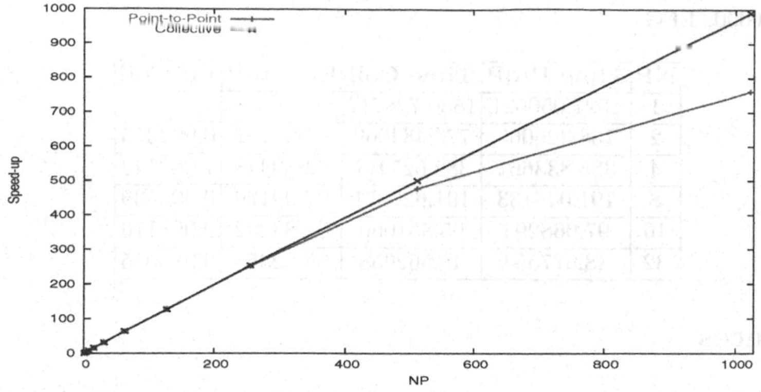

Fig. 2. HPCx, $N = 12800000$

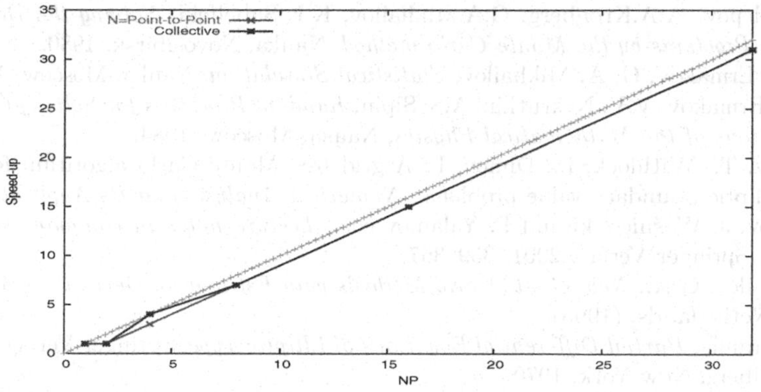

Fig. 3. Lomond, $N = 12800000$

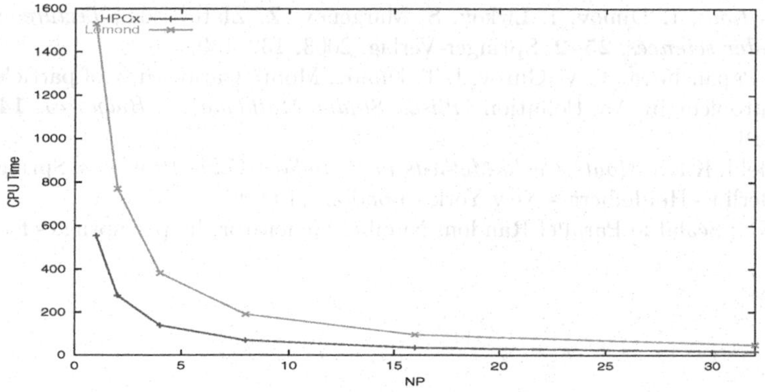

Fig. 4. Comparison of runtime on HPCx with Lomond

Table 4. Comparison between collective and point–to–point communication, Lomond, $N = 12800000$, LFG

NP	Time PtoP	Time Coll	Eff PtoP	Eff Coll
1	1527.050621	1530.728717	1	1
2	768.326065	770.484069	0.993751	0.993355
4	385.833667	382.627477	0.989448	1.000142
8	191.048133	191.620971	0.999126	0.998539
16	97.068294	95.851060	0.983232	0.998116
32	48.577554	48.562958	0.982353	0.985015

References

1. A. V. Bitzadze, *Equations of the Mathematical Physics*, Nauka, Moscow, 1982.
2. Dimov I., Gurov T., Estimates of the computational complexity of iterative Monte Carlo algorithm based on Green's function approach, *Mathematics and Computers in Simulation*, **47** 1998, 183–199.
3. B.S.Elepov, A.A.Kronberg, G.A.Mikhailov, K.K.Sabelfeld, *Solving the Boundary Value Problems by the Monte Carlo method*, Nauka, Novosibirsk, 1980.
4. S. M. Ermakov, G. A. Mikhailov, *Statistical Simulation*, Nauka, Moscow, 1982.
5. S.M. Ermakov, V.V. Nekrutkin, A.S. Sipin, *Random Processes for Solving Classical Equations of the Mathematical Physics*, Nauka, Moscow, 1984.
6. Gurov, T., Withlock, P., Dimov, I.: A grid free Monte Carlo algorithm for solving elliptic boundary value problems, *Numerical Analysis and Its Applications*, L. Vulkov, J. Waśniewski and P. Yalamov eds., *Lecture notes in computer sciences*, **1988**, Springer-Verlag, 2001, 359–367.
7. Mikhailov, G.A., *New Monte Carlo Methods with Estimating Derivatives*, Utrecht, The Netherlands, (1995).
8. C. Miranda, *Partial Differential Equations of Elliptic Type*, Springer-Verlag, Berlin, Heidelberg, New York, 1970.
9. M.Muller, Some continuous Monte Carlo methods for the Dirichlet problem, *Ann Math. Stat.*, **27** 3 (1956) 569–589.
10. R. Y. Papancheva, I. T. Dimov, T. V. Gurov, A New class of grid–free Monte Carlo Algorithms for Elliptic Boundary Value Problems, *Numerical methods and applications*, I. Dimov, I. Lirkov, S. Margenov, Z. Zlatev eds., *Lecture notes in computer sciences*, **2542**, Springer-Verlag, 2003, 132–139.
11. R. Y. Papancheva, T. V. Gurov, I. T. Dimov, Monte Carlo study of particle transport problem in Air Pollution, *Pliska Studia Mathematica Bulgarica*, **14**, 2003, 107–116.
12. Sabelfeld, K.K., *Monte Carlo Methods in Boundary Value Problems*, Springer Verlag, Berlin - Heidelberg - New York - London, (1991).
13. SPRNG: Scalable Parallel Random Number Generator, http://sprng.cs.fsu.edu/

Adaptive Genetic Algorithm and Quasi-parallel Genetic Algorithm: Application to Knapsack Problem

Kwok Yip Szeto[1] and Jian Zhang[1,2]

[1] Department of Physics,
Hong Kong University of Science and Technology,
Clear Water Bay, Hong Kong SAR, China
[2] Department of Mathematics,
Hong Kong University of Science and Technology,
Clear Water Bay, Hong Kong SAR, China
phszeto@ust.hk

Abstract. A new adaptive genetic algorithm using mutation matrix is introduced and implemented in a single computer using the quasi-parallel time sharing algorithm for the solution of the zero/one knapsack problem. The mutation matrix $M(t)$ is constructed using the locus statistics and the fitness distribution in a population $A(t)$ with N rows and L columns, where N is the size of the population and L is the length of the encoded chromosomes. The mutation matrix is parameter free and adaptive as it is time dependent and captures the accumulated information in the past generation. Two strategies of evolution, mutation by row (chromosome), and mutation by column (locus) are discussed. Time sharing experiment on these two strategies is performed on a single computer for solving the knapsack problem. Based on the investment frontier of time allocation, the optimal configuration for solving the knapsack problem is found.

1 Introduction

Parallel computation using the Darwinian principle of survival of the fittest has been implemented quite successfully in the framework of Genetic algorithms [1,2] with many successful application in many areas, such as solving the crypto-arithmetic problem [3], time series forecasting [4], traveling salesman problem [5], function optimization [6], and adaptive agents in stock markets [7,8]. However, the necessity of parameter setting in the application of genetic algorithm is a serious drawback for its practitioners, as its efficiency depends very much on the experience of the user on the problem at hand. One notable example of this drawback concerns the ad-hoc manner in the choice of the selection mechanism. One may need to use different percentage of the population for survival for different problems. Indeed, even for the same problem, the percentage of survivors in the evolution process should be time dependent for higher efficiency. Though some advances in adaptive parameter control on selection have been made, such as in the solution of the financial knapsack problem [9], the need for parameters setting remains. Here we like to address a novel way to do the selection process

I. Lirkov, S. Margenov, and J. Waśniewski (Eds.): LSSC 2005, LNCS 3743, pp. 189–196, 2006.

by the introduction of a mutation matrix that is time dependent but problem *independent.* We call our method mutation only genetic algorithm, or MOGA.

A second issue of parallel computation is the allocation of computer resource. It is desirable to devise a method for locating the optimal parameters in running the bottleneck program in a single computer that satisfies the criteria of both high speed and high confidence. Based on the ideas of Hogg and Huberman and collaborators [10], Szeto and Jiang [11] developed a formalism of quasi-parallel genetic algorithm, which is a method of combining existing algorithms into new ones that are unequivocally preferable to any of the component algorithms using the notion of risk in economics [12]. Here we assume that only one computer is available and the sharing of resource is realized only in the time domain. The concept of optimal usage is defined economically by the "investment frontier", characterized by low risk and high speed to solution. In this paper, we combine the work on mutation only genetic algorithm and time sharing in the framework of quasi-parallel genetic algorithm. We test this approach on the 0/1 knapsack problem with satisfactory results, locating the investment frontier for the knapsack problem.

2 Mutation Matrix

2.1 Mutation Matrix for Traditional Genetic Algorithm

In traditional simple genetic algorithm, the mutation/crossover operators are processed on the chromosome indiscriminately over the loci. The loci statistics is never employed. The recent work of Ma and Szeto [13] on Locus Oriented Adaptive Genetic Algorithm (LOAGA) has demonstrated the importance of the locus specific mutation rate for solving the zero/one knapsack problem. In this paper, we generalize their method and further demonstrate the advantage of using the information on the loci statistics on mutation operator. First let's show that traditional genetic algorithm can be treated as a special case in our formulation. We consider a population of N chromosomes, each of length L and binary encoded. We describe the population by a $N \times L$ matrix, which entry $A_{ij}(t), i = 1, ..., N; j = 1, ..., L$ being the value of the jth locus of the ith chromosome. The convention is to order the rows of A by the fitness of the chromosomes, $f_i(t) \leq f_k(t)$ for $i \geq k$. Next we introduce a mutation matrix with elements $M_{ij} \equiv a_i(t)b_j(t), i = 1, ..., N; j = 1, ..., L$, where $0 \leq a_i(t), b_j(t) \leq 1$ are called the row mutation probability and column mutation probability respectively. Traditionally we divide the population of N chromosomes into three groups: (1) Survivors who are the fit ones. They form the first N_1 rows of the population matrix $A(t+1)$. Here $N_1 = c_1 N$ with the survival selection ratio $0 < c_1 < 1$. (2) The number of children is $N_2 = c_2 N$ and is generated from the fit chromosomes by genetic operators such as mutation. Here $0 < c_2 < 1 - c_1$ is the second parameter of the model. We replace the next N_2 population matrix $A(t+1)$ (3) The remaining $N_3 = N - N_1 - N_2$ rows are the randomly generated chromosomes to ensure the diversity of the population so that the genetic algorithm continuously explores the solution space. In our formalism, traditional genetic algorithm with mutation only corresponds to a time independent mutation matrix with elements $M_{ij} \equiv 0$ for $i = 1, ..., N_1$, $M_{ij} \equiv m \in (0, 1)$ for $i = N_1, ..., N_2$, and finally we have $M_{ij} = 1$ for $i = N_2, ..., N$. Here m is the time independent mutation rate. We see that traditional genetic algorithm with mutation only requires at least three parameters: N_1, N_2 and m .

2.2 Mutation Probability

We first consider the case of mutation on a fit chromosome. We expect to mutate only a few loci so that it keeps most of the information unchanged. This corresponds to "*exploitation*" of the features of fit chromosomes. On the other hand, when an unfit chromosome undergoes mutation, it should change many of its loci so that it can explore more regions of the solution space. This corresponds "*exploration*". Therefore, we require that $M_{ij}(t)$ should be a monotonic increasing function of the row index i since we order the population in descending order of fitness. One simple solution is to use $a_i(t) = (i - 1)/(N - 1)$. Next, we must decide on the choice of loci for mutation once we have selected a chromosome to undergo mutation. This is accomplished by computing the locus mutation probability of changing to $X(X = 0\text{or}1)$ at locus j as p_{jX} by

$$p_{jX} = \frac{\sum_{k=1}^{N}(N + 1 - k) \times \delta_{kj}(X)}{\sum_{m=1}^{N} m} \tag{1}$$

Here k is the rank of the chromosome in the population. $\delta_{kj}(X) = 1$ if the jth locus of the kth chromosome assume the value X, and zero otherwise. The factor in the denominator is for normalization. Note that p_{jX} contains information of both locus and row and the locus statistics is biased so that heavier weight for chromosomes with high fitness is assumed. This is in general better than the original method of Ma and Szeto[13] where there is no bias on the row. After defining p_{jX}, we define the column mutation rate as

$$b_j = \frac{1 - |p_{j0} - 0.5| - |p_{j1} - 0.5|}{\sum_{j'=1}^{N} b_{j'}} \tag{2}$$

For example, if 0 and 1 are randomly distributed, then $p_{j0} = p_{j1} = 0.5$. We have no useful information about the locus, so we should mutate this locus, and $b_j = 1$. When there is definitive information, such as when $p_{j0} = 1 - p_{j1} = 0$ or 1, we should not mutate this column and $b_j = 0$.

3 Mutation Only Genetic Algorithm: MOGA

Once the mutation matrix **M** is obtained, we are ready to discuss the strategy of using **M** to evolve **A**. There are two ways to do Mutation Only Genetic Algorithm. We can first decide which row (chromosome) to mutate, then which column (locus) to mutate, we call this particular method the *Mutation Only Genetic Algorithm by Row* or abbreviated as MOGAR. Alternatively, we can first select the column and then the row to mutate, and we call this the *Mutation Only Genetic Algorithm by Column* or abbreviated as MOGAC.

For MOGAR, we go through the population matrix $A(t)$ by row first. The first step is to order the set of locus mutation probability $b_j(t)$ in descending order. This ordered set will be used for the determining of the set of column position (locus) in the mutation process. Now, for a given row i, we generate a random number x. If $x < a_i(t)$, then we perform mutation on this row, otherwise we proceed to the next row and $A_{ij}(t+1) = A_{ij}(t), j = 1, ..., L$. If row i is to be mutated, we determine the set $R_i(t)$ of loci in row i to be changed by choosing the loci with $b_j(t)$ in descending order, till we obtain $K_i(t) = a_i(t) \times L$ members. Once the set $R_i(t)$ has been constructed, mutation will be performed on these columns of the ith row of the $A(t)$ matrix to obtain the matrix elements $A_{ij}(t+1), j = 1, ..., L$. We then go through all N rows, so that in one generation, we need to sort a list of L probabilities and generate N random numbers for the rows. After we obtained $A(t+1)$, we need to compute the $M_{ij}(t+1) = a_i b_i(t+1)$and proceed to the next generation.

For MOGAC, the operation is similar to MOGAR mathematically except now we rotate the matrix **A** by 90 degrees. Now, for a given column j we generate a random number y. If $y < b_j(t)$, then we mutate this column, otherwise we proceed to the next column and $A_{ij}(t+1) = A_{ij}(t), i = 1, ..., N$. If column j is to be mutated, we determine the set $S_j(t)$ of chromosomes in column j to be changed by choosing the rows with the $a_i(t)$ in descending order, till we obtain $W_j(t) = b_j(t) \times N$ members. Since our matrix **A** is assumed to be row ordered by fitness, we simply need to choose the $N, N-1, ..., N-W_j+1$ rows to have the jth column in these row mutated to obtain the matrix elements $A_{ij}(t+1), i = 1, ..., N$. We then go through all L columns, so that in one generation, we need to sort a list of N fitness values and generate L random numbers for the columns.

4 Quasi-parallel Algorithm

Now we switch our discussion from MOGA to the problem of allocation of computational resource to two algorithms: MOGAR and MOGAC in one single computer when solving a particular optimization problem. The framework for proper mixing of computing algorithms is the quasi-parallel algorithm of Szeto and Jiang [11]. Here we first summarize this algorithm. A simple version of our quasi-parallel genetic algorithm ($QPGA = (M, SubGA, \Gamma, T)$) consists of M independent sub-algorithms $SubGA$. The time sharing of the computing resource is described by the resource allocation vector Γ. If the total computing resource is R, shared by M sub-algorithms $G_i, i = 1, 2, ..., M$, with resource R_i assigned to G_i in unit time, then we introduce $\tau_i = R_i/R$, $0 \leq \tau_i \leq 1$ for any $i = 1, 2, ..., M$, and $\sum_{i=1}^{M} \tau_i = 1$, and the allocation of resource for sub-algorithms is defined by the *resource allocation vector*, $\Gamma = (\tau_1, \tau_2, ..., \tau_M)'$. In our case, we have $M = 2$ and our $SubGA$ are MOGAR and MOGAC. For resource allocation, we only have one parameter $0 < \gamma < 1$ which is the fraction of time the computer is using MOGAR, and the remaining fraction of time $(1 - \gamma)$ we use MOGAC.

The termination criterion T is used to determine whether a $QPGA$ should stop running. Thus, we have a mixture of MOGAR with MOGAC in the framework of quasi-parallel genetic algorithm with a mixing parameter γ. The parallel genetic algorithm described above can be implemented in a serial computer. For a particular generation t, we will generate a random number z. If $z < \gamma$, then we perform MOGAR, otherwise we perform MOGAC to generate the population $A(t+1)$. We now apply this quasi-parallel mutation only genetic algorithm to solve the 0/1 knapsack problem and try to obtain the investment frontier that give the mixing parameter γ that yields the fastest speed for solving the problem while also running with most certainty (minimum risk) of getting the solution.

5 The Zero/One Knapsack Problem

The model problem to test our ideas on mutation only genetic algorithm is the Knapsack problem. We define the 0/1 knapsack problem [14] as follow. Given L items, each with profit P_i, weight w_i and the total capacity limit c, we need to select a subset of L items to maximize the total profit, but its total weight does not exceed the capacity limit. Mathematically, we want to find the set $x_i \in \{0,1\}, i = 1, ..., L$ to

$$\text{Maximize} \sum_{j=1}^{L} P_j x_j \text{ subjected to constraint } c \geq \sum_{j}^{L} w_j x_j. \tag{3}$$

We consider a particular knapsack problem with size $L = 150$ items, $c = 4000$. The set $P_i \in [0, 1000]$ and $w_i \in [0, 100]$ are chosen randomly to define our problem, but afterwards fixed. In order not to violate the constraint of the problem, we use two tricks, *"Punishment"* and *"Repairing"*. *Punishment* reduces the fitness when the constraint is violated, while *Repairing* modifies the chromosome (adding/deleting items) until the constraint is satisfied. We will use a method called Greedy Repair. If a chromosome violates the constraint (total weight is over the constraint in the Knapsack), the repair scheme will find the site k with minimum value of P_k/w_k and reset x_k to zero, i.e., removing the kth item from the knapsack. This process continues till the constraint is satisfied. When the constraint is satisfied, and if some empty space remains, Greedy Repair will tried to fill the knapsack "as full as possible" by picking up the unselected item (those sites m where x_m=0) and fill them in the knapsack in descending order of P_m/w_m. Repair operation stops once the constraint is violated. This scheme can repair all chromosomes into local optimal solution in Hamming space.

6 Results

In the early version of our ideas on mutation matrix [13], we have found that locus oriented adaptive genetic algorithm (LOAGA) outperforms dynamic programming which is the usual method for Knapsack problem. We have found evidence

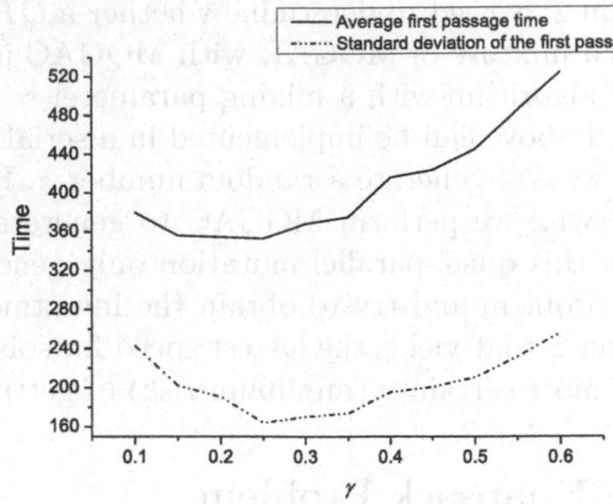

Fig. 1. Mean first passage time to solution and its standard deviation of 1000 runs as a function of time sharing parameter γ

that mixing MOGAR with MOGAC in a time-sharing manner produces superior results compared to (LOAGA) in numerical experiment for the 0/1 knapsack problem. Here we like to find the optimal time sharing parameter by locating the investment frontier of our mutation only genetic algorithm. We first define our MOGAR and MOGAC. We choose the simplest form of $a_i(t) = (i-1)/(N-1)$. Here $N(=100)$ is the size of the population in all genetic algorithms. For a given generation time t, we generate a random number z. If $z < \gamma$, then we perform MOGAR, otherwise we perform MOGAC to generate the population $A(t+1)$. Next we address the stopping criterion. We use exhaustive search to locate the true optimal solution of the knapsack problem. Then, we run our mixed MOGA in *QPGA* formalism 1000 times to collect statistical data. For each run, we define the stopping time to be the generation number when we obtain the optimal solution, or 1500, whichever is smaller. The choice of 1500 as the upper bound is based on our numerical experience since for a wide range of γ, all the runs are able to find the optimal solution within 1500 generations. Only for those extreme cases where γ is near 1 or 0, meaning that we use MOGAR alone or MOGAC alone, a few runs fail to find the optimal solution within 1500 generations. This is expected since we know row mutation only GA has low speed of convergence while column mutation only GA has early convergence problem. These extreme cases turn out to be irrelevant in our search of the investment frontier as demonstrated in Fig.1, where we plot the mean first passage time to solution and its standard deviation of 1000 runs as a function of the time sharing parameter γ. These results demonstrate the power of *QPGA* (i.e., time-sharing of computational resource) based on Mutation Only Genetic Algorithm. In Fig.2, we plot average first passage time to solution versus standard deviation. The curve is parameterized by γ. We see that there is a point on the curve which is closest to the origin. This point is unique in this experiment, corresponding to a value of $\gamma_c = 0.22 \pm 0.02$. The interpretation of this time sharing parameter is that our *QPGA* will be fastest and most reliable (least risky) in finding the optimal

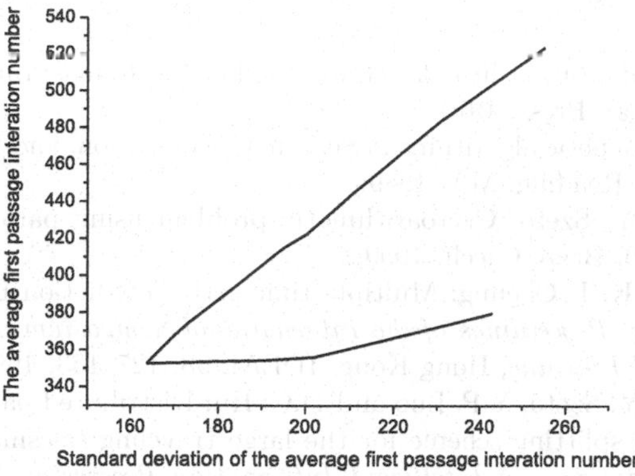

Fig. 2. Average passage time to solution versus standard deviation. The curve is parameterized by γ.

solution of the 0/1 knapsack problem. In another word, the investment frontier of this problem consists of a single critical point at γ_c.

7 Conclusion

Using the simple observation that ordinary genetic algorithm can be considered as a special case of evolutionary computation using a special static form of mutation matrix, we develop a general formalism for mutation matrix that allows adaptive genetic algorithm without the need to preset selection parameters. We further generalize the evolution by making use of the locus statistics and develop MOGAC, *mutation only genetic algorithm by column*. This new algorithm has high speed of convergence. By combining it with MOGAR, *mutation only genetic algorithm by row*, we find a way to combine efficiently two processes: exploration of solution space and exploitation of the features of locus statistics for the fit chromosomes. The method we use is time sharing of MOGAR and MOGAC in the framework of quasi-parallel genetic algorithm. This methodology is tested on the 0/1 knapsack problem. We succeed in locating the critical value of time sharing in the investment frontier of mixing MOGAR and MOGAC to be 0.22, meaning that statistically we should use 22% of the computational resource on mutation by row, and 78% on mutation by column when solving the knapsack problem. Our general formalism can be used to address various types of optimization problems such as Ising model in random fields, Potts model, and traveling salesman problem. Our future work will address the incorporation of crossover in our formulation.

Acknowledgement

K.Y. Szeto acknowledged that this work is supported by RGC grant no. HKUST6157/01P and 603203.

References

1. J.H. Holland: Adaptation in natural and artificial systems. Ann Arbor, MI: University of Michigan Press, 1975.
2. D.E. Goldberg: Genetic algorithms in Search, Optimization, and Machine Learning, Addison-Wesley, Reading, MA, 1989.
3. S. P. Li and K.Y. Szeto: Crytoarithmetic problem using parallel Genetic Algorithms, Mendl'99, Brno, Czech, 1999.
4. K.Y. Szeto and K.H. Cheung: Multiple time series prediction using genetic algorithms optimizer. *Proceedings of the International Symposium on Intelligent Data Engineering and Learning*, Hong Kong, IDEAL'98, 127–133, 1998.
5. R. Jiang and K.Y. Szeto, Y.P. Luo and D.C. Hu: Distributed parallel genetic algorithm with path splitting scheme for the large traveling salesman problems. *Proceedings of Conference on Intelligent Information Processing, 16th World Computer Congress 2000, Aug.21-25, 2000*, Beijing, Ed. Z. Shi, B. Faltings, and M. Musen, Publishing House of Electronic Industry, 478–485, 2000.
6. K.Y. Szeto, K.H. Cheung and S.P. Li: Effects of dimensionality on parallel genetic algorithms. *Proceedings of the 4th International Conference on Information System, Analysis and Synthesis, Orlando, Florida, USA*, **2**, 322–325, 1998.
7. K.Y. Szeto and L.Y. Fong: How adaptive agents in stock market perform in the presence of random news: a genetic algorithm approach, LNCS/LNAI, Vol. 1983, Ed. K. S. Leung et al. Spriger-Verlag, Heidelberg, 2000, IDEAL 2000, 505–510, 2000.
8. Alex L.Y. Fong and K.Y. Szeto: Rule Extraction in Short Memory Time Series using Genetic Algorithms, European Physical Journal B Vol.20, 569–572 (2001).
9. Kwok Yip Szeto and Man Hon Lo: An Application of Adaptive Genetic Algorithm in Financial Knapsack Problem, *The 17th International Conference on Industrial & Engineering Applications of Artificial Intelligenc & Expert Systems, Ed Bob Orchard, et al. May 17-20, 2004*, LNAI **3029**, Springer Verlag, 1220–1227, 2004.
10. B.A. Huberman, R.M. Lukose and T. Hogg: An economics approach to hard computational problems, Science, **275** 3, 51–54, 1997.
11. Kwok Yip Szeto and Jiang Rui: A quasi-parallel realization of the Investment Frontier in Computer Resource Allocation Using Simple Genetic Algorithm on a Single Computer, *LNCS 2367, 6th International Conference, PARA 2002, Espoo, Finland, June 15-18, 2002* 116–126. Springer-Verlag.
12. H. Markowitz: J. of Finance, **7**, 77, 1952.
13. C.W. Ma and K.Y. Szeto: Locus Oriented Adaptive Genetic Algorithm: Application to the Zero/One Knapsack Problem, *Proceeding of The 5th International Conference on Recent Advances in Soft Computing, RASC2004 Nottingham, UK.* 410–415, 2004.
14. V. Gordon, A. Bohm, and D. Whitley: A Note on the Performance of Genetic Algorithms on Zero-One Knapsack Problems, *Proceedings of the 9th Symposium on Applied Computing (SAC'94), Genetic Algorithms and Combinatorial Optimization, Phoenix, Az*, 194–195, 1994.

Part V

Control Systems

Approximate Gradient/Penalty Methods with General Discretization Schemes for Optimal Control Problems

Ion Chryssoverghi

Department of Mathematics, National Technical University of Athens,
Zografou Campus, 15780 Athens, Greece
ichris@central.ntua.gr

Abstract. We consider an optimal control problem described by ordinary differential equations, with control and state constraints. The state equation is first discretized by a general explicit Runge-Kutta scheme and the controls are approximated by piecewise polynomial functions. We then propose approximate gradient and gradient projection methods, and their penalized versions, that construct sequences of discrete controls and progressively refine the discretization during the iterations. Instead of using the exact discrete cost derivative, which usually requires tedious calculations of composite functions, we use here an approximate derivative of the cost defined by discretizing the continuous adjoint equation by the same, but nonmatching, Runge-Kutta scheme backward and the integral involved by a Newton-Cotes integration rule. We show that strong accumulation points in L^2 of sequences constructed by these methods satisfy the weak necessary conditions for optimality for the continuous problem. Finally, numerical examples are given.

1 Introduction

In this paper, we consider an optimal control problem for systems governed by nonlinear ordinary differential equations, with control and state constraints. In order to solve this problem numerically, the state equation is first discretized by an explicit Runge-Kutta scheme of maximal global order m and the controls are approximated by vector functions whose components are piecewise polynomial of degree $l \leq m - 1$, but not necessarily continuous. We then propose approximate gradient and gradient projection methods, and their penalized versions, that construct sequences of discrete controls and progressively refine the discretization during the iterations. Since the matching adjoint of the discrete state equation and the derivative of the cost functional usually involve tedious calculations of partial derivatives of composed functions, we use at each iteration an approximate cost derivative defined by discretizing the continuous adjoint equation by the same, but nonmatching, Runge-Kutta scheme backward and the integral defining the cost derivative by a Newton-Cotes integration rule with nodes equal to the $l + 1$ polynomial interpolation points, both schemes involving maximal

I. Lirkov, S. Margenov, and J. Waśniewski (Eds.): LSSC 2005, LNCS 3743, pp. 199–207, 2006.

global order approximations of intermediate values of states and adjoints. Since the discrete adjoints are nonmatching here, the exact derivative of the discrete cost is not defined in adjoint form, and one must necessarily use a progressive refining procedure, with the adjoint matching only in the limit. This approach has the advantage of reducing computing time and memory. The main result is that strong accumulation points in L^2, if they exist, of sequences generated by this method satisfy the weak necessary conditions for optimality for the continuous problem. Finally, numerical examples are given. For discretization and optimization methods in optimal control problems, see [1, 2, 3, 4, 5, 7, 10, 11]. The results of this paper extend those of [2] by using general discretization schemes (Runge-Kutta, integration rule, polynomial parameterization) in conjunction with a penalized gradient projection method for solving state constrained problems.

2 The Continuous Optimal Control Problem

Consider the following optimal control problem, with state equation

$$y'(t) = f(t, y(t), w(t)) \quad \text{in } I := [0, T], \quad y(0) = y^0,$$

where $y(t) \in \mathrm{R}^p$, constraints on the control w

$$w(t) \in U \text{ in } I,$$

where U is a convex, closed, but not necessarily bounded, subset of R^q, state constraints

$$G_1(w) := g_1(y(T)) = 0, \quad G_2(w) := g_2(y(T)) \leq 0,$$

where $g_1 : \mathrm{R}^p \to \mathrm{R}^{p_1}$, $g_2 : \mathrm{R}^p \to \mathrm{R}^{p_2}$, and cost functional to be minimized

$$G_0(w) := g_0(y(T)).$$

If the problem involves additional integral state constraints or cost functionals, it can classically be transformed into a problem with final cost only by introducing scalar differential equations. The problem could also include pointwise pure state constraints, but this was not done here for simplicity of presentation.

We denote by $\|.\|$ the Euclidean norm in R^n and by $\|.\|_2$ the usual norm in L^2. We define the sets of controls

$$W = \{w \in L^2(I, \mathrm{R}^q) \,|\, w : I \to U\}, \quad W_\infty := W \cap L^\infty,$$

both endowed with the relative norm topology of L^2. We suppose in the sequel at least that f, f_y, f_u are Lipschitz continuous w.r.t. (t, y, u) and g, ∇g are Lipschitz continuous on every bounded set. Then, in particular, for every $w \in W_\infty$, the state equation has a unique absolutely continuous solution $y := y_w$. Moreover, for every given $b_0 \geq 0$, there exists $b_1 \geq 0$ such that $\|y_w\|_\infty \leq b_1$, for every $w \in W_\infty$, with $\|w\|_2 \leq b_0$ (or $\|w\|_\infty \leq b_0$). The two following results are standard, see e.g. [12].

Proposition 1. *Dropping the index r in g_r and G_r, given controls $w, w' \in W_\infty$, the directional derivative of the functional G is given by*

$$DG(w, w' - w) := \lim_{\alpha \to 0+} \frac{G(w + \alpha(w' - w)) - G(w)}{\alpha}$$
$$= \int_I z(t)^T f_u(t, y(t), w(t))[w'(t) - w(t)]dt,$$

where $y := y_w$, and the adjoint state $z := z_w$ is defined by the adjoint equation

$$z'(t) = -f_y(t, y(t), w(t))^T z(t) \text{ in } I, \;\; z(T) = \nabla g(y(T)).$$

Theorem 1. *If the control $w \in W_\infty$ is optimal, then w is extremal, i.e. there exist multipliers $\lambda_0 \in \mathrm{R}$, $\lambda_1 \in \mathrm{R}^{p_1}$, $\lambda_2 \in \mathrm{R}^{p_2}$ (λ_1, λ_2 row vectors), with $\lambda_0 \geq 0$, $\lambda_2 \geq 0$, $\lambda_0, \lambda_1, \lambda_2$ not all zero, such that*

$$\sum_{r=0}^{2} \lambda_r DG_r(w, w' - w) \geq 0, \;\; \textit{for every } w' \in W_\infty,$$

$$\lambda_2 G_2(w) = 0 \quad \text{(Transversality condition)}.$$

The above inequalities are equivalent to the weak pointwise minimum principle

$$z(t)^T f_u(t, y(t), w(t)) w(t) = \min_{u \in U} [z(t)^T f_u(t, y(t), w(t)) u], \;\; \text{a.e. in } I,$$

where z is defined with $g := \sum_{r=0}^{2} \lambda_r g_r$ in the end-point condition of the adjoint equation.

3 Discretizations

Let $(N_n)_{n \geq 1}$ be a sequence of positive integers. We suppose that $N_n \to \infty$ as $n \to \infty$ and that for each n, either $N_{n+1} = N_n$, or $N_{n+1} = MN_n$, where $M \geq 2$ is a positive integer. For each $n \geq 1$, we define the *discretization* Δ_n by setting

$$N := N_n, \; h_n := T/N, \; t_{ni} = ih_n, \; i = 0, ..., N,$$

$$I_{ni} := [t_{n,i-1}, t_{n,i}), \; i = 1, ..., N-1, \; I_{nN} := [t_{n,N-1}, t_{nN}].$$

For given $l+1$ ($l \leq m-1$) interpolation points t_{ni}^k in each I_{ni} of the form

$$t_{ni}^k = t_{n,i-1} + h_n/l, \; k = 0, ..., l.$$

we define the *set of discrete admissible controls*

$$W_n = \{w_n \in W_\infty | \, w_n \in \Pi_l(I_{ni}) \text{ on } I_{ni}, \, w_{ni}(t_{ni}^k) = w_{ni}^k \in U, \, k = 0, ..., l, \; i = 1, ..., N\},$$

where $\Pi_l(I_{ni})$ denotes the set of q-vector functions whose components are polynomials of degree $\leq l$ on I_{ni}, and where it is understood that the values at the possible interpolation jump points $t^k_{ni} = t_{n,i-1}, t_{ni}$ are right/left limit values, on each I_{ni}.

Next, we discretize the state equation by an explicit Runge-Kutta scheme of *maximal global order* m, and with m intermediate points (not necessarily distinct), which can be written in the form

$$\phi^j_{ni} = f(\bar{t}^j_{ni}, y_{n,i-1} + h_n \sum_{s=1}^{j-1} \alpha_{is}\phi^s_{ni}, \bar{w}^s_{ni}),\ \ \bar{w}^j_{ni} = w_{ni}(\bar{t}^j_{ni}),\ \ j = 1, ..., m,$$

$$\bar{t}^j_{ni} = t_{n,i-1} + \bar{\theta}^j h_n,\ \text{ with } \bar{\theta}^j \in [0,1],\ \ j = 1, ..., m,\ \ \bar{\theta}^1 = 0,\ \bar{\theta}^m = 1,$$

$$y_{ni} = y_{n,i-1} + h_n \sum_{j=1}^{m} \beta^j \phi^j_{ni},\ \text{ with } \sum_{j=1}^{m} \beta^j = 1,\quad \beta^j \geq 0,\ j = 1, ..., m,$$

$$i = 1, ..., N,\ \ y_{n0} = y^0.$$

We have chosen here an explicit scheme for simplicity, but the theory below can be extended to implicit ones. For $w_n \in W_n$, with corresponding state y_n, we define the discrete functionals

$$G_{rn}(w_n) := g_r(y_{nN}),\ r = 0, 1, 2.$$

The next Theorems 2,3,4 generalize Theorems 2,4,5 in [2], respectively.

Theorem 2. *(Consistency) Let (w_n) be a sequence with $w_n \in W_n$, $\|w_n\|_\infty \leq \bar{b}'$, for $n \geq 1$, and first order divided differences (DD) of the w_n uniformly bounded. If $w_n \to w$ in L^2, then, $y^-_n \to y$, $y^+_n \to y$ uniformly (y^-_n, y^+_n piecewise constant functions defined by left/right values), and $G_n(w_n) \to G(w)$, as $n \to \infty$.*

Theorem 3. *(Error estimates) For $w_n \in W_n$, with $\|w_n\|_\infty \leq \bar{b}'$ and DD of order $1, ..., l$ of w_n bounded by $\bar{L}$, let y_n be the corresponding discrete state and $\tilde{y}_n$ the corresponding solution of the continuous state equation. If f is sufficiently smooth (e.g. $f \in C^m$) w.r.t. (t, y, u) ((t, y) if $l = 0$), then*

$$\max_{0 \leq i \leq N} \|y_{ni} - \tilde{y}_n(t_{ni})\| \leq ch^m_n,$$

$$\|G_n(w_n) - G(w_n)\| \leq ch^m_n,$$

$$\|G_n(w_n) - G_{n'}(w_n)\| \leq ch^m_n,\ \text{ if } n' > n, \text{ and } N_n \neq N_{n'}.$$

where c denotes various constants, independent of n and w_n.

Next we define, for given $w_n \in W_n$, with corresponding discrete state y_n, the (general) *approximate discrete adjoint* z_n (dropping r in g_r) as the solution of the initial Runge-Kutta scheme applied formally backward to the continuous time adjoint equation

$$\psi^j_{ni} = f_y(\bar{t}^j_{ni}, \bar{y}^j_{ni}, \bar{w}^j_{ni})^T (z_{n,i-1} + h_n \sum_{s=m}^{j+1} \alpha_{is}\psi^s_{ni}),\ \ j = m, ..., 1,$$

$$z_{n,i-1} = z_{ni} + h_n \sum_{j=m}^{1} \beta^j \psi^j_{ni},\ \ i = N, ..., 1,\ \ z_{nN} = \nabla g(y_{nN}),$$

and using, instead of the exact values, *intermediate state approximations* $\tilde{y}^j_{ni}$ at the points $\tilde{t}^j_{ni}$, of *maximal local order* m (hence inducing, at best, a local error $O(h_n^{m+1})$ in the adjoint scheme), which can be computed as linear combinations of the intermediate function evaluations ϕ^j_{ni} of the Runge-Kutta scheme for the state equation, with some additional function evaluations, if $m \geq 5$ (for such Runge-Kutta approximations, see [6, 8]). These evaluations require much less computations than the direct calculation of the matching adjoint of the discrete state equation, which requires the computation of Jacobians w.r.t. y of multi-stage composite functions.

Now let as above y^k_{ni}, z^k_{ni} be approximations of *maximal local order* m of the state and adjoint, and w^k_{ni} the exact control values, at the interpolation points t^k_{ni}. For given $w_n, w'_n \in W_n$, and y_n, z_n corresponding to w_n, the *approximate discrete derivative* of each functional G is defined by applying formally, on each I_{ni}, some Newton-Cotes integration rule, with nodes t^k_{ni} and of *maximal global order* m', to the continuous time cost derivative, and using in this rule the approximate values y^k_{ni}, z^k_{ni}, instead of the exact values, and the exact control interpolation values $w^k_{ni} = w_n(t^k_{ni})$

$$D_n G(w_n, w'_n - w_n) := h_n \sum_{i=1}^{N} \sum_{k=0}^{l} C^k_l (z^k_{ni})^T f_u(t^k_{ni}, y^k_{ni}, w^k_{ni})(w'^k_{ni} - w^k_{ni}),$$

$$\text{with} \sum_{k=0}^{l} C^k_l = 1, \quad C^k_l \geq 0,\ k = 0, ..., l.$$

Define the I_{ni}-piecewise constant functions $w^k_n(t) = w^k_{ni}$ in I_{ni}, $i = 1, ..., N$, and similarly for y^k_n, z^k_n.

Theorem 4. *(i) (Consistency) Let* (w_n), (w'_n) *be sequences with* $w_n, w'_n \in W_n$, $\|w_n\|_\infty \leq \bar{b}'$, $\|w'_n\|_\infty \leq \bar{b}'$, *for* $n \geq 1$, *and first order DD of the* w_n, w'_n *uniformly bounded. If* $w_n \to w$ *and* $w'_n \to w'$ *in* L^2 *strongly. Then* $y^k_n \to y = y_w$, $z^k_n \to z = z_w$, *uniformly, and*

$$D_n G(w_n, w'_n - w_n) \to DG(w, w' - w).$$

(ii) (Error estimates) If f, f_y *are sufficiently smooth (e.g.* $\in C^m$*) w.r.t.* (t, y, u) $((t, y)$ *if* $l = 0)$, *and the DD of order* $1, ..., l$ *of the* w_n *are uniformly bounded, then*

$$\max_{0 \leq i \leq N} \|z_{ni} - \tilde{z}_n(t_{ni})\| \leq c h_n^m,$$

where $\tilde{z}_n$ *is the exact solution of the continuous adjoint equation corresponding to* w_n *and* $\tilde{y}_n$. *If the DD of order* $1, ..., \min(l, m')$ *of the* w_n, w'_n *are uniformly bounded and* f, f_y, f_u *are sufficiently smooth (e.g.* $\in C^m$*), then*

$$|D_n G(w_n, w'_n - w_n) - DG(w_n, w'_n - w_n)| \leq c h_n^{\bar{m}}, \text{ with } \bar{m} = \min(m, m').$$

Let $\bar{W}_n$ denote the set of discrete piecewise constant controls. We have $\bar{W}_n \subset W_n$ for every n. The following result is classical (see [9]).

Proposition 2. *(Approximation by discrete controls) For every* $w \in W$ *(or* $w \in W_\infty$*), there exists a sequence* $(w_n \in \bar{W}_n)$ *such that* $w_n \to w$ *in* L^2.

4 Approximate Gradient/Penalty Methods

For a given constant $L' \geq 0$, define the *projection set of controls*

$$W'_n := \{w_n \in W_n | \ 1^{\text{st}} \text{ order DD of } w_n \text{ bounded by } L'\} \subset W_n,$$

and for $w_n \in W_n$, the *discrete norm*

$$\|w_n\|_{2,n}^2 := h_n \sum_{i=1}^{N} \sum_{k=0}^{l} C_l^k \left\|w_{ni}^k\right\|^2 = \sum_{k=0}^{l} C_l^k \left\|w_n^k\right\|_2^2.$$

Let (M_r^j), $r = 1, 2$, be nonnegative increasing sequences such that $M_r^j \to \infty$ as $j \to \infty$, (ϵ^j) a nonnegative decreasing sequence such that $\epsilon^j \to \infty$, and define the sequence of *penalized discrete functionals*

$$G_n^j(w_n) = G_{0n}(w_n) + \frac{1}{2}\{M_1^j \sum_{s=1}^{p_1} |G_{1n,s}(w_n)|^2 + M_2^j \sum_{s=1}^{p_2} [\max(0, G_{2n,s}(w_n))]^2\}.$$

Algorithm

Step 1. Choose an initial discretization Δ_1, integers $m \geq 1$, $l \in [0, m-1]$, $M \geq 2$ ($U = \mathrm{R}^q$, or $U \neq \mathrm{R}^q$ with $l \leq 1$), or $M = l$ ($U \neq \mathrm{R}^q$ with $l > 1$), $L' \geq 0$, $b, c \in (0,1)$, $s \in (0,1]$ ($s \in (0, +\infty)$ if $U = \mathrm{R}^q$), $\gamma > 0$, $w_1 \in W'_1$, and set $n := 1$, $\kappa := 1$, $j := 1$.

Step 2. Find $v_n \in W'_n$ such that

$$\begin{aligned} e_n &:= D_n G^j(w_n, v_n - w_n) + (\gamma/2)\|v_n - w_n\|_{2,n}^2 \\ &= \min_{v'_n \in W'_n} \left[D_n G^j(w_n, v'_n - w_n) + (\gamma/2)\|v'_n - w_n\|_{2,n}^2\right], \end{aligned}$$

and set $d_n := D_n G^j(w_n, v_n - w_n)$. If there are no state constraints, go to Step 4.

Step 3. If $|d_n| \leq \epsilon^j$, set $n^j := n$, $w_n^j := w_n$, $v_n^j := v_n$, $e_n^j := e_n$, $d_n^j := d_n$, $j := j + 1$, and go to Step 2.

Step 4. (Armijo step search) Set and $\alpha^0 = s$. If the inequality

$$G_n^j(w_n + \alpha^l(v_n - w_n)) - G_n^j(w_n) \leq \alpha^l b d_n,$$

is not satisfied, set successively $\alpha^{l+1} := c\alpha^l$ and find, if it exists, the first $\alpha^l \in (0,1]$, say $\bar{\alpha}$, such that it is satisfied. [*Optional*: Else, set successively $\alpha^{l+1} := \alpha^l/c$ and find the last $\alpha^l \in (0,1]$, say $\bar{\alpha}$, such that this inequality is satisfied.]

If $\bar{\alpha}$ is found, set $\alpha_n := \bar{\alpha}$, $\tilde{w}_n := w_n + \alpha_n(v_n - w_n)$, $n_\kappa := n$, $\kappa := \kappa + 1$. Else, set $\tilde{w}_n := w_n$.

Step 5. Define w_{n+1} by:

(a) Cases $U = \mathrm{R}^q$, or $U \neq \mathrm{R}^q$ with $l \leq 1$: Set $N_{n+1} := N_n$ or $N_{n+1} := M N_n$ (refining). In both cases, set $w_{n+1} := \tilde{w}_n$.

(b) Case $U \neq \mathrm{R}^q$ with $l > 1$: Set $N_{n+1} := N_n$ or $N_{n+1} := l N_n$ (refining). If $N_{n+1} = N_n$, set $w_{n+1} := \tilde{w}_n$. If $N_{n+1} = l N_n$, then, for each $i = 1, ..., N_{n+1}$, compute the multi-vector of interpolation values $\tilde{\mathbf{w}}_{n+1,i} := (\tilde{w}^0_{n+1,i}, ..., \tilde{w}^l_{n+1})$ of $\tilde{w}_n$ on $I_{n+1,i}$ for the discretization Δ_{n+1}, and find the projection $P_{n+1,i}\tilde{\mathbf{w}}_{n+1,i}$ of $\tilde{\mathbf{w}}_{n+1,i}$ onto U^{l+1} subject to the (linear) first order DD_{n+1} constraints (i.e. first order DD_{n+1} bounded by L'). Then define w_{n+1} as the piecewise polynomial function of degree $\leq l$ interpolating all these projection values on I for the discretization Δ_{n+1}.

Step 6. Set $n := n + 1$ and go to Step 2.

Define the *set of successful iterations* $K := (n_\kappa)_{\kappa\in \mathrm{N}}$ (see Step 4), and the *sequences of multipliers* (w.r.t. j)

$\lambda^j_{1n} = M^j_1 G_{1n}(w^j_n)$, $\lambda^j_{2n} = M^j_2$ $\mathbf{max}$ $(0, G_{2n}(w^j_n))$, with $n := n^j$ (see Step 3), where **max** denotes a vector of max values and w^j_n is defined in Step 3.

Theorem 5. *We suppose that f, f_y, f_u are at least Lipschitz continuous w.r.t. (t, y, u) ((t, y) if $l = 0$).*

(i) (No state constraints – Gradient projection method) If K is finite (resp. infinite) and there exists a subsequence $(w_n)_{n\in L\subset \mathrm{N}}$ (resp. $(w_n)_{n\in L\subset K}$) that converges strongly in L^2 to some w and is bounded in L^∞ if U is unbounded, then $d_n \underset{n\in L}{\to} 0$, $e_n \underset{n\in L}{\to} 0$, $w \in W_\infty$, and w is extremal.

(ii) (State constraints – Penalized gradient projection method) If the sequence (w_n) is infinite (i.e. $n \to \infty$) and converges strongly in L^2 to some w, (w_n) is bounded in L^∞ if U is unbounded, and the sequences of multipliers (λ^j_{1n}), (λ^j_{2n}) are bounded, then w is admissible and extremal.

Proof. (*Sketch*). (i) The results are proved here using the techniques of [2], with some modifications due to the more general polynomials used, if $l > 1$, and the discrete projections in Step 5 (b).

(ii) By our assumptions, Step 3, and (i), j cannot remain constant in the Algorithm. Setting $n := n^j$ (see Step 3), we thus have $j \to \infty, d^j_n \to 0$, $e^j_n \to 0$, hence $\left\|v'_n - w^j_n\right\|^2_{2,n} \to 0$. It can then be shown, similarly to (i), that $w \in W_\infty$. Since the sequences (λ^j_{1n}), (λ^j_{2n}) are bounded, we can suppose that (up to subsequences) $\lambda^j_{1n} \to \lambda_1$, $\lambda^j_{2n} \to \lambda_2$. By Theorem 4, we have

$$0 = \lim_{l\to\infty} \frac{\lambda^j_{1n}}{M^j_1} = \lim_{j\to\infty} G_{1n}(w^j_n) = G_1(w),$$

$$0 = \lim_{j\to\infty} \frac{\lambda^j_{2n}}{M^j_2} = \lim_{j\to\infty} [\mathbf{max}(0, G_{2n}(w^j_n))] = \mathbf{max}(0, G_2(w)),$$

which show that w is admissible. Now let any $v' \in W$ and, by Proposition 2, let $(v'^n \in W^n)$ be a sequence of piecewise constant discrete controls that converges to v'. Let $(\lambda^j_{1n}), (\lambda^j_{2n})$ be subsequences converging to some λ_1, λ_2. By Step 2, we have

$$D_nG_0(w_n^j, v_n'^j - w_n^j) + \lambda_{1n}^j D_nG_1(w_n^j, v_n'^j - w_n^j) + \lambda_{2n}^j D_nG_2(w_n^j, v_n'^j - w_n^j) + (\gamma/2)\|v_n' - w_n^j\|_{2,n}^2 \geq d_n^j.$$

Using Theorem 4, we can pass to the limit as $j \to \infty$ in this inequality and obtain

$$DG_0(w, v' - w) + \lambda_1 DG_1(w, v' - w) + \lambda_2 DG_2(w, v' - w) \geq 0, \quad \forall v' \in W.$$

If $G_{2,s}(w) < 0$, for some s, then for sufficiently large j we have $G_{2n,s}(w_n^j) < 0$ and $\lambda_{2n}^j = 0$, hence $\lambda_2 = 0$, i.e. the transversality condition holds. Therefore, w is extremal. □

5 Numerical Examples

Set $I := [0,1]$, and define the control $\bar{w} = (\bar{w}_1, \bar{w}_2)$, where

$$\bar{w}_1(t) := \begin{cases} 0, & t \in [0, 0.5) \\ \frac{e^{t-0.5}-1}{e^{0.5}-1}, & t \in [0.5, 1] \end{cases} \qquad \bar{w}_2(t) := \frac{e^{1-t}-1}{e-1}, \; t \in I,$$

and the state $\bar{y}(t) := (e^{-t},\, e^{-t},\, 0)$, $t \in I$.

a) Consider the following optimal control problem, with state equations

$$\begin{aligned} &y_1' = -y_2 + w_1 - \bar{w}_1, \quad y_2' = -y_1 + w_2 - \bar{w}_2, \\ &y_3' = [(y_1 - \bar{y}_1)^2 + (y_2 - \bar{y}_2)^2 + (w_1 - \bar{w}_1)^2 + (w_2 - \bar{w}_2)^2]/2, \\ &y_1(0) = y_2(0) = 1, \; y_3(0) = 0, \end{aligned}$$

control constraint set $U := [0,1]^2$, and cost $G(w) := y_3(1)$. Clearly, the optimal control here is $w^* = \bar{w}$, with optimal state $y^* = \bar{y}$ and cost $G(w^*) = G(\bar{w}) = 0$.

The Algorithm was applied to this example without penalties, using the 4th order 4-point Runge-Kutta scheme, with $\bar{\theta}^2 = 1/3$, $\bar{\theta}^3 = 2/3$, the 3/8-Newton-Cotes 4th order 4-point integration rule, and piecewise cubic controls ($l = 3$) with interpolation points coinciding with the Runge-Kutta points

$$t_{ni}^k = \bar{t}_{ni}^k = t_{n,i-1},\; t_{n,i-1} + h_n/3,\; t_{n,i-1} + 2h_n/3,\; t_{ni}.$$

With the following successive step sizes

$$h_n = M^{-j}/60, \text{ for } Kj + 1 \leq n \leq K(j+1),\; j = 0, 1, 2,$$

refining factor $M = l = 3$, refining period $K = 12$, first order DD constraints constant $L' = 10$, gradient projection parameter $\gamma = 0.35$, Armijo step search parameters $b = c = 0.5$, option skipped in Step 4, and constant initial control $(0.5, 0.5)$, we obtained after 36 iterations the results

$$\begin{aligned} G_{0n}(w_n) &= 4.727 \cdot 10^{-14}, \quad d_n = -0.723 \cdot 10^{-21}, \\ \zeta_n &= 2.505 \cdot 10^{-11}, \quad \eta_n = 1.404 \cdot 10^{-11}, \end{aligned}$$

where ζ_n (resp. η_n) is the discrete max control (resp. state) error at the t^k_{ni} (resp. t_{ni}).

b) With the same state equations, cost, and step sizes $M = 3$, $K = 27$, control constraint set $U := [0.2, 1]^2$, and additional state constraint $G_1(w) := y_1(1) - 0.5 = 0$, we obtained after 81 iterations in n the results

$$G_{0n}(w_n) = 1.785503596281293 \cdot 10^{-2},$$
$$G_{1n}(w_n) = y_{1n}(1) - 0.5 = -1.489 \cdot 10^{-6}, \quad d_n = -5.733 \cdot 10^{-7}.$$

Finally, the above results with progressive refining were found to be of similar accuracy to those obtained with constant last step size 1/540, but required here about half the computing time. This shows that finer discretizations become progressively more efficient as the control iterate gets closer to the extremal control, while coarser ones in the early iterations have not much influence on the final results. Anyway, the progressive refining is necessary here for the convergence of the method, since this method uses approximate gradients.

References

1. Chryssoverghi, I., Coletsos, J., Kokkinis, B.: Approximate relaxed descent method for optimal control problems. Control Cybernet. **30** (2001) 385–404
2. Chryssoverghi, I., Coletsos, J., Kokkinis, B.: Approximate gradient projection method with Runge-Kutta schemes for optimal control problems. Comput. Optim. Appl. **29** (2004) 91–115
3. Dontchev, A.L.: An a priori estimate for discrete approximations in nonlinear optimal control. SIAM J. Control Optimization **34** (1996) 1315–1328
4. Dontchev, A.L., Hager, W.W., and Veliov, V.M.: Second-order Runge-Kutta approximations in control constrained optimal control. SIAM J. Numer. Anal. **38** (2000) 202–226
5. Dunn, J.C.: On L^2 sufficient conditions and the gradient projection method for optimal control problems. SIAM J. Control Optim. **34** (1996) 1270–1290
6. Enright, W.H., Jackson, K.R., Norsett, S.P., Thomsen, P.G.: Interpolants for Runge-Kutta formulas. ACM TOMS **12** (1986) 193–218
7. Malanowski, K., Buskens, C., Maurer, H.: Convergence of approximations to nonlinear optimal control problems, Mathematical Programming with Data Perturbations. Lect. Notes Pure Appl. Math. **195** Dekker, New York (1998) 253–284
8. Papakostas, S.N., Tsitouras, CH.: Highly continuous interpolants for one-step ODE solvers and their application to Runge-Kutta methods. SIAM J. Numer. Anal. **34** (1997) 22–47
9. Polak, E.: Optimization Algorithms and Consistent Approximations. Springer Berlin (1997)
10. Schwartz, A., Polak, E.: Consistent approximations for optimal control problems based on Runge-Kutta integration. SIAM J. Control Optim. **34** (1996) 1235–1269
11. Veliov, V.M.: On the time-discretization of control systems. SIAM J. Control Optim. **35** (1997) 1470–1486
12. Warga, J.: Optimal control of Differential and Functional Equations. Academic Press New York (1972)

Stabilization of a Nonlinear Anaerobic Wastewater Treatment Model

Neli S. Dimitrova and Mikhail I. Krastanov

Institute of Mathematics and Informatics,
Bulgarian Academy of Sciences,
Acad. G. Bonchev, Bl. 8, 1113 Sofia, Bulgaria
nelid@bio.bas.bg, krast@math.bas.bg

Abstract. A nonlinear anaerobic digester model of wastewater treatment plants is considered. The stabilizability of the dynamic system is studied and a continuous stabilizing feedback, depending only on an on-line measurable variable, is proposed. Computer simulations are carried out in Maple to illustrate the theoretical results.

1 Introduction

The interest to biological wastewater treatment plants has recently highly increased due to the strong necessity of keeping the quantity of the organic matter in industrial and urban effluents up to a critical level. This necessity has led to development of adequate mathematical models and to application of various techniques for monitoring, optimization and control of the processes [2, 6]. The present paper proposes a continuous stabilizing feedback of a model of an anaerobic digestion process, described by the following ODEs [1]

$$\frac{ds_1}{dt} = u(s_1^i - s_1) - k_1\mu_1 x_1 \tag{1}$$

$$\frac{dx_1}{dt} = (\mu_1 - \alpha u)x_1 \tag{2}$$

$$\frac{ds_2}{dt} = u(s_2^i - s_2) + k_2\mu_1 x_1 - k_3\mu_2 x_2 \tag{3}$$

$$\frac{dx_2}{dt} = (\mu_2 - \alpha u)x_2 \tag{4}$$

$$\frac{dc}{dt} = u(c^i - c) + k_4\mu_1 x_1 + k_5\mu_2 x_2 - Q \tag{5}$$

$$\frac{dz}{dt} = u(z^i - z), \tag{6}$$

where

$$\mu_1 = \mu_1(s_1) = \frac{\mu_{\max} s_1}{k_{s_1} + s_1}, \qquad \mu_2 = \mu_2(s_2) = \frac{\mu_0 s_2}{k_{s_2} + s_2 + \left(\dfrac{s_2}{k_I}\right)^2}$$

I. Lirkov, S. Margenov, and J. Waśniewski (Eds.): LSSC 2005, LNCS 3743, pp. 208–215, 2006.

are model functions for the specific growth rates of the microorganisms. The state variables s_1, s_2, x_1 and x_2 are concentrations (measured in g/l) of chemical oxygen demand (COD), volatile fatty acids (VFA), acidogenic and methanogenic bacteria respectively; c and z are concentrations (measured in mmol/l) of the total inorganic carbon and of the strong ions in the medium; s_1^i, s_2^i, c^i and z^i are the corresponding influent concentrations; u is the dilution rate [day^{-1}]; Q is the gaseous CO_2 molar flow rate; k_1 is the yield coefficient for COD degradation; k_2 is the yield coefficient for VFA production of x_1; k_3 is the yield coefficient for VFA consumption of x_2; k_4 is the yield coefficient for CO_2 production due to x_1; k_5 is the yield coefficient for CO_2 production due to x_2; $\mu_{\max}$ is the maximum acidogenic biomass growth rate; μ_0 is the maximum methanogenic biomass growth rate; k_{s_1} is a saturation parameter associated with s_1; k_{s_2} is a saturation parameter associated with s_2; k_I is an inhibition constant associated with s_2; α is the proportion of dilution rate reflecting the process heterogeneity, taking values from the interval $(0, 1]$.

This model has been experimentally validated for an anaerobic up-flow fixed bed reactor used for the treatment of industrial wine distillery wastewater. More details about that can be found in [1] and in the references there.

The dilution rate u is considered as a control input which takes its values in a convex compact subset $\mathcal{U}$ of the set of the positive real numbers. It is assumed that the substrate concentrations s_1 and s_1^i are measurable. Usually, model-based on-line estimations of the variables s_1 and s_1^i are obtained applying the so called adaptive observers (the Luenberger observer [3, 4], the extended Kalman filter [5] etc.).

The paper is organized as follows. Steady state analysis of the dynamic system (1)–(6) is included in Section 2. A continuous feedback, stabilizing asymptotically the control system (1)–(6), is proposed in Section 3. Computer simulations illustrating the robustness of the proposed feedback, are reported in Section 4.

2 The Equilibrium Points

Let α be an arbitrary point from the interval $(0, 1]$. The equilibrium points are solutions of the nonlinear algebraic system, obtained from (1)–(6) by setting the right-hand side functions equal to zero.

Excluding the trivial solutions $x_1 = 0$, $x_2 = 0$, $s_1 = s_1^i$, $s_2 = s_2^i$ (called wash-out steady states), it is straightforward to see that the equilibrium points are presented by

$$s_1(u) = \frac{\alpha u k_{s_1}}{\mu_{\max} - \alpha u}, \qquad x_1(u) = \frac{s_1^i - s_1(u)}{\alpha k_1},$$

$$s_2(u) = \frac{2\alpha u k_{s_2}}{\mu_0 - \alpha u + \sqrt{\Delta(u)}}, \qquad \Delta(u) = \alpha^2 \left(1 - 4\frac{k_{s_2}}{k_I^2}\right) u^2 - 2\alpha\mu_0 u + \mu_0^2,$$

$$x_2(u) = \frac{s_2^i - s_2(u) + k_2\alpha x_1(u)}{\alpha k_3},$$
$$c(u) = c^i + \alpha(k_4 x_1(u) + k_5 x_2(u)) - \frac{Q}{u}, \qquad z(u) = z^i,$$

and are defined for every u from the interval

$$\mathbf{U} = [u_0, u_1] = \left[u_0, \frac{1}{\alpha}\min\left\{\mu_2\left(k_I\sqrt{k_{s_2}}\right), \mu_1(s_1^i)\right\}\right],$$

where u_0 is such that $c(u_0) \equiv 0$ (such a point exists because $\lim_{u\to 0_+} c(u) = -\infty$ and $c(u_1) > 0$). Note also that $\tilde{s}_2 = k_I\sqrt{k_{s_2}}$ is the point where the function $\mu_2(s_2)$ takes its maximum [4].

Let u^* be an arbitrary point from the interior *int* $\mathbf{U}$ of $\mathbf{U}$. Denote $s_1^* = s_1(u^*)$, $x_1^* = x_1(u^*)$, $s_2^* = s_2(u^*)$, $x_2^* = x_2(u^*)$, $c^* = c(u^*)$, $z^* = z(u^*) = z^i$ and $\zeta^* = (s_1^*, x_1^*, s_2^*, x_2^*, c^*, z^*)^{\mathrm{T}}$. In the following, ζ^* is called a rest or a reference point.

3 Stabilization of the Nonlinear Model

We shall study the asymptotic stabilizability of the system (1)–(6) in a suitable compact and convex neighborhood Ω of the point ζ^*. To state the problem, we introduce some assumptions, notions and notations. Let us fix a positive real number $r > 0$. By $B(\zeta^*, r)$ we denote the closed ball with radius r centered at the point ζ^*. Let the set $\mathcal{U}$ of admissible values of the control is a compact interval containing the set $\mathbf{U}$ in its interior. Any continuous function $k : B(\zeta^*, r) \to \mathcal{U}$ is called a continuous feedback.

Now, let us fix an arbitrary point α from the interval $(0, 1]$. We define the following continuous feedback

$$k_\delta(s_1) := \frac{1}{\alpha}\mu_1(s_1) - \delta(s_1 - s_1^*), \tag{7}$$

where $\delta > 0$ is a parameter. It should be noted that the feedback $k_\delta(\cdot)$ depends only on the on-line measurable variable s_1. Under suitable assumptions, we prove that this feedback stabilizes asymptotically the system (1)–(6) to the point ζ^*.

Theorem 1. *There exist a positive δ and a convex and compact neighborhood Ω of the point ζ^* such that for each point $\zeta \in \Omega$ the feedback (7) stabilizes asymptotically the control system (1)–(6) to ζ^*.*

Proof. Since the proof is very technical, we shall present only its basic features and drop many technicalities.

Let us substitute u by the feedback $k_\delta(\cdot)$ defined in (7) in (1)–(6). The value of the positive parameter δ will be determined later. With respect to the obtained closed-loop system Σ, we have that the set

$$\Omega_0 := \left\{(s_1, x_1, s_2, x_2, c, z)^{\mathrm{T}} : 0 < s_1 < s_1^i, x_1 > 0, s_2 > 0, x_2 > 0, c > 0, z > 0\right\}$$

is positively invariant. This means that every trajectory of Σ starting from a point of Ω_0 remains in Ω_0. In particular, the second and the fourth coordinate, x_1 and x_2, of this trajectory will never vanish.

Let us fix the positive constants δ, $b_{s_1}^-$ and $b_{s_1}^+$ in such a way that $b_{s_1}^- < s_1^* < b_{s_1}^+$ and the values of the feedback $k_\delta(\cdot)$ are admissible values for the control at each point of the set

$$\Omega_1 := \{(s_1, x_1, s_2, x_2, c, z)^{\mathrm{T}} \in \Omega_0 : \ b_{s_1}^- \le s_1 \le b_{s_1}^+\},$$

i. e. $k_\delta(s_1) \in \mathcal{U}$ for each point $(s_1, x_1, s_2, x_2, c, z)^{\mathrm{T}} \in \Omega_1$.

Using the fact that $s_1^i - s_1 = s_1^* - s_1 + \alpha k_1 x_1^*$, the first and the second equation of the closed-loop system Σ can be written as follows:

$$\frac{d}{dt}s_1 = -k_\delta(s_1)[s_1 - s_1^* + \alpha k_1(x_1 - x_1^*)] - \alpha\delta k_1 x_1(s_1 - s_1^*) \tag{8}$$

$$\frac{d}{dt}x_1 = \alpha\delta x_1(s_1 - s_1^*). \tag{9}$$

Consider the function

$$V_1(s_1, x_1) = [s_1 - s_1^* + \alpha k_1(x_1 - x_1^*)]^2 + \alpha(1-\alpha)k_1^2(x_1 - x_1^*)^2.$$

Let us denote by $F_1(s_1, x_1)$ the right-hand side of (8)–(9). It can be directly checked that

$$\begin{aligned}\langle \text{grad } V_1(s_1, x_1), F_1(s_1, x_1)\rangle &= -2\left\{[k_\delta(s_1) + \delta k_1\alpha(1-\alpha)x_1](s_1 - s_1^*)^2\right.\\ &\quad \left.+ 2k_1\alpha k_\delta(s_1)(s_1 - s_1^*)(x_1 - x_1^*) + \alpha^2 k_1^2 k_\delta(s_1)(x_1 - x_1^*)^2\right\}.\end{aligned}$$

The discriminant $D_1(s_1, x_1)$ of the last expression, considered as a quadratic function with respect to the variables $s_1 - s_1^*$ and $x_1 - x_1^*$, is

$$D_1(s_1, x_1) = -4\alpha^3(1-\alpha)k_1^3\delta x_1 k_\delta(s_1).$$

Obviously, the value of $D_1(s_1, x_1)$ is negative on the set $\Omega_1 \setminus \{\zeta^*\}$, hence

$$\langle \text{grad } V_1(s_1, x_1), F_1(s_1, x_1)\rangle < 0 \tag{10}$$

for each point $(s_1, x_1, s_2, x_2, c, z)^T \in \Omega_1 \setminus \{\zeta^*\}$.

Consider now the third and the fourth equation of (1)–(6) with x_1 and u substituted by x_1^* and $u^* := k_\delta(s_1^*)$, respectively, i. e. we consider the following system of differential equations:

$$\frac{d}{dt}s_2 = u^*(s_2^i - s_2) + k_2\alpha u^* x_1^* - k_3\mu_2(s_2)x_2 \tag{11}$$

$$\frac{d}{dt}x_2 = (\mu_2(s_2) - \mu_2(s_2^*))x_2. \tag{12}$$

For each $\nu \in (0, \tilde{s}_2)$, we define the set

$$\Omega_2^\nu := \{(s_1, x_1, s_2, x_2, c, z)^{\mathrm{T}} \in \Omega_1 : \ s_2 \le \tilde{s}_2 - \nu\}.$$

Let us consider the function

$$V_2(s_2, x_2) := [s_2 - s_2^* + \alpha k_3(x_2 - x_2^*)]^2 + \alpha(1-\alpha)k_3^2(x_2 - x_2^*)^2.$$

According to the mean-value theorem, there exists a point θ between the numbers s_2 and s_2^* such that

$$\begin{aligned}\langle \text{grad } V_2(s_2, x_2), F_2(s_2, x_2)\rangle &= -2[u^* + k_3(1-\alpha)\mu_2'(\theta)x_2](s_2 - s_2^*)^2 \\ &- 4k_3\mu_2(s_2^*)(x_2 - x_2^*)(s_2 - s_2^*) - 2\alpha k_3^2\mu_2(s_2^*)(x_2 - x_2^*)^2,\end{aligned} \tag{13}$$

where $F_2(s_2, x_2)$ denotes the right-hand side of (11)–(12). Let $D_2(s_2, x_2)$ be the discriminant of the expression (13), considered as a quadratic function with respect to the variables $s_2 - s_2^*$ and $x_2 - x_2^*$. The function $\mu_2(\cdot)$ is increasing and strictly concave. Moreover, its derivative $\mu_2'(\cdot)$ is a decreasing function which vanishes at the point $\tilde{s}_2$. Hence, $\mu_2'(\theta) > \mu_2'(\tilde{s}_2 - \nu) > \mu_2'(\tilde{s}_2) = 0$. This and the definition of the set Ω_2^ν imply that

$$D_2(s_2, x_2) < -4\alpha(1-\alpha)k_3^3\mu_2(s_2^*)\mu_2'(\tilde{s}_2 - \nu)x_2 < 0$$

on the set $\Omega_2^\nu \setminus \{\zeta^*\}$.

For each $\varepsilon > 0$, we define the function

$$V^\varepsilon(\zeta) = V_1(s_1, x_1) + \varepsilon V_2(s_2, x_2) + \varepsilon(c - c^*)^2 + (z - z^*)^2,$$

where $\zeta := (s_1, x_1, s_2, x_2, c, z)^{\mathrm{T}}$. Clearly, $V^\varepsilon(\cdot)$ is a smooth function which is positive on the set $\Omega_2^\nu \setminus \{\zeta^*\}$ and $V^\varepsilon(\zeta^*) = 0$. Let us denote by $F(\cdot)$ the right-hand side of the closed-loop system Σ. Let us fix an arbitrary point ζ from the set $\Omega_2^\nu \setminus \{\zeta^*\}$. Then $\langle \text{grad } V^\varepsilon(\zeta), F(\zeta)\rangle$ can be represented as a quadratic function with respect to the variables $s_1 - s_1^*$, $x_1 - x_1^*$, $s_2 - s_2^*$, $x_2 - x_2^*$, $c - c^*$, $z - z^*$ whose coefficients depend on ζ. To check that it is negative definite, we calculate successively the leading principal minors of the corresponding symmetric matrix generated by the coefficients of the considered quadratic form:

$$\begin{aligned}
&\Delta_1^\varepsilon(\zeta) = -2(k_\delta(s_1) + \delta k_1\alpha(1-\alpha)x_1) < 0 \\
&\Delta_2^\varepsilon(\zeta) = -D_1(s_1, x_1) > 0 \\
&\Delta_3^\varepsilon(\zeta) < -2\varepsilon D_1(s_1, x_1)(u^* + k_3x_2(1-\alpha)\mu_2'(\tilde{s}_2 - \nu)) + o_1(\varepsilon) \\
&\Delta_4^\varepsilon(\zeta) = \varepsilon^2 D_1(s_1, x_1)D_2(s_2, x_2) + o_2(\varepsilon^2) \\
&\Delta_5^\varepsilon(\zeta) = -2\varepsilon^3 k_\delta(s_1)D_1(s_1, x_1)D_2(s_2, x_2) + o_3(\varepsilon^3), \\
&\Delta_6^\varepsilon(\zeta) = 4\varepsilon^3 k_\delta^2(s_1)D_1(s_1, x_1)D_2(s_2, x_2) + o_4(\varepsilon^3).
\end{aligned}$$

Clearly, if we fix $\varepsilon_0 > 0$ to be sufficiently small, then

$$\Delta_3^{\varepsilon_0}(\zeta) < 0,\ \Delta_4^{\varepsilon_0}(\zeta) > 0,\ \Delta_5^{\varepsilon_0}(\zeta) < 0,\ \Delta_6^{\varepsilon_0}(\zeta) > 0.$$

Hence,

$$\langle \text{grad } V^{\varepsilon_0}(\zeta), F(\zeta)\rangle < 0 \tag{14}$$

for each point $\zeta \in \Omega_2^\nu \setminus \{\zeta^*\}$. Let us choose $\beta > 0$ as large as possible and such that the set

$$\Omega := \{\zeta \in R^6 : V^{\varepsilon_0}(\zeta) \leq \beta\}$$

is a subset of the set Ω_2^ν. Clearly, Ω is a compact and convex neighborhood of the point ζ^*. Moreover, the relation (14) implies that $V^{\varepsilon_0}(\cdot)$ is a Lyapunov function of the closed-loop system Σ on the set Ω. So, Σ is asymptotically stable on Ω.

This completes the proof.

4 Numerical Simulation

The theoretical results are illustrated numerically and graphically using the computer algebra system *Maple*.

From [1] we take the following numerical values for the model coefficients: $Q = 20$ [mmol/l], $k_1 = 10.53$ [g COD/g], $k_2 = 28.6$ [mmol VFA/g], $k_3 = 1074$ [mmol VFA/g], $k_4 = 12.42$ [mmol CO_2/g], $k_5 = 1375$ [mmol CO], $\mu_{\max} = 1.2$ [day^{-1}], $\mu_0 = 0.74$ [day^{-1}], $k_{s_1} = 7.1$ [g COD/l], $k_{s_2} = 9.28$ [(mmol VFA/l)$^{1/2}$], $k_I = 16$ [mmol VFA/l].

With $\alpha = 0.5$ and $s_1^i = 6.9$, $s_2^i = 70$, $c^i = 52$, $z^i = 67$, the admissible interval for the control is $\mathbf{U} = [0.1169,\ 1.0718]$. For $u^* = 0.4 \in int\,\mathbf{U}$ (which

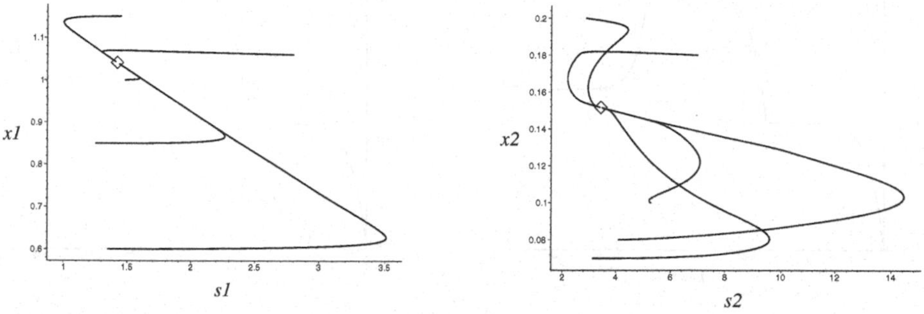

Fig. 1. Phase portraits in (s_1, x_1) (left) and (s_2, x_2) (right) phase planes

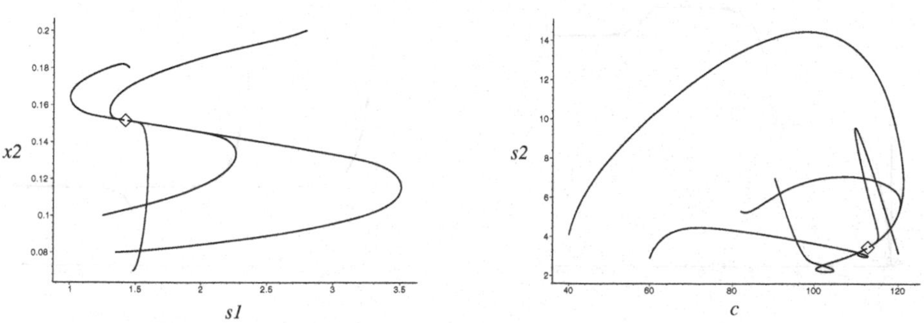

Fig. 2. Phase portraits in (s_1, x_2) (left) and (c, s_2) (right) phase planes

is close to the on-line experimental results in [1]), the rest point ζ^* is $\zeta^* = (1.42,\ 1.04,\ 3.45,\ 0.15,\ 112.71,\ 67)$. Using the feedback $k_\delta(s_1)$ with $\delta = 7$, Figures 1 and 2 present the projections of the phase portraits of the stabilized system on different phase planes. In all plots, the symbol diamond $\diamond$ denotes the corresponding projection of the rest point.

In practice, the influent concentrations s_1^i, s_2^i, c^i and z^i are usually not exactly known. For the computer simulations we assume that these values change stepwise (see [1], Figures 2 to 5):

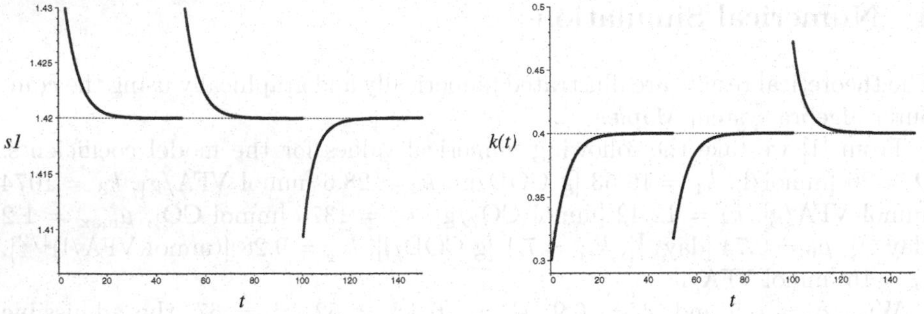

Fig. 3. Time evolution of the state variable $s_1(t)$ (left) and of the feedback $k_\delta(t)$ (right)

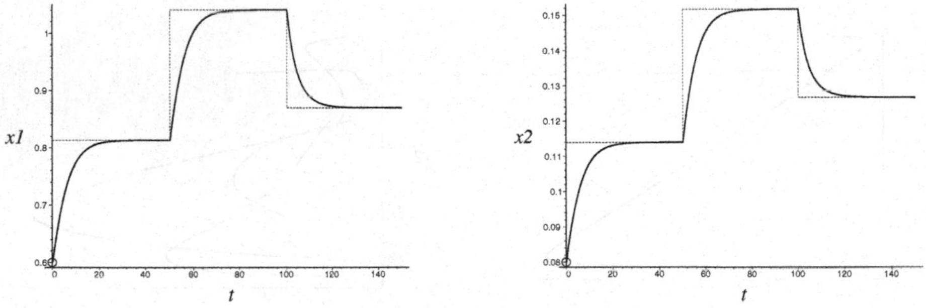

Fig. 4. Time evolution of the state variables $x_1(t)$ (left) and $x_2(t)$ (right)

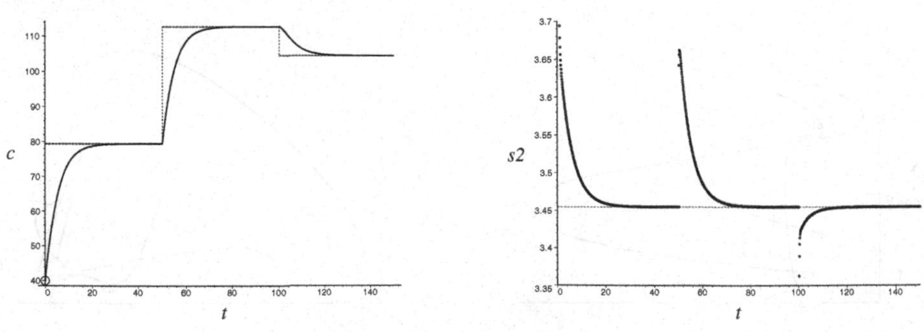

Fig. 5. Time evolution of the state variables $c(t)$ (left) and $s_2(t)$ (right)

$$s_1^i \in \{5.7,\ 6.9,\ 6\},\ s_2^i \in \{53,\ 70,\ 59\},\ c^i \in \{46,\ 52,\ 62\},\ z^i \in \{53,\ 67,\ 58\}.$$

The end-points of the admissible interval $\mathbf{U}$ and the steady states $x_1(u)$, $x_2(u)$, $c(u)$ and $z(u)$ depend explicitly on s_1^i, s_2^i, c^i and z^i, thus they change as well. The three different admissible intervals have an intersection $\mathbf{U} = [0.1461,\ 1.0687]$. The three reference points ζ_1^*, ζ_2^* and ζ_3^*, related to the above step changes in the influent concentrations are $\zeta_1^* = (1.42, 0.81, 3.45, 0.11, 79.36, 53)$, $\zeta_2^* = (1.42,\ 1.04,\ 3.45,\ 0.15,\ 112.7,\ 67)$, $\zeta_3^* = (1.42,\ 0.87,\ 3.45,\ 0.13,\ 104.4,\ 58)$. Starting with the initial point $s_1(0) = 1.39$, $x_1(0) = 0.6$, $s_2(0) = 4.1$, $x_2(0) = 0.08$, $c(0) = 40$, $z(0) = 51$, the numerical results are visualized on Figures 3, 4 and 5. In these plots the symbol circle $\circ$ denotes the corresponding coordinate of the initial point. The horizontal dot-line segments go through the corresponding coordinate of the rest point. The right plot in Figure 3 shows the time profile of the feedback $k_\delta(t)$; the horizontal dot-line segment goes trough the point u^*.

Acknowledgement. This research has been partially supported by the Bulgarian National Science Fund under contract MM-1104/01.

References

1. Alcaraz-González, V., Harmand, J., Rapaport, A., Steyer, J., González–Alvarez, V., Pelayo-Ortiz, C.: Software sensors for highly uncertain WWTPs: a new apprach based on interval observers. Water Research **36** (2002) 2515–2524.
2. Bernard, O., Hadj-Sadok, Z., Dochain, D.: Advanced monitoring and control of anaerobic wastewate treatment plants: dynamic model development and identification. In: Proc. of the Fifth IWA International Symposium WATERMATEX, Gent, Belgium (2000) 3.57–3.64.
3. Ciccarella, C., Dalla, M., Germani, A.: A Luenberger-like observer for nonlinear systems. Int. J. Control **57** (1993) 536–556.
4. Bastin, G., Dochain, D.: On-line estimation and adaptive control of bioreactors. Elsevier, Amsterdam (1990).
5. Petersen, I., Savkin, A.: Robust Kalman filtering for signals and systems with large uncertainties. Birkhäuser, Boston (1999).
6. Simeonov, I., Babary, J., Lubenova, V., Dochain, D.: Linearizing control of continuous anaerobic fermentation processes. In: Proc. Bioprocess Systems'97 **III** (1997) 21–24.

Approximation of the Solution Set of Optimal Control Problems

Tzanko Donchev

Department of Mathematics,
University of Architecture and Civil Engineering,
1 Hr. Smirnenski, 1046 Sofia, Bulgaria
tdd51us@yahoo.com

Abstract. We investigate approximation in $W^{1,2}$ topology of the solution set of a differential inclusion with Kamke Lipschitz right-hand side. The results are then applied to Bolza optimal control problem in form of differential inclusions. Namely it is shown that the optimal solution is the limit of optimal solution of appropriately defined finite dimensional nonlinear programming problems.

1 Introduction

Here we continue recent authors investigations presented in [2]. We study discrete approximations of the following differential inclusion:

$$\dot{x}(t) \in F(t, x(t)),\ x(0) = x_0,\ t \in I = [0, 1] \tag{1}$$

in order to apply the results to the Bolza optimal control problem:

$$\min J[x] := \varphi(x(1)) + \int_0^1 f(t, x(t), \dot{x}(t))\, dt \tag{2}$$

subject to (1). Here and further in the paper $F(\cdot, \cdot)$ is a continuous map from $I \times \mathbb{R}^n$ into $C(\mathbb{R}^n)$ (the set of all nonempty compact subsets of $\mathbb{R}^n$).

The problem of finding optimal Hamiltonian or/and Euler–Lagrange conditions is studied in a large number of papers. We note here only [1, 7, 9, 12] and the references therein. In all these papers the right-hand side is assumed to be locally Lipschitz.

There are also a lot of papers devoted to approximation of the solution and reachable set of (1). We mention only [4, 6], survey papers [5, 8] and the references therein.

We note [9], where refined Euler-Lagrange optimal conditions are obtained with the help of discrete approximations. Although the optimality conditions require some properties of subdifferentials which hold only for locally Lipschitz functions, the approximation in $W^{1,2}$ can be proved for much larger class.

Denote by $\mathbb{B}$ the open unit ball in $\mathbb{R}^n$. Given two closed bounded sets A, B we let $D_H(A, B) := \max\{\max_{b\in B} \min_{a\in A} |b - a|, \max_{a\in A} \min_{b\in B} |b - a|\}$ – Hausdorff distance.

I. Lirkov, S. Margenov, and J. Waśniewski (Eds.): LSSC 2005, LNCS 3743, pp. 216–222, 2006.

Recall that a continuous function $w : I \times \mathbb{R}^+ \to \mathbb{R}^+$ is said to be Kamke function if $w(t,0) \equiv 0$ and the only solution of the differential equation $\dot{s}(t) = w(t, s(t))$, $s(0) = 0$ is $s(t) \equiv 0$.

Standing Hypothesis

1) There exists $\lambda > 0$ such that $|F(t,x)| \le \lambda(1 + |x|)$ for every x.
2) $F(\cdot,\cdot)$ is continuous multifunction with nonempty compact values.

Further we assume that for every bounded set $A \subset \mathbb{R}^n$ there exists a Kamke function $w_A(\cdot,\cdot)$ such that $D_H(F(t,x), F(t,y)) \le w_A(t, |x-y|)$ for every $x, y \in A$, which we call locally Kamke–Lipschitz (KL) condition.

It is shown in [2] that almost all continuous convex compact valued (CCC) multi-functions satisfy the locally KL condition. That means that our results are applicable for almost all CCC differential inclusions. Commonly used locally Lipschitz condition hold on a first Baire category set (in the complete metric space of continuous CCC multi-functions).

When the right-hand side is one sided Lipschitz the approximation of the solution set (in $C(I, \mathbb{R}^n)$) is studied in [3]. Although this condition is weaker than Lipschitz one and it does not follow from the Kamke–Lipschitz condition it is shown in [2] that the set of all locally one sided Lipschitz continuous multi-functions is of first Baire category. Of course one can study locally one sided KL multi-functions, however, it is not clear how to obtain $W^{1,2}$ approximation replacing KL by one sided KL.

The aim of this paper is the extension of the approximation results of Mordukhovich [9] to the case of locally KL right-hand side. Notice that our results hold for almost all continuous differential inclusions (as it was pointed out above). This implies that the approach of [9, 10] is more flexible then the approach in [1, 7, 12] etc.

We mentioned the paper [13], where the approximation of reachable set for locally Kamke–Lipschitz differential inclusions is studied. More for Kamke functions one can find in [11, 14].

Let $\Delta_k = \left\{ t_j = \frac{j}{k} \right\}$, $j = 0, 1, \ldots, k$ be a uniform grid of I. To (1) we juxtapose the following discrete system:

For $j = 0, 1, \ldots, k-1$ we let $x_j^k = \lim_{t \uparrow t_j} x^k(t)$ and for $t \in [t_j, t_{j+1})$:

$$\dot{x}^k(t) = v_j \in F(t_j, x_j^k), \; x^k(t) = x_j^k + v_j(t - t_j). \tag{3}$$

Of course $x^k(0) = x_0$.

2 The Results

In this section we prove (in more general form) two theorems announced in [2]. These theorems are (partial) extensions of Theorem 3.1 and Theorem 3.3 of [9] respectively.

First we study $W^{1,2}$ approximation of the solution set of (1) with discrete trajectories (the solution set of (3)).

Theorem 1. *Assume F satisfies the* **standing hypothesis** *and locally KL condition. Let $x(\cdot)$ be a solution of (1). Then there exists a sequence $\{x^k(\cdot)\}_{k=1}^{\infty}$ of solutions of (3) such that*

$$\lim_{k\to\infty} \max_{t\in I} |x^k(t) - x(t)| = 0, \quad \textit{and}$$

$$\lim_{k\to\infty} \int_I |\dot{x}(t) - \dot{x}^k(t)|^2 \, dt = 0.$$

Proof. First it is easy to see that there exist constants M and P such that $|x(t)| \le M$ and $|F(t, x(t) + \mathbb{B})| \le P$ for every absolutely continuous function $x(\cdot)$ with $x(0) = x_0$ and $\dot{x}(t) \in F(t, x(t) + \mathbb{B})$. Hence without loss of generality one can assume that $F(t, \cdot)$ is KL with respect to Kamke function $w(\cdot, \cdot)$.

On the set $A := I \times (M\mathbb{B})$ the mappings $F(\cdot, \cdot)$ and $w_A(\cdot, \cdot)$ are uniformly continuous. Furthermore every discrete or continuous trajectory is P–Lipschitz.

Since the step functions are dense in $L^2(I)$, one has that there exists a sequence of polygonal functions $y^k(\cdot)$ (functions with constant on $[t_j, t_{j+1})$ derivatives) such that $\lim_{k\to\infty} \int_I |\dot{x}(t) - \dot{y}^k(t)|^2 \, dt = 0$ and $y^k(t) \to x(t)$ uniformly on I. Moreover if $u^k(\cdot) \to \dot{x}(\cdot)$ in $L^2(I)$, then for every $\varepsilon > 0$ there exists a compact $I_\varepsilon \subset I$ with measure greater than $1 - \varepsilon$ such that $u^k(\cdot) \to \dot{x}(\cdot)$ uniformly on I_ε. Since $F(\cdot, \cdot)$ is uniformly continuous, one has that for every $\delta > 0$ there exists $k(\delta)$ such that for $j = 0, 1, \ldots, k-1$ we have $\operatorname{dist}(\dot{y}^k(t), F(t_j, y_j^k)) \le \delta$ for every $t \in [t_j, t_{j+1})$ and every $k > k(\delta)$.

Let $\delta_n \downarrow 0$ as $n \to \infty$. We let δ above to be equal to δ_n For $j = 0, 1, \ldots, k - 1$ we define $x_j^{k_n} = \lim_{t \uparrow t_j} x^{k_n}(t)$, $x^{k_n}(t) = x_j^{k_n} + (t - t_j)v^{k_n}$. Here $t \in [t_j, t_{j+1})$ and v^{k_n} belongs to the projection set of $\dot{y}^{k_n}(t)$ on $F(t_j, x_j^{k_n})$. It is easy to see that $\left|\dot{x}^{k_n}(t) - \dot{y}^{k_n}(t)\right| \le w\left(t, \left|x^{k_n}(t) - y^{k_n}(t)\right| + \delta_n\right) + \delta_n$. Therefore denoting $r_{k_n}(t) = \left|x^{k_n}(t) - y^{k_n}(t)\right|$ we have: $r_{k_n}(0) = 0$ and $\dot{r}_{k_n}(t) \le w(t, r_{k_n}(t) + \delta_n) + \delta_n$.

Since $w(\cdot, \cdot)$ is a Kamke function, one has that $\lim_{n\to\infty} \left|x^{k_n}(t) - y^{k_n}(t)\right| = 0$ uniformly on I. Moreover, $\lim_{n\to\infty} \int_I \left|\dot{x}^{k_n}(t) - \dot{y}^{k_n}(t)\right|^2 dt = 0$. The proof is therefore complete thanks to the triangle inequality. □

Remark 1. Theorem 1 when $F(\cdot, \cdot)$ is also convex valued (provided with very concise and not complete proof) is given in [2].

Notice that for absolutely continuous functions with bounded by P derivatives the $W^{1,1}$ and $W^{1,2}$ norms are equivalent.

The following definition is from Mordukhovich [9]:

Definition 1. *The arc $\bar{x}(\cdot)$ is said to be a local intermediate minimum (LIM) if there exists $\varepsilon > 0$ such that $J[\bar{x}] \le J[x]$ for any other trajectory $x(\cdot)$, satisfying $|\bar{x}(t) - x(t) < \varepsilon$, $\forall t \in I$ and $\int_0^1 |\dot{x}(t) - \dot{\bar{x}}|^2 \, dt < \varepsilon$.*

Let $\bar{x}(\cdot)$ be LIM. To optimal control problem (1) – (2) we juxtapose the following discrete minimization problem:

$$\min\ J_k(x^k) = \varphi(x_k^k) + \frac{1}{k}\sum_{j=0}^{k-1} f\left(t_j^k, x_j^k, k\left(x_{j+1}^k - x_j^k\right)\right)$$
$$+ \sum_{j=0}^{k-1}\int_{t_j}^{t_{j+1}} \left|\left(x_{j+1}^k - x_j^k\right)k - \dot{\bar{x}}(t)\right| dt, \tag{4}$$

for discrete trajectories $x^k(\cdot)$ such that $|x_j^k - \bar{x}(t_j)| \le \frac{\varepsilon}{2}$, $j = 0, 1, \ldots, k$ and

$$\sum_{j=0}^{k-1}\int_{t_j}^{t_{j+1}} \left|k(x_{j+1}^k - x_j^k) - \dot{\bar{x}}(t)\right| dt \le \frac{\varepsilon}{2}. \tag{5}$$

The functions $f(\cdot,\cdot,\cdot)$ and $\varphi(\cdot)$ are assumed to be continuous.

Theorem 2. *Assume all the conditions of Theorem 1 hold. Let $z(\cdot)$ be LIM of the problem (1) – (2). Then every sequence $\{x^n(\cdot)\}_{n=1}^{\infty}$ of solutions of the discrete problem (3) – (4) converges as $n \to \infty$ to $z(\cdot)$ on $W^{1,2}(I)$.*

Proof. We will follow with modifications the proof of Theorem 3.3 of [9]. It follows from Theorem 1 that every optimal solution of (3) – (4) satisfies (5) for n big enough. Due to Weierstrass theorem such an optimal solution, denoted by $x^n(\cdot)$, exists.

First we will show that $\lim_{n\to\infty} J_n[x^n] = J[z]$, where the sequence $x^k(\cdot)$ approximates $z(\cdot)$ in $W^{1,2}$.

Since $\varphi(\cdot)$ is continuous, one has that $\varphi(x^n(1)) \to \varphi(z(1))$. Also

$$\lim_{n\to\infty}\sum_{j=0}^{k-1}\int_{t_j}^{t_{j+1}} \left|\frac{x_{j+1}^k - x_j^k}{h_k} - \dot{z}(t)\right| dt = 0,$$

because $x^n(\cdot) \to z(\cdot)$ in $W^{1,2}$ and hence also in $W^{1,1}$.

We have only to see that

$$\frac{1}{n}\sum_{j=0}^{n-1} f\left(t_j^n, x_j^n, n(x_{j+1}^n - x_j^n)\right) \to \int_0^1 f(t, x(t), \dot{z}(t))\, dt.$$

Since $f(\cdot,\cdot,\cdot)$ is continuous one has that:

$$\lim_{n\to\infty}\frac{1}{n}\sum_{j=0}^{n-1} f\left(t_j, x_j^n, n\left(x_{j+1}^n - x_j^n\right)\right) = \lim_{n\to\infty}\sum_{j=0}^{n-1}\int_{t_j}^{t_{j+1}} f\left(t_j, x_j^n, v^n(t)\right) dt$$
$$= \lim_{n\to\infty}\sum_{j=0}^{n-1}\int_{t_j}^{t_{j+1}} f(t, x^n(t), v^n(t))\, dt = \lim_{n\to\infty}\sum_{j=0}^{n-1}\int_{t_j}^{t_{j+1}} f(t, z(t), \dot{z}(t))\, dt$$

If $\bar{x}^n(\cdot)$ is an optimal solution of (3) – (4) then $\limsup_{n\to\infty} J_n[\bar{x}^n] \le J[z]$.

Now we will use the fact that $z(\cdot)$ is LIM for (1) – (2). We let $c_n := \int_I |\dot{\bar{x}}^n(t) - \dot{z}(t)|^2\,dt$. It remains to prove that $\lim_{n\to\infty} c_n = 0$. Suppose the contrary, i.e. $\limsup_{n\to\infty} c_n = c > 0$. Since $x^n(\cdot)$ are P – Lipschitz and $x^n(0) = x_0$, one has that (passing to subsequences if necessary) $x^n(\cdot) \to x(\cdot)$ uniformly on I and $\dot{x}^n(\cdot) \rightharpoonup x(\cdot)$ weakly $L^2(I)$. Furthermore

$$\lim_{n\to\infty} \left| \frac{1}{n} \sum_{j=0}^{n-1} f\left(t_j, x_j^n, n\left(x_{j+1}^n - x_j^n\right)\right) - \int_I f(t, x^n(t), \dot{x}^n(t))\,dt \right| \to 0$$

thanks to Theorem 1.

Due to Mazur theorem there is a sequence of convex combinations which (passing to subsequences) converges for a.e. $t \in I$ to $\dot{x}(\cdot)$.

We let $f_F(t, x, v) := f(t, x, v) + \chi_{F(t,x)}(v)$, where

$$\chi_A(v) = \begin{cases} 0 & \text{if } v \in A \\ \infty & \text{otherwise.} \end{cases}$$

Let $\hat{f}_F$ be the biconjugate function of f_F. Obviously $x(\cdot)$ is a solution of the relaxed differential inclusion (the right-hand side is closed convex hull of $F(t,x)$). It is easy to scc that

$$\int_I \hat{f}_F(t, x(t), \dot{x}(t))\,dt \le \liminf_{n\to\infty} \frac{1}{n} \sum_{j=0}^{n-1} f\left(t_j, x_j^n, n\left(x_{i+1}^n - x_j^n\right)\right). \tag{6}$$

Furthermore $I[v] := \int_I |v(t) - \dot{x}(t)|^2\,dt$ is lower semicontinuous with respect to the weak topology of $L^2(I)$. Therefore

$$\int_I |\dot{x}(t) - \dot{z}(t)|^2\,dt \le \liminf_{n\to\infty} \frac{1}{n} \sum_{j=0}^{n-1} \int_{t_j}^{t_{j+1}} \left|n\left(x_{i+1}^n - x_j^n\right) - \dot{z}(t)\right|^2 dt. \tag{7}$$

One has $|x(t) - z(t)| \le \frac{\varepsilon}{2}$ on I. Furthermore $\int_I |\dot{x}(t) - \dot{z}(t)|^2 \le \frac{\varepsilon}{2}$ thanks to (7). Due to (6) we have:

$$\hat{J}[x] := \varphi(x(1)) + \int_I \hat{f}_F(t, x(t), \dot{x}(t))\,dt + c \le \liminf_{n\to\infty} J_n[x^n].$$

Thus $\hat{J}[x] < J[z]$ – contradiction. □

3 Concluding Remarks

Remark 2. When $F(\cdot,\cdot)$ is Caratheodory (almost continuous) one can study $W^{1,2}$ approximation of (1) by (3) with the help of the multifunctions:

$$F_h(t,x) = \frac{1}{\operatorname{meas}(I_{t,h})} \int_{I_{t,h}} F(t,x)\,dt.$$

Here $I_{t,h} = I \bigcap [t-h, t+h]$. Obviously $D_H\left(F_h(t,x), F_h(t,y)\right) \le w_h(t, |x-y|)$, where $w_h(\cdot,\cdot)$ is defined as $F_h(\cdot,\cdot)$. The main difficulty is to see when $w_h(\cdot,\cdot)$ is a Kamke function.

These questions are interesting, but are not considered here.

Notice that our results may be extended to the case of functional differential systems having the form:

$$\dot{x}(t) \in F(t, x_t), \quad x_0 = \phi.$$

Here $x \in \mathbb{R}^n$, $t \in I = [0,1]$ and F is a compact valued map from $I \times X$ into $\mathbb{R}^n$. $X = C([-\tau, 0], \mathbb{R}^n)$ is the space of the continuous maps from $[-\tau, 0]$ into $\mathbb{R}^n$ and $x_t(s) = x(t-s)$ for $s \in [-\tau, 0]$. The minimization criterium is:

$$\min J[x] := \varphi(x(1)) + \int_0^1 f(t, x(t), x_t, \dot{x}(t))\,dt.$$

This problem requires some other techniques and will be studied in another paper.

Acknowledgement

This work was supported by the Australian Research Council Discovery-Project Grant DP0346099.

References

1. Clarke F., Ledyaev Yu., Stern R., Wolenski P.: *Nonsmooth Analysis and Control Theory*, Springer, New York, 1998.
2. Donchev T.: Generic properties of differential inclusions and control problems. *Numerical analysis and its applications*, Zhilin Li, Lubin Vulkov, Jerzy Wasniewski eds., *Lecture notes in computer sciences*, **3401**, Springer Verlag, Berlin Heidelberg, 2005, 266–271.
3. Donchev T., Farkhi E.: Euler approximation of discontinuous one-sided Lipschitz convex differential inclusions, *Calculus of Variations and Differential Equations* A. Ioffe, S. Reich and I. Shafrir eds., *Chapman & Hall/CRC* Boca Raton, New York, 1999, 101–118.
4. Dontchev A., Farkhi E.: Error estimates for discretized differential inclusions. *Computing*, **41** (1989), 349–358.

5. Dontchev A., Lempio F.: Difference methods for differential inclusions: a survey. *SIAM Rev.*, **34** (1992), 263–294.
6. Grammel G.: Towards fully discretized differential inclusions. *Set Valued Analysis*, **11** (2003), 1–8.
7. Ioffe A.: Euler–Lagrange and Hamiltonian formalisms in dynamical optimization. *Trans. Amer. Math. Soc.*, **349** (1997), 2871–2900.
8. Lempio F., Veliov V.: Discrete approximations of differential inclusions, *Bayreuter Mathematische Schiften*, Heft **54** (1998), 149–232.
9. Mordukhovich B.: Discrete approximations and refined Euler-Lagrange conditions for differential inclusions. *SIAM J. Control. Optim.*, **33** (1995), 882–915.
10. Mordukhovich B.: *Approximation Methods in Problems of Optimization and Control*, Nauka, Moskow, 1988 (in Russian)
11. Pianigiani G.: On the fundamental theory of multivalued differential equations. *J. Differential Equations*, **25** (1977), 30–38.
12. Pshenichniy B.: *Convex Analysis and Extremal Problems*, Nauka, Moskow, 1980. (in Russian).
13. Rurenko E.: Approximate calculation of the attainability set for a differential inclusions. *Vestn. Mosk. Univ. Ser*, **15**, Vych. Math. Kibern. (1989), 81–83.
14. Tolstonogov A.: *Differential Inclusions in a Banach Space*, Kluwer, Dordrecht, 2000.

Trajectory Tubes to Impulsive Control Systems

Tatiana F. Filippova

Institute of Mathematics and Mechanics, Russian Academy of Sciences,
16 S. Kovalevskaya Str., 620219 Ekaterinburg GSP-384, Russia

Abstract. The properties of set-valued solutions (trajectory tubes) for measure driven (impulsive) differential control systems are considered. Numerical simulation results related to the procedures of set-valued approximations of trajectory tubes of linear impulsive systems are also given.

1 Introduction

In many applied problems the evolution of a dynamical system depends not only on the current system states but also on uncertain disturbances or errors in modelling. There are many publications devoted to different aspects of treatment of uncertain dynamical systems (e.g., [1, 2, 3, 4, 5]).

The model of uncertainty considered here is deterministic, with set — membership description of uncertain items which are taken to be unknown but bounded with given bounds.

In the paper we consider the properties of set-valued solutions (trajectory tubes) for measure driven (impulsive) differential control systems. The solution to the impulsive differential system is introduced and studied here in the framework of the theory of uncertain dynamical systems through the techniques of informational states (or set-valued state vectors) of the impulsive system:

$$dx(t) = f(x(t), u(t))dt + B(x(t), u(t))dv(t), \tag{1}$$

$$x \in R^n, \quad t_0 \le t \le T,$$

with unknown but bounded initial condition

$$x(t_0 - 0) = x_0, \quad x_0 \in X_0. \tag{2}$$

Here $u(t)$ is a usual (measurable) control, with constraint: $u(t) \in U$, $U \subset R^m$, and $v(t)$ is an impulsive control function which is continuous from the right, with the bounded variation on $[t_0, T]$. The trajectory tube to the system (1) - (2) is the set

$$X[\cdot] = \{x[\cdot] = x(\cdot, t_0, x_0) | x_0 \in X_0\} \tag{3}$$

of solutions to (1) – (2) with its t-cross-section $X[t]$ being the reachable set (the informational set) at instant t of the system (1)–(2) which is found under above given assumptions on the uncertainty constraints.

I. Lirkov, S. Margenov, and J. Waśniewski (Eds.): LSSC 2005, LNCS 3743, pp. 223–230, 2006.

In such problems the trajectories $x(t)$ are discontinuous and belong to a space of functions with bounded variation. Among many problems related to treatment of dynamic systems of this kind let us mention the results devoted to a precise definition of a solution to (1) especially for the case $B = B(t,x)$ ([6]) and a long list of publications concerning the optimality conditions (e.g., [7, 8, 9]).

In this paper we consider the properties of generalized trajectory tubes $X[\cdot]$ (such as compactness, sensitivity or continuity of these tubes on some groups of parameters that define the restrictions on uncertain data). Numerical results using examples related to procedures of set-valued approximations of $X[\cdot]$ are also presented.

2 Nonlinear Systems: Qualitative Results

Basing on the techniques of approximation of the discontinuous generalized trajectory tubes to (1)–(2) by the solutions of usual differential systems without measure terms [10, 11] it is possible to study the dependence of generalized trajectory tubes and their cross-sections (reachable sets) on parameters that define the restrictions on uncertain values (initial data, a variation of impulses, constraints on measurable controls).

2.1 Uncertain Impulsive Systems with Parameter Disturbance

In this section we consider a dynamic system described by the following differential equation with measures

$$dx(t) = f(t, x(t), u(t))dt \; + \; B(t)dv(t), \tag{4}$$

with unknown but bounded initial condition

$$x(t_0 - 0) = x^0, \quad x^0 \in X^0_\lambda. \tag{5}$$

Here $x \in R^n$, $t_0 \le t \le T$ and we assume that matrix function B in (1) depends on time only and $u(t)$ is a usual (measurable) control with the parameterized constraint

$$u(t) \in U_\lambda, \quad U_\lambda \subset R^m,$$

and $v(t)$ is an impulsive control function which is continuous from the right, with

$$\mathrm{Var}_{t\in[t_0,T]}\, v(t) \le \mu_\lambda.$$

Here λ is a finite dimensional parameter ($\lambda \in R^k$) that will be taken as tending to some fixed value λ_0 (obviously without loss of generality we may take $\lambda_0 = 0$).

We study the dependence of trajectory tubes $X[\cdot]$ of the impulsive differential system on system parameters λ that define the constraints on initial data, a variation of impulses, restrictions on measurable controls. These parameter disturbances may be treated, e. g., as errors of the system modelling or as hard bounds on admissible system noises.

So the main problem of the section is to study the sensitivity of the considered differential system (and its set-valued solutions) with respect to model errors.

Along with the system (4)–(5) let us consider a new system, namely, a differential inclusion of the following type

$$dx(t) \in F_\lambda(t, x(t))dt + B(t)dv(t), \tag{6}$$

with the initial condition

$$x(t_0 - 0) = x^0, \quad x^0 \in X^0_\lambda. \tag{7}$$

Here we use the notation

$$F_\lambda(t, x) = f(t, x, U_\lambda) = \cup\{ f(t, x, u) \mid u \in U_\lambda \}$$

for the set-valued map F_λ in (6).

The introduction of this differential inclusion (6) which we will study further may be motivated by the well known results given by the control theory [12] and also by results of the theory of differential inclusions [13, 14].

Let $X_\lambda(\cdot, t_0, X^0_\lambda)$ be the set of all solutions to the inclusion (6) that emerge from X^0_λ (the trajectory tube related to all initial state vectors x^0 and all admissible impulse controls $v(t)$ is defined in (3)). Denote $X_\lambda[t] = X_\lambda(t, t_0, X^0_\lambda)$ to be its cross-section at instant t. The set $X_\lambda[t]$ is actually the reachable set of the impulsive differential inclusion (6) - (7) (or, equivalently, of the impulsive control system (4) – (5)) from the initial set X^0_λ taken at instant t.

So the main problem may be reformulated in terms of the differential inclusions theory, i.e. in order to answer the main questions we need to find first the type of the dependance of set-valued solutions $X_\lambda[\cdot]$ of (6) on the variation of parameter vectors λ.

2.2 Preliminaries

Assume that F_λ is a continuous multivalued map ($F_\lambda : [t_0, T] \times R^n \to \mathrm{conv}R^n$) that satisfies the Lipschitz condition with constants $L_1, L_2 > 0$, namely for all $x, y \in R^n$ and $t_1, t_2 \in [t_0, T]$ we have

$$h(F_\lambda(t_1, x), F_\lambda(t_2, y)) \leq L_1|t_1 - t_2| + L_2 \parallel x - y \parallel,$$

where $\mathrm{conv}R^n$ denotes the space of all compact and convex subsets of R^n and $h(A, B)$ is the Hausdorff distance [2, 13, 14] for A, $B \subseteq R^n$.

Assume also the Lipschitz continuity of the matrix function $B(t)$

$$\parallel B(t_1) - B(t_2) \parallel \leq L_3|t_1 - t_2|, \ \forall t_1, t_2 \in [t_0, T]$$

and also the so-called extendability condition [12]

$$F_\lambda(t, x) \subset c(1 + ||x||)S, \quad S = \{ x \in R^n \mid ||x|| \leq 1 \}.$$

Definition 1. *A function* $x[t] = x(t, t_0, x^0)$ $(x^0 \in X^0_\lambda,\ t \in [t_0, T])$ *will be called a solution (a trajectory) of the differential inclusion (6) if for all* $t \in [t_0, T]$

$$x[t] = x^0 + \int_{t_0}^{t} \psi(t)dt + \int_{t_0}^{t} B(t)dv(t), \tag{8}$$

where $\psi(\cdot) \in L_1^n[t_0, T]$ *is a selector of* F_λ*, i.e.* $\psi(t) \in F_\lambda(t, x[t])$ *a.e.*

The last integral in (8) is taken as the Riemann–Stieltjes one. Following the scheme of the proof of the well-known Caratheodory theorem we can prove the existence of solutions $x[\cdot] = x(\cdot, t_0, x^0) \in BV^n[t_0, T]$ for all $x^0 \in X^0$ where $BV^n[t_0, T]$ is the space of n-vector functions with bounded variation at $[t_0, T]$.

Let us introduce a new time variable η and a new state coordinate τ ([6, 7, 8]):

$$\eta(t) = t + \int_{t_0}^{t} dv(t), \quad \tau(\eta) = \inf \{ \, t \mid \eta(t) \geq \eta \, \}.$$

Consider the following auxiliary differential inclusion

$$\frac{d}{d\eta}\begin{pmatrix} z \\ \tau \end{pmatrix} \in G_\lambda(\tau, z) \tag{9}$$

with the initial condition

$$z(t_0) = x^0, \quad \tau(t_0) = t_0, \quad t_0 \leq \eta \leq T + \mu_\lambda.$$

Here

$$G_\lambda(\tau, z) = \bigcup_{0\ \leq \nu \leq\ 1} \left\{ (1-\nu)\begin{pmatrix} F_\lambda(\tau, z) \\ 1 \end{pmatrix} + \nu \begin{pmatrix} B(\tau) \\ 0 \end{pmatrix} \right\}. \tag{10}$$

Let us mention here two auxiliary results connected with two properties of the system (9) [15].

Lemma 1. *The map* $G_\lambda(\tau, z)$ *is convex and compact valued*

$$G_\lambda : [t_0, T + \mu_\lambda] \times R^n \rightarrow convR^{n+1}$$

and $G_\lambda(\tau, z)$ *is Lipschitz continuous in both variables* τ, z.

In addition to the above assumptions we will assume further that the initial problem constraints depend continuously on a parameter λ $(\lambda \in R^k)$ in such a way that

$$\lim_{\lambda \to 0} h(X^0_\lambda, X^0) = 0, \ \lim_{\lambda \to 0} h(U_\lambda, U) = 0, \ \lim_{\lambda \to 0} \mu_\lambda = \mu.$$

The next auxiliary property provides the continuous dependance of the set-valued right-hand side $G_\lambda(\tau, z)$ of the differential inclusion (9) on a parameter λ.

Lemma 2. *Under the above assumptions we have*

$$\lim_{\lambda \to 0} h(G_\lambda(\tau, z), G_0(\tau, z)) = 0, \quad \forall (\tau, z) \in R^{n+1}.$$

2.3 Sensitivity Properties

Denote by $w = \{z, \tau\}$ the extended state vector of the system (9) and consider trajectory tube of this differential inclusion (which has no measure or impulse components):

$$W_\lambda[\eta] = \bigcup_{w^0 \in X_\lambda^0 \times \{t_0\}} w(\eta, t_0, w^0), \quad t_0 \leq \eta \leq T + \mu_\lambda.$$

From Lemmas 1–2 and from the properties of trajectory tubes of ordinary differential inclusions [12, 5] we can conclude that the following result is valid.

Theorem 1. *The limit equality*

$$\lim_{\lambda \to 0} h(W_\lambda[T + \mu_\lambda], W_0[T + \mu]) = 0.$$

is true.

The next lemma explains the construction of the auxiliary differential inclusion (9).

Lemma 3. *The set $X_\lambda[T]$ is the projection of $W_\lambda[T + \mu_\lambda]$ at the subspace of variables z:*

$$X_\lambda[T] = \pi_z W_\lambda[T + \mu_\lambda].$$

The proof of this Lemma follows from the structure of the auxiliary system (9).

Combining Theorem 1 and Lemma 3 we get the property of the continuity of time cross-sections of trajectory tubes with respect to system parameters.

Theorem 2. *The following equality*

$$\lim_{\lambda \to 0} h(X_\lambda[T], X_0[T]) = 0$$

is true.

3 Linear Impulsive Systems: Ellipsoidal Estimation Algorithm

3.1 Linear Impulsive Systems with Ellipsoidal Constraints

Let us consider a linear control system

$$\begin{aligned} dx &= A(t)xdt + B(t)du, \\ x(0) &\in X_0, \quad 0 \leq t \leq T \end{aligned} \tag{11}$$

with impulsive control $u(\cdot)$ restricted by a set U that will be defined further, X_0 is convex and compact in R^n (in particular, X_0 may be an ellipsoid in R^n).

Let $E_0 = \{l \in R^m \mid l'Ml \leq 1\}$ be an ellipsoid in R^m where M is a symmetric positive definite matrix.

Denote by $C^m[0,T]$ the space of all continuous m-vector functions defined on $[0,T]$. It is well known that the conjugate space to $C^m[0,T]$ coincides with the space $BV^m[0,T]$ of m-vector functions with bounded variation at $[0,T]$: $(\ C^m[0,T]\ )^* = BV^m[0,T]$. Denote also $E = \{y(\cdot) \in C^m[0,T] \mid y(t) \in E_0 \ \forall t \in [0,T]\}$ and let us take $U = E^*$ where E^* is the conjugate ellipsoid [2,14] to E $(U = E^* \subset BV^m[0,T])$ that is

$$E^* = \{u(\cdot) \in BV^m[0,T] \mid \int\limits_{[0,T]} y'(t)du(t) \le 1 \ \forall \ y(\cdot) \in E\}.$$

We assume in this section that admissible controls u satisfy the restriction $u(\cdot) \in U = E^*$. In particular it follows from the structure of the set U of admissible controls $u(\cdot)$ that their jumps $\Delta u(t_i) = u(t_{i+1}) - u(t_i)$ have to belong to the conjugate ellipsoid $E_0^* = \{l \in R^m \mid l'M^{-1}l \le 1\}$ [17,18].

The following theorem gives the structure of the cross-section of trajectory tubes [17,18].

Theorem 3. *The reachable set $X(t,t_0,X_0)$ is convex and compact for all $t \in [0,T]$. Every state vector $x \in X(t,t_0,X_0)$ may be generated by a solution $x(\cdot)$ of (11) with $x(t) = x$ and corresponding to a piecewise control $u(\cdot)$ with $(n+1)$ jumps belonging to E_0^*.*

3.2 Ellipsoidal Estimation Approach

The method of constructing the external (or "upper" with respect to inclusion) estimates of trajectory tubes $X(\cdot,t_0,X_0)$ of a differential system with uncertainty is based on the ellipsoidal calculus [16,4] and on the new procedures of external approximation of a convex hull of the union of a variety of some ellipsoids [17,18]. Each of these ellipsoids corresponds to the reachable set of the system (11) when only unique impulse "jump" of the admissible control function is allowed [18]. The convex hull operation which we need to take additionally over the union of all these auxiliary ellipsoids in order to get the final reachable set $X[t]$ is motivated by the above Theorem 3.

The following example shows how to find the reachable set $\mathcal{X}(T;\mathcal{X}_0)$ basing on the above theorems.

*Example 1 (*External Ellipsoidal Estimates*).* Consider the following control system:

$$\begin{cases} dx_1(t) &=& x_2(t)dt + du_1(t), \\ dx_2(t) &=& du_2(t), \end{cases} \tag{12}$$

Here we take $\mathcal{X}_0 = \{0\}$ and the set U generated by the ellipsoid

$$\mathcal{E}_0 = \{l \in R^2 \mid \ l'Q_0 l \le 1\}, \ \ Q_0 = \begin{pmatrix} a^2 & 0 \\ 0 & b^2 \end{pmatrix}, \ \ a,b \in R, \ \ a,b > 0.$$

From Theorems 1-3 we find the reachable set $\mathcal{X}(T;0)$ given at Fig. 1.

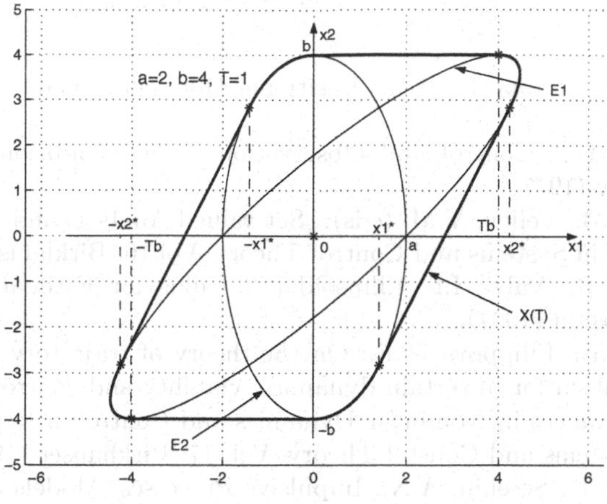

Fig. 1. The reachable set $\mathcal{X}(T) = \mathcal{X}(T; \mathcal{X}_0)$

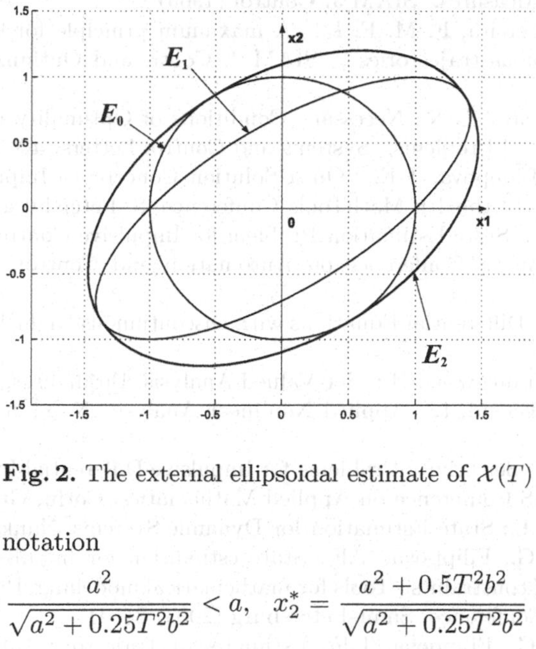

Fig. 2. The external ellipsoidal estimate of $\mathcal{X}(T)$

Here we use the notation

$$x_1^* = \frac{a^2}{\sqrt{a^2 + 0.25T^2b^2}} < a, \quad x_2^* = \frac{a^2 + 0.5T^2b^2}{\sqrt{a^2 + 0.25T^2b^2}} > Tb,$$

$$\mathcal{E}_1 = \{x \in R^2 \mid \frac{1}{a^2}(x_1 - x_2T)^2 + \frac{x_2^2}{b^2} \leq 1\}, \quad \mathcal{E}_2 = \{x| \ \ x'Q_0^{-1}x \leq 1\}.$$

The external ellipsoidal estimate $\mathcal{E}_2$ is shown at Fig. 2. Comparing with Fig.1 we note that $\mathcal{X}(T) \subset \mathcal{E}_2$.

Acknowledgments

The research was supported by the Russian Foundation for Basic Research (RFBR) under Project 03-01-00528.

References

1. Krasovskii, N.N., Subbotin, A.I.: Positional Differential Games. Nauka, Moscow (1974)
2. Kurzhanski, A.B.: Control and Observation under Conditions of Uncertainty. Nauka, Moscow (1977)
3. Kurzhanski, A.B., Veliov, V.M. (eds): Set-valued Analysis and Differential Inclusions. Progress in Systems and Control Theory, Vol.16. Birkhauser, Boston (1990)
4. Kurzhanski, A.B., Valyi, I.: Ellipsoidal Calculus for Estimation and Control. Birkhauser, Boston (1997)
5. Kurzhanski, A.B., Filippova, T.F.: On the theory of trajectory tubes — a mathematical formalism for uncertain dynamics, viability and control. In: Kurzhanski, A.B. (ed.): Advances in Nonlinear Dynamics and Control: a Report from Russia. Progress in Systems and Control Theory, Vol. 17. Birkhauser (1993) 122-188
6. Zavalishchin, S.T., Sesekin, A.N.: Impulsive Processes. Models and Applications. Nauka, Moscow (1991)
7. Rishel, R.: An Extended Pontryagin Principle for Control System whose Control Laws Contain Measures. SIAM J. Control (1965)
8. Vinter, R.B., Pereira, F. M. F. L.: A maximum principle for optimal processes with discontinuous trajectories . SIAM J. Contr. and Optimization, **26** (1988) 155-167
9. Pereira, F.L., Silva, G.N.: Necessary Conditions of Optimality for Vector-Valued Impulsive Control Problems. Systems and Control Letters, **40** (2000) 205–215
10. Pereira, F.L., Filippova, T.F.: On a Solution Concept to Impulsive Differential Systems. Proc. of the 4th MathTools Conference, S.-Petersburg, Russia (2003)
11. Filippova, T.F.: State Estimation Problem for Impulsive Control Systems. Proc. 10th Mediterranean Conference on Automation and Control, Lisbon, Portugal (2002)
12. Filippov, A.F.: Differential Equations with Discontinuous Right-hand Side. Nauka, Moscow (1985)
13. Aubin, J.-P., Frankowska, H.: Set-Valued Analysis. Birkhauser, Boston (1990)
14. Aubin, J.P., Ekeland, I.: Applied Nonlinear Analysis. N.-Y. etc., John Wiley & Sons (1984)
15. Filippova, T.F.: Sensitivity Problems for Impulsive Differential Inclusions. Proc. of the 6th WSEAS Conference on Applied Mathematics. Corfu, Greece (2004)
16. Chernousko, F.L.: State Estimation for Dynamic Systems. Nauka, Moscow (1988)
17. Vzdornova, O.G., Filippova, T.F.: State estimation for impulse control systems with ellipsoidal constraints. Tools for mathematical modelling, Proc. of The Fourth International Conference. Saint-Petersburg (2003) 34–41
18. Vzdornova, O.G., Filippova, T.F.: Estimates of Trajectory Tubes of Differential Systems of Impulsive Type. Proc. of the 3th Conference on Mathematics, Informatics and Control. Irkutsk, Russia (2004) 1–12

Limit Shapes of Reachable Sets for Linear Control Systems

Elena Goncharova[1] and Alexander Ovseevich[2]

[1] Institute of System Dynamics and Control Theory,
Siberian Branch of Russian Academy of Sciences,
134, Lermontov st., 664033 Irkutsk, Russia
goncha@icc.ru

[2] Institute for Problems in Mechanics,
Russian Academy of Sciences,
101, Vernadsky av., 119526 Moscow, Russia
ovseev@ipmnet.ru

Abstract. We study the limit behavior of reachable sets for time-invariant linear control systems under two types of the control bounds: the geometric bounds, and the bound for the total impulse.

Our main results consist in the description of the arising (as time tends to ∞) attractors in the space of shapes of the reachable sets, shape being the totality of sets obtained from a fixed one by an invertible affine transformation.

1 Introduction

In this paper we compare fundamental asymptotic properties of reachable sets of linear control systems under two types of the control bounds: the geometric bounds, and the bound for the total impulse. The investigation of the asymptotic behavior is quite nontrivial and far from completion for the non-autonomous linear systems. It is still nontrivial in the time-invariant case which we will deal with here, but in this case a more or less complete description of the asymptotics can be found.

The governing equations are traditional

$$\dot{x} = Ax + Bu, \quad x \in \mathbb{V} = \mathbb{R}^n, \; u \in \mathbb{W} = \mathbb{R}^m, \tag{1}$$

and should be augmented with the initial condition

$$x(0) \in M, \tag{2}$$

and the control constraints. We consider two types of the constraints: geometric one meaning

$$u = u(t) \in U, \tag{3}$$

where U is a fixed set, and the following bound for the total impulse:

$$\left| \int_0^T f(t)\, u(t)\, dt \right| \leq 1 \tag{4}$$

I. Lirkov, S. Margenov, and J. Waśniewski (Eds.): LSSC 2005, LNCS 3743, pp. 231–238, 2006.

for all continuous vector functions f such that $|f(t)| \le 1$, $t \in [0, T]$. In the geometric case, the admissible control is a measurable vector function subject to (3), while in the impulse case, it is a vector-valued measure satisfying (4).

Here, we are going to compare two families

$$\mathcal{D}_{\text{geom}}(t) = \mathcal{D}(t; M) = \{x(t);\ x \text{ satisfies (1), (2), (3)}\},$$

and

$$\mathcal{D}_{\text{impulse}}(t) = \mathcal{D}(t; M) = \{x(t);\ x \text{ satisfies (1), (2), (4)}\}$$

of the reachable sets as $t \to +\infty$.

It was discovered in [5] that there is a great advantage in studying the *shapes* of the reachable sets instead of the reachable sets themselves. Here, the shape $\operatorname{Sh} \Omega$ is the entity of the sets which can be obtained from one set Ω by means of invertible linear (or affine) maps.

A brief reflection reveals that for *stable* systems, (both families of) the reachable sets have a limit as the time $t \to +\infty$. By the stable systems we mean the systems such that the spectrum $\operatorname{Spec} A$ (set of eigenvalues) of the matrix A is contained in the left half-plane. In particular, it was found in [5] that $\operatorname{Sh} \mathcal{D}_{\text{geom}}(t)$ has a limit as $t \to +\infty$ irrespective of whether the matrix A is stable or not. This is a proper generalization of the mentioned property of stable systems.

In a more general setup, say, for time-dependent linear systems , there is no reason to expect the similar simplest possible asymptotic behavior even at the level of shapes of the reachable sets. Furthermore, it was found recently [8] that, in general, $\operatorname{Sh} \mathcal{D}_{\text{impulse}}(t)$ has no limit as $t \to +\infty$ as well. Instead, there arises an attractor $\mathcal{A}$ of positive dimension.

Generally speaking, if $t \mapsto \gamma(t)$ is a curve in a compact K, it defines an obvious attractor $\mathcal{A} = \bigcap_{T>0} \overline{\{\gamma(T + s),\ s \ge 0\}}$ also known as the ω-limit set of the curve γ. Here, $\overline{X}$ stands for the closure of a set X in K. The attractor $\mathcal{A}$ consists of all limits of sequences $\gamma(t_i)$, $i = 0, 1, \dots$, where $t_i \to +\infty$. In our cases of interest, $\gamma(t)$ is either $\operatorname{Sh} \mathcal{D}_{\text{geom}}(t)$ or $\operatorname{Sh} \mathcal{D}_{\text{impulse}}(t)$, while K is a suitable space of shapes.

If γ is an integral curve of a dynamic system, the maps $\phi(\tau) : \gamma \to \gamma$ given by $\phi(\tau)(\gamma(t)) = \gamma(t + \tau)$ can be extended by continuity on the attractor $\mathcal{A}$, i.e., to a dynamic system $\phi(\tau) : \mathcal{A} \to \mathcal{A}$, $\tau \in \mathbb{R}$. Note that although $\mathcal{D}_{\text{geom}}(t)$ and $\mathcal{D}_{\text{impulse}}(t)$ are integral curves of the dynamic system, the curves $\operatorname{Sh} \mathcal{D}_{\text{geom}}(t)$ or $\operatorname{Sh} \mathcal{D}_{\text{impulse}}(t)$ are not. Nevertheless, one can define by continuity a natural dynamic system $\phi(\tau)$, $\tau \in \mathbb{R}$, acting on the corresponding attractor.

Now, the problem we deal with is to give an explicit description of the attractor $\mathcal{A}$ together with the dynamic system $\phi(\tau) : \mathcal{A} \to \mathcal{A}$. It should be emphasized that here we do not recede too far from the original problem on the asymptotics of the reachable sets. The final results can be translated into the asymptotic equality:

$$\mathcal{D}(t; M) \sim C(t)\phi(t)\Omega \quad \text{as } t \to +\infty,$$

where $C(t)$ is a matrix multiplier, and $\Omega \in \mathcal{A}$ is a fixed set. Note that the structures of the multiplier $C(t)$ and the operators $\phi(t)$ are, in fact, very simple.

2 Preliminaries on Convex Bodies

Let the geometric bound U on control (see (3)) be a central symmetric compact convex body, and the system (1) satisfy the Kalman controllability condition which can be equivalently expressed by any of the following ones:

- The composite matrix $(B, AB, \dots, A^{n-1}B)$ has the maximal rank equal to $n = \dim \mathbb{V}$.
- If the set U of admissible control vectors is replaced by the vector space $[U] = \operatorname{Span} U$, the system is completely controllable.
- The minimal A-invariant vector space containing U coincides with $\mathbb{V}$.
- For any $T > 0$, the reachable set $\mathcal{D}(T)$ is a convex body (i.e., has a nonempty interior).

We also assume that the initial set M is a central symmetric convex compact. Under these assumptions, the set $\mathcal{D}(T)$, where $\mathcal{D}(T)$ stands for either $\mathcal{D}_{\text{geom}}(T)$ or $\mathcal{D}_{\text{impulse}}(T)$, is a central symmetric convex body with the center at $0 \in \mathbb{V}$.

The space $\mathbb{B}$ of central symmetric convex bodies possesses a distinguished metric ρ invariant under natural action of the general linear group $GL(\mathbb{V})$ of nonsingular matrices ($\rho(\Omega_1, \Omega_2) = \rho(C\Omega_1, C\Omega_2)$, where C is any nonsingular matrix).

Definition 1. *Let $\Omega_1, \Omega_2 \subset \mathbb{V}$ be central symmetric convex bodies,*

$$t(\Omega_1, \Omega_2) = \inf\{t \geq 1;\ t\Omega_1 \supset \Omega_2\}, \qquad \textit{and}$$

$$\rho(\Omega_1, \Omega_2) = \log(t(\Omega_1, \Omega_2)t(\Omega_2, \Omega_1)).$$

Then the function ρ is called the Banach–Mazur distance in the space $\mathbb{B}$.

Example 1. Let $\mathbb{V} = \mathbb{R}^2$ and $\mathcal{L}$ be the subspace of B consisting of ellipses $E(Q)$ with $\det Q = 1$. Then the metric space $(\mathcal{L}, \rho)$ coincides with the Lobachevsky plane. (Indeed, $(\mathcal{L}, \rho)$ is a symmetric space with involution $E(Q) \mapsto E(Q^{-1})$ corresponding to the unit circle as a base point).

Definition 2. *The space $\mathbb{S}$ of shapes of central symmetric convex bodies is the factor space $\mathbb{S} = \mathbb{B}/GL(\mathbb{V})$, while the shape $\operatorname{Sh}\Omega \in \mathbb{S}$ of a convex body $\Omega \in \mathbb{B}$ is the orbit*

$$\operatorname{Sh}\Omega = \{C\Omega;\ \det C \neq 0\} \qquad \textit{of } \Omega \textit{ under } GL(\mathbb{V})\textit{-action.}$$

The space $\mathbb{S}$ can be equivalently characterized as the space of Banach structures (norms) on $\mathbb{R}^n$ up to an isomorphism.

Since the Banach–Mazur distance is $GL(\mathbb{V})$-invariant, it defines a good metric ρ on $\mathbb{S}$ making it a metric space.

The convergence of the reachable sets $\mathcal{D}(T)$ and their shapes $\operatorname{Sh}\mathcal{D}(T)$ is to be understood in the Banach–Mazur metric. In particular, two $\mathbb{B}$-valued functions are said to be asymptotically equal $\Omega_1(T) \sim \Omega_2(T)$ if $\rho(\Omega_1(T), \Omega_2(T)) \to 0$ as $T \to \infty$.

Remark. From the homotopy-theoretic point of view the space $\mathbb{S}$ is the classifying space BG of the group $G = GL_n(\mathbb{R})/\pm 1$ since $\mathbb{S} = \mathbb{B}/G$, $\mathbb{B}$ is contractible, and the action of G on $\mathbb{B}$ is free outside a subset of infinite codimension (cf. [1]).

3 Attractors and Related Dynamics

Suppose we are given a continuous curve $\Gamma : [0, \infty) \to K$ in a compact K. It defines the ω-limit set, or attractor $\mathcal{A}$, as follows:

$$\mathcal{A} = \bigcap_{T>0} \overline{\Gamma_{\geq T}},$$

where $\Gamma_{\geq T}$ is the set $\{\Gamma(T + s),\, s \geq 0\}$. Denote by $\phi(\tau) : \Gamma_{-\tau} \to \Gamma$ the map given by $\phi(\tau)(\Gamma(t)) = \Gamma(t + \tau)$.

We will speak about dynamics on the attractor in the event that the maps $\phi(\tau)$ can be extended to the attractor $\mathcal{A}$ by continuity, i.e., if $t_i \to \infty$, and $\Gamma(t_i) \to a \in \mathcal{A}$, then $\phi(\tau)\Gamma(t_i) = \Gamma(t_i + \tau) \to b \in \mathcal{A}$. In this case, we write $b = \phi(\tau)a$.

Theorem 1. *Consider either the curve* $\operatorname{Sh}\mathcal{D}_{\text{geom}}(t)$ *or* $\operatorname{Sh}\mathcal{D}_{\text{impulse}}(t)$ *in the space* $\mathbb{S}$ *of shapes. Then, one can define by continuity a natural dynamic system* $\phi(\tau)$, $\tau \in \mathbb{R}$, *on the corresponding attractor.*

Remark. In fact, the $\operatorname{Sh}\mathcal{D}_{\text{geom}}(t)$-part of the above theorem is trivial. As is shown in the next section (see Theorem 2), the corresponding attractor is reduced to one point.

4 Splitting Principle and Main Results

We will use below the following well-known constructions of convex geometry. Let $\Omega_i \subset \mathbb{V}_i$ be convex bodies in the vector spaces $\mathbb{V}_i$, and the index i run over a finite set. Then, $\oplus\Omega_i \subset \oplus\mathbb{V}_i$ is by definition the cartesian product of Ω_i in the cartesian product (also known as the direct sum) of $\mathbb{V}_i$. Suppose now that $\Omega_i \subset \mathbb{V}$ are convex compacts in a vector space $\mathbb{V}$, then we define the join $\Omega = *_i\Omega_i$ as

$$\Omega = \{x \in \mathbb{V} :\, x = \sum_i t_i x_i, \text{ where } t_i \geq 0, \sum_i t_i = 1, \text{ and } x_i \in \Omega_i\}.$$

By a slight abuse of the language, we will use the similar notations for the corresponding shapes: $\oplus \operatorname{Sh}\Omega_i := \operatorname{Sh}\oplus\Omega_i$, and $*_i \operatorname{Sh}\Omega_i := \operatorname{Sh} *_i\Omega_i$.

Consider the canonical decomposition of the matrix A from (1)

$$A = A_+ \oplus A_0 \oplus A_- \tag{5}$$

into unstable, neutral and stable components (according to the sign of the real part of eigenvalues), and the corresponding decomposition of the phase space

$$\mathbb{V} = \mathbb{V}_+ \oplus \mathbb{V}_0 \oplus \mathbb{V}_- . \tag{6}$$

Denote by $\mathbb{V}_{ij}$, where $i,\ j \in \{+,0,-\}$, and $i \neq j$, the direct sums $\mathbb{V}_{ij} = \mathbb{V}_i \oplus \mathbb{V}_j$, and by $P_i, P_{ij} : \mathbb{V} \to \mathbb{V}_i, \mathbb{V}_{ij}$ the corresponding canonical projectors. For each index α s.t. $\alpha = i$ or $\alpha = ij$, $i \neq j \in \{+,0,-\}$, we obtain a new dynamic system governed by

$$\dot{x}_\alpha = A_\alpha x_\alpha + B_\alpha u, \quad x_\alpha \in \mathbb{V}_\alpha,\ u \in \mathbb{W} = \mathbb{R}^m, \tag{7}$$

where $A_\alpha = P_\alpha A$, $B_\alpha = P_\alpha B$. The initial conditions are

$$x_\alpha(0) \in M_\alpha = P_\alpha M, \tag{8}$$

while the control constraints (3) or (4) remain unchanged.

Theorem 2. *The limit shape $\bar{\mathcal{D}}(\infty) = \lim\limits_{T\to\infty} \operatorname{Sh} \mathcal{D}_{\text{geom}}(T)$ of the reachable set $\mathcal{D}_{\text{geom}}(T)$ does exist (in the space $\mathbb{S}$) and splits as*

$$\bar{\mathcal{D}}(\infty) = \bar{\mathcal{D}}_+(\infty) \oplus \bar{\mathcal{D}}_0(\infty) \oplus \bar{\mathcal{D}}_-(\infty)$$

in accordance with decomposition (5), (6). Here, $\bar{\mathcal{D}}_\alpha(\infty)$, $\alpha = i \in \{+,0,-\}$, is the limit shape of the reachable set corresponding to the system (7), (8), (3).

In this case, the attractor $\mathcal{A}$ is reduced, in fact, to a single point $\bar{\mathcal{D}}(\infty)$, and the corresponding limit shape admits a further reduction to the independent sum of three limit shapes.

In order to state the result for impulsive control systems, we consider the Jordan decomposition $A_0 = D + N$ of the neutral component A_0 of the matrix A. Here, the matrix D is diagonalizable, N is nilpotent, and $DN = ND$.

Define $\mathcal{T} \subset \mathrm{GL}(\mathbb{V}_0)$ as the torus generated by the one-parameter group of operators $\{e^{Dt} : \mathbb{V}_0 \to \mathbb{V}_0\}$ in the neutral canonical subspace $\mathbb{V}_0 \subset \mathbb{V}$.

Theorem 3. *The following asymptotic formula holds as $T \to \infty$:*

$$\operatorname{Sh} \mathcal{D}_{\text{impulse}}(T) \sim \operatorname{Sh}\left[\left(e^{DT}\Omega_{+0}\right) * \Omega_{-0} + e^{DT}\mathcal{M}_{+0}\right],$$

where Ω_{+0}, Ω_{-0} are certain limit convex bodies in the spaces $\mathbb{V}_{+0}, \mathbb{V}_{-0}$, and $\mathcal{M}_{+0} = M_+ \oplus F_\infty M_0$, where F_∞ is a projector of $\mathbb{V}_0$ upon $\ker N$.

In other words, the attractor $\mathcal{A}$ is an image of the torus $\mathcal{T}$ under a continuous map Φ such that for $\mathbf{t} \in \mathcal{T}$ we have $\Phi(\mathbf{t}) = \operatorname{Sh}[(\mathbf{t}\Omega_{+0}) * \Omega_{-0} + \mathbf{t}\mathcal{M}_{+0}]$. Furthermore, the canonical dynamic system $\phi(\tau)$ on the attractor $\mathcal{A}$ comes from multiplication by $e^{D\tau}$ on the torus $\mathcal{T}$: $\phi(\tau)\Phi(\mathbf{t}) = \Phi(e^{D\tau}\mathbf{t})$.

5 Examples

Assume that A is a "generic" skew-symmetric matrix, and the system (1) in the case of geometric bounds takes the form

$$\dot{x} = Ax + bu, \quad b \in \mathbb{V}, \quad |u| \leq 1. \tag{9}$$

Here, the genericity condition is that

$$\operatorname{Spec} A = \{\pm\sqrt{-1}\,\omega_j,\ \omega_j > 0,\ j = 1, ..., n = \dim \mathbb{V}/2\},$$

and there are no nontrivial relations of the form

$$\sum_{j=1}^{n} m_j \omega_j = 0, \quad m_j \in \mathbb{Z}.$$

This means that (9) is a non-resonant oscillatory system. Then the limit shape $\bar{\mathcal{D}}(\infty)$ of the reachable set can be described by the following support function:

$$H(\xi) = \int_0^{2\pi} \dots \int_0^{2\pi} \left| \sum |\xi_i| \cos \phi_i \right| d\phi_1 \dots d\phi_n, \tag{10}$$

where $\xi = (\xi_1, \dots, \xi_n) \in \mathbb{C}^n$. This means that, for any body Ω of the shape $\bar{\mathcal{D}}(\infty)$, there exists an isomorphism $P : \mathbb{V}^* \to \mathbb{C}^n$ such that $H_\Omega(\eta) = H(P\eta)$, where $H_\Omega(\eta)$ is the support function of Ω. Note that according to (10) the limit shape $\bar{\mathcal{D}}(\infty)$ depends neither on the matrix A nor the vector b provided that these data are "generic".

Note that, in the case of the bounded total impulse, there still exists a limit shape and even the limit reachable set $\mathcal{D}(\infty)$. Moreover, its support function can be obtained from (10) when substituting integration by taking supremum, and takes a simple form:

$$H(\xi) = \sum |\xi_i|,$$

which reveals that $\mathcal{D}(\infty)$ is a polydisk, i.e., $\{z \in \mathbb{C}^n;\ |z_i| \leq 1\}$.

Reminder. By definition $H_\Omega(\eta) = \sup_{x \in \Omega} \langle x, \eta \rangle$, where $\eta \in \mathbb{V}^*$ is a vector of a dual space, and $\langle x, \eta \rangle$ is the canonical pairing. The support function H_Ω determines a closed convex set Ω uniquely.

One can give an expression for the support function (10) via the Bessel functions

$$H(\xi) = \int_0^{\infty} \left(\prod_{k=1}^{n} J_0(|\xi_k|\rho) - 1 \right) \frac{d\rho}{\rho^2} \tag{11}$$

and without multiple integrals. Here,

$$J_0(x) = \frac{1}{\pi} \int_0^{\pi} e^{ix\cos\phi}\, d\phi = \sum_{k=0}^{\infty} \frac{(-1)^k}{k!^2} \left(\frac{x}{2}\right)^{2k}$$

is the zero-index Bessel function.

More generally, consider the *generic neutral* control system with geometric bounds on control:

$$\dot{x} = Ax + Bu, \quad x \in \mathbb{R}^{2n} = \mathbb{C}^n,\ u \in U \subset \mathbb{R}^m, \tag{12}$$

where A is the same as in (9), and $U \subset \mathbb{R}^m$ is a central symmetric convex body with the support function h.

Then, the limit shape of the relevant reachable sets can be described by a formula similar to (10):

$$H(\xi) = \int_{\mathcal{T}} h(B^* e^{\phi} \xi)\, d\phi. \tag{13}$$

Here, $\mathcal{T}$ is the n-dimensional torus $\mathcal{T} = (\mathbb{R}/2\pi\mathbb{Z})^n$, and

$$e^{\phi}\xi = (e^{i\phi_1}\xi_1, \ldots, e^{i\phi_n}\xi_n),$$

where $\phi = (\phi_1, \ldots, \phi_n) \in (\mathbb{R}/2\pi\mathbb{Z})^n$, $d\phi = d\phi_1 \ldots d\phi_n$, and $\xi = (\xi_1, \ldots, \xi_n) \in \mathbb{C}^n$. In this case, it is still possible to rewrite the multiple integral (13) as an integral over $\mathbb{R}^m$ involving the Bessel functions.

In fact,

$$H(\xi) = \int_{\mathbf{R}^m} \hat{h}(\lambda) \prod_{k=1}^{n} J_0(|\xi_k \otimes B\lambda|_k)\, d\lambda, \tag{14}$$

where $\hat{h}$ is the Fourier transform of h, and $|\cdot|_k$ stands for a Euclidean norm on the space $\mathbb{R}^{2m} = \mathbb{C} \otimes_{\mathbb{R}} \mathbb{R}^m$.

Formula (11) corresponds to

$$\hat{h}(\lambda) = \frac{1 - \delta(\lambda)}{|\lambda|^{m+1}},$$

which is true for any dimension m if U is the unit ball.

6 Conclusion

Our investigation of the asymptotic properties of the reachable sets to linear control systems reveals a hidden rich geometric structure. Although we can describe to a great extent the asymptotic behavior of the reachable sets in the time-invariant case with either geometric or impulse bounds, some basic questions remains obscure. In particular, it is not clear yet how to unify the results for the cases of geometric and impulse bounds. Much less clear is the corresponding issue in time-dependent case.

Acknowledgments

The work was supported by Russian Foundation for Basic Research, grants No. 05-01-00477, 05-01-00647, and 05-08-50226, and the Scientific Schools Support Program, grant 1627.2003.01.

References

1. Atiayh, M.F., Bott, R.: The Yang-Mills equations over Riemann surfaces. Proc. R. Soc. London A. **308** (1982) 523-615
2. Arnold, V.I.: Mathematical methods of classical mechanics. Moscow: Nauka (1967)
3. Chernousko, F. L.: State estimation for dynamic systems. Boca Raton. Florida: CRC Press (1994)
4. Grüne, L.: Asymptotic behavior of dynamical and control systems under perturbation and discretization. Springer-Verlag. Lecture Notes in Math. **1783** (2002)
5. Ovseevich, A.I.: Asymptotic behavior of attainable and superattainable sets. Proc. of the Conference on Modelling, Estimation and Filtering of Systems with Uncertainty. Sopron. Hungary. Basel, Switzerland: Birkhaüser. (1991) 324–333
6. Figurina, T.Yu., Ovseevich, A.I.: Asymptotic behavior of attainable sets of linear periodic control systems. J. Optim. Theory and Appl. **100** No. 2 (1999) 349–364
7. Goncharova, E.V., Ovseevich, A.I.: Asymptotic behavior of attainable sets of a linear impulse control system. Proc. of the IFAC Workshop on Generalized Solutions in Control Problems. Pereslavl-Zalesski. Russia. Moscow: Fizmatlit (2004) 72–76
8. Goncharova, E.V., Ovseevich, A.I.: Limit shapes of reachable sets for linear impulse control systems. Proc. of the 16th IFAC World Congress. Prague (2005) (to appear)

On the Synthesis of a Stabilizing Feedback Control

Mikhail I. Krastanov

Institute of Mathematics and Informatics, Bulgarian Academy of Sciences,
1113 Sofia, Bulgaria
krast@bas.bg
http://www.math.bas.bg/~krast/

Abstract. Starting from states near to a closed set S we want to steer S and to stay always close to S. Unfortunately, open-loop controls are very sensitive to disturbances and can lead to very bad practical results. For that reason, we propose an approach for constructing a discontinuous feedback control law that asymptotically stabilizes the system in a neighborhood of the set S.

1 Introduction

Let S be a closed subset of R^n, Ω be a neighborhood of S and U be a closed subset of R^m. We consider the following control system:

$$\dot{x}(t) \;=\; f(x(t), u(t)),\; u(t) \in U \text{ a.e.}, \tag{1}$$

where $x(t) \in \Omega$ is the state, $u(t) \in U$ is the control and $f : R^n \times R^m \to R^n$ is a smooth map.

Starting from states close to the set S, we want to steer S and to stay always close to S. To explain the difficulties related to this problem, let us consider the case when the set S is a point: Let $S = \{0\}$ and let us assume that the control system (1) is small-time local controllable at the origin, i.e. for any $T > 0$ there exists a neighbourhood Ω of the origin such that for any point $x_0 \in \Omega$ there exists an open-loop control $u_{x_0}(.)$ that steers the point x_0 to the origin in time not greater than T. Unfortunately, open-loop controls are very sensitive to disturbances and can lead to very bad practical results (cf., e.g., [14], Chapter 1, §4). Taking this into account, one can try to find a feedback control law $k : \Omega \to U$ such that all solutions of the ordinary differential equation:

$$\dot{x}(t) = f(x(t), k(x(t))) \tag{2}$$

asymptotically approach the origin. Such a feedback has the advantage of compensating automatically all random perturbations when they are sufficiently small.

A minimal condition for the existence of classical solutions to the ordinary differential equation (2) is that the feedback law $k(\cdot)$ be continuous on $R^n \setminus \{0\}$ (cf. [1] where several concepts of generalized solutions of ODE are considered). A

I. Lirkov, S. Margenov, and J. Waśniewski (Eds.): LSSC 2005, LNCS 3743, pp. 239–246, 2006.

classical result (cf., e.g., [14], Theorem 7, p.134) shows that the small-time local controllability of a linear control system implies that this system can be assymptotically stabilized by means of a stationary continuous feedback. However, as was shown by Sontag and Sussmann [15], even when $m = n = 1$, such a continuous feedback $k(\cdot)$ need not exist. Moreover, Brockett proves in [2] a topological condition, which is necessary for the existence of a continuous stabilizing feedback law. It can be directly checked that the following analytic three-dimensional control system

$$\dot{x} = u,\ \dot{y} = v,\ \dot{z} = vx - uy \tag{3}$$

is small-time locally (and even globally) controllable at the origin but does not satisfy the Brockett necessary condition, and hence can not be asymptotically stabilized by means of a continuous feedback law. Brockett's necessary condition holds true even when Filippov solutions are considered.

To get around the problem of impossibility to stabilize nonlinear systems by a continuous autonomous feedback, some alternative approaches are proposed. The approach proposed by Coron [5, 6] and [7] deals with continuous periodic time-varying feedbacks. Clarke, Ledyaev, Sontag and Subbotin [3] constructed a discontinuous stabilizing feedback. Their approach is based on the so called "closed loop system sampling" solution concept for (2) employed by Krasovskii and Subbotin [10] in the differential game theory. Unfortunately, the above mentioned two approaches have mainly theoretical importance. The reason is that it is not clear how to construct explicitly stabilizing controls in the general case (when only a control system is given, but we do not know a Lyapunov function or any other feedback design tool). A different approach is proposed by Hermes in [8] (cf. also [13]). The basic idea is to compute "stabilizing feedback controls" when a suitable sufficient condition for small-time local controllability holds true.

In this paper we propose an approach similar to those of Hermes. This approach uses a suitable class of high-order control variations variations with respect to a closed set. The control system (3) is considered as an illustrative example. Some simulation results, carried out in Maple, are presented at the end of the paper.

2 The Main Result

Let S be an arbitrary closed subset of R^n. Let $\delta_0 > 0$ and S_{δ_0} be the closed neighbourhood of the set S consisting of all points x such that $\mathrm{dist}_S(x) \leq \delta_0$ (here $\mathrm{dist}_S\,(x)$ denotes the distance between the point x and the set S). If x is an arbitrary point of R^n, we set

$$\pi_S(x) := \{y \in S :\ \|x - y\| = \mathrm{dist}_S(x)\}\,,$$

i.e. $\pi_S(x)$ is the set of all metric projections of the point x on the set S. Let y belong to the boundary ∂S of the set S. A vector $\xi \in R^n$ is called a proximal normal to S at y provided there exists $r > 0$ so that the point $y + r\xi$ has closest point y in S. The set of all proximal normals at a point y is a cone, and is denoted by $N_S^p(y)$ (for a more detailed treatment of proximal analysis, cf. [4]).

We consider the following control system in S_{δ_0}:

$$\dot{x}(t) = f(x(t), u(t)), \tag{4}$$

where $x(t)$ is the state and $u(t) \in U \subset R^m$ is the control. We assume that $f : S_{\delta_0} \times U \to R^n$ is a smooth map and for each point s from the boundary of S there exists $u_s \in U$ such that $f(s, u_s) = 0$.

Let $u(\cdot)$ be an integrable function defined on the interval $[0, T]$ with values from U. An absolutely continuous function $x(\cdot)$ satisfying (4) for almost every t from $[0, T]$ is called an admissible trajectory of (4) defined on $[0, T]$, starting from the point $x(0)$ and corresponding to the control $u(\cdot)$. By $\mathcal{R}(x, t)$ we denote all points of R^n reachable from the point x by means of admissible trajectories of (4) defined on $[0, T]$ and starting from the point x.

We assume that measurements can be made at discrete moments of time. Suppose that the first measured state is x_1. We give an constructive algorithm to generate an admissible control $u(t, x_1)$, $t \in [0, t_{x_1}]$. Let x_2 denote the value of the solution of (4) at time $t = t_{x_1}$ starting from the point x_1 at time $t = 0$ and corresponding to the control $u(\cdot, x_1)$. The algorithm may then be used to generate a new control $u(t, x_2)$, $t \in [0, t_{x_2}]$, etc. The controls generated in this way are such that if $\delta \in (0, \delta_0]$ is sufficiently small and whenever the starting point belongs to the neighborhood S_δ of the set S, then the sequence $x_1, x_2, \ldots$ converges to the set S. In this sense, the proposed algorithm may be considered as generating a stabilizing feedback control.

To explain our approach, we need

Definition 1. *Let S be an arbitrary closed subset of R^n, x_0 be a point of R^n, α be a positive real number and A be a locally Lipschitz continuous vector-field defined on a neighborhood of x_0. It is said that A is a control variation of order α of the control system (4) at the point x_0 with respect to the set S iff there exist a convex compact neighborhood K of the point x_0 and some positive numbers T, M, N, θ, $\beta > \alpha$, $p_i, i = 1, \ldots, k$, and $1 \leq q_1 < q_2 < \cdots < q_k$ such that for each $t \in [0, T]$ and each $x \in K$ the following inclusion holds true*

$$x + t^\alpha A(x) + b(x, t) \in \mathcal{R}(x, p(t)), \tag{5}$$

where $p(t) = \sum_{i=1}^{k} p_i t^{q_i}$ *and the continuous function* $b(\cdot, \cdot) : K \times [0, T] \to R^n$ *satisfies the estimate* $\|b(x, t)\| \leq M.t^\theta.dist_S(x) + N.t^\beta$. *By* $\mathcal{V}_{x_0,S}^\alpha$ *we denote the set of all control variations of order α of the control system (4) at the point x_0 with respect to the set S.*

Remark 1. By setting $t := t^{\tilde{\alpha}/\alpha}$, it can be proved that the relation $A \in \mathcal{V}_{x,S}^\alpha$ implies the relation $A \in \mathcal{V}_{x,S}^{\tilde{\alpha}}$ whenever $\tilde{\alpha} > \alpha$.

Remark 2. For the smooth case, the results in [9, 11, 12, 16] and [17] can be useful for constructing of high-order control variations when a control system is given in the form of differential equation, differential inclusion and etc.

Let $\mathcal{V}$ be a subset of $\bigcup_{x \in S_{\delta_0}} \mathcal{V}^{\alpha}_{x,S}$. To formulate our main result, we need some regularity properties of the set of control variations $\mathcal{V}$. Roughly speaking, it is said that $\mathcal{V}$ is a regular subset of $\bigcup_{x \in S_{\delta_0}} \mathcal{V}^{\alpha}_{x,S}$ iff all elements of $\mathcal{V}$ are Lipschitz continuous functions with one and the same Lipschitz constant as well as all functions (involved according to Definition 1) are uniformly bounded. This assumption is technical and guaranteed the existence of suitable uniformly bounded trajectories of (4) on some fixed interval $[0, T]$. This is especially important for the case of unbounded closed set S.

Now we can formulate the main result.

Theorem 1. *Let S be a closed subset of R^n, $\mu > 0$, $T_0 > 0$, $\delta_0 > 0$ and $\mathcal{A}$ be a regular subset of $\bigcup_{x \in S_{\delta_0} \setminus S} \mathcal{V}^{\alpha}_{x,S}$ consisting of control variations defined on the interval $[0, T_0]$. Let us assume that whenever $x \in S_{\delta_0} \setminus S$ there exist $y \in \pi_S(x)$, $\xi \in N^p_S(y)$ and $A \in \mathcal{V}$ such that*

$$< \xi, A(y) > \leq -\mu . \|\xi\|. \tag{6}$$

Then there exist real numbers $q \in (0, 1)$ and $\delta > 0$ such that for each point $x \in S_\delta$ there exist $t_x > 0$ and a piece-wise constant control function $u_x : [0, t_x] \to U$ such that the solution $z(\cdot, x, u_x)$ of (4) starting from x and corresponding to the control u_x is well defined on $[0, t_x]$ and satisfies the inequality $dist_S(z(t_x, x, u_x)) \leq q.dist_S(x)$.

Remark 3. Let all assumptions of Theorem 1 hold true and let x_1 be an arbitrary point of S_δ. Applying Theorem 1, we obtain the existence of a number $q \in (0, 1)$ such that the point $x_2 := z(t_{x_1}, x_1, u_{x_1})$ satisfies the inequality $\mathrm{dist}_S(x_2)) \leq q.\mathrm{dist}_S(x_1)$. Then, Theorem 1 may be applied again to generate a new control $u_{x_2}(t)$, $t \in [0, t_{x_2}]$. The point $x_3 := z(t_{x_2}, x_2, u_{x_2})$ satisfies the inequality $\mathrm{dist}_S(x_3)) \leq q.\mathrm{dist}_S(x_2)$, etc. Hence, the obtained sequence $x_1, x_2, \ldots,$ of points of S_δ converges to the set S. In this sense, Theorem 1 may be applied to generate a stabilizing feedback control.

3 Numerical Results

In this section we present some illustrative examples. All computations are performed using the computer algebra system Maple V.

Example 1. Let us consider the following three-dimensional control system

$$\begin{aligned}
\dot{x}_1 &= u, & x_1(0) &= 0,\ u \in [-1, 1], \\
\dot{x}_2 &= v, & x_2(0) &= 0,\ v \in [-1, 1], \\
\dot{x}_3 &= v x_1 - u x_2, & x_3(0) &= 0,
\end{aligned}$$

which does not satisfy the Brockett necessary condition, and hence can not be assymptotic stabilized by means of a continuous feedback law.

Let us denote $x = (x_1, x_2, x_3)^{\mathrm{T}}$, $B(x) = (1, 0, -x_2)^{\mathrm{T}}$, $C(x) = (0, 1, x_1)^{\mathrm{T}}$. Then $[B, C](x) = (0, 0, 2)^{\mathrm{T}}$. It could be directly checked that B, C and $[B, C]$ are the only nonvanishing Lie brackets of the Lie algebra $\mathcal{L}$ generated by the vector fields B and C. The considered control system belongs to the class of the so called "symmetric" control systems. To realize B, C and $[B, C]$ as control variations of high order, we use the following construction: Let $(\beta_1, \beta_2, \beta_3)^{\mathrm{T}}$ be an arbitrary vector whose components belong to the interval $[-1, 1]$. We define the following admissible control:

$$u_\beta := \begin{cases} u = \beta_3,\ v = 0, & \text{for } t \in [0, t) \\ u = 0,\ v = 1, & \text{for } t \in [t, 2t) \\ u = -\beta_3,\ v = 0, & \text{for } t \in [2t, 3t) \\ u = 0,\ v = -1, & \text{for } t \in [3t, 4t) \\ u = \beta_1,\ v = \beta_2, & \text{for } t \in [4t, 4t + t^2) \end{cases}.$$

It could be directly verified that the corresponding trajectory $z(\cdot, x, u_\beta)$ is defined on $[0, 4t + t^2]$ and

$$z(x, u_\beta, 4t + t^2) = x + t^2(\beta_1 B(x) + \beta_2 C(x) + \beta_3 [B, C](x)).$$

We set

$$\Omega := \{x = (x_1, x_2, x_3)^{\mathrm{T}} : |x_1| \leq 1,\ |x_2| \leq 1,\ |x_3| \leq 2\} \ \text{ and } \ t_x := \sqrt{\|x\|}.$$

Then, every point $x \in \Omega$ can be present as follows:

$$x = \alpha_1(x)B(x) + \alpha_2(x)C(x) + \alpha_3(x)[BC](x), \ \text{ with } \alpha_i(x) \in [-1, 1],\ i = 1, 2, 3\}.$$

We set

$$\beta_i(x) := -\frac{\alpha_i(x)}{\|x\|},\ i = 1, 2, 3,$$

Table 1. Simulation results for Example 1 for four different starting points $x = (x_1, x_2, x_3)^{\mathrm{T}}$. Each end-point $z = (z_1, z_2, z_3)^{\mathrm{T}}$ is reached from the corresponding starting point x in time t_x.

x_1	x_2	x_3	t_x	z_1	z_2	z_3
1	-2	5	$\sqrt{30} + 4\sqrt[4]{30}$	$-3.219647.10^{-14}$	$9.992007.10^{-15}$	$-5.457833.10^{-14}$
1	2	-5	$\sqrt{30} + 4\sqrt[4]{30}$	$-3.219647.10^{-14}$	$-9.992007.10^{-15}$	$-4.534174.10^{-14}$
1	1	-2	$\sqrt{6} + 4\sqrt[4]{6}$	$7.394085.10^{-14}$	$-2.597922.10^{-14}$	$-1.562844.10^{-13}$
1	-1	2	$\sqrt{6} + 4\sqrt[4]{6}$	$-2.59792188.10^{-14}$	$2.597922.10^{-14}$	0.0

and use the feedback control $u_{\beta(x)}$ on the interval $[0, 4t_x + t_x^2]$. The condition (6) holds true with $\mu = -1$. Moreover, it could be directly verified that the corresponding trajectory $z(\cdot, x, u_{\beta(x)})$ is defined on $[0, 4t_x + t_x^2]$ and $z(4t_x + t_x^2, x, u_{\beta(x)}) = 0$. We apply this feedback control for different starting points x. Some of the corresponding numerical results are presented in Table 1.

Example 2. Let us consider the following closed subset of R^2: $S = \{(x, y)^{\mathrm{T}} : 0 \leq x \leq 1,\ y \leq 0\}$ and the following control system

$$\begin{array}{ll} \dot{x} = u, & u \in [-1, 1], \\ \dot{y} = x_2 + 1 - x_1 + v, & v \in [-1, 1]. \end{array} \tag{7}$$

To move in direction $(0, -1)^{\mathrm{T}}$, we construct control variation of second order as follows: Let $T \in [0, 1]$, $t \in (0, T/2]$ and $z = (x, y)^{\mathrm{T}}$ with $x, y \in [0, 1]$. We set $v_t(s) = -1$ for each $s \in [0, t]$ and $u_t(s) := \begin{cases} 1, & \text{if } s \in [0, t]; \\ -1, & \text{if } s \in [t, 2t]; \end{cases}$

Then the trajectory $z_t(\cdot) = (x_t(\cdot), y_t(\cdot))^{\mathrm{T}}$ of (7) starting from the point z and corresponding to the controls $u_t(\cdot)$ and $v_t(\cdot)$ is well defined on the interval $[0, 2t]$, $x_t(2t) = 0$ and

$$y_t(2t) = e^{2t}\,(y - 1 - x) + 2e^t + x - 1.$$

Hence, we can represent $z_t(2t)$ as follows:

$$z_t(2t) = z + t^2 A(z) + a^1(t, z) + a^2(t, z), \tag{8}$$

where $A(z) = (0, -1)^{\mathrm{T}}$, $a^1(t, z) = (0,\ 2t(1 + t)(y - x))^{\mathrm{T}}$ and $a^2(t, z)$ is determined by (8). Taking into account that $d_S(z) = y$, we obtain that $\|a^1(t, z)\| \leq 4t.\mathrm{dist}_S(z)$ and $\|a^1(t, z)\| \leq C.t^3$ for some positive constant C . This implies that $A(z) = (0, -1)^{\mathrm{T}}$ is a control variation of second order. Some simulation results, obtained for different starting points, are presented in Table 2.

Table 2. Simulation results for Example 2 for four different starting points $x = (x_1, x_2)^{\mathrm{T}}$. Each end-point $z = (z_1, z_2)^{\mathrm{T}}$ is reached from the corresponding starting point x in time t_x.

x_1	x_2	t_x	z_1	z_2
1.5	0.5	1.284211638	1.0	$-2.949259209.10^{-5}$
−0.2	0.5	5.371304140	$3.0087043967.10^{-14}$	$-1.891439399.10^{-4}$
0.8	0.5	7.166944604	0.80000000000000154	$-2.454711917.10^{-5}$
0.1	0.5	2.057154154	0.09999999999999976	$-3.737032781.10^{-5}$

Example 3. Let us consider the following three-dimensional control system (cf. [8]):

$$\begin{aligned} \dot{x}_1 &= x_2 + \quad u, \; x_1(0) = 0, \; u \in [-1, 1], \\ \dot{x}_2 &= x_1 x_3 + \; u, \; x_2(0) = 0, \\ \dot{x}_3 &= \sin(x_1) \quad , \; x_3(0) = 0. \end{aligned}$$

We take $X(x) = (x_2, x_1 x_3, \sin\, x_1)^{\mathrm{T}}$, $Y = (1, 1, 0)^{\mathrm{T}}$. These vector fields generate an infinite-dimensional Lie algebra. So, theoretically, we can not expect to reach the origin in finite steps. It can be directly calculated the following Lie brackets: $[Y, X](x) = (1, x_3, \cos\, x_1)^{\mathrm{T}}$ and $[X, [X, Y]](x) = (x_3, x_1 \cos\, x_1 + x_3 - \sin\, x_1, \cos\, x_1 + x_2 \sin\, x_1)^{\mathrm{T}}$. Then Y, $[Y, X]$ and $[X, [X, Y]]$ are linearly independent at every point sufficiently close to the origin. The results of [9, 16, 17] and [12] imply that Y, $[X, Y]$ and $[X, [X, Y]]$ are control variations of third order. For the sample run, shown in Table 3, the computations end after 11 steps.

Table 3. Simulation results for Example 3. At each step, starting from the point $x = (x_1, x_2, x_3)^{\mathrm{T}}$ we move along a suitable chosen trajectory time t_x. The calculated end-point is a starting point for the next iteration.

step	x_1	x_2	x_3	t_x
1	−0.7	0.6	−0.7	5.964160037
3	$6.46243645.10^{-1}$	$2.68204626.10^{-1}$	$8.17140796.10^{-1}$	4.977525366
5	$-1.38120390.10^{-1}$	$-7.25770966.10^{-2}$	$-1.11551718.10^{-1}$	1.896470928
7	$1.65818311.10^{-5}$	$-1.86612233.10^{-5}$	$-1.23494724.10^{-5}$	$1.26774871.10^{-1}$
9	$4.99196520.10^{-11}$	$1.12568962.10^{-9}$	$-6.05404117.10^{-10}$	$2.53457987.10^{-3}$
11	$4.47233396.10^{-19}$	$1.91497208.10^{-17}$	$-3.13242048.10^{-18}$	$7.93634219.10^{-6}$

References

1. Bacciotti, A.:Some remarks on generalized solutions of discontinuous differential equations, Int. J. Pure Appl. Math. **10** (2004) 257–266.
2. Brockett, R.: Asymptotic stability and feedback stabilization, in Differential Geometric Control Theory, Progr. Math. **27** (1983), R. Brockett, R. Millmann, H. Sussmann eds., Birkhäuser, Basel-Boston, 181–191.
3. Clarke, F., Ledyaev, Yu., Sontag, E., Subbotin, A.: Asymptotic controllability omplies feedbackstabilizatuion, IEEE Trans. Automat. Control **42** (1997) 1394–1407.

4. Clarke, F., Ledyaev, Yu., Stern R., Wolenski, P.: Nonsmooth analysis and control theory, Graduate Text in Mathematics **178** Springer-Verlag, New York, 1998.
5. Coron, J.-M., Global asymptotic stabilization for controllable systems without drift, Math. Control Signals Systems **5** (1992) 295–312.
6. Coron, J.-M.: Links between local controllability and local continuous stabilization, IFAC Nonlinear Control Systems Design (1992), M. Fliess ed., Bordeaux, 165–171.
7. Coron, J.-M.: Stabilization in finite time of locally controllable systems by means of continuous time-varying feedback laws, SIAM J. Control Optim. **33** (1995) 804–833.
8. Hermes, H.,: On the synthesis of a stabilizing feedback control via Lie algebraic methods, SIAM Journal on Control and Optimization **16** (1978) 715–727.
9. Hermes, H.: Lie algebras of vector fields and local approximation of attainable sets, SIAM Journal on Control and Optimization **18** (1980) 352–361.
10. Krasovskii, N., Subbotin, A.: Game-Theoretical Control Problems, Springer Verlag, New York, 1988.
11. Krastanov, M., Quincampoix, M.: Local small-time controllability and attainability of a set for nonlinear control systems, ESAIM: Control. Optim. Calc. Var. **6** (2001) 499–516.
12. Krastanov, M. I., Veliov, V. M.: On the controllability of switching linear systems, Automatica **41** (2005) 663–668.
13. Michalska, H., Torres-Torriti, M.: Feedback stabilization of strongly nonlinear systems using the CBH formula, Int. J. Control **77** (2004) 562–571.
14. Sontag, E.: Mathematical Control Theory: Deterministic Finite Dimensional Systems, Texts in Applied Mathematics **6**, Springer-Verlag, New York, Berlin, Heidelberg, London, Paris, Tokyo, Hong Kong (1990).
15. Sontag, E., Sussmann, H.: Remarks on continuous feedback, Proc. of the 19-th IEEE Conf. Decision and Control (Albuquerque) (1980), IEEE Press, Piscataway, 916–921.
16. Veliov, V., M. Krastanov, M. I.: Controllability of piecewise linear systems, Systems & Control Letters **7** (1986) 335–341.
17. Veliov, V.: On the controllability of control constrained linear systems, Math. Balk., New Ser. **2** (1988) 147–155.

Robust Attainability of a Closed Set for Nonlinear Systems with Imperfect Initial State Information

Sylvain Rigal

Laboratoire de Mathématiques, Université de Bretagne Occidentale,
6 Avenue Victor Le Gorgeu, BP 809, 29285 Brest, France
Laboratoire de Recherches Balistiques et Aérodynamiques,
Forêt de Vernon, B.P. 914, 27207 Vernon cédex, France
sylvain.rigal@dga.defense.gouv.fr

Abstract. In this paper, we investigate the existence of controls which allow to reach a given target through trajectories of a nonlinear control system in the case of a non exactly known initial state. For doing this, we use the key concept of weakly invariant tubes and we give a new compactness property for weakly invariant tubes with values in a prescribed collection of sets. We give some consequences of this property on the minimal time function to reach the target, and we prove a sufficient condition for the attainability of the target by weakly invariant tubes of the considered system. If the attainability property is not satisfied, we characterize a subcollection of initial sets from which the attainability property holds true, and we provide an algorithm to compute it.

1 Introduction

We consider a control system defined by

$$x'(t) = f(x(t), u(t)), \quad a.e.\ t \geq 0, \tag{1}$$

where $x \in \mathbb{R}^n$ is the state vector, $u : [0, +\infty) \to U$ is the control, $U \subset \mathbb{R}^l$ is compact, $f : \mathbb{R}^n \times U \to \mathbb{R}^n$ Lipschitz. Given an initial condition $x(0) = x_0$ and a measurable control function $u(\cdot)$, we denote by $t \mapsto x[x_0, u(\cdot)](t)$ the trajectory of system (1) starting from x_0 associated with control $u(\cdot)$.

Let us first recall that the problem of small time local attainability of a given closed set C by trajectories of (1) consists in answering the following question:
– For any given $T > 0$, does there exist a closed neighborhood K of C such that for any point $x_0 \in K \setminus C$, there exists a measurable control $u(\cdot)$ such that the corresponding solution to (1) reaches the set C before T:

$$\forall T > 0, \exists r > 0, \forall x_0 \in K \setminus C \text{ where } K := C + r\mathbf{B}, \exists u(\cdot), \exists t < T, x[x_0, u(\cdot)](t) \in C.$$

A comprehensive answer to this question can be found in [3, 6, 10, 14]. The specificity of the problem considered in the present paper lies in the fact that the initial state $x(0)$ is not supposed to be exactly known but to belong to a given

I. Lirkov, S. Margenov, and J. Waśniewski (Eds.): LSSC 2005, LNCS 3743, pp. 247–254, 2006.

set: $x(0) = e \in E_0 \subset \mathbb{R}^n$. We denote by $x[E_0, u(\cdot)](t) := \{x[e, u(\cdot)](t) \mid e \in E_0\}$ the reachable state of system (1) at time t for all the trajectories starting from E_0 associated with control $u(\cdot)$. The set-valued map $t \mapsto x[E_0, u(\cdot)](t)$ is called the *solution tube of (1) in* $\mathbb{R}^n$ starting from $E_0 \subset \mathbb{R}^n$.

Then, the attainability property with uncertain initial state could be reformulated as follows: Given an initial set E_0 and a measurable control $u(\cdot)$, we say that the corresponding solution tube $x[E_0, u(\cdot)](\cdot)$ of (1) in $\mathbb{R}^n$ "reaches" C at time T when we have $x[E_0, u(\cdot)](T) \subset C$. In other words, if the tube $x[E_0, u(\cdot)](\cdot)$ reaches C at time T, we guarantee that any corresponding trajectory $x[x_0, u(\cdot)](\cdot)$ starting from any $x_0 \in E_0$ reaches C before T. Roughly speaking, our attainability property is "robust" with respect to variations of the initial condition x_0 in E_0. Thus, the small time local attainability question becomes:

- *For any given $T > 0$, does there exist a neighborhood $K := C + r\mathbf{B}$ of C such that for any initial set $E_0 \subset K$, there exists a measurable control $u(\cdot)$ such that a corresponding solution tube of (1) in $\mathbb{R}^n$ starting from E_0 reaches the set C before T:*

$$\forall T > 0,\ \exists r > 0, \forall E_0 \subset K := C + r\mathbf{B}, E_0 \not\subset C, \exists u(\cdot), \exists t < T, x[E_0, u(\cdot)](t) \subset C.$$

In the control context, further numerical applications of these theories require to deal with "simple" prescribed collections of sets $\mathcal{E}$, rather than general classes of subsets of $\mathbb{R}^n$. Therefore, it is important to consider evolutions of outer approximations of the solution tube $t \mapsto x[E_0, u(\cdot)](t)$. So, following [15], we will use evolutions of sets in a given collection of compact sets $\mathcal{E} \subset \mathrm{comp}(\mathbb{R}^n)$. Of course when $\mathcal{E} = \mathrm{comp}(\mathbb{R}^n)$, it reduces to the previous problem, but interesting classes of compact sets $\mathcal{E}$ will be formed of simple sets - for instance polyhedrons, ellipsoids, etc.

To describe the evolution of such outer approximations, we can find a set-valued map $E(\cdot) : [0, +\infty] \to \mathcal{E}$ satisfying some invariance properties with respect to (1), and such that $x[E_0, u(\cdot)](t) \subset E(t) \in \mathcal{E}$ for all $t \geq 0$. We might refer to $E(\cdot)$ as a *weakly invariant tube of system (1) with values in* $\mathcal{E}$ starting from E_0.

Now, let $\mathcal{E}$ be given. The main question we address in the present paper is that of the small time local attainability of C by weakly invariant tubes of system (1) with values in $\mathcal{E}$. In other words:

- *For any given $T > 0$, does there exist a neighborhood $K := C + r\mathbf{B}$ of C such that for any initial set $E_0 \in \mathcal{E}_K := \mathcal{E} \cap \mathrm{comp}(K)$ with $E_0 \not\subset C$, there exists a measurable control $u(\cdot)$ such that a corresponding weakly invariant tube of (1) with values in $\mathcal{E}$, starting from E_0, reaches the set C in a time not greater than T:*

$$\forall T > 0,\ \exists r > 0,\ K := C + r\mathbf{B},\ \forall E_0 \in \mathcal{E}_K := \mathcal{E} \cap \mathrm{comp}(K), E_0 \not\subset C,$$
$$\exists u(\cdot),\ \exists t < T,\ E(t) \subset C.$$

Before answering this question, let us give a few important consequences of a possible positive answer to the previous question:

- The set C is viable for system (1). Namely, starting from any initial set $E_0 \subset C$ there exists a weakly invariant tube $E(\cdot)$ of system (1) with values in $\mathcal{E}$ such that $E(t) \subset C$ for all $t \geq 0$.

– The *minimal time function*

$$\Theta(E_0) := \inf\{t > 0 \mid \exists E(\cdot) \text{ weakly invariant tube of system (1) with values in } \mathcal{E}, \text{ starting from } E_0, \text{ such that } E(t) \subset C\} \quad (2)$$

takes finite values when $E_0 \in \mathcal{E}_K$.

– When $\mathcal{E} = \mathrm{comp}(\mathbb{R}^n)$, we have in particular: $\forall T > 0, \exists r > 0, \exists u, \exists t < T, x[C+rB,u](t) \subset C$ which requires that, starting from $C+rB$, there exist some "decreasing" tubes $x[C+rB,u](\cdot)$ of system (1), the latter meaning that $\forall t_1,\ t_2 \in [0,T],\ t_1 < t_2 \Rightarrow x[C+rB,u](t_2) \subset x[C+rB,u](t_1)$.

In order to describe the evolution of set-dynamics, one can consider two related approaches: The approach of the "funnel equations" [7,8,9,11,15] and the framework of the mutational equations [2,4,5] which is based on a generalization of the notion of derivative. Within the control context, it is necessary to use prescribed collections of simple sets rather than general classes of subsets of the state space. Moreover, it is crucial to use here the Hausdorff semidistance which is suitable to characterize set inclusions. Therefore, our work is also related to the work of [12].

Let us explain how the paper is organized. In Section 2, we introduce some preliminaries on collections of sets, weakly invariant tubes with values in a prescribed collection, and viability for tubes. Then, we state in Section 3 a new compactness result on the space of weakly invariant tubes in $\mathcal{E}$ and we give some consequences on the minimal time function $\Theta(\cdot)$. Section 4 is devoted to our main result: Using the key concept of viability for tubes introduced in [12], we prove a sufficient condition for the attainability of a closed set C by weakly invariant tubes of system (1) with values in a prescribed collection $\mathcal{E}$. Finally, using viability kernels, we characterize those collections of initial sets from which starts at least one trajectory reaching C in finite time.

2 Preliminaries

2.1 Hypothesis

We denote by $\mathrm{comp}(\mathbb{R}^n)$ the collection of all compact subsets of $\mathbb{R}^n$ and by $\mathbf{B}$ the unit ball in $\mathbb{R}^n$ with respect to Euclidean norm. For two sets $A, B \in \mathrm{comp}(\mathbb{R}^n)$, we denote by $H^+(A,B) := \min\{\varepsilon \geq 0 \mid A \subset B + \varepsilon\mathbf{B}\}$ the Hausdorff semidistance from A to B and by $H(A,B) := \max\{H^+(A,B), H^+(B,A)\}$ the Hausdorff distance between A and B. Let $C \in \mathrm{comp}(\mathbb{R}^n)$ be a given set. In order to make the reading easier, we denote by $H_C^+(A) := H^+(A,C)$ the Hausdorff semidistance to C. Throughout the paper, we denote by $\mathcal{E} \subset \mathrm{comp}(\mathbb{R}^n)$ a nonempty collection of compact subsets of $\mathbb{R}^n$ satisfying the following condition:

Condition 1.

1. *$\mathcal{E}$ consists of nonempty compact sets and is closed in the Hausdorff topology,*
2. *For every $Z \in \mathrm{comp}(\mathbb{R}^n)$ there is some $E \in \mathcal{E}$ containing Z,*

3. *There exist*[1] $\bar{\epsilon} > 0$ *and* $L_\epsilon > 0$ *such that for each* $\epsilon \in [0, \bar{\epsilon})$ *and each* $E \in \mathcal{E}$ *there exists* $E' \in \mathcal{E}$ *for which* $E + \epsilon\mathbf{B} \subset E' \subset E + \epsilon L_\epsilon \mathbf{B}$.

Equipped with the Hausdorff distance, the set comp($\mathbb{R}^n$) is a complete metric space. Therefore the notions of closedness and compactness of a collection of sets $\mathcal{E} \subset$ comp($\mathbb{R}^n$) make sense. Some of our results also require the following condition which is related to Condition 1.2:

Condition 2. *For every* $Z \in$ comp($\mathbb{R}^n$), *there is a unique minimal (with respect to inclusion) element of* $\mathcal{E}$ *containing* Z.

In the sequel, we consider system (1) under the following standing condition:

Condition 3.

1. $f(\cdot,\cdot) : \mathbb{R}^n \times U \to \mathbb{R}^n$ *has the form* $f(x,u) = f_0(x) + f_1(x)u$, *where* $f_0(\cdot) : \mathbb{R}^n \to \mathbb{R}^n$ *and* $f_1(\cdot) : \mathbb{R}^n \to \mathbb{R}^{n\times l}$ *are locally Lipschitz mappings and* U *is a convex compact subset of* $\mathbb{R}^l$ *bounded by* $M > 0$ ($U \subset M\mathbf{B}_{\mathbb{R}^l}$, $\mathbf{B}_{\mathbb{R}^l}$ *being the unit ball in* $\mathbb{R}^l$)
2. f *satisfies the linear growth condition:* $\exists c > 0, \forall x \in \mathbb{R}^n, \forall u \in U, \|f(x,u)\| \leq c(1 + \|x\|)$.

2.2 Tubes in a Prescribed Collection

Definition 1. *A set-valued map* $E(\cdot) : [0,T] \to$ comp($\mathbb{R}^n$) *is called a* tube *if it is nonempty compact valued and has closed graph.*

All tubes in this paper are supposed to fulfill the following condition:

Condition 4. *The tube* $E(\cdot)$ *is upper semicontinuous and locally Lipschitz from the right:*

$$\forall T > 0,\ \exists L > 0, \quad E(s) \subset E(t) + L(t-s)\mathbf{B}, \qquad \forall\, 0 \leq s \leq t \leq T.$$

For example, let $u(\cdot)$ be given and $E_0 \in$ comp($\mathbb{R}^n$). The solution tube $x[E_0, u(\cdot)](\cdot) : [0,T] \to \mathbb{R}^n$ is a tube. An important property of tubes is the *weak invariance* which is defined as follows:

Definition 2. *Let a measurable control function* $u(\cdot)$ *be given. A tube* $E(\cdot)$ *is said to be* weakly invariant *with respect to (1) in* $[0,T]$ *if, for all* $t \in [0,T)$ *and for all* $x(\cdot)$ *solution to (1) on* $[t,T]$ *with* $x(t) \in E(t)$, *we have*

$$x(t+s) \in E(t+s), \quad \forall s \in [0, T-t].$$

The weak invariance property is also characterized in the following way:

$$\forall t \in [0,T),\ \forall s \in [0, T-t], \quad x[E(t), u(t+\cdot)](s) \subset E(t+s). \tag{3}$$

[1] Condition 1.3 is equivalent to the following one: There exist real $\epsilon > 0$ and L_ϵ such that for each $Z \in$ comp($\mathbb{R}^n$) and each $E \in \mathcal{E}$ for which $H(Z,E) \leq \epsilon$ there exists a minimal element E' of $\mathcal{E}$ containing Z and satisfying $H^+(E', Z) \leq L_\epsilon H(Z,E)$.

Thus, a weakly invariant tube of (1) with values in $\mathcal{E}$ *is a tube* $E(\cdot):[0,T]\to\mathcal{E}$ *satisfying property (3)*[2].

This property, even though it is noticeably different from the invariance property introduced by the Viability Theory (see [1]), is called "weak invariance" because, for a given control $u(\cdot)$, any trajectory of system (1) starting from a point inside the tube remains in the tube. We describe the evolution of a weakly invariant tube $E(t)$ of system (1) with values in $\mathcal{E}$, starting from $E_0 \in \mathcal{E}$, by the following set-dynamic equation:

$$\lim_{h\to 0+}\frac{1}{h}H^+\left((Id+hf(\cdot,u(t)))(E(t)),E(t+h)\right)=0, \tag{4}$$

with initial condition

$$E(t_0)=E_0. \tag{5}$$

System (4) is similar to the funnel equations (see [8]), but it applies for nonlinear systems. All of the solutions of (4) are weakly invariant with respect to (1). We now introduce the *solution tube with values in* $\mathcal{E}$ which is a minimal weakly invariant tube of (1) with values in $\mathcal{E}$:

Definition 3. *Let* $\mathcal{E}$ *satisfy Condition 1.*
A solution tube of (1) with values in $\mathcal{E}$ *is a weakly invariant tube* $E(\cdot)$ *with values in* $\mathcal{E}$*, such that* $E(\cdot)$ *is minimal (with respect to inclusion).*

Moreover, if $\mathcal{E}$ satisfies Condition 2, the solution tube with values in $\mathcal{E}$ starting from a given E_0 is unique. For example, when $\mathcal{E}=\mathrm{comp}(\mathbb{R}^n)$, under Condition 3, the solution tube is $E(t)=x[E_0,u(\cdot)](t)$. It is the unique *minimal* tube that starts from E_0 and is *weakly invariant* with respect to (1). But when $\mathcal{E}\subset\mathrm{comp}(\mathbb{R}^n)$, we have $E(t)\supset x[E_0,u(\cdot)](t)$. More examples of "tractable" collections are given in [13].

2.3 Viability

In this section, we recall some basic material from Viability theory with uncertain initial state information. Let $\mathcal{L}$ be a convex subset of the space $\mathrm{C}(\mathbb{R}^n,\mathbb{R}^n)$ of continuous mappings $\mathbb{R}^n\to\mathbb{R}^n$ such that the restrictions of the functions from $\mathcal{L}$ to any compact subset of $\mathbb{R}^n$ are equi-Lipschitz and uniformly bounded. By definition (see [12]), the collection $\mathcal{E}\subset\mathrm{comp}(\mathbb{R}^n)$ is a *viability domain* for $\mathcal{L}$ if and only if $\forall E\in\mathrm{cl}(\mathcal{E})$, $\mathcal{T}_{\mathcal{E}}(E)\cap\mathcal{L}_{|E}\neq\emptyset$, with

$$\mathcal{T}_{\mathcal{E}}(E)=\{l(\cdot):E\to\mathbb{R}^n \mid \liminf_{h\to 0+}\frac{1}{h}\inf_{\tilde{E}\in\mathcal{E}}H^+\left(I_d+hl(\cdot))(E),\tilde{E}\right)=0\},$$

where $l(\cdot)$ is called a *contingent field.* In other words,

$$l(\cdot):E\to\mathbb{R}^n\in\mathcal{T}_{\mathcal{E}}(E)\Longleftrightarrow\begin{cases}\exists h_k\to 0+\\ \exists\gamma_k\to 0\\ \exists\tilde{E}_k\in\mathcal{E}\end{cases},\ \forall x\in E,\quad x+h_kl(x)\in\tilde{E}_k+h_k\gamma_k\mathbf{B}.$$

[2] As a consequence of property (3), we obviously have: $\forall t\in[0,T]$, $x[E_0,u(\cdot)](t)\subset E(t)\in\mathcal{E}$.

A contingent field $l(\cdot)$ does not need to be necessarily continuous but, in this paper, we restrict our considerations to continuous contingent fields for the sake of simplicity. Moreover, in the case of a collection of single points, the notion of contingent field coincides with that of the Bouligand contingent vector. We now recall a viability result for tubes adapted to our framework.

Theorem 1 (Theorem 1.2 in [12]). *Suppose that control system (1) satisfies Condition 3. Let collection $\mathcal{E}$ satisfy Condition 1 and let K be a given bounded set.*

If $\mathcal{E}_K := \mathcal{E} \cap \mathrm{comp}(K)$ is a closed viability domain, then for every $E_0 \in \mathcal{E}_K$, there exist a time τ and a measurable $u(\cdot) : [0, \tau] \to U$ such that there exists a tube $E(\cdot)$ solution to (4) on $[0, \tau]$ with values in $\mathcal{E}_K$ and initial condition (5). This tube is a weakly invariant tube of (1) with values in $\mathcal{E}$, starting from E_0.

Under Condition 3 and according to Proposition 2 of [12], given a collection $\mathcal{E}$ which satisfies Condition 1, given a bounded set K and a subset $\mathcal{L} = \{f(\cdot, u) \mid u \in U\}$ of $\mathrm{C}(\mathbb{R}^n, \mathbb{R}^n)$, there exists a maximal viability domain for $\mathcal{L}$ that is contained in $\mathcal{E}_K$. This set, denoted by $\mathrm{Viab}_{\mathcal{L}}(\mathcal{E}_K)$, is called the *viability kernel* of $\mathcal{E}_K$ for $\mathcal{L}$.

3 Compactness of the Set of Solutions

Proposition 1. *Let $\mathcal{E}$ satisfy Conditions 1 and 2, and let $f : \mathbb{R}^n \times U \to \mathbb{R}^n$ satisfy Condition 3. Fix $K \in \mathrm{comp}(\mathbb{R}^n)$ and $T > 0$. Let $(E_0^k)_{k\in\mathbb{N}}$ be a sequence of compact sets from collection $\mathcal{E}_K$ and let $(E_k(\cdot))_{k\in\mathbb{N}}$ be a sequence of weakly invariant tubes of system (1) with values in $\mathcal{E}$ associated with controls $(u_k(\cdot))_{k\in\mathbb{N}}$, where for every $k \in \mathbb{N}$, tube $E_k(\cdot)$ starts from E_0^k.*

Then, there exists a sequence of solution tubes $(\tilde{E}_k(\cdot))_{k\in\mathbb{N}}$ of system (1) with values in $\mathcal{E}$ (where for every $k \in \mathbb{N}$, tube $\tilde{E}_k(\cdot)$ starts from E_0^k) such that

$$\forall k \in \mathbb{N},\ \forall t \in [0, T],\ \tilde{E}_k(t) \subset E_k(t) \in \mathcal{E},$$

and there exists a compact subset $\overline{E}_0$ of $\mathcal{E}$ and a continuous weakly invariant tube $\overline{E}(\cdot)$ of system (1) with values in $\mathcal{E}$ starting from $\overline{E}_0$, such that

$$\liminf_{k\to+\infty} H\left(\tilde{E}_k(t), \overline{E}(t)\right) = 0, \quad \forall t \in [0, T].$$

We deduce the following from Proposition 1.

Corollary 1. *Let Conditions 1, 2 and 3 be satisfied.*

1. *If $\Theta(E_0) < +\infty$ there exists an optimal weakly invariant tube of (1) with values in $\mathcal{E}$ starting from E_0. Namely, the infimum in (2) is attained.*
2. *The minimal time function $\Theta(\cdot)$ is lower semicontinuous.*

4 Small Time Local Attainability

Here is the main result of this paper:

Proposition 2. *Suppose that Conditions 1 and 3 are fulfilled. Let γ be a positive constant, and define $\mathcal{L}^* := \{(f(\cdot, u), -\gamma) \mid u \in U\}$. Fix a nonempty compact set*

C and a constant $r > 0$. Let $K := C + r\mathbf{B}$ and $\mathcal{E}_K := \mathcal{E} \cap \mathrm{comp}(K)$. We denote by $\mathcal{E}_K^*$ the epigraph of $H_C^+(\cdot)$ restricted to $\mathcal{E}_K$, defined as follows:

$$\mathcal{E}_K^* := \mathrm{Epi}(H_C^+(\cdot)_{|\mathcal{E}_K}) = \{(E, y) \in \mathcal{E}_K \times \mathbb{R}^+ \mid H_C^+(E) \le y\}.$$

If the following condition is satisfied

$$\forall E \in \mathcal{E}_K \text{ such that } H_C^+(E) > 0,\ \mathcal{T}_{\mathcal{E}_K^*}(E, H_C^+(E)) \cap \mathcal{L}^*_{|(E, H_C^+(E))} \neq \emptyset, \tag{6}$$

then for any set of initial conditions $E_0 \in \mathcal{E}_K$, there exists at least one tube of system (4) starting from E_0 and reaching C in finite time.

From the proof of Proposition 2, we deduce the following.

Corollary 2. *Suppose that Condition 2 is satisfied. Under the assumptions made in Proposition 2, condition (6) yields $\Theta(E_0) \le \mathcal{V}\left(H_C^+(E_0)\right) = \frac{H_C^+(E_0)}{\gamma}$.*

If a given collection $\mathcal{E}$ satisfies condition (6) then, for any given initial state $E_0 \in \mathcal{E}$, there exists at least one weakly invariant tube $E(\cdot)$ of system (1) in $\mathcal{E}$ reaching C before time $T = \mathcal{V}\left(H_C^+(E_0)\right)$. Roughly speaking, we say that the tube $E(\cdot)$ reaches C *with a speed greater than* γ. If condition (6) is not satisfied, we want to characterize the collection of initial sets E_0 from which starts at least one weakly invariant tube of system (1) with values in $\mathcal{E}$ reaching C with a speed greater than γ. This collection of initial sets is known as the *capture basin* of C for system (1). It is denoted by $\mathrm{Capt}_{\mathcal{L}}(C, \mathcal{E}_K)$ and can be characterized through a viability kernel of $\mathcal{E}^*$ for an extended system:

Proposition 3. *Suppose that Conditions 1 and 3 are fulfilled and let $\Pi_{\mathcal{E}} : \mathcal{E}_K^* \to \mathcal{E}$ be the projection onto $\mathcal{E}$ such that $\Pi_{\mathcal{E}}(E, y) := E$. For $h > 0$, we denote by $\mathcal{E}_K^h$ the limit of the following algorithm when $k \to +\infty$:*

$$\begin{cases} \mathcal{E}_0^h = \mathcal{E}_K^* \\ \mathcal{E}_{k+1}^h = \Big\{(E, y) \in \mathcal{E}_k^h \mid E \subset C \text{ or } \exists u \in U,\ \exists(\tilde{E}, \tilde{y}) \in \mathcal{E}_k^h,\ y - \gamma h \ge \tilde{y} \\ \qquad \text{and } (Id + hf(\cdot, u))(E) \subset \tilde{E} + \dfrac{LM_f}{2} h^2 \mathbf{B}\Big\},\ k = 0, 1, \ldots \end{cases} \tag{7}$$

where M_f is a bound of $f(\cdot, u)$ uniformly in $u \in U$. Then we have

$$\mathrm{Capt}_{\mathcal{L}}(C, \mathcal{E}_K) = \Pi_{\mathcal{E}}\left(\lim_{h \to 0_+} \mathcal{E}_K^h\right).$$

where the limit is with respect to the Hausdorff distance between compact subsets in the space $\mathrm{comp}(\mathbb{R}^n)$.

This algorithm is a first step towards numerical analysis. Even though it is essentially a tool for providing a collection of initial sets satisfying some attainability properties, we can imagine to use it in a constructive way: When implemented, it could save all of the "good" values of u in order to find a selection $u(\cdot)$ such that a corresponding weakly invariant tube of (1) with values in $\mathcal{E}$ reaches C with a speed greater than γ.

References

1. J.-P. Aubin. *Viability Theory.* Birkhäuser Boston, 1991.
2. J.-P. Aubin. *Mutational and Morphological Analysis.* Birkhäuser Boston, 1999.
3. R. Bianchini and G. Stefani. Time Optimal Problem and Time Optimal Map. *Rend. Sem. Mat. Univ. Politec. Tonrino*, 48:401–429, 1990.
4. L. Doyen. Filippov and Invariance Theorems for Mutational Equations for Tubes. *Set-Valued Analysis*, 1:289–303, 1993.
5. A. Gorre. Evolutions of Tubes under Operability Constraints. *J. Math. Anal. Appl.*, 216(1):1–22, 1997.
6. M. Krastanov and M. Quincampoix. Local Small Time Controllability and Attainability of a Set for Nonlinear Control System. *ESAIM Control Optimiz. and Calculus of Variations*, 6:499–516, 2001.
7. A. B. Kurzhanski and T. F. Filippova. On the theory of trajectory tubes - A Mathematical Formalism for Uncertain Dynamics, viability and control. In *Advances in Nonlinear Dynamics and Control: A Report from Russia*, Progress in Systems and Control, 17:122–188. Birkhäuser, 1993.
8. A. B. Kurzhanski and O.I. Nikonov. Funnel equations and multivalued integration problems for control synthesis. In B. Jakubczyk and K. Malanowski and W. Respondek, editor, *Perspectives in Control Theory*, Progress in Systems and Control, 2:143–153. Birkhäuser Basel, 1990.
9. A. B. Kurzhanski and I. Vályi. *Ellipsoidal Calculus for Estimation and Control.* Birkhäuser Boston, 1997.
10. P. Nistri and M. Quincampoix. On open-loop and feedback attainability of a closed set for nonlinear control systems. *J. Math. Anal. Appl.*, 270(2):474–487, 2002.
11. A. I. Panasyuk. Equations for attainable set dynamics, Part 1: Integral Funnel Equations. *J. Optim. Theory Appl.*, 64(2):349–366, 1990.
12. M. Quincampoix and V. M. Veliov. Open-Loop Viable Control Under Uncertain Initial State Information. *Set-Valued Analysis*, 7(1):55–87, 1999.
13. M. Quincampoix and V. M. Veliov. Solution Tubes to Differential Inclusions within a Collection of Sets. *Control and Cybernetics*, 31(3):849–862, 2002.
14. P. Soravia. Hölder Continuity of the Minimum-Time Function for C^1-Manifold Targets. *JOTA*, 75, 1992.
15. V. M. Veliov. Funnel equations and regulation of uncertain systems. In A. B. Kurzhanski and V. M. Veliov, editor, *Set-Valued Analysis and Differential Inclusions*, 183–199. Birkhäuser Basel, 1993.

Numerical Methods for Optimal Control Problems with ODE or Integral Equations

Werner H. Schmidt

Ernst–Moritz–Arndt–Universität Greifswald, Germany
Institut für Mathematik und Informatik,
F.-L.-Jahnstr. 15 a, D-17487 Greifswald, Germany
wschmidt@uni-greifswald.de
http://www.math-inf.uni-greifswald.de

Abstract. An overview of numerical methods for solving optimal control problems described by ODE and integral equations is presented. We consider direct and indirect methods. The finer indirect methods use necessary optimality conditions. Direct methods transform the control problem after discretization to an optimization problem. The nonlinear optimization problem can be solved by means of SQP–methods or gradient methods. Known variants of this method are GESOP and DIRCOL. Then a wave–method is mentioned, in which the state variables are varied at first. The direct methods apply the maximum principle, often it is possible to eliminate the control with the help of the necessary condition. The control problem is transformed to a boundary value problem for the state and the adjoint variable, which is solved by multiple shooting. The iterative procedures of Krylow/Chernousko and Sakawa, respectively, are based on the maximum principle, too. It is referred to the gradient methods described in the monograph of Pytlak and to prox–methods.

1 Introduction

The control theory (a generalization of the calculus of variations) was developed in the last fifty years as a powerful tool to describe real processes in economics, armed forces, industry and life sciences and to obtain optimal solutions in the mathematical model. Numerical methods were searched from the very beginning to solve such mathematical problems, particularly such coming from military aircraft and astronautics. Most of the papers dealing with such methods were top secret for a long time and some are still unknown. The numerical ideas were restricted by the slowness of the computers and the strong limitedness of their results until the 80th. Necessary optimality conditions are the most well-known results in optimal control theory for engineers. Pontryagin and his team (see [9]) proposed to calculate the optimal control using the maximum principle.

Here we consider some direct and indirect numerical methods for solving optimal control problems. In the direct methods the control problem is approximated by a (finite dimensional) optimization problem. The finer indirect methods are based on necessary optimality conditions. Pytlak [10] studied feasible direction algorithms, in his opinion they are the most promising first order algorithms. In [1] we proposed a combination of gradient and prox–methods.

I. Lirkov, S. Margenov, and J. Waśniewski (Eds.): LSSC 2005, LNCS 3743, pp. 255–262, 2006.

2 The Problem

Assume $f : [0,T] \times \mathbb{R}^n \times \mathbb{R}^r \longrightarrow \mathbb{R}^n$, $f_0 : [0,T] \times \mathbb{R}^n \times \mathbb{R}^r \longrightarrow \mathbb{R}$ and $r : \mathbb{R}^n \times \mathbb{R}^n \longrightarrow \mathbb{R}^p$ and $g : \mathbb{R}^n \longrightarrow \mathbb{R}^s$ to be continuous functions and $U \subseteq \mathbb{R}^r$, $G \subseteq \mathbb{R}^n$ given sets. G may be described by $g(x(t)) \geq 0$ for all $0 \leq t < T$, where $[0,T]$ is a fixed time–interval.

To minimize is the functional

$$I(u(\cdot)) = J(x(\cdot), u(\cdot)) = \int_0^T f_0(t, x(t), u(t))dt \tag{1}$$

with respect to

$$\dot{x} = f(t, x(t), u(t)) \tag{2}$$
$$r(x(0), x(T)) \geq 0 \tag{3}$$
$$x(t) \in G \text{ a.e.} \tag{4}$$
$$u(t) \in U \text{ a.e.} \tag{5}$$

$u(\cdot)$ piecewise continuous, $x(\cdot)$ absolutely continuous.

Two points of view are possible: J can be considered as a functional of $x(\cdot)$ and $u(\cdot)$, where the state equation (2) and the conditions (3) – (5) are restrictions of the control problem. On the other hand, (1) is a functional I of $u(\cdot)$, where $u(\cdot)$ determines $x(\cdot)$ by means of (2) – (5).

To apply direct methods we put a grid $\{t_i\}$ on $[0,T]$, for instance an equidistant grid with mesh points $t_i = ih$, $h = T/N$, $N \in \mathbb{N}$. Discretize the trajectory and the control: $x(\cdot) \longrightarrow (x_0, x_1, \ldots, x_N)$, $u(\cdot) \longrightarrow (u_0, u_1, \ldots, u_{N-1})$, where $x_i \approx x(t_i)$ and $u_i \approx u(t_i)$, $i = 0, \ldots, N-1$. A very simple discretization of (1) – (5) is obtained by means of the Euler scheme and the rectangle integration formula

$$\sum_{i=0}^{N-1} f(t_i, x_i, u_i) = \min! \tag{6}$$

subject to

$$x_{i+1} = x_i - hf(t_i, x_i, u_i) \tag{7}$$
$$r(x_0, x_N) \geq 0 \tag{8}$$
$$x_i \in G, u_i \in U \text{ for all } i. \tag{9}$$

The following method deals with (6) as a functional of x_i with corresponding controls u_i.

3 Variations in States

Reinbach [11] tested some well-known methods in applications. The idea is to reduce the dimension of the finite dimensional optimization problem (6) – (9), the price is to deal with a big number of optimization problems. Let us approximate the functional (1) by the trapeze rule. Start with an admissible vector

$(x_0, x_1, \dots, x_N) \in \mathbb{R}^{n(N+1)}$. Let $e_k \in \mathbb{R}^n$ denote the k–th unit vector. Compare the initial vector $(x_0, \dots x_N)$ for every $i = 1, \dots, N-1$, starting with $i = 1$, with $2n$ disturbed trajectories $(x_0, \dots, x_{i-1}, y_i^{\pm k}, x_{i+1}, \dots, x_N)$, where $y_i^{\pm k} = x_i \pm h_k e_k$, $h_k \in \mathbb{R}$ fixed, such that $y_i^{\pm k} \in G$. That means, compare the vector with all trajectories connecting x_{i-1} with x_{i+1} via $y_i^{\pm k}$ and find that with a minimal cost functional: For every $y_i^{\pm k}$ minimize

$$J_{ik}^{\pm} = f_0(t_{i-1}, x_{i-1}, u_{i-1}) + f_0(t_i, x_i \pm h_k e_k, u_i)$$

with respect to u_{i-1}, $u_i \in U$. Compare the best way with the cost functional of the trajectory $(x_0, \dots, x_N)$. If all $J_{ik}^{\pm}$ are bigger or equal to $f_0(t_{i-1}, x_{i-1}, u_i) + f_0(t_i, x_i, u_i)$ then put $\tilde{x}_i = x_i$ $\tilde{u}_{i-1} = u_{i-1}$, $\tilde{u}_i = u_i$; otherwise define $\tilde{x}_i$, $\tilde{u}_{i-1}$, $\tilde{u}_i$ with the solution of the problem with the smallest value $J_{ik}^{\pm}$: $\tilde{x}_i = x_i + h_e e_e$ or $\tilde{x}_i = x_i - h_e e_e$, respectively, and $\tilde{u}_{i-1}$, $\tilde{u}_i \in U$ are the corresponding controls. In the next step we repeat the algorithm with $i+1$. $(N-1)$ steps give $(x_0, \tilde{x}_1, \dots, \tilde{x}_{N-1}, x_N)$ and $(\tilde{u}_0, \dots, \tilde{u}_{N-1})$. Go to the beginning and vary $\tilde{x}_1$ once more, and so on. Stop if the process is numerically convergent. Have in mind, that there must not exist solutions $\tilde{u}_{i-1}$, $\tilde{u}_i \in U$ minimizing the "small" optimization problems. We are not able to prove convergence. There are some modifications known as wave method, method of traveling tubes, see [6]. We mention, that Bellman's dynamic programming is a tool to solve (6) – (9), too.

4 The Finite Dimensional Problem

The problem (6) – (9) obtained by simple discretization is to solve as a (nonlinear) optimization problem. Therefore consider $u_0, \dots u_{N-1} \in \mathbb{R}^r$ and $x_0, \dots x_N \in \mathbb{R}^n$ to be free variables between which certain constraints (equations and inequalities) (7) – (9) are to be fulfilled. We used the solver LANCELOT by Conn, Gould and Toint [3], that is a FORTRAN–package for Large–Scale Nonlinear Optimization. LANCELOT is based on SQP–methods. The user has to declare all variables and can cluster some groups of variables and functions into groups which are also to declare. In a certain sense, LANCELOT is a special computer language for nonlinear optimization.

5 Direct Collocation Method

The discretization of the control problem is done in a better and in a "more effective" manner. There are software packages TROPIC (by Well, Schnepper, Jänsch) and DIRCOL (by v. Stryk [15]).

The idea is to choose a (collocation) grid $0 = t_0 < \dots < t_N = T$ of the time interval. The control $u(\cdot)$ is approximated by a constant or a linear function in every subinterval $[t_i, t_{i+1}]$, the state $x(\cdot)$ is taken as a cubic polynomial $S(t)$ in each $[t_i, t_{i+1}]$. For simplicity we take $n = 1$. Then

$$S_i(t) = a_i + b_i(t - t_i) + c_i(t - t_i)^2 + d_i(t - t_i)^3,$$

$t_i \le t \le t_{i+1}$, is the state with certain coefficients a_i, b_i, c_i, d_i. Of course, $a_i = S_i(t_i) = x_i$, $S(t_{i+1}) = x_{i+1}$. From the state equation (2) it follows

$$S_i'(t_i) = f(t_i, x_i, u_i) \text{ and } S_i'(t_{i+1}) = f(t_{i+1}, x_{i+1}, u_{i+1}).$$

Therefore the unknown parameters a_i, b_i, c_i, d_i can be calculated as functions of $x_i, x_{i+1}, u_i, u_{i+1}$.

The piecewise cubic polynomial $S(t)$ is at least once continuously differentiable in $[0, T]$ and satisfies the state equation at the nodes t_i. The computer–software DIRCOL demands the state equation in the points $t_i^{1/2} = \frac{1}{2}(t_{i+1} - t_i)$, too: The derivative of $S_i(t)$ is calculated as $S_i'(t) = b_i + 2c_i(t - t_i) + 3d_i(t - t_i)^2$ and the condition

$$S_i'(t_i^{1/2}) = f(t_i^{1/2}, S(t_i^{1/2}), u(t_i^{1/2}))$$

is to take into consideration. Because $u(\cdot)$ is constant or linear, the new conditions are equations of type

$$K_i(x_i, x_{i+1}, u_i, u_{i+1}) = 0, \quad i = 0, \dots, N-1.$$

The constraints (9) are to be fulfilled at the middle points of each interval:

$$g(t_i^{1/2}, S_i(t_i^{1/2}), u(t_i^{1/2})).$$

The original control problem is replaced by the (nonlinear) programming problem in the variables $x_0, \dots, x_N, u_0, \dots u_N$

$$\sum_i \gamma_i f_0(t_i, x_i, u_i) = \min!$$

subject to constraints

$$K_i(x_i, x_{i+1}, u_i, u_{i+1}) = 0$$

and phase constraints

$$g(t_i^{1/2}, x_i, x_{i+1}, u_i, u_{i+1}) \geq 0, \quad i = 0, \dots, N-1.$$

6 Direct Multiple Shooting

This method is similar to that described in the next section, but it does not exploit necessary optimality conditions. We choose a grid (for multiple shooting) with respect to the states: $0 = t_0 < t_1 \cdots < t_N = T$ and finer grid for the controls $t_i = \tau_1^i < \tau_2^i < \cdots < \tau_{k_i}^i = t_{i+1}$, $i = 0, \dots, N_1$. Each component of the control vector is approximated within a shooting interval $[t_i, t_{i+1}]$ by polynomials (constant, linear, quadratic, cubic) or splines, respectively, which depend on unknown controls u_{ij}, $j = 1, \dots, k_i$, $i = 1, \dots, N-1$. We suppose (unknown) initial values s_i for $x(t_i)$ in every interval $[t_i, t_{i+1}]$. Then we integrate the process equation in $[t_i, t_{i+1}]$ numerically and obtain a solution $x(t, ; t_i, s_i)$ with the initial value $x(t_i; t_i, s_i) = s_i$. The trajectory is an admissible one if $x(t_{i+1}; t_i, s_i) = s_{i+1}$ for all $i = 0, \dots, N-1$. Of course, the function $x(t; t_i, s_i)$ depends on the controls u_{ij}, $j = 1, \dots, k_i$, $i = 0, \dots, N-1$.

Here the control vectors u_{ij} and the initial values s_i are the variables which determine $x(t; s_i, u_{ij})$ as the (numerical) solution of an ODE. The programming problem is to minimize

$$\int_0^T f_0(t, x(t, u_{ij}, s_i), u(t))dt$$

subject to

$$x(t_i; t_i, s_i) = s_i$$
$$x(t_{i+1}; t_i, s_i) = s_{i+1}$$

and

$$g(\tau_j^i, x_i(\tau_j^i), u_i(\tau_j^i)) \geq 0, \quad i = 0, \ldots, N-1, \quad j = 1, \ldots, k_i$$

in the variables u_{ij} and the initial values s_i. This can be done by a combination of a SQP–method and an ODE–solver.

7 Multiple Shooting – An Indirect Method

It was the idea of Pontryagin and his team [9] in 1961 and Morrison (1962) to find the optimal control by means of the maximum principle as a function of the state and the adjoint variable. We get a boundary value problem of ($2n$) ODE which can be solved by a shooting method. This method for solving boundary value problems depends very sensitively on the initial values which are to calculate. An effective way is to solve the boundary value problem by the multiple shooting method. It is very well described in the textbook of Stoer and Bulirsch [13], a FORTRAN program can be found in Oberle and Grimm [8]. A similar method is used by Korytowski and Szymkat [4] to find an optimal control of a fedbatch fermentation process.

The H–function corresponding to (1) – (5) is

$$H(t, x, u, \lambda_0, \lambda, \psi) = \psi f(t, x, u) + \lambda_0 f_0(t, x, u) + \lambda g(t, x, u).$$

If $\hat{u}(\cdot)$, $\hat{x}(\cdot)$ is optimal, then the adjoint equation is

$$\dot{\psi}(t) = -\psi(t) f_x(t, \hat{x}(t), \hat{u}(t)) - \lambda_0 f_{0x}(t, \hat{x}(t), \hat{u}(t)) - \lambda(t) g_x(t, \hat{x}(t), \hat{u}(t)),$$
$$\psi(T) = 0.$$

From the maximum principle we often get $\hat{u}(t) = u(t, x(t), \psi(t))$. If the process is singular in the sense $g(t, x(t), u(t)) = 0$ in a certain interval, we use this boundary condition to find $\hat{u}(t) = u(t, x(t))$. It is necessary to know the structure of the optimal solution. The state equation and the adjoint equation are

$$\dot{x} = f(t, x(t), u(t, x(t), \psi(t)) = F(t, x(t), \psi(t))$$
$$\dot{\psi}(t) = -\psi(t) f_x(t, x(t), u(t, x(t), \psi(t)) - \lambda_0 f_{0x}(t, x(t), u(t, x(t), \psi(t)) -$$
$$-\lambda(t) g_x(t, x(t), u(t, x(t), \psi(t)) = \Psi(t, x(t), \psi(t))$$

and it is $\lambda(t) g(t, x(t), u(t)) = 0$ a.e.. We want to find initial values x_0, ψ_0 such that the (numerical) solution of

$$\dot{x} = F(t, x(t), \psi(t)), \quad x(0) = x_0$$
$$\dot{\psi} = \Phi(t, x(t), \psi(t)), \quad \psi(0) = \psi_0$$

fulfills the boundary conditions $r(x_0, x(T)) = 0$ and $\psi(T) = 0$. Therefore we take a grid in $[0, T]$: $0 = t_0 < t_1 < \cdots < t_n = T$ and we solve the ODE–system in every $[t_i, t_{t+1}]$ with an initial vector s_i. Then $z = (x, \psi)$ depends on s_i and it is $z(t) = z(t; t_i, s_i)$ with $z(t_i, t_i, s_i) = s_i$. We have to find $(s_0, \ldots, s_{N-1})$ such that $z(t_{i+1}, t_i, s_i) = s_{i+1}$, $i = 0, \ldots, N-1$ and $r(x_0, x(T); t_{N-1}, s_{N-1}) = 0$, $\psi(T) = 0$. The derived system of (nonlinear) equations can be solved by Newton's method. There are variants of this multiple shooting methods for parallel computing. The collocation method of v. Stryk gives informations on the values of the adjoint variables at the mesh–points.

8 The Method of Krylov and Chernousko

This is an iterative procedure. Start with a dispatcher control $u^0(\cdot)$ and calculate the corresponding solution $x^0(\cdot)$ of the state equation and then $\psi^0(\cdot)$ of the adjoint equation. From the maximum condition we find

$$u^1(t) = \arg \max_{u \in U} H(t, x^0(t), u, \psi^0(t)), \text{ a.e.}$$

and repeat the procedure with u^1 instead of u^0. For simplicity we did not consider phase constraints. Unfortunately, in many cases the method is not convergent. Chernousko and Lyubushin [5] presented some modifications with better properties.

The algorithm of Sakawa is applicable to a much broader class of control problems. Sakava [12] introduced an augmented Hamiltonian of the form

$$H(t, x, u, v, \lambda_0, \lambda) + \beta ||u - v||^2.$$

Then

$$u^1(t) = \arg \max_{u \in U} \{H(t, x^0(t), u, \psi^0(t)) + \beta(u - u^0(t))\}, \text{ a.e.}$$

9 Control Process Described by Integral Equations

The functional

$$\mathcal{F} = \int_0^T f_0(t, x(t), u(t))dt \tag{10}$$

is to be minimized subject to

$$x(s) = \int_0^T f(s, t, x(t), u(t))dt, \quad s \in [0, T] \tag{11}$$

$$u(t) \in U, \quad t \in [0, T] \tag{12}$$

$$g(t, x(t), u(t)) \geq 0 \quad t \in [0, T] \tag{13}$$

A direct method is to solve the programming problem

$$\sum_i f_0(t_i, x_i, u_i)$$

subject to

$$x_i = \sum_{j=1}^{N} \gamma_j f(t_i, t_j, x_j, u_j),$$
$$u_i \in U, \quad g(t_i, x_i, u_i) \geq 0, \quad i = 0, \dots, N.$$

$0 = t_0 < t_1 < \cdots < t_N = T$ is a grid, the integrals are approximately replaced by an integration formula with modes t_i and weights γ_i. We apply SQP–methods which do not need the knowledge of admissible starting vectors x_i, u_i, $i = 0, \dots, N$.

A variant of the Chernousko–method is described in [13] as an indirect method. Define

$$H(t, x, \psi(\cdot), u) = \int_0^T \psi(s) f(s, t, x, u) ds - f_0(t, x, u), \quad 0 \leq t \leq T$$

and the adjoint equation

$$\dot{\psi}(t) = H_x(t, \hat{x}(t), \hat{u}(t), \psi(\cdot)), \quad 0 \leq t \leq T. \tag{14}$$

Phase constraints (13) are to be taken into account as penalty terms. Then the algorithm is:

(i) Input $N_{\max}$ as the maximal number of subintervals taken into consideration.
(ii) $k := 0$. Start with an admissible control $u_0(\cdot)$.
(iii) Compute a solution x_k of (11) and ψ_k of (14) with respect to the arguments x_k, u_k. Find a control v_k as a solution of the optimization problems

$$H(t, x_k(t), \psi_k(\cdot), v_k(t)) = \max_{v \in U} H(t, x_k(t), \psi_k(\cdot), v), \quad t \in J.$$

(iv) Set $\Delta H(t) = H(t, x_k(t), \psi_k(\cdot), v_k(t)) - H(t, x_k(t), \psi_k(\cdot), u_k(t))$, $t \in J$ and calculate $\mu_k = \int_0^T \Delta H(t) dt$.
(v) $N := 1$.
(vi) $h := 2^{-N}T$. Find $\tau \in \{h, 3h, \dots, (2^N - 1)h\}$ such that

$$\frac{1}{2h} \int_{r-h}^{r+h} \Delta H(t) dt \geq \frac{1}{T} \mu_k.$$

(vii) Define $u_{r,h}(t) = v_k(t)$ for $\tau - h < t \leq \tau + h$ and $u_{r,h}(t) = u_k(t)$ else.
(viii) Compute $x_{r,h}$ corresponding to $u_{r,h}$ as solution of (11) and the value $\mathcal{F}(u_{r,h})$ of the functional (10).
(ix) If $\mathcal{F}(u_{r,h}) \leq \mathcal{F}(u_k) - \frac{h}{T}\mu_k$ then {put $u_{k+1} := u_{r,h}$; $k := k + 1$ and goto (iii)} else goto (x).
(x) $N := N + 1$.
(xi) If $N \leq N_{\max}$ goto (vi) else stop.

Under some regularity conditions the cost functional of the constructed controls and states converge to the optimal one. The Sakawa–algorithm (see [11]) can be applied to control processes with a Volterra integral equation.

Integral processes can be solved by gradient methods, too.

References

1. Azmyakov, V., Schmidt, W.: Stable Methods for Convex Optimal Control Problems with Constraints Preprint 4/2005. Greifswald.
2. Conn, A. R., Gould, N. I. M., Toint, Ph. L.: *LANCELOT – A Fortran Package for Large Scale Nonlinear Optimization.* Springer 1991.
3. Korytowski, A., Szymkat, M.: Optimal control of a fedbatch fermentation process. Preprint. AGH University of Krakow. 2004.
4. Lyubushin, A. A.: Modifications and studies of convergence of the method of successive approximations in optimal control. (in Russ.). *J. Num. Math./Math. Phys.*, **19** 6, 1979, 1414–1421.
5. Michalewiszis, W. S.: Iterative algorithms of optimization and their application. (in Russ.) Kybernetika. 1965, Nr. 1, 2.
6. Moskalenko, A. I.: *New methods of improvement in optimal control.* (in Russ.) Nauka. Novosibirsk. 1987.
7. Oberle, J. H., Grimm, W.: BNDSO – A program for the numerical solution of optimal control problems. Internal Report 515 – 89/22 Oberpfaffenhofen. 1989.
8. Pontryagin, L. S., Boltjanski, V. G., Gamkrelidze, R. V., Miscenko, E. F.: *The Mathematical Theory of Optimal Processes.* Wiley, New York, 1962.
9. Pytlak, R.: *Numerical methods for optimal control problems with state constraints.* Springer. 1999.
10. Reinbach, I.: Methoden der Variation im Phasenraum. Preprint 1/1995. Greifswald.
11. Sakawa, Y., Shindo, Y.: Local convergence of an algorithm for solving optimal control problems. *JOTA* **46** 3, 1982, 265–293.
12. Schmidt, W. H.: Iterative methods for optimal control processes governed by integral equations. *ISNM* **111**, 1993, 69–82.
13. Stoer, J., Bulirsch, R.: *Introduction to Numerical Analysis.* Springer. 1980, 1993.
14. v. Stryk, O.: *Numerische Lösung optimaler Steuerungsprobleme: Diskretisierung, Parameteroptimierung und Berechnung der adjungierten Variablen.* VDI–Verlag, 1995.

Approximations with Error Estimates for Optimal Control Problems for Linear Systems*

Vladimir M. Veliov

Institute of Mathematical Methods in Economics, Vienna University of Technology, Argentinierstrasse 8, A-1040 Vienna, Austria
Institute of Mathematics and Informatics, Bulgarian Academy of Sciences, Acad. G. Bonchev str. bl.8, 1113 Sofia, Bulgaria
vveliov@eos.tuwien.ac.at

Abstract. The paper presents a class of time-discretization schemes for terminal optimal control problems for linear systems. An error estimate is obtained for the optimal control and for the optimal performance, although the optimal control is typically discontinuous, and neither Lipschitz nor structurally stable with respect to perturbations.

1 Introduction

A key point in the error/sensitivity analysis in optimal control is the sufficiency and Lipschitz stability of the system of necessary optimality conditions. In the context of discrete approximations, also certain regularity (differentiability) of the optimal control is required in order to ensure consistency of the discretization (see e.g. Dontchev at al. [3] for a bibliographic account). For problems that are linear with respect to the control neither of the above requirements is satisfied. Sufficient conditions for optimality and for structural stability in the case of non-coercive problems have been obtained only recently (Osmolovskii [10], Agrachev et al. [1], Noble and Schaettler [9], Kostyukova and Kostina [7], Felgenhauer [4, 5], Maurer and Osmolovskii [8]) and the research in this direction is still in progress.

In this paper we consider a terminal optimal control problem for a linear system. To ensure consistency of the discrete approximation (also for discontinuous controls) we use as control variables in the discrete problem the integral moments of the continuous-time control. This is in contrast to the traditional approach, where the control in the discrete problem takes values in the same set in which the continuous-time control takes values (c.f. [14, 3]). Using integral moments as discrete-time controls is proposed in [16] for a specific Runge-Kutta scheme, and developed in [6] for general Runge-Kutta schemes and linear systems. To obtain error estimates for the optimal control problem in the lack of Lipschitz stability of the system of necessary conditions we use an indirect approach involving four main ingredients: (i) estimation of the error in the reachable set caused by the time-discretization; (ii) a sensitivity estimate for convex problems, depending on

* This research was partially supported by the Austrian Science Found P18161.

I. Lirkov, S. Margenov, and J. Waśniewski (Eds.): LSSC 2005, LNCS 3743, pp. 263–270, 2006.

the convexity index of the objective function and that of the constraining set; (iii) estimation of the convexity index of the reachable set; (iv) estimation of the sensitivity of the optimal control with respect to perturbations in the adjoint equation. Our main result extends for higher order approximations that in [17], which, to our knowledge, provides the first error estimate for discretizations of bang-bang control systems. Using higher order schemes is especially advantageous for large scale problems arising from space-discretizations of distributed control systems. Among the main motivations are the "compartmental" versions of first order PDEs arising in population dynamics and convection-reaction equations.

The paper is organized as follows: Section 2 presents the problem and its discrete approximation, Section 3 contains the main result – the error estimate, Section 4 sketches the proof, and Section 5 discusses the accuracy of the estimation.

2 The Problem and Its Discrete Approximation

We consider the following optimal control problem (P)

$$g(x(T)) \longrightarrow \min \tag{1}$$

$$\dot{x} = Ax + Bu, \quad x(0) = x^0, \tag{2}$$

$$u \in U \subset \mathbf{R}^m, \tag{3}$$

where $x \in \mathbf{R}^n$ is the state, $u \in \mathbf{R}^m$ is the control, A and B are matrices of respective dimensions, assumed constant just for the sake of brevity.

We start with a list of notations that will be used later on:

$\mathbf{R}^n$ (resp. $\mathbf{R}^m$, $\mathbf{R}$) is the Euclidean space with the respective dimension;
$|\cdot|$ and $\langle \cdot, \cdot \rangle$ are the norm and the scalar product;
∂R is the boundary of the set $R \subset \mathbf{R}^n$;
$\mathrm{co}\, X$ is the convex hull of the set $X \subset \mathbf{R}^n$;
$N_R(x)$ is the (external) normal cone to the convex closed set R at $x \in R$;
$N_R^1(x)$ is the set of all external unit normal vectors;
$H(X,Y)$ is the Hausdorff distance between two compact subsets $X, Y \subset \mathbf{R}^n$;
$\mathrm{meas}(\Delta)$ is the Lebesgue measure of $\Delta \subset \mathbf{R}$;
* means transposition.

Definition 1. A function $g : R \mapsto \mathbf{R}$ (where R is a convex subset of $\mathbf{R}^n$) is locally κ-convex at $x \in R$ if there exists a constant $\rho > 0$ and a neighborhood Z of x, such that for every $y \in R \cap Z$ it holds that $g(\frac{1}{2}(x+y)) \le \frac{1}{2}(g(x)+g(y)) - \rho|x-y|^\kappa$.

We shall formally admit also the case $\kappa = +\infty$, where the inequality in Definition 1 is required with $\rho = 0$.

Let $(\hat{x}(\cdot), \hat{u}(\cdot))$ be a solution of problem (P).

Standing Assumptions: The function $g : \mathbf{R}^n \mapsto \mathbf{R}$ is convex, differentiable with locally Lipschitz derivative, and is locally κ-convex at $\hat{x}(T)$ with $\kappa \in [2, +\infty]$; moreover, $g'(\hat{x}(T)) \neq 0$. The set U is a convex compact polyhedron with finite number of vertices, for which the general position hypothesis [13] holds: for every edge l of U

$$\operatorname{rank}\{Bl, \dots, A^{n-1}Bl\} = n.$$

Suppose that U be represented as a product of $q \geq 1$ polyhedral sets[1]: $U = U_1 \times \ldots \times U_q$. Denote by $\nu(i)$ the maximal number such that U_i has $\nu(i)$ edges, each two non-parallel to each other. For example, if U is a box-like set in $\mathbf{R}^m$, then $q = m$ and $\nu(i) = 1$ for all $i = 1, \dots, q$. If U_i is a nondecomposable equilateral polygon in $\mathbf{R}^2$ with k vertices we have

$$\nu(i) = \begin{cases} 3 & \text{if } k = 3, \\ 2 & \text{if } k = 4, \\ 5 & \text{if } k = 5, \\ 3 & \text{if } k = 6, \\ \dots & \end{cases}$$

Also we denote

$$\mathcal{U}_j(U_i) = \{u(\cdot) : [0,1] \to U_i : u(\cdot) \text{ is piece-wise constant, with at most } j \text{ jumps}\}.$$

Then for the natural numbers $r \geq 1$ and $p \geq 0$ we define the sets

$$W_{r,p} = \left\{ (v^0, \dots, v^r) = \left(\int_0^1 u(t)\,\mathrm{d}t, \dots, \int_0^1 (1-t)^r u(t)\,\mathrm{d}t \right) : \right. \tag{4}$$
$$\left. u(\cdot) \in \mathcal{U}_{p\nu(1)}(U_1) \times \ldots \times \mathcal{U}_{p\nu(q)}(U_q) \right\}.$$

The lower indexes of $\mathcal{U}_{\cdot}(U_i)$, indicating the number of jumps, is the product $p\nu(j)$, therefore they depend on the geometry of the Cartesian components of U. Notice that the integration is taken on the normalized interval $[0,1]$ rather than on $[0,h]$.

Given N, we introduce the N-stage problem with state trajectories $(y_0, \dots, y_N)$ and control sequences $(v_0, \dots, v_{N-1})$, where $y_k \in \mathbf{R}^n$ and $v_k = (v_k^0, \dots, v_k^r) \in \operatorname{co} W_{r,r-1}$. The discrete-time problem (P_N) is defined as

$$g(y_N) \longrightarrow \min, \tag{5}$$

$$y_{k+1} = y_k + hA_r(h)y_k + h\sum_{i=0}^{r} B_i(h)v_k^i, \quad y_0 = x^0, \tag{6}$$

$$(v_k^0, \dots, v_k^r) \in \operatorname{co} W_{r,r-1}, \tag{7}$$

where $h = T/N$ and

$$A_r(h) = A + \ldots + \frac{h^r}{(r+1)!}A^{r+1}, \qquad B_i(h) = \frac{h^i}{i!}A^i B.$$

[1] As it will become clear below, it is preferable to decompose U in form of a Cartesian product of as many as possible components. For this purpose it may be useful to perform an appropriate change of the control variable in advance.

Notice that (P_N) has the same state dimension as (P), while the dimension of the control is, in general, higher. There is only one exception: the case $p = 0$, where

$$\mathrm{co}\, W_{r,0} = W_{r,0} = \{(u, \frac{u}{2}, \ldots, \frac{u}{r+1}) :\; u \in U\},$$

therefore $\dim(W_{r,0}) = \dim(U)$. In general, $\dim(\mathrm{co}\, W_{r,p}) = (r+1)\dim(U)$. One can prove that $\mathrm{co}\, W_{r,r-1} \subset W_{r,r} = \mathrm{co}\, W_{r,r}$, and that all the results below hold (with simpler proofs, in fact) if $\mathrm{co}\, W_{r,r-1}$ is replaced with $W_{r,r}$ in (7). Using $W_{r,r}$ instead of $W_{r,r-1}$ is essential in the context of Lagrange problems, where a functional $\int_0^1 G(x(t))\, dt$ is to be minimized. The error estimate in this case is, however, somewhat worse than that in Theorem 1 in the next section. Due to the space limitation we do not consider Lagrange problems in this paper.

Notice that the set $\mathrm{co}\, W_{r,p}$ does not depend on the data of problem (P) excepting the control constraining set U. Therefore it can be calculated in advance for some typical sets U, in particular for $U = [-1, 1]$ (which encompasses all box-like sets $U = [a_1, b_1] \times \ldots \times [a_m, b_m]$ after a Cartesian decomposition).

3 The Error Estimate

Let R denote the reachable set of (2), (3) on $[0, T]$, which consists of all end points $x(T)$ of trajectories of (2) corresponding to measurable selections of U.

Let V be the set of all vertices of U, and E be the set of all edges[2] of U. Moreover, given $p \in \mathbf{R}^n$ we denote by $\lambda[p](\cdot)$ the backward solution of the adjoint equation

$$\dot{\lambda} = -A^* \lambda, \quad \lambda(T) = p. \tag{8}$$

For $l \in \mathbf{R}^n$ we define

$$V(l) = \{v \in V :\; \langle l, v\rangle = \max_{u \in U} \langle l, u\rangle\},$$

and

$$E(l) = \{[v, w] \in E :\; v, w \in V(l)\}.$$

That is, $E(l)$ consists of all "extremal" edges with respect to the direction l. Clearly, $E(l) = \emptyset$ if $V(l)$ is a singleton.

For $x \in \partial R$ we define $\sigma(x)$ as the minimal natural number $\sigma \geq 2$ for which

$$\sum_{i=1}^{\sigma-1} |\langle \lambda[p](t), A^{i-1} B e\rangle| > 0 \quad \forall t \in [0, T], \;\; \forall p \in N_R^1(x), \;\; \forall e \in E(B^* \lambda[p](t)). \tag{9}$$

The last part of Standing Assumptions ensures that $\lambda[p](t) \neq 0$ for every $p \in N_R^1(x)$ and t, which easily implies that the number $\sigma(x)$ exists for every $x \in \partial R$, and $\sigma(x) \leq n + 1$. Further we denote $\sigma = \sigma(\hat{x}(T))$.

[2] The edges will be interpreted either as sets (segments) $[u, v]$, or as vectors $v - u$. In both cases $u, v \in V$, and $[u, v]$ is an extremal subset of U.

By the definition of $W_{r,r-1}$, for every sequence $v = \{(v_k^0, \ldots, v_k^r)\}_{k=0}^{N-1}$ of elements of $W_{r,r-1}$ there is a measurable selection $u = u(v)$ of U such that $\tau \longrightarrow u(t_k + h\tau)$ belongs to $\mathcal{U}_{p\nu(1)}(U_1) \times \ldots \times \mathcal{U}_{p\nu(q)}(U_q)$ for $p = r-1$, and corresponds to $(v_k^0, \ldots, v_k^r)$ in (4). For each boundary point v of $W_{r,r-1}$ the corresponding function $u(v)$ is unique, piece-wise constant with values in V, and can be determined explicitly if $W_{r,r-1}$ is determined explicitly (i.e. for small r).

Theorem 1. *There exist numbers C and N_0 such that for every $N \geq N_0$ and for every optimal control $\hat{v}^N$ of (P_N), the corresponding control $\hat{u}^N(\cdot) = u(\hat{v}^N)$ is unique, and the following estimation holds:*

$$\text{meas}\{t \in [0,T] : \hat{u}^N(t) \neq \hat{u}(t)\} \leq C\left(h^{\frac{r}{\sigma-1}} + L_{g'} h^{\frac{r+1}{s(\sigma-1)}}\right), \tag{10}$$

where $L_{g'}$ is the Lipschitz constant of g' at $\hat{x}(T)$, and $s = \min\{\kappa, \sigma\}$. Moreover, if the control $\hat{u}^N(\cdot)$ is plugged in (2), the corresponding objective value satisfies

$$g(\hat{x}^N(T)) \leq g(\hat{x}(T)) + Ch^{r+1}.$$

The above theorem does not include the Euler scheme, which corresponds to the case $r = 0$, $p = 0$. The right-hand side of (10) becomes in this case

$$C\left(h^{\frac{1}{\sigma-1}} + L_{g'} h^{\frac{1}{s(\sigma-1)}}\right).$$

4 Sketch of the Proof

1. The first step in the proof is to estimate the Hausdorff distance between the reachable sets R and R_N of the continuous and the discrete-time system. The latter consists of all points y_N generated by system (6), (7). Clearly, both R and R_N are convex compact sets. Below c and C denote real numbers which are independent of N but may be different in different places.

Having in mind the definition of the numbers $\nu(i)$, one can deduce the following result from Theorem 1 in [2].

Proposition 1. *For every $r \geq 1$*

$$H(R, R_N) \leq Ch^{r+1}.$$

2. The second step is to estimate the difference between the solutions $\hat{x}(T)$ and $\hat{y}_N$ of problems (P) and (P_N), respectively. These problems can be reformulated as

$$\min_{x \in R} g(x), \qquad \min_{y \in R_N} g(y).$$

Definition 2. The set $R \subset \mathbf{R}^n$ is locally μ-convex at the point $x \in R$ if there exists a constant $\gamma > 0$ and a neighborhood Z of x, such that for every $y \in R \cap Z$ the ball of radius $\gamma|x-y|^\mu$ centered at $\frac{1}{2}(x+y)$ is contained in R.

For $\gamma = 0$ this is merely the local star-shape property, which formally corresponds to the case $\mu = +\infty$. If the above property is fulfilled at every point of R with $\mu = 2$ and with the same γ, then R it is also called "strong convexity" (c.f [11, 12]).

Proposition 2. *Assume that R is locally μ-convex at $\hat{x}$ (with $\mu \in [2, +\infty]$). Assume also that $s = \min\{\kappa, \mu\} < +\infty$. Then there exist numbers $\varepsilon > 0$ and c such that for every compact set $\tilde{R} \subset \mathbf{R}^n$ with $H(\tilde{R}, R) \le \varepsilon$ and for every minimizer $\tilde{x}$ of g on $\tilde{R}$ it holds that*

$$|\tilde{x} - \hat{x}| \le c(H(\tilde{R}, R))^{\frac{1}{s}}. \tag{11}$$

Combining the above two propositions we obtain the estimation

$$|\hat{x}(T) - \hat{y}_N| \le ch^{\frac{r+1}{s}}. \tag{12}$$

3. The next step is to estimate the convexity index μ at $\hat{x}(T)$. Here we use a result from [15] in the stronger form presented in Proposition 2 in [17].

Proposition 3. *The reachable set is locally $\sigma(\hat{x}(T))$-convex at $\hat{x}(T)$.*

Hence (12) holds with the definition of s in the formulation of Theorem 1.

4. The next step involves the maximum principle for (P) and (P_N). For problem (P) it claims that the optimal control $\hat{u}(\cdot)$ satisfies for a.e. t

$$B^* \hat{\lambda}(t) \in N_U(\hat{u}(t)) \text{ where } \quad \hat{\lambda}(t) = \lambda[-g'(\hat{x}(T))](t),$$

or equivalently,

$$\langle B^* \hat{\lambda}(t), u - \hat{u}(t) \rangle \le 0 \quad \forall u \in U. \tag{13}$$

The maximum principle for (P_N) claims that

$$\begin{pmatrix} B_0^*(h)\hat{\lambda}_{k+1} \\ \vdots \\ B_r^*(h)\hat{\lambda}_{k+1} \end{pmatrix} \in N_{W_{r,r-1}}(\hat{v}_k^0, \dots, \hat{v}_k^r), \tag{14}$$

where $\hat{\lambda}_k$ is the backward solution of the discrete system

$$\frac{\lambda_k - \lambda_{k+1}}{h} = A_r^*(h)\lambda_{k+1}, \quad k = N-1, \dots, 1, \quad \lambda_N = g'(\hat{y}_N).$$

According to (12), we obtain

$$|\hat{\lambda}(T) - \hat{\lambda}_N| \le cL_{g'} h^{\frac{r+1}{s}}. \tag{15}$$

For every sequence $v = \{(v_k^0, \dots, v_k^r)\}_{k=0}^{N-1}$ we defined in Section 3 a control $u = u(v)$ that generates v according to the definition of the set $W_{r,r-1}$ in (4). As before, let $\hat{u}^N(\cdot) = v(\hat{v}^N)$ correspond to an optimal control of (P_N).

Denote

$$\lambda^N(t) = \sum_{i=0}^{r} \frac{(t_{k+1} - t)^i}{i!} (A^*)^i \hat{\lambda}_{k+1} \quad \text{for } \ t \in (t_k, t_{k+1}].$$

One can verify that (14) is equivalent to the variational inequality

$$\sum_{0}^{N-1} \int_{t_k}^{t_{k+1}} \langle B^* \lambda^N(t), \tilde{u}(t) - u^N(t) \rangle \leq 0,$$

which holds for every piece-wise constant selection $\tilde{u}$ of U, for which the j-th component in its Cartesian decomposition has at most $(r-1)\nu(j)$ jumps.

Lemma 1. $|\lambda^N(t) - \lambda[\hat{\lambda}_N](t)| \leq ch^r$.

Combining the above inequality with (15) we obtain

$$|\lambda^N(t) - \hat{\lambda}(t)| \leq ch^r + cL_{g'}h^{\frac{r+1}{s}}.$$

5. The first claim of the theorem follows from the last estimation by arguments similar to those in Lemma 6 in [17].

The second claim of the theorem follows directly from Proposition 1.

5 Discussion

The proof of Theorem 1 involves several steps and some of them may bring redundancy in the order of the estimation. Below we discuss this issue.

1. To analyze the sharpness of step 5 of the proof we assume that g is linear, therefore $L_{g'} = 0$ and the estimation (10) is of order $r/(\sigma - 1)$. This estimation is certainly sharp if $\sigma = 2$. If $\sigma > 2$, however, the numerical experiments with $\sigma = 3, 4, 5$ suggest an error estimate of order $r/(\sigma - 2)$. The reason is a sort of a super-convergence of the discrete approximation to the switching function around its zeros. The issue is not studied theoretically yet.

2. Another important issue is that of the definition of the number $\sigma = \sigma(\hat{x}(T))$. Our definition differs from that in [4] (where only the case $\sigma = 2$ is considered), since (9) does not involve only $p = \hat{\lambda}(T)$, rather, the whole normal set $N_R^1(\hat{x}(T))$. This makes the definition of σ non-constructive even a posteriori, in contrast to the following definition: $\sigma \geq 2$ is the minimal natural number such that

$$\sum_{i=1}^{\sigma-1} |\langle \hat{\lambda}(t), A^{i-1}Be \rangle| > 0 \quad \forall t \in [0, T], \;\; \forall e \in E(B^* \hat{\lambda}(t)). \tag{16}$$

In fact, σ is involved in two different steps of the proof, and the definition as given in (9) is needed only to estimate the convexity index of R at $\hat{x}(T)$, while definition (16) is good enough in part 5 of the proof. Having in mind that $s \leq \kappa$, and assuming that κ is finite, we may replace estimation (10) with

$$\text{meas}\{t \in [0, T] : \hat{u}^N(t) \neq \hat{u}(t)\} \;\leq\; Ch^{\frac{r+1}{\kappa(\sigma-1)}},$$

where now σ is defined by (16), therefore the estimation is a posteriori constructive. In the typical case $\kappa = 2$, $\sigma = 2$ the estimation is of order $(r+1)/2$, that is of first order if $r = 1$, which is sharp.

3. Another source of redundancy in the estimation (9) comes from the numbers $\nu(i)$ which are involved in the definition of $W_{r,p}$. The value of ν can, in fact, be slightly decreased for some non-decomposable sets. For example, for a triangular we have $\nu = 3$, while it may be shown that Theorem 1 is still true if one takes $\nu = 2$ in the definition of $W_{r,r-1}$.

References

1. A.A. Agrachev, G. Stefani, and P. Zezza. Strong optimality for a bang-bang trajectory. *SIAM J. Control Optim.*, **41**(4):991–1014, 2002.
2. B.D. Doitchinov and V.M. Veliov. Parametrisations of integrals of set-valued mappings and applications. *J. of Math. Anal. and Appl.*, **179**(2):483–499, 1993.
3. A.L. Dontchev, W.W. Hager, and V.M. Veliov. Second-order Runge-Kutta approximations in control constrained optimal control, *SIAM J. Numerical Anal.*, **38**(1):202–226, 2000.
4. U. Felgenhauer. On stability of bang-bang type controls. *SIAM J. Control Optim.*, **41**(6):1843–1867, 2003.
5. U. Felgenhauer. On the optimality of optimal bang-bang controls for linear and semilinear systems. *Control & Cybernetics.* To appear.
6. R. Ferretti. High-order approximations of linear control systems via Runge-Kutta schemes. *Computing*, **58**:351–364, 1997.
7. O. Kostyukova and E. Kostina. Analysis of properties of the solutions to parametric time-optimal problems. *Computational Optimization and Applications* **26**:285–326, 2003.
8. H. Maurer and N. Osmolovskii. Second order sufficient conditions for time-optimal bang-bang control. *SIAM J. Control Optim.*, **42**(6):2239–2263, 2004.
9. J. Noble and H. Schättler. Sufficient conditions for relative minima of broken extremals in optimal control theory. *J. Math. Anal. Appl.*, **269**(1):98–128, 2002.
10. N.P. Osmolovskii. Second-order conditions for broken extremal. In *Calculus of variations and optimal control* (Haifa, 1998), pp. 198–216, Chapman & Hall/CRC Res. Notes Math., **411**, Boca Raton, FL, 2000.
11. A. Pliś. Accessible sets in control theory. *International Conference on Differential Equations* (Calif., 1974), pp. 646–650, Academic Press, 1975.
12. E. Polovinkin. Strongly convex analysis. *Mat. Sb.* **187**(2):103–130, 1996; translation in *Sb. Math.*, **187**(2):259–286, 1996.
13. L.S. Pontryagin, V.G. Boltyanskii, R.V. Gamkrelidze, and E.F. Mishchenko. *The mathematical theory of optimal processes.* John Wiley & Sons, 1962.
14. A. Schwartz, and E. Polak. Consistent approximations for optimal control problems based on Runge-Kutta integration, *SIAM J. Control Optim.*, **34**:1235–1269, 1996.
15. V.M. Veliov. On the convexity of integrals of multivalued mappings: applications in control theory. *J. Optim. Theory Appl.*, **54**(3):541–563, 1987.
16. V.M. Veliov. *Approximations to Differential Inclusions by Discrete Inclusions.* WP–89–017, IIASA, Laxenburg, Austria, 1989.
17. V.M. Veliov. Error analysis of discrete approximations to bang-bang optimal control problems: the linear case. *Control&Cybernetics*, to appear.

Part VI

Uncertain Systems

Numerical Study of Algebraic Solutions to Linear Problems Involving Stochastic Parameters

Rene Alt[1], Jean-Luc Lamotte[1], and Svetoslav Markov[2]

[1] Laboratoire Informatique Paris 6, U. Pierre et Marie Curie, 4 place Jussieu, 75252 Paris cedex 05, France
{Rene.Alt, Jean-Luc.Lamotte}@lip6.fr

[2] Institute of Mathematics and Informatics, BAS, "Acad. G. Bonchev" st., block 8, 1113 Sofia, Bulgaria
smarkov@bio.bas.bg

Abstract. We formulate certain numerical problems with stochastic numbers and compare algebraically obtained results with experimental results provided by the CESTAC method. Such comparisons give additional information related to the stochastic behavior of random roundings in the course of numerical computations. The good coincidence between theoretical and experimental results confirms the adequacy of our algebraic model and its possible application in the numerical practice.

Keywords: stochastic numbers, stochastic arithmetic, standard deviations, s-space, stochastic linear system.

AMS Subject Classification: 65C99, 65G99, 93L03.

1 Introduction

Stochastic numbers are gaussian random variables with a known mean value and a known standard deviation. Some fundamental properties of stochastic numbers are considered in [9]. The mean values of the stochastic numbers satisfy the usual real arithmetic, whereas standard deviations are added and multiplied by scalars in a specific way. As regard to addition standard deviations form an abelian monoid with cancellation law. This monoid can be embedded in an additive group and after a suitable extension of multiplication by scalars one obtains a so-called s-space, which is in fact a vector space with a specifically defined multiplication by scalar [2], [4]. This allows us to introduce in s-spaces concepts like linear combination, basis, dimension etc. Thus, in theory, computations in s-spaces are reduced to computations in vector spaces. This opens the road to finding explicit expressions for the solution of certain algebraic problems involving stochastic numbers.

Alternatively, stochastic numbers can be computed experimentally using the CESTAC method, which is a Monte-Carlo method consisting in performing each arithmetic operation several times using an arithmetic with a random rounding mode [3], [7], [8]. For a survey of methods using Monte-Carlo arithmetic see [6].

I. Lirkov, S. Margenov, and J. Waśniewski (Eds.): LSSC 2005, LNCS 3743, pp. 273–280, 2006.

In Sections 2 we briefly present the main results of our theory of s-spaces as regard to the arithmetic operations for addition and multiplication by scalars needed for the purposes of this study; for a detailed presentation of the theory, see [4]. Section 3 considers the algebraic solution of linear systems of equations with right-hand sides involving stochastic numbers. In Section 4 we extend further our idea from [5] to compare the theoretic solution of an algebraic problem involving stochastic numbers with the solution obtained numerically by the CESTAC method. Numerical experiments are reported and a good coincidence between theoretical and experimental results is observed.

2 Stochastic Numbers and Stochastic Arithmetic

By $\mathbb{R}$ we denote the set of reals; the same notation is used for the linearly ordered field of reals $\mathbb{R} = (\mathbb{R}, +, \cdot, \leq)$. For any integer $n \geq 1$ we denote by $\mathbb{R}^n$ the set of all n-tuples $(\alpha_1, \alpha_2, ..., \alpha_n)$, $\alpha_i \in \mathbb{R}$. The set $\mathbb{R}^n$ forms a vector space under the familiar operations of addition and multiplication by scalars denoted by $\mathbf{V}^n = (\mathbb{R}^n, +, \mathbb{R}, \cdot)$, $n \geq 1$. By $\mathbb{R}^+$ we denote the set of nonnegative reals.

2.1 The Arithmetic for Stochastic Numbers

A *stochastic number* $X = (m; s)$ is a gaussian random variable with mean value $m \in \mathbb{R}$ and (nonnegative) standard deviation $s \in \mathbb{R}^+$. The set of all stochastic numbers is $\mathbb{S} = \{(m; s) \mid m \in \mathbb{R}, s \in \mathbb{R}^+\}$. Let $X_1 = (m_1; s_1)$, $X_2 = (m_2; s_2) \in \mathbb{S}$. Addition and multiplication by scalars are defined by:

$$X_1 + X_2 = (m_1; s_1) + (m_2; s_2) \stackrel{def}{=} \left(m_1 + m_2;\ \sqrt{s_1^2 + s_2^2}\right),$$

$$\gamma * X = \gamma * (m; s) \stackrel{def}{=} \left(\gamma m;\ |\gamma| s\right),\ \gamma \in \mathbb{R}.$$

It has to be noticed that the operations on stochastic numbers are error free and are only used for theory. In this approach stochastic numbers are only used as a model for computation on data containing errors.

A stochastic number of the form $(0; s)$, $s \in \mathbb{R}^+$, is called *(centrally) symmetric.* If X_1, X_2 are symmetric stochastic numbers, then $X_1 + X_2$ and $\lambda * X_1$, $\lambda \in \mathbb{R}$, are also symmetric stochastic numbers. Thus there is a 1–1 correspondence between the set of symmetric stochastic numbers and the set $\mathbb{R}^+$. We shall use special symbols "$\oplus$", "$*$" for the arithmetic operations over standard deviations, as these operations are different from the corresponding ones for numbers. The operations "$\oplus$", "$*$" induce a special arithmetic on the set $\mathbb{R}^+$. Consider the system $(\mathbb{R}^+, \oplus, \mathbb{R}, *)$, such that for $s, t \in \mathbb{R}^+$, $\gamma \in \mathbb{R}$:

$$s \oplus t = \sqrt{s^2 + t^2}, \quad \gamma * s = |\gamma| s. \tag{1}$$

Proposition 1. *[4] The system $(\mathbb{R}^+, \oplus, \mathbb{R}, *)$ is an abelian additive monoid with cancellation, such that for $s, t \in \mathbb{R}^+$, $\alpha, \beta \in \mathbb{R}$:*

$$\alpha * (s \oplus t) = \alpha * s \oplus \alpha * t, \tag{2}$$

$$\alpha * (\beta * s) = (\alpha\beta) * s, \tag{3}$$

$$1 * s = s, \tag{4}$$

$$(-1) * s = s, \tag{5}$$

$$\sqrt{\alpha^2 + \beta^2} * s = \alpha * s \oplus \beta * s, \ \alpha, \beta \geq 0. \tag{6}$$

More generally, we can extend componentwise operations (1) for n-tuples $s = (s_1, ..., s_n)$, $s_i, \in \mathbb{R}^+$, that is,

$$(s_1, ..., s_k) \oplus (t_1, ..., t_k) = (s_1 \oplus t_1, ..., s_k \oplus t_k), \tag{7}$$

$$\gamma * (s_1, s_2, ..., s_k) = (|\gamma| s_1, |\gamma| s_2, ..., |\gamma| s_k), \ \gamma \in \mathbb{R}. \tag{8}$$

The corresponding system $((\mathbb{R}^+)^n, \oplus, \mathbb{R}, *)$ again satisfies the conditions of Proposition 1. A system satisfying the conditions of Proposition 1 is called an *s-space of monoid structure*. Such a structure can be naturally embedded into a group, obtaining thus an *s-space of group structure*, as shown below.

2.2 The S-Space of Group Structure

For $s \in \mathbb{R}$ denote $\tau(s) = \{+, \text{ if } s \geq 0; \ -, \text{ if } s < 0\}$. We extend the operation addition "$\oplus$" for all $s, t \in \mathbb{R}$, admitting thus negative reals, corresponding to *improper* standard deviations:

$$s \oplus t \stackrel{def}{=} \tau(s + t)\sqrt{|\tau(s)s^2 + \tau(t)t^2|}. \tag{9}$$

We note that $\tau(s + t) = \tau(\tau(s)s^2 + \tau(t)t^2) = \tau(s \oplus t)$ for $s, t \in \mathbb{R}$. Using (9) we embed isomorphically the monoid $(\mathbb{R}^+, \oplus)$ into the system $(\mathbb{R}, \oplus)$, which is an abelian group with null 0 and opposite element $\text{opp}(s) = -s$, i. e. $s \oplus (-s) = 0$. Indeed, from (9) we have $s \oplus (-s) = \tau(s - s)\sqrt{|\tau(s)s^2 - \tau(s)s^2|} = \tau(0)\sqrt{0} = 0$. Here are some examples of addition in the system $(\mathbb{R}, \oplus)$: $1 \oplus 1 = \sqrt{2}, 1 \oplus 2 = \sqrt{5}$, $3 \oplus 4 = 5$, $4 \oplus (-3) = \sqrt{7}$, $3 \oplus (-4) = -\sqrt{7}$, $5 \oplus (-4) = 3$, $4 \oplus (-5) = -3$, $(-3) \oplus (-4) = -5$, $1 \oplus 2 \oplus 3 = \sqrt{14}$, $1 \oplus 2 \oplus (-3) = -2$.

Using (9) and $\tau(s_1 \oplus ... \oplus s_n) = \tau(s_1 + ... + s_n)$ we obtain for $n \geq 2$

$$s_1 \oplus s_2 \oplus ... \oplus s_n = \tau(s_1 + ... + s_n)\sqrt{|\tau(s_1)s_1^2 + ... + \tau(s_n)s_n^2|}. \tag{10}$$

Proposition 2. *For $s_1, s_2, ..., s_n, t \in \mathbb{R}$ the equation $s_1 \oplus s_2 \oplus ... \oplus s_n = t$ is equivalent to $\tau(s_1)s_1^2 + ... + \tau(s_n)s_n^2 = \tau(t)t^2$.*

The proof follows immediately from the fact that the equation $\tau(s)\sqrt{|s|} = t$ implies $s = \tau(t)t^2$, and, in particular, $\tau(t) = \tau(s)$.

Multiplication by scalars is naturally extended on the set $\mathbb{R}$ of generalized standard deviations by: $\gamma * s = |\gamma|s$, $s \in \mathbb{R}$. Multiplication by -1 (negation) is $(-1) * s = |-1|s = s$, $s \in \mathbb{R}$, in accordance with (4)–(5). To avoid confusion we shall write the scalars always to the left side of the standard deviation. Under this convention we have, e. g. $(-2) * 2 = 4$, whereas $2 * (-2) = -4$.

Note that if s is a standard deviation, then we have $\gamma * s = (-\gamma) * s$ for any $\gamma \in \mathbb{R}$; thus multiplication by negative scalar does not change the type of s (proper/improper).

It is easy to check that all conditions (2)–(6) of Proposition 1 hold true for generalized standard deviations. This justifies the following definition:

Definition 1. *A system $(\mathcal{S}, \oplus, \mathbb{R}, *)$, such that: i) $(\mathcal{S}, \oplus)$ is an abelian additive group, and, ii) for any $s, t \in \mathcal{S}$, $\alpha, \beta \in \mathbb{R}$ relations (2)–(6) hold, is called an s-space over $\mathbb{R}$ (with group structure).*

3 Linear Systems with Stochastic Right-Hand Side

3.1 Canonical S-Spaces and Dot Product

For any integer $k \geq 1$ the set $\mathcal{S} = \mathbb{R}^k$ of all k-tuples $(s_1, s_2, ..., s_k)$ forms an s-space over $\mathbb{R}$ under the operations (7)–(8), whenever the sums $s_i \oplus t_i$ in (7) are defined by (9). The s-space $\mathbf{S}^k = (\mathbb{R}^k, \oplus, \mathbb{R}, *)$ is the *canonical s-space (of standard deviations)*. In the s-space $\mathbf{S}^k$ we introduce a scalar (dot) product. Namely, for $\alpha = (\alpha_1, \alpha_2, ..., \alpha_k) \in \mathbb{R}^k, s = (s_1, s_2, ..., s_k) \in \mathbf{S}^k$ we define $\alpha * s = \alpha_1 * s_1 \oplus \alpha_2 * s_2 \oplus ... \oplus \alpha_k * s_n$.

Using (10) we obtain for $\alpha = (\alpha_1, \alpha_2, ..., \alpha_k) \in \mathbb{R}^k,\ s = (s_1, s_2, ..., s_k) \in \mathbf{S}^k$

$$\alpha * s = \alpha_1 * s_1 \oplus ... \oplus \alpha_k * s_k = \tau(\alpha * s)\sqrt{|\alpha_1^2 \tau(s_1) s_1^2 + ... + \alpha_k^2 \tau(s_k) s_k^2|}.$$

Example 1. Let $\alpha_i = 1, s_i = s, i = 1, ..., k$. Then $\alpha * s = s \oplus ..._{(k\ times)} \oplus s = \tau(s)\sqrt{ks^2} = s\sqrt{k}$. This fact has been already known for long [1].

Proposition 3. *For $\alpha = (\alpha_1, \alpha_2, ..., \alpha_k) \in \mathbb{R}^k, (s_1, s_2, ..., s_k) \in \mathbf{S}^k$ the equation $\alpha * s = t$ is equivalent to $\alpha_1^2 \tau(s_1) s_1^2 + ... + \alpha_k^2 \tau(s_k) s_k^2 = \tau(t) t^2$.*

Remark. It is used in the proof that $\tau(\alpha_i * s_i) = \tau(s_i)$.

3.2 S-Spaces and Their Relation to Vector Spaces

Proposition 4. *Let $(\mathcal{S}, +, \mathbb{R}, *)$ be an s-space over $\mathbb{R}$. Then the system $(\mathcal{S}, +, \mathbb{R}, \cdot)$ where the operation "$\cdot$": $\mathbb{R} \times \mathcal{S} \longrightarrow \mathcal{S}$ is defined by*

$$\alpha \cdot c = \begin{cases} \sqrt{|\alpha|} * c, & if\ \ \alpha \geq 0; \\ \sqrt{|\alpha|} * (-c), & if\ \ \alpha < 0, \end{cases} \tag{11}$$

is a vector space over $\mathbb{R}$. Conversely, let $(\mathcal{S}, +, \mathbb{R}, \cdot)$ be a vector space over $\mathbb{R}$. The system $(\mathcal{S}, +, \mathbb{R}, *)$ is an s-space over $\mathbb{R}$ whenever "$*$" is defined by

$$\alpha * c = \alpha^2 \cdot c. \tag{12}$$

Proposition 4 shows that each one of the two *associted* spaces $(\mathcal{S}, +, \mathbb{R}, *)$ and $(\mathcal{S}, +, \mathbb{R}, \cdot)$ can be obtained from the other one by a redefinition of the operation multiplication by scalars using (11), resp. (12). Assume that $\mathcal{S} = (\mathcal{S}, +, \mathbb{R}, *)$ is an s-space over $\mathbb{R}$ and $(\mathcal{S}, +, \mathbb{R}, \cdot)$ is the associated vector space. All vector space concepts from the vector space $(\mathcal{S}, +, \mathbb{R}, \cdot)$, such as linear combination, linear dependence, basis etc., apply to the s-space $(\mathcal{S}, +, \mathbb{R}, *)$ [4].

Theoretically stochastic numbers are defined as elements of the direct sum $\mathcal{V} \oplus \mathcal{S}$ of a vector space $\mathcal{V}$ and a s-space $\mathcal{S}$ both of same dimension k. Namely, let $\mathcal{V} = \mathbf{V}^k$ be a k-dimensional vector space with a basis $(v^{(1)}, ..., v^{(k)})$ and let $\mathcal{S} = \mathbf{S}^k$ be a k-dimensional s-space having a basis $(s^{(1)}, ..., s^{(k)})$. Then $(v^{(1)}, ..., v^{(k)}; s^{(1)}, ..., s^{(k)})$ is a basis of the *k-dimensional* space $\mathbf{V}^k \oplus \mathbf{S}^k$. Such a setting allows us to consider numerical problems involving vectors and matrices, wherein certain numeric variables have been substituted by stochastic ones.

3.3 Stochastic Linear Systems

We consider a linear system $Ax = b$, such that A is a real $n \times n$-matrix and the right-hand side b is a vector of stochastic numbers. Then the solution x also consists of stochastic numbers, and, respectively, all arithmetic operations (additions and multiplications by scalars) in the expression Ax involve stochastic numbers; we denote this by writing $A * x$ instead of Ax.

Problem. Assume that $A = (\alpha_{ij})_{i,j=1}^n$, $\alpha_{ij} \in \mathbb{R}$, is a real $n \times n$-matrix, and $b = (b'; b'')$ is a n-tuple of (generalized) stochastic numbers, such that $b', b'' \in \mathbb{R}^n$, $b' = (b'_1, ..., b'_n)$, $b'' = (b''_1, ..., b''_n)$. We look for a (generalized) stochastic vector $x = (x'; x'')$, $x', x'' \in \mathbb{R}^n$, satisfying the system $A * x = b$.

Solution. Due to $A * x = A * (x'; x'') = (Ax'; A * x'')$ the system $A * x = b$ reduces to a linear system $Ax' = b'$ for the vector $x' = (x'_1, ..., x'_n)$ of mean values and a system $A * x'' = b''$ for the standard deviations $x'' = (x''_1, ..., x''_n)$. If $A = (\alpha_{ij})$ is nonsingular, then $x' = A^{-1}b'$. We shall next concentrate on the solution of the system $A * x'' = b''$ for the standard deviations.

The i-th equation of the system $A * x'' = b''$ reads $\alpha_{i1} * x''_1 \oplus ... \oplus \alpha_{in} * x''_n = b''_i$. According to Proposition 3, this is equivalent to

$$\alpha_{i1}^2 \tau(x''_1) x_1''^2 + ... + \alpha_{in}^2 \tau(x''_n) x_n''^2 = \tau(b''_i) b_i''^2, \quad i = 1, ..., n.$$

Setting $\tau(x''_i)(x''_i)^2 = y_i$, $\tau(b''_i)(b''_i)^2 = c_i$, we obtain a linear $n \times n$ system $Dy = c$ for $y = (y_i)$, where $D = (\alpha_{ij}^2)$, $c = (c_i)$. If D is nonsingular we can solve the system $Dy = c$ for the vector y, $y = D^{-1}c$, and then obtain the standard deviation vector x'' by means of $x''_i = \tau(y_i)\sqrt{|y_i|}$. Thus for the solution of the original problem it is necessary and sufficient that both matrices $A = (\alpha_{ij})$ and $D = (\alpha_{ij}^2)$ are nonsingular.

Summarizing, to solve $A * x = b$ we perform the following steps:

i) check the matrices $A = (\alpha_{ij})$ and $D = (\alpha_{ij}^2)$ for nonsingularity;
ii) find the solution $x' = A^{-1}b'$ of the linear system $Ax' = b'$;
iii) find the solution $y = D^{-1}c$ of the linear system $Dy = c$, where $c = (c_i)$, $c_i = \tau(b''_i)(b_i''^2)$. Compute $x''_i = \tau(y_i)\sqrt{|y_i|}$; then the solution of $A * x = b$ is $x = (x'; x'')$.

4 Numerical Experiments

Numerical experiments have been performed in order to compare the theoretical results with numerical results obtained by means of the CESTAC method for imprecise stochastic data.

Scalar Product. Let α be a real vector of size N with $\alpha_i = i$, $i = 1, ..., N$. Assume that b is a stochastic vector of size N. All samples for the components of b have been generated with a gaussian generator with mean value $m = 1$ and standard deviation $\sigma = 0.001$.

Theoretically, the standard deviation of the dot product $\alpha * b$ is equal to $\sigma\sqrt{N(N+1)(2N+1)/6}$. On the other hand, according to the theory of the CESTAC method, a stochastic number can be represented by an n-tuple of random values with a known mean value m and a known standard deviation σ. In our examples $n = 3$ as implemented in the CADNA software [3].

With the above conditions ($m = 1, \sigma = 0.001$) the scalar product $\alpha * b$ has been computed k times for various sizes $N = 10, 100, ..., 10000$. For each size N the mean value $\overline{\delta}$ of the standard deviation δ_i of the result ($i = 1, 2, ..., k$) has been computed.

This provides samples of size k whose mean values approximate the theoretical standard deviation.

Table 1 reports the percentages of cases where the theoretical standard deviation $\sigma\sqrt{N(N+1)(2N+1)/6}$ is outside the computed confidence interval. These percentages have been computed with 1000 runs.

Comments: From Table 1, it is clear a posteriori that the distribution of the scalar product is effectively gaussian, as a size of 4 to 5 for the samples is enough to approximate the theoretical value, whereas if this were not the case, then the samples should have rather be of size 30.

4.1 Solution of a Linear System $A * x = b$

In this numerical example A is a real matrix such that $a_{ij} = i$, if $i = j$, else $a_{ij} = 10^{-|i-j|}$, $i, j = 1, .., N$, $N = 10$. Assume that b is a stochastic vector such that the component b_i is generated with a gaussian generator with a mean value $\sum_{j=1}^{n} a_{ij}$ and a standard deviation equal to $1.e-4 = 10^{-4}$. With such kind of system, the solutions x_i are around 1.

The theoretical standard deviations on each component of the solution are obtained according the method described in the previous section. First the matrix D is computed from matrix A. Then the system $y = D^{-1}c$ is solved, and

Table 1. Percentages of theoretical standard deviation outside the confidence interval

N \ k	3	4	5	6	7	10
10	12.1	6.3	3.3	2.1	1.5	0.3
100	12.6	5.3	3.8	2.3	1.0	0.3
1000	13.2	4.6	3.9	1.6	1.4	0.2
10000	11.6	5.4	2.9	1.9	1.4	0.2

Table 2. Theoretical and computed standard deviations

Component i	Theoretical standard deviation x''	mean value of the computed standard deviations
1	9.98e-05	10.4e-05
2	4.97e-05	4.06e-05
3	3.32e-05	3.21e-05
4	2.49e-05	2.02e-05
5	1.99e-05	1.81e-05
6	1.66e-05	1.50e-05
7	1.42e-05	1.54e-05
8	1.24e-05	1.02e-05
9	1.11e-05	0.778e-05
10	0.999e-05	0.806e-05

the standard deviations are computed with the formula $x_i'' = \tau(y_i)\sqrt{|y_i|}$. The values x_i'' are given in the first column in Table 2.

The experimental results only concern the standard deviations on the components of the solution. They are obtained in the following way: 30 different vectors $b^{(k)}, k = 1, ..., 30$ and thus 30 systems $A * x = b^{(k)}$ are generated as above. Then they are solved using the CADNA software using Gaussian elimination. This CADNA software provides the standard deviation of each component the solution. In the end, the mean value of the standard deviations of the 30 samples are computed for the $N = 10$ components and printed in Table 2. As we can see in Table 2, the theoretical standard deviations and the computed values are very close to each other.

The influence of the variation of the error σ on the right hand side b on the results is studied as follows: The same as above procedure is performed but here only the first component of the solution is considered. As before, 30 solutions for

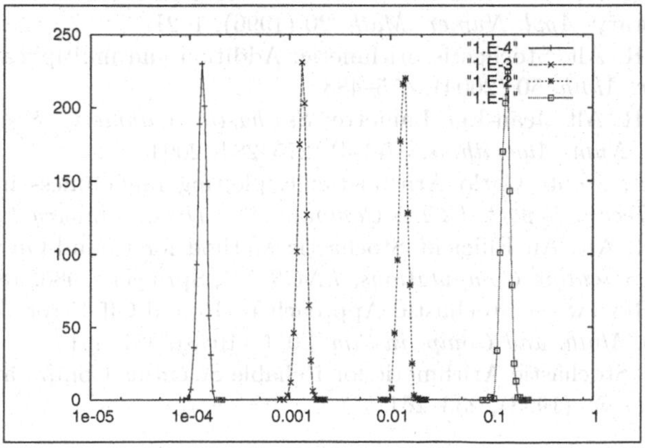

Fig. 1.

x are obtained and the standard deviations of the 30 first components $x^{(1)}$ and their corresponding mean value named $\bar{\delta}_i(1)$ are computed. This procedure has been performed 1000 times. Therefore, we obtained 1000 values for $\bar{\delta}_i(1)$ which are classified in 20 classes from $0.5x_i''$ to $1.5x_i''$. The graphs of the number of elements in each class obtained with the 4 values of $\sigma = 1.e-4, 1.e-3, 1.e-2, 1.e-1$ are reported in Fig. 1.

5 Conclusion

The theoretic study of the properties of stochastic numbers with respect to the operations addition and multiplication by scalars allows the solution of certain algebraic problems involving stochastic numbers. This gives us a possibility to compare algebraically obtained results with practical applications of stochastic numbers, such as the ones provided by the CESTAC method [3]. Such comparisons give additional information related to the stochastic behaviour of random roundings in the course of numerical computations. It may be expected that the proposed theory can be used in the computational practice.

Acknowledgments

This work was partially supported by Bulgarian NSF grant MM-1104/01.

References

1. Alt, R., Error Propagation in Fourier Transforms, Mathematics and Computers in Simulation XX (1978) 37–43.
2. Alt, R., S. Markov, On the Algebraic Properties of Stochastic Arithmetic. Comparison to Interval Arithmetic, In: W. Kraemer et al.(Eds.) Scientific Computing, Validated Numerics, Interval Methods, Kluwer, 2001, 331–341.
3. Alt, R., J. Vignes, Validation of Results of Collocation Methods for ODEs with the CADNA Library. *Appl. Numer. Math.* 20 (1996), 1–21.
4. Markov, S., R. Alt, Stochastic arithmetic: Addition and multiplication by scalars, *Appl. Numer. Math.* 50 (2004), 475–488.
5. Markov, S., R. Alt, Jean-Luc Lamotte; *Stochastic Arithmetic: S-spaces and Some Applications*, *Num. Algorithms*, 37(1-4), 275–284, 2004.
6. Parker, D. S., Monte Carlo Arithmetic: Exploiting randomness in floating point arithmetic, *Techn. Report, UCLA Computer Sci. Dept., January 1997.*
7. Vignes, J., R. Alt, An Efficient Stochastic Method for Round-Off Error Analysis, in: *Accurate Scientific Computations, LNCS 235*, Springer, 1985, 183–205.
8. Vignes, J., Review on Stochastic Approach to Round-Off Error Analysis and its Applications. *Math. and Comp. in Sim.* 30, 6 (1988), 481–491.
9. Vignes, J., A Stochastic Arithmetic for Reliable Scientific Computation, *Math. and Comp. in Sim.* 35 (1993), 233–261.

On the Normed Linear Space of Hausdorff Continuous Functions

Roumen Anguelov[1], Svetoslav Markov[2], and Blagovest Sendov[3]

[1] University of Pretoria, Pretoria 0002, South Africa
roumen.anguelov@up.ac.za
[2] Institute of Mathematics and Informatics, BAS,
Acad. G. Bonchev, block 8, 1113 Sofia, Bulgaria
smarkov@bio.bas.bg
[3] Bulgarian Embassy, 36-3, Yoyogi 5-chome,
Shibuia-ku, Tokyo, 151-0053 Japan
sendov2003@yahoo.com

Abstract. In the present work we show that the linear operations in the space of Hausdorff continuous functions are generated by an extension property of these functions. We show that the supremum norm can be defined for Hausdorff continuous functions in a similar manner as for real functions, and that the space of all bounded Hausdorff continuous functions on an open set is a normed linear space. Some issues related to approximations in the space of Hausdorff continuous functions by subspaces are also discussed.

1 Introduction

The concept of Hausdorff continuity generalizes the concept of continuity of real functions to interval-valued functions [4, 8]. Due to a minimality condition with respect to inclusion of graphs, Hausdorff continuous (H-continuous) functions retain some important properties of continuous functions, e.g. they are completely determined by their values on a dense subset of their domain. It is well-known that the operations (addition and multiplication by scalars) associated with interval structures typically do not infer a linear space [5]. In this regard the set of H-continuous functions is a notable exception. It is shown in [9] that one can define addition and multiplication by scalars on the set $\mathbb{H}(\Omega)$ of all H-continuous functions on an open subset Ω of $\mathbb{R}^n$ in such a way that $\mathbb{H}(\Omega)$ is a linear space. Naturally, these operations are not defined in a point-wise manner. In Sections 3 and 4 of the present work we show that the linear space operations on $\mathbb{H}(\Omega)$ are a direct consequence of an extension property of H-continuous functions. In Section 5 we show that the supremum norm can be defined for H-continuous functions in a similar way as for real functions, and that the space $\mathbb{H}_b(\Omega)$ of all bounded H-continuous functions on the open set Ω is a normed linear space. However, due to the involvement of discontinuous functions, a natural metric to be associated with the space $\mathbb{H}_b(\Omega)$ is the Hausdorff metric considered in Section 6. Issues related to approximations in $\mathbb{H}(\Omega)$ by subspaces are also discussed.

I. Lirkov, S. Margenov, and J. Waśniewski (Eds.): LSSC 2005, LNCS 3743, pp. 281–288, 2006.

2 General Setting

The real line is denoted by $\mathbb{R}$ and the set of all finite real intervals by $\mathbb{IR} = \{[\underline{a}, \overline{a}] : \underline{a}, \overline{a} \in \mathbb{R},\ \underline{a} \leq \overline{a}\}$. Given an interval $a = [\underline{a}, \overline{a}] = \{x : \underline{a} \leq x \leq \overline{a}\} \in \mathbb{IR}$, $w(a) = \overline{a} - \underline{a}$ is the width of a, while $|a| = \max\{|\underline{a}|, |\overline{a}|\}$ is the modulus of a. An interval a is called proper interval, if $w(a) > 0$ and point interval, if $w(a) = 0$. Identifying $a \in \mathbb{R}$ with the point interval $[a, a] \in \mathbb{IR}$, we consider $\mathbb{R}$ as a subset of $\mathbb{IR}$. We denote by $\mathbb{A}(\Omega)$ the set of all locally bounded interval-valued functions defined on an arbitrary set $\Omega \subseteq \mathbb{R}^n$. The set $\mathbb{A}(\Omega)$ contains the set $\mathcal{A}(\Omega)$ of all locally bounded real functions defined on Ω. Recall that a real function or an interval-valued function f defined on Ω is called locally bounded if for every $x \in \Omega$ there exist $\delta > 0$ and $M \in \mathbb{R}$ such that $|f(y)| < M$, $y \in B_\delta(x)$, where $B_\delta(x) = \{y \in \Omega : ||x - y|| < \delta\}$ denotes the open δ-neighborhood of x in Ω.

Let D be a dense subset of Ω. The mappings $I(D, \Omega, \cdot), S(D, \Omega, \cdot) : \mathbb{A}(D) \longrightarrow \mathcal{A}(\Omega)$ defined for $f \in \mathbb{A}(D)$ and $x \in \Omega$ by

$$I(D, \Omega, f)(x) = \sup_{\delta > 0} \inf\{f(y) : y \in B_\delta(x) \cap D\},$$
$$S(D, \Omega, f)(x) = \inf_{\delta > 0} \sup\{f(y) : y \in B_\delta(x) \cap D\},$$

are called lower and upper Baire operators, respectively. The mapping $F : \mathbb{A}(D) \longrightarrow \mathbb{A}(\Omega)$, called *graph completion operator*, is defined by

$$F(D, \Omega, f)(x) = [I(D, \Omega, f)(x), S(D, \Omega, f)(x)],\ x \in \Omega,\ f \in \mathbb{A}(D).$$

In the case when $D = \Omega$ the sets D and Ω will be omitted, thus we write $I(f) = I(\Omega, \Omega, f),\ S(f) = S(\Omega, \Omega, f),\ F(f) = F(\Omega, \Omega, f)$.

Definition 1. *A function $f \in \mathbb{A}(\Omega)$ is S-continuous, if $F(f) = f$.*

Definition 2. *A function $f \in \mathbb{A}(\Omega)$ is Hausdorff continuous (H-continuous), if $g \in \mathbb{A}(\Omega)$ with $g(x) \subseteq f(x)$, $x \in \Omega$, implies $F(g)(x) = f(x)$, $x \in \Omega$.*

Theorem 1. *[1, 8] For every $f \in \mathbb{A}(\Omega)$ the functions $F(I(S(f)))$ and $F(S(I(f)))$ are H-continuous.*

H-continuous functions are similar to usual continuous real functions in that they assume point values everywhere on Ω except for a set of first Baire category. More precisely, it is shown in [1] that for every $f \in \mathbb{H}(\Omega)$ the set

$$W_f = \{x \in \Omega : w(f(x)) > 0\} \tag{1}$$

is of first Baire category and f is continuous on $\Omega \setminus W_f$. Since a finite or countable union of sets of first Baire category is also a set of first Baire category we have:

Theorem 2. *Let the set Ω be open and let $\mathcal{F}$ be a finite or countable set of H-continuous functions. Then the set $D_{\mathcal{F}} = \{x \in \Omega : w(f(x)) = 0,\ f \in \mathcal{F}\} = \Omega \setminus \bigcup_{f \in \mathcal{F}} W_f$ is dense in Ω and all functions $f \in \mathcal{F}$ are continuous on $D_{\mathcal{F}}$.*

The graph completion operator is inclusion isotone i) w. r. t. the functional argument, that is, if $f, g \in \mathbb{A}(D)$, where D is dense in Ω, then

$$f(x) \subseteq g(x),\ x \in D \implies F(D,\Omega,f)(x) \subseteq F(D,\Omega,g)(x),\ x \in \Omega, \tag{2}$$

and, ii) w. r. t. the set D in the sense that if D_1 and D_2 are dense subsets of Ω and $f \in \mathbb{A}(D_1 \cup D_2)$ then

$$D_1 \subseteq D_2 \implies F(D_1,\Omega,f)(x) \subseteq F(D_2,\Omega,f)(x),\ x \in \Omega. \tag{3}$$

In particular, (3) implies that for any dense subset D of Ω and $f \in \mathbb{A}(\Omega)$ we have $F(D,\Omega,f)(x) \subseteq F(f)(x),\ x \in \Omega$. The graph completion operator is idempotent. Moreover [2], if the sets D_1 and D_2 are both dense in Ω and $D_1 \subseteq D_2$ then

$$F(D_2,\Omega,\cdot) \circ F(D_1,\Omega,\cdot) = F(D_1,\Omega,\cdot). \tag{4}$$

Let $f \in \mathbb{A}(\Omega)$. For every $x \in \Omega$ the value of f is an interval $[\underline{f}(x), \overline{f}(x)] \in \mathbb{IR}$. Hence, f can be written in the form $f = [\underline{f}, \overline{f}]$ where $\underline{f}, \overline{f} \in \mathcal{A}(\Omega)$ and $\underline{f}(x) \leq \overline{f}(x),\ x \in \Omega$. The lower and upper Baire operators as well as the graph completion operator of an interval-valued function f can be represented in terms of $\underline{f}$ and $\overline{f}$, namely, for every dense subset D of Ω: $I(D,\Omega,f) = I(D,\Omega,\underline{f})$, $S(D,\Omega,f) = S(D,\Omega,\overline{f})$, $F(D,\Omega,f) = [I(D,\Omega,\underline{f}), S(D,\Omega,\overline{f})]$.

3 Extension and Restriction Properties

Let $\Omega \subseteq \mathbb{R}^n$ and let D be dense in Ω. Extending a function f defined on D to Ω while preserving its properties (e.g. linearity, continuity) is an important issue in functional analysis. Recall that if f is continuous on D it does not necessarily have a continuous extension on Ω. The next theorem shows that an H-continuous function on D has a unique H-continuous extension on Ω.

Theorem 3. *Let $\varphi \in \mathbb{H}(D)$, where D is dense subset of Ω. Then there exists unique $f \in \mathbb{H}(\Omega)$, such that $f(x) = \varphi(x)$, $x \in D$. Namely, $f = F(D,\Omega,\varphi)$.*

Proof. Let $f = F(D,\Omega,\varphi)$. From the fact that φ is H-continuous on D it follows that $F(D,D,\varphi) = \varphi$. Therefore, for every $x \in D$ we have $f(x) = F(D,\Omega,\varphi)(x) = F(D,D,\varphi)(x) = \varphi(x)$. Hence f is an extension of φ over Ω. We show next that f is H-continuous on Ω. Using property (4) we obtain

$$F(f) = F(\Omega,\Omega,F(D,\Omega,\varphi)) = F(D,\Omega,\varphi) = f. \tag{5}$$

Let $g \in \mathbb{A}(\Omega)$ satisfy the inclusion

$$g(x) \subseteq f(x),\ x \in \Omega. \tag{6}$$

Then using the inclusion isotone property (2) of the operator F we have

$$F(g)(x) \subseteq F(f)(x) = f(x),\ x \in \Omega. \tag{7}$$

Relation (6) implies $g(x) \subseteq f(x) = \varphi(x),\ x \in D$. Using again the H-continuity of φ on D we obtain $F(D,D,g)(x) = \varphi(x),\ x \in D$. From this equality and properties (2) and (3) of F we obtain

$$f(x) = F(D,\Omega,F(D,D,g))(x) \subseteq F(D,\Omega,F(g))(x) \subseteq F(g)(x),\ x \in \Omega. \quad (8)$$

Inclusions (7) and (8) give $F(g) = f$ which implies that f is H-continuous on Ω.

Finally, we prove uniqueness. Let $h \in \mathbb{H}(\Omega)$ be another extension of φ, that is, $h(x) = \varphi(x),\ x \in D$. We have $f(x) = F(D,\Omega,\varphi)(x) = F(D,\Omega,h)(x) \subseteq F(h)(x) = h(x),\ x \in \Omega$. Then the H-continuity of h implies that $F(f) = h$. Using (5) we obtain $f = h$, which completes the proof of the theorem.

Corollary 1. *Let $f,g \in \mathbb{H}(\Omega)$ and let D be a dense subset of Ω. Then*
a) $f(x) \le g(x),\ x \in D \implies f(x) \le g(x),\ x \in \Omega$,
b) $f(x) = g(x),\ x \in D \implies f(x) = g(x),\ x \in \Omega$.

Let $D \subseteq \Omega$. For $f \in \mathbb{A}(\Omega)$ denote by $f|_D$ the restriction of f on D, i. e. $f|_D \in \mathbb{A}(D)$ and $f|_D(x) = f(x),\ x \in D$. The next theorem shows that the restriction of an H-continuous function on an open subset is H-continuous.

Theorem 4. *Let D be an open subset of Ω. If $f \in \mathbb{H}(\Omega)$ then $f|_D \in \mathbb{H}(D)$.*

Proof. Since D is open for every $x \in D$ we have $B_\delta(x) \subseteq D$ for $\delta > 0$ small enough. Hence, for $x \in D$ we have $S(D,D,f|_D)(x) = S(f)(x), I(D,D,f|_D)(x) = I(f)(x), F(D,D,f|_D)(x) = F(f)(x)$. Then the theorem follows from Theorem 1.

4 The Linear Space of Hausdorff Continuous Functions

In the sequel we assume that the set Ω is open. For every two functions $f,g \in \mathbb{H}(\Omega)$ denote $D_{fg} = \Omega \setminus (W_f \cup W_g)$, where W_f and W_g are defined by (1). Using addition of intervals the point-wise sum of $f = [\underline{f}, \overline{f}] \in \mathbb{H}(\Omega)$ and $g = [\underline{g}, \overline{g}] \in \mathbb{H}(\Omega)$ is given by $(f+g)(x) = f(x) + g(x) = [\underline{f}(x) + \underline{g}(x), \overline{f}(x) + \overline{g}(x)],\ x \in \Omega$. It is easy to see that the point-wise sum of H-continuous functions is not always H-continuous [9]. However, the restrictions of f, g and $f+g$ on the set D_{fg} which is dense in Ω, see Theorem 2, are continuous real functions. This suggests the definition of a new operation addition "$\oplus$" on $\mathbb{H}(\Omega)$ as follows.

Definition 3. *Let $f,g \in \mathbb{H}(\Omega)$. Then $f \oplus g$ is the unique H-continuous extension of $(f+g)|_{D_{fg}}$ on Ω given by Theorem 3, that is, $f \oplus g = F(D_{fg},\Omega,f+g)$.*

Multiplication by scalars on $\mathbb{H}(\Omega)$ is defined point-wise; for $f \in \mathbb{H}(\Omega)$, $\alpha \in \mathbb{R}$

$$(\alpha * f)(x) = \alpha f(x) = \begin{cases} [\alpha \underline{f}(x), \alpha \overline{f}(x)] \text{ if } \alpha \ge 0, \\ [\alpha \overline{f}(x), \alpha \underline{f}(x)] \text{ if } \alpha < 0. \end{cases}$$

It can be verified that operations "$\oplus$" and "$*$" satisfy on $\mathbb{H}(\Omega)$ the axioms of a linear space [9]. In particular, the second distributive law, which is usually violated in interval structures, holds true; thus $(\mathbb{H}(\Omega), \oplus, *)$ is a linear space.

Denote by $\mathbb{H}_b(\Omega)$ the set of all bounded H-continuous functions on Ω. Clearly $\mathbb{H}_b(\Omega)$ is a linear subspace of $\mathbb{H}(\Omega)$. Note that the assumption that Ω is open, made in the beginning of the section, is not a significant restriction with regard to $\mathbb{H}_b(\Omega)$. One can easily see that the sets $\mathbb{H}_b(\Omega)$ and $\mathbb{H}(\overline{\Omega})$, where $\overline{\Omega}$ is the closure of Ω are identical. Indeed, according to Theorem 3 every function $f \in \mathbb{H}_b(\Omega)$ has a unique H-continuous extension $e(f)$ on $\overline{\Omega}$, that is, $e(f) \in \mathbb{H}(\overline{\Omega})$. Conversely, the restriction of every function $\mathbb{H}(\overline{\Omega})$ on Ω belongs to $\mathbb{H}_b(\Omega)$, see Theorem 4. Then the mapping $e : \mathbb{H}_b(\Omega) \longrightarrow \mathbb{H}(\overline{\Omega})$ is a bijection. Identifying f with $e(f)$ gives $\mathbb{H}_b(\Omega) = \mathbb{H}(\overline{\Omega})$. Hence by considering $\mathbb{H}_b(\Omega)$ we deal implicitly with the case when the domain is a closure of an open set. It is also easily seen that the supremum norm and the Hausdorff metric discussed in the sequel are preserved by e. Further we prefer to work with $\mathbb{H}_b(\Omega)$ rather then $\mathbb{H}(\overline{\Omega})$ since the linear operations are defined for Ω open and working with Ω closed or compact requires an extension of these definitions. We remark that similar approach is not possible for sets of continuous functions, since the set $C_b(\Omega)$ of all bounded continuous functions on Ω satisfies the inclusion $C(\overline{\Omega}) \subseteq C_b(\Omega)$ but the inverse inclusion is generally not true.

5 Supremum Norm and Approximations

The supremum norm on $\mathbb{H}_b(\Omega)$ can be defined as usually by

$$||f|| = \sup_{x\in\Omega} |f(x)|, \; f \in \mathbb{H}_b(\Omega). \tag{9}$$

Lemma 1. *If D is dense in Ω, then for $f \in \mathbb{H}_b(\Omega)$ we have $||f|| = \sup_{x\in D} |f(x)|$.*

Proof. The inequality $\sup_{x\in D} |f(x)| \leq ||f||$ is obvious. To prove the inverse inequality denote $m = \sup_{x\in D} |f(x)|$. From $-m \leq f(x) \leq m$, $x \in D$, it follows $-m \leq F(D,\Omega,f)(x) \leq m$, $x \in \Omega$. Since f is H-continuous the inclusion $F(D,\Omega,f)(x) \subseteq F(f)(x) = f(x)$, $x \in \Omega$, implies $f = F(D,\Omega,f)$. Therefore $|f(x)| \leq m$ $x \in \Omega$, which gives $||f|| \leq m$. This completes the proof of Lemma 1.

Theorem 5. *The mapping $||\cdot|| : H_b(\Omega) \longrightarrow \mathbb{R}$ given in (9) is a norm on the linear space $\mathbb{H}_b(\Omega)$.*

Proof. Let $f, g \in \mathbb{H}_b(\Omega)$. According to Definition 3 for every $x \in D_{fg}$ we have $(f \oplus g)(x) = f(x) + g(x)$. Hence,

$$\sup_{x\in D_{fg}} |(f+g)(x)| = \sup_{x\in D_{fg}} |f(x)+g(x)| \leq \sup_{x\in D_{fg}} |f(x)| + \sup_{x\in D_{fg}} |g(x)|.$$

Using Lemma 1 the above inequality implies $||f+g|| \leq ||f|| + ||g||$. The remaining properties of the norm are trivially satisfied.

Theorem 5 shows that $\mathbb{H}_b(\Omega)$ considered with the operations "$\oplus$", "$*$" and the supremum norm is a normed linear space. Clearly the supremum norm on $\mathbb{H}_b(\Omega)$ is an extension of the supremum norm on the set of usual bounded continuous functions which is a subset of $\mathbb{H}_b(\Omega)$. Thus the familiar normed linear space $C_b(\Omega)$ is a subspace of $\mathbb{H}_b(\Omega)$.

It is well-known that the supremum norm has limited applications in the approximation of discontinuous functions. It is easy to construct examples of approximations in $\mathbb{H}_b(\Omega)$ by subspaces where the error of the approximation remains bounded away from zero irrespective of the dimension of the subspace. However, approximations with respect to the supremum norm work, in the case when the approximated function and/or some of its derivatives have only "jump" type of discontinuities at a finite number of points which are known. This is a situation which may arise e. g. in the solution of PDE's where discontinuities of the given boundary conditions are propagated in a predictable way within the interior of the domain of the solution [3].

6 Hausdorff Distance and Approximations by Finite Dimensional Subspaces

A natural metric to be associated with H-continuous functions is the Hausdorff metric denoted here by ρ. Let us recall that for $f, g \in \mathbb{H}_b(\Omega)$ the distance $\rho(f,g)$ is defined as the Hausdorff distance between the graphs of f and g considered as subsets of $\mathbb{R}^{n+1}$ [8]. It should be noted that the operation "$\oplus$" is not continuous with respect to the Hausdorff metric as can be shown by easy examples. Hence $\mathbb{H}_b(\Omega)$ is not a linear metric space in the sense of [6]. However, the next theorem shows that the operation "$\oplus$" satisfies a condition rather close to continuity. In the sequel "convergence" is meant in the sense of Hausdorff metric.

Theorem 6. *If the sequences $(f_k)_{k\in\mathbb{N}} \subseteq \mathbb{H}_b(\Omega)$ and $(g_k)_{k\in\mathbb{N}} \subseteq \mathbb{H}_b(\Omega)$ converge respectively to $f, g \in \mathbb{H}_b(\Omega)$, then the sequence $(f_k \oplus g_k)_{k\in\mathbb{N}}$ converges to an S-continuous function h, s. t. the only H-continuous function satisfying the inclusion $\phi(x) \subseteq h(x)$, $x \in \Omega$, is $\phi = f \oplus g$. Moreover, if $h \in \mathbb{H}_b(\Omega)$ then $h = f \oplus g$.*

The proof is rather technical and will be omitted.

We next illustrate the ideas of approximation in $\mathbb{H}_b(\Omega)$ by elements of finite dimensional linear subspaces in the case of functions of one variable, that is, $\Omega = (a,b) \subseteq \mathbb{R}$. Denote by φ the Π-form function:

$$\varphi(x) = \begin{cases} 1 & \text{if } 0 < x < 1; \\ [0,1] & \text{if } x \in \{0,1\}; \\ 0 & \text{if } x < 0 \text{ or } x > 1. \end{cases}$$

For every $j \in \mathbb{N}$ we consider the following set of linearly independent functions

$$\{\phi_{jk} : k \in \mathbb{Z}\},\ \phi_{jk}(x) = \varphi(2^j x - k),\ j,k \in \mathbb{Z}. \tag{10}$$

It is easy to see that $\phi_{jk} \in \mathbb{H}(\mathbb{R})$, $j,k \in \mathbb{Z}$. Therefore every linear combination of functions from the set (10) is also in $\mathbb{H}(\mathbb{R})$.

We now discuss approximation of H-continuous functions by linear combinations of functions from the set (10). To simplify matters we consider approximations on the interval $(0,1)$, that is, in the set $\mathbb{H}_b(0,1)$. It follows from Theorem 4 that the restrictions of the functions (10) to the interval $(0,1)$ belong to $\mathbb{H}_b(0,1)$.

In the sequel ϕ_{jk} denotes the restriction to the interval $(0,1)$ of the function ϕ_{jk} given in (10). Clearly in $\mathbb{H}_b(0,1)$ for every $j \in \mathbb{N}$ it is enough to consider the set

$$\{\phi_{jk} : k = 0, 1, ..., 2^j - 1\}. \tag{11}$$

Denote by V_j the linear subspace of $\mathbb{H}(0,1)$ spanned by the set of functions (11). Using that $\phi_{j-1,k}(x) = \phi_{j,2k}(x) + \phi_{j,2k+1}(x)$, $j, k \in \mathbb{Z}$, one can see that the inclusions $V_0 \subset V_1 \subset ... \subset V_j \subset ... \subset \mathbb{H}(0,1)$ hold true. Hence we have a similar situation to the adaptive multiresolution analysis discussed in [7].

Consider the operators $I_\delta : \mathbb{A}(0,1) \longrightarrow \mathcal{A}(0,1)$ and $S_\delta : \mathbb{A}(0,1) \longrightarrow \mathcal{A}(0,1)$ where $\delta > 0$ and for every $f \in \mathbb{A}(0,1)$

$$I_\delta(f)(x) = \inf\{z \in f(y) : y \in (0,1), |y - x| < \delta\},\ x \in (0,1),$$
$$S_\delta(f)(x) = \sup\{z \in f(y) : y \in (0,1), |y - x| < \delta\},\ x \in (0,1).$$

For a given $\delta > 0$ the modulus of H-continuity $\tau(f, \delta)$ of a function $f \in \mathbb{A}(0,1)$ is the Hausdorff distance between the completed graphs of $I_{\delta/2}(f)$ and $S_{\delta/2}(f)$, that is, $\tau(f,\delta) = \rho(F(I_{\delta/2}(f)), F(S_{\delta/2}(f)))$. It is shown in [8] that a function $f \in \mathbb{A}(0,1)$ is H-continuous if and only if $\lim_{\delta\to 0} \tau(f,\delta) = 0$.

Let $f = [\underline{f}, \overline{f}] \in \mathbb{H}(0,1)$ and let $j \in \mathbb{N}$. Using the operators $I_\delta(f)$ and $S_\delta(f)$ we can construct in V_j a lower approximation $L(f,j)$ of f and an upper approximation $U(f,j)$ of f as follows:

$$L(f,j) = \sum_{k=0}^{2^j-1} I_h(f)((2k+1)h)\phi_{jk},\quad U(f,j) = \sum_{k=0}^{2^j-1} S_h(f)((2k+1)h)\phi_{jk}, \tag{12}$$

where $h = 2^{-j-1}$ and the sums are in terms of the addition "$\oplus$". The inequality

$$L(f,j)(x) \le f(x) \le U(f,j)(x),\quad x \in (0,1), \tag{13}$$

can be easily verified. Indeed, if we have $x \in (2^{-j}k, 2^{-j}(k+1))$ for some $k \in \{0, 1, ..., 2^j - 1\}$ then $L(f,j)(x) = I_h(f)(2^{-j}(k + \frac{1}{2}))\phi_{jk}(x) \le f(x)$. Similarly, $U(f,j)(x) \ge f(x)$. Hence (13) holds on the set $\bigcup_{k=0}^{2^j-1} (2^{-j}k, 2^{-j}(k+1))$ which is dense on $(0,1)$. Using that the functions involved in (13) are all H-continuous, we obtain that (13) holds for all $x \in (0,1)$, see Corollary 1.

Theorem 7. *For every $f \in \mathbb{H}(0,1)$ and $j \in \mathbb{N}$ we have*

$$\rho(L(f,j), f) \le \tau(f, 2^{-j+1}),\quad \rho(U(f,j), f) \le \tau(f, 2^{-j+1}).$$

Proof. Let $h = 2^{-j-1}$ as in (12). From the inequalities

$$I_{2h}(f)(x) \le L(f,j)(x) \le f(x) \le U(f,j) \le S_{2h}(f)(x),\quad x \in (0,1),$$

it follows that $\rho(L(f,j), f) \le \rho(F(I_{2h}(f)), F(S_{2h}(f))) = \tau(f, 4h) = \tau(f, 2^{-j+1})$. Similarly, $\rho(U(f,j), f) \le \tau(f, 2^{-j+1})$.

It follows from Theorem 7 that for every $f \in \mathbb{H}_b(0,1)$ both sequences $(L(f,j))_{j\in\mathbb{N}}$ and $(U(f,j))_{j\in\mathbb{N}}$ converge to f with respect to the Hausdorff distance ρ. Hence $\bigcup_{j=1}^{\infty} V_j$ is a dense subspace of $\mathbb{H}_b(0,1)$ considered as a metric space w. r. t. ρ.

7 Conclusion

H-continuous functions have a number of interesting and rather unique properties due to the fact that they share characteristics of both real-valued and interval-valued functions. The extension property discussed in Section 3 is in this category as it is typical neither for classes of real functions usually considered in Functional Analysis nor for classes of interval functions considered in Interval Analysis. We show that this extension property generates the linear space operations in $\mathbb{H}(\Omega)$ proposed in our previous work [9]. Our further discussion is devoted to issues of norm, metric and approximations of H-continuous functions. We introduce the supremum norm for H-continuous functions and prove that the set $\mathbb{H}_b(\Omega)$ of all bounded H-continuous functions is a normed linear space w. r. t. this norm. Recognizing the limitations of the supremum norm when discontinuous functions are involved we consider the Hausdorff metric on $\mathbb{H}_b(\Omega)$ and establish a strong connection between the metric and the linear space operations. The considered approximations by a subspace show that the Hausdorff metric is a natural metric to be associated with H-continuous functions.

Acknowledgments

This work is partially supported by Bulgarian NSF grant MM-1104/01 as well as by a research grant from the National Research Foundation of South Africa.

References

1. Anguelov, R.: Dedekind Order Completion of C(X) by Hausdorff Continuous Functions. Quaestiones Mathematicae **27**(2004) 153–170.
2. Anguelov, R.: An Introduction to some Spaces of Interval Functions, Technical Report UPWT2004/3, University of Pretoria, 2004.
3. Anguelov, R.: Spline-Fourie Approximations of Discontinuous Waves, J. UCS **4** (1998) 110–113.
4. Anguelov, R., Markov, S.: Extended Segment Analysis. Freiburger Intervall-Berichte **10** (1981) 1–63.
5. Markov, S.: On Quasilinear Spaces of Convex Bodies and Intervals, J. Comput. Appl. Math. 162 (2004), 93–112.
6. Rolewicz, S.: Metric Linear Spaces, PWN–Polish Scientific Publishers, Warsaw, D. Reidel Publ., Dordrecht, 1984.
7. Sendov, Bl.: Adapted Multiresolution Analysis, In: L. Leindler, F. Schipp, J. Szabados (Eds.), Functions, Series, Operators, Janos Bolyai Math. Soc., Budapest, 2002, 23–38.
8. Sendov, Bl.: Hausdorff Approximations, Kluwer, 1990.
9. Sendov, Bl., Anguelov, R., Markov, S.: The Linear Space of Hausdorff Continuous Functions, submitted to JCAM.

Comparison of Interval and Ellipsoidal Bounds for the Errors of Vector Operations

Felix L. Chernousko, Alexander J. Ovseevich, and Yuri V. Tarabanko

Institute for Problems in Mechanics RAS,
Pr. Vernadskogo 101, Moscow, Russia
chern@ipmnet.ru, ovseev@ipmnet.ru

Abstract. We compare two approaches, the interval and the ellipsoidal ones, to the guaranteed estimation of errors of vector operations by considering the problem of multiplication of a vector by a matrix. It is shown that for a large class of linear operators the ellipsoidal estimates are more precise than the interval ones, even if the initial vector has interval error bounds.

1 Introduction

These days the interval analysis is the most common approach to the guaranteed estimation of computational errors [1]. In this approach, we deal with the intervals containing the values known up to a bounded error. For instance, instead of a scalar a one considers an interval $[a^-, a^+]$, where a is known to belong. Similarly, instead of an n–dimensional vector x with components x_i, $i = 1, \ldots, n$, one considers a box defined by the inequalities

$$x_i^- \le x_i \le x_i^+, \quad i = 1, \ldots, n, \tag{1}$$

such that its edges are parallel to the coordinate axes. We will call these parallelepipeds boxes. The boxes (1), when regarded as indeterminacy sets for unknown vectors, has some important properties which make them suitable for computations. Namely, a box is characterized by a relatively small amount of defining parameters ($2n$ if $x \in \mathbf{R}^n$), simple shape of the boundary, and easy visualization. However, this class of indeterminacy sets is not invariant with respect to affine transformations; thus, each transform of this kind makes the estimate of the indeterminacy set worse, since we have to substitute a box for the transformed one. This accumulation of errors might make the final estimate worthless.

To avoid this loss of precision, it is reasonable to use an affine invariant class of indeterminacy sets. For instance, one might use ellipsoidal estimators. The optimal algebraic operations with ellipsoids are designed in [2, 3]. These operations are applied to estimation of reachable sets of dynamic systems. Ellipsoids require more defining parameters (n coordinates of the center and $n(n+1)/2$ elements of the symmetric matrix, if $x \in \mathbf{R}^n$). This number is, however, almost two times less then the number $n(n+1)$ of defining parameters of an arbitrary

I. Lirkov, S. Margenov, and J. Waśniewski (Eds.): LSSC 2005, LNCS 3743, pp. 289–296, 2006.

parallelepiped. The class of ellipsoids is invariant under affine transformations. Therefore, there is no need to make estimator more worse after this transformation. We note that both methods (the interval and the ellipsoidal ones) are the same when applied to scalars, because in the scalar case both ellipsoids and parallelepipeds degenerate into intervals.

2 Statement of the Problem

We consider the problem of multiplication of a vector x, given by interval bounds (1), by a known square $n \times n$ matrix A. We need to estimate the extent of error of this operation, i.e., indeterminacy of the vector $y = Ax \in \mathbf{R}^n$. By coordinate scaling and translation of the origin one might bring the bounds (1) to the form:

$$-\frac{1}{2} \leq x_i \leq \frac{1}{2}, \quad i = 1, \ldots, n. \tag{2}$$

Thus, the initial problem is reduced to that of multiplication of a unit n-dimensional cube, centered at the origin, by a given matrix A. We use two approaches in order to solve the problem.

The first approach is to apply the interval analysis, so that the final bound for the vector $y = Ax$ takes a form of a box:

$$y_i^- \leq y_i \leq y_i^+, \quad i = 1, \ldots, n. \tag{3}$$

In the second approach, we begin with the substitution of the minimal volume ellipsoid containing the cube for the initial unit cube. Clearly, this ellipsoid is a ball. Then we transform the ball by the matrix A and declare the resulting ellipsoid the final estimate for y. We conduct an analysis which shows that for most matrices A the second approach is better, at least if our criterion is the volume of the set of indeterminacy. From now on, we assume that the matrix A is nonsingular. We notice that if A is singular, then the transformed ellipsoid has zero volume, which is the best possible outcome not reachable in general by the interval approach.

3 Interval Approach

The transformed unit cube (2) is not a box in general. Therefore, we have to make a weaker estimate by substituting minimal box for the exact transform. To do this, we find the maximal value of the projection of the vector y on the ith coordinate axis. In other words, we have to find the maximum over all x containing in the unit cube (2) of the expression

$$y_i = \sum_{k=1}^{n} a_{ik} x_k, \quad |x_k| \leq 1/2, \quad i, k = 1, \ldots, n,$$

where a_{ik} are elements of A. It is clear that the maximum is attained at $x_k = \frac{1}{2}\,\mathrm{sign}\,(a_{ik})$. This gives us the following interval estimate:

$$-\frac{1}{2}\sum_{k=1}^{n}|a_{ik}| \leq y_i \leq \frac{1}{2}\sum_{k=1}^{n}|a_{ik}|, \quad i = 1, \dots, n.$$

The volume of the above box equals to:

$$V_{\Pi} = \prod_{i=1}^{n}\sum_{k=1}^{n}|a_{ik}|. \tag{4}$$

Now we turn to the structure of the matrix A. One can represent it in the form of a product (the Iwasava decomposition) $A = SO$, where S is an upper triangular matrix such that its diagonal element are nonnegative, O is an orthogonal matrix. We can obtain a lower estimate for V_{Π} which does not involve the orthogonal part of the matrix A. To do this, we inscribe the maximal volume ellipsoid into the initial unit cube (2). Clearly, this is the unit ball. In notations of [2, 3] for an n-dimensional ellipsoid

$$E(a, Q) = \{x : \ (Q^{-1}(x-a), (x-a)) \leq 1\}, \tag{5}$$

where a is n-dimensional vector of the center of the ellipsoid, Q is a positive-definite $n \times n$ matrix, this ball takes the form

$$E_{\text{inn}} = E\left(0, \frac{1}{4}I\right),$$

where I is a unit matrix. Then, we transform the ball by the matrix A and substitute a minimal box for the obtained ellipsoid. The volume of this ellipsoid is less than or equal to the volume (4). It follows from the identity $AE(a, Q) = E\left(Aa, AQA^T\right)$, cf. [2,3], and from the orthogonality of the matrix O that the ball E_{inn} is transformed into the ellipsoid

$$E'_{\text{inn}} = E\left(0, \frac{1}{4}SS^T\right). \tag{6}$$

In order to estimate the ellipsoid E'_{inn} from outside by a box, we find its maximal projection on the kth coordinate axis. According to the well-known formula for the support function of the ellipsoid $E(a, Q)$, its maximal projection on the direction η equals to

$$(Q\eta, \eta)^{\frac{1}{2}} + (a, \eta).$$

By substituting the parameters of ellipsoid (6) for the above formula and choosing $\eta = e_k$, where e_k is the kth coordinate vector, we obtain the parameters of the parallelepiped we are looking for:

$$-\frac{1}{2}\left(SS^T e_k, e_k\right)^{\frac{1}{2}} \leq y_k \leq \frac{1}{2}\left(SS^T e_k, e_k\right)^{\frac{1}{2}}, \quad k = 1, \dots, n. \tag{7}$$

The triangular matrix S can be represented in the form:

$$S = DU = \begin{pmatrix} \rho_1 & 0 & 0 & \dots & 0 \\ 0 & \rho_2 & 0 & \dots & 0 \\ 0 & 0 & \rho_3 & \dots & 0 \\ \vdots & \vdots & \vdots & \ddots & \vdots \\ 0 & 0 & 0 & \dots & \rho_n \end{pmatrix} \begin{pmatrix} 1 & \delta_{11} & \delta_{12} & \dots & \delta_{1,n-1} \\ 0 & 1 & \delta_{21} & \dots & \delta_{2,n-2} \\ 0 & 0 & 1 & \dots & \delta_{3,n-3} \\ \vdots & \vdots & \vdots & \ddots & \vdots \\ 0 & 0 & 0 & \dots & 1 \end{pmatrix}, \tag{8}$$

where $\rho_i > 0$, $i = 1, \dots, n$, since $A = SO$ is nonsingular. It is clear that $(SS^T e_k, e_k)$ (in other words, the kth diagonal element d_k of the matrix SS^T) is equal to

$$d_k = \rho_k^2 (1 + \sum_{i=1}^{n-k} \delta_{ki}^2), \quad k = 1, \dots, n-1; \quad d_n = \rho_n^2.$$

By introducing the notation

$$\sum_{i=1}^{n-k} \delta_{ki}^2 = \Delta_k^2, \quad k = 1, \dots, n-1, \tag{9}$$

we can present the parallelepiped (7) in the form

$$\begin{aligned} -\tfrac{1}{2}\rho_k \sqrt{1 + \Delta_k^2} \le y_k \le \tfrac{1}{2}\rho_k \sqrt{1 + \Delta_k^2}, \ k = 1, \dots, n-1, \\ -\tfrac{1}{2}\rho_n \le y_n \le \tfrac{1}{2}\rho_n. \end{aligned} \tag{10}$$

From (10) we obtain the following lower bound for the volume of the parallelepiped which is provided by the interval analysis:

$$V_\Pi \ge \prod_{k=1}^{n} \rho_k \sqrt{\prod_{k=1}^{n-1} (1 + \Delta_k^2)}. \tag{11}$$

4 Ellipsoidal Approach

Now, we find the volume of the ellipsoid resulting from the ellipsoidal analysis of the problem. To this end, we replace the initial unit cube with the minimal volume bounding ellipsoid. This ellipsoid is, clearly, the ball centered at the origin and having radius equal to half the diagonal of the unit n-cube, i.e., $E_{\text{out}} = E\left(0, \frac{n}{4} I\right)$. Transforming it by A we come to ellipsoid

$$E'_{\text{out}} = E\left(0, \frac{n}{4} AA^T\right) = E\left(0, \frac{n}{4} SS^T\right).$$

The volume of the transformed ellipsoid E'_{out} is

$$\text{Vol}\,(E'_{\text{out}}) = c_n |\det A| = c_n \prod_{k=1}^{n} \rho_k, \tag{12}$$

where ρ_k are diagonal elements (12) of the matrix S, and

$$c_n = \frac{(\pi n)^{\frac{n}{2}}}{2^n \Gamma\left(\frac{n}{2}+1\right)} \tag{13}$$

is the volume of n-dimensional ball of radius $\sqrt{n}/2$.

5 Comparison

From (11) and (12) we obtain an inequality describing a domain in the parameter space of matrices S, where the ellipsoidal estimates are evidently better than the interval ones at least in the sense of volume. Namely,

$$\sqrt{\prod_{k=1}^{n-1}(1+\Delta_k^2)} > c_n. \tag{14}$$

Note, that c_n exponentially increases with n. The sequence

$$\mu_n = \sqrt[n-1]{c_n} \tag{15}$$

is bounded and monotone increasing with the limit

$$\mu_\infty = \lim_{n\to\infty} \mu_n = \frac{\pi e}{2} \approx 4.27.$$

Getting back to notation (9) we obtain a sufficient condition (in terms of the elements of the matrix S) for the advantage of the ellipsoidal approach vs. the interval one. Namely,

$$\Delta_k^2 = \sum_{i=1}^{n-k} \delta_{ki}^2 > \left(\mu_n^2 - 1\right), \quad k = 1, \ldots, n-1. \tag{16}$$

Every inequality (16) gives the complement of a ball in the space of elements of kth row of the matrix U. Thus, the domain, where the interval estimates are better than the ellipsoidal ones, is bounded by a ball such that its radius never (for any dimension of the initial vector) exceeds $\sqrt{\mu_\infty^2 - 1} \approx 4.151$. This result might be interpreted geometrically as follows: the diagonal matrix A retains the class of boxes, and, therefore, there is no loss of precision due to noninvariance under affine transformations. If, on the other hand, the matrix A is far from diagonal, the ellipsoidal analysis gives better estimate than the interval one.

6 Two-Dimensional Case

Estimate (16) of the domain, where ellipsoids have an advantage over the intervals, might be improved greatly if the orthogonal part O of the matrix A is taken to account. Consider the case $n = 2$ in more detail. Any invertible 2×2 matrix A can be represented uniquely in the form $A = SO$, where

$$S = \begin{pmatrix} \rho_1 & 0 \\ 0 & \rho_2 \end{pmatrix} \begin{pmatrix} 1 & \xi \\ 0 & 1 \end{pmatrix}, O = \begin{pmatrix} \cos\varphi & -\sin\varphi \\ \sin\varphi & \cos\varphi \end{pmatrix}$$

and $\rho_i > 0$, $i = 1, 2$.

We can find the volume (area) of the rectangle by (4):

$$V_{\Pi} = \rho_1 \rho_2 \left(|\sin\varphi| + |\cos\varphi| \right) \left(|\cos\varphi + \xi \sin\varphi| + |-\sin\varphi + \xi\cos\varphi| \right).$$

The area of the ellipse is given by (12): $V_E = \pi\rho_1\rho_2/2$. Then the domain, where ellipsoids has an advantage over rectangles, is given by the following inequality in the space of parameters ξ, φ:

$$\left(|\sin\varphi| + |\cos\varphi| \right) \left(|\cos\varphi + \xi \sin\varphi| + |-\sin\varphi + \xi\cos\varphi| \right) > \frac{\pi}{2}. \qquad (17)$$

Denote the left-hand side of the inequality (17) by $v(\xi, \varphi)$ and describe the domain, where

$$v(\xi, \varphi) > \frac{\pi}{2}. \qquad (18)$$

One can easily check, that

$$v\left(\xi, \varphi + \frac{\pi}{2}\right) = v(\xi, \varphi), \quad v(-\xi, -\varphi) = v(\xi, \varphi). \qquad (19)$$

The first relation (19) implies, that it suffices to study (18) in the domain $|\varphi| \le \pi/4$, and the second relation (19) implies, that one might confine himself with the study of the domain $0 \le \varphi \le \pi/4$, where $\cos\varphi > \sin\varphi > 0$.

There are three cases:

1) $\xi < -\cot\varphi$. Inequality (18) reduces to

$$v(\xi,\ \varphi) = -\cos 2\varphi - \xi(1 + \sin 2\varphi) > \pi/2. \qquad (20)$$

2) $-\cot\varphi \le \xi \le \tan\varphi$. From (18) we obtain

$$v(\xi,\ \varphi) = 1 + \sin 2\varphi - \xi \cos 2\varphi > \pi/2. \qquad (21)$$

3) $\xi > \tan\varphi$. In this case, it follows from (18) that

$$v(\xi,\ \varphi) = \cos 2\varphi + \xi(1 + \sin 2\varphi) > \pi/2. \qquad (22)$$

Inequality (20) always holds in the domain $\xi < -\cot\varphi$, $\varphi \in [0,\ \pi/4]$ because in this set

$$v(\xi,\ \varphi) \ge -\cos 2\varphi + \cot\varphi\,(1 + \sin 2\varphi) = \cot\varphi + 1 > 2 > \pi/2.$$

Inequalities (21) and (22) imply that

$$\xi < \xi_1(\varphi) = \frac{2 + 2\sin 2\varphi - \pi}{2\cos 2\varphi}, \quad \xi > \xi_2(\varphi) = \frac{\pi - 2\cos 2\varphi}{2 + 2\sin 2\varphi}.$$

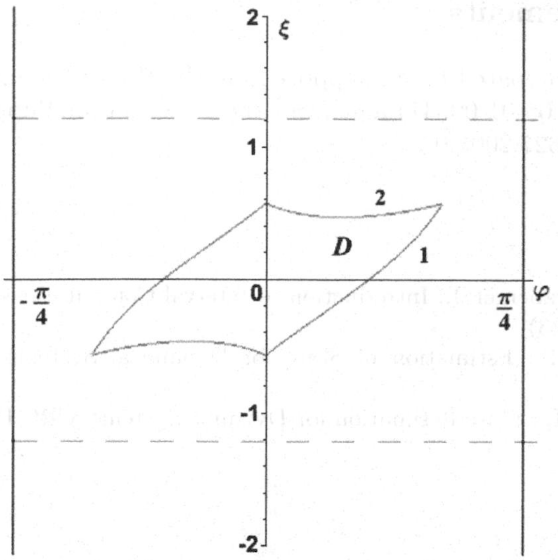

Fig. 1.

The curves $\xi_1(\varphi)$ and $\xi_2(\varphi)$ are drawn in Fig. 1 and marked with numbers 1, 2. These curves are defined in the interval $[0, \varphi^*]$, where

$$\varphi^* = \arctan[(\pi - 2)/2] = 0.5187 < \pi/4,$$

and they cross at the point (φ^*, ξ^*), where $\xi^* = (\pi - 2)/2$. Outside the domain **D**, bounded by these curves and curves obtained from these by symmetry with respect to the origin, the interval estimates has an advantage over the ellipsoidal ones. One can see from Fig. 1 that the domain **D** is a relatively small part of the strip $|\varphi| < \pi/4, \ -\infty < \xi < \infty$ of parameters of an arbitrary matrix A.

Note, that estimate (16) together with (15) and (13) gives

$$|\xi| > \sqrt{\mu_2^2 - 1} = \sqrt{\pi^2/4 - 1} \approx 1.211.$$

This is an upper estimate of the set of parameters, where interval estimates are better than the ellipsoidal ones. In Fig. 1 the boundary of the corresponding domain is drawn as a dotted line. One can see easily, that this bound is much worse than the exact estimate given by the domain **D**.

7 Conclusion

In many cases the ellipsoidal estimates are better than the interval ones, even if the initial vector is estimated by means of intervals.

Acknowledgements

This paper was prepared under support from the Russian Foundation for Basic Research (grant 05-01-00647) and from the Program for Support of Scientific Schools (grant 1627.2003.01).

References

1. Alefeld G., Herzberger J.: Introduction to Interval Computations. Academic Press, New York, (1983)
2. Chernousko F.L.: Estimation of State for Dynamical Systems. Moscow: Nauka, (1988)
3. Chernousko F.L.: State Estimation for Dynamic Systems. CRC Press, Boca Raton, (1994)

On the Time-Discretization of Singularly Perturbed Uncertain Systems

G. Grammel

Center for Mathematics, TU Munich, Boltzmann St. 3, 85747 Garching, Germany
grammel@ma.tum.de
http:www-m6.ma.tum.de/~grammel

Abstract. Uncertain dynamical systems with two time scales are under consideration. The ratio of the magnitudes of the multi-valued vector fields related to the slow and fast subsystems is given as a singular perturbation parameter. The averaging method is employed in order to construct a limiting system for the slow subsystem, representing the case of a vanishing perturbation parameter. This method is a classical one in connection with continuous time systems, but works as well for uncertain systems in discrete time. However, the relation between a continuous time system and its time-discretized version along with the limiting behavior as the perturbation tends to zero has not yet been elaborated. In the present work it is shown that both limiting procedures, the vanishing singular perturbation parameter and the vanishing discretization step, are commutable.

1 Introduction

We consider the continuous time singularly perturbed system $\mathbf{\Sigma}_\epsilon$ in $\mathbb{R}^m \times \mathbb{R}^n$ of the form

$$\dot{z}(t) = \epsilon f(z(t), y(t)), \quad z(0) = z^0,$$
$$\dot{y}(t) \in G(z(t), y(t)), \quad y(0) = y^0, \quad t \in \left[0, \frac{1}{\epsilon}\right].$$

Here, the perturbation parameter $\epsilon > 0$ is considered to be small and hence reflects the presence of two time scales. Via averaging we construct a limiting system $\mathbf{\Sigma}_{\text{aver}}$ describing the dynamics of the $\mathbb{R}^m$-components for vanishing perturbation parameter. We simultaneously examine the corresponding discrete time singularly perturbed system $\mathbf{\Sigma}_\epsilon^{\mathbf{h}}$ in $\mathbb{R}^m \times \mathbb{R}^n$ given by

$$z_{k+1} - z_k = h\epsilon f(z_k, y_k), \quad z_0 = z^0,$$
$$y_{k+1} - y_k \in hG(z_k, y_k), \quad y_0 = y^0, \quad k \in \left\{0, 1, \ldots, \left[\frac{1}{h\epsilon}\right]\right\}.$$

Naturally, for any fixed perturbation parameter $\epsilon > 0$ the system $\mathbf{\Sigma}_\epsilon^{\mathbf{h}}$ is just the Euler discretization of $\mathbf{\Sigma}_\epsilon$. The approximating properties of Euler discretizations

I. Lirkov, S. Margenov, and J. Waśniewski (Eds.): LSSC 2005, LNCS 3743, pp. 297–304, 2006.

for differential inclusions have been investigated in a variety of articles, see for instance [1, 2, 6, 7]. In particular in [7] it is shown that under stability type suppositions on the $\mathbb{R}^n$-components, the rate of convergence, as the stepsize $h > 0$ tends to zero, does not depend on the perturbation parameter $\epsilon > 0$. Still, the limiting behavior of $\mathbf{\Sigma}_\epsilon^\mathbf{h}$ as $(\epsilon, h) \to (0, 0)$ has not yet been investigated. It is the main purpose of this paper to show that the $\mathbb{R}^m$-components of $\mathbf{\Sigma}_\epsilon^\mathbf{h}$ converge to the trajectories of the continuous time averaged system $\mathbf{\Sigma}_{\text{aver}}$, as $(\epsilon, h) \to (0, 0)$. This convergence implies that both limiting processes, $\epsilon \to 0$ and $h \to 0$, are commutable. If we first let $\epsilon \to 0$, then the $\mathbb{R}^m$-components of $\mathbf{\Sigma}_\epsilon^\mathbf{h}$ converge to the $\mathbb{R}^m$-components of a discrete time averaged system $\mathbf{\Sigma}_{\text{aver}}^\mathbf{h}$. In particular it turns out that the rate of convergence is independent of the step size $h > 0$. If, in a second step, we let $h \to 0$, then the $\mathbb{R}^m$-components of $\mathbf{\Sigma}_{\text{aver}}^\mathbf{h}$ tend to the trajectories of the continuous time averaged system $\mathbf{\Sigma}_{\text{aver}}$. This interplay of discretization and perturbation can be expressed in the following commutative diagram.

$$\begin{array}{ccccc} & & \textit{Perturbation} & & \\ & \mathbf{\Sigma}_{\text{aver}} & \longrightarrow & \mathbf{\Sigma}_\epsilon & \\ \textit{Discretization} & \downarrow & & \downarrow & \textit{Discretization} \\ & \mathbf{\Sigma}_{\text{aver}}^\mathbf{h} & \longrightarrow & \mathbf{\Sigma}_\epsilon^\mathbf{h} & \\ & & \textit{Perturbation} & & \end{array}$$

We present the regularity suppositions on the mappings involved.

Assumption 1. The mapping $f : \mathbb{R}^m \times \mathbb{R}^n \to \mathbb{R}^m$ is Lipschitz continuous with Lipschitz constant $L \geq 0$.

Assumption 2. The multi-valued mapping $G : \mathbb{R}^m \times \mathbb{R}^n \to \mathcal{CC}(\mathbb{R}^n)$ is Lipschitz continuous with Lipschitz constant $L \geq 0$. Furthermore, there are real constants $0 < \alpha \leq \beta$ such that for all $z \in \mathbb{R}^m$, $y_1, y_2 \in \mathbb{R}^n$, $v_1 \in G(z, y_1)$ there is a $v_2 \in G(z, y_2)$ with

$$\langle v_2 - v_1, y_2 - y_1 \rangle \leq -\alpha \|y_2 - y_1\|^2, \quad \|v_2 - v_1\| \leq \beta \|y_2 - y_1\|.$$

These assumptions guarantee that for any compact set $M^0 \times N^0 \subset \mathbb{R}^m \times \mathbb{R}^n$ of initial conditions, the trajectories of the singularly perturbed systems involved stay in a compact set $M \times N \subset \mathbb{R}^m \times \mathbb{R}^n$ and that certain averages constructed below have limits.

Organization of the paper. In Section 2, we focus on continuous time systems and recall some known results on the relation between $\mathbf{\Sigma}_\epsilon$ and $\mathbf{\Sigma}_{\text{aver}}$. In particular the construction of the continuous time averaged system $\mathbf{\Sigma}_{\text{aver}}$ is displayed.

Section 3 provides a discrete time version of the averaging result. Here it is shown that the relation between $\mathbf{\Sigma}_\epsilon^\mathbf{h}$ and $\mathbf{\Sigma}_{\text{aver}}^\mathbf{h}$ is independent of the step size $h > 0$.

In Section 4, the discrete time averaged system $\mathbf{\Sigma}_{\text{aver}}^\mathbf{h}$ is considered as a Euler discretization of the continuous time averaged system $\mathbf{\Sigma}_{\text{aver}}$.

Finally, in Section 5 we recall some known results on the relation between $\mathbf{\Sigma}_\epsilon$ and its Euler discretization $\mathbf{\Sigma}_\epsilon^\mathbf{h}$.

Notation. Given a normed space X, we denote by $\mathcal{C}(X)$ the family of compact, nonvoid subsets of X. By $\mathcal{CC}(X)$, we denote the family of compact, convex, nonvoid subsets of X. For two subsets $A, B \in \mathcal{C}(X)$, the distance of A to B is given by $\text{dist}_{\text{H}}(A, B) := \sup_{a \in A} \text{dist}(a, B)$, where, as usual we set $\text{dist}(x, B) := \inf_{b \in B} \|x - b\|$, for $x \in X$. The Hausdorff metric is defined by $\text{d}_{\text{H}}(A, B) := \max\{\text{dist}_{\text{H}}(A, B), \text{dist}_{\text{H}}(B, A)\}$. For the comparison of trajectory sets, throughout the maximum norm is used. For the comparison of a continuous time trajectory with a discrete time trajectory, we linearly interpolate the discrete time trajectory.

2 Continuous Time Averaging

Definition 1. *For $z \in \mathbb{R}^m$, $y^0 \in \mathbb{R}^n$, $S > 0$, we define the individual continuous time average as the set*

$$F_S(z, y^0) := \bigcup_{y(\cdot)} \frac{1}{S} \int_0^S f(z, y(t))dt,$$

where the union is taken over all solutions to the multi-valued differential equation

$$\dot{y}(t) \in G(z, y(t)), \quad y(0) = y^0, \quad t \in [0, S]. \tag{1}$$

Lemma 1. *Let Assumptions 1 and 2 be effective. Let $M \times N^0 \subset \mathbb{R}^m \times \mathbb{R}^n$ be a compact subset. Then there is a Lipschitz continuous multi-valued mapping $F_{\text{aver}} : M \to \mathcal{CC}(\mathbb{R}^m)$ such that uniformly in $(z, y^0) \in M \times N^0$ the individual continuous time averages satisfy the estimate*

$$\text{d}_{\text{H}}\left(F_S(z, y^0), F_{\text{aver}}(z)\right) = O(S^{-\frac{1}{2}}), \quad \text{as} \quad S \to \infty.$$

Proof. This statement can be shown in the same way as Lemma 3 in [5]. A complete proof of the convergence is contained in [4]. □

The continuous time averaged system $\mathbf{\Sigma}_{\text{aver}}$ in $\mathbb{R}^m$ is given by

$$\dot{z}(t) \in \epsilon F_{\text{aver}}(z(t)), \quad z(0) = z^0, \quad t \in \left[0, \frac{1}{\epsilon}\right].$$

Let $\mathcal{S}_\epsilon(z^0, y^0)$ be the solution set to $\mathbf{\Sigma}_\epsilon$ (projected to the $\mathbb{R}^m$-components) and let $\mathcal{S}_{\text{aver}}(z^0)$ be the solution set to $\mathbf{\Sigma}_{\text{aver}}$. The following theorem states that $\mathbf{\Sigma}_{\text{aver}}$ adequately describes the limiting behavior of $\mathbf{\Sigma}_\epsilon$, as the perturbation parameter $\epsilon > 0$ is vanishing.

Theorem 1. *Let Assumptions 1 and 2 be effective. For any compact sets $M^0 \subset \mathbb{R}^m$, $N^0 \subset \mathbb{R}^n$, we can estimate*

$$\text{d}_{\text{H}}\left(\mathcal{S}_\epsilon(z^0, y^0), \mathcal{S}_{\text{aver}}(z^0)\right) = O(\epsilon^{\frac{1}{3}}), \quad \text{as} \quad \epsilon \to 0,$$

uniformly in $(z^0, y^0) \in M^0 \times N^0$.

Proof. This statement can be shown in the same way as Theorem 1 in [5]. A complete proof of the convergence is contained in [4]. □

3 Discrete Time Averaging

Definition 2. *For $z \in \mathbb{R}^m$, $y^0 \in \mathbb{R}^n$, $K > 0$, we define the individual discrete time average as the set*

$$F_K^h(z, y^0) := \bigcup_{y(\cdot)} \frac{1}{K} \sum_{k=0}^{K-1} f(z, y_k),$$

where the union is taken over all solutions to the multi-valued discrete time system

$$y_{k+1} - y_k \in hG(z, y_k), \quad y_0 = y^0, \quad k \in \{0, \dots, K-1\}. \tag{2}$$

Lemma 2. *Let Assumptions 1 and 2 be effective. Let $M \times N^0 \subset \mathbb{R}^m \times \mathbb{R}^n$ be a compact subset. Then there is a multi-valued mapping $F_{\text{aver}}^h : M \to CC(\mathbb{R}^m)$ such that uniformly in $(z, y^0) \in M \times N^0$ and in $h \in \left(0, \frac{\alpha}{\beta^2}\right]$ the individual discrete time averages satisfy the estimate*

$$h^{\frac{1}{2}} d_H\left(F_K^h(z, y^0), F_{\text{aver}}^h(z)\right) = O(K^{-\frac{1}{2}}), \quad \text{as} \quad K \to \infty.$$

Furthermore, the mapping $F_{\text{aver}}^h : M \to CC(\mathbb{R}^m)$ is Lipschitz continuous with Lipschitz constant $\Lambda \geq 0$.

Proof. Let $(y_k^1)_{k\in\{0,\dots,K-1\}}$ be a solution to the multi-valued discrete system

$$y_{k+1} - y_k \in hG(z, y_k), \quad k \in \{0, \dots, K-1\}, \tag{3}$$

with initial value $y_0^1 = \psi_1$. We construct a solution y_k^2 to the multi-valued discrete system (3) with initial condition $y_0^2 = \psi_2$ according to Assumption 2. Then we can achieve that

$$\begin{aligned}
\|y_{k+1}^2 - y_{k+1}^1\|^2 &= \langle y_k^2 - y_k^1 + h(v_2 - v_1), y_k^2 - y_k^1 + h(v_2 - v_1)\rangle \\
&= \|y_k^2 - y_k^1\|^2 + h^2\|v_2 - v_1\|^2 + 2h\langle v_2 - v_1, y_k^2 - y_k^1\rangle \\
&\leq (1 + \beta^2 h^2 - 2\alpha h)\|y_k^2 - y_k^1\|^2,
\end{aligned}$$

which yields for $0 < h \leq \frac{\alpha}{\beta^2}$ the estimation

$$\|y_{k+1}^2 - y_{k+1}^1\| \leq \left(1 - \frac{\alpha h}{2}\right) \|y_k^2 - y_k^1\|.$$

Obviously, the estimation above implies exponential stability type properties of the multi-valued discrete system (3). Accordingly we have for the individual discrete time averages

$$d_H\left(F_K^h(z, \psi_2), F_K^h(z, \psi_1)\right) \leq \frac{L}{K} \sum_{k=0}^{K-1} \left(1 - \frac{\alpha h}{2}\right)^k \|\psi_2 - \psi_1\| \leq \frac{2L}{\alpha h K} \|\psi_2 - \psi_1\|.$$

Note that in general the individual discrete time averages are not convex, but according to the estimation above their convex hulls form a Cauchy sequence in $\mathcal{CC}(\mathbb{R}^m)$, as $(hK) \to \infty$. The claim follows from the fact that the convex combinations of any set $A \in \mathcal{C}(\mathbb{R}^m)$ approximate its convex hull via

$$\mathrm{d_H}\left(\frac{1}{k}\sum_{i=1}^{k} A, \mathrm{conv}(A)\right) = O(k^{-1}), \quad \text{as} \quad k \to \infty,$$

see [3], Lemma 4. As for the Lipschitz continuity of the multi-valued mapping $F^h_{\text{aver}} : M \to \mathcal{CC}(\mathbb{R}^m)$ we argue as follows. The exponential stability type properties of the multi-valued discrete system (3) imply that there is a constant $C \geq 0$ such that for any sequences $(z^1_k), (z^2_k) \subset M$, any initial value $y^0 \in N^0$, any $K > 0$, any step size $0 < h \leq \frac{\alpha}{\beta^2}$ and any solution y^1_k to

$$y_{k+1} - y_k \in hG(z^1, y_k), \quad y_0 = y^0, \quad k \in \{0, \ldots, K-1\},$$

there is a solution y^2_k to

$$y_{k+1} - y_k \in hG(z^2, y_k), \quad y_0 = y^0, \quad k \in \{0, \ldots, K-1\},$$

with

$$\max_{k \in \{0,\ldots,K-1\}} \|y^1_k - y^2_k\| \leq C \max_{k \in \{0,\ldots,K-1\}} \|z^1_k - z^2_k\|.$$

Hence, the individual discrete time averages have a common Lipschitz constant in the first argument, which is transferred to the limit set and the proof is finished. □

The discrete time averaged system $\mathbf{\Sigma}^{\mathbf{h}}_{\text{aver}}$ in $\mathbb{R}^m$ is given by

$$z_{k+1} - z_k \in h\epsilon F^h_{\text{aver}}(z_k), \quad z(0) = z^0, \quad k \in \left\{0, 1, \ldots, \left[\frac{1}{h\epsilon}\right]\right\}.$$

Let $\mathcal{S}^h_\epsilon(z^0, y^0)$ be the solution set to $\mathbf{\Sigma}^{\mathbf{h}}_\epsilon$ (projected to the $\mathbb{R}^m$-component) and let $\mathcal{S}^h_{\text{aver}}(z^0)$ be the solution set to $\mathbf{\Sigma}^{\mathbf{h}}_{\text{aver}}$. The following theorem is the main result of this paper. It states in particular that the approximation via averaging is independent of the step size.

Theorem 2. *Let Assumptions 1 and 2 be effective. For any compact sets $M^0 \subset \mathbb{R}^m$, $N^0 \subset \mathbb{R}^n$ we can estimate*

$$\mathrm{d_H}\left(\mathcal{S}^h_\epsilon(z^0, y^0), \mathcal{S}^h_{\text{aver}}(z^0)\right) = O(\epsilon^{\frac{1}{3}}), \quad \text{as} \quad \epsilon \to 0,$$

uniformly in the initial states $(z^0, y^0) \in M^0 \times N^0$ and in the stepsize $h \in \left(0, \frac{\alpha}{\beta^2}\right]$.

Proof. We define the mapping $n : (0,1] \times \left(0, \frac{\alpha}{\beta^2}\right] \to \mathbb{N}$, $(\epsilon, h) \mapsto n^h_\epsilon := \left[\frac{\epsilon^{-\frac{2}{3}}}{h}\right]$. We set

$$l^h_\epsilon := \left[\frac{1}{\epsilon h n^h_\epsilon}\right].$$

Let $(z_k)_{k\in\{0,..,[1/\epsilon]\}} \subset \mathbb{R}^m$ be a trajectory of the discrete time singularly perturbed system $\mathbf{\Sigma}_\epsilon^\mathbf{h}$. Then we have $z_0 = z^0$ and

$$z_{(l+1)n_\epsilon^h} = z_{ln_\epsilon^h} + \epsilon \sum_{k=ln_\epsilon^h}^{(l+1)n_\epsilon^h-1} hf(z_k, y_k)$$

for $l \in \{0,..,l_\epsilon^h\}$. We define a family $(\xi_l)_{l\in\{0,..,l_\epsilon^h\}} \subset \mathbb{R}^m$ by $\xi_0 := z^0$ and

$$\xi_{l+1} := \xi_l + \epsilon \sum_{k=ln_\epsilon^h}^{(l+1)n_\epsilon^h-1} hf(\xi_l, y_k^l),$$

where $(y_k^l)_{k\in\{ln_\epsilon^h,\dots,(l+1)n_\epsilon^h-1\}} \subset \mathbb{R}^n$ denotes the trajectory of the system

$$q_{k+1} - q_k \in hG(\xi_l, q_k), \quad q_{ln_\epsilon^h} = y_{ln_\epsilon^h}, \quad k \in \{ln_\epsilon^h, \dots, (l+1)n_\epsilon^h - 1\}.$$

We define $\Delta_l := \|z_{ln_\epsilon^h} - \xi_l\|$ and obtain

$$\Delta_{l+1} \le \Delta_l + \epsilon \sum_{k=ln_\epsilon^h}^{(l+1)n_\epsilon^h-1} h(L(\Delta_l + \epsilon n_\epsilon^h hP) + L\|y_k - y_k^l\|).$$

By Assumption 1 we can choose y_k^l in a way that for $k \in \{ln_\epsilon^h, \dots, (l+1)n_\epsilon^h - 1\}$ we have

$$\|y_k - y_k^l\| \le C \max_{k\in\{ln_\epsilon^h,\dots,(l+1)n_\epsilon^h-1\}} \|z_k - \xi_l\| \le C\left(\Delta_l + \epsilon n_\epsilon^h hP\right),$$

where the constant $C \ge 0$ only depends on the one-sided Lipschitz constant $\alpha > 0$ and on the Lipschitz constant $L \ge 0$. We conclude that

$$\Delta_{l+1} \le \Delta_l(1 + \epsilon h n_\epsilon^h L(1 + C)) + (\epsilon h n_\epsilon^h)^2(LP + LCP).$$

Since $l_\epsilon^h \le 1/(\epsilon h n_\epsilon^h)$ we have

$$\Delta_l \le \epsilon h n_\epsilon^h (LP + LCP)e^{L(1+C)}$$

for all $l \in \{0,..,l_\epsilon^h\}$. We choose $v_l \in F_{\text{aver}}(\xi_l)$ such that

$$\sqrt{h}\left\| v_l - \frac{1}{n_\epsilon^h} \sum_{k=ln_\epsilon^h}^{(l+1)n_\epsilon^h-1} f(\xi_l, y_k^l)\right\| \le O\left(\frac{1}{\sqrt{n_\epsilon^h}}\right)$$

and define a family $(\eta_l)_{l\in\{0,..,l_\epsilon^h\}} \subset \mathbb{R}^m$ by $\eta_0 := z^0$ and

$$\eta_{l+1} := \eta_l + \epsilon h n_\epsilon^h v_l.$$

Then we can estimate

$$\sqrt{h}\|\eta_l - \xi_l\| \leq O\left(\frac{1}{\sqrt{n_\epsilon^h}}\right)$$

for all $l \in \{0,..,l_\epsilon^h\}$. Note that η_l is not yet a trajectory of the discrete time averaged system $\Sigma_{\text{aver}}^{\mathbf{h}}$. For this reason we finally define a trajectory $(x_k)_{k\in\{0,..,[1/\epsilon]\}} \subset \mathbb{R}^m$ of the discrete time averaged system $\Sigma_{\text{aver}}^{\mathbf{h}}$ by $x_0 = z^0$ and

$$x_{k+1} - x_k = \epsilon w_k,$$

where $w_k \in F_{\text{aver}}(x_k)$ is chosen such that for $k \in \{ln_\epsilon^h, \ldots, (l+1)n_\epsilon^h - 1\}$ we have

$$\|w_k - v_l\| \leq \Lambda\|x_k - \eta_l\|.$$

We define $D_l := \|x_{ln_\epsilon^h} - \eta_l\|$ and obtain

$$D_{l+1} \leq D_l + \epsilon h n_\epsilon^h \Lambda(\epsilon h n_\epsilon^h P + D_l)$$

which yields

$$D_l \leq \epsilon h n_\epsilon^h \Lambda P e^K$$

for all $l \in \{0,..,l_\epsilon^h\}$. Overall we have the estimation

$$\|z_k - x_k\| \leq O(\epsilon h n_\epsilon^h) + O\left(\frac{1}{\sqrt{h n_\epsilon^h}}\right)$$

and by the choice of n_ϵ^h we obtain the estimation

$$\text{dist}_{\text{H}}\left(\mathcal{S}_\epsilon^h(z^0, y^0), \mathcal{S}_{\text{aver}}^h(z^0)\right) = O(\epsilon^{\frac{1}{3}}),$$

as $\epsilon \to 0$. The proof of the converse estimation is similar and we omit it. □

4 Discretization of the Averaged System

Theorem 3. *Let Assumptions 1 and 2 be effective. For any compact sets $M^0 \subset \mathbb{R}^m$ we can estimate*

$$\text{d}_{\text{H}}\left(\mathcal{S}_{\text{aver}}^h(z^0), \mathcal{S}_{\text{aver}}(z^0)\right) = O(h), \quad \text{as} \quad h \to 0,$$

uniformly in the initial states $z^0 \in M^0$.

Proof. Let us note that the individual discrete time averages $F_K^h(z, y^0) \subset \mathbb{R}^m$ converge to the limit set $F^h(z) \subset \mathbb{R}^m$ for $0 < h < \frac{\alpha}{\beta^2}$. The construction of the limit set $F^h(z) \subset \mathbb{R}^m$ is based on exponential stability properties of the multi-valued discrete system (2). By the convexity of the sets $G(z, y) \subset \mathbb{R}^n$ the solutions to the multi-valued discrete system (2) are Euler approximations of the multi-valued differential equation (1) and the approximation is of order $O(h)$ on bounded time intervals., see for instance [1, 6, 7]. By the exponential stability properties of (2) and (1), the Euler approximation can be extended to the interval $[0,\infty)$, where it is of order $O(h)$ as well. Hence, we have the estimation

$$\mathrm{d_H}\left(F^h_{\mathrm{aver}}(z), F_{\mathrm{aver}}(z)\right) = O(h),$$

as $h \to 0$, uniformly in $z \in M^0$. On the other hand, the trajectories of the discrete time averaged system $\mathbf{\Sigma}^{\mathbf{h}}_{\mathrm{aver}}$ form Euler approximations of the solutions to the multi-valued differential equation given by $F^h_{\mathrm{aver}}(z)$ and the order of approximation is $O(h)$ as well. This observation finishes the proof.

5 Discretization of the Singularly Perturbed System

Theorem 4. *Let Assumptions 1 and 2 be effective. For any compact sets* $M^0 \subset \mathbb{R}^m$, $N^0 \subset \mathbb{R}^n$ *and any upper bound* $\epsilon^0 > 0$ *we can estimate*

$$\mathrm{d_H}\left(\mathcal{S}^h_\epsilon(z^0,y^0), \mathcal{S}_\epsilon(z^0,y^0)\right) = O(h), \quad \text{as} \quad h \to 0,$$

uniformly in the initial states $(z^0,y^0) \in M^0 \times N^0$ *and in the perturbation parameter* $\epsilon \in (0,\epsilon^0]$.

Proof. The estimation

$$\mathrm{dist_H}\left(\mathcal{S}^h_\epsilon(z^0,y^0), \mathcal{S}_\epsilon(z^0,y^0)\right) = O(h),$$

as $h \to 0$, can be deduced from Theorem 3 in [7]. The proof of the converse estimation is similar and we omit it. □

References

1. Donchev, Tz., Farkhi, E.: Stability and Euler Approximation of One-Sided Lipschitz Differential Inclusions, SIAM Journal on Control and Optimization, **36** (1998), 780–796
2. Donchev, T., Farkhi, E., Reich, S.: Fixed Set Iterations for Relaxed Lipschitz Multimaps, Nonlinear Analysis TMA, **53** (2003), 997–1015
3. Grammel, G.: Towards Fully Discretized Differential Inclusions, Set-Valued Analysis, **11** (2003), 1–8
4. Grammel, G.: Singularly Perturbed Differential Inclusions: An Averaging Approach, Set-Valued Analysis, **4** (1996), 361–374
5. Grammel, G.: Order Reduction of Multi-scale Differential Inclusions, *Numerical Analysis and Its Applications* Z. Li, L. Vulkov, J. Wasniewski (eds.), *Lecture Notes in Computer Science*, **3401**, Springer Verlag, (2005), 296–303
6. Lempio, F., Veliov, V.: Discrete Approximations of Differential Inclusions, Bayreuther Mathematische Schriften, **54** (1998), 149–232
7. Veliov, V.M.: Differential Inclusions with Stable Subinclusions, Nonlinear Analysis TMA, **23** (1994), 1027–1038

Improved Solution Enclosures for Over- and Underdetermined Interval Linear Systems*

Evgenija D. Popova

Inst. of Mathematics & Informatics, Bulgarian Academy of Sciences,
Acad. G. Bonchev, block 8, BG-1113 Sofia, Bulgaria
epopova@bio.bas.bg

Abstract. In this paper we discuss an inclusion method for solving rectangular (over- and under-determined) dense linear systems where the input data are uncertain and vary within given intervals. An improvement of the quality of the solution enclosures is described for both independent and parameter dependent input intervals. A fixed-point algorithm with result verification that exploits the structure of the problems to be solved is given. *Mathematica* functions for solving the discussed rectangular problems are developed and presented. Numerical examples illustrate the advantages of the proposed improved approach.

1 Introduction

Consider linear systems (1) where $A \in \mathbb{S}^{m\times n}$, $b \in \mathbb{S}^m$ and $\mathbb{S} \in \{\mathbb{R}, \mathbb{C}\}$,

$$A \cdot x = b. \tag{1}$$

For $m > n$, the linear system (1) is overdetermined and has no solution in general. In this case we are interested in a vector $\tilde{x} \in \mathbb{S}^m$ which minimizes the Euclidean norm $||b - A\tilde{x}||$ of the residual vector. If $m < n$ we have an underdetermined system. In general, there are infinitely many solutions and we look for a vector $\tilde{y} \in \mathbb{S}^m$ for which $A\tilde{y} = b$ and $||\tilde{y}||$ is minimal. If the rank of A is maximal, the solution for both problems is uniquely determined. It is well known [9], that if

$$m > n \text{ and } rank(A) = n \text{ then } \tilde{x} \text{ is the solution of } A^H Ax = A^H b \tag{2}$$

$$m < n \text{ and } rank(A) = m \text{ then } \tilde{y} = A^H x, \text{ where } AA^H x = b. \tag{3}$$

However, $A^H A$ and AA^H are in general ill-conditioned and furthermore not representable in the computer. In order to find guaranteed enclosures of the solutions to the above non-square problems, S. Rump [6] proposes to consider the following augmented square linear systems

$$\begin{pmatrix} A & -I \\ 0 & A^H \end{pmatrix} \cdot \begin{pmatrix} x \\ y \end{pmatrix} = \begin{pmatrix} b \\ 0 \end{pmatrix} \qquad \text{for } m > n, \tag{4}$$

$$\begin{pmatrix} A^H & -I \\ 0 & A \end{pmatrix} \cdot \begin{pmatrix} x \\ y \end{pmatrix} = \begin{pmatrix} 0 \\ b \end{pmatrix} \qquad \text{for } m < n, \tag{5}$$

* This work was supported by the Bulgarian National Science Fund under grant No. MM-1301/03.

I. Lirkov, S. Margenov, and J. Waśniewski (Eds.): LSSC 2005, LNCS 3743, pp. 305–312, 2006.

instead of solving (2), (3). It is shown in [6] as a direct consequence that Rump's fixed-point inclusion method for square systems can be used to verify solvability of the over-/underdetermined problem, the maximum rank $= \min(m, n)$ of the matrix A, and to produce guaranteed bounds for the uniquely determined solution. The method is naturally extendable to interval and complex interval matrices $[A] \in \mathbb{IS}^{m\times n}$ and vectors $[b] \in \mathbb{IS}^{m}$.

Following this approach, over-/underdetermined linear system solvers with result verification were implemented in the environments of PASCAL-XSC [2] and C-XSC [1]. For problems with interval (real or complex) input data these implementations are based on solving square interval linear systems (4), (5) assuming that all the elements vary independently in their intervals. A close look at the structure of the matrices in the systems (4), (5) shows that each element of the matrix A appears twice in the augmented square matrix which means that this matrix involves dependencies. As is well-known "the dependency problem" in interval analysis causes severe overestimation of the corresponding solution set and may lead to an interval iteration matrix involving singular ones.

In this paper we derive an algorithm for sharp enclosure of the solution set to the square linear system used as a basis for the solution of an over-/underdetermined interval (real or complex) linear problem (2), (3). The proposed algorithm improves the approach used in [1, 2] by taking into account the dependencies in the augmented matrix in order to improve the sharpness of the solution enclosure, and by using block matrix computations, related to the special structure of the matrix, to reduce the number of floating-point operations.

The following notations are used. $\mathbb{IR}, \mathbb{IC}$ are the sets of real, resp. complex intervals. We assume that the reader is familiar with basic results of interval analysis (cf. [3]). For an interval $[a] := [a^-, a^+] \in \mathbb{IR}$, $\mathrm{mid}([a]) := (a^- + a^+)/2$ is the mid point of $[a]$. By $\mathbb{IS}^n$, $\mathbb{IS}^{m\times n}$ denote the sets of interval (real or complex) n-vectors, resp. $m \times n$ matrices. For $\Sigma \subseteq \mathbb{R}^n$ the interval hull $\Diamond(\Sigma)$ is defined by $\Diamond(\Sigma) := \bigcap\{[w] \in \mathbb{IR}^n \mid \Sigma \in [w]\}$. A^H denotes the Hermitian matrix of a matrix A (i.e. the transposed matrix in the real, resp. real interval case). I is the identity matrix. An interval matrix $[A]$ has rank n if every $A \in [A]$ has rank n.

2 Theory and Algorithm

Let $[A] \in \mathbb{IS}^{m\times n}$, $[b] \in \mathbb{IS}^m$, where $\mathbb{S} \in \{\mathbb{R}, \mathbb{C}\}$, and $m \neq n$.

If $m > n$ and $\mathrm{rank}([A]) = n$, define the set

$$\begin{aligned}\Sigma_{m>n} &:= \{\tilde{x} \in \mathbb{S}^n \mid \exists\, A \in [A], \exists\, b \in [b],\ \tilde{x} = \mathrm{argmin}_{x\in\mathbb{S}^n} \|b - Ax\|\} \\ &= \left\{x \in \mathbb{S}^n \mid \exists\, A \in [A], \exists\, b \in [b],\ A^H A x = A^H b\right\}. \end{aligned} \tag{6}$$

If $m < n$ and $\mathrm{rank}([A]) = m$, define the set

$$\begin{aligned}\Sigma_{m<n} &:= \left\{\tilde{y} \in \mathbb{S}^n \mid (\exists\, A \in [A])(\exists\, b \in [b])(\tilde{y} = \mathrm{argmin}_{y\in\mathbb{S}^n}\|y\|,\ Ay = b)\right\} \\ &= \left\{y \in \mathbb{S}^n \mid (\exists\, A \in [A])(\exists\, b \in [b])(y = A^H x,\ AA^H x = b)\right\}. \end{aligned} \tag{7}$$

We are interested in finding verified enclosures of the solution sets (6), (7) whenever they are bounded, that is interval vectors $[x], [y] \in \mathbb{IS}^n$ such that

$[x] \supseteq \lozenge(\Sigma_{m>n})$ and $[y] \supseteq \lozenge(\Sigma_{m<n})$. The following theorems permit to calculate very sharp bounds to the solution sets (6), (7) and to estimate the sharpness of the calculated bounds. Define

$$(o) \qquad [B] := \begin{pmatrix} [A] & -I \\ 0 & [A]^H \end{pmatrix}, \qquad [v] := ([b], 0)^\top; \tag{8}$$

$$(u) \qquad [B] := \begin{pmatrix} [A]^H & -I \\ 0 & [A] \end{pmatrix}, \qquad [v] := (0, [b])^\top. \tag{9}$$

Theorem 1. *Let $[A] \in \mathbb{IS}^{m\times n}$, $[b] \in \mathbb{IS}^m$, $m > n$. Define $[B] \in \mathbb{IS}^{(m+n)\times(m+n)}$, $[v] \in \mathbb{IS}^{m+n}$ to be the square matrix, resp. the vector in (8) and let $R \in \mathbb{S}^{(m+n)\times(m+n)}$, $\tilde{u} \in \mathbb{S}^{m+n}$.*
1) Let $[z] \in \mathbb{IS}^{m+n}$ be defined by

$$[z_i]|_{i=1}^{m+n} := \sum_{\mu=1}^{m} r_{i\mu}([b_\mu] + \tilde{u}_{n+\mu}) - \sum_{\mu=1}^{m}\sum_{\nu=1}^{n}(r_{i\mu}\tilde{u}_\nu + r_{i,m+\nu}\tilde{u}_{n+\mu})Re([a_{\mu\nu}])$$
$$- \sum_{\mu=1}^{m}\sum_{\nu=1}^{n}(r_{i\mu}\tilde{u}_\nu - r_{i,m+\nu}\tilde{u}_{n+\mu})Im([a_{\mu\nu}]) \tag{10}$$

then $[z] = \lozenge(\{R\cdot(v - B\tilde{u}) \mid B \in [B], v \in [v]\})$.
2) Let $[C] \in \mathbb{IS}^{(m+n)\times(m+n)}$ is defined by $[C] := I - R\cdot[B]$ and let $[u] \in \mathbb{IS}^{m+n}$. Define $[w] \in \mathbb{IS}^{m+n}$ by

$$[w_i]|_{i=1}^{m+n} := \{[z] + [C]\cdot[uu]\}_i, \quad \text{where } [uu] := ([w_1], \ldots, [w_{i-1}], [u_i], \ldots, [u_{m+n}])^\top$$

If $[w] \subsetneqq [u]$, then every matrix $A \in [A]$ has a full rank n, and for every $A \in [A]$, $b \in [b]$ with $\hat{x}$ minimizing $||b - Ax||$, the unique solution $\hat{x}$ satisfies $\hat{x} \in \tilde{x} + [x]$, where $\tilde{x}$ and $[x]$ are the first n components of $\tilde{u}$, resp. $[w]$.
3) With $[\Delta] := [C]\cdot[w] \in \mathbb{IS}^{m+n}$ and $[xx] \in \mathbb{IS}^n$ defined by

$$[xx_i] := [\inf([z_i]) + \sup([\Delta_i]),\ \sup([z_i]) + \inf([\Delta_i])], \quad 1 \leqq i \leqq n,$$

the following inner and outer estimations hold true

$$\tilde{x} + [xx] \subseteqq \lozenge(\Sigma_{m>n}) \subseteqq \tilde{x} + [x].$$

Proof. 1) Let us represent R by a block structure $R = (R^1, R^2)$, where

$$R^1 = (r^1_{i\mu}) := (r_{i\mu}),\ \ R^2 = (r^2_{i\nu}) := (r_{i,m+\nu}),\ \ 1 \le \mu \le m,\ 1 \le \nu \le n, \quad 1 \le i \le m+n. \tag{11}$$

Since for arbitrary $A \in [A]$, $b \in [b]$ we have

$$z := R(v - B\cdot\tilde{u}) = R^1\cdot b - (R^1, R^2)\cdot\begin{pmatrix} A\tilde{u}^1 - \tilde{u}^2 \\ A^H\tilde{u}^2 \end{pmatrix}$$
$$= R^1\cdot b - (R^1, R^2)\cdot\begin{pmatrix} A\tilde{u}^1 - \tilde{u}^2 \\ (\tilde{u}^{2\top}\overline{A})^\top \end{pmatrix}$$

$$= R^1 \cdot b - R^1 \cdot \begin{pmatrix} \sum_{\nu=1}^{n} a_{\mu\nu}\tilde{u}^1_\nu - \tilde{u}^2_\mu \\ \vdots \\ \mu = 1,\ldots,m \end{pmatrix} - R^2 \cdot \begin{pmatrix} \sum_{\mu=1}^{m} \overline{a}_{\mu\nu}\tilde{u}^2_\mu \\ \vdots \\ \nu = 1,\ldots,n \end{pmatrix},$$

$$\begin{aligned} z_i|_{i=1}^{m+n} &= \sum_{\mu=1}^{m} r_{i\mu} b_\mu - \sum_{\mu=1}^{m} r^1_{i\mu} \sum_{\nu=1}^{n} a_{\mu\nu}\tilde{u}^1_\nu + \sum_{\mu=1}^{m} r^1_{i\mu}\tilde{u}^2_\mu - \sum_{\nu=1}^{n} r^2_{i\nu} \sum_{\mu=1}^{m} \tilde{u}^2_\mu \overline{a}_{\mu\nu} \\ &= \sum_{\mu=1}^{m} r_{i\mu}(b_\mu + \tilde{u}_{n+\mu}) - \sum_{\mu=1}^{m}\sum_{\nu=1}^{n} (r_{i\mu}\tilde{u}_\nu + r_{i,m+\nu}\tilde{u}_{n+\mu}) \mathrm{Re}(a_{\mu\nu}) \\ &\quad - \sum_{\mu=1}^{m}\sum_{\nu=1}^{n} (r_{i\mu}\tilde{u}_\nu - r_{i,m+\nu}\tilde{u}_{n+\mu}) \mathrm{Im}(a_{\mu\nu}) \end{aligned}$$

then 1) follows by a Theorem by Moore [3] because in (10) each interval variable occurs only once and to the first power. 2) follows from the fixed-point verification theory for non-square systems [6] and for parametric systems [8]. 3) is a consequence of the theory for estimating the quality of an outer enclosure [7].

Define another block structure for $R = (R^3, R^4)$, where $1 \le i \le m+n$ and

$$R^3 = (r^3_{i\nu}) := (r_{i\nu}), \quad R^4 = (r^4_{i\mu}) := (r_{i,n+\mu}), \; 1 \le \nu \le n, \; 1 \le \mu \le m. \quad (12)$$

Deriving analogously the expression which gives the exact range for the residuum

$$\begin{aligned} z := R(v - B \cdot \tilde{u}) &= R^4 \cdot b - (R^3, R^4) \cdot \begin{pmatrix} A^H \tilde{u}^1 - \tilde{u}^2 \\ A\tilde{u}^2 \end{pmatrix} \\ &= R^4 \cdot b - R^3 \cdot \begin{pmatrix} \sum_{\mu=1}^{m} \overline{a}_{\mu\nu}\tilde{u}^1_\mu - \tilde{u}^2_\nu \\ \vdots \\ \nu = 1,\ldots,n \end{pmatrix} - R^4 \cdot \begin{pmatrix} \sum_{\nu=1}^{n} \overline{a}_{\mu\nu}\tilde{u}^2_\nu \\ \vdots \\ \mu = 1,\ldots,m \end{pmatrix}, \end{aligned}$$

$$\begin{aligned} z_i|_{i=1}^{m+n} &= \sum_{\mu=1}^{m} r^4_{i\mu} b_\mu - \sum_{\nu=1}^{n} r^3_{i\nu} \sum_{\mu=1}^{m} \overline{a}_{\mu\nu}\tilde{u}^1_\mu + \sum_{\nu=1}^{n} r^3_{i\nu}\tilde{u}^2_\nu - \sum_{\mu=1}^{m} r^4_{i\mu} \sum_{\nu=1}^{n} a_{\mu\nu}\tilde{u}^2_\nu \\ &= \sum_{\mu=1}^{m} r_{i,n+\mu} b_\mu - \sum_{\mu=1}^{m}\sum_{\nu=1}^{n} (r_{i\nu}\tilde{u}_\mu + r_{i,n+\mu}\tilde{u}_{m+\nu}) \,\mathrm{Re}(a_{\mu\nu}) \\ &\quad + \sum_{\nu=1}^{n} r_{i\nu}\tilde{u}_{m+\nu} + \sum_{\mu=1}^{m}\sum_{\nu=1}^{n} (r_{i\nu}\tilde{u}_\mu - r_{i,n+\mu}\tilde{u}_{m+\nu}) \,\mathrm{Im}(a_{\mu\nu}), \end{aligned}$$

we proof the following theorem for the underdetermined case.

Theorem 2. *Let* $[A] \in \mathbb{IS}^{m\times n}$, $[b] \in \mathbb{IS}^m$, $m < n$ *and define* $[B] \in \mathbb{IS}^{(m+n)\times(m+n)}$, $[v] \in \mathbb{IS}^{m+n}$ *to be the square matrix, resp. the vector in (9) and let* $R \in \mathbb{S}^{(m+n)\times(m+n)}$, $\tilde{u} \in \mathbb{S}^{m+n}$.

1) Let $[z] \in \mathbb{IS}^{m+n}$ *be defined by*

$$[z_i]|_{i=1}^{m+n} := \sum_{\mu=1}^{m} r_{i,n+\mu} b_\mu + \sum_{\nu=1}^{n} r_{i\nu} \tilde{u}_{m+\nu} - \sum_{\mu=1}^{m}\sum_{\nu=1}^{n} (r_{i\nu}\tilde{u}_\mu + r_{i,n+\mu}\tilde{u}_{m+\nu})\, Re(a_{\mu\nu})$$
$$+ \sum_{\mu=1}^{m}\sum_{\nu=1}^{n} (r_{i\nu}\tilde{u}_\mu - r_{i,n+\mu}\tilde{u}_{m+\nu})\, Im(a_{\mu\nu}), \quad (13)$$

then $[z] = \Diamond\left(\{R \cdot (v - B\tilde{u}) \mid B \in [B], v \in [v]\}\right)$.
2) Let $[C] \in \mathbb{IS}^{(m+n)\times(m+n)}$ *is defined by* $[C] := I - R \cdot [B]$ *and let* $[u] \in \mathbb{IS}^{m+n}$. *Define* $[w] \in \mathbb{IS}^{m+n}$ *by*

$$[w_i]|_{i=1}^{m+n} := \{[z] + [C] \cdot [uu]\}_i, \text{ where } [uu] := ([w_1], \ldots, [w_{i-1}], [u_i], \ldots, [u_{m+n}])^\top$$

If $[w] \subsetneqq [u]$, *then every matrix* $A \in [A]$ *has a full rank* m, *and for every* $A \in [A]$, $b \in [b]$ *with* $A\hat{y} = b$ *and* $\hat{y} = \min ||y||$, *the unique solution* $\hat{y}$ *satisfies* $\hat{y} \in \tilde{y} + [y]$, *where* $\tilde{y}$ *and* $[y]$ *are the last* n *components of* $\tilde{u}$, *resp.* $[w]$.
3) With $[\Delta] := [C] \cdot [w] \in \mathbb{IS}^{m+n}$ *and* $[yy] \in \mathbb{IS}^n$ *defined by*

$$[yy_i] := [\inf([z_i]) + \sup([\Delta_i]),\ \sup([z_i]) + \inf([\Delta_i])], \quad m+1 \leqq i \leqq m+n,$$

the following inner and outer estimations hold true

$$\tilde{y} + [yy] \subseteqq \Diamond(\Sigma_{m<n}) \subseteqq \tilde{y} + [y].$$

Now, the following algorithm computes guaranteed enclosures of the solution sets (6), (7), for $[A] \in \mathbb{IR}^{m\times n}$, $[b] \in \mathbb{IR}^m$, $m \neq n$.

1. *Initialization.* $\check{b} := \text{mid}([b])$; $\check{A} := \text{mid}([A])$;
   Compose $\check{B}$ corresponding to (4), (5); Compute $R \approx \check{B}^{-1} \in \mathbb{R}^{(m+n)\times(m+n)}$.
   Decompose R into blocks defined by (11), (12).
2. *Compute the approximate mid-point solution* $\tilde{u} = R\check{v}$, where

   $$(o) \quad \tilde{u} = R^1\check{b}, \qquad\qquad (u) \quad \tilde{u} = R^4\check{b}.$$

3. *Enclosure for the residuum,* where for $1 \leq i \leq m+n$

   $$(o)\ \ [z_i] = \sum_{\mu=1}^{m} r_{i\mu}([b_\mu] + \tilde{u}_{n+\mu}) - \sum_{\mu=1}^{m}\sum_{\nu=1}^{n}(r_{i\mu}\tilde{u}_\nu + r_{i,m+\nu}\tilde{u}_{n+\mu})[a_{\mu\nu}];$$

   $$(u)\ \ [z_i] = \sum_{\mu=1}^{m} r_{i,n+\mu}[b_\mu] + \sum_{\nu=1}^{n} r_{i\nu}\tilde{u}_{m+\nu} - \sum_{\mu=1}^{m}\sum_{\nu=1}^{n}(r_{i\nu}\tilde{u}_\mu + r_{i,n+\mu}\tilde{u}_{m+\nu})[a_{\mu\nu}].$$

4. *Enclosure for the iteration matrix:*
   (o) $[C] = I - (R^1[A],\ R^2[A]^\top - R^1)$, (u) $[C] = I - (R^3[A]^\top,\ R^4[A] - R^3)$.
5. *Verification*
   $[uu] := [z]$;
   repeat
   $\quad [u_{\text{new}}] := [uu] := \text{blow}([uu], \varepsilon)$
   $\quad$ **for** $i = 1$ to $m+n$ **do** $[uu_i] := [z_i] + [C_i] \cdot [uu]$
   until $[uu] \subsetneqq [u_{\text{new}}]$ or max iteration exceeded

6. **If** $[uu] \subsetneqq [u_{\text{new}}]$ **then**
   a unique solution u exists and $u \in \tilde{u} + [uu]$, hence
   $\lozenge(\Sigma_{m>n}) \subseteq \tilde{x}+[x]$, where $\tilde{x}$ and $[x]$ are the first n components of $\tilde{u}$, resp. $[uu]$;
   $\lozenge(\Sigma_{m<n}) \subseteq \tilde{y}+[y]$, where $\tilde{y}$ and $[y]$ are the last n components of $\tilde{u}$, resp. $[uu]$.
   else algorithm fails, matrix $[B]$ is ill conditioned or singular.

The above algorithm can be rigorously implemented on a computer. Function blow($[uu], \varepsilon$) (for the definition see [1, 4, 2]) does the so-called ε-inflation which blows up the intervals somewhat, in order to catch the nearby fixed point. Some modifications of the above algorithm can be done in order to make it more robust in the case of almost singular matrices [2]. A cheap method for computing an estimation of the sharpness of the calculated outer bounds is presented in [4].

3 Parameter Dependence

Let us suppose now that the elements of the matrix A and the vector b depend affine-linearly on a parameter vector $p = (p_1, \ldots, p_k)^\top$, $p \in [p] \in \mathbb{IR}^k$. In this case $A(p)$ and $b(p)$ have the following factored representation

$$A(p) = A^{(0)} + A^{(1)}p_1 + \cdots + A^{(k)}p_k, \qquad A^{(j)} \in \mathbb{R}^{m\times n}, \tag{14}$$

$$b(p) = b^{(0)} + b^{(1)}p_1 + \cdots + b^{(k)}p_k, \qquad b^{(j)} \in \mathbb{R}^{m},\ j = 0, 1, \ldots, k. \tag{15}$$

For $1 \le i \le m+n$ we have the following exact ranges for the residuum

$$(o) \qquad [z_i] = \sum_{\mu=1}^{m}\left(\tilde{u}_{n+\mu} + b_\mu^{(0)} + \sum_{\nu=1}^{n}(r_{i\mu}\tilde{u}_\nu + r_{i,m+\nu}\tilde{u}_{n+\mu})a_{\mu\nu}^{(0)}\right)$$
$$+\sum_{j=1}^{k}[p_j]\left(\sum_{\mu=1}^{m} r_{i\mu}b_\mu^{(j)} - \sum_{\mu=1}^{m}\sum_{\nu=1}^{n}(r_{i\mu}\tilde{u}_\nu + r_{i,m+\nu}\tilde{u}_{n+\mu})a_{\mu\nu}^{(j)}\right),$$

$$(u) \qquad [z_i] = \sum_{\mu=1}^{m} r_{i,n+\mu}b_\mu^{(0)} + \sum_{\nu=1}^{n} r_{i\nu}\tilde{u}_{m+\nu} - \sum_{\mu=1}^{m}\sum_{\nu=1}^{n}(r_{i\nu}\tilde{u}_\mu + r_{i,n+\mu}\tilde{u}_{m+\nu})a_{\mu\nu}^{(0)}$$
$$+\sum_{j=1}^{k}\left(\sum_{\mu=1}^{m} r_{i,n+\mu}b_\mu^{(j)} - \sum_{\mu=1}^{m}\sum_{\nu=1}^{n}(r_{i\nu}\tilde{u}_\mu + r_{i,n+\mu}\tilde{u}_{m+\nu})a_{\mu\nu}^{(j)}\right)[p_j].$$

The improved enclosure of the iteration matrix C, for $m > n$ e.g., is

$$[C] = I - \left(R^1A^{(0)} + \sum_{j=1}^{k}[p_j]\left(R^1A^{(j)}\right),\ R^2A^{(0)\top} - R^1 + \sum_{j=1}^{k}[p_j]\left(R^2A^{(j)\top}\right)\right).$$

An implementation of the methods/algorithm, presented in Section 2, which exploits the above given expressions for the exact ranges of the residual and the iteration matrix will produce very sharp bounds for the solution sets to over- and underdetermined linear systems which elements depend on parameters varying within given intervals.

4 Mathematica Software, Numerical Examples

Methods and algorithms discussed in this paper are implemented in the environment of *Mathematica* [10]. The *Mathematica* package `IntervalComputations 'LinearSystems'` contains a collection of functions which compute guaranteed inclusions for the solution set of a square interval linear system [5]. The particular solvers differ upon the type of the linear system to be solved and the implemented solution method. Now, this package is extended by functions that find guaranteed enclosures to the solutions of over- and underdetermined real linear systems which do or do not involve parameter dependencies.

`ILinearSolve[A, b]` computes guaranteed bounds for the solution sets of a square interval linear system and for the solution sets (6), (7), where all elements vary independently in their intervals. The input elements can be either numerical intervals or elements from the domain `Real`.

`ParametricSSolve[Ap, bp, parLst]` computes guaranteed enclosures for the solutions of over- and underdetermined real linear systems whose elements depend on interval parameters. The parameters and their interval values should be specified by a list of rules `parLst`. *Mathematica* [10] allows convenient symbolic description of the parametric matrix and the right hand side vector.

All the iterative solvers can take options affecting the computational process and/or the output of the particular function. The three options, associated with everyone of the iterative solvers are `InnerEstimation`, `Refinement`, and `Statistics`. `InnerEstimation`, when set to `True`, specifies the computing of component-wise inner approximation of the solution set in addition to the outer enclosure. Inner estimations allow to obtain the very important measure for the degree of sharpness of an outer solution set enclosure [7]. `Refinement` set to `True` implies the application of an iterative refinement procedure for the outer solution set approximation. `Statistics` set to `True` implies output of the number of iterations and the relative improvement by the refinement procedure. The three options are set to `False` by default.

A web access to the above solvers for interval linear systems can be found at http://cose.math.bas.bg/webComputing/.

Example 1. Consider the (6×5)-matrix M_3, taken from [11–p.143], with rank= 4 and depending on a real parameter a. The following matrix $A(a)$ with full rank is obtained by deleting the fourth column in M_3. The right-hand side vector b is constructed so that for every a the exact solution of $A(a)^\top A(a)x = A(a)^\top b$ is $ex = (1, -1, 1, -1)^\top$ (see [12]), thus

$$A(a) = \begin{pmatrix} a & a+1 & a+2 & a \\ a & a+2 & a+3 & a+1 \\ a+1 & a+2 & a+3 & a+2 \\ a+2 & a+3 & a+4 & a+3 \\ a+3 & a+4 & a+5 & a+5 \\ a+5 & a+5 & a+6 & a+7 \end{pmatrix}, \quad b = \begin{pmatrix} 3 \\ 1 \\ -2 \\ -3 \\ 0 \\ 0 \end{pmatrix}, \quad ex = \begin{pmatrix} 1 \\ -1 \\ 1 \\ -1 \end{pmatrix}.$$

Matrix $A(a)$ is ill-conditioned with $\mathrm{cond}_\infty(A) \approx 2602$ for $a = 1.0$. We solve the system $[A]x = b$, where $[A] = \lozenge(\{A(a) \mid a \in [0.999, 1.001]\})$ by the improved algorithm and get the following inclusion intervals

$$\lozenge(\Sigma_{m>n}) \subseteq \tag{16}$$
$$([0.8629, 1.1371], [-1.8273, -0.1727], [0.4924, 1.5076], [-1.2742, -0.7257])^\top.$$

The linear solver for the corresponding augmented square system which does not accounts for the particular matrix structure gives a solution enclosure overestimating the enclosure (16) by 6–10%.

Example 2. We solve an overdetermined parametric system $A(a)x = b$, where $A(a)$ and b are defined in Example 1, and $a \in [0.99, 1.01]$. Our parametric solver found in 1 iteration a sharp enclosure (the numbers have 16-digits mantissa)

$$([0.9\ldots97440, 1.0\ldots02573], [-1.0\ldots016600, -0.9\ldots983394],$$
$$[0.9\ldots989643, 1.0\ldots010354], [-1.0\ldots05294, -0.9\ldots94715])^\top$$

of the exact solution $ex = (1, -1, 1, -1)^\top$ which actually does not depend on a.

The main application of the improved methods and algorithms presented in this paper is for parameter identification in approximation and interpolation when there are errors (uncertainties) in both the output and the variables.

References

1. Hölbig, C.; Krämer W.: Selfverifying Solvers for Dense Systems of Linear Equations Realized in C-XSC. Universität Wuppertal, Preprint BUGHW-WRSWT 2003/1, 2003. (`http://www.math.uni-wuppertal.de/wrswt/literatur.html`)
2. Krämer W.; Kulisch, U.; Lohner, R.: Numerical Toolbox for Verified Computing II – Advanced Numerical Problems. Universität Karlsruhe, 1994. (`http://www.uni-karlsruhe.de~Rudolf.Lohner/papers/tb2.ps.gz`)
3. Moore, R.: Methods and Applications of Interval Analysis. SIAM, Philadelphia, 1979.
4. Popova, E.D.; Krämer, W.: Parametric Fixed-Point Iteration Implemented in C-XSC, Preprint 2003/03, WRSWT, Universität Wuppertal, 2003.
5. Popova, E.: Parametric Interval Linear Solver. Numerical Algorithms, **37** (2004) 1–4, 345-356.
6. Rump, S.: Solving Algebraic Problems with High Accuracy. In: Kulisch, U. and Miranker, W. (eds.): A New Approach in Scientific Computation. Academic Press (1983) 51-120.
7. Rump, S. M.: Rigorous Sensistivity Analysis for Systems of Linear and Nonlinear Equations, Mathematics of Computation **54**, 190, (1990), 721-736.
8. Rump, S. M.: Verification Methods for Dense and Sparse Systems of Equations, in Topics in Validated Computations, J. Herzberger, ed., Elsevier Science B. V. (1994), 63-135.
9. Stewart G.H.: Introduction to Matrix Computations. Academic Press, N.Y., 1973.
10. Wolfram, S.: The Mathematica Book, Wolfram Media/Cambridge U. Press, 1999.
11. Zielke, G.: Report on test matrices for generalized inverses, Computing **36** (1986) 105-162.
12. Zielke, G.; Drygalla, V.: Genaue Lösung linearer Gleichungssyteme, Mitteilungen der GAMM **26** (2003) Heft 1/2, 7-107.

The Implicit Midpoint Rule for a Class of Convex Differential Inclusions

Nedka Pulova[1] and Vladimir Pulov[2]

[1] Department of Mathematics, Technical University-Varna
[2] Department of Physics, Technical University-Varna, Studentska str. 1, Varna 9010, Bulgaria
vpulov@ieee.bg

The ideas used here belong to Radostin Ivanov.
We wish to dedicate this work to his memory.

Abstract. A multivalued version of the implicit midpoint rule is applied to one type of convex differential inclusions. Local as well as global estimates are obtained. For numerical implementations of the method certain selection strategy algorithm is suggested and its accuracy estimated.

1 Introduction

We consider the differential inclusion

$$\frac{dx(t)}{dt} \in \text{co} \bigcup_{i=1}^{k} a_i(x(t)), \quad x(0) = x_0, \quad t \in [0,T], \tag{1}$$

where co means the convex hull and $a_i(\cdot) : R^n \to R^n$. We address the problem of approximation of this differential inclusion by discrete-time inclusions by means of a specific Runge-Kutta scheme. We apply two versions (multivalued) of the implicit midpoint rule when the functions $a_i(\cdot)$ satisfy conditions for linear growth and continuous differentiability. Our analysis is based on the representation of the solutions set of (1) by the solutions of some family of differential equations. The set-valued modification of the implicit midpont rule, known from the field of differential equations, requires a solution of certain implicit algebraic equation in order some approximate solution of (1) to be obtained. The second version of the implicit midpoint rule, suggested here, allows one to obtain a set of approximate solutions as convex combinations of the solutions of a finite number k of algebraic implicit equations, which makes the method more useful for practical purposes. Some of the estimates obtained here for either of the considered schemes are of second order, including the case of nonsmooth approximated solutions. To the authors best knowledge, approximation order better than linear for Runge-Kutta schemes is obtained for strongly convex differential inclusions [1], and when the approximated trajectory has continuous derivatives with bounded variation [2].

I. Lirkov, S. Margenov, and J. Waśniewski (Eds.): LSSC 2005, LNCS 3743, pp. 313–320, 2006.

2 Preliminary Material

Let us assume:

A1: $a_i(\cdot)$ are continuously differentiable, and
A2: $a_i(\cdot)$ are with linear growth: $\|a_i(x)\| \le \vartheta(1+\|x\|)$ for some positive ϑ, where $\|\cdot\|$ is the Euclidean norm.

The assumptions A1 and A2 imply that the solutions set $X[0,T]$ of (1) is nonempty and compact in the space of the continuous functions $C[0,T]$ and the set of all values of the solutions is compact in R^n [3].

For any sequence of k elements $(f_1, f_2, \dots, f_k)$ we use the notation $\{f_i\}_{i=1}^k$. Denote $\mathcal{A} = \left\{ \{\alpha_i\}_{i=1}^k \,\middle|\, \alpha_i \in R,\ \sum_{i=1}^k \alpha_i = 1,\ \alpha_i \ge 0 \right\}$. The set of all measurable (single-valued) selections of the mapping $t \mapsto \mathcal{A}$, defined on $[0,T]$ is given by $\mathcal{M}[0,T] = \left\{ \{\mu_i(\cdot)\}_{i=1}^k \,\middle|\, \mu_i(\cdot) \in L_1[0,T],\ \sum_{i=1}^k \mu_i(t) = 1,\ \mu_i(t) \ge 0 \right\}$. It follows from the lemma of Filippov in [4] that, for a given solution $x(t)$ of the differential inclusion (1), there exists a measurable selection $\{\mu_i(\cdot)\}_{i=1}^k \in \mathcal{M}[0,T]$, such that

$$\frac{dx(t)}{dt} = \sum_{i=1}^k \mu_i(t) a_i(x(t)) \quad \text{for almost all} \quad t \in [0,T], \quad x(0) = x_0. \tag{2}$$

Conversely, thanks to the conditions A1 and A2, for a given sequence $\{\mu_i(\cdot)\}_{i=1}^k$ in $\mathcal{M}[0,T]$, (2) has an unique solution, which is a solution of the differential inclusion (1). We come to the conclusion that $X[0,T]$ can be represented by the solutions of the differential equations (2), defined by the elements of $\mathcal{M}[0,T]$, namely it holds

$$X[0,T] = \left\{ x(\cdot) \middle| x(t) = x_0 + \int_0^t \sum_{i=1}^k \mu_i(s) a_i(x(s)) ds \ \text{ on } [0,T], \right.$$
$$\left. \{\mu_i(\cdot)\}_{i=1}^k \in \mathcal{M}[0,T] \right\}. \tag{3}$$

The reachable set of (1) at $t \in [0,T]$ is defined by $R(t) = \left\{ x(t) \middle| \, x(\cdot) \in X[0,T] \right\}$.

Implicit Midpoint Rule. For some $h = \frac{T}{m}$ (m is a given natural number) and solutions set $Z(h, x_{jh})$ of the implicit inclusion $z \in \left\{ \sum_{i=1}^k \alpha_i a_i(x_{jh} + \frac{h}{2} z) \middle| \{\alpha_i\}_{i=1}^k \in \mathcal{A} \right\}$ calculate the vector $x_{(j+1)h} \in \left\{ x_{jh} + hz \middle| \, z \in Z(h, x_{jh}) \right\}$.

Convex Combinations Selection Scheme. Select any z that belongs to the set $\bar{Z}(h, x_{jh}) \equiv \left\{ \sum_{i=1}^k \alpha_i z_i \middle| \{\alpha_i\}_{i=1}^k \in \mathcal{A} \right\}$, where z_i is a solution of the equation $z_i = a_i(x_{jh} + \frac{h}{2} z_i)$ and compute $x_{(j+1)h} = x_{jh} + hz$.

Provided h is sufficiently small, $Z(h, x_{jh})$ and $\bar{Z}(h, x_{jh})$ are not empty [5]. The approximate solution $(x_0, x_h, \ldots, x_{mh})$ is denoted, either by $\mathcal{X}_h$, or by $\bar{\mathcal{X}}_h$, depending on whether it is associated with the implicit midpoit rule, or the convex combinations selection scheme, respectively. The corresponding approximate reachable sets are defined by $R_{jh} = \{x | x = x_{jh} \text{ for some } \mathcal{X}_h\}$ and $\bar{R}_{jh} = \{x | x = x_{jh} \text{ for some } \bar{\mathcal{X}}_h\}$. Obviously, $\bar{R}_{jh} \subset \text{co}\, R_{jh}$. By induction it can be proven that the set $\{x | x = x_{jh} \text{ for some } \mathcal{X}_h \bigcup \bar{\mathcal{X}}_h\}$ belongs to some closed ball $\widehat{B} \subset R^n$ together with all exact solutions in $X[0,T]$. Via the assumption A1 we take advantage of the obvious Lipschitz continuity of $a_i(\cdot)$ and the Jacoby matrixes $a_i'(\cdot)$ on $\widehat{B}$ denoting their Lipschitz constant by M. $H(A, B)$ denotes the excess of A from B and haus(A, B) is used for the Hausdorff distance between A and B, where $A, B \subset R^n$.

3 Main Results

Theorem 1 (local estimation). *Consider the problem (1), its reachable set $R(h)$, and the approximate reachable set R_h obtained by the implicit midpoint rule. If the conditions A1 and A2 are assumed, then for all sufficiently small h the following estimates hold true:*

1. *For any x_h in R_h, there exists $x(h)$ in $R(h)$ such that $\|x(h) - x_h\| = \mathcal{O}(h^3)$, where $\mathcal{O}(h^3)$ does not depend on x_h;*
2. *For any $x(\cdot) \in X[0,T]$, there exists $x_h \in R_h$ such that $\|x(h) - x_h\| = \mathcal{O}(h^2)$, where $\mathcal{O}(h^2)$ does not depend on $x(\cdot)$;*

Proof. Let us consider the difference

$$x(h) - x_h = \int_0^h \sum_{i=1}^k \left[\mu_i(s) a_i(x(s)) - \alpha_i a_i(x_0 + \frac{h}{2} z(h, x_0)) \right] ds,$$

(see the representation (3) of $x(\cdot)$) and apply the Taylor's formula to the functions $a_i(\cdot)$

$$x(h) - x_h = \int_0^h \sum_{i=1}^k (\mu_i(s) - \alpha_i) a_i(x_0) ds +$$

$$+ \int_0^h \sum_{i=1}^k \mu_i(s) a_i'(x_0)(x(s) - x_0) - \int_0^h \sum_{i=1}^k \alpha_i a_i'(x_0) s z(h, x_0) ds + \mathcal{O}(h^3). \quad (4)$$

The term $\mathcal{O}(h^3)$ depends on the Lipschitz constant M and the radius of the ball $\widehat{B}$. Interpolating by $x_s = x_0 + s z(h, x_0)$ we can write

$$x(h) - x_h = \int_0^h \sum_{i=1}^{k} (\mu_i(s) - \alpha_i) a_i(x_0) ds +$$

$$\int_0^h \sum_{i=1}^{k} (\mu_i(s) - \alpha_i) a_i'(x_0) s z(h, x_0) ds + \int_0^h \sum_{i=1}^{k} \mu_i(s) a_i'(x_0)(x(s) - x_s) ds + \mathcal{O}(h^3).$$

Finally, we arrive at

$$\|x(h) - x_h\| \le \left\| \sum_{i=1}^{k} a_i(x_0) \int_0^h (\mu_i(s) - \alpha_i)\, ds \right\| +$$

$$\|z(h, x_0)\| \sum_{i=1}^{k} \|a_i'(x_0)\| \left| \int_0^h (\mu_i(s) - \alpha_i) s ds \right| + M_0 \int_0^h \|x(s) - x_s\|\, ds + \mathcal{O}(h^3);$$

$M_0 = \sum_{i=1}^{k} \|a_i'(x_0)\|$. Applying the Gronwall's inequality we obtain

$$\|x(h) - x_h\| \le c_1(h) + c_2(h) + \mathcal{O}(h^3) + M_0 e^{M_0 h} \int_0^h (c_1(h) + c_2(h) + \mathcal{O}(h^3))\, ds,$$

where

$$c_1(h) = \left\| \sum_{i=1}^{k} a_i(x_0) \int_0^h (\mu_i(s) - \alpha_i)\, ds \right\|$$

$$c_2(h) = \|z(h, x_0)\| \sum_{i=1}^{k} \|a_i'(x_0)\| \left| \int_0^h (\mu_i(s) - \alpha_i) s ds \right| .$$

1. Let us fix an element in R_h associated with the sequence $\{\alpha_i\}_{i=1}^k$ in $\mathcal{A}$. Hence, by the sequence of functions $\{\mu_i(\cdot)\}_{i=1}^k$ in $\mathcal{M}[0, T]$ defined on $[0, h]$ by the substitutions $\mu_i(s) = \alpha_i$, we choose a point $x(h)$ in $R(h)$, for which the last inequality implies the first estimate.

2. Let us fix a sequence $\{\mu_i(\cdot)\}_{i=1}^k$ in $\mathcal{M}[0, T]$, that is equivalent to choose a solution $x(\cdot)$ in $X[0, T]$. Set

$$\alpha_i(h) = \frac{1}{h} \int_0^h \mu_i(s)\, ds, \quad i = 1, 2, \cdots, k. \tag{5}$$

It is easy to check that

$$\left| \int_0^h (s - \frac{h}{2}) \mu_i(s) ds \right| \le \frac{h^2}{8}, \tag{6}$$

As a result we obtain $c_1(h) = 0$ and $c_2(h) = \mathcal{O}(h^2)$. For the latter estimate we make use of the boundedness of $\|z(h, x_0)\|$.

The local error estimation can be apparently expressed in terms of the Hausdorff distance.

Corollary. *On the assumptions A1 and A2 for all sufficiently small h it holds* $\mathrm{haus}(R_h, R(h)) = \mathcal{O}(h^2)$.

Remark. *For continuous* $\{\mu_i(\cdot)\}_{i=1}^k$, *by applying the L'Hopital rule, we have* $\lim_{h\to 0} \frac{1}{h^2} \int_0^h (s-\frac{h}{2})\mu_i(s)\,ds = 0$, *thereby* $c_2(h) = o(h^2)$. *As a result we obtain: for any continuously differentiable* $x(\cdot) \in X[0,T]$, *produced by continuous selections* $\{\mu_i(\cdot)\}_{i=1}^k$, *there exists* $x_h \in R_h$, *such that* $\|x(h) - x_h\| = o(h^2)$, *where* $o(h^2)$ *depends on* $x(\cdot)$.

The next example shows that the estimate of the Hausdorff distance between R_h and $R(h)$ can not be better than $\mathcal{O}(h^2)$.

Example. Consider the differential inclusion

$$\frac{dx_1(t)}{dt} = x_2(t), \quad \frac{dx_2(t)}{dt} \in [-1, 1], \quad x(0) = 0, \quad t \in [0, T] \tag{7}$$

and the family of its solutions

$$x_1(\tau;\, t) = \frac{t^2}{2}, \; x_2(\tau;\, t) = t, \; t \in [0, \tau]$$

$$x_1(\tau;\, t) = -\frac{t^2}{2} + 2\tau t - \tau^2, \; x_2(\tau;\, t) = -t + 2\tau, \; t \in [\tau, T], \; \tau \in [0, T].$$

It is easily obtained that $R_h = \left\{(x_1, \frac{2}{h}x_1) \,\middle|\, -\frac{h^2}{2} \le x_1 \le \frac{h^2}{2}\right\}$. Let us introduce the function $\rho(\tau;\, h) = \frac{1}{h^2} d(x(\tau; h), R_h)$. We have $\rho(\tau;\, h) = 0$ for $h \in [0, \tau]$ and $\rho(\tau; h) = \frac{2\tau(h-\tau)}{h^2\sqrt{h^2+4}}$ for $h \in [\tau, T]$.

It is obvious, in fact, that for a given solution from the family (for each $\tau \in [0, T]$) we get $\lim_{h\to 0} \rho(\tau; h) = 0$, i.e. we obtain $d(x(\tau; h), R_h) = o(h^2)$. But, we claim that this latter estimate is, actually, not uniform with respect to the parameter τ. Indeed, take $h \in (0, T]$ and note the limit $\rho(\frac{h}{2^n}; \frac{h}{2^{n-1}}) \to \frac{1}{4}$ for $n \to \infty$, implying: for each ϵ in the interval $(0, \frac{1}{4})$ there exists a positive number $N(\epsilon)$, such that for all integer $n > N(\epsilon)$ it holds $\rho(\frac{h}{2^n}; \frac{h}{2^{n-1}}) > \epsilon$. This proves our claim.

As it is shown in [6], the reachable set of (7) is

$$R(h) = \left\{(x_1, x_2) \,\middle|\, \frac{(x_2 + h)^2}{4} - \frac{h^2}{2} \le x_1 \le \frac{h^2}{2} - \frac{(x_2 - h)^2}{4}, \; |x_2| \le h\right\}.$$

Hence, for the excess from R_h to R(h) we have

$$H(R(h), R_h) = \sup_{(x_1,x_2)\in R(h)} \frac{h\,|\frac{2}{h}x_1 - x_2|}{\sqrt{h^2+4}} = \sup_{x_2\in[-h,h]} \frac{|\pm x_2^2 \mp h^2|}{2\sqrt{h^2+4}},$$

and thus $H(R(h), R_h) = \frac{h^2}{2\sqrt{h^2+4}}$, implying the estimate of $H(R(h), R_h)$ is $\mathcal{O}(h^2)$ but not $o(h^2)$. Taking into account $R_h \subset R(h)$, we come to the conclusion that $\mathrm{haus}(R_h, R(h))$ is not better than $\mathcal{O}(h^2)$.

Theorem 2 (global estimation). *Consider the problem (1) and the implicit midpoint rule with an uniform stepsize $h = \frac{T}{m}$. Assume A1 and A2. Then for all sufficiently small h, the following estimates hold true:*

1. *For every approximate solution $\mathcal{X}_h$, there exists a solution $x(\cdot)$ of the differential inclusion (1) such that $\max\limits_{0\le j\le m} \|x(jh) - x_{jh}\| = \mathcal{O}(h^2)$, where $\mathcal{O}(h^2)$ does not depend on $\mathcal{X}_h$;*
2. *Given a solution $x(\cdot)$ of (1), there exists an approximate solution $\mathcal{X}_h$ such that $\max\limits_{0\le j\le m} \|x(jh) - x_{jh}\| = \mathcal{O}(h)$, where $\mathcal{O}(h)$ does not depends on $x(\cdot)$.*

The proof is obtained in a standard way by using the previous result.

Theorem 3 (convex combinations selection scheme). *Consider the problem (1), its reachable set $R(h)$, and the approximate reachable set $\bar{R}_h$ obtained by the convex combinations selection scheme. If the conditions A1 and A2 are assumed, then for all sufficiently small h the following estimates hold true:*

1. *For any $x(\cdot) \in X[0,T]$, there exists $x_h \in \bar{R}_h$ such that $\|x(h) - x_h\| = \mathcal{O}(h^2)$, where $\mathcal{O}(h^2)$ does not depend on $x(\cdot)$;*
2. *For any x_h in $\bar{R}_h$, there exists $x(h) \in R(h)$ such that $\|x(h) - x_h\| = \mathcal{O}(h^2)$, where $\mathcal{O}(h^2)$ does not depend on x_h;*
3. *Consider, either $n = 1$, or $n > 1$ with $a_i(\cdot)$ of the form*

$$a_i(x) = a_k(x) + A_i x + P_i, \quad i = 1, \dots, k-1, \tag{8}$$

where $A_i \in R$, $P_i \in R^n$ are arbitrary constants. Then for any x_h in $\bar{R}_h$, there exists a $x(h)$ in $R(h)$ such that $\|x(h) - x_h\| = \mathcal{O}(h^3)$, where $\mathcal{O}(h^3)$ does not depend on x_h.

Proof. The proofs of the estimates 1 and 2 are obtained by using (4), where z is replaced by z_i and by choosing $\{\mu_i(\cdot)\}_{i=1}^k$ and $\{\alpha_i\}_{i=1}^k$ as it is done in the respective parts 1. and 2. in the Theorem 1.

3. Let us fix a $x_h \in \bar{R}_h$ defined by $\{\alpha_i\}_{i=1}^k \in \mathcal{A}$ and consider the difference

$$x(h) - x_h = \int_0^h \sum_{i=1}^k \left[\mu_i(s) a_i(x(s)) - \alpha_i a_i(x_0 + \frac{h}{2} z_i)\right] ds,$$

where $\{\mu_i(\cdot)\}_{i=1}^k \in \mathcal{M}[0,T]$ are twice continuously differentiable and $x_h \in \bar{R}_h$. Applying the Taylor's formula we obtain

$$x(h) - x_h = h \sum_{i=1}^k (\mu_i(0) - \alpha_i) a_i(x_0) + \frac{h^2}{2} \sum_{i=1}^k \left[\frac{d\mu_i(0)}{dt} a_i(x_0) + \right.$$

$$\left. + \mu_i(0) a_i'(x_0) \frac{dx(0)}{dt} - \alpha_i a_i'(x_0) z_i \right] + \mathcal{O}(h^3).$$

Then, taking into account that $z_i = a_i(x_0 + \frac{h}{2} z_i) = a_i(x_0) + \mathcal{O}(h)$ and

$$\frac{dx(0)}{dt} = \sum_{i=1}^{k} \mu_i(0) a_i(x_0)$$

the last expression is transformed to

$$x(h) - x_h = h \sum_{i=1}^{k} (\mu_i(0) - \alpha_i) a_i(x_0) + \frac{h^2}{2} \sum_{i=1}^{k} \Bigg[\frac{d\mu_i(0)}{dt} E + \mu_i(0) \sum_{j=1}^{k} \mu_j(0) a_j'(x_0) - \alpha_i a_i'(x_0) \Bigg] a_i(x_0) + \mathcal{O}(h^3), \tag{9}$$

where E is $n \times n$ identity matrix. Then we find a sequence of functions $\{\mu_i(\cdot)\}_{i=1}^k \in \mathcal{M}[0,T]$ that satisfy on $[0,h]$ the equation

$$\sum_{i=1}^{k} \left(\frac{d\mu_i(s)}{dt} E + \mu_i(s) \sum_{j=1}^{k} \mu_j(s) a_j'(x_0) - \mu_i(s) a_i'(x_0) \right) a_i(x_0) = 0 \tag{10}$$

with the initial conditions $\mu_i(0) = \alpha_i$, $i = 1, 2, \ldots, k$. Imposing each one of the terms in (10) to be equal to zero, after some equivalent transformations we obtain the system

$$\left(\frac{1}{\mu_i(s)} \frac{d\mu_i(s)}{dt} - \frac{1}{\mu_k(s)} \frac{d\mu_k(s)}{dt} \right) E = a_i'(x_0) - a_k'(x_0),$$
$$\sum_{j=1}^{k} \mu_j(s) = 1, \; i = 1, 2 \ldots, k-1 \tag{11}$$

that must be satisfied by the unknown functions $\mu_i(s)$ on the interval $[0,h]$, under the initial conditions $\mu_i(0) = \alpha_i$, $i = 1, 2, \ldots, k$.

The system (11) has the solution:

- for the one-dimensional case $n = 1$

$$\mu_i(s) = \frac{\alpha_i}{\alpha_k} \mu_k(s) e^{(a_i'(x_0) - a_k'(x_0))s} \;, i = 1, 2, \ldots, k-1$$
$$\mu_k(s) = \left(1 + \sum_{i=1}^{k-1} \frac{\alpha_i}{\alpha_k} e^{(a_i'(x_0) - a_k'(x_0))s} \right)^{-1}, \quad s \in [0,h]; \tag{12}$$

- for $n > 1$ with $a_i(\cdot)$ given by (8):

$$\mu_i(s) = \frac{\alpha_i}{\alpha_k} \mu_k(s) e^{A_i(s-t)}, \quad i = 1, 2, \ldots, k-1$$
$$\mu_k(s) = \left(1 + \sum_{i=1}^{k-1} \frac{\alpha_i}{\alpha_k} e^{A_i(s-t)} \right)^{-1}, \quad s \in [0,h]. \tag{13}$$

Finally, for any $x_h \in \bar{R}_h$ we deduce from (9): each sequence $\{\mu_i(\cdot)\}_{i=1}^k$ in $\mathcal{M}[0,T]$, defined on $[0,h]$ either by (12) or by (13), determines a point $x(h)$ in $R(h)$, which satisfies the desired estimate for the respective considered case.

References

1. Veliov, V.: Second order discrete approximations to stronly convex differential inclusions, Systems and Control Letters. **13** (1989), 263-269.
2. Lempio, F., Veliov V.: Discrete approximations of differential inclusions, Bayreuter Mathematische Schriften **54** (1998), 149-232.
3. Clarke, F.H. et al.: Nonsmoth Analysis and Control Theory, Springer-Verlag, Berlin-Heidelberg-New York,1988
4. Filippov, A.: On some problems of optimal control theory, Vestnik Moskowskogo Universiteta, Math. no.**2** (1959), 25-32 (in Russian)
5. Hairer E., Norsett S. P. and Wanner G.: Solving Ordinary Differential Equations I (nonstiff problems), Springer-Verlag, Berlin-Heidelberg-New York
6. Pontryagin, L. S. et al.: The Mathematical Theory of Optimal Processes, transl. Wiley Interscience, New York, 1962

Synchronization of Chaotic Systems with Diagonal Coupling

Andrzej Stefański, Jerzy Wojewoda, and Tomasz Kapitaniak

Technical University of Lodz, 90-924 Lodz, ul. Stefanowskiego 1/15, Poland
steve@p.lodz.pl

Abstract. In this paper we define a simple criterion of the synchronization threshold in the set of coupled chaotic systems (flows or maps) with diagonal diffusive coupling. The condition of chaotic synchronization is determined only by two "parameters of order", i.e. the largest Lyapunov exponent and the coupling coefficient. Our approach can be applied for both regular chaotic networks and arrays or lattices of chaotic oscillators with irregular, arbitrarily assumed structure of coupling.

1 Introduction

Chaotic synchronization in networks of coupled dynamical systems has been intensively investigated in recent years. It has been demonstrated that two or more chaotic systems can synchronize by linking them with mutual coupling or with a common signal or signals (see e.g. [1, 3]). The first analytical condition for complete synchronization of the sets of symmetrically coupled identical continuous-time dynamic systems has been formulated in [3]. The complete synchronization [7] takes place when all trajectories converge to the same value and remain in step during further evolution in phase space. Next this condition has been developed for discrete-time systems [10]. Many approaches have been applied for describing the synchronization problem for particular coupling configurations as well as for more general cases [2, 4, 5, 8, 9, 10, 11, 12]. Most of the existing works on networks synchronization refer to regular, symmetrical structure of coupling. However, nonsymmetrical and random coupling configuration have been also considered in some papers [2, 4, 5, 8, 9, 11, 12]. Especially noteworthy is a concept called *Master Stability Function* (MSF) introduced by Pecora & Carroll [8, 9] which allows to solve the networks synchronization problem for any set of coupling weights and connections and any number of coupled oscillators. Other interesting solutions are the applications of graph theory to configurations of oscillators [12] and the concept of the so-called small-world networks [11] which connect the properties of regular and random networks.

In this paper, we present how to exploit the properties of diagonal diffusive coupling for the estimation of network synchronization threshold. Our approach can be successfully applied both for flows and for maps with arbitrarily assumed structure of coupling.

I. Lirkov, S. Margenov, and J. Waśniewski (Eds.): LSSC 2005, LNCS 3743, pp. 321–328, 2006.

2 Stability of Synchronization

Consider a set of n identical dynamical systems with diagonal diffusive coupling of arbitrary configuration between the oscillators. The equations of motion for the system are

$$\dot{\mathbf{x}}_i = \mathbf{f}(\mathbf{x}_i) + \sum_{j=1}^{n} \mathbf{D}_{ij}(\mathbf{x}_j - \mathbf{x}_i) \tag{1}$$

where $\mathbf{x}_i \in \mathbf{R^k}$ ($k \in \mathrm{N} \geq \mathbf{3}$), $\mathbf{f}(\mathbf{x}_i)$ is a function which governs the dynamics of each individual oscillator and $\mathbf{D}_{ij}$=diag$[d_{ij}, d_{ij}, \ldots, d_{ij}] \in \mathbf{R^k}$ are diagonal coupling matrices defining rates of coupling between each pair of the subsystems in network $(i, j = 1, 2, \ldots, n)$. For $\mathbf{D}_{ij}$=$\mathbf{0}$ (absence of coupling) each of subsystems given by Eq.(1) evolves on the asymptotically stable chaotic attractor $\boldsymbol{A}$. Since these oscillators are identical, it can be assumed that the solutions of equation $\dot{\mathbf{x}}_i = \mathbf{f}(\mathbf{x}_i)$ starting from different initial points of the same basin of attraction, represent the set of n uncorrelated trajectories evolving on the attractor $\boldsymbol{A}$ (after a period of transient motion). Let us introduce a new variable

$$\mathbf{x}_{ij} = \mathbf{x}_i - \mathbf{x}_j \tag{2}$$

representing the *trajectories separation* between any pair of oscillators. Complete synchronization of all subsystems requires a fulfillment of the expression

$$\lim_{t\to\infty} \|\mathbf{x}_{ij}(t)\| = 0, \; \forall i, j. \tag{3}$$

As it results from the definition of Lyapunov exponents, the average distance between nearby trajectories diverges with the rate determined by LLE. On the other hand, non-zero diffusive coupling causes mutual convergence of these trajectories. In several prior works [4,11,14,16] it has been confirmed that diffusive interaction of identical strange attractors leads to the direct dependence between LLE and coupling coefficient, which can be used for the estimation of synchronization threshold. In our analysis we have assumed that the analogous effect occurs in the system under consideration (Eq.(1)). According to this approach, for sufficiently small initial *trajectory separation* distance $\mathbf{x}_{ij}(0)$ (where linear effects are dominant) synchronization process is a product of two independent factors:

(i) exponential divergence of nearby trajectories with mean rate being proportional to the positive LLE,
(ii) exponential convergence caused by introduced diffusive coupling with a rate being proportional to effective coupling.

An exact determination of synchronization condition can be done analytically only in some simple cases of coupling configurations, e.g. symmetrical or global coupling. More complex structure of network requires an application of advanced mathematical and numerical techniques [2, 4, 8, 9, 12]. As follows from the Master Stability Function approach [8, 9], in the case of diagonal coupling, only these two parameters of order are important for the complete synchronization. Thus, we can substitute the node by any other system characterized by the same value of LLE without the influence on the process of network synchronization and the level of synchronization threshold. This property can be used to simplify the mathematical description of complete synchronization of chaotic networks. Namely, we can reduce the system under consideration (Eq.(1)) to the linear case with $\mathrm{x}_i \in \mathbf{R}^1$ and determine the synchronization threshold on the basis of linear stability analysis of the simplified system. In order to preserve necessary properties, two conditions have to be fulfilled in the simplified system:

(i) the substituted system in $\mathbf{R}^1$ is characterized by the same value of LLE as the original one,
(ii) original and simplified systems have identical configuration of coupling.

The presented approach can be applied for continuous-time systems as well as for discrete-time systems.

2.1 Continuous-Time Systems

In order to construct a linear model of the system (1) with one-dimensional nodes we use the substitutions:

$$\mathbf{f}(\mathbf{x}) = \lambda_1 x, \tag{4}$$

$$\mathbf{D}_{ij} = d_{ij}. \tag{5}$$

Substituting (4) and (5) into (1) we obtain a model for the network of one-dimensional systems

$$\dot{x}_i = \lambda_1 x_i + \sum_{j=1}^{n} d_{ij}(x_j - x_i). \tag{6}$$

This simplified model can be rewritten in the vector form:

$$\dot{\mathbf{X}} = \lambda_1 \mathbf{X} + \mathbf{G}\mathbf{X}, \tag{7}$$

where $\mathbf{X}=[x_1,x_2, \ldots ,x_n]^T$, and

$$\mathbf{G} = \begin{bmatrix} -\sum\limits_{j=1}^{n} d_{1j} \; d_{12} \; \cdots & & d_{1n} \\ d_{21} & \ddots & \vdots \\ \vdots & \ddots & d_{n-1n} \\ d_{n1} & \cdots \; d_{nn-1} & -\sum\limits_{j=1}^{n} d_{nj} \end{bmatrix} \tag{8}$$

is the coupling configuration matrix (note that d_{ii}=0).

Let us now introduce the *trajectories separation* between arbitrarily chosen base subsystem and any other j-th oscillator of the network. If we mark the base subsystem by subscript "1" we obtain

$$\begin{aligned} x_{1j} &= x_1 - x_j , \\ x_j - x_r &= x_{1r} - x_{1j} , \end{aligned} \tag{9}$$

where $(j, r = 2, 3, ..., n)$. Subtracting the remaining subsystems from the base node and applying the introduced substitutions (9) we can rewrite the simplified system in $(n-1)$-dimensional form where *trajectories separation* variables are given clearly:

$$\dot{\mathbf{Y}} = \mathbf{SY}, \tag{10}$$

where $\mathbf{Y} = [x_{12}, x_{13}, \ldots, x_{1n}]^T \in \mathbf{R^{n-1}}$ and $(n-1)\times(n-1)$ matrix $\mathbf{S}$ assumes the form

$$\mathbf{S} = \begin{bmatrix} \lambda_1 - \left(d_{12} + \sum_{j=1}^{n} d_{2j} \right) & \cdots \; d_{2k} - d_{1k} & \cdots \; d_{2n} - d_{1n} \\ \vdots & \ddots \vdots & \iddots \vdots \\ d_{i2} - d_{12} & \cdots \; \lambda_1 - \left(d_{1k} + \sum_{j=1}^{n} d_{ij} \right) & \cdots \; d_{in} - d_{1n} \\ \vdots & \iddots \vdots & \ddots \vdots \\ d_{n2} - d_{12} & \cdots \; d_{nk} - d_{1k} & \cdots \; \lambda_1 - \left(d_{1n} + \sum_{j=1}^{n} d_{nj} \right) \end{bmatrix}, \tag{11}$$

where indices i and k enumerate rows and columns, respectively. System (9) now incorporates only transverse dynamics to the synchronization hyperplane. Therefore complete synchronization of all subsystems of the system (6) takes place if the critical point of *trajectories separation* $\mathbf{Y}=\mathbf{0}$ is a stable attractor. Such situation occurs if real parts of all eigenvalues of the matrix (10) are negative. Thus, in agreement with the above assumptions, we can formulate the synchronization condition for general case of network of chaotic time-continuous systems (Eq.(1)) in the following form:

$$Re(s_i) < 0, \tag{12}$$

where s_i (i=1, 2, ... , $n-1$) are eigenvalues of matrix $\mathbf{S}$, which we named *Synchronization Stability Matrix* (SSM) due to its universal character, i.e. the form of SSM depends only on the network coupling configuration and LLE of the dynamical system considered as a network node. The SSM can be constructed directly from the coupling matrix (8) according to the model formula given by (10). In general case we can choose any node of the network as the base to define the SSM, because it is of no significance for the results of synchronization stability analysis.

2.2 Discrete-Time Systems

The system analogous to Eq.(1) but consisting of n diffusively coupled identical maps is described as follows:

$$\mathbf{x}_i(m+1) = \mathbf{f}\left(\mathbf{x}_i(m)\right) + \sum_{j=1}^{n} d_{ij}\mathbf{I}_k \left[\mathbf{f}\left(\mathbf{x}_j(m)\right) - \mathbf{f}\left(\mathbf{x}_i(m)\right)\right], \qquad (13)$$

where $\mathbf{x}_i(m) \in \mathbf{R^k}$, $k \in \mathbf{N} \geq \mathbf{1})$ and $\mathbf{I}_k$ represents $k \times k$ unit matrix. We obtain the simplified version of Eq.(13) applying the simplest discrete-time system

$$x(m+1) = \exp(\lambda_1)x(m), \qquad (14)$$

which fulfills the first condition of the system simplification, i.e. LLE of the map given by Eq.(13) is equal to λ_1. Using Eqs(5) and (13), the system under consideration (Eq.(12)) is reduced to the form analogous to Eq.(6) but described by the following set of difference equations:

$$x_i(m+1) = \exp(\lambda_1)x_i(m) + \sum_{j=1}^{n} d_{ij}\left[\exp(\lambda_1)x_j(m) - \exp(\lambda_1)x_i(m)\right], \qquad (15)$$

or in the vector form

$$\mathbf{X}_{m+1} = \exp(\lambda_1)\left[\mathbf{X}_m + G\mathbf{X}_m\right]. \qquad (16)$$

Substituting Eq.(9) into Eq.(14) and proceeding in way shown in section 2.1 we formulate the difference equations of *trajectories separation* evolution:

$$\mathbf{Y}(m+1) = \mathbf{M}\mathbf{Y}(m), \qquad (17)$$

and a version of SSM for maps:

$$\mathbf{M} = \exp(\lambda_1)\begin{bmatrix} 1-\left(d_{12}+\sum\limits_{k=1}^{n} d_{2k}\right) & \cdots\ d_{2j}-d_{1j} & \cdots\ d_{2n}-d_{1n} \\ \vdots & \ddots & \vdots \\ d_{i2}-d_{12} & \cdots\ 1-\left(d_{1j}+\sum\limits_{k=1}^{n} d_{ik}\right) & \cdots\ d_{in}-d_{1n} \\ \vdots & \vdots & \ddots \\ d_{n2}-d_{12} & \cdots\ d_{nj}-d_{1j} & \cdots\ 1-\left(d_{1n}+\sum\limits_{k=1}^{n} d_{nk}\right) \end{bmatrix}. \qquad (18)$$

Hence, the synchronization threshold for the ensembles of chaotic maps with regular or random configuration of coupling is defined by the inequality

$$|\mu_i| < 1,\ \forall i, \qquad (19)$$

where μ_i (i=1, 2, ... , n−1) are eigenvalues of SSM (Eq.(17)).

3 Numerical Example

In numerical simulations the examples of classical dynamical systems (Rössler system and Henon map) have been applied as the network nodes. Table 1 presents form of detailed equations which describe these examples with their corresponding LLE's. The network consists of four randomly coupled chaotic oscillators according to the scheme shown in Fig.1. The corresponding coupling configuration matrix has the following form

$$\mathbf{G} = d\begin{bmatrix} -3 & 1 & 2 & 0 \\ 2 & -2 & 0 & 0 \\ 1 & 0 & -1 & 0 \\ 3 & 2 & 0 & -5 \end{bmatrix} \tag{20}$$

Irregular coupling configuration causes non-symmetrical, random structure of SSM's:

$$\mathbf{S} = \begin{bmatrix} \lambda_1 - 3d & -2d & 0 \\ -d & \lambda_1 - 3d & 0 \\ 0 & -2d & \lambda_1 - 4d \end{bmatrix} \tag{21}$$

and

$$\mathbf{M} = \exp(\lambda_1)\begin{bmatrix} 1 - 3d & -2d & 0 \\ -d & 1 - 3d & 0 \\ 0 & -2d & 1 - 4d \end{bmatrix}. \tag{22}$$

Eigenvalues of the above matrices (20) and (21) can be calculated analytically:

$$s_1 = \lambda_1 - 4d,\ s_{2,3} = \lambda_1 - \left(3 \pm \sqrt{2}\right) d$$

or

$$\mu_1 = \exp(\lambda_1)(1 - 4d),\ \mu_{2,3} = \exp(\lambda_1)\left[1 - \left(3 \pm \sqrt{2}\right) d\right].$$

Substituting these eigenvalues into inequalities (11) and (18) we obtain the synchronization ranges of parameter d for the network shown in Fig.2:

$$d > \frac{\lambda_1}{3 - \sqrt{2}} \tag{23}$$

for flows (Rössler systems) and

$$\frac{1 - \exp(-\lambda_1)}{3 - \sqrt{2}} < d < \frac{1 + \exp(-\lambda_1)}{3 + \sqrt{2}} \tag{24}$$

Table 1. Dynamical systems used in numerical simulations

Dynamical system	Equations of motion	LLE – λ_1
Rössler system	$\dot{x} = -y - z$ $\dot{y} = x + 0.15y$ $\dot{z} = 0.20 + z(x - 10.0)$	0.085
Henon map	$x_{m+1} = 1 - 1.40x_m^2 + y_m$ $y_{m+1} = 0.30x_m$	0.419

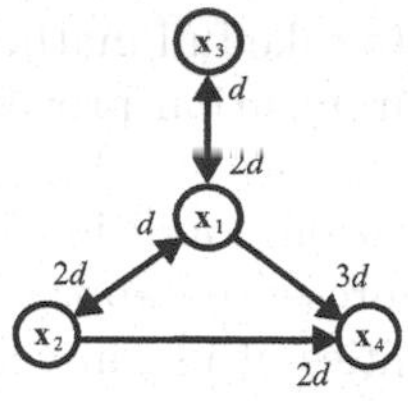

Fig. 1. Four oscillators with irregular configuration of coupling

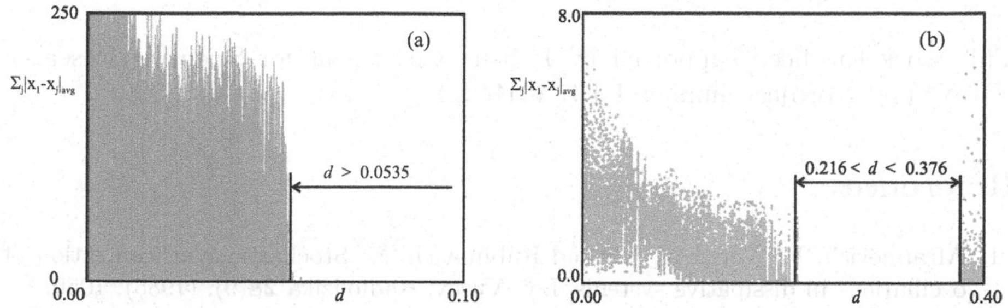

Fig. 2. Bifurcation diagrams (the sum of average *trajectories separations* vs. coupling coefficient) representing the comparison of the synchronization ranges in the ensembles of dynamical systems ((a) set of Rössler systems, (b) set of Henon maps) connected as in Fig. 1. Analytically obtained ranges according to (22) and (23) marked in black.

for maps (Henon maps). The confirmation of complete synchronization stability regions given by inequalities (22) and (23) is shown in Figs 2(a) and 2(b).

4 Conclusions

The presented theoretical analysis supported by numerical simulations leads to main conclusion that chaotic synchronization in the networks composed of identical oscillators with diagonal, diffusive-type interaction between them can be considered as simple, linear dynamical process. Two "parameters of order", i.e. the largest Lyapunov exponent of the network node system and effective coupling rate between the nodes, play the dominant role in this process. This property of diagonal coupling allows us to estimate the synchronization threshold for arbitrary configuration of coupling. Such method bases on the simplified, linear model of the network. The advantage of this approach is a simplicity of its application for both continuous-time and discrete-time systems. In order to examine the stability of synchronization state we introduce the concept of *Synchronization Stability Matrix*. The SSM is constructed directly from coupling configuration matrix and allows the linear stability analysis. However, one should note that our approach can be realized only in the case of diagonal coupling because only in such case we can substitute the coupling matrices for coupling coefficients according to (5). Non-diagonal coupling (realized by not all system coordinates for each pair of nodes) forces us to take full mathematical form of node system in considerations of network synchronization process, that makes impossible simplification of the network given by (4) and (5). In such cases other techniques for determination of synchronization condition have to be used, eg. MSF. We would like to point out that the presented approach can be qualified as a version of MSF method, but its possibilities of use in very different systems (maps and flows) makes it widely useful.

Acknowledgement

This work has been supported by Polish Department for Scientific Research (DBN) under project number 4 T07A 044 28.

References

1. Afraimovich, VS Verichev, NN and Rabinovich, M: Stochastic synchronization of oscillations in dissipative systems, Izv. Vusov. Radiofizika **28**(9), (1985), 1050.
2. Dmitriev, AS, Shirokov, M and Starkov, SO: Chaotic synchronization in ensembles of coupled maps, IEEE Trans. Circuits Syst. I. Fund. Th. Appl. **44**(9), (1997), 918.
3. Fujisaka, H and Yamada, T: Stability theory of synchronized motion in coupled oscillator systems, Prog. Theor. Phys. **69**, (1983), 32.
4. Gade, PM, Cerdeira, H and Ramaswamy, R: Coupled maps on trees, Phys. Rev. E **52**, (1995), 2478.
5. Heagy, JF, Carroll, TL and Pecora, LM: Synchronous chaos in coupled oscillators systems, Phys. Rev. **50**(3), (1994), 1874.
6. Kapitaniak, T, Sekieta, M, Ogorzałek, M: Monotone synchronization of chaos, Int. J. Bifurcation Chaos **6**, (1996), 211.
7. Pecora, LM, Carroll, TL: Synchronization in chaotic systems, Phys. Rev. Lett. **64**, (1990), 821.
8. Pecora, LM and Carroll, TL: Master stability functions for synchronized coupled systems, Phys. Rev. Lett. **80**(9), (1998), 2109.
9. Pecora, LM and Carroll, TL, Johnson, G, Mar, D, Fink, K: Synchronization stability in coupled oscillator arrays: solution for arbitrary configurations, Int. J. Bifurcation Chaos **10**(2), (2000), 273.
10. Pikovsky, A: On the interaction of strange attractors, Zeitschrift Phys. B **55**, (1984), 149.
11. Wang, XF, Chen, G: Synchronization in small-world dynamical networks, Int. J. Bifurcation Chaos **12**(1), (2002), 273.
12. Wu, CW, Chua, LO: Application of graph theory to the synchronization in an array of coupled nonlinear oscillators, IEEE Trans. Circuits Syst. **42**(8), (1995), 430.

Part VII

Operator Splitting, Their Application and Realization

Operator Splitting Procedures for Air Pollution Transport Models*

Petra Csomós

Eötvös Loránd University, Department of Applied Analysis,
H-1117 Budapest, Pázmány Péter st. 1/C
csomos@ludens.elte.hu

Abstract. In the present paper a simple two dimensional air pollution transport model is investigated applying a sequential operator splitting procedure. A comparison is done between the cases when Eulerian and semi-Lagrangian schemes are used for the advection sub-problem.

1 Introduction

Air pollution transport models forecast the spatial distribution of the concentrations of air pollutants. Changes in the concentrations are caused by different physical processes in the atmosphere: advection, diffusion, emission, deposition, and chemical reactions. The effect of these processes lead to the time-evolution of the concentrations, which is mathematically modelled by a system of partial differential equations. Due to the complexity of these equations, usually, it is hard to find appropriate numerical methods which could be efficiently used to solve the problem numerically.

In order to simplify the system, *operator splitting procedures* have been introduced. Application of an operator splitting procedure means that instead of the original problem, a sequence of certain sub-problems is solved. In the case of air pollution transport models each sub-model describes mathematically one of the above mentioned physical processes. Applying a splitting procedure, however, gives rise to a certain amount of error.

In the present paper this error is investigated through a numerical study of solving a simple air pollution transport model applying sequential splitting and two different numerical methods for solving the advection sub-problem.

2 Air Pollution Transport Models

Time-evolution of concentrations of air pollutants is caused by the combined effect of the following atmospheric physical processes. *Advection* describes the transportation due to the wind field. *Diffusion* occurs due to the termic motion of the gas particles in the air. *Deposition* means the purification of the atmosphere

* This research was supported by Hungarian Scientific Research Fund under the grant OTKA T043765, and NATO Collaborative Linkage Grant EST CLG 980505.

I. Lirkov, S. Margenov, and J. Waśniewski (Eds.): LSSC 2005, LNCS 3743, pp. 331–338, 2006.

due to the gravity and the rain. *Chemical reactions* between different species cause reduction and increase of certain concentrations. *Emission* is the source of the air pollutants.

Let $\mathbf{c} = \mathbf{c}(\mathbf{x}, t) \in \mathbb{R}^r$ denote the concentrations of r species of air pollutants, where $\mathbf{c}$ is a function of the location ($x \in \mathbb{R}^3$) and the time ($t \in [0, T]$). The time-evolution of the concentration vector can be mathematically described by a system of the following partial differential equations referring to the c_l component of $\mathbf{c}$ ([1]):

$$\left.\begin{aligned} \frac{\partial c_l}{\partial t} &= -\nabla(\mathbf{u}c_l) + \nabla(K\nabla c_l) + \mathbf{R}_l(\mathbf{c}) + E - \sigma c_l, \quad t \in (0, T], \\ c_l(\mathbf{x}, 0) &= c_{0_l}(\mathbf{x}) \end{aligned}\right\} \tag{1}$$

for $l = 1, \ldots, r$, where $\mathbf{u} = \mathbf{u}(\mathbf{x}, t)$ describes the wind velocity, $K = K(\mathbf{x}, t)$ is the diffusion coefficient, the usually non-linear function $\mathbf{R}_l(\mathbf{c}) = \mathbf{R}_l(\mathbf{x}, t, \mathbf{c})$ describes the chemical reactions between the investigated species, $E = E(\mathbf{x}, t)$ is the emission function, and $\sigma = \sigma(\mathbf{x}, t)$ describes the deposition. Initial and boundary conditions are also attached to the system (1).

We note that the operator on the right-hand side of system (1) is a sum of different sub-operators, among those there are also first- and second-order spatial differential operators. Each of these sub-operators describes one of the physical processes. The idea of the operator splitting comes from the observation that system (1) cannot be solved efficiently by using a numerical method which treats together all the sub-operators. However, solving the sub-problems having only the corresponding sub-operators on the right-hand side, leads to an easier problem.

3 Operator Splitting Procedures

Although, there exist several splitting techniques (see, e.g. [2, 3, 4]), we will focus on the *sequential splitting*, because usually this is applied in the air pollution transport models.

Let B denote a Banach space, and X is a properly chosen space of functions of type $[0, T] \to B$. We consider the following abstract Cauchy problem:

$$\left.\begin{aligned} \frac{dw(t)}{dt} &= Aw(t), \quad t \in (0, T] \\ w(0) &= w_0, \end{aligned}\right\} \tag{2}$$

where $w \in X$ is the unknown function, $w_0 \in B$ is a given element, and A is an operator of type $B \to B$. Let us assume that A can be written formally as a sum of s sub-operators A_i ($i = 1, \ldots, s$). We may assume that the sub-operators A_i have simpler structure than operator A, and problem (2) has a unique solution in X. Then let us divide the time interval $[0, T]$ into $m \in \mathbb{N}$ sub-intervals with length τ, where τ is called the *splitting timestep*, i.e. $T = m\tau$.

Application of the sequential splitting means that first we solve the sub-problem related to the sub-operator A_1 on the time interval $[0, \tau]$, using the original initial condition w_0 (thus, initial condition is needed also in the case of (1)). Then we solve the sub-problem related to A_2 also on the time interval $[0, \tau]$, using the solution of the previous sub-problem as an initial condition. We continue this process until A_s (always using the previous solution as initial condition). Then we start again with A_1, using the solution of the previous sub-problem (i.e. related to A_s) as an initial condition, but on the time interval $[\tau, 2\tau]$. Finally, we arrive to the last time interval $[(m-1)\tau, m\tau]$. The solution of problem (2) at time $t = T$ applying sequential splitting is the solution of the sub-problem related to A_s at the time $t = m\tau$. We note that the approximation obtained after the treatment of the last sub-model is accepted as an approximation to the solution at the splitting timestep under consideration. If $s = 2$, i.e. there are only two sub-operators, the above process can be written as follows:

$$\left.\begin{aligned} &\frac{dw_1^{(k)}(t)}{dt} = A_1 w_1^{(k)}(t), \qquad t \in ((k-1)\tau, k\tau] \\ &w_1^{(k)}((k-1)\tau) = w_{spl}((k-1)\tau) \end{aligned}\right\}$$

$$\left.\begin{aligned} &\frac{dw_2^{(k)}(t)}{dt} = A_2 w_2^{(k)}(t), \qquad t \in ((k-1)\tau, k\tau] \\ &w_2^{(k)}((k-1)\tau) = w_1(k\tau) \\ &w_{spl}(k\tau) := w_2^{(k)}(k\tau) \end{aligned}\right\}$$

with $k = 1, \ldots, m$, and $w_{spl}(0) = w_0$, where $w_{spl}(k\tau)$ is the solution of the split problem defined in the points $\{k\tau : \; k = 0, 1, \ldots, m\}$.

We remark that the application of a splitting procedure for solving (2) gives rise to a certain amount of error.

4 Numerical Methods

In real situations, the exact solution of system (1) is not known, therefore, certain numerical methods have to be used in order to determine the forecast values of the concentrations $\mathbf{c}$ at time $t = T$. Applying a splitting procedure, instead of one problem more sub-problems have to be solved individually. Therefore, the advantage of applying a splitting procedure is that different numerical methods can be used for solving each sub-problem. For each numerical method different time step Δt can be chosen. The only important restriction is that the solutions have to "meet" at time $t = k\tau$ $(k = 1, \ldots, m)$. This means that each step size (for fixed grid size) can be chosen to satisfy the stability condition only of the corresponding numerical method. Taking the advantage of this possibility the integration of the model can take shorter time. In the following, we summarize the numerical methods which we use for the different sub-problems in our test air pollution transport model introduced in the next section.

We use finite difference methods for discretizing the diffusion equation. The emission and the deposition equations are discretized by using first-order finite difference schemes. Chemical reactions are not considered. Since in our investigation we apply two different numerical schemes for the advection sub-model, in what follows, we describe them in more details.

4.1 Advection Equation

For the sake of simplicity, let us assume that $x \in \mathbb{R}$. Let us define the temporal and the spatial meshes ω_t and ω_x, respectively, as

$$\omega_t := \{i \cdot \Delta t, \quad i = 0, 1, \ldots, I\} \qquad \text{and} \qquad \omega_x := \{j \cdot \Delta x, \quad j = 0, 1, \ldots, J\},$$

where Δt is the numerical step size, and Δx denotes the spatial grid size. Then the advection equation has the following form:

$$\frac{\partial c}{\partial t} = -\frac{\partial (uc)}{\partial x}, \tag{3}$$

where c is the concentration of a certain species, and u is the wind velocity. For solving the advection equation (3), two methods are introduced.

Eulerian aspect. When measurements in certain temporal and spatial points are available, it is useful to discretize equation (3) directly on the meshes ω_t and ω_x, applying a finite difference method. The simplest one is the so-called *up-wind scheme*, which can be written in the following way:

$$\frac{c_j^{i+1} - c_j^i}{\Delta t} = -\frac{u_{j+1}^i c_{j+1}^i - u_j^i c_j^i}{\Delta x}, \tag{4}$$

where c_j^i and u_j^i denote the approximations of c and u, respectively, at time $i \cdot \Delta t \in \omega_t$ at the gridpoint $j \cdot \Delta x \in \omega_x$.

The stability condition of a numerical scheme has a crucial importance. Scheme (4) is stable only if the *Courant-Friedrichs-Levy condition* is satisfied, namely,

$$\frac{\Delta t \cdot u_{\max}}{\Delta x} \leq 1, \tag{5}$$

where $u_{\max}$ is the maximal wind velocity appearing during the integration of the model. Hence, applying this scheme the step size has to be chosen carefully.

Semi-Lagrangian aspect. Another aspect of solving the advection equation (3) is from the point of view of the flowing medium. We assume that each fluid domain contains a certain amount of pollution which is constant during the integration. When solving the equation in the semi-Lagrangian manner, we follow all fluid domains back in time, and examine their pollution-contents. The same amount of pollution they had at time t, they have at time $t + \Delta t$. In order to determine the path of each fluid domain, for first guess we say that a certain fluid domain comes from the dimensionless distance $\alpha := u_j^i \cdot \frac{\Delta t}{\Delta x}$, where $\alpha = p + \hat{\alpha}$ with $p \in \mathbb{Z}$ and $\hat{\alpha} \in (0, 1)$. Hence, it may happen that a fluid domain starts between two gridpoints. To avoid this phenomenon, an interpolation onto the mesh ω_x is made in each time step. Applying linear interpolation, the value of c_j^{i+1} (the forecast value of the concentration at the gridpoint $j \cdot \Delta x \in \omega_x$) can be determined as

$$c_j^{i+1} = (1 - \hat{\alpha})\, c_{j-p}^i + \hat{\alpha}\; c_{j-p-1}^i. \tag{6}$$

When the wind velocity u depends on the location and the time, then usually an iteration is done.

Because of the interpolation (we do not follow the same domains for all time), this scheme is called *semi-Lagrangian* scheme (see, e.g. [5]). The great advantage of this scheme is that it is *unconditionally stable* if the flow is laminar. This means that the numerical step size Δt can be chosen independently from the grid size Δx.

5 Errors Appearing in the Numerical Solution

If we apply splitting procedure and numerical methods for solving system (1), the following kinds of solutions can be defined (as before, τ is the splitting timestep):

- $c(\mathbf{x}, t)$ denotes the exact solution of problem (1) at time t;
- $\tilde{c}(\mathbf{x}, k\tau)$ denotes the numerical solution of problem (1) at time $k\tau$;
- $\tilde{c}_{\text{spl}}(\mathbf{x}, k\tau)$ denotes the numerical solution of the split problem at time $k\tau$,

for $k = 1, \ldots, m$. Using the above notations, we can define the following two errors:

- *total error:* $E_{\text{tot}}(\mathbf{x}, k\tau) := |c(\mathbf{x}, k\tau) - \tilde{c}_{spl}(\mathbf{x}, k\tau)|$
- *practical error:* $E_{\text{prac}}(\mathbf{x}, k\tau) := |\tilde{c}(\mathbf{x}, k\tau) - \tilde{c}_{\text{spl}}(\mathbf{x}, k\tau)|$

for $k = 1, \ldots, m$. The difference between the two errors is that the total error can only be computed when the exact solution is known. However, we are interested in this error because this tells us how much the numerical solution differs from the exact solution. In cases when (1) describes a real physical situation, the exact solution is not known, therefore, the practical error has been introduced. In this case the numerical solution obtained without applying splitting plays the role of the exact solution.

6 A Test Problem

In our test model the time-evolution of the concentration of one chemical species is investigated in two dimensions (i.e. $\mathbf{x} = (x, y)$), without taking into account the chemical reactions. The physical coefficients and the wind velocity are chosen to be constant in space and time. Hence, we study the following equation:

$$\left.\begin{aligned} &\frac{\partial c}{\partial t} = -u_{0_x}\frac{\partial c}{\partial x} - u_{0_y}\frac{\partial c}{\partial y} + K_0\left(\frac{\partial^2 c}{\partial x^2} + \frac{\partial^2 c}{\partial y^2}\right) + E - \sigma_0 c, \quad t \in (0, T], \\ &c(x, y, 0) = c_0(x, y), \end{aligned}\right\} \tag{7}$$

where $\mathbf{u}(x,y,t) = \big(u_{0_x}, u_{0_y})\big)$. In order to measure the above defined errors caused by the application of the splitting procedure and the numerical methods, two kinds of emission functions are defined.

One can derive an emission function, for which the exact solution can be defined as a Gaussian surface in two dimensions growing in time:

$$c_G(x,y,t) = E_0 \; e^t e^{-\beta\left((x-x_0)^2+(y-y_0)^2\right)}. \tag{8}$$

From this solution not only the practical but the total error can be computed. In what follows we refer this case as *Gaussian emission.*

The *emission of a town* describes a real physical phenomenon: we consider the uniform emission field of a circle-shaped town in the following form:

$$E_{\mathrm{ch}}(x,y,t) = \begin{cases} E_0 & \left(x-\frac{1}{2}\right)^2 + \left(y-\frac{1}{2}\right)^2 \le R^2 \\ 0 & \text{anywhere else,} \end{cases} \tag{9}$$

where E_0 is constant in space and time, and R is the radius of the field. For this case $c_0(x,y) = 0$. The exact solution is not known, therefore, only the practical error can be computed.

6.1 Solution of the Test Problem

For solving numerically equation (7), we run the model in the following three ways: (i) without applying splitting procedure, and using the up-wind scheme for the advection equation; (ii) applying sequential splitting procedure, and using the up-wind scheme for the advection equation – we will refer to this case as Eulerian case; (iii) applying sequential splitting procedure, and using the semi-Lagrangian scheme for the advection equation – we will refer to this case as semi-Lagrangian case.

Our forecast domain is 50 $km \times 50$ km large, the grid sizes are equal in both directions: $\Delta x = \Delta y = 0.5$ km. We forecast the concentration for $T = 8$ $hour$ in the case of the Gaussian emission, and $T = 1$ day in the case of the town, using a step size $\Delta t = 1$ $minute$. The splitting timestep is chosen as $\tau = 1$ $hour$. The most important difference between the Eulerian and the semi-Lagrangian case is that the latter method is unconditionally stable. Therefore, in this case the integration of the advection equation does not need the small step size Δt, therefore, we can use the larger timestep τ. Hence, the integration of the model takes shorter time in the semi-Lagrangian case.

The parameters in (7) are chosen as follows:

$$u_{0_x} = u_{0_y} = 0.5\; km/h,\; K_0 = 0.5\; km^2/h,\; E_0 = 10^{-4}\; kg/km^3/h,\; \sigma_0 = 10^{-5}\; 1/h,$$

the radius of the emission field of the town equals $R = 7$ km, which is located in the center of the domain (i.e. $x_0 = y_0 = 25$ km).

In *Figure 1* the numerical solutions of equation (7) can be seen using the two emission functions. We have very similar results without applying splitting, and also for the Eulerian and the semi-Lagrangian case.

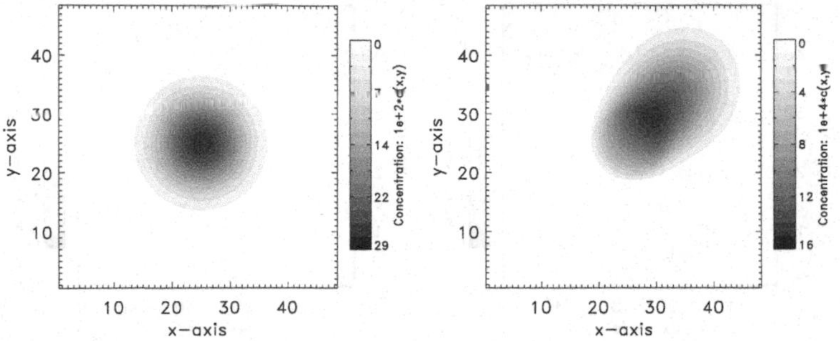

Fig. 1. Numerical solution of (7) using the Gaussian emission (left panel), and the emission field of a town (right panel) (similar results without splitting, and for Eulerian and semi-Lagrangian case)

6.2 Errors Appearing in the Numerical Solution

The spatial distribution of the practical error in the Eulerian (left panel) and semi-Lagrangian case (right panel) can be seen in *Figure 2* and *3* for the Gaussian emission and for the town, respectively. We note that the spatial distribution of the total error for the Gaussian emission is similar to the case of the practical error.

One can see that the structures of the errors reflect the physical processes, i.e. one can distinguish the effect of the emission (the error field has a typical center), the advection (the error field is transported "top-right"), and the diffusion (the error field is spread).

In the case of the Gaussian emission there is not a significant difference between the shapes of the error fields for the Eulerian and the semi-Lagrangian methods. This case, however, is very simple, because the solution always has the same shape, the concentration is only growing in time. The case of the town is more complex. In this case the practical errors also behave similarly, however,

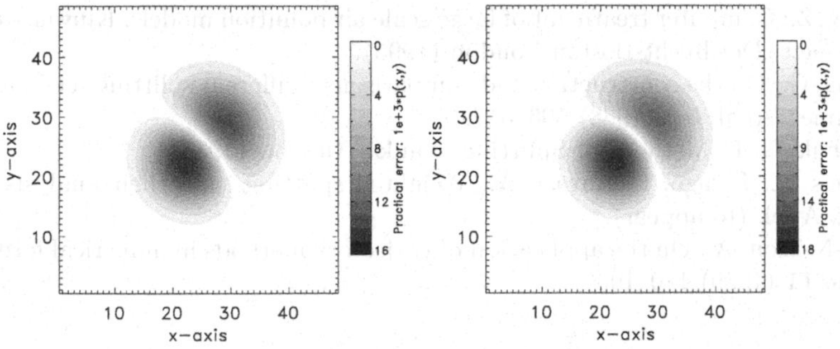

Fig. 2. Spatial distribution of practical error for Gaussian emission in the Eulerian (left panel) and semi-Lagrangian case (right panel) at time $t = 8$ *hour*

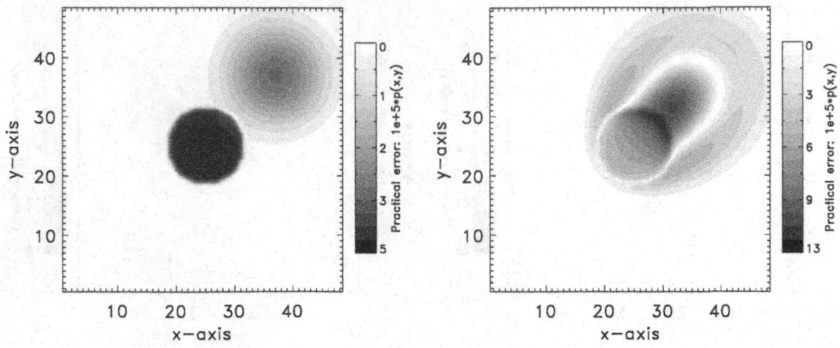

Fig. 3. Spatial distribution of practical error for the town in the Eulerian (left panel) and semi-Lagrangian case (right panel) at time $t = 1$ *day*

their shapes differ significantly. It is important that in both cases the practical error of the Eulerian method is smaller than it is for the semi-Lagrangian method.

7 Conclusion

To summarize our paper we can say that applying a splitting procedure for solving a partial differential equation numerically, gives us the possibility to use different numerical method for each sub-problem. Application of semi-Lagrangian instead of Eulerian scheme for solving the advection equation results shorter integration time, however, it is less accurate.

After these preliminary results we plan to continue our investigations regarding to the sequential splitting using physical parameters changes in space and time. Then the same investigations should be done applying Strang – Marchuk and weighted splittings.

References

1. Zlatev, Z.: Computer treatment of large scale air pollution models. Kluwer Academic Publishers, Dordrecht-Boston-London (1995)
2. Strang, G.: On the construction and comparison of different splitting schemes. SIAM J. Numer. Anal. **5** 3 (1968) 506–517.
3. Marchuk, G. I.: Methods of Splitting. Nauka, Moscow (1988)
4. Csomós, P., Faragó, I., Havasi, Á.: Weighted splittings and their analysis. Comp. Math. Appl. (to appear).
5. Wiin-Nielson, A.: On the application of trajectory methods in numerical forecasting. Tellus, **11** (1959) 180–196.

Adjoint Computations in Data Assimilation Problems Using a 4-Stage Rosenbrock Method

Gabriel Dimitriu

University of Medicine and Pharmacy, Faculty of Pharmacy,
Department of Mathematics and Informatics, 700115 Iaşi, Romania
dimitriu@umfiasi.ro

Abstract. In the past decade the variational methods (3D-var, 4D-var) have been successfully applied in meteorological data assimilation and showed promising results for atmospheric chemistry models. In this paper we describe adjoint computations in data assimilation problems using a 4-stage Rosenbrock method. The order 4 formula is L-stable and has an embedded order 3 formula for error control.

1 Introduction

In the 4D-var data assimilation approach a cost function is defined as the weighted least squares distance between model predictions and observations over the assimilation window. A minimization procedure is then used to find the set of control variables that minimizes the cost function. Most of the powerful minimization methods require the evaluation of the gradient of the cost function which for large scale models is a very intensive computational process. Using the adjoint technique, the gradient of the cost funtion can be computed at the expense of few evaluations, making the optimization algorithm very efficient.

In this paper we describe adjoint computations in data assimilation problems using a 4-stage Rosenbrock method. The order 4 formula is L-stable and has an embedded order 3 formula for error control. The number of function evaluations for the 4 stage method is kept to 3, i.e., one function evaluation is saved.

We consider a 3D atmospheric transport-chemistry model given by the system of differential equations ([3]):

$$\frac{\partial}{\partial t} c_i = -\nabla \cdot (u c_i) + \nabla \cdot (\mathbb{K} \cdot \nabla c_i) + f_i(c) + E_i, \qquad i = 1, \ldots, S. \tag{1}$$

The initial condition is $c(t_0) = c_0$ and appropriate boundary values are prescribed. The solution $c(t, x, y, z) \in \mathbb{R}^S$ of problem (1) represents the concentration vector of the chemical species in the model, u is the wind field and $\mathbb{K}$ is the eddy diffusivity tensor. The chemical reactions are governed by the nonlinear stiff terms $f_i(c) = P_i(c) - D_i(c)c_i$, with $P_i(c)$, $D_i(c)$, the chemical production and destruction terms; E_i represents the source term, and depositions are modeled as a boundary condition at the surface of the earth: $-n_i \cdot (\mathbb{K} \cdot \nabla c_i) = Q_i - \nu_i c_i$ with n_i the inward vector normal to the surface, Q and ν_i, the surface emission

I. Lirkov, S. Margenov, and J. Waśniewski (Eds.): LSSC 2005, LNCS 3743, pp. 339–346, 2006.

rate and deposition velocity of species i, respectively. We will refer to problem (1) as the forward model and $c(t)$, $t \in [t_0, T]$ will represent the "forward trajectory". Typical choices for the set of control variables in data assimilation are the boundary values, initial concentrations, emissions and deposition rates.

2 4D Variational Data Assimilation

We consider here the 4D variational data assimilation problem associated to (1) with the set of control variables given by the initial state of the model, c_0. Under suitable assumptions, forward model (1) has a unique solution, viewed as a function of the initial conditions, $c = c(x, y, z, c_0)$.

If space discretization is applied to forward model (1) on a grid (N_x, N_y, N_z), the resulting ODE system of dimension $N = S \times N_x \times N_y \times N_z$ is:

$$\frac{dc}{dt} = F_A(c) + F_D(c) + F_R(c), \qquad c(t_0) = c_0, \tag{2}$$

where F_A represents the advection and horizontal diffusion, F_D is the vertical diffusion, and the reaction terms are given by F_R. We assume that a "background estimate" c^b of c_0 with the error covariances matrix B, has been generated from the output of a previous analysis, using some assumptions of consistency in time of the model state, like stationarity (hypothesis of persistence) or the evolution predicted by a forecast model. The measurements c_k^0, $k = 1, \ldots, m$ of the concentrations at moments t_k are scattered over the interval $[t_0, T]$. The errors in measurements and model representativeness are given by the covariances matrices R_k, $k = 1, \ldots, m$. The covariance matrices B and R_k are symmetric and positive definite such that B^{-1}, R_k^{-1} are well defined. In practice, B and R_k are often taken diagonal, which corresponds to the assumption that there is no spatial and chemical correlation in the background errors, and measurement and model errors are uncorrelated in space and time.

The 4D-var data assimilation determines an initial state c_0 that minimizes the distance between the model predictions and observations expressed by the cost function:

$$F(c_0) = \frac{1}{2}(c_0 - c^b)^T B^{-1}(c_0 - c^b) + \frac{1}{2}\sum_{k=1}^{m}(c_k - c_k^0)^T R_k^{-1}(c_k - c_k^0)\,. \tag{3}$$

Most of the powerful optimization techniques require the evaluation of the gradient $\nabla_{c_0} F$ of the cost function. In a comprehensive atmospheric chemistry model the dimension of the vector c_0 can easily be of order 10^6, which makes the optimization a very expensive computational process.

In the variational approach, the gradient of the functional F is computed by using the "adjoint method". The general theory of adjoint equations and the derivation of the adjoint model are given in many studies (see, for example,

[5, 7, 11] and [15]). Below we sketch the basic ideas following [3]. The gradient of the cost function is:

$$\nabla_{c_0} F(c_0) = B^{-1}(c_0 - c^b) + \sum_{k=1}^{m} \left(\frac{\partial c_k}{\partial c_0}\right)^T R_k^{-1}(c_k - c_k^0) \,. \tag{4}$$

Using the chain rule $(\partial c_k)/\partial c_0)^T = (\partial c_{k-1}/\partial c_0)^T (\partial c_k/\partial c_{k-1})^T$ in its transpose form, we can construct the algorithm to compute the gradient:

Step 1. Initialize *gradient* $= 0$.
Step 2. for $k = m, 1, -1$ do

$$gradient = (\partial c_k/\partial c_{k-1})^T \left[R_k^{-1}(c_k - c_k^0) + gradient\right].$$

Step 3. $gradient = B^{-1}(c_0 - c^b) + gradient$.

The main advantage of the adjoint method is that explicit computation of the Jacobian matrices $\partial c_k/\partial c_{k-1}$ is avoided and the matrix-vector products can be computed directly at Step 2 ([7]).

Problem (2) is usually solved by using the operator splitting ([1, 2, 6]). This method has the advantage that processes such as advection, vertical diffusion and chemical reactions can be treated with different numerical schemes. In a second order accurate Strang splitting approach ([14]) with the time step $h = t_{n+1} - t_n$ the solution c_{n+1} is obtained from c_n as follows:

$$c_{n+1} = \overline{F}_A\left(t_{n+\frac{1}{2}}, \frac{h}{2}\right) \overline{F}_D\left(t_{n+\frac{1}{2}}, \frac{h}{2}\right) \overline{F}_R(t_n, h) \overline{F}_D\left(t_n, \frac{h}{2}\right) \overline{F}_A\left(t_n, \frac{h}{2}\right) c_n \,, \tag{5}$$

where the operators $\overline{F}$ are defined by the numerical method used to solve the corresponding processes. If $\overline{J}$ denotes the Jacobian matrix associated to $\overline{F}$, the adjoint algorithm to compute the gradient (4) of the cost function requires products of the form $\overline{J}^T u$, with u an arbitrary seed vector. For large systems, constructing the adjoint code by hand can be a frustrating process and raises questions of technical implementation. In this respect, automatic tools have been developed ([4, 7]).

3 Adjoint Computations for a 4-Stage Rosenbrock Method

Following the basic ideas in [3] applied to a 2-stage Rosenbrock method, this section develops the adjoint formulas for a general 4-stage Rosenbrock method leading to an efficient implementation, suitable for automatization. We consider the following problem:

$$\frac{dc}{dt} = f(c), \qquad c(t_0) = c_0 \,, \tag{6}$$

with $c(t)$, $c_0 \in \mathbb{R}^n$ and $f : \mathbb{R}^n \to \mathbb{R}^n$, $f = f_1, f_2, \ldots, f_n)^T$. One step from t_0 to t_1 with $h = t_1 - t_0$ of a 4-stage Rosenbrock method reads:

$$\begin{aligned}
(\frac{1}{\gamma_{11}h}I - J_0)k_1 &= f(c_0) \\
(\frac{1}{\gamma_{22}h}I - J_0)k_2 &= f(c_0 + \alpha_{21}k_1) + \frac{\beta_{21}}{h}k_1 \\
(\frac{1}{\gamma_{33}h}I - J_0)k_3 &= f(c_0 + \alpha_{31}k_1 + \alpha_{32}k_2) + \frac{\beta_{31}}{h}k_1 + \frac{\beta_{32}}{h}k_2 \qquad (7) \\
(\frac{1}{\gamma_{44}h}I - J_0)k_4 &= f(c_0 + \alpha_{41}k_1 + \alpha_{42}k_2 + \alpha_{43}k_3) + \frac{\beta_{41}}{h}k_1 + \frac{\beta_{42}}{h}k_2 + \frac{\beta_{43}}{h}k_3 \\
c_1 &= c_0 + m_1k_1 + m_2k_2 + m_3k_3 + m_4k_4 \\
\hat{c}_1 &= c_0 + \hat{m}_1k_1 + \hat{m}_2k_2 + \hat{m}_3k_3 + \hat{m}_4k_4\,,
\end{aligned}$$

where J_0 is the Jacobian matrix of f evaluated at c_0, $J_0 = (\partial f/\partial c_j)_{ij}|_{c=c_0}$ and the coefficients γ_{ii}, $i = \overline{1,4}$, α_{21}, α_{3j}, β_{3j}, $j = 1, 2$, α_{4s}, β_{4s}, $s = 1, 2, 3$. The coefficients m_1, m_2, m_3 and m_4 are chosen to obtain a desired order of consistency and numerical stability (see Table 1, [9]).

Table 1. Set of coefficients for the 4-stage Rosenbrock method

α_{21} = 1.14563212	β_{21} = 2.34199312711201394917052 0
α_{31} = 0.52092078913062902932851 6	β_{31} = -0.027333746543489836196505
α_{32} = 2.34199312711201394917052 0	β_{32} = 0.213811650836699689867472
α_{41} = 0.52092078913062902932851 6	β_{41} = -0.259083837785510222112641
α_{42} = 0.134294186842504800149232	β_{42} = -0.190595807732311751616358
α_{43} = 0.	β_{43} = -0.228031035973133829477744
m_1 = 0.324534707891734513474196	$\hat{m}_1$ = 0.520920789130629029328516
m_2 = 0.049086544787523308684633	$\hat{m}_2$ = 0.144549714665364599584681
m_3 = 0.	$\hat{m}_3$ = 0.124559686414702049774897
m_4 = 0.626378747320742177841171	$\hat{m}_4$ = 0.209969809789304321311906

We consider the case when $\gamma_{11} = \gamma_{22} = \gamma_{33} = \gamma_{44} = \gamma$, since of special interest are the methods that require only one LU decomposition of $(1/\gamma_{ii})I - J_0$ per step. The value of the parameter γ determines the stability properties of the Rosenbrock method. In this context, the diagonal entry of the Rosenbrock formula is suggested in [9] as $\gamma = 0.57281606$, which is used for the proposed algorithm. The coefficients $\hat{m}_i$ belong to the order 3 embedded formula. The difference $c_1 - \hat{c}_1$ can be used as a local error estimator. For the adjoint computations we have from (7):

$$\left(\frac{\partial k_1}{\partial c_0}\right)^T = \left(J_0^T + (\frac{\partial J_0}{\partial c_0} \times k_1)^T\right)\left((\frac{1}{\gamma h}I - J_0)^T\right)^{-1}$$

$$\left(\frac{\partial k_2}{\partial c_0}\right)^T = \left(\left(I+\alpha_{21}\frac{\partial k_1}{\partial c_0}\right)^T J_1^T + \frac{\beta_{21}}{h}(\frac{\partial k_1}{\partial c_0})^T + \left(\frac{\partial J_0}{\partial c_0}\times k_2\right)^T\right)$$
$$\cdot\left((\frac{1}{\gamma h}I - J_0)^T\right)^{-1}$$
$$\left(\frac{\partial k_3}{\partial c_0}\right)^T = \left(\left(I+\alpha_{31}\frac{\partial k_1}{\partial c_0}+\alpha_{32}\frac{\partial k_2}{\partial c_0}\right)^T J_2^T + \frac{\beta_{31}}{h}(\frac{\partial k_1}{\partial c_0})\right.$$
$$\left.+\frac{\beta_{32}}{h}(\frac{\partial k_2}{\partial c_0})^T + \left(\frac{\partial J_0}{\partial c_0}\times k_3\right)^T\right)\cdot\left((\frac{1}{\gamma h}I - J_0)^T\right)^{-1}$$
$$\left(\frac{\partial k_4}{\partial c_0}\right)^T = \left(\left(I+\alpha_{41}\frac{\partial k_1}{\partial c_0}+\alpha_{42}\frac{\partial k_2}{\partial c_0}\right)^T J_3^T + \frac{\beta_{41}}{h}(\frac{\partial k_1}{\partial c_0})^T\right.$$
$$\left.+\frac{\beta_{42}}{h}(\frac{\partial k_2}{\partial c_0})^T + \frac{\beta_{43}}{h}(\frac{\partial k_3}{\partial c_0})^T + \left(\frac{\partial J_0}{\partial c_0}\times k_4\right)^T\right)\cdot\left((\frac{1}{\gamma h}I - J_0)^T\right)^{-1},$$

where J_1 is the Jacobian of f evaluated at $c_0+\alpha_{21}k_1$, J_2 is the Jacobian evaluated at $c_0+\alpha_{31}k_1+\alpha_{32}k_2$, J_3 is the Jacobian evaluated at $c_0+\alpha_{41}k_1+\alpha_{42}k_2+\alpha_{43}k_3$ and the terms $(\partial J_0/\partial c_0)\times k_i$, $i=1,2,3,4$ are $n\times n$ matrices whose j column is $(\partial J_0/\partial c_0^j)k_i$, $i=1,2,3,4$. We took into account in the evaluation of the last equality above that $\alpha_{43}=0$ (see Table 1). We want to point out here the fact that these matrices are not symmetric. From (7) and using the notation $(\frac{1}{\gamma h}I-J_0)^T v = u$ and the equalities stated above, for an arbitrary seed vector $u\in \mathbb{R}^n$, one obtains:

$$\left(\frac{\partial c_1}{\partial c_0}\right)^T u = u + m_1\left(J_0^T + \left(\frac{\partial J_0}{\partial c_0}\times k_1\right)^T\right)v + m_2\left(\left(I+\alpha_{21}\frac{\partial k_1}{\partial c_0}\right)^T J_1^T\right.$$
$$\left.+\frac{\beta_{21}}{h}\left(\frac{\partial k_1}{\partial c_0}\right)^T + \left(\frac{\partial J_0}{\partial c_0}\times k_2\right)^T\right)v + m_3\left(\left(I+\alpha_{31}\frac{\partial k_1}{\partial c_0}+\alpha_{32}\frac{\partial k_2}{\partial c_0}\right)^T J_2^T\right.$$
$$\left.+\frac{\beta_{31}}{h}\left(\frac{\partial k_1}{\partial c_0}\right)^T + \frac{\beta_{32}}{h}\left(\frac{\partial k_2}{\partial c_0}\right)^T + \left(\frac{\partial J_0}{\partial c_0}\times k_3\right)^T\right)v + m_4\left(\left(I+\alpha_{41}\frac{\partial k_1}{\partial c_0}+\alpha_{42}\frac{\partial k_2}{\partial c_0}\right)^T J_3^T\right.$$
$$\left.+\frac{\beta_{41}}{h}\left(\frac{\partial k_1}{\partial c_0}\right)^T + \frac{\beta_{42}}{h}\left(\frac{\partial k_2}{\partial c_0}\right)^T + \frac{\beta_{43}}{h}\left(\frac{\partial k_3}{\partial c_0}\right)^T + \left(\frac{\partial J_0}{\partial c_0}\times k_4\right)^T\right)v\,.$$

We are going to show that the computations can in fact be arranged in a much efficient way. In order to avoid frequent recomputations and to exploit the particular properties of the method, the order of the operations in the formula above becomes important. Below we describe the following algorithm:

Step 1. Solve for v the linear system $(\frac{1}{\gamma h}I - J_0)^T v = u$. Then, taking into account that $m_3 = 0$ (see Table 1), we have

$$\begin{aligned}\left(\frac{\partial c_1}{\partial c_0}\right)^T u &= u + m_1\left(J_0^T + \left(\frac{\partial J_0}{\partial c_0}\times k_1\right)^T\right)v + m_2 J_1^T v + m_2\left(\frac{\partial J_0}{\partial c_0}\times k_2\right)^T v\\ &+ m_2\left(\frac{\partial k_1}{\partial c_0}\right)^T\left(\alpha_{21}J_1^T + \frac{\beta_{21}}{h}I\right)v + m_4\left(\frac{\partial k_1}{\partial c_0}\right)^T\left(\alpha_{41}J_3^T + \frac{\beta_{41}}{h}I\right)v\\ &+ m_4 J_3^T v + m_4\left(\frac{\partial k_2}{\partial c_0}\right)^T\left(\alpha_{42}J_3^T + \frac{\beta_{42}}{h}I\right)v \qquad (8)\\ &+ m_4\frac{\beta_{43}}{h}\left(\frac{\partial k_3}{\partial c_0}\right)^T v + m_4\left(\frac{\partial J_0}{\partial c_0}\times k_4\right)^T v\,.\end{aligned}$$

Step 2. Compute

$$\begin{aligned}\omega_1 &= J_1^T(m_2 v); & \omega_{11} &= \alpha_{21}\omega_1 + (m_2\beta_{21}/h)v;\\ \omega_2 &= J_3^T(m_4 v); & \omega_{21} &= \alpha_{41}\omega_2 + (m_4\beta_{41}/h)v;\\ & & \omega_{22} &= \alpha_{42}\omega_2 + (m_4\beta_{42}/h)v;\\ \omega_3 &= J_2^T(m_4 v); & \omega_{33} &= (m_4\beta_{43}/h)v\,.\end{aligned}$$

Step 3. Solve for θ_1, θ_2, θ_3 and θ_4 the linear systems:

$$(\frac{1}{\gamma h}I - J_0)^T\theta_1 = \omega_{11}\,, \qquad (\frac{1}{\gamma h}I - J_0)^T\theta_2 = \omega_{21}\,,$$

$$(\frac{1}{\gamma h}I - J_0)^T\theta_3 = \omega_{22}\,, \quad \text{and} \quad (\frac{1}{\gamma h}I - J_0)^T\theta_4 = \omega_{33}\,,$$

respectively. Then, compute $\theta = \sum_{i=1}^4 \theta_i$. After replacing in (8), one obtains:

$$\begin{aligned}\left(\frac{\partial c_1}{\partial c_0}\right)^T u &= u + J_0^T(m_1 v) + J_1^T(m_2 v) + J_3^T(m_4 v)\\ &+ \left(\frac{\partial J_0}{\partial c_0}\times k_1\right)^T(m_1 v) + \left(\frac{\partial J_0}{\partial c_0}\times k_2\right)^T(m_2 v) + \left(\frac{\partial J_0}{\partial c_0}\times k_4\right)^T(m_4 v)\\ &+ \left(J_0^T + \left(\frac{\partial J_0}{\partial c_0}\times k_1\right)^T\right)\theta_1 + \left(J_0^T + \left(\frac{\partial J_0}{\partial c_0}\times k_1\right)^T\right)\theta_2 \qquad (9)\\ &+ \left(\frac{\partial k_2}{\partial c_0}\right)^T\theta_3 + \left(\frac{\partial k_3}{\partial c_0}\right)^T\theta_4\,,\end{aligned}$$

and after arranging the terms in (9) we have:

Step 4. Compute

$$\left(\frac{\partial c_1}{\partial c_0}\right)^T u = u + \omega_1 + \omega_2 + J_0^T(m_1 v + \theta) + J_1^T(\theta_3 + \theta_4) + J_2^T(\theta_4)$$
$$+ \left(\frac{\partial J_0}{\partial c_0} \times k_1\right)^T (m_1 v + \theta) + \left(\frac{\partial J_0}{\partial c_0} \times k_2\right)^T (m_2 v + \theta_3 + \theta_4)$$
$$+ \left(\frac{\partial J_0}{\partial c_0} \times k_4\right)^T (m_4 v) \,. \tag{10}$$

4 Computational Issues

From relation (10) we see that it is enough to call the routine to compute the product $J_0^T s$ (s a seed vector) once, with the seed vector $m_1 v + \theta$. The same remark is made for the product $[(\partial J_0/\partial c_0) \times k_1]^T (m_1 v + \theta)$ and for the last two terms of (10).

We now concentrate on the terms $[(\partial J_0/\partial c_0) \times k]^T v$ whose evaluation dominate the computational cost of the algorithm given by Steps 1-4. Here $k, v \in I\!R^n$ are arbitrary constant vectors. For the ith component we have:

$$\left((\frac{\partial J_0}{\partial c_0} \times k)^T v\right)_i = \left(\frac{\partial J_0}{\partial c_0^i} k\right)^T v = k^T \left(\frac{\partial (J_0^T v)}{\partial c_0^i}\right) = \left(\frac{\partial (J_0^T v)}{\partial c_0^i}\right)^T k \,. \tag{11}$$

Consider now the function $g : I\!R^n \to I\!R^n$, $g(c_0) = J_0^T v$. Using (7) results in:

$$[(\partial J_0/\partial c_0) \times k]^T v = [\partial g(c_0)/\partial c_0] k \,.$$

Notice that the Jacobian matrix of g is symmetric (for details, see [3]). The symmetry of the Jacobian matrix of the function g plays a significant role in the implementation of the adjoint code.

The integration of the forward model (1) using implicit methods ([8, 9]), together with the performance analysis was carried out in [13]. Implementation of the Rosenbrock methods ([3, 10]) can be done in the symbolic kinetic preprocessor KPP environment ([4]), which generates the sparse matrix factorization LU required in (2), (3) and the routine to forward-backward solve the linear systems without indirect addressing. The computations are independent, allowing parallel implementation ([12]).

5 Concluding Remarks

In this study we described adjoint computations in data assimilation problems using a 4-stage Rosenbrock method. We focused on constructing an efficient algorithm which exploits the particular properties of the method and avoids frequent recomputations. The algorithm presented in Section 3 has the great benefit that the adjoint of the chemistry integration can be generated completely automatically, taking full advantage of the sparsity of the system. Moreover, rounding errors are avoided since symbolic computations can be used, and the accuracy of the results goes up to the machine precision.

Acknowledgement

This work was supported by the NATO Collaborative Linkage Grant under reference EST.CLG.980505.

References

1. J. Bartholy, I. Faragó, and Á. Havasi. Splitting Method and its Application in Air Pollution Modeling. *Időjárás*, 105 (2001) 39–58.
2. M. Botchev, I. Faragó, and Á. Havasi. Testing Weighted Splitting Schemes on a One-Column Transport-Chemstry Model. In: *Large scale scientific computing*, I. Lirkov, S. Margenov, J. Waśniewski, P. Yalamov eds., *Lecture notes in computer sciences*, **2907**, Springer-Verlag, (2004) 295–302.
3. D. Daescu, G.R. Carmichael, and A. Sandu. Adjoint Implementation of Rosenbrock Methods Applied to Variational Data Assimilation Problems. *J. Comput. Phys.*, **165** 2 (2000) 496–510.
4. V. Damian-Iordache and A. Sandu. *KPP - A symbolic preprocessor for chemistry kinetics - User's guide.* Tech. Rep., Univ. of Iowa, Department of Mathematics, 1995.
5. G. Dimitriu. *Parameter identification for some classes of nonlinear systems.* PhD thesis, 1999 (in Romanian).
6. I. Dimov, I. Faragó, Á. Havasi, and Z. Zlatev. *L*-commutativity of the operators in splitting methods for air pollution models. *Annales Univ. Sci. Sec. Math.*, **44** (2001) 127–148.
7. R. Giering and T. Kaminski. Recipes for Adjoint Code Construction. *ACM Trans. Math. Software*, 1998.
8. E. Hairer, S.P. Norsett, and G. Wanner. *Solving Ordinary Differential Equations I. Nonstiff Problems.* Springer-Verlag, Berlin Heidelberg New York, 1996.
9. E. Hairer and G. Wanner. *Solving Ordinary Differential Equations II. Stiff and Differential-Algebraic Problems.* Springer-Verlag, Berlin Heidelberg New York, 1996.
10. J. Lang, and J. Verwer. ROS3P – An accurate third-order Rosenbrock solver designed for parabolic problems. *BIT*, **41** 4 (2001) 731–738.
11. G.I. Marchuk, I.V. Agoshkov, and P.V. Shutyaev. *Adjoint Equations and Perturbation Algorithms in Nonlinear Problems.* CRC Press, 1996.
12. T. Ostromsky and Z. Zlatev. Flexible Two-Level Parallel Implementation of a Large Air Pollution Model. In: *Numerical methods and applications*, I. Dimov, I. Lirkov, S. Margenov, Z. Zlatev eds., *Lecture notes in computer sciences*, **2542**, Springer Verlag, (2003) 545–554.
13. A. Sandu, J.G. Verwer, J.G. Blom, E.J. Spee, and G.R. Carmichael. Benchmarking stiff ODE solvers for atmospheric chemistry problems II: Rosenbrock solvers. *Atmos. Environ.*, **31** (1997) 3459–3472.
14. G. Strang. On the construction and comparison of difference schemes. *SIAM Journal on Numerical Analysis*, **5** (1968) 506–517.
15. K.Y. Wang, D.J. Lary, D.E. Shallcross, S.M. Hall, and J.A. Pyle. A review on the use of the adjoint method in four-dimensional atmospheric-chemistry data assimilation. *Q.J.R. Meteorol. Soc.*, **127** (576) Part B (2001) 2181–2204.

Operator Splittings and Numerical Methods*

I. Faragó

Eotvos Lorand University, Department of Applied Analysis,
1117 Budapest, Pazmany P. setany 1/c, Hungary
faragois@cs.elte.hu

Abstract. The operator splitting method is a widely used technique which is frequently applied to the solution of complex problems. However, its application is not enough to the practical solution of the problems. The split sub-problems still require some numerical method. In this paper we give a unified investigation of the operator splitting and the numerical discretization. Moreover, we consider the interaction of the operator splitting method and the applied numerical methods to the solution of the different sub-processes. We show that many well-known fully-discretized numerical schemes to solving the Cauchy problems can be obtained in this manner. We investigate the convergence of these methods, too.

1 Introduction

Many problems of the mathematical modelling lead to the solution of the Cauchy problem for the system of ordinary equations in the form

$$\left.\begin{array}{c}\dfrac{dw(t)}{dt} = \mathbf{A}w(t) \equiv \displaystyle\sum_{i=1}^{d} \mathbf{A}_i w(t), \quad t \in (0,T) \\ w(0) = w_0,\end{array}\right\} \tag{1}$$

where $w : [0,T) \to \mathbb{R}^N$ is the unknown function, $w_0 \in \mathbb{R}^N$ is a given element, $\mathbf{A}_i \in \mathbb{R}^{N\times N}$, $(i = 1, 2, \dots d)$ are given matrices. (This approach has a lot of application on different fields, like in the parameter identification [3] and the numerical solution of the Maxwell equation [5].) We approximate the above problem with the one-step iterative method of the form

$$y^{n+1} = \mathrm{r}(\tau\mathbf{A})y^n, \tag{2}$$

where $\tau \ll T$ is some given parameter. Here the problem is the determination of $\mathrm{r}(\tau\mathbf{A})$ being a suitable approximation of the matrix exponential $\exp\left(\tau\sum_{i=1}^{d} A_i\right)$.

In fact this means that using the notation $\mathbf{a} = (a_1, a_2, \dots a_n)$, $\mathrm{r}(\mathbf{a})$ should approximate the exponential function $\exp(\sum_{i=1}^{d} a_i)$. Hence, the approximation method

* Supported by Hungarian National Research Founds (OTKA) under grant N. T043765 and NATO Collaborative Linkage Grant N. 980505.

I. Lirkov, S. Margenov, and J. Waśniewski (Eds.): LSSC 2005, LNCS 3743, pp. 347–354, 2006.

is defined by the choice of the function $r(\tau\mathbf{A})$. For $d = 1$ the problem is widely investigated and has a waste of literature. (For an overview see, e.g. [2, 6].) The different operator splitting methods are based on the special form of the approximation for the case $d > 1$. (E.g. [4, 6].) This means that these methods aren't defined by the formal substitution the sum $A = \sum_{i=1}^{d} \mathbf{A}_i$ into the function $r(\tau\mathbf{A})$. However, the classical operator splittings are elaborated mainly for $d = 2$. In this paper we generalize the concept for any d.

The paper is organised as follows. In Section 2 we consider the possible approximations of the exponential function with different d. We introduce the sequential, Strang-Marchuk and the weighted sequential splittings. In Section 3 we define special numerical methods, the so called θ-scheme to the numerical solution of the sub-problems. We prove the convergence of these fully discretizations for different operator splittings. In Section 4 we re-call some well-known discretization schemes which can be obtained from the general setting.

2 Operator Splitting Methods

We recall that our aim is a suitable approximation of the exponential function $\exp(\sum_{i=1}^{d} a_i)$.

The first choice of the approximation function $r(\mathbf{a})$ is the following

$$r_{ss}(\mathbf{a}) = \prod_{i=1}^{d} \exp(a_{d+1-i}). \tag{3}$$

Then the iteration (2) reads as

$$y_{ss}^{n+1} = \prod_{i=1}^{d} \exp(\tau \mathbf{A}_{d+1-i}) y_{ss}^{n}. \tag{4}$$

In order to put the choice (3) in a new light, let us consider the problem (1) for $d = 2$ and we investigate the following sequence of the initial value sub-problems

$$\left.\begin{aligned} &\frac{dw_1^n}{dt}(t) = \mathbf{A}_1 w_1^n(t), \qquad (n-1)\tau < t \le n\tau, \\ &w_1^n((n-1)\tau) = w_2^{n-1}((n-1)\tau), \end{aligned}\right\} \tag{5}$$

and

$$\left.\begin{aligned} &\frac{dw_2^n}{dt}(t) = \mathbf{A}_2 w_2^n(t), \qquad (n-1)\tau < t \le n\tau, \\ &w_2^n((n-1)\tau) = w_1^n(n\tau), \end{aligned}\right\} \tag{6}$$

for $n = 1, 2, \ldots K$, where K denotes the supremum of the integers K_1 such that $K_1\tau \le T$ and $w_2^0(0) = u_0$. This is called *sequential operator splitting* and the function $w_{sp}(n\tau) = w_2^n(n\tau)$, defined at the points $t_n = n\tau$ is called splitting solution of the problem. Obviously, in (4), for $d = 2$, the approximation y_{ss}^n

corresponds to the exact solution at $t = n\tau$ of the above sequential operator splitting problem, i.e. $w_{sp}(n\tau) = y_{ss}^n$.

The second choice is

$$\mathrm{r}_{str}(\mathbf{a}) = \left(\prod_{i=1}^{d-1} \exp(\frac{a_i}{2})\right) \exp(a_d) \left(\prod_{i=1}^{d-1} \exp(\frac{a_{d+1-i}}{2})\right). \tag{7}$$

Then the iteration (2) means

$$y_{str}^{n+1} = \left(\prod_{i=1}^{d-1} \exp(\frac{\tau}{2}\mathbf{A}_i)\right) \exp(\tau \mathbf{A}_d) \left(\prod_{i=1}^{d-1} \exp(\frac{\tau}{2}\mathbf{A}_{d+1-i})\right) y_{str}^n. \tag{8}$$

Similarly to the sequential operator splitting, for $d = 2$ we establish contact between the iteration and some sequence of initial value problems. Let us consider the sub-problems

$$\left.\begin{aligned} &\frac{dw_1^n}{dt}(t) = \mathbf{A}_1 w_1^n(t), \qquad (n-1)\tau < t \leq (n-0.5)\tau, \\ &w_1^n((n-1)\tau) = w_3^{n-1}((n-1)\tau), \end{aligned}\right\} \tag{9}$$

$$\left.\begin{aligned} &\frac{dw_2^n}{dt}(t) = \mathbf{A}_2 w_2^n(t), \qquad (n-1)\tau < t \leq n\tau, \\ &w_2^n((n-1)\tau) = w_1^n((n-0.5)\tau), \end{aligned}\right\} \tag{10}$$

and

$$\left.\begin{aligned} &\frac{dw_3^n}{dt}(t) = \mathbf{A}_1 w_3^n(t), \qquad (n-0.5)\tau < t \leq n\tau, \\ &w_3^n((n-0.5)\tau) = w_2^n(n\tau), \end{aligned}\right\} \tag{11}$$

for $n = 1, 2, \ldots K$, where $w_3^0(0) = u_0$, and the function $w_{sp}(n\tau) = w_3^n(n\tau)$, defined at the points $t_n = n\tau$. Clearly, $w_{sp}(n\tau)$ is the corresponding splitting solution of the problem. This kind of operator splitting is called *Strang-Marchuk operator splitting* [11, 8]. As before, y_{str}^n in (8) corresponds to the exact solution at $t = n\tau$ of the above Strang-Marchuk operator splitting process, i.e. $w_{sp}(n\tau) = y_{str}^n$.

The generalization of the sequential and Strang operator splittings to d operators ($d \geq 3$) is straightforward.

Remark 1. There are several other choice of the operator splitting functions. In order to symmetrize the sequential operator splitting, we can set

$$\mathrm{r}_{wss}(\mathbf{a}) = \frac{1}{2}\left(\prod_{i=1}^{d} \exp(a_{d+1-i}) + \prod_{i=1}^{d} \exp(a_i)\right). \tag{12}$$

This method is called weighted sequential operator splitting, introduced for $d = 2$ in [10] and analyzed in [1].

Remark 2. Important question is the accuracy of the approximation of $\exp(\tau\mathbf{A})$ by $\mathrm{r}(\tau\mathbf{A})$. Clearly, for the scalar case we have equality. However, when the sub-operators $\mathbf{A}_i$ are not commuting then usually we don't have equality. The expression $Err_{spl} = \exp(\tau\mathbf{A}) - \mathrm{r}(\tau\mathbf{A})$ is called local splitting error. If the order of the local splitting error is $\mathcal{O}(\tau^{p+1})$ then the given splitting is called of p-th order. One can check that the sequential operator splitting is of first order, while the Strang and the weighted sequential operator splittings are of second order [1, 6]. When the sub-operators in the sum of the operators $\mathbf{A}$ are commuting then the local splitting error is vanishing, i.e. the operator splitting is accurate.

However, the practical use of the operator splitting method has a basic drawback: each splitting requires the computation of the matrix exponentials for the sub-operators in the sum of $\mathbf{A}$. Whereas these matrices usually have simpler structure, the exact computation of their exponentials is seldom possible. So, the operator splitting is not directly useful for practical purposes. The main benefit is that this method leads the original complex problem to the sequence of simpler sub-problems. Hence, the rational approximation of the exponential function (i.e. application of some numerical method) is not only the necessary step in the solution process but its suitable choice becomes easier.

3 Numerical Methods Based on the Splittings

The typical choice of the rational approximation (which defines the numerical method) to the exponential function, called θ-approximation, is

$$\mathrm{r}_\theta(z) = (1 - \theta z)^{-1}(1 + (1 - \theta)z), \tag{13}$$

where $\theta \in [0, 1]$ is a given parameter. In this section we will use this approximation to the different split sub-problems.

When we approximate in the sequential splitting (3) each exponential by the θ-approximations (13), then we get the approximation function

$$\mathrm{r}_{ss,\theta}(\mathbf{a}) = \prod_{i=1}^{d}(1 - \theta_i a_i)^{-1}(1 + (1 - \theta_i)a_i), \tag{14}$$

where $\theta_i \in [0, 1]$ $(i = 1, 2, \ldots d)$ are given constants. Then the iteration (2) means

$$y_{ss,\theta}^{n+1} = \prod_{i=1}^{d}(\mathbf{I} - \theta_i\tau\mathbf{A}_i)^{-1}(\mathbf{I} + \tau(1 - \theta_i)\mathbf{A}_i)y_{ss,\theta}^{n}. \tag{15}$$

This approximation has second order when $\theta_i = 0.5$ for each $i = 1, 2, \ldots d$, and first order otherwise. Hence (15) yields a consistent numerical scheme to the problem (1). In the following we analyze its absolute stability and contractivity in the Hilbert space $\mathbf{H} = (\mathbb{R}^N, (\cdot, \cdot)_H)$, where $(\cdot, \cdot)_H$ denotes a scalar product in $\mathbb{R}^N$.

Theorem 1. *Assume that the matrices* $\mathbf{A}_i$ *are negative definite and* $\theta_i \in [0.5, 1]$ *for each* $i = 1, 2, \ldots d$. *Then the numerical method (15) is absolute stable and contractive in the norm* $\|x\|_H^2 = (x, x)_H$.

Proof. Obviously it is enough to show that under the conditions of the lemma the estimation

$$\|(\mathbf{I} - \theta_i \tau \mathbf{A}_i)^{-1}(\mathbf{I} + \tau(1-\theta_i)\mathbf{A}_i)\|_H^2 \leq 1 \tag{16}$$

holds for each $i = 1, 2, \ldots d$. We have

$$\begin{aligned}
&\|(\mathbf{I} - \theta_i \tau \mathbf{A}_i)^{-1}(\mathbf{I} + \tau(1-\theta_i)\mathbf{A}_i)\|_H^2 = \\
&\sup_{\mathbf{x} \in \mathbf{H}} \frac{\left((\mathbf{I} - \theta_i \tau \mathbf{A}_i)^{-1}(\mathbf{I} + \tau(1-\theta_i)\mathbf{A}_i)\mathbf{x}, (\mathbf{I} - \theta_i \tau \mathbf{A}_i)^{-1}(\mathbf{I} + \tau(1-\theta_i)\mathbf{A}_i)\mathbf{x}\right)_H}{(\mathbf{x}, \mathbf{x})_H} \\
&= \sup_{\mathbf{y} \in \mathbf{H}} \frac{((\mathbf{I} + \tau(1-\theta_i)\mathbf{A}_i)\mathbf{y}, (\mathbf{I} + \tau(1-\theta_i)\mathbf{A}_i)\mathbf{y})_H}{((\mathbf{I} - \theta_i \tau \mathbf{A}_i)\mathbf{y}, (\mathbf{I} - \theta_i \tau \mathbf{A}_i)\mathbf{y})_H} = \\
&= \sup_{\mathbf{y} \in \mathbf{H}} \frac{(\mathbf{y}, \mathbf{y})_H + 2\tau(1-\theta_i)(\mathbf{A}_i\mathbf{y}, \mathbf{y})_H + \tau^2(1-\theta_i)^2(\mathbf{A}_i\mathbf{y}, \mathbf{A}_i\mathbf{y})_H}{(\mathbf{y}, \mathbf{y})_H - 2\tau\theta_i(\mathbf{A}_i\mathbf{y}, \mathbf{y})_H + \tau^2(1-\theta_i)^2(\mathbf{A}_i\mathbf{y}, \mathbf{A}_i\mathbf{y})_H}.
\end{aligned}$$

Since $(\mathbf{A}_i\mathbf{y}, \mathbf{y})_H \leq 0$, therefore in case $\theta_i \in [0.5, 1]$ the expression in the above supremum is not greater than one for any $\mathbf{y} \in \mathbf{H}$, which proves the statement.

Remark 3. The statement (16) is the generalization of the Kellogg lemma [7].

When we use the θ-approximations (13) for each exponential in the Strang-Marchuk splitting (7) then the approximation function is

$$\begin{aligned}
\mathrm{r}_{str,\theta}(\mathbf{a}) = &\prod_{i=1}^{d-1}(1 - \theta_i^1 \frac{a_i}{2})^{-1}(1 + (1-\theta_i^1)\frac{a_i}{2}) \\
&(1 - \theta_d^1 a_d)^{-1}(1 + (1-\theta_d^1)a_d)\prod_{i=1}^{d-1}(1 - \theta_i^2 \frac{a_i}{2})^{-1}(1 + (1-\theta_i^2)\frac{a_i}{2}),
\end{aligned} \tag{17}$$

where $\theta_i^k \in [0, 1]$ for $k = 1, 2$ and $i = 1, 2, \ldots d$. Then the iteration (2) means

$$y_{str,\theta}^{n+1} = \mathrm{r}_{str,\theta}(\tau\mathbf{A})y_{str,\theta}^n. \tag{18}$$

When $d = 2$ then (18) results in the following numerical scheme:

$$\begin{aligned}
y_{str,\theta}^{n+1} = &(\mathbf{I} - \theta_1^1 \frac{\tau}{2}\mathbf{A}_1)^{-1}(\mathbf{I} + (1-\theta_1^1)\frac{\tau}{2}\mathbf{A}_1)(\mathbf{I} - \theta_2^1 \tau \mathbf{A}_2)^{-1} \\
&(\mathbf{I} + (1-\theta_2^1)\tau\mathbf{A}_2)(\mathbf{I} - \theta_1^2 \frac{\tau}{2}\mathbf{A}_1)^{-1}(\mathbf{I} + (1-\theta_1^2)\frac{\tau}{2}\mathbf{A}_1)y_{str,\theta}^n.
\end{aligned} \tag{19}$$

On the base of the proof of the Theorem 1 we can check the validity of the following

Theorem 2. *Assume that the matrices* $\mathbf{A}_i$ *are negative definite and* $\theta_i \in [0.5, 1]$ *for each* $i = 1, 2, \ldots d$. *Then the numerical method (18) is absolute stable and contractive in the norm* $\|x\|_H^2 = (x, x)_H$.

Remark 4. For the weighted sequential splitting the definition of the $\mathrm{r}_{wss,\theta}(z)$ from the function $\mathrm{r}_{ss,\theta}(z)$ is straightforward. Namely, for two operators ($d=2$) we have

$$y^{n+1}_{wss,\theta} = \mathrm{r}_{wss,\theta}(\tau\mathbf{A})y^n_{wss,\theta}, \tag{20}$$

where

$$\mathrm{r}_{wss,\theta}(\tau\mathbf{A}) = 0.5[\mathrm{r}^{(1,2)}_{ss,\theta}(\tau\mathbf{A}) + \mathrm{r}^{(2,1)}_{ss,\theta}(\tau\mathbf{A})]. \tag{21}$$

(Here $\mathrm{r}^{(1,2)}_{ss,\theta}(\tau\mathbf{A})$ and $\mathrm{r}^{(2,1)}_{ss,\theta}(\tau\mathbf{A})$ denote the operators $\mathrm{r}_{ss,\theta}(\tau\mathbf{A})$ in different ordering w.r.t. the sub-operators $\mathbf{A}_1$ and $\mathbf{A}_2$.) As an obvious consequence of Theorem 1, we get that for the negative definite matrices $\mathbf{A}_1$ and $\mathbf{A}_2$ the numerical method (20) is also absolute stable and contractive in the norm $\|\cdot\|_H$ by the choices of the parameters $\theta_i \in [0.5, 1]$ for each $i = 1, 2, 3, 4$.

Obviously, in numerical methods for the different sub-problems not only the weight parameters θ^k_i can be chosen differently, but the step-sizes, too. Let us denote by Δt the step-size of the numerical method. Then we can choose $\Delta t = \tau/M$ for some M integer. This results in that the numerical method chosen should be used M-times in order to solve the split sub-problem in one split step. For instance, for the sequential splitting, by the choice of the step-size $\Delta t_i = \tau/M_i$ for the i-th sub-problem ($i = 1, 2, \ldots d$), we have

$$y^{n+1}_{ss,\theta} = \prod_{i=1}^{d}[(\mathbf{I} - \frac{\theta_i\tau}{M_i}\mathbf{A}_i)^{-1}(\mathbf{I} + \frac{\tau(1-\theta_i)}{M_i}\mathbf{A}_i)]^{M_i} y^n_{ss,\theta}. \tag{22}$$

Clearly, Theorems 1 and 2 remain valid for such case, too. We remark that, even in the case of commutativity of the operators, the different choices of M result in different numerical methods. E.g. for $M = 1$ (i.e. $\tau = \Delta t$) for one operator the explicit Euler method is

$$y^{n+1} = (\mathbf{I} + \tau\mathbf{A})y^n, \tag{23}$$

while for $M = 2$ (i.e. $\tau = \Delta t/2$) the method is

$$y^{n+1} = (\mathbf{I} + \frac{\tau}{2}\mathbf{A})(\mathbf{I} + \frac{\tau}{2}\mathbf{A})y^n = (\mathbf{I} + \tau\mathbf{A} + \frac{\tau^2}{4}\mathbf{A}^2)y^n. \tag{24}$$

However the difference of the methods coincides with the order of the method.

4 Some Splitting Schemes

In this section we give some known splitting schemes to the problem (1) and we point out their relation to the above operator splitting-numerical schemes.

4.1 The Splitting Method Based on Implicit Scheme ([9], p.231)

Let us consider the following scheme

$$\begin{aligned}\frac{y^{n+\frac{1}{d}}-y^n}{\tau} &= \mathbf{A}_1 y^{n+\frac{1}{d}}\\ &\vdots\\ \frac{y^{n+1}-y^{n+\frac{d-1}{d}}}{\tau} &= \mathbf{A}_d y^{n+1}\end{aligned} \tag{25}$$

where $y^0 = w_0$ and $n = 0, 1, 2, \ldots$.

Obviously, this method corresponds to the sequential splitting with implicit Euler method to each sub-problems, i.e. (25) is a special case of (15) with $\theta_i = 1$ for each $i = 1, 2, \ldots d$.

4.2 Componentwise Splitting Based on Crank-Nicolson Schemes ([9], p.235)

Let us consider the following scheme

$$\begin{aligned}\frac{y^{n+\frac{1}{d}}-y^n}{\tau} &= \frac{1}{2}\mathbf{A}_1(y^n + y^{n+\frac{1}{d}})\\ &\vdots\\ \frac{y^{n+1}-y^{n+\frac{d-1}{d}}}{\tau} &= \frac{1}{2}\mathbf{A}_d(y^{n+\frac{d-1}{d}} + y^{n+1})\end{aligned} \tag{26}$$

where $y^0 = w_0$ and $n = 0, 1, 2, \ldots$.

Obviously, this method corresponds to the sequential splitting with middle point method to each subproblems, i.e. (26) is a special case of (15) with $\theta_i = 0.5$ for each $i = 1, 2, \ldots d$. This method has of second order accuracy.

4.3 Two-Cycle Multi-component Splitting ([9], p.247)

Let us consider the following scheme

$$\begin{aligned}\frac{y^{n+\frac{1}{2d}}-y^n}{\tau} &= \frac{1}{4}\mathbf{A}_1(y^n + y^{n+\frac{1}{2d}})\\ &\vdots\\ \frac{y^{n+\frac{d}{2d}}-y^{n+\frac{d-1}{2d}}}{\tau} &= \frac{1}{4}\mathbf{A}_d(y^{n+\frac{d-1}{2d}} + y^{n+\frac{d}{2d}})\\ &\vdots\\ \frac{y^{n+1}-y^{n+\frac{2d-1}{2d}}}{\tau} &= \frac{1}{4}\mathbf{A}_1(y^{n+\frac{2d-1}{2d}} + y^{n+1})\end{aligned} \tag{27}$$

where $y^0 = w_0$ and $n = 0, 1, 2, \ldots$. This method corresponds to the Strang-Marchuk splitting with the middle point integration rule and $\Delta t = \tau$ for the problems with the operators $\mathbf{A}_i$ $(i = 1, 2, \ldots d-1)$ and $\Delta t = \tau/2$ for the $\mathbf{A}_d$. We note that the method (27) is connected with the method (26) in the following way. Let us cut in half the splitting interval. If we use the method (26) on the first half on the interval in the indicated order of the operators and then we use the same method on the second half but in the reverse order of the operators, then we obtain (27).

Finally we remark that many others numerical schemes can be also obtained, see [4].

5 Conclusion

In this paper we analyzed the combination of the operator splitting method with the numerical methods applied to the split sub-problems. We proved that both methods can be considered as approximation to the exponential function. Therefore, their common investigation is possible and the well-known numerical schemes can be obtained.

This approach perhaps can be used successfully for the general case, i.e. when we consider (1) as an abstract Cauchy problem in Banach space under the assumption that the linear operator A is a generator of the C_0-semigroup. However, this approach needs some further investigations.

References

1. Csomós, P., Faragó, I., and Havasi, Á., Weighted sequential splittings and their analysis, *Comput. Math. Appl.*, **50** (2005) 1017–1031.
2. Dekker, K. and Verwer, J. G. *Stability of Runge-Kutta methods for stiff nonlinear differential equations*, North-Holland, Amsterdam (1984).
3. Dimitriu, G., *Parameter identification in a two-dimensional parabolic equation using an ADI based solver*, Lect. Notes Comp.Sci. 2179, Springer Verlag, Berlin, (2001) 479–486.
4. Faragó, I., Splitting methods for abstract Cauchy problems, *Lect. Notes Comp.Sci.*, **3401**, Springer Verlag, Berlin, (2005) 35–45.
5. Horváth, R., Uniform Treatment of the Numerical Time-Integration of the Maxwell Equations, *Lecture Notes in Computational Science and Engineering*, Springer Verlag, Berlin, (2003) 231–239.
6. Hundsdorfer, W. and Verwer, J., *Numerical solution of time-dependent advection-diffusion-reaction equations*, Springer, Berlin, (2003).
7. Kellogg, R. B., Another alternating-direction-implicit method, *J. Soc. Indust. Appl. Math.*, **11** (1963) 976–979.
8. Marchuk, G., Some applicatons of splitting-up methods to the solution of problems in mathematical physics, *Aplikace Matematiky*, **1** (1968) 103–132.
9. Marchuk, G., *Splitting and alternating direction methods*, North Holland, Amsterdam, (1990).
10. Strang, G., Accurate partial difference methods I: Linear Cauchy problems. *Archive for Rational Mechanics and Analysis*, **12** (1963) 392–402.
11. Strang, G., On the construction and comparison of difference schemes, *SIAM J. Num. Anal.* **5** (1968) 506–517.

Dispersion Analysis of Operator Splittings in the Linearized Shallow Water Equations

Ágnes Havasi

Eötvös Loránd University, H-1117 Budapest, Pázmány P.s. 1/A
hagi@nimbus.elte.hu

Abstract. The shallow water equations describe motions in a shallow, incompressible, non-viscous fluid layer on the rotating Earth. Due to their relative simplicity, they are widely used for testing and analysing new numerical methods developed for weather predicition models. In this paper we apply different operator splittings to the linearized form of the shallow water equations obtained by the method of small perturbations. This system has three harmonic wave solutions with known dispersion relations and phase velocities. We investigate how the application of operator splitting modifies these important characteristics, and compare the performance of different splitting methods from this point of view.

1 Introduction

Operator splitting is widely applied in different fields of applied mathematics, where huge systems of partial differential equations are to be solved. This procedure allows us to lead the solution of the original problem back to the solution of a sequence of simpler sub-problems.

An important area where splitting is often used is large-scale air pollution modelling. Results on the application of splitting in transport-chemistry models can be found e.g. in [9, 1, 2]. Splitting can also be applied in dynamical (weather prediction) models, where the hydro- and thermodynamic variables of the atmophere are predicted. In [4] the Strang splitting and approximate matrix factorization are applied to the linearized shallow water equations. We would like to supplement the results of that paper by examining and comparing different splitting methods in the test problem fomulated there.

The structure of the paper is as follows. In Section 2 the shallow water equations and their linearized form are presented. In Section 3 we apply different splitting methods to this problem and give the formulae of the corresponding numerical frequencies. In Section 4 we show the results of our numerical comparisons between the sequential, symmetrically weighted sequential (SWS) and additive splittings. Two characteristic properties of the wave solutions are examined: their artificial amplification or damping and their phase velocities. The conclusions are summarized in Section 5.

I. Lirkov, S. Margenov, and J. Waśniewski (Eds.): LSSC 2005, LNCS 3743, pp. 355–362, 2006.

2 The Shallow Water Equations and Their Dispersion Relations

The shallow water equations describe motions in a shallow, incompressible, nonviscous fluid layer on the rotating Earth. The derivation of the shallow water equations can be found e.g. in [5]. Let u and v denote the horizontal velocity components, h the height of the top boundary of the fluid, g the gravitational acceleration and f the Coriolis parameter $2\Omega \sin\phi$, where Ω is the angular velocity of the Earth and ϕ the latitudinal degree. Then the shallow water equations read as

$$\frac{\partial u}{\partial t} + u\frac{\partial u}{\partial x} + v\frac{\partial u}{\partial y} + g\frac{\partial h}{\partial x} - fv = 0, \tag{1}$$

$$\frac{\partial v}{\partial t} + u\frac{\partial v}{\partial x} + v\frac{\partial v}{\partial y} + g\frac{\partial h}{\partial y} + fu = 0, \tag{2}$$

$$\frac{\partial h}{\partial t} + u\frac{\partial h}{\partial x} + v\frac{\partial h}{\partial y} + h(\frac{\partial u}{\partial x} + \frac{\partial v}{\partial y}) = 0. \tag{3}$$

We restrict our investigation to the so-called normal modes of this system. To this aim we linearize the equations by means of the method of small perturbations. Assume that changes of the model variables around their average fields are small compared to these average fields. Then, referring to the average values by upper bars and to the small perturbations by apostrophies, the instantaneous values of u, v and h can be written as $u = \overline{u} + u'$, $v = \overline{v} + v'$ and $h = \overline{h} + h'$. Let us substitute these sums into equations (1)-(3). Exploiting the fact that the average fields satisfy the nonlinear equations and that the terms containing products of perturbation quantities can be neglected, we obtain the following linearized form of the shallow water equations

$$\frac{\partial u'}{\partial t} + \overline{u}\frac{\partial u'}{\partial x} + \overline{v}\frac{\partial u'}{\partial y} + g\frac{\partial h'}{\partial x} - fv' = 0, \tag{4}$$

$$\frac{\partial v'}{\partial t} + \overline{u}\frac{\partial v'}{\partial x} + \overline{v}\frac{\partial v'}{\partial y} + g\frac{\partial h'}{\partial y} + fu' = 0, \tag{5}$$

$$\frac{\partial h'}{\partial t} + \overline{u}\frac{\partial h'}{\partial x} + \overline{v}\frac{\partial h'}{\partial y} + \overline{h}(\frac{\partial u'}{\partial x} + \frac{\partial v'}{\partial y}) = 0. \tag{6}$$

with the corresponding initial and periodic boundary conditions which are motivated by the physics of the problem.

As usual in dispersion analysis, we apply Fourier's method, i.e., we seek the solution in the form

$$u'(x, y, t) = Ue^{-i\omega t}e^{i(k_1 x + k_2 y)}, \tag{7}$$

$$v'(x, y, t) = Ve^{-i\omega t}e^{i(k_1 x + k_2 y)}, \tag{8}$$

$$h'(x, y, t) = He^{-i\omega t}e^{i(k_1 x + k_2 y)}. \tag{9}$$

Here U, V and H are the amplitudes of the waves, and

$$k_1 = \frac{2\pi}{L_x}, \quad k_2 = \frac{2\pi}{L_y}, \quad \omega = \frac{2\pi}{T} \tag{10}$$

are the wave numbers and the frequency. Substituting (7), (8) and (9) into system (4)-(6), we are led to the so-called characteristic system of homogeneous, linear algebraic equations for the determination of the amplitudes U, V and H. This has a non-trivial solution only if its determinant is zero, which means the following condition:

$$\hat{\omega}^3 + \hat{\omega}(f^2 + g\overline{h}(k_1^2 + k_2^2)) = 0, \tag{11}$$

where $\hat{\omega} = -i\omega + \overline{u}ik_1 + \overline{v}ik_2$. For fixed values of k_1 and k_2 this equation gives three different solutions for ω, namely:

$$\omega_1 = \overline{u}k_1 + \overline{v}k_2, \tag{12}$$

$$\omega_2 = \overline{u}k_1 + \overline{v}k_2 - \sqrt{f^2 + g\overline{h}(k_1^2 + k_2^2)}, \tag{13}$$

$$\omega_3 = \overline{u}k_1 + \overline{v}k_2 + \sqrt{f^2 + g\overline{h}(k_1^2 + k_2^2)}. \tag{14}$$

The first value corresponds to a slow, so-called advective wave, while the second and third to fast gravity waves. Note that none of these frequencies has an imaginary part, which means that the wave solutions are neither amplified, nor damped. The propagation (or phase) velocities of these waves can be obtained by dividing the frequencies by $|\underline{k}| = \sqrt{(k_1^2 + k_2^2)}$.

3 Operator Splitting Techniques

In order to apply operator splitting in the linearized shallow water equations, first we transform (4)-(6) into a more convenient from. For solutions of the form $\hat{q}(t)e^{i(k_1x+k_2y)}$ system (4)-(6) is equivalent to the following system of ODE's:

$$\frac{d\hat{q}}{dt} = A\hat{q}(t), \tag{15}$$

where

$$A = \begin{pmatrix} -\overline{u}k_1i - \overline{v}k_2i & f & -gk_1i \\ -f & -\overline{u}k_1i - \overline{v}k_2i & -gk_2i \\ -\overline{h}k_1i & -\overline{h}k_2i & -\overline{u}k_1i - \overline{v}k_2i \end{pmatrix}. \tag{16}$$

We split the right-hand side of (15) into the sum $A_1\hat{q}(t) + A_2\hat{q}(t)$, where

$$A_1 = -ik_1 \begin{pmatrix} \overline{u} & i\frac{f}{k_1} & g \\ 0 & \overline{u} & 0 \\ \overline{h} & 0 & \overline{u} \end{pmatrix}, \quad A_2 = -ik_2 \begin{pmatrix} \overline{v} & 0 & 0 \\ -i\frac{f}{k_2} & \overline{v} & g \\ 0 & \overline{h} & \overline{v} \end{pmatrix}. \tag{17}$$

This decomposition is convenient to apply in large-scale computations since it involves directional separation of all derivatives and the Coriolis terms in system (4)-(6). For the application of splitting we first divide the time axis into sub-intervals of length τ, the so-called splitting time step. Denote the n-th time level by t_n, and let $\hat{q}(t_n)$ be arbitrary. The simplest splitting method is the sequential splitting, which means that we solve the following sequence of problems at the sub-intervals $[t_n, t_n + \tau]$:

$$\begin{cases} \dfrac{d\hat{q}^{(1)}}{dt} = A_1 \hat{q}^{(1)}, \\ \hat{q}^{(1)}(t_n) = \hat{q}(t_n) \end{cases} \tag{18}$$

$$\begin{cases} \dfrac{d\hat{q}^{(2)}}{dt} = A_2 \hat{q}^{(2)}, \\ \hat{q}^{(2)}(t_n) = \hat{q}^{(1)}(t_n + \tau), \end{cases} \tag{19}$$

and the splitting solution at t_{n+1} is defined as $\hat{q}^{(2)}(t_{n+1}) = \hat{q}^{(2)}(t_n + \tau)$.

If a function $\hat{q}_{\text{sp}}(t_n)$, $n = 1, 2, \ldots$ is a solution to problem (18), (19) then

$$\hat{q}_{\text{sp}}(t_n + \tau) = e^{A_2 \tau} e^{A_1 \tau} \hat{q}_{\text{sp}}(t_n). \tag{20}$$

Let us look for the solution in the form

$$\hat{q}_{\text{sp}}(t_n) = q e^{-i\omega_{\text{sp}} t_n}, \tag{21}$$

where $q = const \in I\!R^3$. By substituting (21) into (20), we are led to the equality

$$q e^{-i\omega_{\text{sp}}(t_n + \tau)} = e^{A_2 \tau} e^{A_1 \tau} q e^{-i\omega_{\text{sp}} t_n}. \tag{22}$$

Dividing the two sides by $e^{-i\omega_{\text{sp}} t_n}$ we obtain that

$$e^{-i\omega_{\text{sp}} \tau} q = e^{A_2 \tau} e^{A_1 \tau} q. \tag{23}$$

Consequently, $q e^{-i\omega_{\text{sp}} t_n}$ is a solution of (20) if and only if $e^{-i\omega_{\text{sp}} \tau}$ is an eigenvalue of the matrix $e^{A_2 \tau} e^{A_1 \tau}$, and the values of ω_{sp} are

$$\omega_{\text{sp},j} = \frac{\ln \lambda(M_{\text{seq}})_j}{\tau} i, \quad j = 1, 2, 3, \tag{24}$$

where M_{seq} stands for the matrix $e^{A_2 \tau} e^{A_1 \tau}$ and λ for its eigenvalues.

Other traditional splitting methods are the Marchuk–Strang splitting [7] and the symmetrically weighted sequential (SWS) splitting [6, 1]. A recently developed splitting method is the additive splitting [3]. Here we give the corresponding matrices, by which M_{seq} should be replaced when the latter three splitting methods are applied:

- Marchuk–Strang (MS) splitting: $M_{\text{MS}} = e^{A_1 \frac{\tau}{2}} e^{A_2 \tau} e^{A_1 \frac{\tau}{2}}$,
- SWS splitting: $M_{\text{SWS}} = \frac{1}{2} e^{A_2 \tau} e^{A_1 \tau} + \frac{1}{2} e^{A_1 \tau} e^{A_2 \tau}$,
- additive splitting: $M_{\text{add}} = e^{A_2 \tau} + e^{A_1 \tau} - I$.

It is well-known that in terms of the local splitting error the MS and SWS splittings have second order, while the sequential and additive splittings only first order. Therefore, one would expect the former two methods to perform better in the comparisons. However, it can be proved that in our case the MS splitting and the sequential splitting will give the same result.

First, it is easy to show that the sequential splitting in the order $A_1 \to A_2$ will result in the same numerical frequencies as in the order $A_2 \to A_1$. This follows from the fact that for any matrices $F, G \in I\!R^{n\times n}$, $n \in I\!N$, $\sigma(FG) = \sigma(GF)$, where σ denotes the spectrum of a matrix. (For an elegant proof see [8], p. 63.) Applying this theorem to $F = e^{A_1\tau}$ and $G = e^{A_2\tau}$, we obtain that $\sigma(e^{A_1\tau}e^{A_2\tau}) = \sigma(e^{A_2\tau}e^{A_1\tau})$.

The following statement deals with the relation between the spectrum of M_{seq} and M_{MS}.

Proposition 1. Let $\lambda \in \sigma(e^F e^G)$ with an eigenvector $\underline{x} \in I\!R^n$. Then $\lambda \in \sigma(e^{G/2}e^F e^{G/2})$ with the eigenvector $\hat{\underline{x}} = e^{G/2}\underline{x}$.

Proof. Due to the assumption, we have

$$e^F e^G \underline{x} = \lambda \underline{x}. \tag{25}$$

Hence, the following relations are valid:

$$e^{G/2}e^F e^{G/2}\hat{\underline{x}} = e^{G/2}e^F e^G \underline{x} = e^{G/2}(\lambda \underline{x}) = \lambda e^{G/2}\underline{x} = \lambda \hat{\underline{x}}, \tag{26}$$

which proves our statement.

Remark. It is easy to prove that the converse inclusion is also true.

From the above consideration it follows that, from the viewpoint of the dispersion analysis, the MS splitting is not worth applying in our model problem, because the application of the sequential splitting gives the same result with less computation. Therefore, we will not deal with this splitting scheme in the sequel. (We remark that this equality of the eigenvalues is already not true concerning the SWS and additive splitting methods.) Our aim is to compare the numerical frequencies $\omega_{\text{sp},j}$, $j = 1, 2, 3$ obtained in the sequential, SWS and additive splitting methods with their exact values under (12)-(14).

4 Numerical Comparisons

In this part we investigate both the real and imaginary parts of the numerical frequencies $\omega_{\text{sp},j}$, $j = 1, 2, 3$ obtained by the different splittings. Positive imaginary parts express amplification, while negative imaginary parts damping of the wave solutions. Let us reiterate here that the exact normal modes propagate with a constant amplitude. The real parts divided by $|\underline{k}|$ give the phase velocity of the waves.

We only investigate waves with wave vectors $\underline{k}$ for which $k_1 = \cos\beta_l$ and $k_2 = \sin\beta_l$ with $\beta_l = l\frac{2\pi}{36}, l = 1, 2, \ldots 36$. (Increasing the resolution did not affect the results considerably.) In the comparisons the splitting time steps τ are chosen as 10^{-4}, 10^{-3} and $10^{-2}s$. (The time period of the fastest wave of the exact solution for $|\underline{k}| = 1$ is approximately $2 \cdot 10^{-2}s$.) We used the same parameter set as that used in [4], namely: $\overline{u} = \overline{v} = 10m/s$, $\overline{h} = 10000m$, $g = 9.8m/s^2$ and $\phi = 45^o$. The computations were done in Matlab.

4.1 Artificial Amplification or Damping

We computed the minimum and maximum values of the imaginary parts of the frequencies $\omega_{\text{sp},j}$ for the three different wave solutions and for the 36 wave vectors. The results obtained for the sequential, SWS and additive splittings are given in Table 1.

Table 1. Minimum and maximum values of the imaginary parts of the frequencies for the different splitting schemes. The first value in each cell corresponds to the advective wave, and the second and third values to the gravity waves.

τ	Sequential		SWS		Additive	
	Min	Max	Min	Max	Min	Max
10^{-4}	-2.08E-9	2.07E-9	-2.91E-4	-1.11E-12	-2.38E+0	-2.22E-12
	-1.04E-9	1.04E-9	-1.46E-4	-2.22E-12	-7.77E-12	1.41E+0
	-1.04E-9	1.04E-9	-1.46E-4	0	-7.77E-12	1.41E+0
10^{-3}	-2.07E-7	2.07E-7	-2.88E-1	-2.22E-12	-2.40E+1	-2.44E-12
	-1.04E-7	1.04E-7	-1.44E-1	-8.88E-13	3.77E-12	1.38E+1
	-1.04E-7	1.04E-7	-1.44E-1	-7.77E-13	3.55E-12	1.38E+1
10^{-2}	-2.56E-5	2.56E-5	-2.98E+2	-4.39E-12	-1.88E+2	4.72E+1
	-1.28E-5	1.28E-5	-2.88E+2	1.87E-12	-5.98E+1	8.44E-1
	-1.28E-5	1.28E-5	-1.43E+2	1.89E-12	-4.32E-2	4.72E+1

According to our expectations, all the methods produce decreasing errors with decreasing time step. In the case of the sequential splitting the amplification and damping are equally important. This splitting causes the smallest damping among the three methods. In the SWS splitting there is only negligible amplification for some waves, but the damping is a few magnitudes stronger than for the sequential splitting, and especially strong for the biggest τ. The additive splitting shows strong amplification for the gravity waves and strong damping for the advective waves even for the smaller values of τ, and for the biggest time step both effects are significant.

4.2 Numerical Phase Velocities

We computed the real parts of the frequencies $\omega_{\text{sp},j}$ for the 36 wave vectors under consideration. Since for these wave number vectors $|\underline{k}| = 1$, therefore the real parts of the frequencies $\omega_{\text{sp},j}$ are equal to the phase velocities of the corresponding wave solutions.

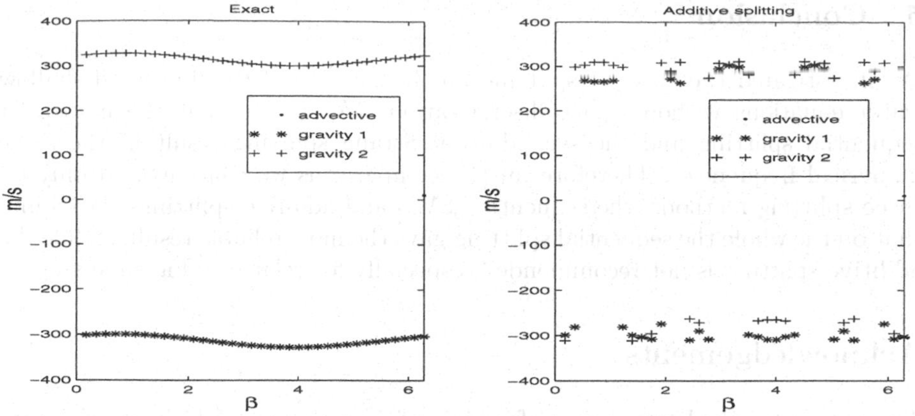

Fig. 1. The exact phase velocities (left panel) and the numerical phase velocities obtained by the additive splitting (right panel)

We found that for $\tau < 10^{-2}s$ all the three methods give excellent results. For $\tau = 10^{-2}s$ the solutions are already different. The left panel of Fig. 1 shows the exact phase velocities. The additive splitting produces unacceptably big errors, especially for the advective wave, as shown in the right panel of Fig. 1. The other two methods give realistic velocity values for all wave components. However, the directions of propagation of the gravity waves are not always simulated properly. Note that the exact gravity waves propagate always in the opposite directions (left panel of Fig. 1). In the case of the sequential splitting we can already find some wave vectors for which the direction of propagation is the same for the two gravity waves, see the left panel of Fig. 2. From this point of view, the SWS splitting gave even worse results: according to the right panel of Fig. 2, here the two gravity wave solutions propagate always in the same direction!

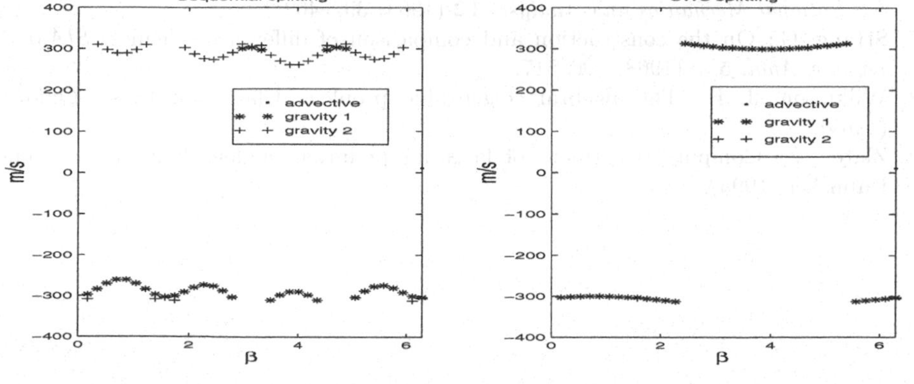

Fig. 2. The numerical phase velocities obtained by the sequential splitting (left panel) and the SWS splitting (right panel)

5 Conclusion

We investigated the effect of splitting on the solution of the linearized shallow water equations without space discretization. We showed that the first-order sequential splitting and the second-order Strang splitting result in the same numerical frequencies. Therefore, in the comparisons we concentrated only on three splitting methods: the sequential, SWS and additive splittings. We found that on the whole the sequential splitting gave the most reliable results, while the additive splitting is not recommended, especially for relatively big time steps.

Acknowledgements

The author thanks István Faragó for his useful suggestions. This work was supported by Hungarian National Research Founds (OTKA) N. T043765 and NATO Collaborative Linkage Grant N. 980505.

References

1. Botchev, M., Faragó, I., Havasi, Á.: Testing weighted splitting schemes on a one-column transport-chemistry model, *Int. J. Env. Pol.* **22**, Nos. 1/2, (2004) 3–16.
2. Dimov, I., Faragó, I., Havasi, Á., Zlatev, Z.: Operator splitting and commutativity analysis in the Danish Eulerian Model, *Math. Comp. Sim.* **67** (2004) 217–233.
3. Gnandt, B.: A new operator splitting method and its numerical investigation. In: *Advances in air pollution modeling for environmental security*, Faragó, I., Georgiev, K, Havasi, Á. (Eds.), Nato Science Series: IV: Earth and Environmental Sciences, **54** (2005) 229–241.
4. Lanser, D.: A comparison of operator splitting and approximate matrix factorization for the shallow water equations in spherical geometry. Technical Report MAS-R0115, CWI, Amsterdam (2001).
5. Pedlosky, J.: Geophysical fluid dynamics. Springer (1998).
6. Strang, G.: Accurate partial difference methods I: Linear Cauchy problems. *Archive for Rational Mechanics and Analysis* **12** (1963) 392–402.
7. Strang, G.: On the construction and comparison of difference schemes. *SIAM J. Numer. Anal.* **5** 3, (1968) 506–517.
8. Wilkinson, J. H.: The algebraic eigenvalue problem. Clarendon Press, Oxford, (1965).
9. Zlatev, Z.: Computer treatment of large air pollution models. Kluwer Academic Publisher (1995).

Operator Splittings for the Numerical Solution of the Maxwell's Equations

Róbert Horváth

University of West-Hungary,
Erzsébet u. 9, 9400 Sopron, Hungary
rhorvath@ktk.nyme.hu

Abstract. In this paper, the operator splitting techniques are applied for the semi-discretized Maxwell's equations. The semi-discretization is performed on a staggered grid structure like other frequently used methods (YEE, NZCZ, KFR). We show how these methods fit into the framework of the splitting methods. We construct a new unconditionally stable solution method, which possesses all favourable properties of the NZCZ-method, and additionally it conserves the energy density of the electromagnetic field. We compare the new method with the NZCZ-method presenting a 2D numerical example.

Keywords: Maxwell's equations, FDTD-Method, Unconditional Stability.

1 Introduction

The mathematical model of electromagnetic problems can be formulated in the form of the so-called Maxwell's equations (in source-free case)

$$\varepsilon\partial_t \mathbf{E} = \nabla \times \mathbf{H}, \quad \mu\partial_t \mathbf{H} = -\nabla \times \mathbf{E} \tag{1}$$

$$\nabla(\varepsilon\mathbf{E}) = 0, \quad \nabla(\mu\mathbf{H}) = 0, \tag{2}$$

where $\mathbf{E} = (E_x(t,x,y,z), E_y(t,x,y,z), E_z(t,x,y,z))$ is the electric field strength, $\mathbf{H} = (H_x(t,x,y,z), H_y(t,x,y,z), H_z(t,x,y,z))$ is the magnetic field strength, $\varepsilon = \varepsilon(x,y,z)$ is the electric permittivity and $\mu = \mu(x,y,z)$ is the magnetic permeability. It is well-known that the divergence equations in (2) follow from the curl equations in (1) when we suppose that the fields $\mathbf{E}$ and $\mathbf{H}$ were divergence-free at the initial point of time (see e.g. [10]). This means that we must solve only the curl equations applying divergence-free initial conditions for $\mathbf{E}$ and $\mathbf{H}$. In real-life problems, the exact solution of this system is very complicated or even impossible. This is why numerical methods are generally applied.

The most frequently used method is the YEE-method (or Finite-Difference Time-Domain method), which was introduced for the Maxwell's equations in 1966 ([13]). This method starts with the definition of a generally rectangular mesh (with the choice of the step-sizes Δx, Δy and Δz) for the electric field

I. Lirkov, S. Margenov, and J. Waśniewski (Eds.): LSSC 2005, LNCS 3743, pp. 363–371, 2006.

and another staggered (by $\Delta x/2, \Delta y/2$ and $\Delta z/2$) grid for the magnetic field in the computational domain. In the time domain, the YEE-method applies the leap-frog scheme, ensuring second order accuracy both in time and spatial variables. Albeit investigations show (see e.g. [8]) that comparing the non-staggered, the colocated staggered and the non-colocated staggered grids the last one has the most favourable properties, the YEE-method suffers from a strict stability condition. Namely, the method is stable if and only if the condition

$$\Delta t < \frac{1}{c\sqrt{(1/\Delta x)^2 + (1/\Delta y)^2 + (1/\Delta z)^2}} \tag{3}$$

is fulfilled, where $c = 1/\sqrt{\varepsilon\mu}$ is the maximal speed of light in the computational domain. A comprehensive survey of the YEE-method can be found in [10].

A lot of efforts has been invested during the last decades to bridge the stability problem of the YEE-method. The main goal was to construct methods, where Δt can be chosen based on accuracy considerations instead of stability reason. Papers [9] and [11] came up with an unconditionally stable method in 1999 (NZCZ-method). The NZCZ-method is an alternating direction implicit (ADI) method, where the alteration is applied in the terms of the curl operator. In the above papers, the stability was shown only on test-problems or with the extensive use of computer algebra (see [12]). Pure mathematical proofs can be found, for instance, in [1, 3, 6]. The paper [7], which was appeared in 2001, also discusses an unconditionally stable numerical scheme (KFR-method).

In this paper, we shed light on the way the above methods fit into the framework of operator splitting techniques and apply operator splitting to construct and investigate a new unconditionally stable numerical scheme.

The operator splitting method is an efficient tool for the solution of differential equations. The basic idea of the method is the splitting of the original problem into sub-problems according to the physical and/or chemical processes involved. Then these sub-problems, which are supposed to be handled more easily, are solved sequentially using some appropriate methods. The solution of the original problem is approximated by the solutions of the sub-problems. As we will apply the splitting methods for the semi-discretized Maxwell's equations, we give a short mathematical description of the method for linear systems of first order ordinary differential equations. (For more general description consult e.g. [2] and [4].)

Let us consider the initial value problem

$$w'(t) = \mathbf{A}w(t), \quad t \in (0, T], \quad w(0) = w_0, \tag{4}$$

where $w : [0, T] \to \mathbb{R}^d$ is the unknown function and $\mathbf{A} \in \mathbb{R}^{d \times d}$ is an arbitrary matrix. As it is well-known, the solution of the problem can be written in the form $w(t) = \exp(t\mathbf{A})w_0$. Unfortunately, in the general case, the matrix exponential cannot be computed explicitly for large matrices even though they are sparse, what typically occurs in the numerical solutions of the Maxwell's equations.

Now assume that the matrix $\mathbf{A}$ is split into the sum of the matrices $\mathbf{A}_1$ and $\mathbf{A}_2$. (Naturally the procedure what follows can be directly extended to more than two split matrices.) We introduce a parameter $\tau > 0$, which is much less than T and, instead of the original problem, we consider the sequence of initial value problems of the form

$$\tilde{w}_k'(t) = \mathbf{A}_1\tilde{w}_k(t),\ t \in ((k-1)\tau, k\tau],\ \tilde{w}_k((k-1)\tau) = \hat{w}_{k-1}((k-1)\tau), \quad (5)$$

$$\hat{w}_k'(t) = \mathbf{A}_2\hat{w}_k(t),\ t \in ((k-1)\tau, k\tau],\ \hat{w}_k((k-1)\tau) = \tilde{w}_k(k\tau), \quad (6)$$

for $k = 1, 2, \ldots, T/\tau$ and $\hat{w}_0(0) = w_0$. After the kth cycle, we approximate the exact solution $w(k\tau) \equiv \exp(k\tau\mathbf{A})w_0$ by the function

$$u(k\tau) = (\exp(\tau\mathbf{A}_2)\exp(\tau\mathbf{A}_1))^k w_0. \quad (7)$$

We will refer to this splitting method as sequential splitting (S-splitting).

Another well-known splitting procedure is the so-called Strang-splitting:

$$\tilde{w}_k'(t) = \mathbf{A}_1\tilde{w}_k(t),\ t \in ((k-1)\tau, (k-1/2)\tau],\ \tilde{w}_k((k-1)\tau) = \bar{w}_{k-1}((k-1)\tau), \quad (8)$$

$$\hat{w}_k'(t) = \mathbf{A}_2\hat{w}_k(t),\ t \in ((k-1)\tau, k\tau],\ \hat{w}_k((k-1)\tau) = \tilde{w}_k((k-1/2)\tau), \quad (9)$$

$$\bar{w}_k'(t) = \mathbf{A}_1\bar{w}_k(t),\ t \in ((k-1/2)\tau, k\tau],\ \bar{w}_k((k-1/2)\tau) = \hat{w}_k(k\tau), \quad (10)$$

with $k = 1, 2, \ldots, T/\tau$ and $\bar{w}_0(0) = w_0$. In this case the exact solution $w(k\tau)$ is approximated by the function

$$u(k\tau) = (\exp((\tau/2)\mathbf{A}_1)\exp(\tau\mathbf{A}_2)\exp((\tau/2)\mathbf{A}_1))^k w_0. \quad (11)$$

Obviously, in the above splitting methods some approximation error occurs unless the matrices $\mathbf{A}_1$ and $\mathbf{A}_2$ commute. The so-called splitting error, defined as $Err_{Sp} = w(\tau) - u(\tau)$ shows that the S-splitting is a first order splitting scheme, while the Strang-splitting has second order accuracy.

2 Semi-discretized Maxwell's Equations

Discretizing the Maxwell's equations in spatial coordinates on a non-colocated staggered grid (applying N Yee-cells) we arrive at a system of first order linear ordinary differential equations of the form (4). Namely, let us introduce the set of indices

$$\mathcal{I} = \{(i/2, j/2, k/2) \mid i, j, k \in \mathbb{N},\ \text{not all odd and not all even}, \quad (12)$$

$$(i\Delta x/2, j\Delta y/2, k\Delta z/2)\ \text{is in the computational domain}\},$$

and define the functions $\Psi_{i/2,j/2,k/2} : \mathbb{R} \to \mathbb{R}$ $((i/2, j/2, k/2) \in \mathcal{I})$ as

$$\Psi_{i/2,j/2,k/2}(t) = \begin{cases} \sqrt{\varepsilon_{i/2,j/2,k/2}} E_x(t, i\Delta x/2, j\Delta y/2, k\Delta z/2), & \text{if } 2 \not| i,\ 2|j, k, \\ \sqrt{\varepsilon_{i/2,j/2,k/2}} E_y(t, i\Delta x/2, j\Delta y/2, k\Delta z/2), & \text{if } 2 \not| j,\ 2|i, k, \\ \sqrt{\varepsilon_{i/2,j/2,k/2}} E_z(t, i\Delta x/2, j\Delta y/2, k\Delta z/2), & \text{if } 2 \not| k,\ 2|i, j, \\ \sqrt{\mu_{i/2,j/2,k/2}} H_x(t, i\Delta x/2, j\Delta y/2, k\Delta z/2), & \text{if } 2|j, k,\ 2 \not| i, \\ \sqrt{\mu_{i/2,j/2,k/2}} H_y(t, i\Delta x/2, j\Delta y/2, k\Delta z/2), & \text{if } 2|i, k,\ 2 \not| j, \\ \sqrt{\mu_{i/2,j/2,k/2}} H_z(t, i\Delta x/2, j\Delta y/2, k\Delta z/2), & \text{if } 2|i, j,\ 2 \not| k. \end{cases} \tag{13}$$

Here $\varepsilon_{i/2,j/2,k/2}$ and $\mu_{i/2,j/2,k/2}$ denote the electric permittivity and magnetic permeability at the point $(i\Delta x/2, j\Delta y/2, k\Delta z/2)$, respectively. Ordering this functions into a vector we get a vector-scalar function $\Psi : \mathbb{R} \to \mathbb{R}^{6N}$, for which we have a system of first order linear ordinary differential equations

$$\Psi'(t) = \mathbf{A}\Psi(t), \ \Psi(0) \text{ is given,} \tag{14}$$

where the matrix $\mathbf{A} \in \mathbb{R}^{6N \times 6N}$ is a skew-symmetric matrix. Furthermore, it can be seen easily that every row of $\mathbf{A}$ consists at most four nonzero elements in the form

$$\frac{1}{\sqrt{\varepsilon_{.,.,.}\mu_{.,.,.}}\Delta.} \tag{15}$$

(see e.g. [6] and [7]).

Owing to the earlier mentioned reasons, (14) can be solved only applying some numerical methods. The direct numerical solution of (14) is problematic, because the explicit Euler method is unstable and implicit methods require a solution of a system of linear equations, which is very expensive considering the huge number of unknowns. This is why splitting methods are applied for (14) in order to split the original system into several sub-problems. Solving these sub-problems exactly or with a suitable combination of explicit and implicit Euler methods we are able to define relatively cheap and stable numerical methods. In the YEE-, NZCZ- and KFR-methods the above approach was employed, although, this is not always clear from the formulation of the methods.

3 Splitting in the YEE-, NZCZ- and KFR-Methods

First let us consider the YEE-method. Here we split the matrix $\mathbf{A}$ into the form $\mathbf{A} = \mathbf{A}_{1Y} + \mathbf{A}_{2Y}$, where the splitting is done according to the magnetic and electric fields. Namely, the two matrices $\mathbf{A}_{1Y}$ and $\mathbf{A}_{2Y}$ are defined as follows. The matrix $\mathbf{A}_{1Y}$ is composed from the matrix $\mathbf{A}$ changing the rows belonging to the electric field variables to zero rows. $\mathbf{A}_{2Y}$ can be derived in similar manner, zeroing the rows belonging to the magnetic field variables. Employing S-splitting with these matrices we can notice that the sub-problems (5)–(6) can be solved exactly due to the equalities $\mathbf{A}_{1Y}^2 = \mathbf{0}$ and $\mathbf{A}_{2Y}^2 = \mathbf{0}$. Thus, we arrive at the iteration

$$\Psi^{n+1} = (\mathbf{I} + \tau\mathbf{A}_{2Y})(\mathbf{I} + \tau\mathbf{A}_{1Y})\Psi^n, \ \Psi^0 \text{ is given,} \tag{16}$$

where Ψ^n stands for the approximation of Ψ at time-level $n\tau$ and $\mathbf{I}$ is the unit matrix. We remark that the YEE-method starts with a vector Ψ^0 that consists the approximations of the magnetic field at the time-level $\tau/2$. To keep the YEE-method stable the time-step must be bounded (see (3) with $\tau = \Delta t$).

The NZCZ-method employs the splitting $\mathbf{A} = \mathbf{A}_{1N} + \mathbf{A}_{2N}$, where $\mathbf{A}_{1N}$ comes from the discretization of the first items in the curl operator, and $\mathbf{A}_{2N}$ comes from the second ones. It is not difficult to see that $\mathbf{A}_{1N}$ and $\mathbf{A}_{2N}$ are skew-symmetric matrices. Applying the Strang-splitting, the sub-systems (8)–(10) cannot be solved exactly, numerical methods are needed. We denote the time-step in the numerical methods by Δt, which is chosen to be equal to τ. We arrive at the NZCZ-method when (8) is solved by the explicit Euler method, (9) is solved by the Crank-Nicolson method, and (10) is solved by the implicit Euler method. In this way we obtain the iteration

$$\Psi^{n+1} = (\mathbf{I} - (\Delta t/2)\mathbf{A}_{1N})^{-1}(\mathbf{I} + (\Delta t/2)\mathbf{A}_{2N})(\mathbf{I} - (\Delta t/2)\mathbf{A}_{2N})^{-1}(\mathbf{I} + (\Delta t/2)\mathbf{A}_{1N}) \tag{17}$$

(the second and third matrices commute). In practice, the above procedure can be simplified to the solution of two systems of linear equations with symmetric tridiagonal matrices in each iteration step. The NZCZ-method is unconditionally stable.

In the KFR-method, the matrix $\mathbf{A}$ is split into six skew-symmetric matrices and the S-splitting, the Strang-splitting or other higher order splittings are used. The sub-systems can be solved exactly again, and the unconditional stability is guaranteed by the fact that the exponential of a skew-symmetric matrix is an orthogonal matrix, and as such it has unit second norm. For example, for the system $\Psi'(t) = \mathbf{A}\Psi(t)$ with the matrix $\mathbf{A} = \mathrm{tridiag}[-1, 0, 1] \in \mathbb{R}^{5\times 5}$, the splitting $\mathbf{A} = \mathbf{A}_{1K} + \mathbf{A}_{2K}$ and the iteration (in the case of the S-splitting)

$$\Psi^{n+1} = \exp(\tau\mathbf{A}_{1K})\exp(\tau\mathbf{A}_{2K})\Psi^n =$$

$$= \exp\left(\begin{bmatrix} 0 & \tau & 0 & 0 & 0 \\ -\tau & 0 & 0 & 0 & 0 \\ 0 & 0 & 0 & \tau & 0 \\ 0 & 0 & -\tau & 0 & 0 \\ 0 & 0 & 0 & 0 & 0 \end{bmatrix}\right) \cdot \exp\left(\begin{bmatrix} 0 & 0 & 0 & 0 & 0 \\ 0 & 0 & \tau & 0 & 0 \\ 0 & -\tau & 0 & 0 & 0 \\ 0 & 0 & 0 & 0 & \tau \\ 0 & 0 & 0 & -\tau & 0 \end{bmatrix}\right) \Psi^n \tag{18}$$

can be applied, where the exponentials can be computed with the equality

$$\exp\left(\begin{bmatrix} 0 & \tau \\ -\tau & 0 \end{bmatrix}\right) = \begin{bmatrix} \cos\tau & \sin\tau \\ -\sin\tau & \cos\tau \end{bmatrix}. \tag{19}$$

Both the NZCZ- and KFR-methods possess the nice properties of the YEE-method. Some advantages of the methods are: easy understandability, solution of a wide frequency range with one simulation (time domain method), animation displays, specification of the material properties at all points within the computational domain and the computation of the electric and magnetic fields directly.

The NZCZ and KFR methods are unconditionally stable, which means that the time-step can be chosen arbitrarily in the numerical calculations. Naturally, the increase in the time-step necessitates decrease in the accuracy. It is observed for example in [5, 6, 12], that although the NZCZ-method is slower with a factor about five than the YEE-method, choosing greater time-steps than the stability bound of the YEE-method, it computes the solution faster in the long run. In the KFR-method, we experienced that the method is relatively inaccurate, and to make it much more accurate (for example choosing a fourth order splitting method or higher) costs a lot of computational time. This phenomenon shows the importance of a proper splitting.

4 A New Efficient Scheme Based on Splittings

In this section, we construct a new numerical scheme for the Maxwell's equations. As we have seen in the previous section it is worth to keep in view the following issues in the construction.

1. The method must be based on some splitting of the matrix **A**.
2. The sub-problems must be exactly solvable or at least easily solvable with computationally cheap numerical methods.
3. The new method must be unconditionally stable.

Let us consider the splitting $\mathbf{A} = \mathbf{A}_{1N} + \mathbf{A}_{2N}$, which splitting was applied in the NZCZ-method. Instead of the Strang-splitting, we apply now the S-splitting, and solve the sub-systems (5)–(6) by the Crank-Nicolson method. This method will be called sequential NZCZ-method (shortly SNZCZ) in the sequel. The temporal discretization results in the equations

$$\frac{\hat{\Psi} - \Psi^n}{\Delta t} = \frac{1}{2}\mathbf{A}_{1N}\hat{\Psi} + \frac{1}{2}\mathbf{A}_{1N}\Psi^n, \tag{20}$$

$$\frac{\Psi^{n+1} - \hat{\Psi}}{\Delta t} = \frac{1}{2}\mathbf{A}_{2N}\Psi^{n+1} + \frac{1}{2}\mathbf{A}_{2N}\hat{\Psi}, \tag{21}$$

where getting rid off the intermediate vector $\hat{\Psi}$ we obtain the iteration

$$\Psi^{n+1} = (\mathbf{I} - \frac{\Delta t}{2}\mathbf{A}_{2N})^{-1}(\mathbf{I} + \frac{\Delta t}{2}\mathbf{A}_{2N})(\mathbf{I} - \frac{\Delta t}{2}\mathbf{A}_{1N})^{-1}(\mathbf{I} + \frac{\Delta t}{2}\mathbf{A}_{1N})\Psi^n. \tag{22}$$

It is easy to see (see e.g. [6]) that the product of the first two matrices, and the product of the second two ones are orthogonal matrices owing to the skew-symmetry of the matrices $\mathbf{A}_{1N}$ and $\mathbf{A}_{2N}$. Thus we obtain the relation $\|\Psi^n\|_2 = \|\Psi^0\|_2$ $(n = 1, 2, \dots)$, which indicates the unconditional stability of the iteration. What is more, the second norm of the iteration vector does not change during the iteration process, that is the energy density of the electromagnetic field is conserved in the computations.

We can notice that the iteration (22) actually is a reordering of (17). This suggests that the number of operations of the two methods to achieve a certain time-level are close to each other. Indeed, the inverses of the matrices in the SNZCZ-method can be computed similarly to the NZCZ-method, that is solving a system of linear equations with a symmetric tridiagonal matrix. In one time-step of the SNZCZ-method, we have to store the old values of one of the field components, which increases the memory consumption and slows down the computations with 5-10%.

5 Numerical Example

We demonstrate the applicability of the SNZCZ-method on a 2D example. Let us suppose that the Maxwell's equations are solved on the square $[0,1] \times [0,1]$ with perfect conductor boundaries and only the components E_z, H_x and H_y change with the spatial coordinates. In this case, an exact solution can be written in the form

$$E_z(t,x,y) = -\sin(\pi x)\sin(\pi y)\sin(\sqrt{2}\pi t), \tag{23}$$

$$H_x(t,x,y) = -\frac{1}{\sqrt{2}}\sin(\pi x)\cos(\pi y)\cos(\sqrt{2}\pi t), \tag{24}$$

$$H_y(t,x,y) = \frac{1}{\sqrt{2}}\cos(\pi x)\sin(\pi y)\cos(\sqrt{2}\pi t) \tag{25}$$

We set the values $\varepsilon = \mu = 1$, $\Delta x = \Delta y = 1/30$ and compute the error in l_2-norm at the time-level $t = 1.7067 = 2^{10}\Delta x/20$. The errors and the CPU times for the NZCZ- and SNZCZ-methods are listed in Table 1.

The classical YEE-method is not stable for time-steps greater than $\Delta t_{\max} = 0.7071\Delta x = \Delta x/\sqrt{2} = 0.0236$. The considered two methods, however, result in acceptable errors even for the time-step $\Delta t = 10\Delta t_{\max}$, that is in the long run they compute the solution two times faster than the YEE-method. (The YEE-method computes one iteration step five times faster than the other two methods.)

Table 1. Computational results with the NZCZ- and SNZCZ-methods. The times-step is measured compared to Δx. The first value is the l_2 error, the CPU time can be found in parenthesis, in seconds.

Δt ($\times\Delta x$)	NZCZ	SNZCZ
0.05	1.20×10^{-3} (289.24)	5.57×10^{-4} (329.16)
0.10	1.22×10^{-3} (104.82)	5.04×10^{-5} (117.98)
0.20	1.36×10^{-3} (52.40)	1.25×10^{-3} (59.04)
0.40	1.71×10^{-3} (36.31)	3.41×10^{-3} (29.49)
0.80	3.33×10^{-3} (13.13)	7.11×10^{-3} (20.67)
1.60	9.59×10^{-3} (6.59)	1.13×10^{-2} (7.41)
3.20	3.35×10^{-2} (4.56)	8.50×10^{-3} (4.74)
6.40	1.15×10^{-1} (1.70)	3.94×10^{-2} (1.87)
12.80	2.54×10^{-1} (0.88)	2.24×10^{-1} (0.93)

The values in Table 1 show that the SNZCZ-method is a little bit slower, but it computes the solution with a slightly smaller error especially for large time-steps. This is a favourable property of the method.

6 Conclusion

Based on the splitting methods for systems of first order linear ordinary differential equations, we have constructed a new numerical method for the time-dependent Maxwell's equations. The SNZCZ-method employs the same decomposition of **A** like the NZCZ-method, but instead of the Strang-splitting it applies the S-splitting in the time coordinate. The SNZCZ-method possesses all the favourable properties of the NZCZ-method, and what is more, it is unconditionally stable by construction conserving the energy density of the electromagnetic field. This can make the method very useful in computations of density of states.

Acknowledgment

The author of the paper was supported by the National Scientific Research Found (OTKA) N. T043765.

References

1. M. Darms, R. Schuhmann, H. Spachmann, T. Weiland, Asymmetry Effects in the ADI-FDTD Algorithm, *IEEE Microwave Guided Wave Lett.* 12 (2002), 491–493.
2. I. Faragó, Á. Havasi, The Mathematical Background of Operator Splitting and the Effect of Non-Commutativity, *Large-Scale Scientific Computing*, S. Margenov, J. Waśniewski, P. Yalamov, eds., *Lecture notes in computer sciences*, **2179**, Springer Verlag, 2001, 264–271.
3. B. Fornberg, Some Numerical Techniques for Maxwell's Equations in Different Type of Geometries, in: *Topics in Computational Wave Propagation*, M. Ainsworth, P. Davies, D. Duncan, P. Martin, B. Rynne (Eds.), , *Lecture Notes in Computational Wave Propagation*, **31**, Springer, Berlin, 2003, 265–299.
4. A. R. Gourlay, Splitting Methods for Time-Dependent Partial Differential Equations, in The State of Art in Numerical Analysis, Proc. Conf. Univ. York, Heslington, 1976, Acad. Press, London 1977.
5. C.C.-P. Chen, Tae-Woo Lee, N. Murugesan, S.C. Hagness, *Generalized FDTD-ADI: An Unconditionally Stable Full-Wave Maxwell's Equations Solver for VLSI Interconnect Modeling Computer Aided Design*, ICCAD-2000. IEEE/ACM International Conference, (2000), 156–163.
6. R. Horváth, Uniform Treatment of the Numerical Time-Integration of the Maxwell Equations, *Lecture Notes in Computational Science and Engineering*, Proceedings Scientific Computing in Electrical Engineering (SCEE-2002, June 23-28, 2002, Eindhoven, The Netherlands), 2003, 231–239.
7. J. S. Kole, M. T. Figge, H. De Raedt, Unconditionally Stable Algorithms to Solve the Time-Dependent Maxwell Equations, *Phys. Rev. E*, **64**, 066705, 2001.

8. Y. Liu, Fourier Analysis of Numerical Algorithms for the Maxwell Equations, *Journal of Comp. Phys.* **124** (1996), 396–416.
9. T. Namiki, A New FDTD Algorithm Based on Alternating-Direction Implicit Method, *IEEE Transactions on Microwave Theory and Techniques*, **47** 10, Oct. 1999, 2003–2007.
10. A. Taflove, S. C. Hagness, *Computational Electrodynamics: The Finite-Difference Time-Domain Method*, 2 ed., Artech House, Boston, MA, 2000.
11. F. Zheng, Z. Chen, J. Zhang, A Finite-Difference Time-Domain Method Without the Courant Stability Conditions, *IEEE Microwave Guided Wave Lett.* **9** (1999), 441–443.
12. F. Zheng, Z. Chen, J. Zhang, Toward the Development of a Three-Dimensional Unconditionally Stable Finite-Difference Time-Domain Method, *IEEE Trans. Microwave Theory and Techniques*, **48** 9, September 2000, 1550–1558.
13. K. S. Yee, Numerical Solution of Initial Boundary Value Problems Involving Maxwell's Equations in Isotropic Media, *IEEE Transactions on Antennas and Propagation*, **14** 3, 302–307, March 1966.

Coefficient Identification in Elliptic Partial Differential Equation

Tchavdar T. Marinov[1], Rossitza S. Marinova[2], and Christo I. Christov[3]

[1] Dept. of Math. Sci., University of Alberta, Edmonton, AB, Canada T6G 2G1
[2] Dept. of Math. & Computing Sci., Concordia Univ. College of Alberta, 7128 Ada Boul., Edmonton, AB, Canada T5B 4E4
[3] Dept. of Math., University of Louisiana at Lafayette, LA 70504-1010, USA

Abstract. We consider the inverse problem for identification of the coefficient in an elliptic partial differential equation inside of the unit square $\mathcal{D}$, when over-posed boundary data are available. Following the main idea of the Method of Variational Imbedding (MVI), we "imbed" the inverse problem into a fourth-order elliptic boundary value problem for the Euler-Lagrange equation being the necessary condition for minimization of the quadratic functional of the original equation. The fourth-order boundary value problem becomes well-posed with the two boundary conditions considered here. The Euler-Lagrange equation for the unknown coefficient provides an explicit equation for the coefficient. A featuring example is elaborated numerically.

1 Introduction

Consider the elliptic partial differential equation

$$\Delta u(x,y) + n(x,y)u(x,y) = 0 \tag{1}$$

inside the unit square $\mathcal{D} = (x,y) : 0 < x < 1; 0 < y < 1$. The equation (1) has a unique solution under the boundary condition

$$u|_{\partial\mathcal{D}} = \varphi\,, \tag{2}$$

provided that the coefficient $n(x,y)$ is a known negative function(see [3]).

Suppose that the coefficient n is unknown. In order to identify it, we need more information. There can be different sources of such information, e.g., one or more of the solution functions; additional data at the boundaries, etc.

Let us consider the case when over–posed boundary data are available:

$$\left.\frac{\partial u}{\partial \nu}\right|_{\partial\mathcal{D}} = \psi\,. \tag{3}$$

If the coefficient $n = n(x,y)$ is given, then the problem (1)–(3) for $u(x,y)$ is over-determined; i.e., for arbitrary φ and ψ there may be no solution $u(x,y)$ satisfying all of the conditions (2) and (3). On the other hand, when $n(x,y)$ is

I. Lirkov, S. Margenov, and J. Waśniewski (Eds.): LSSC 2005, LNCS 3743, pp. 372–379, 2006.

not known a priori, then under certain conditions it may be possible to find a coefficient $n(x,y)$ such that the problem (1) has a unique solution $u(x,y)$, and this solution also satisfies (2) and (3). In this case we say that the functions (u,n) constitute a solution to the problem (1), (2), (3).

To assure the uniqueness we will consider the following two cases:

Case 1. The coefficient $n(x,y)$ is a piecewise constant function in $\mathcal{D}$, i.e., $n(x,y) = c_\gamma = \text{const}$, when $(x,y) \in \mathcal{D}_\gamma$, where $\mathcal{D}$ is divided into Γ disjoint regions $\mathcal{D}_\gamma$ and $\mathcal{D} = \bigcup_{\gamma=1}^{\Gamma} \mathcal{D}_\gamma$.
Case 2. Another special case is present when the coefficient is a function of a single variable, i.e., $n(x,y) = n(\omega)$, where $\omega = \omega(x,y)$ is known a priori. For example, $\omega = x$, or $\omega = x^2 + y^2$, etc.

Generally speaking, the problem (1)–(3) is over-determined under the assumptions above. We assume that the boundary conditions (2), (3) are self-consistent and the solution of the problem (1), (2), (3) exists. The problem for identification (u,n) from (1), (2), (3) is of an inverse nature and it is similar to the problem of identification of the heat-conductivity coefficient from over-posed data [1, 2]. For more information on the existing methods for solving inverse problems, we refer the reader to Isakov [5], Engl et al [4], Tikhonov et al [6], and the references therein.

2 Variational Imbedding

Following the previous authors' works on the coefficient identification in parabolic equations, we replace the original problem by the problem of the minimization of the following functional

$$\mathcal{J}(u,n) = \iint_{\mathcal{D}} (\Delta u + nu)^2 \, dx dy \,, \tag{4}$$

where u must satisfy the conditions (2) and (3). The functional $\mathcal{J}$ is a quadratic and homogeneous function of $(\Delta u + nu)$, and hence it attains its minimum if and only if $\Delta u + nu \equiv 0$. In this sense, there is a one–to–one correspondence between the original equation (1) and the minimization problem (4).

2.1 The Imbedding Boundary-Value Problem for $u(x,y)$

The necessary conditions for the minimization of (4) are the Euler-Lagrange equations for the functions $u(x,y)$ and $n(x,y)$. The equation for u reads

$$\Delta\Delta u + \Delta[n(x,y)u] + n(x,y)\Delta u + n^2(x,y)u = 0 \,. \tag{5}$$

This equation is of the fourth order and its solution can satisfy the two boundary conditions (2) and (3). In this sense the problem (5), (2) and (3) is well–posed if the function $n(x,y)$ is considered as known.

2.2 Equation for the Coefficient $n(x, y)$

The Euler-Lagrange equation for the unknown coefficient $n(x, y)$ provides an explicit equation of following form

$$n(x,y) = -\frac{\mathcal{A}_1(x,y)}{\mathcal{A}_2(x,y)} . \tag{6}$$

The calculation of the functions $\mathcal{A}_1(x, y)$ and $\mathcal{A}_2(x, y)$ is slightly different for the two cases discussed above. In *Case 1*, the functional $\mathcal{J}(u, n)$ can be rewritten as a sum

$$\mathcal{J}(u,n) = \iint\limits_{\mathcal{D}} (\Delta u + nu)^2 \, dxdy = \sum_{\gamma=1}^{\Gamma} \iint\limits_{\mathcal{D}_\gamma} (\Delta u + nu)^2 \, dxdy \, , \tag{7}$$

and, taking into account that $n(x, y) = c_\gamma = \text{const}$, if $(x, y) \in \mathcal{D}_\gamma$, after some manipulations

$$\mathcal{A}_1 = \iint\limits_{\mathcal{D}_\gamma} u \Delta u dxdy, \qquad \mathcal{A}_2 = \iint\limits_{\mathcal{D}_\gamma} u^2 dxdy \, . \tag{8}$$

In *Case 2*, after the change of variables $\omega = \omega(x, y)$ and $\eta = \eta(x, y)$, the functional $\mathcal{J}(u, n)$ can be rewritten as

$$\mathcal{J}(u,n) = \iint\limits_{\mathcal{D}^*} (\Delta u(\omega,\eta) + n(\omega) u(\omega,\eta))^2 \left| \frac{\partial(x,y)}{\partial(\omega,\eta)} \right| d\omega d\eta \, . \tag{9}$$

Here $\eta = \eta(x, y)$ is the function for which $\left|\frac{\partial(x,y)}{\partial(\omega,\eta)}\right| \neq 0$. Taking into account that $n(x, y) = n(\omega)$, the functions $\mathcal{A}_1(x, y)$ and $\mathcal{A}_2(x, y)$ adopt the form

$$\mathcal{A}_1 = \int\limits_{\Omega_c} u \Delta u \left| \frac{\partial(x,y)}{\partial(\omega,\eta)} \right| d\eta, \qquad \mathcal{A}_2 = \int\limits_{\Omega_c} u^2 \left| \frac{\partial(x,y)}{\partial(\omega,\eta)} \right| d\eta \, , \tag{10}$$

where $\Omega_c := \{(x, y) \in \mathcal{D}; \omega(x, y) = c\}$,

3 Difference Scheme

Consider the model problem in the unit square $\mathcal{D}$. Let us introduce an axis–parallel orthogonal mesh with a total number of grid lines in the $x-$ and $y-$direction, equal to M and N, respectively. A straightforward approximation of equation (1) with boundary condition (2) gives the correct number of equations for identifying the discrete function u. The number of additional equations obtained from the boundary condition (3) is $2(M + N - 1)$. Our problem lies in finding a way to transfer this additional information to the unknown function $n(x, y)$. The number of unknown values of the function $n(x, y)$ must be equal to the number of equations we have available for this function, i.e., $2(M + N - 1)$. This means that it is not possible to use the same mesh for the coefficient—the number of grid nodes for the function $n(x, y)$ must be less or equal to $2(M + N - 1)$. For this reason we have constructed a special numerical scheme.

3.1 Grid Pattern and Approximations

In order to have second–order approximations of the derivatives that enter the boundary conditions, we use staggered grids in both directions. For the grid spacings we have $h_x \equiv 1/(M-2)$ and $h_y \equiv 1/(N-2)$, where M is the total number of grid lines in the x–direction and N—in the y–direction. Then the grid lines are defined as follows (see Figure 1): $x_i = (i-1.5)h_x$ for $i = 1,\ldots,M$ and $y_j = (j-1.5)h_y$ for $j = 1,\ldots,N$.

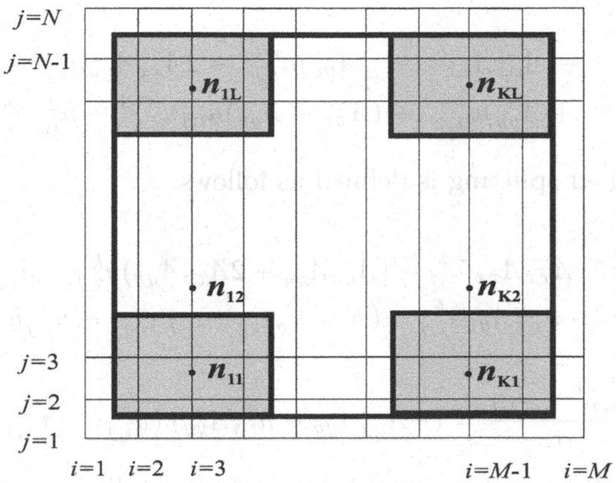

Fig. 1. Grid pattern

The grid for the coefficient n is also staggered, the total number of grid lines in the x–direction is equal to K, and the total number of grid lines in the y–direction—to L. Then the grid spacings for n are: $H_x \equiv 1/K$ and $H_y \equiv 1/L$, and the grid lines are defined as follows (see Figure 1): $X_l = (l-0.5)H_x$ for $l = 1,\ldots,L$, and $Y_k = (j-0.5)H_y$ for $k = 1,\ldots,K$. Let us introduce the mesh functions (notations):

$$u_{ij} = u(x_i, y_j) \quad \text{for} \quad i = 1,\ldots,M; \quad j = 1,\ldots,N; \tag{11}$$

$$n_{kl} = n(x_k, y_l) \quad \text{for} \quad k = 1,\ldots,K; \quad l = 1,\ldots,L. \tag{12}$$

It is convenient to use an additional piecewise mesh function $\tilde{n}_{ij}$ defined as:

$$\tilde{n}_{ij} = n_{kl} \quad \text{when} \quad X_{k-1/2} \le x_i \le X_{k+1/2}; \quad Y_{l-1/2} \le y_j \le Y_{l+1/2}\,, \tag{13}$$

where $X_{k\pm1/2} = X_k \pm H_x/2$ and $Y_{l\pm1/2} = Y_l \pm H_y/2$. However, there are relationships between the mesh-steps for the functions u and n as $H_x = ph_x$ and $H_y = qh_y$, where p and q are integers.

3.2 Scheme for the Fourth-Order Elliptic Equation

We use the iterative procedure based on the coordinate–splitting method similar to the method in [7], because of its computational efficiency. Let us introduce the notations Λ_{xx} and Λ_{yy} for the central difference approximation of the operators $\frac{\partial^2}{\partial x^2}$ and $\frac{\partial^2}{\partial y^2}$, respectively. The most straightforward approximation for the imbedding fourth-order problem is the following

$$(\Lambda_{xx}+\Lambda_{yy})\left[(\Lambda_{xx}+\Lambda_{yy})+\tilde{n}_{i,j}\right]u_{i,j}+\tilde{n}_{i,j}(\Lambda_{xx}+\Lambda_{yy})u_{i,j}+\tilde{n}^2_{i,j}u_{i,j}=0 \quad (14)$$

for $i=3,\ldots,M-2$ and $j=3,\ldots,N-2$, respectively.

The equation (14) adopts the form of a parabolic difference equation after introducing the fictitious time:

$$\begin{aligned}\frac{u^{k+1}_{i,j}-u^k_{i,j}}{\sigma} &= -(\Lambda_{xx}\Lambda_{xx}+\Lambda_{yy}\Lambda_{yy})u^{k+1}_{i,j}-2\Lambda_{xx}\Lambda_{yy}u^k_{i,j}-\tilde{n}_{i,j}(\Lambda_{xx}\\ &\quad +\Lambda_{yy})u^{k+1}_{i,j}-(\Lambda_{xx}+\Lambda_{yy})\tilde{n}_{i,j}u^{k+1}_{i,j}-\tilde{n}^2_{i,j}u^{k+1}_{i,j}\,.\end{aligned} \quad (15)$$

Then the applied splitting is defined as follows:

$$\begin{aligned}\frac{\tilde{u}_{i,j}-u^k_{i,j}}{\sigma} &= -\Lambda_{xx}\Lambda_{xx}\tilde{u}_{i,j}-(\Lambda_{yy}\Lambda_{yy}+2\Lambda_{xx}\Lambda_{yy})u^k_{i,j}-\tilde{n}_{i,j}\Lambda_{xx}\tilde{u}_{i,j}\\ &\quad -\tilde{n}_{i,j}\Lambda_{yy}u^k_{i,j}-(\Lambda_{xx}+\Lambda_{yy})(\tilde{n}_{i,j}u^k_{i,j})-\tilde{n}^2_{i,j}\tilde{u}_{i,j}\end{aligned} \quad (16)$$

$$\frac{u^{k+1}_{i,j}-\tilde{u}_{i,j}}{\sigma}=\left(-\Lambda_{yy}\Lambda_{yy}-\tilde{n}_{i,j}\Lambda_{yy}\right)\left(u^{k+1}_{i,j}-u^k_{i,j}\right), \quad (17)$$

where $\tilde{u}_{i,j}$ is a half-time-step variable, and $u^k_{i,j}$ is a full-time-step one.

The fractional-step scheme (16), (17) has a total approximation in full steps for equation (15).

The staggered in both directions grid for u allows one to use central differences with a second-order of approximation on two-point stencils for the boundary conditions.

3.3 Scheme for the Coefficient

In *Case 1*, taking into account that the function n is piecewise, the Euler-Lagrange equation (6) for the piecewise function $\tilde{n}(x,y)$ is approximated by the following second order difference scheme:

$$n_{k,l}=-\frac{\sum\limits_{X_{k-1/2}\le x_i\le X_{k+1/2}}\ \sum\limits_{Y_{l-1/2}\le y_j\le Y_{l+1/2}} u_{i,j}(\Lambda_{xx}+\Lambda_{yy})u_{i,j}}{\sum\limits_{X_{k-1/2}\le x_i\le X_{k+1/2}}\ \sum\limits_{Y_{l-1/2}\le y_j\le Y_{l+1/2}} u^2_{i,j}}, \quad (18)$$

for $k=1,\ldots K$ and $l=1,\ldots L$.

In *Case 2* the idea is similar, although the calculations are more complicated.

3.4 Algorithm

(I) With a given initial guess for $n^{\text{old}}_{k,l}<0$, the fourth-order boundary value problem (14), (2) and (3) is solved for the function $u_{i,j}$.

(II) With the newly computed values of $u_{i,j}$, the function $n_{k,l}^{\text{new}}$ is evaluated. If the difference between the new and the old field for n is less than ε, i.e.

$$\max_{i,j} |n_{k,l}^{\text{new}} - n_{k,l}^{\text{old}}| < \varepsilon,$$

then the calculations are terminated, otherwise go to step **(I)**.

4 Numerical Experiments

The first experiment is performed to verify the fact that for a given coefficient and boundary data, the solution of the "imbedding" problem does coincide with the solution of the "direct" problem. Our calculations confirmed this fact in order of the round-off error in double precision arithmetics.

The accuracy of the developed difference scheme is checked with the mandatory tests involving different grid spacing h_x, h_y, H_x and H_y. We conducted calculations with different values of mesh parameters and compared them in order to verify the practical convergence, approximation, and consistency of the difference scheme.

4.1 Constant Coefficient

The truncation error term is $O(h_x^2+h_y^2+H_x^2+H_y^2)$, and the total error also should be $O(h_x^2+h_y^2+H_x^2+H_y^2)$. In order to verify the $O(h_x^2+h_y^2)$ approximation of the scheme, let us consider an example where the coefficient $n(x,y)$ is a constant, because in such a case one can use $K = L = 1$, i.e. $H_x = H_y = 1$ and

$$\tilde{n}_{k,l} = n = -\left[\sum_{i=2}^{M-1}\sum_{j=2}^{N-1}(\Lambda_{xx} + \Lambda_{yy})u_{i,j}\right]\left[\sum_{i=2}^{M-1}\sum_{j=2}^{N-1} u_{i,j}^2\right]^{-1}, \tag{19}$$

for $k = 2, \ldots M-1$ and $l = 2, \ldots N-1$. In this case, the problem is strongly over-determined, but for the boundary data, which are self-consistent, the MVI converges to the solution, because it provides the global minimum of the functional.

Table 1. Obtained values of the coefficient $n = const$ and the rate of convergence for five different mesh size

h	n	rate of convergence
exact	-2.0	—
1/8	-2.00260551699214	—
1/16	-2.00065111900421	2.000576511
1/32	-2.00016275800837	2.000192715
1/64	-2.00004068266175	2.000242553
1/128	-2.00001016475638	2.000838435

The following numerical results demonstrate clearly the error orders and the convergence. We have computed a test problem with an exact solution

$$u_{\text{direct}}(x,y) = e^{x+y}, \qquad n_{\text{direct}}(x,y) = -2 \tag{20}$$

with five different sizes of the mesh-steps $h_x = h_y = h$. The values of the identified coefficient n with five steps are given in the Table 1. The rate of convergence, calculated as

$$\text{rate} = \log_2 \left| \frac{n_{2h} - n_{\text{direct}}}{n_h - n_{\text{direct}}} \right| , \tag{21}$$

is shown also in Table 1.

In this case, the developed scheme shows the second order of approximation. Even more: when we introduce the difference solution into the difference scheme for the "direct" problem, we achieve point-wise satisfaction of the latter better than 10^{-8}, i.e., the variational functional is of the order of 10^{-15}.

4.2 The Coefficient Is a Function of Single Variable

Another test example was constructed in order to illustrate the *Case 2* from Section 1 and to check the accuracy of the developed difference scheme and algorithm with tests involving different grid spacings H_x and H_y. For this reason, we conducted a number of calculations with different values of the mesh parameters, and compared them in order to verify the practical convergence and approximation of the difference scheme.

A test problem with a known exact solution

$$u_{\text{direct}}(x,y) = e^{(x^2+y)}, \qquad n_{\text{direct}}(x,y) = -(4x^2+3), \tag{22}$$

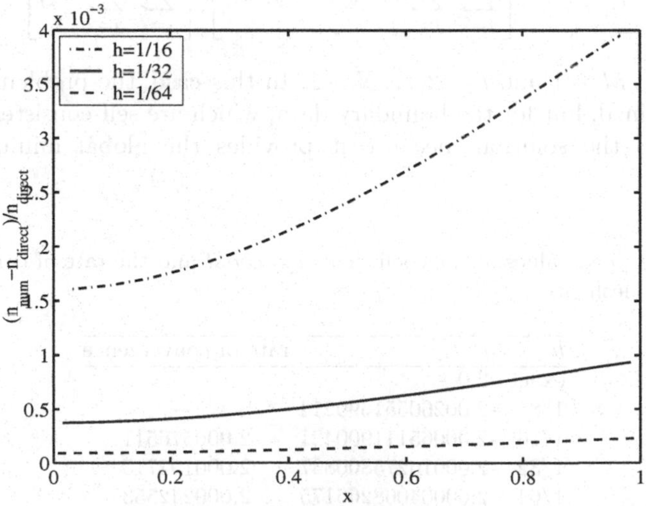

Fig. 2. The discretization error for three different grid steps $h = 1/16; 1/32; 1/64$

is used to illustrate the $O(h_x^2 + h_y^2 + H_x^2 + H_y^2)$. For these calculations

$$\tilde{n}_{i,j} = n_i = -\left[\sum_{j=2}^{N-1} (\Lambda_{xx} + \Lambda_{yy}) u_{i,j}\right] \left[\sum_{j=2}^{N-1} u_{i,j}^2\right]^{-1}, \tag{23}$$

for $i = 2, \ldots M-1$ and $j = 2, \ldots N-1$, since $H_y = 1$.

The shape of the discretization error $(n_{\text{inverse}} - n_{\text{direct}})/n_{\text{direct}}$ for the obtained values of the coefficient $n_{i,j}$ with three different grid steps $h_x = h_y = H_x = h$ is shown in Figure 2. It is clearly seen that the numerical solution approximates the analytical one with $O(h^2)$.

5 Conclusions

In the present paper we have displayed the performance of the technique called Method of Variational Imbedding for solving the inverse problem of coefficient identification in elliptic partial differential equation. The inverse problem for identification $n(x, y)$ from over–posed boundary data is replaced by a minimization problem for the quadratic functional of the original equation. The Euler–Lagrange equations for minimization comprise a fourth–order elliptic equation for the function $u = u(x, y)$, and an explicit equation for the unknown coefficient $n = n(x, y)$. For this system, the boundary data are not over–posed. Thus the inverse problem is imbedded into a higher–order but well posed for the given boundary data elliptic boundary value problem. Two examples are considered. The numerical results confirm the theoretical statement that the two problems are equivalent, giving a solution for the inverse problem.

References

1. Christov C, Marinov T, Method of Variational Imbedding for the Inverse Problem of Boundary-Layer-Thickness Identification, M^3AS, **7** 7, 1005–1022, 1997.
2. Christov C, Marinov T, Identification of Heat-Conduction Coefficient via Method of Variational Imbedding, *Mathematical and Computer Modelling* **27** 3, 109–116, 1998.
3. D. Colton, R. Kress. Integral Equation Methods in Scattering Theory. *A Wiley—Interscience Publication,* New York, 1983.
4. Engl H W, Hanke M, Neubauer A, *Regularization of Inverse Problems*, Kluwer Academic Publishers, Dordrecht, 1996.
5. Isakov V, *Inverse Problems for Partial Differential Equations*, Springer, NY, 1998.
6. Tikhonov A N, Leonov A S, Yagola A G, *Non-linear Ill-Posed Problems* **1, 2**, Chapman and Hall, London, 1998.
7. Marinova, R, Christov C, Marinov T, A Fully Coupled Solver for Incompressible Navier-Stokes Equations using Operator Splitting. *International Journal of Computational Fluid Dynamics,* **17** 5, 371–385, 2003.

Integrating Factor Methods as Exponential Integrators

Borislav V. Minchev

Department of Mathematical Science,
NTNU, 7491 Trondheim, Norway
Borko.Minchev@ii.uib.no

Abstract. Recently a lot of effort has been made in the construction and implementation of a class of methods called exponential integrators. These methods are preferable when one has to deal with stiff and highly oscillatory semilinear problems, which often arise after spatial discretization of Partial Differential Equations (PDEs). The main idea behind the methods is to use the exponential and some closely related functions inside the numerical scheme. In this note we show that the integrating factor methods, introduced by Lawson in 1967, are also examples of exponential integrators with very special structure for the related exponential functions. In order to prove this relation, we use the approach based on bi-coloured rooted trees and B-series. We also show under what conditions every bi-coloured rooted tree can be expressed as a linear combination of standard non-coloured rooted trees.

1 Introduction

Realistic models of many physical processes require effective numerical solvers for a special class of partial differential equations, which after semi-discretization in space, can be written in the following form

$$u' = Lu + N(u(t)), \qquad u(t_0) = u_0, \tag{1}$$

where $u : \mathbb{R} \to \mathbb{R}^d$, $L \in \mathbb{R}^{d\times d}$, $N : \mathbb{R}^d \to \mathbb{R}^d$ and d is a discretization parameter equal to the number of spatial grid points. Several interesting problems can be brought to this form. Examples are the Allen-Cahn, Burgers, Cahn-Hilliard, Kuramoto-Sivashinsky, Navier-Stokes, Swift-Hohenberg and nonlinear Scrödinger equations. Typically the linear part of the problem will be stiff and the nonlinear part will be nonstiff. Many numerical integrators have been developed to overcome the phenomenon of *stiffness*. Exponential integrators were introduced in the early sixties as an alternative approach for solving stiff systems. The main idea behind these methods is to integrate exactly the linear part of the problem and then use an appropriate approximation of the nonlinear part. Thus the exponential function, and functions which are closely related to the exponential function, appear in the format of the method. This was the reason why, until recently, these methods have not been regarded as practical. The latest achievements in the field of computing approximations to the matrix

I. Lirkov, S. Margenov, and J. Waśniewski (Eds.): LSSC 2005, LNCS 3743, pp. 380–386, 2006.

exponential, have raised a new interest in the construction and implementation of exponential integrators [2, 3, 7, 6, 9].

The main requirements imposed on the functions which appear in the format of an exponential integrator are: to be analytic, map the spectrum of L into a bounded region in $\mathbb{C}$ and can be computed exactly or up to arbitrarily high order cheaply. Suppose that, for all $l \in \mathbb{N}$ and $\lambda \in \mathbb{R}$, the operators $\phi^{[l]}(\lambda) : \mathbb{R}^{d\times d} \to \mathbb{R}^{d\times d}$ satisfy the above conditions and can be expanded in the form

$$\phi^{[l]}(\lambda)(hL) = \sum_{j\geq 0} \phi_j^{[l]}(\lambda)(hL)^j.$$

The $\phi^{[l]}$ functions, which are used in practice, are associated with the so called Exponential Time Differencing methods [3, 4, 10, 11], and can be written explicitly as

$$\phi^{[l]}(\lambda)(hL) = (\lambda hL)^{-l}\left(e^{\lambda hL} - \sum_{k=0}^{l-1} \frac{(\lambda hL)^k}{k!}\right). \tag{2}$$

If h represents the stepsize and U_i denotes the internal stage approximation of the exact solution for $i = 1, 2, \ldots, s$, then the computations performed in step number n of an exponential Runge–Kutta (RK) method are related by the equations

$$\begin{aligned} U_i &= \sum_{j=1}^{s}\sum_{l=1}^{s} \alpha_{ij}^{[l]}\, \phi^{[l]}(c_i)(hL)\, hN(U_j) + e^{c_i hL} u_{n-1}, \\ u_n &= \sum_{j=1}^{s}\sum_{l=1}^{s} \beta_j^{[l]}\, \phi^{[l]}(1)(hL)\, hN(U_j) + e^{hL} u_{n-1}, \end{aligned} \tag{3}$$

where $\alpha_{ij}^{[l]}$ and $\beta_j^{[l]}$ are the parameters of the method and the vector $c = (c_1, c_2, \ldots, c_s)^T$ is the abscissa vector. If $\alpha_{ij}^{[l]} = 0$ for all $j \geq i$, then the method is explicit, and implicit otherwise. Alternatively the computations performed in step number n can be represented in a more Runge–Kutta type formulation as follows

$$\begin{array}{c|c|c|c|c} c & \alpha^{[1]} & \alpha^{[2]} & \cdots & \alpha^{[s]} \\ \hline & \beta^{[1]^T} & \beta^{[2]^T} & \cdots & \beta^{[s]^T} \end{array},$$

where each element in row number i of the matrix $\alpha^{[l]}$ is multiplied by $\phi^{[l]}(c_i)(hL)$ and each element in the vector $\beta^{[l]^T}$ is multiplied by $\phi^{[l]}(1)(hL)$. The resulting matrices are then added componentwise.

An other important class of methods which are also used for solving the semilinear problem (1) are the Integrating Factor (IF) methods. The idea behind these methods goes back to the work of Lawson [8]. He proposes to ameliorate the effect of the stiff linear part of equation (1) by using the change of variables (also known as Lawson transformation)

$$v(t) = e^{-tL}u(t).$$

The initial value problem (1) written in the new variable is

$$v'(t) = e^{-tL}N(e^{tL}v(t)) \qquad v(t_0) = v_0, \tag{4}$$

where $v_0 = e^{-t_0L}u_0$. The approach now is to apply an arbitrary s-stage Runge–Kutta method to the transformed equation (4) and then to transform the result back into the original variable. Thus, a method which satisfy just the nonstiff order conditions will not suffer from severe order reduction when it is applied to stiff problems.

The aim of this paper is to show that the IF methods are examples of exponential RK methods with special choices for the $\phi^{[l]}$ functions and the parameters $\alpha^{[l]}$ and $\beta^{[l]^T}$. We also prove that in this special case the nonstiff order theory for the exponential RK methods reduces to the classical Runge–Kutta order theory, which explains why the IF methods exhibit the expected order.

The paper is organized as follows: We briefly survey the nonstiff order theory for the exponential RK methods in Section 2. Next, in Section 3, we define the structure of the matrices $\alpha^{[l]}$ and the vectors $\beta^{[l]^T}$ as well as the form of the functions $\phi^{[l]}$, which correspond to the IF methods. Finally, in Section 4, we conclude with several remarks and questions of future interest.

2 Nonstiff Order Conditions

The nonstiff order theory for the exponential RK methods was first constructed in [4] and later developed in [11]. Here we follow the approach suggested in [9]. It is based on bi-coloured rooted trees and B-series. For those not familiar with these concepts we suggest the monographs [1, 5] for a complete treatment.

Let 2T* denote the set of all bi-coloured (black and white) rooted trees with the requirement that the valency of the white nodes is always one. This correspods to the fact that the first term on the right hand side of (1) is linear. Let $\emptyset$ represents the empty set, which remains if the root of the one node tree . or . is removed. The order of the tree $\tau \in 2\mathrm{T}^*$ is defined as the number of vertices in the tree, and it is denoted by $|\tau|$. The density γ of the tree is defined as the product over all vertices of the order of the subtree rooted at that vertex. An exponential Runge–Kutta method with elementary weight function $a : 2\mathrm{T}^* \to \mathbb{R}$ has nonstiff order p, if for all $\tau \in 2\mathrm{T}^*$, such that $|\tau| \leq p$, $a(\tau) = 1/\gamma(\tau)$.

In order to give a practical representation of the elementary weight function a of the numerical solution, it is convenient to introduce some notations. Let for $l = 0, 1, \ldots$ and $k = 1, \ldots, m$ the $s \times s$ matrix $\phi_l^{[k]}(c) = \mathrm{diag}\left(\phi_l^{[k]}(c_1), \ldots, \phi_l^{[k]}(c_s)\right)$ and $C = \mathrm{diag}(c_1, \ldots, c_s)$.

Define

$$A^{[l]} = \sum_{k=1}^{s} \phi_l^{[k]}(c)\alpha^{[k]}, \quad b^{[l]^T} = \sum_{k=1}^{s} \phi_l^{[k]}(1)\beta^{[k]^T}, \qquad C^{[l]} = \tfrac{1}{(l+1)!}C^{l+1}. \tag{5}$$

The elementary weight function a of the numerical solution can be computed using the following non-recursive rule:

- Attach $b^{[j]^T}$ to the root black node.
- Attach $A^{[j]}$ to all remaining nonterminal black nodes.
- Attach $A^{[j]}e$ to all terminal black nodes.
- Attach $C^{[j]}e$ to all terminal white nodes.
- Attach I to all remaining white nodes.

The value j is the number of white nodes directly below the corresponding node, I is the $s \times s$ identity matrix and $e = (1, \dots, 1)^T$. Now for each tree multiply from the root to the leaf as in the case for Runge–Kutta methods, then multiply these expressions in a component by component sense.

3 IF Methods as a Special Case

Applying a standard s-stage Runge–Kutta method to the transformed equation (4) and then transforming back the result into the original variable, leads us to the following $\phi^{[l]}$ functions

$$\phi^{[l]}(\lambda)(hL) = e^{(\lambda - c_l)hL}. \tag{6}$$

Every IF method can be represented in the form (3) with $\phi^{[l]}$ functions given by (6) and with a special choice of the coefficient matrices $\alpha^{[l]}$ and the coefficient vectors $\beta^{[l]^T}$. This choice reduces the set of all order conditions to a set which consists only of the order conditions corresponding to the black trees. To prove this fact we need the following lemma.

Lemma 1. *Let $t \in \mathbb{R}\backslash\{0, -1, -2, \dots\}$, then for $j = 0, 1, 2, \dots$*

$$\sum_{k=0}^{j} \frac{(-1)^k}{k!(j-k)!} \frac{1}{(k+t)} = \frac{1}{t(t+1)\cdots(t+j)}.$$

Proof. The proof of this statement is by induction on j.

The following theorem defines the structure of the matrices $\alpha^{[l]}$ and the vectors $\beta^{[l]^T}$ for the IF methods. With this structure of the coefficients, to achieve certain nonstiff order, it is sufficient to satisfy only the black trees. This implies that the transformed differential equation (4) is solved using a Runge–Kutta method.

Theorem 1. *Let all the non-zero coefficients of an exponential Runge–Kutta method (3), with $\phi^{[l]}$ functions given by (6) be located in column number l of the matrix $\alpha^{[l]}$ and in position number l of the vector $\beta^{[l]^T}$ for $l = 1, 2, \dots, s$. The method has nonstiff order p iff all order conditions corresponding to the black trees are satisfied.*

Proof. It follows directly that if the exponential Runge–Kutta method has non-stiff order p then all order conditions corresponding to the black trees are satisfied. Let us assume that all the order conditions corresponding to the black trees are satisfied. We need to prove that all the remaining order conditions are also satisfied. From the definition of the $\phi^{[l]}$ functions (6), it follows that for $j = 0, 1, 2, \ldots$,

$$\phi_j^{[l]}(1) = \frac{(1-c_l)^j}{j!}, \quad \phi_j^{[l]}(c) = \frac{1}{j!}\mathrm{diag}((c_1-c_l)^j, \ldots, (c_s-c_l)^j). \tag{7}$$

Since all order conditions corresponding to the black trees involve only the coefficients $A^{[0]}$, $b^{[0]^T}$ and c, we need to express every other order condition in terms of these coefficients. Having in mind the special structure of the matrices $\alpha^{[l]}$ and the vectors $\beta^{[l]^T}$, after substituting (7) into (5), we obtain for $j = 1, 2, \ldots$,

$$\begin{aligned} A^{[j]} &= \sum_{k=0}^{j} \frac{(-1)^k}{k!(j-k)!} C^{[0]^{j-k}} A^{[0]} C^{[0]^k}, \\ b^{[j]^T} &= \sum_{k=0}^{j} \frac{(-1)^k}{k!(j-k)!} b^{[0]^T} C^{[0]^k}. \end{aligned} \tag{8}$$

From the fact that all order conditions corresponding to the black trees are satisfied, it follows that $A^{[0]}$, $b^{[0]^T}$ and c form a Runge–Kutta method. Therefore,

$$\begin{aligned} C^{[0]}e = A^{[0]}e, \quad C^{[0]}\zeta = (A^{[0]}e)\zeta, \\ C^{[0]^k}\zeta = (A^{[0]}e)\ldots(A^{[0]}e)\zeta, \end{aligned} \tag{9}$$

where ζ is an arbitrary vector and the multiplications between the elements in the brackets are in a component by component sense.

Now, we are in the position to define a procedure which transforms every coloured tree τ into a linear combination of black trees of order at most $|\tau|$. Each tree τ can be decomposed as $\tau = (\tau_b, \tau_j, \tau_t)$, where τ_b is a coloured tree on the bottom with fewer white nodes than τ; τ_j is a tall white tree with $j \geq 1$ white nodes and τ_t is a black tree on the top. First applying formula (8) and then (9), for the order condition corresponding to a tree τ, we obtain the following three representations in terms of black trees or trees with fewer white vertices. If $\tau_t = \emptyset$, then τ reduces to

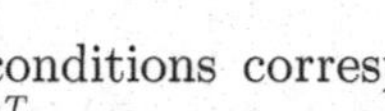

If $\tau_b = \emptyset$, then τ reduces to

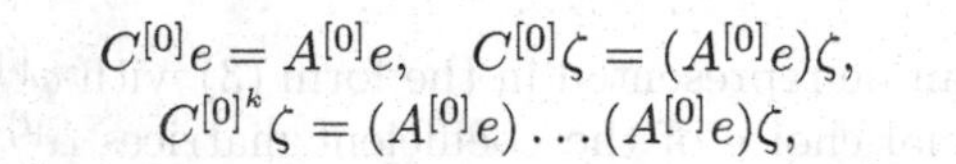

where $\delta_k = \frac{(-1)^k}{k!(j-k)!}$ for $k = 0, 1, 2, \ldots, j$. In the general case when $\tau_{\{t,b\}} \neq \emptyset$, τ can be represented as

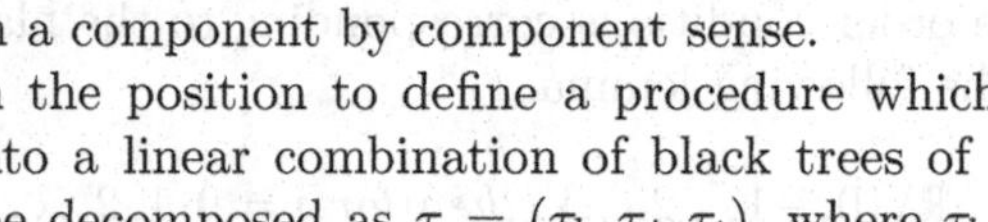

(10)

For each of the trees in the linear combination we apply the same procedure. Thus, after a finite number of steps all the trees in the combination will be black. From (5) it is clear that the order of every single black tree cannot exceed the order of a coloured tree. To complete the proof we need to show that $a(\tau) = 1/\gamma(\tau)$ for all coloured trees τ, where $|\tau| \le p$. We prove this by induction on the number of steps θ in the transformation process. Let $\theta = 1$. Every coloured tree τ has representation $\tau = \sum_{k=0}^{j} \delta_k \tau_k$, where all τ_k are black trees. If $\gamma(\tau_t) = x_1|\tau_t|x_2$ then $a(\tau_k) = \frac{1}{\gamma(\tau_k)} = \frac{1}{x_1(|\tau_t|+k)x_2}$ and by Lemma 1 for $t = |\tau_t|$ it follows that

$$a(\tau) = \sum_{k=0}^{j} \frac{(-1)^k}{k!(j-k)!} a(\tau_k) = \sum_{k=0}^{j} \frac{(-1)^k}{k!(j-k)!} \frac{1}{x_1(|\tau_t|+k)x_2}$$
$$= \frac{1}{x_1|\tau_t|(|\tau_t|+1)\cdots(|\tau_t|+j)x_2} = \frac{1}{\gamma(\tau)}.$$

Assume that $a(\tau) = 1/\gamma(\tau)$ for all coloured trees τ with θ steps in the transformation process. Let τ be a tree with $\theta + 1$ steps in the transformation process. From (10) it follows that $\tau = \sum_{k=0}^{j} \delta_k \tau_k$, where τ_k are coloured trees with θ steps in the transformation process and hence $a(\tau_k) = 1/\gamma(\tau_k)$. If $\gamma(\tau_t) = x_1|\tau_t|x_2$ then $a(\tau_k) = \frac{1}{\gamma(\tau_k)} = \frac{1}{x_1(|\tau_t|+k)x_2}$ and by Lemma 1 for $t = |\tau_t|$ it again follows that $a(\tau) = 1/\gamma(\tau)$.

4 Conclusions

We have shown that the IF methods are examples of exponential Runge–Kutta methods with special structure of the coefficient matrices and the related $\phi^{[l]}$ functions. We have also proven that, in this special case, the nonstiff order theory for the exponential RK methods reduces to the classical Runge–Kutta order theory. This explains why the IF methods exhibit the expected order. Examples of the $\phi^{[l]}$ functions, other than (2) and (6), arise from the framework of Lie group methods, see [9]. The question of how to find the best set of $\phi^{[l]}$ functions is open and needs further investigation.

Acknowledgment

This work was partially supported by NTNU and Shumen University.

References

1. J. C. Butcher. *Numerical methods for ordinary differential equations.* John Wiley & Sons, 2003.
2. M. Calvo and C. Palencia. A class of explicit multistep exponential integrators for semilinear problems. preprint, 2005.
3. P. M. Cox and P.C. Matthews. Exponential time differencing for stiff systems. *J. Comput. Phys.*, 176:430–455, 2002.

4. A. Friedli. Verallgemeinerte Runge–Kutta verfahren zur löesung steifer differentialgleichungssysteme. *Lect. Notes Math.*, 631, 1978.
5. E. Hairer, C. Lubich, and G. Wanner. *Geometric Numerical Integration.* Springer, 2003. Number 31 in Springer Series in Computational Mathematics.
6. M. Hochbruck and A. Ostermann. Exponential Runge–Kutta methods for parabolic problems. *Appl. Numer. Math.*, 53:323–339, 2005.
7. M. Hochbruck and A. Ostermann. Explicit exponential Runge–Kutta methods for semilinear parabolic problems. *SIAM J. Numer. Anal.*, 43 (3):1069–1090, 2005.
8. J. Lawson. Generalized Runge–Kutta processes for stable systems with large Lipschitz constants. *SIAM J. Numer. Anal.*, 4:372–390, 1969.
9. B. Minchev. *Exponential Integrators for Semilinear Problems.* PhD thesis, University of Bergen, 2004.
10. S. Nørsett. An A-stable modification of the Adams-Bashforth methods. *Lect. Notes Math.*, 109:214–219, 1969.
11. K. Strehmel and R. Weiner. B-convergence results for linearly implicit one step methods. *BIT*, 27:264–281, 1987.

An Operator Splitting Scheme for Biharmonic Equation with Accelerated Convergence

X.H. Tang and Ch. I. Christov

Dept. of Mathematics, University of Louisiana at Lafayette, LA, 70504-1010

Abstract. We consider the acceleration of operator splitting schemes for Dirichlet problem for biharmonic equation. The two fractional steps are organized in a single iteration unit where the explicit operators are arranged differently for the second step. Using an *a-priori* estimate for the spectral radius of the operator, we show that there exists an optimal value for the acceleration parameter which speeds up the convergence from two to three times. An algorithm is devised implementing the scheme and the optimal range is verified through numerical experiments.

1 Introduction

Biharmonic boundary value problems arise in many different areas of mechanics of continua such as the stream function formulation for stationary Navier-Stokes equations and the equations for deformation of elastic plates. Constructing efficient numerical algorithms is of prime importance in these cases. A well established approach to the problem is to introduce an artificial time in the elliptic equation and to use operator splitting technique for the resulting parabolic equation. The technique is summarily known as Alternative Directions Implicit (ADI) method. To the authors' knowledge, the first work in which an ADI scheme was applied to biharmonic equation is [5]. The CD scheme exhibits the best of the world of ADI schemes, as being absolutely stable and low cost per iterations, but its rate of convergence has been shown to be rather slow in some cases (see, [7, 6]). In order to accelerate the convergence we consider an iteration unit consisting of a CD scheme and a modified scheme, the latter dependent on a parameter. The two schemes damp different part of the spectrum differently and we show in the present paper that the combination of them yields a faster convergence than the original ingredients.

2 Differential Equations and Conte-Dames ADI Scheme

Consider the following two-dimensional higher-order parabolic equation with Dirichlet boundary value problem

$$u_t = -\Delta^2 u(x,y) + F(x,y), \qquad (x,y) \in D; \tag{1}$$

$$u(0,y) = f_1(y), u(1,y) = f_2(y), u(x,0) = f_3(x), u(x,1) = f_4(x), \tag{2}$$

$$u_x(0,y) = g_1(y), u_x(1,y) = g_2(y), u_y(x,0) = g_3(x), u_y(x,1) = g_4(x), \tag{3}$$

I. Lirkov, S. Margenov, and J. Waśniewski (Eds.): LSSC 2005, LNCS 3743, pp. 387–394, 2006.

where D is a square region $\{(x,y)|0 < x < 1, 0 < y < 1\}$ and $\bar{D} = D \cup \partial D$ is its closure. Δ^2 is the biharmonic operator. Equations (1)-(3) are also called "clamped plate problem" in elasticity.

In order to obtain second order of approximation of the difference scheme we assume that function u possesses derivatives up to sixth order.

We employ an uniformly spaced mesh in $\bar{D}$ with spacings $h_x = h_y = h = 1/N$, where N is the grid size and replace D and ∂D by sets of grid points $D_h = \{(ih, jh)|i = 1, 2, \cdots, N-1; j = 1, 2, \cdots, N-1\}$ and ∂D_h. Respectively, $u_{i,j}^n$ is the difference approximation to u at the grid point $x = ih$, $y = jh$ and time stage $t = n\tau$, where τ is the time increment. The differential operators are approximated by the usual central difference operators, denoted by δ_x^4, δ_y^4 and $\delta_x^2\delta_y^2$. The scheme of Conte and Dames [5] (called in what follows "CD") consists of following two sweeps in x and y directions

$$\tilde{u}_{i,j} = u_{i,j}^n - \tau(\delta_x^4\tilde{u} + 2\delta_x^2\delta_y^2 u_{i,j}^n + \delta_y^4 u_{i,j}^n - F_{i,j}), \quad (4)$$

$$u_{i,j}^{n+1} = \tilde{u}_{i,j} - \tau(\delta_y^4 u_{i,j}^{n+1} - \delta_y^4 u_{i,j}^n), \quad (5)$$

where $u_{i,j}^0 = 0$ is an (arbitrary) initial condition and the time increment τ plays the role of an iteration parameter and can be chosen to accelerate convergence. Second-order approximations for the boundary conditions are obtained by means of central differences on the grid that overflows the actual region, e.g.

$$\tilde{u}_{i,j} = u_{i,j}^{n+1} = f_{i,j}, \qquad \text{for } (ih, jh) \in \partial D_h, \quad (6)$$

$$\phi_{-1,j} = \phi_{1,j} - 2hg_1(jh), \qquad \phi_{N+1,j} = \phi_{N-1,j} + 2hg_2(jh), \quad (7)$$

$$\phi_{i,-1} = \phi_{i,1} - 2hg_3(ih), \qquad \phi_{i,N+1} = \phi_{i,N-1} + 2hg_4(ih), \quad (8)$$

for $i = 1, 2, \cdots, N-1$ and $j = 1, 2, \cdots, N-1$. Here ϕ stands for the different time stages $\tilde{u}$, and u^{n+1}.

Eliminating the intermediate variable $\tilde{u}_{i,j}$ between Eqs. (4) and (5) one gets

$$(E + \tau\delta_x^4 + \tau\delta_y^4 + \tau^2\delta_x^4\delta_y^4)u^{n+1} = (E - 2\tau\delta_x^2\delta_y^2 + \tau^2\delta_x^4\delta_y^4)u^n + \tau F, \quad (9)$$

where F stands for the grid function of $F(x,y)$ on D_h, E denotes the identity matrix and the difference operators are considered as the corresponding matrices for grid function u^n. The subscripts i,j are omitted for brevity of notation. The transition matrix T from one time step to another is

$$T = (E + \tau\delta_x^4 + \tau\delta_y^4 + \tau^2\delta_x^4\delta_y^4)^{-1}(E - 2\tau\delta_x^2\delta_y^2 + \tau^2\delta_x^4\delta_y^4). \quad (10)$$

The convergence of CD scheme for arbitrary positive iteration parameter τ is shown in [5] by demonstrating that $\|T\| < 1$ for any $\tau > 0$ for the case when the second-order derivatives are specified at the boundary. However,the dependence of spectral radius of T on the iteration parameter τ has not been investigated so far, because of the difficulty in obtaining appropriate eigenvectors of the difference operator for biharmonic problem with Dirichlet boundary conditions. If an arbitrary τ is chosen, the convergence rate of CD scheme can be quite slow (see, discussions in[7] and [6]).

Therefore we have three essential objectives to achieve in this work: (1) to reformulate the CD scheme in a manner that allows one to accelerate its convergence rate depending on an iteration parameter; (2) to prove that acceleration is possible and to find an estimate for the iteration parameter introduced; (3) to find optimal choice of the iteration parameter through numerical experiment.

Suppose that w is the grid function that is the solution of the stationary difference problem with the same Dirichlet boundary conditions. Define the error vector for the n-th iteration as $\xi^n = u^n - w$. By (9) and b.c. (6)-(8), we obtain the system of equations with homogeneous boundary conditions for ξ^n

$$\xi^{n+1} = T\xi^n \quad \text{on} \quad D_h, \tag{11}$$

$$\xi^{n+1} = 0 \quad \text{on} \quad \partial D_h, \tag{12}$$

$$\xi^{n+1}_{-1,j} = \xi^{n+1}_{1,j} \qquad \xi^{n+1}_{N+1,j} = \xi^{n+1}_{N-1,j}; \tag{13}$$

$$\xi^{n+1}_{i,-1} = \xi^{n+1}_{i,1} \qquad \xi^{n+1}_{i,N+1} = \xi^{n+1}_{i,N-1}. \tag{14}$$

Using Courant's theorem [2], Conte and Dames [5] showed that for the boundary conditions (12)-(14) the transition matrix T has a complete set of eigenvectors v_k for the vector space Φ defined on D_h and the corresponding eigenvalues are $0 < \alpha(T) = \lambda_1 \leq \lambda_2 \leq \cdots \leq \lambda_{m-1} \leq \lambda_m = \beta(T) < 1$. This is the only a-priori information about transition matrix T needed for us to improve the original CD method. For convenience, we denote the matrices involved in the scheme as

$$B_x = E + \tau\delta_x^4, \qquad B_y = E + \tau\delta_y^4. \qquad R = E - 2\tau\delta_x^2\delta_y^2 + \tau^2\delta_x^4\delta_y^4. \tag{15}$$

Then T can be written as $T = B_y^{-1}B_x^{-1}R$.

3 The Modified Splitting Scheme

In order to accelerate the convergence of the iterations for a given time increment τ we try to find a modification of the CD scheme in which the transition operator has smaller norm than the original scheme.

The gist of present paper is that we introduce an iteration unit consisting of two CD iterations with different arrangements of the explicit terms by means of an auxiliary parameter θ as follows

$$\text{(n+1)-th unit} = \begin{cases} B_xB_yu^{n+1} = Ru^n + \tau F & (a) \\ B_xB_yu^{n+2} = R[(\theta+1)u^{n+1} - \theta u^n] + \tau F & (b) \end{cases}, \tag{16}$$

where the intermediate variable $\tilde{u}$ is eliminated and $(\theta+1)u^{n+1} - \theta u^n$ is considered as the input vector to compute u^{n+2} using CD scheme in (b). The boundary conditions are the same for the two steps and are omitted for the sake of brevity. Clearly, there is no extra cost for implementing of the new scheme.

To study the convergence of the iteration units we observe that one can recast the equations of the full iteration unit (16) as equations for the error ξ^n, namely

$$\xi^{n+1} = T\xi^n, \quad \xi^{n+2} = [(\theta+1)T^2 - \theta T]\xi^n. \tag{17}$$

Begin with a trivial initial condition $u^0 = 0$, which means that the initial condition for the error is $\xi^0 = -w$. The Fourier expansion of ξ with respect to the complete set of eigenvectors $\{v_k\}$ of T reads

$$\xi^0 = -w = \sum_{k=1}^{m} c_k v_k, \qquad \xi^n = \begin{cases} \sum_{k=1}^{m} \lambda_k \rho_\theta^l(\lambda_k) c_k v_k & \text{if } n = 2l+1 \\ \sum_{k=1}^{m} \rho_\theta^l(\lambda_k) c_k v_k & \text{if } n = 2l \end{cases} \tag{18}$$

where c_k is the corresponding Fourier coefficient for $l = 0, 1, 2, \cdots$; where we define quadratic function $\rho_\theta(\lambda_k) = (1+\theta)\lambda_k^2 - \theta\lambda_k$ and λ_k is the corresponding eigenvalue for T. In order to show the convergence of the iteration units, it is sufficient for us to show that for each λ_k, $|\rho_\theta(\lambda_k)| < 1$, since $0 < \lambda_k < 1$. Indeed, we have

$$\max_{\lambda_k} |\rho_\theta(\lambda_k)| \le \max_{0<x<1} |\rho_\theta(x)| \le \max\{|\rho_\theta(0)|, |\rho_\theta(1)|, |\rho_\theta\left(\frac{\theta}{2(\theta+1)}\right)|\}, \tag{19}$$

where we consider only $\theta > 0$ because only positive choice of θ can accelerate the iterations. Since $\rho_\theta(0) = 0$ and $\rho_\theta(1) = 1$, Eq.(19) will hold when

$$|\rho_\theta\left(\frac{\theta}{2(\theta+1)}\right)| = \frac{\theta^2}{4(\theta+1)} < 1, \tag{20}$$

whence it follows that $0 < \theta < 2 + 2\sqrt{2}$ must hold in order to ensure that we have $\max_{\lambda_k} |\rho_\theta(\lambda_k)| < 1$ needed for the convergence of the iteration unit (16).

Now we analyze the acceleration of the convergence for the iteration units. Consider the error ξ_{cd}^n for the CD scheme after n iterations. Using (11) and (18) we can represent ξ_{cd}^n in terms of Fourier expansion with respect to the eigenvectors v_k of the transition matrix T, namely

$$\xi_{cd}^n = \sum_{k=1}^{m} \lambda_k^n (c_k v_k). \tag{21}$$

The coefficients of v_k in the error vector ξ_{cd}^n decrease in absolute value by the multiplicative factor of λ_k. The least affected is the coefficient of v_m, which corresponds to the largest eigenvalue λ_m. Repeating CD iterations leads us to the asymptotic case, because for $n \gg 1$ all other coefficients become negligibly small compared to the coefficient of v_m, and hence we have the error for CD scheme given by

$$\xi_{cd}^n = \lambda_m \xi_{cd}^{n-1} = \beta(T)\xi^{n-1}. \tag{22}$$

From (22) follows for the standard norm $\|\xi_{cd}^n\|^2 = (\xi_{cd}^n, \xi_{cd}^n)$ that

$$\|\xi_{cd}^n\| = \beta(T)\|\xi_{cd}^{n-1}\|. \tag{23}$$

Although for small n the amplifier of the norm depends on the iteration, the performance of the iterative process is usually judged by the asymptotic rate of convergence s $(n \gg 1)$ which is defined as

$$s = -\ln|\beta(T)| = -\ln\frac{\|\xi^n\|}{\|\xi^{n-1}\|}. \tag{24}$$

By the recursive relation (23) we obtain the asymptotic equation

$$\|\xi^n\| = e^{n \ln \beta(T)} \|\xi^0\| = e^{-ns} \|\xi^0\|. \tag{25}$$

In this way, the asymptotic rate of convergence s characterizes the rate of the exponential error decrease. The cause of slow convergence of CD scheme is that $\beta(T)$ is close to unity. In such a case we use the notation $\beta = 1 - p$ where $0 < p \ll 1$. Using the Taylor expansion, the asymptotic rate of convergence s of CD scheme is given by $s = -\ln(1-p) \approx p \ll 1$.

In the same manner, we investigate the asymptotic rate of convergence s of our iteration units. In order not to obscure the main idea we limit the discussion here to some typical values of θ, as $\theta = 3$, which is in the range $0 < \theta < 2 + 2\sqrt{2}$. The coefficient of eigenvector v_k in (18) decreases in absolute value by the multiplicative factor of $|\rho_3(\lambda_k)|$. Since $\rho_3(x) = 4x^2 - 3x$ is a quadratic function, it is easy to show that its minimum is at $x = \frac{3}{8}$ and has the magnitude of $\frac{9}{16}$. Then

$$\max_{0<\lambda_k \leq 1-p} |\rho_3(\lambda_k)| \leq \max\{\tfrac{9}{16}, |\rho_3(1-p)|\}, \tag{26}$$

which mean that if $9/16 > |\rho_3(1-p)|$, we have already obtained a very fast convergence of the iteration units which will reduce the error to 10^{-5} within 20 iteration units.

On the other hand, for $p \ll 1$ we have $|\rho_3(1-p)| = |(1-p)(1-4p)| \approx (1-5p)$ and using (17) asymptotically for $n \gg 1$, we can write $\|\xi^{n+2}\| \approx (1-5p)\|\xi^n\|$. Therefore, the corresponding asymptotic rate of convergence of our iteration units is

$$s = -\tfrac{1}{2}\ln(1-5p) \approx 2.5p, \tag{27}$$

where we compare one iteration unit consisted of two iterations with two original CD iterations for which the reduction factor would be $(1-p)^2 \approx 1-2p$. All this means that the introduction of the iteration unit can speed up the convergence rate at least 2.5 times. Note that the actual factor of acceleration depends on the value of time increment τ and is discussed in the next Section. Similarly, we can verify that for $0 < \theta < 2 + \sqrt{2}$, the asymptotic rate of convergence is

$$s = -\tfrac{1}{2}\ln \rho_\theta(1-p) = -\tfrac{1}{2}\ln\{(1-p)[(\theta+1)(1-p)-\theta]\} \approx \tfrac{1}{2}(\theta+2)p. \tag{28}$$

But this conclusion depends on the assumption that $1 - \beta(T) = p \ll 1$ and

$$\left|\rho_\theta\left(\frac{\theta}{2(1+\theta)}\right)\right| = \frac{\theta^2}{4(1+\theta)} \leq \rho_\theta(1-p). \tag{29}$$

Actually, the maximum eigenvalue $\beta(T)$ of the transition matrix can be estimated in the numerical experiment, based on which we can choose a proper value of the auxiliary parameter θ to maximize the acceleration of the devised iteration units. By (9) and (10), we have the recursive relation

$$(u^{n+1} - u^n) = T(u^n - u^{n-1}). \tag{30}$$

Since $\|T\| = \beta(T) < 1$, after a few CD iterations the largest eigenvalue $\beta(T)$ becomes the dominant multiplicative factor in the iterations while other smaller eigenvalues become negligible. Therefore we can compute the numerical quantity

$$q = \|u^{n+1} - u^n\| / \|u^n - u^{n-1}\| \tag{31}$$

in each iteration and when q varies little between two consecutive iterations, an estimate of $\beta(T)$ is obtained. Based on the a-posteriori numerical estimate $q \approx \beta(T)$, we can determine the optimal choice of θ to maximize the acceleration. At first, we notice that $\rho_\theta(q)$ is a linear function of θ when q is fixed

$$h_1(\theta) = \rho_\theta(q) = (q^2 - q)\theta + q^2. \tag{32}$$

since $q^2 - q < 0$, $h_1(\theta)$ is a monotone decreasing function. Next, we consider another function h_2 of θ

$$h_2(\theta) = \left|\rho_\theta\left(\frac{\theta}{2(1+\theta)}\right)\right| = \frac{\theta^2}{4(1+\theta)}, \tag{33}$$

which is a monotone increasing for $\theta \in (0, 2\sqrt{2})$ because

$$h_2'(\theta) = \frac{\theta^2 + 2\theta}{4(\theta+1)} > 0. \tag{34}$$

To maximize the acceleration, by (19) we need to find θ from

$$\min_{0<\theta<2+\sqrt{2}} \max\{|h_1(\theta)|, h_2(\theta)\}. \tag{35}$$

By the monotonicity of $h_1(\theta)$ and $h_2(\theta)$, it is easy to verify that the optimal θ has to be the positive solution of the following equation and automatically in the range $(0, 2 + 2\sqrt{2})$,

$$(q^2 - q)\theta + q^2 = \tfrac{1}{4}\theta^2(1+\theta)^{-1}. \tag{36}$$

Therefore, we obtain the optimal θ by solving (36)

$$\theta_{opt}(q) = \frac{2q^2 + (\sqrt{2} - 1)q}{-2q^2 + 2q + \frac{1}{2}}. \tag{37}$$

Since $h_1(0) = q^2$ and $h_1(\theta)$ is a decreasing function, then we have

$$0 < h_1(\theta_{opt}) = \frac{\theta_{opt}^2}{4(1+\theta_{opt})} < q^2 \tag{38}$$

where q^2 stands for the convergent effect of two CD iterations. Hence, we have shown that for different choices of the iteration parameter τ in the CD method, which give us different transition matrices T (i.e. different q), we are always able to select a θ_{opt} to accelerate the original CD iterations.

By (28) and (37), we can see that when $\beta(T)$ is closer to unity, that is to say, the original CD method has a slower convergence, larger auxiliary parameter θ can be chosen in the iteration units which leads us to a more significant acceleration over the CD method. Since the value of θ is bounded by $2 + 2\sqrt{2} \approx 4.8$, by (28) the best acceleration that our method can reach is 3.4 times faster than CD method.

4 Results and Discussion

We begin with numerical verification of the performance of a single iteration-unit as introduced above. In implementation of the algorithm we follow [4] where a splitting scheme of type of CD was applied to lid-driven cavity flow of viscous liquid. Later on a similar algorithm was used in [3] for another kind of higher-order diffusion equation.

We select two different problems for which analytical solutions are available

$$\hat{u}(x,y) = \sin^2(\pi x)\sinh^2 y, \ (a) \quad \text{and} \quad \hat{u}(x,y) = 2350x^4(x-1)^2y^4(y-1)^2 \ (b). \tag{39}$$

The first of these was created by us, and the secon is from [1].

The operator to be inverted is the same in both cases and does not depend on the actual solution, and hence on the right hand side. Yet, the convergence rate depends on the specific solution because of the fact that different eigenfunctions decay differently with the iterations. Since different analytical solutions have different content of eigenfunctions, the rate could be rather different which warrants checking the performance for two radically different analytical solutions.

It is important to demonstrate that an iteration unit does give the same truncation error as a single CD iteration. For this particular test we fix $\theta = 1, 2$ and $\tau = \frac{1}{4}h$ and calculate the solution with three different grids. Here, we fix the value of θ and vary the number of grid points both along x and y directions simultaneously. We define the computed order of approximation as $R = \log_2 [\|u_N - \hat{u}\|/\|u_{2N} - \hat{u}\|]$. For second order schemes $O(h^2)$, the value of $R = 2$. Table 1 shows that in both cases, the computed convergence rate is very close to two. Having confirmed the second-order accuracy of the scheme we can address the issue of computational efficiency. The pertinent parameter here is the number of iterations, say N_{iter}, needed to reduce the norm of the difference between two iterations to 10^{-6}. Clearly, the rate of convergence is a function of the time increment, τ, and the optimization parameter θ. For different θ we performed calculations with several different τ. In Fig. 1 we present N_{iter} for which the norm between two iterations for Eq. (39)(b) go down to 10^{-6}. For $\theta \leq 2$ there is a conspicuous minimum of N_{iter} for the original CD scheme, as well as for the scheme with one iteration unit. In this range of θ, our N_{iter} is at least twice smaller than CD scheme. In the range of non-optimal τ, the acceleration is even larger. The result in Fig. 1 is in very good agreement with the theoretical estimate, Eq. (37), which gives us $\theta_{optimal} \approx 1.8$ when $q \approx 0.77$. This is the fastest result one can get for Eq. (39)(b) with a splitting scheme of the type of CD.

Table 1. Order of approximation R for different grids

case Eq. (39)(a)

N	$\theta = 1$	R	$\theta = 2$	R
256	1.578E-5	-	1.582E-5	-
512	3.797E-6	2.05	3.810e-6	2.05
1024	9.687E-7	1.97	9.747E-7	1.967

case Eq. (39)(b)

N	$\theta = 1)$	R	$\theta = 2$	R
256	3.093E-4	-	3.080E-4	-
512	7.981E-5	1.95	8.063e-5	1.93
1024	1.823E-5	2.13	1.884E-5	2.09

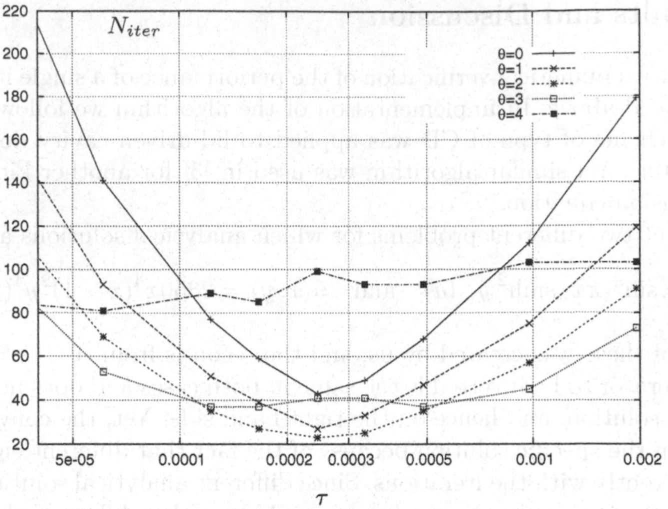

Fig. 1. Number of iterations, as function of time increment for different values of θ and grid size 1024×1024

It is interesting to mention here the nonmonotone behavior of N_{iter} with the increase of θ, which means that there is an optimum for θ, but only in the vicinity of optimal τ. If one cannot chose *a-priori* an optimal τ, then the scheme proposed here will overperform the CD scheme even for a wider range of θ. We obtained a similar result for grid size 512×512, for which the values of N_{iter} consistently lower by 10% from the presented case. This is natural for iterative algorithms since the eigenvalues of difference operators depend on spacing h.

References

1. M. Arad, A. Yakhot, and G BenDor. High-order-accurate discretization stencil for an elliptic equation. *Int. J. Num. Meth. Fluids*, 23:367–377, 1996.
2. E. Bodewig. *Matrix Calculus* **4**. Interscience Publishers Inc., New York, 1956.
3. C. I. Christov and J. Pontes. Numerical scheme for Swift-Hohenberg equation with strict implementation of Lyapunov functional. *Mathematical and Computer Modelling*, 35:87–99, 2002.
4. C. I. Christov and A. Ridha. Splitting scheme for iterative solution of biharmonic equation. application to 2D Navier-Stokes problems. In *Advances in Numerical Methods and Applications*, I. Dimov, Bl. Sendov, and P. Vasilevski, editors, 341–352, Singapore, 1994. World Scientific.
5. S. D. Conte and R. T. Dames. On an alternating direction method for solving the plate problem with mixed boundary conditions. *J. ACM*, 7:264–273, 1960.
6. L. W. Ehrlich and M. M. Gupta. Some difference schemes for the biharmonic equation. *SIAM J. NUMER. ANAL.*, 12:773–790, 1975.
7. D. Greenspan and D. Schultz. Fast finite-difference solution of biharmonic problems. *Communications of ACM*, 15:347–350, 1972.

Testing Variational Data Assimilation Modules

Zahari Zlatev and Jørgen Brandt

National Environmental Research Institute,
Frederiksborgvej 399, P. O. Box 358, DK-4000 Roskilde, Denmark

Abstract. The use of the variational data assimilation approach is becoming more and more popular in the attempts to improve the accuracy of the results obtained by different air pollution models. Different tests were carried out in order both to check the ability of the data assimilation procedures to improve the initial concentrations and to start building up a benchmark for testing the performance of these procedures on different modern high-speed computers.

1 Statement of the Problem

The variational data assimilation approach could be viewed as an attempt to adjust globally the results obtained by a given model to a set of available observations. The idea was probably applied for the first time in 1971 in a paper written by Morel et al. [3]. They defined and implemented a procedure in which the assimilating model is repeatedly integrated forward and backward in time, the observations being constantly introduced in the model. The heuristic idea behind that procedure, supported by a number of numerical results, was that repeated calculations and corrections of model results would converge to a solution which would be compatible, at least to a certain degree of accuracy, with the observations.

2 Basic Ideas

Assume that observations are available at time-points t_p, where $p \in \{0, 1, 2, \ldots, P\}$. These observations can be taken into account in an attempt to improve the results obtained by a given model. This can be done by minimizing the value of the following functional (see, for example, Lewis and Derber [1]):

$$J\{\bar{c}_0\} = \frac{1}{2} \sum_{p=0}^{P} < W(t_p) \, (\bar{c}_p - \bar{c}_p^{obs}) \, , \, \bar{c}_p - \bar{c}_p^{obs} >, \tag{1}$$

where the functional $J\{\bar{c}_0\}$ depends on the initial value $\bar{c}_0$ of the vector of the concentrations, $W(t_p)$ is a matrix containing some weights and $<,>$ is an inner product in an appropriately defined Hilbert space (it will be assumed here that the usual vector space is used, i.e. it is assumed that $\bar{c} \in \mathbf{R}^s$). $J\{\bar{c}_0\}$ is expressed by the weights and the differences between calculated by the model

I. Lirkov, S. Margenov, and J. Waśniewski (Eds.): LSSC 2005, LNCS 3743, pp. 395–402, 2006.

concentrations $\bar{c}_p$ and observations $\bar{c}_p^{obs}$ at the time-levels t_p at which observations are available, but it will be assumed that $W(t_p)$ is the identity matrix I in this study. In general weights are to be defined in some way.

The task is to find an improved initial field $\bar{c}_0$, which minimizes the functional $J\{\bar{c}_0\}$. This can be achieved by using some optimization algorithm. Most of the optimization algorithms are based on the application of the gradient of the functional $J\{\bar{c}_0\}$. The adjoint equation has to be defined and used in the calculation of the gradient of the functional $J\{\bar{c}_0\}$.

It is assumed here that data assimilation techniques are applied to improve an initial field of concentrations, but data assimilation can be applied for other purposes too. Other applications are (a) improving emission fields, (b) checking boundary conditions and, in a more general context, (c) checking the sensitivity of the concentrations to variation of different parameters.

3 Calculating the Gradient of the Functional

It is convenient to explain the basic ideas that are used when the gradient of the functional $J\{\bar{c}_0\}$ is calculated by the following very simple example. Assume that observations are available at five time-points: t_0, t_1, t_2, t_3 and t_5. The calculations have to be performed in five steps.

- **Step 1.** Use the *model* to calculate $\bar{c}_1$ (performing integration, in a forward mode, from time-point t_0 to time-point t_1). Calculate the adjoint variable $\bar{q}_1 = \bar{c}_1 - \bar{c}_1^{obs}$. Form the adjoint equation (corresponding to the model used in the forward mode). Perform backward integration (by applying the adjoint equation) from time-point t_1 to time-point t_0 to calculate the vector $\bar{q}_0^1$, where the lower index shows that $\bar{q}_0^1$ is calculated at time-point t_0 , while the upper index shows that $\bar{q}_0^1$ is obtained by using $\bar{q}_1 = \bar{c}_1 - \bar{c}_1^{obs}$ as an initial vector in the backward integration.
- **Step 2 - Step 4.** Perform the same type of calculations, as those in Step 1 to obtain $\bar{q}_0^2$, $\bar{q}_0^3$ and $\bar{q}_0^4$.
- **Step 5.** The sum of the vectors $\bar{q}_0^1$, $\bar{q}_0^2$, $\bar{q}_0^3$, $\bar{q}_0^4$ obtained in Step 1 - Step 4 and vector $\bar{q}_0^0 = \bar{q}_0 = \bar{c}_0 - \bar{c}_0^{obs}$ gives the required gradient of the functional $J\{\bar{c}_0\}$.

It is clear that this approach can also be used for an arbitrary value of P (instead of $P = 4$). In the general case, the gradient of the functional $J\{\bar{c}_0\}$ can be calculated by performing one forward step from time-point t_0 to time-point t_P and P backward steps from time-points t_p, $p = 1, 2, \ldots, P$, to time-point t_0. This explains the main idea in a very clear way, but it is expensive when p is large. In fact, the computational work can be reduced by performing only one backward step (see, again, [1]).

4 Forming the Adjoint Equations

Assume that a linear model is written in the following general form:

$$\frac{\partial \bar{c}}{\partial t} = A\bar{c}. \tag{2}$$

Let $\bar{q} = \bar{c} - \bar{c}^{obs}$ be the adjoint variable (it is assumed here that some interpolation rules are used in order to get the continuous variable $\bar{c}^{obs}$ from the available discrete values of the observations). Then the adjoint equation is defined by

$$\frac{\partial \bar{q}}{\partial t} = -A^T \bar{q}, \tag{3}$$

where the superscript T means that if we form the matrix by which the adjoint operator is represented after some kind of discretization, then this matrix is transposed to the matrix representing operator A (again after some kind of discretization). Normally, the notation A^* is used instead of A^T. The operator A^* is the conjugate operator of operator A. The notation A^T is used in order to facilitate the understanding of the linear algebra operations which have to be carried out when the data assimilation procedure is used in practice.

If the model is non-linear, then it is first necessary to produce a linearized version. In other words, the non-linear model

$$\frac{\partial \bar{c}}{\partial t} = B(\bar{c}), \tag{4}$$

is rewritten as

$$\frac{\partial (\delta \bar{c})}{\partial t} = B'(\bar{c})\, \delta \bar{c}, \tag{5}$$

where $\delta \bar{c}$ is some small variation of $\bar{c}$ and B' is a linear operator obtained by differentiation of B. The adjoint equation of (4) can be formed as follows:

$$\frac{\partial \bar{q}}{\partial t} = -\left[B'(\bar{c})\right]^T \bar{q}. \tag{6}$$

The adjoint equation is always linear, see (3) and (6). In the non-linear case the adjoint operator depends on the values of $\bar{c}$. This causes difficulties (all values of $\bar{c}$ calculated during the forward mode have to be saved because they are needed in the backward mode).

More details about adjoint equations can be found in Marchuk [2].

5 Algorithmic Representation

A data assimilation algorithm can be represented by applying the procedure described in Fig. 1. An optimization procedure is needed for the calculations that are to be carried out in the loop "DO ITERATIONS". In many optimization subroutines, the direction of the steepest descent is to be found and then the value of parameter ρ that gives the largest decrease in the direction found is to be used to improve the current solution. In practice, however, it is only necessary here to find a good standard minimization subroutine. In our experiments we used the subroutine E04DGF from the NAG Numerical Library [4]. It should be relatively easy to call another appropriate subroutine.

Nearly all optimization subroutines need the value of the functional $J\{\bar{c}_0\}$ and its gradient. The calculation of these values is performed in the loop "DO

LARGE_STEPS" in Fig. 1. P_STEP is the number of time-points at which observations are available. P_LENGTH is the number of time-steps carried out between two time-points at which observations are available. This is a major part of the computational work and is based on scheme described in Section 3. Let us reiterate here that multiple backward steps can be avoided. The algorithm with multiple backward modes is given here only because it facilitates the understanding of the main idea behind the data assimilation procedure.

```
        Initialize scalar variables, vectors and arrays; set the initial gradient
C
        DO ITERATIONS = 1, MAX_ITERATIONS
C
           DO LARGE_STEPS = 1, P_STEP
C
              F_START=(LARGE_STEPS-1)*P_LENGTH+1
              F_END=LARGE_STEPS*P_LENGTH
              DO FORWARD_STEPS = F_START,F_END
                 Perform forward steps by using the model
              END DO FORWARD_STEPS
C
              Form the adjoint variable and update the value of the functional
              DO BACKWARD_STEPS = F_END,1,-1
                 Perform backward steps by using the adjoint equation
              END DO BACKWARD_STEPS
              Update the value of the gradient of the functional
C
           END DO LARGE_STEPS
C
           Compute an approximation of the optimization parameter ρ
           Update the initial values field:
           NEW_FIELD=OLD_FIELD + ρ*GRADIENT
           Exit from loop DO ITERATIONS if the stopping criteria are satisfied
C
        END DO ITERATIONS
C
        Perform output operations and stop the computations
```

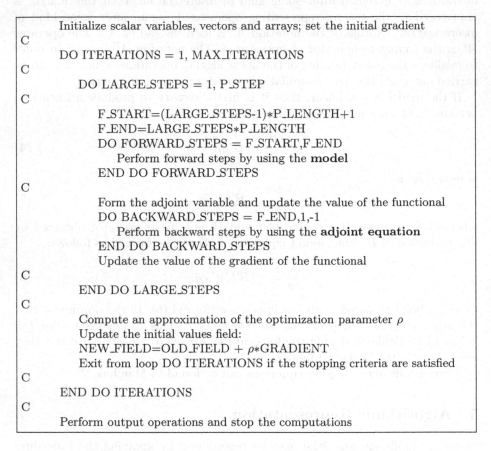

Fig. 1. An algorithm for performing variational data assimilation by carrying out multiple backward calculations

6 Using Splitting Techniques

It is well-known (see, for example, Zlatev [5]) that splitting procedures are used in all operational large-scale air pollution models. The algorithm shown in Fig. 1 can easily be modified for the case where some splitting procedure is implemented. Assume that the original model (the model used in Fig. 1) is split to two sub-models: "sub-model-1" and "sub-model-2". Then in the body of the loop DO

FORWARD STEPS in Fig. 1 one should perform successively at each time-step calculations with "sub-model-1" and "sub-model-2" (instead of the calculations carried out by "model"). Denote the adjoint equations corresponding to "sub-model-1" and "sub-model-2" by "adjoint-equation-1" and "adjoint-equation-2". Then in the body of the loop DO BACKWARD STEPS in Fig. 1 one should perform successively at each time-step calculations with "adjoint-equation-1" and "adjoint-equation-2" (instead of the calculations carried out by the adjoint equation of the original model).

The use of splitting procedures simplifies considerably the computations related to the data assimilation algorithm. It is clear that one should prepare modules for performing forward and backward steps for each sub-model obtained by the splitting procedure and, after that, combine these modules in an appropriate way. This observation was used to prepare modules for the two most important processes in an air pollution model, the advection and the chemistry. Some tests obtained with these two modules will be shown.

It should be emphasized, however, that it is also necessary to check carefully the performance of the splitting procedure. Such checks will be carried out in the near future.

7 Numerical Examples

It is necessary to build up a large benchmark of relevant examples by which different properties of the variational data assimilation procedures can be studied. Two examples will be used in this paper. The first example is a linear first-oder PDE, which is in some sense similar to the problems arising in the advection sub-model when splitting procedures are implemented. The second example is a non-linear system of ODEs, which is similar to the problems arising in the chemical sub-model.

7.1 Example 1

One-dimensional transport of a single pollutant in the atmosphere can be described mathematically (see [1]) by

$$\frac{\partial \bar{c}}{\partial t} = -V \frac{\partial \bar{c}}{\partial x}, \quad x \in [a,b], \quad t \in [0,T], \quad \bar{c}(x,0) = f(x), \tag{7}$$

where

$$[a,b] = [0,2\pi], \quad f(x) = \sin(x), \quad V(x) = \frac{6}{(2\pi)^2} x(2\pi - x). \tag{8}$$

The exact solution of the problem defined by (7) and (8) is given by

$$\bar{c}(x,t) = \sin\left(\frac{2\pi x}{x + (2\pi - x)\exp\left(\frac{3t}{\pi}\right)}\right). \tag{9}$$

The problem is solved numerically by a method based on finite differences and described in [1]. In the experiments the initial values of the discretized problem are perturbed and an attempt to improve the accuracy of the perturbed initial field is carried out by using data assimilation in which the role of the observations is played by the exact solution of the problem at the prescribed time-points.

7.2 Example 2

A non-linear system of ODEs (the chemical reactions in an air pollution model are also described mathematically by non-linear systems of ODEs) of the type $y' = f(t, y)$ has been solved. The exact solution is known. Five numerical methods (the Backward Euler Method, the Implicit Mid-point Rule, the Trapezoidal Rule, a Runge-Kutta Method of order two and a Runge-Kutta Method of order six) were used in the experiments. The same approach as in Example 1 has been used in the experiments with Example 2 (i.e. again the initial values of the problem are perturbed and an attempt to improve the accuracy of the perturbed initial field is carried out by using the data assimilation procedure, in which the role of the observations is played by the exact solution of the problems at the prescribed time-points).

8 Some Results from the Runs

The following issues were tested during the experiments: (a) the correct implementation of the forward module, (b) the effects of using different grids, (c) the application of unbiased perturbations of the initial values, (d) the insertion of biased perturbations in the initial values, (e) the effects of reducing the number of observations and (f) the variation of the number of analyses (or, in other words, the variation of the number of time-points at which observations are available). Only a few results are given in this section. The influence of the number of available observation on the accuracy of the final results for Example 1 is shown in Table 1. $\mathbf{N}_x$ and $\mathbf{N}_t$ are the numbers of grid-points along the x and t axes. Errors of about 5% were inserted in the initial field by using the perturbation parameter $\alpha = 0.1$. $\mathbf{E}RR_{global}^{impr}$ is the global error obtained with the improved by the data assimilation procedure initial field. It is clearly seen that the results do not depend too much on the number of time-points at which observations are available. However the number of observations is very important (the accuracy becomes very poor when the number of observations is small).

Table 1. Values of $\mathbf{E}RR_{global}^{impr}$ obtained by using the variational data assimilation algorithm with $\mathbf{N}_x = 1001$, $\mathbf{N}_t = 10000$, $\alpha = 0.1$, five values of **N_OBS** and different numbers of analyses **P_STEP**

P_STEP	N_OBS=1000	N_OBS=500	N_OBS=125	N_OBS=20	N_OBS=5
1000	$2.64 * 10^{-5}$	$3.21 * 10^{-5}$	$3.39 * 10^{-4}$	$1.34 * 10^{-2}$	$2.05 * 10^{-1}$
100	$2.66 * 10^{-5}$	$3.20 * 10^{-5}$	$3.48 * 10^{-4}$	$1.34 * 10^{-2}$	$2.06 * 10^{-1}$
10	$2.90 * 10^{-5}$	$3.04 * 10^{-5}$	$3.91 * 10^{-4}$	$1.50 * 10^{-2}$	$2.15 * 10^{-1}$
1	$5.27 * 10^{-5}$	$5.27 * 10^{-5}$	$5.87 * 10^{-4}$	$2.75 * 10^{-2}$	$3.55 * 10^{-1}$

Results obtained when Example 2 is run on different grids by using the Backward Euler Method (the order of this method is one) and the Runge-Kutta Method of order six are given in Table 2 and Table 3 respectively. It is seen that

Table 2. Running Example 2 on different grids by using the Backward Euler Method. The initial values are perturbed by using $\alpha = 1.0$ (which means that the size of perturbations is about 50% of the values of the initial concentrations). The times are given in seconds. $\mathbf{E}RR_{global}^{impr}$ is the global error obtained with the improved by the data assimilation procedure initial field. RATIO is the ratio of the errors obtained in two successive runs.

N_t	Time	$\mathbf{E}RR_{global}^{impr}$	RATIO
40	0.02	3.286E-01	
80	0.04	3.416E-01	0.962
160	0.07	2.189E-01	1.561
320	0.17	1.233E-01	1.774
640	0.26	6.426E-02	1.919
1280	0.82	3.317E-02	1.937
2560	0.77	1.690E-02	1.962
5120	1.57	8.514E-03	1.985
10240	3.95	4.270E-03	1.994
20480	4.32	2.140E-03	1.996
40960	12.30	1.071E-03	1.999
81920	28.09	5.357E-04	1.999
163840	67.51	2.679E-04	1.999
327680	110.80	1.340E-04	2.000
655360	257.54	6.699E-05	2.000
1310720	630.42	3.349E-05	2.000
2621440	966.52	1.675E-05	2.000
5242880	1855.20	8.342E-06	2.008
10485760	3539.76	4.164E-06	2.003
20971520	8830.68	2.309E-06	1.803
41943040	14021.17	9.550E-07	2.418
83886080	31744.91	5.246E-07	1.820

Table 3. Running Example 2 on different grids by using the Runge-Kutta Method of order six. The initial values are perturbed by using $\alpha = 1.0$ (which means that the size of perturbations is about 50% of the values of the initial concentrations). The times are given in seconds. $\mathbf{E}RR_{global}^{impr}$ is the global error obtained with the improved by the data assimilation procedure initial field. RATIO is the ratio of the errors obtained in two successive runs.

N_t	Time	$\mathbf{E}RR_{global}^{impr}$	RATIO
40	0.03	6.002E-04	
80	0.04	1.031E-05	58.234
160	0.08	1.586E-07	64.979
320	0.18	1.979E-08	8.015

if the problem solved is smooth (Example 2 is such a problem) then the use of a more accurate numerical method is very profitable. The accuracy achieved by the Runge-Kutta method by using only 320 time-steps is greater than the accuracy achieved when the Backward Euler Method is used with more than 83 million time-steps (although the improvement of the accuracy achieved by the Backward Euler Method is in some sense optimal, i.e. reducing the time-stepsize by a factor of two is leading to an improvement of the accuracy also by a factor of two, which should be expected when the order of the method is one).

Acknowledgements

Two grants (CPU-1002-27 and CPU-1101-17) from the Danish Centre for Scientific Computing (DCSC) gave us access to the IBM computers at the Århus University and the SUN computers at the Technical University of Denmark. The members of the staff of DCSC helped us to resolve some difficult problems related to the efficient exploitation of the grids on IBM and SUN computers.

The work on this project was partly supported by the NATO Scientific Programme (Grant No. CLG 980505).

References

1. Lewis, J. M. and Derber, J. C.: The use of adjoint equations to solve a variational adjustment problem with advective constraints. Tellus, **37A** (1985), 309–322.
2. Marchuk, G. I. (1995): *Adjoint equations and analysis of complex systems.* Kluwer Academic Publishers, Dordrecht.
3. Morel, P., Lefevre, G. and Rabreau, G.: On initialization and non-synoptic data assimilation. Tellus, **23** (1971), 197–206.
4. NAG Library Fortran Manual (2004). *E04 - minimizing and maximizing a function.* http://www.nag.co.uk. Numerical Algorithms Group (NAG), Banbury Road 7, Oxford, England.
5. Zlatev, Z. (1995). *Computer treatment of large air pollution models.* Kluwer Academic Publishers, Dordrecht-Boston-London.

Part VIII

Distributed Numerical Methods and Algorithms for Grid Computing

Ant Algorithm for Grid Scheduling Problem

Stefka Fidanova and Mariya Durchova

IPP – BAS, Acad. G. Bonchev, bl.25A, 1113 Sofia, Bulgaria
stefka@parallel.bas.bg, mabs@parallel.bas.bg

Abstract. Grid computing is a form of distributed computing that involves coordinating and sharing computing, application, data storage or network resources across dynamic and geographically dispersed organizations. The goal of grid task scheduling is to achieve high system throughput and to match the application needed with the available computing resources. This is matching of resources in a non-deterministically shared heterogeneous environment. The complexity of scheduling problem increases with the size of the grid and becomes highly difficult to solve effectively. To obtain good methods to solve this problem a new area of research is implemented. This area is based on developed heuristic techniques that provide an optimal or near optimal solution for large grids. In this paper we introduce a tasks scheduling algorithm for grid computing. The algorithm is based on Ant Colony Optimization (ACO) which is a Monte Carlo method. The paper shows how to search for the best tasks scheduling for grid computing.

1 Introduction

Computational Grids are a new trend in distributed computing systems. They allow the sharing of geographically distributed resources in an efficient way, extending the boundaries of what we perceive as distributed computing. Various sciences can benefit from the use of grids to solve CPU-intensive problems, creating potential benefits to the entire society. With further development of grid technology, it is very likely that corporations, universities and public institutions will exploit grids to enhance their computing infrastructure. In recent years there has been a large increase in grid technologies research, which has produced some reference grid implementations.

Task scheduling is an integrated part of parallel and distributed computing. Intensive research has been done in this area and many results have been widely accepted. With the emergence of the computational grid, new scheduling algorithms are in demand for addressing new concerns arising in the grid environment. In this environment the scheduling problem is to schedule a stream of applications from different users to a set of computing resources to maximize system utilization. This scheduling involves matching of applications needs with resource availability.

There are three main phases of scheduling on a grid [10]. Phase one is resource discovery, which generates a list of potential resources. Phase two involves gathering information about those resources and choosing the best set to match the

I. Lirkov, S. Margenov, and J. Waśniewski (Eds.): LSSC 2005, LNCS 3743, pp. 405–412, 2006.

application requirements. In the phase three the job is executed, which includes file staging and cleanup. In the second phase the choice of the best pairs of jobs and resources is NP-complete problem [4].

A related scheduling algorithm for the traditional scheduling problem is Dynamic Level Scheduling (DLS) algorithm [11]. DLS aims at selecting the best subtask-machine pair for the next scheduling. To select the best subtask-machine pair, it provides a model to calculate the dynamic level of the task-machine pair. The overall goal is to minimize the computational time of the application. In the grid environment the scheduling algorithm no longer focuses on the subtasks of an application within a computational host or a virtual organization (clusters, network of workstations, etc.). The goal is to schedule all the incoming applications to the available computational power. In [1, 7] some simple heuristics for dynamic matching and scheduling of a class of independent tasks onto a heterogeneous computing system have been presented.

There are two different goals for task scheduling: high performance computing and high throughput computing. The former aims is minimizing the execution time of each application and later aims is scheduling a set of independent tasks to increase the processing capacity of the systems over a long period of time.

Our approach is to develop a high throughput computing scheduling algorithm based on ACO. ACO algorithm can be interpreted as parallel replicated Monte Carlo (MC) systems [12]. MC systems [9] are general stochastic simulation systems, that is, techniques performing repeated sampling experiments on the model of the system under consideration by making use of a stochastic component in the state sampling and/or transition rules. Experimental results are used to update some statistical knowledge about the problem, as well as the estimate of the variables the researcher is interested in. In turn, this knowledge can be also iteratively used to reduce the variance in the estimation of the described variables, directing the simulation process toward the most interesting state space regions. Analogously, in ACO algorithms the ants sample the problem's solution space by repeatedly applying a stochastic decision policy until a feasible solution of the considered problem is built. The sampling is realized concurrently by a collection of differently instantiated replicas of the same ant type. Each ant "experiment" allows to adaptively modify the local statistical knowledge on the problem structure. The recursive retransmission of such knowledge determines a reduction in the variance of the whole search process the so far most interesting explored transitions probabilistically bias future search, preventing ants to waste resources in not promising regions of the search.

The organization of the paper is as follows. In section 2 the ACO method is discussed. In section 3 grid scheduling algorithm is introduced. We make some experimental testing and conclude this study in sections 4 and 5.

2 Ant Colony Optimization

Real ants foraging for food lay down quantities of pheromone (chemical cues) marking the path that they follow. An isolated ant moves essentially at random but an ant encountering a previously laid pheromone will detect it and decide to follow it with high probability and thereby reinforce it with a further quantity of pheromone. The repetition of the above mechanism represents the auto catalytic behavior of real ant colony where the more the ants follow a trail, the more attractive that trail becomes.

The ACO algorithms were inspired by the observation of real ant colonies [2, 3]. An interesting behavior is how ants can find the shortest paths between food sources and their nest. While walking from a food source to the nest and vice-versa, ants deposit on the ground a substance called pheromone. Ants can smell pheromone and then they tend to choose, in probability, paths marked by strong pheromone concentrations. The pheromone trail allows the ants to find their way back to the food source (or to the nest).

The above behavior of real ants has inspired ACO algorithm. ACO algorithm, which is a population-based approach, has been successfully applied to many NP-hard optimization problems [2, 3]. One of its main ideas is the indirect communication among the individuals of ant colony. This mechanism is based on an analogy with trails of pheromone which real ants use for communication. The pheromone trails are a kind of distributed numerical information which is modified by the ants to reflect their experience accumulated while solving a particular problem.

The ACO algorithm uses a colony of artificial ants that behave as co-operative agents in a mathematical space were they are allowed to search and reinforce pathways (solutions) in order to find the optimal ones. Solution that satisfies the constraints is feasible. After initialization of the pheromone trails, ants construct feasible solutions, starting from random nodes, then the pheromone trails are updated. At each step ants compute a set of feasible moves and select the best one (according to some probabilistic rules) to carry out the rest of the tour. The transition probability is based on the heuristic information and pheromone trail level of the move. The higher value of the pheromone and the heuristic information, the more profitable it is to select this move and resume the search. In the beginning, the initial pheromone level is set to a small positive constant value τ_0 and then ants update this value after completing the construction stage.

```
procedure ACO
begin
    Initialize the pheromone
    while stopping criterion not satisfied do
        Position each ant in a starting node
        repeat
            for each ant do
                Chose next node by applying the state transition rate
            end for
        until every ant has build a solution
        Update the pheromone
    end while
end
```

All ACO algorithms adopt specific algorithmic scheme as is shown above. After the initialization of the pheromone trails and control parameters, a main loop is repeated until the stopping criteria are met. The stopping criteria can be a certain number of iterations or a given CPU time limit or time limit without improving the result. In the main loop the ants construct feasible solutions and then the pheromone trails are updated. More precisely, partial problem solutions are seen as states: each ant starts from random state and moves from state i to another state j of the partial solution. At each step, ant k computes a set of feasible solutions to its current state and moves to one of these expansions, according to a probability distribution specified as follows. For ant k the probability p_{ij}^k to move from a state i to a state j depends on the combination of two values:

$$p_{ij}^k = \begin{cases} \frac{\tau_{ij}.\eta_{ij}}{\sum_{l \in allowed_k} \tau_{il}.\eta_{il}} & if\ j \in allowed_k \\ 0 & otherwise \end{cases} \tag{1}$$

where

- η_{ij} is the attractiveness of the move as computed by some heuristic information indicating a prior desirability of that move;
- τ_{ij} is the pheromone trail level of the move, indicating how profitable it has been in the past to make that particular move (it represents therefore a posterior indication of the desirability of that move);
- $allowed_k$ is the set of remaining feasible states.

Thus, the higher the value of the pheromone and the heuristic information, the more profitable it is to include state j in the partial solution. In the beginning, the initial pheromone level is set to τ_0, which is a small positive constant. In the nature there is not any pheromone on the ground at the beginning, or the initial pheromone in the nature is $\tau_0 = 0$. If in ACO algorithm the initial pheromone is zero, than the probability to chose next state will be $p_{ij}^k = 0$ and the search process will stop from the beginning. Thus it is important the initial pheromone to be positive value.

The pheromone level of the elements of the solutions is changed by applying following updating rule:

$$\tau_{ij} \leftarrow \rho.\tau_{ij} + \Delta\tau_{ij}, \tag{2}$$

where the rule $0 < \rho < 1$ models evaporation and $\Delta\tau ij$ is an additional pheromone and it is different for different ACO algorithms. Normally the quantity of the added pheromone depends on the quality of the solution.

3 Grid Scheduling Model

Our scheduling algorithm is designed for distributed systems shared asynchronously by both remote and local users.

3.1 Grid Model

The grid considered in this study is composed of a number of hosts send, each host is composed of several computational resources, which may be homogeneous or heterogeneous. The grid scheduler does not own the local hosts, therefore does not have control over them. The grid scheduler must make best effort decisions and then submit the jobs to the hosts selected, generally as a user. Furthermore, the grid scheduler does not have control over the set of jobs submitted to the grid, or local jobs submitted to the computing hosts directly. This lack of ownership and control is the source of many of the problems yet to be solved in this area. The grid scheduling is a particular case of tasks scheduling on machines problem. In the grid scheduling every machine can execute any task, but for different time.

3.2 Grid Scheduling Algorithm

While there are scheduling request from applications, the scheduler allocates the application to the host by selecting the best match from the pool of applications and pool of the available hosts. The selecting strategy can be based on the prediction of the computing power of the host [6]. We will review some terms and definitions [7, 8].

The expected execution time ET_{ij} of task t_i on machine m_j is defined as the amount of time taken by m_j to execute t_i given that m_j has no load when t_i is assigned. The expected completion time CT_{ij} of the task t_i on machine m_j is defined as the wall-clock time at which m_j completes t_i (after having finished any previously assigned tasks). Let M be the total number of the machines. Let S be the set containing the tasks. Let the beginning time of t_i be b_i. From the above definitions, $CT_{ij} = b_i + ET_{ij}$. The makespan for the complete schedule is then defined as $\max_{t_i \in S}(CT_{ij})$. Makespan is a measure of the throughput of the heterogeneous computing system. The objective of the grid scheduling algorithm is to minimize the makespan. It is well known that the problem of deciding on an optimal assignment of jobs to resources is NP-complete. We develop heuristic algorithm based on ACO to solve this problem.

Existing mapping heuristics can be divided into two categories: on-line mode and batch mode. In the on-line mode, a task is mapped onto a machine as soon as it arrives at the mapper. In the batch mode, tasks are not mapped onto the machines as they arrive, instead they are collected in a set that is examined for mapping at pre-scheduled times called mapping events. This independent set of tasks that is considered for mapping at mapping events is called meta-task. In the on-line mode, each task is considered only once for matching and scheduling. The minimum completion time heuristic assigns each task to the machine so that the task will have the earliest computation time [5]. The minimum execution time heuristic assigns each task to the machine that performs that tasks' computation in the least amount of execution time. In batch mode, the scheduler consider a meta-task for matching and scheduling at each mapping event. This enable the mapping heuristics to possibly make better decision, because the heuristics have the resource requirement information for the meta-task and known the actual execution time of a larger number of tasks. Our heuristic algorithm is for batch mode.

Let the number of the tasks in the set of tasks is greater than the number of machines in the grid. The result will be triples $(task, machine, startingtime)$. The function $free(j)$ - shows when the machine m_j will be free. If the task t_i is executed on the machine m_j then the starting time of t_i becomes $b_i = free(j)+1$ and the new value of the function $free(j)$ becomes $free(j) = b_i + ET_{ij} = CT_{ij}$.

An important part of implementation of ACO algorithm is the graph of the problem. We need to decide which elements of the problem to correspond to the nodes and which ones to the arcs. Let $M = \{m_1, m_2, \ldots, m_m\}$ is the set of the machines and $t = \{t_1, t_2, \ldots, t_s\}$ is the set of the tasks and $s > m$. Let $\{T_{ij}\}_{s \times m}$ is the set of the nodes of the graph and to machine $m_j \in M$ corresponds a set of nodes $\{T_{kj}\}_{k=1}^{s}$. The graph is fully connected. The problem is to choose s nodes of the graph thus to minimize the function $F = max(free(j))$, where $[b_i, CT_{ij}] \cap [b_k, CT_{kj}] = \oslash$ for all i, j, k. We will use several ants and every ant starts from random node to create their solution. There is a tabu list corresponding to every ant. When a node T_{ij} is chosen by the ant, the nodes $\{T_{ik}\}_{k=1}^{m}$ is included in tabu list. Thus we prevent the possibility the task t_i to be executed more than ones. An ant add new nodes in the solution till all nodes are in the tabu list. Like heuristic information we use:

$$\eta_{ij} = \frac{1}{free(j)}.$$

Thus if a machine is free earlier, the corresponding node will be more desirable. At the end of every iteration we calculate the objective function $F_k = max(free(j))$ over the solution constructed by ant k and the added pheromone by the ant k is:

$$\Delta\tau_{ij} = \frac{(1-\rho)}{F_k}.$$

Hence in the next iterations the elements of the solution with less value of the objective function will be more desirable. Our ACO implementation is different from ACO implementation on traditional tasks machines scheduling problem. The new of our implementation is using of multiple node corresponding to one machine. It is possible because in grid scheduling problem every machine can execute any task.

Two kind of sets of tasks are needed: set of scheduled tasks and set of arrived and unscheduled tasks. When the set of scheduled tasks becomes empty the scheduled algorithm is started over the tasks from the set of unscheduled tasks. Thus is guaranteed that the machines will be fully loaded.

4 Experimental Testing

We have developed 3 simulated grid examples to evaluate the newly proposed ACO algorithm for grid scheduling. In our experimental testing we use 5 heterogeneous machines and 20 tasks. The initial parameters are set as follows:

$\tau_0 = 0.01$ and $\rho = 0.5$ and we use 1 ant. We compare achieved by ACO algorithm result with often used online-mode.

The results are in minutes. We observe the outperform of ACO algorithm and the improvement of the result with. In online-mode the arriving order is very important. In ACO algorithm the most important is the execution time of the separate task.

Table 1. Makespan for the execution on first free machine and ACO algorithm

online-mode	ACO	improvement
80	67	16%
174	128	26.4%
95	80	15.8%

5 Conclusion

To confront new challenges in tasks scheduling in a grid environment, we present in this study heuristic scheduling algorithm. The proposed scheduling algorithm is designed to achieve high throughput computing in a grid environment. This is a NP-problem and to be solved needs an exponential time. Therefore the heuristic algorithm which finds a good solution in a reasonable time is developed. In this paper heuristic algorithm based on ACO method is discussed and it basic strategies for a grid scheduling are formulated. This algorithm guarantee good load balancing of the machines. In ACO technique it is very important how the graph of the problem is created. Another research direction is to create different heuristic based algorithms for problems arising in grid computing.

Acknowledgments

Stefka Fidanova was supported by the European Community program "Structuring the European Research Area" contract No MERG-CT-2004-510714. Mariya Durchova was supported by the Bulgarian IST Center of Competence in 21st century — BIS-21++ funded by European Commission in FP6 INCO via grant 016639/2005.

References

1. Braun T. D., Siegel H. J., Beck N., Bolony L., Maheswaram M., Reuther A. I., Robertson J.P., Theys M. D., Jao B.: *A taxonomy for describing matching and scheduling heuristics for mixed-machine heterogeneous computing systems*, IEEE Workshop on Advances in Parallel and Distributed Systems, (1998) 330–335.
2. Dorigo, M., Di Caro, G.: The Ant Colony Optimization metaheuristic. In: *New Idea in Optimization*, Corne, D., Dorigo, M., Glover, F. eds., McGrow-Hill (1999) 11–32.

3. Dorigo, M., Gambardella, L.M.: Ant Colony System: A cooperative Learning Approach to the Traveling Salesman Problem. *IEEE Transaction on Evolutionary Computation*, **1** (1999) 53–66.
4. Fernandez-Baca D.: Allocating Modules to Processors in a Distributed System, *IEEE Transactions on Software Engineering*, **15** 11 (1989) 1427–1436.
5. Freund R. F., Gherrity M., Ambrosius S., Camp-Bell M., Halderman M., Hensgen D., Keith E., Kidd T., Kussow M., Lima J.D., Mirabile F., Moore L., Rust B., Siegel H.J.: Scheduling Resources in Multi-User Heterogeneous Computing Environments with SmartNet, IEEE Heterogeneous Computing Workshop, (1998) 184–199.
6. Gong L., Sun X.H., Waston E.: Performance Modeling and Prediction of Non-Dedicated Network Computing, *IEEE Transaction on Computer*, **51** 9 (2002) 1041–1055.
7. Maheswaran M., Ali S., Siegel H.J., Hensgen D., Freund R.: *Dynamic Mapping of a Class of Independent Tasks onto Heterogeneous Computing Systems*, 8th IEEE Heterogeneous Computing Workshop (HCW'99), San Juan, Puerto Rico, (1999) 30–44.
8. Pinedo M.:*Scheduling: Theory, Algorithms and Systems*, Prentice Hall, Englewood Clifts, NJ, (1995).
9. Rubinstein, R. Y.:*Simulation and the Monte Carlo Method.* John Wiley &Sons, (1981).
10. Schopf J.M: *A General Architecture for Scheduling on the Grid*, special issue of JPDC on Grid Computing, (2002).
11. Sih G.C., Lee E.A.: A Compile-Time Scheduling Heuristic for Inter Connection-Constrained Heterogeneous Processor Architectures, *IEEE Transactions Parallel and Distributed Systems*, **4** (1993) 175–187.
12. Strelsov, S., Vakili, P.: Variance Reduction Algorithms for Parallel Replicated Simulation of Uniformized Markov Chains. *J. of Discrete Event Dynamic Systems: Theory and Applications*, **6** (1996) 159–180.

Automatic Tuning Technique Exploring Within the Hardware-Specific Constrained Parameters

Toshiyuki Imamura[1] and Ken Naono[2]

[1] Department of Computer Science, The University of Electro-Communications, 1-5-1 Chofugaoka, Chofu-shi, Tokyo 182-8585, Japan
imamura@im.uec.ac.jp

[2] Central Research Laboratory, Hitachi Ltd., 1-280 Higashi-Koigakubo, Kokubunji-shi, Tokyo 185-8601, Japan
naono@crl.hitachi.co.jp

Abstract. This paper covers an efficient strategy for exploring the sampling parameters on auto-tuning processes. Byte/flop is considered as a performance indicator, and finding the best parameter is interpreted as an optimisation problem with some hardware-specific constrained conditions. In this work, we also evaluate the performance of various unrolled loops both in a rank-update operation and a matrix-vector multiplication which appear in a significant operation of an eigensolver. The tuned routines running on a single processor of a Hitachi SR8000 and a Fujitsu VPP5000 record 1080 MFLOPS and 8342 MFLOPS respectively.

1 Introduction

In the recent development of a numerical library, it is strongly required to reduce the maintenance cost and keep higher and more stable performance. Because of the rapid change of the hardware specification, it becomes quite harder to develop and implement new optimisation techniques. For that difficulty, the auto-tuning mechanism is proposed, and facilitated as the integration of empirical optimisation techniques on a high performance numerical library. The auto-tuning mechanism selects an optimal code fragment at the installation phase, and then generates more effective executables. The most successful projects, which adopt this technique, are ATLAS[7] at UTK, PHiPAC[2] at UCB, I-Lib[4] at Tokyo University, and so forth.

In terms of the network-based computation, the concept of server portals has matured, and the Ninf project[5] at AIST and the Netsolve project[1] at UTK are going on. They enable us to handle a thousand of computational resources geographically spread over the world without users' awareness. Key issues of constructing a numerical portal are 'processing speed', 'accuracy', and 'installation cost' as well as the underlying middleware supports, and they sometimes conflict with the QoS and each other. Therefore, performance stability and the installation cost be deeply discussed. However, the existing performance evaluation reports emphasise mainly a pinpoint result, and they have not considered the stability on the whole range of problems.

I. Lirkov, S. Margenov, and J. Waśniewski (Eds.): LSSC 2005, LNCS 3743, pp. 413–421, 2006.

In this paper, we focus on performance stability and installation costs, and concentrate to exploit full and particular evaluation of various unrolled loops at the auto-tuning process. Specifically, 'Byte/Flops', that is related to the dense matrix computation [6], is used for one of the performance metrics. We propose a novel algorithm to explore the degree of loop-unrolling. Applying the exploring algorithm, the parameters to be swept can be dramatically reduced, and we expect selected loop fragments provide us with the highest performance and performance stability.

2 Householder Transform and Auto-tuning Method

The numerical algorithm objected in this work is block Householder transform, and it is known as a significant operation in an eigenvalue calculation. The algorithm is presented as follows. Here, N and M mean dimension of the matrix and the block width respectively, and subscripts indicate the range of submatrices.

for $j = N, \ldots, 1$ **step** $-M$
  $U \leftarrow \emptyset, V \leftarrow \emptyset, W \leftarrow A_{(*,j-M+1:j)}$
  for $k = 0, \ldots, M-1$
    (1) Householder reflector: $u^{(k)} = H(W_{(*,j-k)})$
    (2) Matrix-Vector multiplication
      $v^{(k-\frac{2}{3})} \leftarrow A_{(1:j-k-1,1:j-k-1)} u^{(k)}$
    (3) $v^{(k-\frac{1}{3})} \leftarrow v^{(k-\frac{2}{3})} - (UV^T + VU^T)u^{(k)}$
    (4) $v^{(k)} \leftarrow v^{(k-\frac{1}{3})} - \left(\left(u^{(k)}, v^{(k-\frac{1}{3})}\right) / 2|u^{(k)}|^2\right) u^{(k)}$
      $U \leftarrow [U, u^{(k)}], V \leftarrow [V, v^{(k)}]$.
    (5) $W_{(*,j-k:j)} \leftarrow W_{(*,j-k:j)} - \left(u^{(k)}v^{(k)^T} + v^{(k)}u^{(k)^T}\right)_{(*,j-k:j)}$
  endfor
  $A_{(*,j-M+1:j)} \leftarrow W$
  (6) $2M$ rank-update
      $A_{(1:j-M,1:j-M)} \leftarrow A_{(j-M:j-M)} - (UV^T + VU^T)_{(j-M,j-M)}$
endfor

```
do M=1,MB
  do J=1,N, 2
    do I=1,J+1
      A(I,J+0)=A(I,J+0)-V(I,M)*U(J+0,M)
                       -U(I,M)*V(J+0,M)
      A(I,J+1)=A(I,J+1)-V(I,M)*U(J+1,M)
                       -U(I,M)*V(J+1,M)
    enddo
  enddo
enddo
```

```
do J=1,N,2
  do I=1,J+1
    V(J+0)=V(J+0)+A(I,J+0)*U(I)
    V(J+1)=V(J+1)+A(I,J+1)*U(I)
    V(I)=V(I)+A(I,J+0)*U(J+0) &
             +A(I,J+1)*U(J+1)
  enddo
enddo
do J=1,N
  V(J)=V(J)+D(J)*U(J)
enddo
```

Fig. 1. An example of two degree of loop unrolling, the rank-update procedure (left), and the matrix-vector multiplication procedure (right)

The core parts of Householder transform are the rank-update process and the matrix-vector multiplication. Former has three-fold loop structure, and latter has two-fold one as illustrated in Figure 1. In the figure, example codes of loop unrolling (the degree of unrolling is two) which appear in an auto-tuning process are presented. The degrees of unrolling to these loops are denoted by (X_r, Y_r, Z_r) and (X_m, Y_m), respectively, in order from the outer loop to the inner one.

3 Exploring Algorithm

General requirements to the numerical library with an auto-tuning facility are

1. Performance, high speed calculation,
2. Stability, which does not depend on problems, and
3. Smaller installation cost.

The first issue is naturally known as the main objective of auto-tuning. We apply loop unrolling to a source code, and investigate a better auto-tuning strategy on general processors. The second issue is removed by a cache stabilising algorithm[3], which authors proposed as a fundamental technique for an auto-tuning facility, and we do not pay special attentions to instability caused by cache conflict. The third issue is intended to reduce the total cost at the installation, while a hundred patterns of parameter sets are sampled. It means a lot loss of the brute force searching must be improved by using more intelligent and reasonable algorithms.

We take advantage of the performance indicator 'Byte/Flops' which is defined by required data amounts per a single floating instruction in a loop structure[6]. Since Byte/Flops can be formulated by a function of the tuning parameters,

$$\begin{aligned} \text{Object function to be minimised}: &\quad F(X_1, X_2, \ldots, X_n) \\ \text{Constrained condition (I)}: &\quad C(X_1, X_2, \ldots, X_n) \le L \\ &\quad X_1 > 0, X_2 > 0, \ldots, X_n > 0. \end{aligned}$$

1. Find the point x^* which minimise the object function F under the constrained condition (I) by Lagrange's method with indeterminate coefficients.
2. Compute the gradient of the object function (∇F) at x^*, and determine an anchor point x^+ by moving along each axis direction by one from x^* toward increasing direction of the object function.
3. Calculate the value $f^+ = F(x^+)$, and add new constrained condition (II), $F(X_1, X_2, ..., X_n) \ge f^+$.
4. Explore the grid points which satisfy the conditions (I) and (II), and evaluate unrolled-loop codes specified by parameters $(X_1, X_2, \ldots, X_n)$ successively.
5. If no grid point satisfies the both conditions, replace the minimising point by the anchor as $x^* := x^+$, and return step 2.

Fig. 2. Exploring algorithm of the degree of loop unrolling

assuming that 'the number of registers required for loop-unrolling cannot exceed the number of physical registers L', immediately, leads searching the best parameter set to the optimisation problem to find the parameter set which minimises Byte/Flops. Formal optimisation, however, involves performance fluctuation. Thus we examine the performance at each parameter, and should expand the range of parameter set. The exploring algorithm is summarised as Figure 2.

3.1 Tuning on a Hitachi SR8000

Rank Update. In case of an SR8000, that has scalar processor element, software pipelining optimisation prevents reuse of registers, though an SR8000 would have 128 floating point registers. We have the assignment strategy that; all the variables to registers, however, they do not waste up the physical registers. According to this strategy, the optimisation problem is formulated as follows.

$$\text{minimise} \quad 4(X_r + Y_r)/X_rY_r \tag{1}$$
$$\text{where} \quad 2(X_rY_r + Z_rX_r) + Y_rZ_r \leq L_s \tag{2}$$
$$X_r > 0, Y_r > 0, Z_r > 0 \tag{3}$$

If we consider the fact that two-word load/store instruction is available on an SR8000, and the degree of loop unrolling to the most-inner loop must be chosen as $Z_r = \{2, 4\}$, then the constrained conditions are

$$X_rY_r + 2X_r + Y_r \leq L_s/2 \quad (Z_r = 2) \tag{4}$$
$$X_rY_r + 4X_r + 2Y_r \leq L_s/2 \quad (Z_r = 4). \tag{5}$$

Though an SR8000 has 128 floating point register, several registers are reserved by a compiler as intermediates, thus we assume $L_s = 110$ in this paper.

Figure 3 illustrates two conditions (4) and (5), and inequalities are held below the lines. Dotted lines are isometric plot of the object function F, and three lines

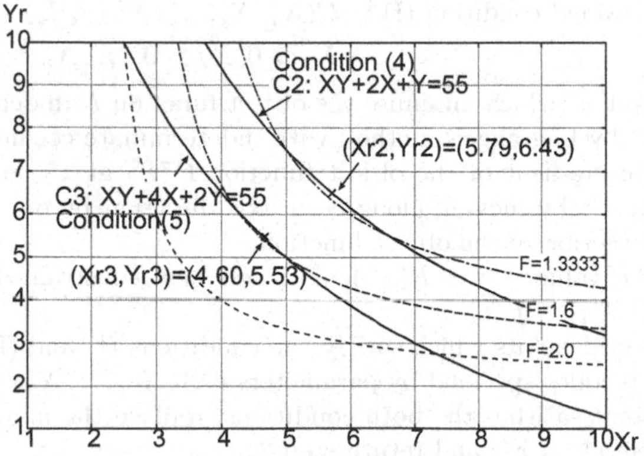

Fig. 3. Isometric plot of the object function and constrained conditions for the Rank-update procedure on an SR8000 (2 lines the constrained conditions (4) and (5))

satisfy $F(X_r, Y_r) = \{1.33, 1.60, 2.00\}$. The minimising point of the constraint (4) is $x_2^* = (5.79, 6.43)$, and the minimum value of F is 1.316. The other constraint (5) leads $x_3^* = (4.60, 5.53)$, and the minimum value is 1.590. According to the step two, anchor points are $x_{2 \text{ or } 3}^+ = x_{2 \text{ or } 3}^* - 1$, and then new constrained conditions are added. Finally we search all the grid points enclosed by $F(X_r, Y_r)$, the original constraints, and the additional constraints $F(X_r, Y_r) = F(x_{2 \text{ or } 3}^+)$. The number of all the grid points contained in the region is totally 16.

Matrix-Vector Product. Optimisation problem for the matrix-vector multiplication is formulated as follows.

$$\text{minimise} \quad 2(X_m + 3)/X_m \tag{6}$$

$$\text{where} \quad X_m Y_m + 3X_m + 2Y_m \leq 110 \tag{7}$$

$$X_m > 0, Y_m > 0 \tag{8}$$

If we fix the parameter Y_m 2 or 4, constrained conditions become $X_m \leq 21.2$ ($Y_m = 2$), $X_m \leq 14.5$ ($Y_m = 4$). Obviously searching range of X_m is $\{1, 2, ..., 21\}$.

3.2 Tuning on a Fujitsu VPP5000

Rank Update. In case of a vector computer, the most inner loop is target of vectorisation, and we do not take the most inner loop as an object to unrolling. Thus we fix formally $X_r = 1$ on a vector computer. When vector pipelines work fully, vector registers for load–store should be separate. Considering this, constrained conditions become as follows.

$$2X_r + 2Y_r \leq L_v \tag{9}$$

$$X_r Y_r \leq L_f \tag{10}$$

L_v and L_f mean the number of available vector registers and floating point registers respectively. In case of a VPP5000, the number of vector registers is varied

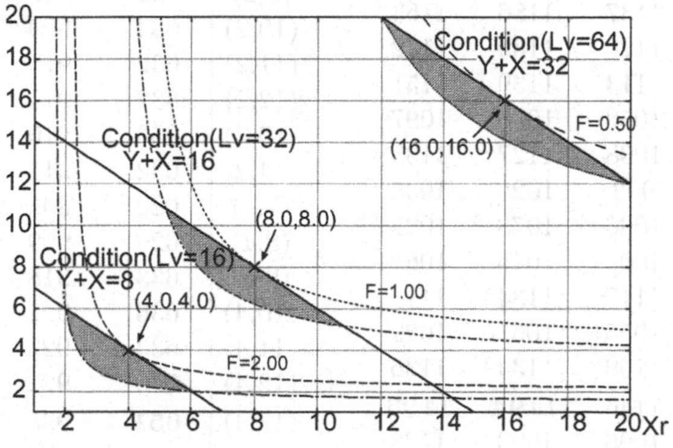

Fig. 4. Isometric plot of the object function and constrained conditions for the Rank-update procedure on a VPP5000

according to vector length, thus $L_v = 2^d (4 \le d \le 8)$ is possible configuration (16, 32, or 64 are practical). The region painted in gray on Figure 4 shows the candidates of exploring, and the number of all the grids is counted up to 32.

Matrix-Vector Product. As the optimisation problem of an SR8000 is formulated, constraints of a VPP5000 are also shown as,

$$X_m + 3 \le L_v = \{16, 32, 64\} \quad (11)$$

$$X_m \le L_f = 64 \quad (12)$$

$$X_m > 0. \quad (13)$$

4 Numerical Experiment and Evaluation

In this section, we examine our exploring algorithm on a single processor of two types of supercomputer systems. By using the exploring algorithm, the number of candidate parameters is reduced, however, it is hard to sweep the entire dimension on all the parameter sets because of the CPU time limitation. Thus we also reduce the sampling points as every 128 dimension, 1 to 4000 on an SR8000, and 1 to 6000 on a VPP5000. The test matrix used in a tridiagonalisation is the Frank matrix. Sampling is carried on tree times on each dimension and the best performance is adopted. Table 1 to 3 show the best result of our codes.

On an SR8000, the best parameters of the rank-update are (4,8,2), (5,5,4), (4,6,4) on less than 1000, 1000 to 2000, and other dimensions, respectively. The

Table 1. Results of the rank-update routine (left), and the matrix-vector multiplication routine (right) on a Hitachi SR8000, (the value shown in a column is the mean value of corresponding dimension range, and unit is in MFLOPS)

Unroll	~ 1000	~ 2000	~ 3000	~ 4000
(4,8,2)	**1006**	1145	1109	1158
(4,9,2)	969	1137	1156	1163
(5,6,2)	984	1120	1139	1142
(5,7,2)	961	1113	1139	1151
(6,5,2)	869	1052	1088	1097
(6,6,2)	917	1098	1127	1135
(7,4,2)	869	992	1026	1035
(7,5,2)	888	1036	1073	1095
(8,4,2)	857	1003	1046	1061
(3,7,4)	1001	1117	1134	1114
(3,8,4)	962	1055	1061	1029
(4,5,4)	942	1109	1134	1136
(4,6,4)	969	1146	**1194**	**1193**
(5,4,4)	959	1096	1093	1128
(5,5,4)	973	**1147**	1152	1188
(7,3,4)	875	1029	1062	1120

Unroll	~ 1000	~ 2000	~ 3000	~ 4000
(9,2)	632	910	959	989
(10,2)	634	909	962	992
(11,2)	632	910	966	994
(12,2)	632	911	964	991
(13,2)	633	910	967	994
(14,2)	632	911	961	988
(7,4)	622	896	915	944
(8,4)	629	909	948	961
(9,4)	633	918	962	979
(10,4)	636	922	965	983
(11,4)	633	923	973	994
(12,4)	645	922	968	990
(13,4)	**651**	935	987	1008
(14,4)	648	**937**	**991**	**1015**
(15,4)	643	931	988	1014

Table 2. Results of the rank-update routine on a Fujitsu VPP5000, (the value shown in a column is the mean value of corresponding dimension range, and unit is in MFLOPS)

Unroll	~1000	~2000	~3000	~4000	~5000	~6000
(2,6,1)	5304	7452	7437	7552	7555	7386
(3,3,1)	4644	7076	7317	7489	7531	7454
(3,4,1)	5099	7383	7611	7805	7871	7816
(3,5,1)	5144	7501	7819	7991	8056	8044
(4,3,1)	5023	7507	7751	7905	7939	7875
(4,4,1)	5405	7700	8018	8165	8204	8115
(5,3,1)	5111	7640	8006	8217	8290	8306
(6,2,1)	4603	7445	7845	8044	8103	8114
(5,11,1)	5287	7697	8161	8375	8484	8453
(6,10,1)	5358	7783	8271	8500	8641	8647
(6,9,1)	5309	7721	8226	8458	8590	8585
(7,7,1)	**5425**	7865	8348	8624	8748	8768
(7,8,1)	5328	7869	8331	8561	8684	8669
(7,9,1)	5334	7886	8357	8591	8729	8714
(8,7,1)	5229	7850	8372	8649	8783	8812
(8,8,1)	5353	**7915**	8414	8629	8751	8750
(9,6,1)	5263	7807	8359	8651	8810	8848
(9,7,1)	5261	7856	8392	8703	**8861**	8902
(10,6,1)	5189	7823	8400	8694	8827	8875
(11,5,1)	4984	7707	8307	8640	8753	8831
(13,18,1)	4854	7658	8306	8652	8741	8891
(13,19,1)	4834	7647	8343	8643	8741	8896
(14,17,1)	4793	7636	8329	8629	8715	8874
(14,18,1)	4841	7674	8360	8672	8767	8920
(15,15,1)	4892	7630	8315	8603	8698	8861
(15,16,1)	4775	7652	8354	8633	8763	8915
(16,15,1)	4701	7580	8269	8579	8666	8832
(16,16,1)	4974	7774	**8449**	**8728**	8781	**8965**
(17,14,1)	4617	7559	8271	8567	8640	8824
(18,13,1)	4554	7499	8189	8502	8580	8777
(18,14,1)	4572	7526	8212	8531	8609	8792
(19,13,1)	4453	7430	8160	8494	8552	8766

performance on 3000 to 4000 dimension records 80% to theoretical peak. As the authors have tuned this code up so far, and the best parameter choice was (5,5,4) [6]. The exploring algorithm reveals that much better parameter exists on unexamined range. The best parameter sets of the matrix-vector multiplication are found on the interval $X_m = \{7, 8, \ldots, 15\}$. (13,4) and (14,4) should be chosen for less than 1000 dimension and for other dimension, respectively.

On a VPP5000, larger parameter space is explored than that of a SR8000. The best parameter sets are (7,7,1), (8,8,1), (16,16,1), (16,16,1), (9,7,1), (16,16,1) in ascending order, for every 1000 dimensional interval respectively. Differences between performance of the best parameter sets and of others are not so large, but the performance depends on the dimension of the input matrices. Thus to

Table 3. Results of the matrix-vector multiplication routine on a Fujitsu VPP5000, (the value shown in a column is the mean value of corresponding dimension range, and unit is in MFLOPS)

Unroll	~1000	~2000	~3000	~4000	~5000	~6000
(10,1)	2449	5283	6284	6806	7067	7272
(11,1)	2792	5846	6794	7272	7472	7641
(12,1)	2857	5982	6939	7395	7598	7749
(13,1)	2922	6088	7037	7478	7673	7829
(20,1)	3032	6289	7264	7726	7932	8088
(21,1)	3005	6285	7274	7749	7961	8104
(22,1)	**3091**	**6392**	**7342**	**7798**	**8013**	**8150**
(23,1)	2941	5806	6650	7118	7338	7488
(24,1)	3017	5908	6715	7192	7392	7546
(25,1)	2770	5384	6161	6619	6834	6985
(26,1)	2938	5705	6526	7015	7221	7390
(27,1)	2828	5528	6339	6832	7054	7216
(28,1)	2976	5819	6675	7166	7384	7546
(29,1)	2836	5586	6448	6949	7173	7342
(30,1)	2958	5849	6718	7210	7438	7603

Table 4. Performance results of our Householder transform routines on an SR8000F1 and a VPP5000, 1000 to 4000 on an SR8000 (left), 1000 to 6000 on a VPP5000 (right)

	1000	2000	3000	4000
MFLOPS	880	1018	1074	1080
time[sec]	1.51	10.4	33.5	78.9

	1000	2000	3000	4000	5000	6000
MFLOPS	5273	7068	7739	8115	8279	8342
time[sec]	0.25	1.51	4.65	10.5	20.1	34.5

select a code fragment, which corresponds to the optimal parameter set, is an effective strategy to develop a stable library. Additionally, it found that $X_m = 22$ is the best parameter of all the dimensions on the matrix-vector multiplication.

Finally, we compare our tuned code with vendor-tuned numerical libraries. Table 4 shows the result of our tuned routines. Peak performance of our codes on two machines record 1080 MFLOPS and 8342 MFLOPS, respectively. On an SR8000, vendor-tuned LAPACK achieves 746 MFLOPS, and on a VPP5000, the function DTRD1 in SSL/II VP does 3720 MFLOPS. This concludes that our tuning strategy is quite effective, and the tuned codes perform stable.

5 Summary

In this paper, we mainly focused on performance stability, and detailed performance evaluation for the loop unrolling technique was carried out. We introduced a novel exploring algorithm for the degree of loop unrolling, in which 'Byte/Flops' plays a role of performance indicator, a cost function of optimisation problems. The exploring algorithm works excellently, and it reduces the number of parameter sets rather than existing strategies like the brute force searching. Our tuned codes also perform stable, and they record 1080 MFLOPS and 8342 MFLOPS on a Hitachi SR8000 and a Fujitsu VPP5000, respectively.

Finally, the authors would like to thank Prof. Yuba, and Dr. Katagiri of the Univ. of Electro-Communications, Prof. Suda of the Univ. of Tokyo, and Dr. Yamamoto of Nagoya University for their fruitful discussions on auto-tuning mechanisms and numerical algebra, and to Japan Atomic Energy Research Institute for their sincere support of the supercomputer systems.

References

1. Arnold D., Agrawal S., Blackford S., Dongarra J.J., Miller M., Seymour K., Sagi K., Shi Z., and Vadhiyar S., Users' Guide to NetSolve V1.4.1, Technical Note of University of Tennessee, ICL-UT-02-05 2002.
2. Bilmes J., Ananoic K., Chin C.-W., and Demmel J., Optimizing Matrix Multiply using PHiPAC: a Portable, High-Performance, ANSI C Coding Methodology, Proceedings of International Conference on Supercomputing 97, 340–347, 1997.
3. Imamura T., C-Stab: Cache stabilizing algorithm for a numerical library, Proceedings of IASTED International conference on Parallel and Distributed Computing and Networks, PDCN2005, 241–245, 2005.
4. Katagiri T., Kuroda H., Ohsawa K., Kudoh M., and Kanada Y., Impact of Autotuning facilities for parallel numerical library, IPSJ Trans. on High Performance Computing Systems, **42**, SIG 12 (HPS 4), 60–76, 2001. (*Japanese*)
5. Nakada H., Sato M., and Sekiguchi S., Design and Implementations of Ninf: towards a Global Computing Infrastructure, Future Generation Computing Systems, Metacomputing Issue, **15**, 5-6, 649–658, 1999.
6. Naono K., and Imamura T., An Evaluation towards an Automatic Tuning Eigensolver, IPSJ SIG Notes, High Performance Computing (HPC), **2002**, 91, 49–54, 2002. (*Japanese*)
7. Whaley R.C., Petitet A., and Dongarra J.J., Automated Empirical Optimization of Software and the ATLAS Project, LAPACK Working Note 147, 2000.

An Evaluation Towards Automatically Tuned Eigensolvers

Ken Naono[1] and Toshiyuki Imamura[2]

[1] Central Research Laboratory, Hitachi Ltd.,
1-280 Higashi-Koigakubo, Kokubunji-shi, Tokyo 185-8601, Japan
naono@crl.hitachi.co.jp

[2] Department of Computer Science, The University of Electro-Communications,
1-5-1 Chofugaoka, Chofu-shi, Tokyo 182-8585, Japan
imamura@im.uec.ac.jp

Abstract. We investigate an automatic tuning method for an eigensolver of a dense symmetric matrix. The aim of this paper is to investigate how to select the unrolling depth. To do this, we evaluate the performance of various unrolled reduction loops of the eigensolver for every matrix size from 3000 to 4000 on the Hitachi SR8000/F1 and on the IBM RS/6000 SP3. We also analyze the trend between Byte/Flop and performance for various patterns of loop unrolling. The result shows that the performance is degraded with higher depth of unrolling in some matrix sizes, where it does not occur with lower depth of unrolling. The result also shows that selection of the unrolling depth should be examined in the case of several matrix sizes.

1 Introduction

The aim of this paper is to investigate the automatic tuning functions for the implementation of eigensolvers on different types of supercomputers. This paper focuses on the loop unrolling tuning for the main loop of eigenvalue computations and presents a numerical experiment based on research [9], which found performance degradation by unrolling of a loop in our eigensolver.

There are two primary background issues that led to doing this study. First, in the past decade, network libraries such as Ninf [11] and NetSolve [10] have been developed. These are used from any PC without being conscious of the location of a supercomputer in use. They enable us to use several types of matrix libraries on such supercomputers without installing them on our PCs. Second, libraries with automatic tuning functions, such as PHiPAC [1], FFTW [4], ATLAS [12], I-LIB [7] [6], and frameworks for automatic tuning, such as SANS [2], FIBER [5], and SIMPLE [8], have recently been developed. These are used to achieve higher performance on the computers where the libraries are installed. For example, ATLAS tunes LAPACK automatically, while I-LIB tunes a parallel library automatically.

In light of such trends, we assume that automatic tuning functions will be regarded as more important for network libraries, because more access to the

I. Lirkov, S. Margenov, and J. Waśniewski (Eds.): LSSC 2005, LNCS 3743, pp. 422–429, 2006.

various kinds of network libraries will lead to a greater need for high quality software installed on each platform.

However, the previous studies on automatic tuning remain inadequate for two reasons. First, it is possible to select an unsuitable loop unrolling. The methods used to determine the loop unrolling in these studies measured the performance on limited pinpoint parameters. For example, they measured the performances with matrix size N = 1000, 2000, and 4000. An inherent instability sometimes appears because of cache conflicts, but this small number of measurements does not reveal the instability; therefore, an unsuitable loop unrolling is sometimes selected. Second, the previous studies had to treat a wide range of loop unrolling to determine loop unrolling parameters when the automatic tuning libraries are installed on the target platform.

Therefore, this paper evaluates the performance of loops for every matrix size, with multiple patterns of loop unrolling, to reveal the inherent instability of loop unrolling. Moreover, from these detailed measurements, evaluations are made on the relationship between a loop performance indicator and the measured performance in order to develop an effective method of selecting the optimal pattern of loop unrolling.

2 Target of the Evaluation

This paper targets the reduction loop in the tridiagonalization of eigenvalue computations in the evaluation. The number of floating operations in reduction loop counts about $2/3N^3$, where N is the size of the matrix. Therefore, the reduction loop is critical in performance. The reduction loop targeted in this paper is based on the block tridiagonalization method [3] and described as follows;

```
do M = 1, MB
   do J = 1, N
      do I = 1, J
         A(I, J) = A(I, J) |( V(I, M)*U(J, M) + U(I, M)*V(J,M) )
      enddo
   enddo
enddo
```

where A is the matrix with the size N by N, V and U are the bundle of vectors with the size NB by N, and NB is fixed at 20 or 40, for example.

This paper evaluates loop unrolling method for the reduction loop. The method is often used for the reduction loop to achieve higher performance.

The evaluation is executed on the range of matrix size N = 3000 to 4000 because of the following two reasons. First, performance evaluations for matrix computations find performance fluctuation problems occurred by cache conflicts. Therefore, the evaluation executes on the consecutive points of the matrix size to clarify the fluctuation. Second, performance evaluations also find lower performance problems occurred by startup overhead in the case of small size matrix computation. On the platforms used in this paper, performance of the reduction loop with matrix size N from 3000 to 4000 does not dramatically change.

```
      subroutine tunecore222(
     &     i$1, i$5, i$3, m0_mod6, m0, i$nnod, i$inod, l1$a, u, v,  nm
     &     )
      implicit double precision (a-h,o-z),integer(i-n)
      double precision l1$a(nm,*), u(nm,*), v(nm,*)
*voption vec
*soption noloopreroll
        do i$1=i$5,i$3,2
*soption noloopreroll
           do i0=m0_mod6+1,m0,2
              j$0=(i$1-1+0)*i$nnod+i$inod; j$1=j$0+i$nnod
              u0_0 = u(j$0,i0+0);  u1_0 = u(j$1,i0+0)
              u0_1 = u(j$0,i0+1);  u1_1 = u(j$1,i0+1)
              v0_0 = v(j$0,i0+0);  v1_0 = v(j$1,i0+0)
              v0_1 = v(j$0,i0+1);  v1_1 = v(j$1,i0+1)
*soption unroll(2)
*soption noloopreroll
*voption vec
              do k=1,j$1
                 l1$a(k,i$1+0)=l1$a(k,i$1+0)
     1             - u0_0*v(k,i0+0) - v0_0*u(k,i0+0)
     1             - u0_1*v(k,i0+1) - v0_1*u(k,i0+1)
                 l1$a(k,i$1+1)=l1$a(k,i$1+1)
     1             - u1_0*v(k,i0+0) - v1_0*u(k,i0+0)
     1             - u1_1*v(k,i0+1) - v1_1*u(k,i0+1)
              enddo
           enddo
        enddo
      return
      end
```

Fig. 1. Programming source code of loop unrolling (2-2-2)

The evaluation also limits the patterns of loop unrolling. This paper denotes the pattern of loop unrolling as (X-Y-Z), where X is the depth of the most outer loop, Y is the depth of the next outer loop, and Z is the depth of the innermost loop. For example, the pattern of unrolling (4-3-2) means the unrolling of M with the depth of four, J with the depth of three, and I with the depth of two.

The patterns of the evaluation on the SR8000/F1 are the following thirteen; (1-1-1), (1-1-2), (2-2-1), (2-2-2), (3-3-1), (3-3-2), (4-4-1), (4-4-2), (5-5-1), (5-5-2), (5-5-4), (6-6-1), (6-6-2),

while the patterns of the evaluation on the RS6000/SP3 are the following eight; (1-1-1), (2-2-1), (2-2-2), (2-3-2), (3-2-2), (3-3-1), (3-3-2), (4-4-1).

The example of the programming source code with the pattern of (2-2-2) loop unrolling on the SR8000/F1 is described in Figure 1. As shown in the figure, in the two outer loops, the hand unrolling is used. On the other hand, in the innermost loop, the directives '*soption unroll(2)' is used. The directive means the compiler should unroll the loop in the depth of two. The calculation is executed only in the upper triangular part of the matrix in the original source code, while the unrolled code executes the lower triangular part in the block diagonal part. Although the amount of the calculations is limited, we can neglect the degradation of performance compared with the total amount of the reduction computation.

3 Detailed Performance Measurements of the Loop Unrolling

This section details the performance measurement results of the patterns of loop unrollings defined in Section 2, in order to reveal the instability of loop unrolling at higher depths.

The performance results for every matrix size N from 3000 to 4000, with the patterns of unrolling at the depth of (*-*-1), where * = 1,2,3,4,5, and 6, on the SR8000/F1 are plotted in Figure 2. It was confirmed that the performance generally improves as the unrolling depth increases from (1-1-1) to (6-6-1). With some matrix sizes, however, the performance does not increase even when the unrolling depth increases. Furthermore, it was found that the performance degrades on some other points when the unrolling depth increases. For example, the performance of (4-4-1) was superior to that of (6-6-1) for the matrix size of 3280, while the situation was opposite for most matrix sizes from 3000 to 4000. This suggests that there is a pitfall in evaluating the effectiveness of unrolling on just a few points.

From Figure 2, we can see that there were also steep degradations in performance. It was confirmed that the unrolling performance of (1-1-1) did not degrade very much from the average performance in the measured range, but the performances of the others degraded remarkably. The performance of (6-6-1), for example, fell to about 1020 Mflop/s for the matrix size from 3057 to 3080,

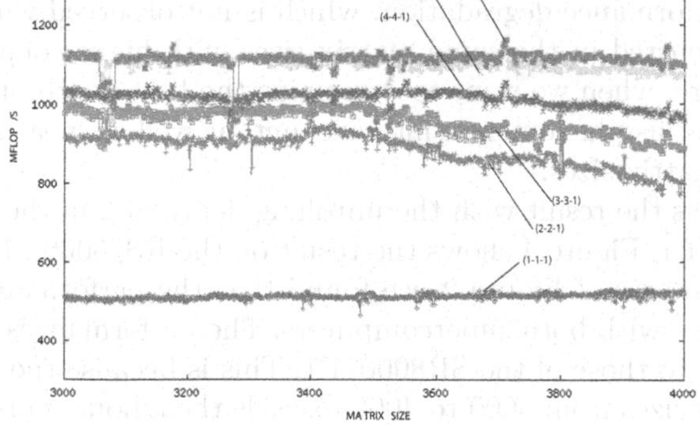

Fig. 2. Evaluation results of (*-*-1) unrolling (SR8000/F1, 1CPU)

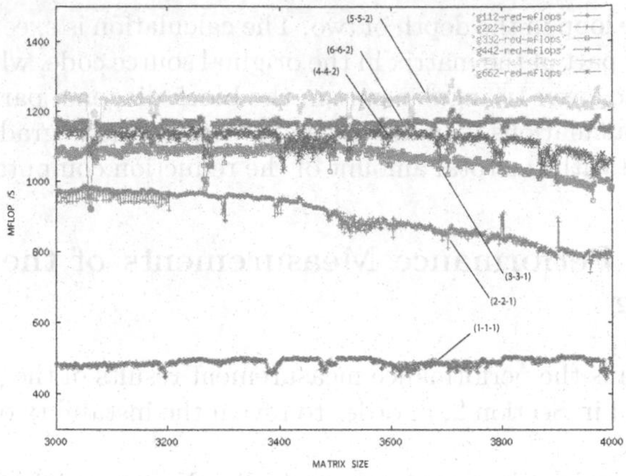

Fig. 3. Evaluation results of (*-*-2) unrolling (SR8000/F1, 1CPU)

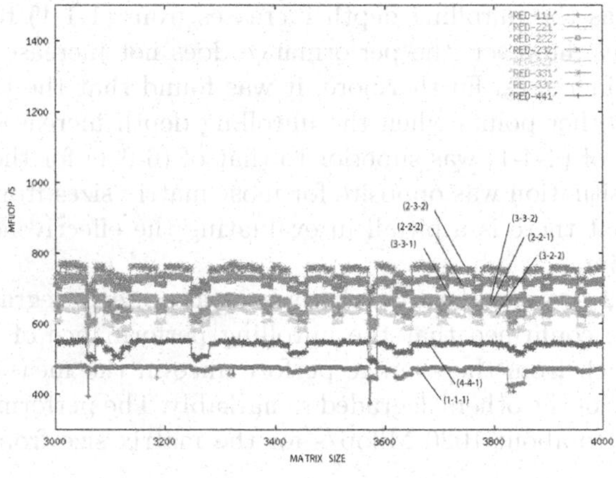

Fig. 4. Evaluation results of unrolling (RS/6000 SP3, 1CPU)

while the performance for the other matrix sizes was about 1110 Mflop/s. This implies that performance degradation, which is not observed with lower depths of unrolling, occurred in the same matrix sizes with higher depths of loop unrolling. Therefore, when we want to determine the best depth of loop unrolling, we have to take this phenomenon into account, as well as measure performance with multiple matrix sizes.

Figure 3 shows the result with the unrolling depth of 2 in the innermost loop on the SR8000/F1. Figure 4 shows the result on the RS/6000 SP3 (Power3, 375 MHz). As in the case of Figure 2, we found that the performances degrade for some matrix sizes with both supercomputers. The performances of the RS/6000 SP3 are inferior to those of the SR8000/F1. This is because the amount of data with the matrix sizes from 3000 to 4000 exceeds the amounts of L2 cache of the RS/6000 SP3. The performances around the matrix sizes of 1000 on the RS/6000 SP3 achieve about 830 Mflop/s.

4 Evaluating Performance Trends in Loop Unrolling

To quantitatively clarify the effectiveness of loop unrolling, the following three steps are used. First, the indicator Byte/Flop is defined. For each loop, the Byte/Flop means the required load and store data per one floating point operation. Second, the values of Byte/Flop for unrolled loops are calculated. When a loop's unrolling is (X-Y-Z), then the required load and store data is $8Byte * (2X+2Y)*Z$, and the number of floating point operations is $4*XYZ$. Therefore, the Byte/Flop is $4(X+Y)/XY$. Using this calculation, the values of Byte/Flop for each unrolled loop in section 3 are presented in Table 1. Third, using the detailed measurements of section 3, the average performances from N = 3000 to 4000 are calculated.

The relationship between the Byte/Flop and the average performance for each unrolled loop is presented in Figure 5. "SR-unroll1.d" consists of the points of unrolling for (*-*-1), where *=1, 2, 3, 4, 5, and 6. "SR-unroll2.d" consists of the points of unrolling for (*-*-2), where *=1, 2, 3, 4, 5, and 6. "SR-unroll4.d" consists of one point of unrolling for (5-5-4). "SP3-unroll1.d" consists of the points of unrolling for (*-*-1), where *=1, 2, 3, and 4. "SP3-unroll2.d" consists of the points of unrolling for (2-2-2), (2-3-2), (3-2-2), and (3-3-2). Both (2-3-2) and (3-2-2) are about 3.3 Byte/Flop, and the above is the data for (2-3-2).

Table 1. Byte/Flop of each unrolled loop

Unrolling	(6-6-1) (6-6-2)	(5-5-1) (5-5-2) (5-5-4)	(4-4-1) (4-4-2)	(3-3-1) (3-3-2)	(2-3-2) (3-2-2)	(2-2-1) (2-2-2)	(1-1-1) (1-1-2)
Byte/Flop	1.33333	1.6	2.0	2.6666	3.33333	4.0	8.0

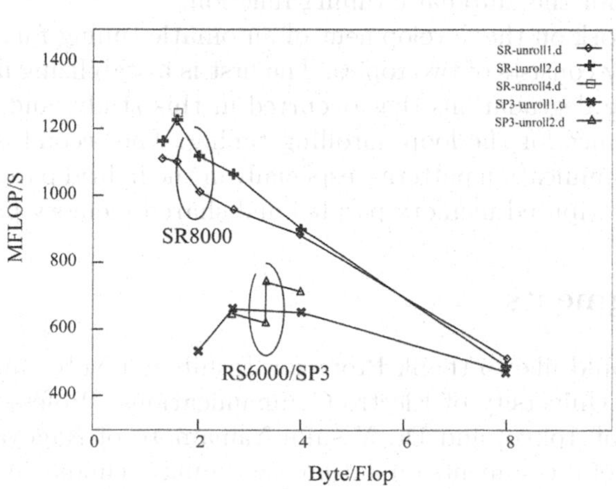

Fig. 5. Relationship between Byte/Flop and average performance for each unrolled loop

We can see from Figure 5 that a trend is revealed in which the average performance declines as the Byte/Flop increases. Especially on the SR8000/F1, almost linear performance degradation is confirmed except for the case of (6-6-2). Of the eleven unrolled loops on the SR8000/F1, (5-5-4) is the best on average. Of the eight unrolled loops on the RS/6000 SP3, (2-3-2) is the best on average. The case of (6-6-2) on the SR8000/F1, and of (4-4-1), (3-3-1), (3-2-2), and (3-3-2) on the RS/6000 SP3 show that a higher depth of loop unrolling at some point causes a degradation of performance. This is probably due to a register spill. The result of the trend clarified that the Byte/Flop is an effective means to predict the performance of unrolled loops, and that it is one possible factor for the automatic tuning function. These results also show that obtaining an average of measurements is important to describe the performances of the unrolled loops.

5 Conclusion and Future Work

To establish the automatic tuning function for eigensolvers, the quantitative effectiveness of unrolling was discussed. The performances of various unrolled loops were evaluated on the SR8000/F1 and the RS/6000 SP3 with every matrix size from N = 3000 to 4000. From the evaluation, the following two results were obtained.

First, the results showed that the performance degradations are found at higher depths of unrolling, while degradation occurs in the same matrix size with lower depths of unrolling. This suggests that the selection of unrolling with pinpoint measurements can degrade performance.

Second, the results showed a connection between the Byte/Flop and the average performance for N=3000 to 4000. In particular, the result on the SR8000/F1 shows an almost linear trend, where the average performance degrades as the Byte/Flop increases. This implies that the Byte/Flop is an effective means of predicting the performance of unrolled loops, and that the Byte/Flop is a possible candidate for the automatic tuning function.

Our future work on the development of automatic tuning functions for eigensolvers primarily consists of two topics. The first is a stabilizing method to avoid the performance degradations that occurred in this study, and to present performance assurance for the loop unrolling tuning. The second is a function for optimizing communication patterns, especially in the hybrid parallel architecture consisting of distributed memory parallel and shared memory parallel.

Acknowledgments

The authors would like to thank Professor Toshitsugu Yuba, and Dr. Takahiro Katagiri of the University of Electro-Communications, Professor Reiji Suda of the University of Tokyo, and Dr. Yusaku Yamamoto of Nagoya University for sharing their useful comments on automatic tuning technologies. Our gratitude is also expressed to JAERI, Japan Atomic Energy Research Institute, for the use of their supercomputers. The authors also express their sincere appreciation to Mr. Nobuhiro Ioki and Mr. Shin'ichi Tanaka of Software Division, Hitachi, Ltd., and to Mr. Hiroaki Fujii and Mr. Naoki Hamanaka of Central Research Laboratory, Hitachi, Ltd., for their continuous support of our research.

References

1. Blimes, J., Asanovic, K., Chin, C.-W., and Demmel, J.: Optimizing matrix multiply using PHiPAC: a portable, high-performance, ANSI C coding methodology, Proceedings of International Conference on Supercomputing 97, 340–347, 1997.
2. Dongarra, J. J. and Eijkhout, V.: Self-adapting numerical software for next generation applications, *The International Journal of High Performance Computing Applications*, **17** 2, Summer 2003, 125–131.
3. Dongarra, J. J. and van de Geijn, R.A.: Reduction to condensed form for the eigenvalue problem on distributed architectures, *Parallel Computing*, **18** 9, 973–982, 1992.
4. Frigo, M.: A fast Fourier transform compiler, Proceedings of the 1999 ACM SIGPLAN Conference on Programming Language Design and Implementation, 169–180, Atlanta, Georgia, May, 1999.
5. Katagiri, T., Kise, K., Honda, H., and Yuba, T.: FIBER: A General Framework for Auto-Tuning Software, The Fifth International Symposium on High Performance Computing (ISHPC-V), *Lecture Notes in Computer Science*, **2858**, Springer, 146–159, 2003.
6. Kudoh, M., Kuroda, H., and Kanada, Y.: Parallel Blocked Sparse Matrix-Vector Multiplication with Dynamic Parameter Selection Method, ICCS2003 (International Conference on Computational Science 2003), International Conference, Melbourne, Australia and St. Petersburg, Russia, June 2-4, 2003. Proceedings, Part III, *Lecture Notes in Computer Science*, **2659**, Springer, 581–591, 2003.
7. Kuroda, H., Katagiri, T., and Kanada, Y.: Knowledge Discovery in Auto-tuning Parallel Numerical Library, Progress in Discovery Science, Final Report of the Japanese Discovery Science Project, *Lecture Notes in Computer Science*, **2281**, Springer, 628–639, 2002.
8. Naono, K.: A framework for development of the library for massively parallel processors with auto-tuning function and with the single memory interface, SIAM Conference on Parallel Processing for Scientific Computing, San Francisco, 2004.
9. Naono, K. and Imamura, T.: An Evaluation towards an Automatic Tuning Eigensolver, IPSJ SIG Notes (2002-HPC-91): 49–54, 1992 (in Japanese).
10. NetSolve project: http://www.cs.utk.edu/netsolve/.
11. Ninf Project: http://ninf.apgrid.org/.
12. Whaley, R., Petitet, A. and Dongarra, J. J.: Automated empirical optimizations of software and the ATLAS project, Parallel Computing, 27, 3–35, 2001.

Peer to Peer Large Scale Linear Algebra Programming and Experimentations

Serge G. Petiton, Lamine M. Aouad, and Laurent Choy

Laboratoire d'Informatique Fondamentale de Lille, and INRIA Futurs,
University of Science and Technology of Lille, France
{serge.petiton, lamine.aouad, laurent.choy}@inria.fr

Abstract. In this paper we discuss the deployment of large scale numerical applications on a Grid. Currently, Grid deployments have coarse-grained and massively parallel nature and are suited for applications that are not communication intensive. For our numerical applications, we minimize the communication needs by a good choice of data placement, in particular by using persistent storage of data. We also introduce out-of-core programming for the task farming paradigm in order to achieve interesting degree of scalability. We discuss the performances of the bisection method to compute the eigenpair of a real symmetric tridiagonal matrix and a block-based matrix-vector product. As experimental middleware we use the XtremWeb system in different configurations; a local one and a WAN based platform with heterogeneous resources distributed on two geographic sites: the university of Lille I and Paris-XI university in Orsay.

1 Introduction

The availability of powerful computers and high-speed network technologies has changed the way of using computers in the last decade. A number of scientific applications that have traditionally performed on supercomputers or on cluster of workstations with traditional tools will be running on a variety of heterogeneous resources geographically distributed.

Grid computing platforms aim to distribute and share computing resources for deploying large scale applications. Typically, they consist in a heterogeneous collection of computational resources, the characteristics of each resource may vary greatly, including processor speed, processor load, disk space, hardware configurations, variance in network bandwidths, latencies, etc. Several realizations have already been proposed, including the Boinc projects[1], the Globus Toolkit[2], XtremWeb[3] and so on. Their aim is to harness heterogeneous resources in some collaborative way in order to share and increase the computing power. Grid computing presents many major challenges, in particular, resources availability, security, scheduling, data confidentiality, etc. We focus here on the effectiveness

[1] http://boinc.ssl.berkeley.edu

[2] http://www.globus.org

[3] http://www.xtremweb.net

I. Lirkov, S. Margenov, and J. Waśniewski (Eds.): LSSC 2005, LNCS 3743, pp. 430–437, 2006.

to execute numerical applications using a Grid middleware such as XtremWeb. Currently, this system needs a central authority. We also present a new programming paradigm based on out-of-core technique to treat the memory constraints on peers, which is a very restrictive factor, and to enhance scalability of distributed numerical applications.

This paper presents Grid implementations of a bisection method and a block-based matrix-vector product. We present the experimental middleware in the next section. In section 3, we present the tested numerical applications. The section 4 shows and discuss the results. In section 5, we introduce the out-of-core programming paradigm for numerical applications in global computing environments. We conclude and present research perspectives in section 6.

2 Experimental Middleware

As an experimental platform, we used the XtremWeb system [2]. XtremWeb is an experimental global computing project intended to distribute applications over dynamic resources according to their availability and implements its own security and fault tolerance policies.

XtremWeb manages tasks following the coordinator-worker paradigm. The coordinator masters the tasks management process, especially results storage such there is, currently, no true notions of tasks scheduling (tasks are scheduled to workers according to their demand in FIFO mode). Workers are distributed volunteer entities which use a part of their CPU time to compute tasks provided by the coordinator. They are not under the control of the coordinator and are volatile. Each action and all connections are then initiated by workers only (the pull model). Every worker connection is registered by the coordinator, and it requests task to compute accordingly to its own local policy. The workers download task software and all expected objects (input files and arguments) in zipped format, store them and start computing. When a task is completed, the workers send back the result to the coordinator.

Several fault tolerance mechanisms are used in XtremWeb to handle resources failures. The main purpose is to enable the system to restart properly after any failure (workers and coordinator). The coordinator manages its tasks using transactions and stores them in reliable media (disks) so that the full system integrity is preserved even if the coordinator shuts down for any reason. At starting time, the coordinator reads the information stored on disk to set up its proper state. The client submits tasks and the worker fetches tasks using transactions; this ensures a consistent state when the coordinator restarts from fault and the client/worker have not failed. Workers failures are detected by the alive signal, so if this signal is not received after a time out, the coordinator considers the worker as lost and reschedules the same task on another available worker. Also, to avoid redundant task and result overwriting, a worker can be brought to stop its current task if it has been disconnected for a long time. XtremWeb also ensures user authentication, workers integrity, application and results protection and user execution logging. It mainly relies on three mechanisms:

- a list of authorized users : after registration, the coordinator provides to each user a key to be used for each subsequent connection. All communications between the user and the coordinator are also encrypted using SSL.
- authentication of the coordinator by workers and clients and,
- sand-boxing system utility: if a malicious user succeeds on launching an aggressive application, XtremWeb workers still protect their host by implementing sand-boxing functionality. Workers run any task inside a sand-box which is customizable, from memory usage to file system operations. This can be disabled on secured systems.

3 Linear Algebra Applications Distributed on a Grid

In this section we describe the experimented applications. We consider in this paper homogeneous data distribution among peers.

3.1 The Bisection Method

Let T be a real symmetric tridiagonal matrix of dimension N. Our goal is to compute $\lambda \in I$ (I is an interval of $\mathbb{R}$) such that $TU = \lambda U$, with $U \in \mathbb{R}^N$. The interval of research can be the domain of Gerschgorin which contains all the eigenvalues (see [3]). As a major hypothesis, we assume that each volunteer peer should be able to store T in its main memory (to avoid any memory paging). The principle of the bisection algorithm consists in sharing I into several subintervals which are smaller and smaller until each eigenvalue is closely surrounded by the low and the high ranges of an interval. Then, we compute the associated eigenvectors with the inverse iteration method.

The task farming paradigm can be easily found here as follow; we split the start interval in several subintervals, then, each volunteer peer is responsible for computing the eigenvalues (and/or the associated eigenvectors) of a set of subintervals instead of only one (otherwise it gives a poor efficiency). We also focus on the strategy used to split the start interval. We must not share it into subintervals of same length because it can lead to an unbalanced distribution of work (when the eigenvalues are clustered). We had rather take into account the distribution of the eigenvalues in order to split dynamically the start domain. First of all, we choose a threshold which corresponds to the maximum number of eigenvalues contained by a subinterval. Let us assume that we have already computed a subinterval. On the one hand, if it has more eigenvalues than the threshold, we split it into two equal subintervals and the same process is done on both. On the other hand, if it has not too many eigenvalues, we keep it. We must notice that a subinterval may contain no eigenvalue. In that case, we delete it.

3.2 The Block-Based Matrix-Vector Product

The matrix-vector product (MVP), $Ax = b$, is a simple but fundamental operation in matrix computing. It forms the core of many linear algebra iterative methods. Historically, the MVP is often the primary operation around which

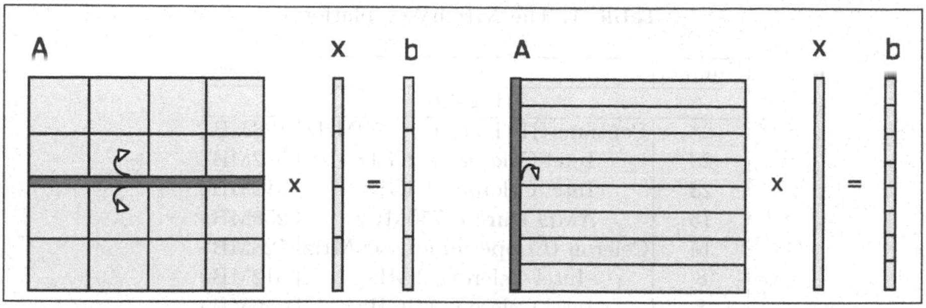

Fig. 1. The two matrix-vector product versions

all supercomputers were benchmarked. We describe here two algorithms of the block-based MVP. In the iterative methods, the matrix A remains unchanged, this must enables us to fasten the data, we call that "persistent data storage", this is discussed in [1] and [4].

A is a matrix of dimension N, x is a vector of dimension N. In the first version, rows of blocks of A are communicated horizontally with P vector blocks across P^2 peers (blocks of size $n_1 \times n_1$, $n_1 = \frac{N}{P}$). In the second version, P^2 blocks of A of dimension $n_2 \times N$ ($n_2 = \frac{N}{P^2}$) and the vector x are communicated vertically across P^2 peers. Figure 1 shows these two versions. We also consider two ways for the blocks transfer; firstly, all the matrix blocks are sent through the network. Secondly, we simulate persistent storage of A to optimize data placement by generating the matrix blocks on peers.

4 Performance Evaluation

The coordinator of the XtremWeb networks runs on a dedicated machine[4] in Lille. For the local configuration, Workers (128 PCs, largely non-dedicated, under Linux) run on undergraduate teaching computers at the Polytech'Lille engineer school. The WAN configuration uses the local configuration in Lille and 36 PCs at the university of Paris-XI, also non-dedicated and under Linux. Table 1 gives a description. All communications are forwarded by the coordinator, it initiates all computations and manages dependencies. The coordinator initializes the runtime environment, submits tasks, retrieves results and makes synchronizations.

For the bisection method, we use two matrix generation tools from Matrix Market to build tridiagonal matrix; the Clement's and Dorr's generators[5]. The matrices size is $N = 10000$. For both matrices, we choose a maximum threshold of 500 eigenvalues per subintervals. About the Clement matrix, 10 sets made of 3 subintervals are built and scheduled among volunteer peers. For the Dorr matrix, 8 sets are made of 3 subintervals and the others are made of 2.

For the MVP, the dense matrix size is $N = 30000$ (double floating point). We have experimented the product with various sizes of blocks. For the version 1, $n_1 =$

[4] http://193.48.57.140/XtremWeb/
[5] http://math.nist.gov/MatrixMarket/

Table 1. The XtremWeb platform

Number	CPU	Memory
At Lille		
28	Pentium III (Katmai), 450MHz	128MB
28	Intel Celeron, 2.2GHz	512MB
23	Intel Celeron, 2.4GHz	512MB
15	AMD Duron, 750MHz	256MB
14	Celeron (Coppermine), 600MHz	128MB
8	Intel Celeron, 2GHz	512MB
8	Intel Celeron, 1.4GHz	256MB
4	Pentium 4, 2.4GHz	512MB
At Orsay		
30	AMD Athlon(tm), 2.8GHz	1GB
6	AMD Athlon(tm), 1.8GHz	1GB

1500, 3000 and 3750. In the version 2, $n_2 = 300$ (blocks size 300×30000). At each execution, we introduced between 72 and 420 tasks into the XtremWeb network.

The computing times vary greatly since the differences in the characteristics of each resource (peers and the network) and the availability of peers across the XtremWeb platforms changes frequently. Indeed, considering the dynamic nature of the Grid the results are nondeterministic. The average bandwidth used is 9.14Mb/s in a local configuration and 6.22Mb/s between Lille and Orsay. The average computing times are shown in tables 2 and 3.

For the bisection method, in table 2, we clearly see that the amount of work is well balanced and, above all, that task farming is a simple way to reduce the time of computations, for bisection and inverse iteration. In fact, let's consider the Dorr matrix, the average time for the computation of the eigenvalues of a set is 18 seconds and we need about 148 seconds to compute the associated eigenvectors. If 10 peers are free when the customer submits its request, all times shown in the table occur in parallel whereas, in a sequential context, a common peer works about 1660 seconds.

For the MVP, in table 3, the results show that the use of persistent storage performs better than the standard version, which manages all communications,

Table 2. The bisection results

The Clement matrix: times picked up on the peers' side (in seconds)

Id of sets	0	1	2	3	4	5	6	7	8	9
Number of couple (λ, U)	875	1031	1031	1031	1032	875	1031	1031	1031	1032
Computing eigenvalues	15	17	18	16	18	15	18	18	18	19
Computing eigenvector	98	121	133	131	128	98	125	123	116	126

The Dorr matrix: times picked up on the peers' side (in seconds)

Id of sets	0	1	2	3	4	5	6	7	8	9
Number of couple (λ, U)	1096	1007	940	1154	1080	1080	1175	950	765	753
Computing eigenvalues	25	16	21	20	20	18	19	15	14	12
Computing eigenvector	230	131	140	180	160	146	144	129	113	103

Table 3. The MVP results

blocks size	With managing all communications	With persistent storage	Number of peers required
In a LAN configuration			
Version 1			
1500	1072 s	663 s	400
3000	239 s	110 s	100
3750	726 s	94 s	64
Version 2			
300×30000	240 s	-	100
In a WAN configuration			
Version 1			
3000	436 s	283 s	100
3750	995 s	201 s	64
Version 2			
300×30000	526 s	-	100

by a factor of 1.5 to 7.7. In this case, efficient data placement is very important factor to maximize performances. Also, the version 1 performs better than the version 2 because we increase the size of communications without increasing the number of computation associated. That means that the classical task farming paradigm is not well adapted in this case and it is necessary to increase the number of operations for each communication.

Efficient data placement is an important factor for numerical applications in a Grid. This reduces the communication needs and gives an important benefit to maximize application performance. This is relatively simple to set up because each numerical application has a knowledge of its data locality and can perform an efficient data placement. We also propose to use out-of-core technique on peers (using explicit I/O disk accesses) to enhance applications scalability and treat memory restrictions. We discuss that in the next section.

5 Out-of-Core Programming on a Grid Environment

Many large problems that could not be solved on the largest supercomputer twenty years ago, can now be solved on a standard laptop computer. However, data structure remains too large to fit in the main memory and must be stored on disks. Out-of-core algorithms are designed to achieve acceptable performance when data are stored on disks. They access data in large contiguous blocks and reuse data that is stored in main memory several times. The ordering of independent operations must be chosen so as to minimize I/O, so that all or most of data in the main memory is used before it is evicted. We avoid then the use of the trivial virtual memory paging which has poor performances due to slow disk accesses. The I/O characteristics of peers can also be used to derive the parameters for our algorithms.

The MVP can be scheduled out-of-core efficiently using a very simple technique that we describe here. We denote the size of main memory by M. We consider a $p \times N$ submatrix of A with $p = (M - N)/(N + 1)$. The algorithm is made up in N/p steps. At each step we perform the product of the submatrix $A_{(p \times N)}$ and the vector x, which gives the subvector $b_{(p)}$. In this schedule, the words that are accessed in the inner iteration need to be read from disk only once because all of them fit within main memory and are written back once.

5.1 Results

We executed this out-of-core product on a laptop computer (Pentium 4, 2.4 GHz and 512 MB), under Linux. The average computing time is 907 seconds with $p = 3000$ ($N = 30000$). This result is more interesting than the results of the product under XtremWeb with blocks size of 1500. This shows that it is possible to solve much larger problems (theoretically multiplied by the number of workers on the platform). The idea is then to use out-of-core technique on peers to deal with their limited memory sizes and to increase the size of tasks to compensate task scheduling and enhance scalability [1].

In these experiments, all communications are forwarded by the coordinator which initiates all computations and manages dependencies. We tested these programs in the LAN XtremWeb platform. We used the matrix-vector out-of-core product shown below on peers to solve the MVP with the same size and $p = 1000$ (i.e. block sizes 1000×30000, version 2). The average computing time is 39 seconds by using persistent data placement. This is the best result obtained for the MVP of sizc 30000 on all considered configurations and versions. We increase the size of the matrix A, $N = 300000$. For $p = 1000$, the average computing time is 575 seconds. This is a very promising result since the traditional task farming paradigm can't achieve this degree of scalability. We can also expect that with an efficient scheduler, the result would have been better.

Also, comparing these results with those obtained in section 4, we see that the combination of explicit I/O disk accesses and persistent storage increase the performances by a factor up to 6 (by using three times less peers). This is a simple and efficient way to program numerical applications on Grid systems based on a better exploitation of data locality, by choosing an adapted placement of data and by redefining the algorithms using explicit I/O accesses.

6 Conclusion

In this work we proposed a performance evaluation for numerical applications on a Grid platform. We have experimented the bisection method for real symmetric tridiagonal matrices and the block-based matrix-vector product for full and general matrices. We used the XtremWeb system for managing non-dedicated resources. Our execution include a classical task farming and a new approach using out-of-core programming and a scheduling strategy which uses persistent data placement.

We showed that persistent storage of data is a very important factor which reduces the communication needs and then increase the computation/communication ratio. We also presented a Grid out-of-core implementation based on explicit input/output disk accesses on peers. We feel that both persistent data storage and an efficient memory management are of prime importance for any distributed application running on a global Grid, and specially for numerical applications. This enhances applications scalability and reaches a good level of performance.

Acknowledgments

This research was supported by the INRIA "Grand Large" Project and the ACI "Masse de données" of the French Ministry of Research. We wish to thank the LRI laboratory in Orsay and the Polytech'Lille engineer school for providing experimentation platforms.

References

1. L. Aouad and S. Petiton. Experimentations and Programming Paradigm For Matrix Computing on Peer to Peer Grid. *GRID 2004, fifth international workshop on Grid Computing.* Pittsburgh, 2004.
2. F. Cappello, S. Djilali, G. Fedak, T. Herault, F. Magniette, V. Néri, and O. Lodygensky. Computing on Large Scale Distributed Systems: XtremWeb Architecture, Programming Models, Security, Tests and Convergence with Grid. In FGCS (Future Generation Computer Science) 2004.
3. J. Demmel, I. Dhillon, and H. Ren. *On the correctness of parallel bisection in floating point.* UCB/CSD 94/805, Berkeley, CA, USA, 1994.
4. S. Petiton and L. Aouad. Large scale peer to peer performance evaluations, with Gauss-Jordan method as an example. *Proceedings of the PPAM2003, fifth international conference on parallel processing and applied mathematics.* Poland, Sep. 2003. *Lecture Notes in Computer Science*, **3019**, 938–945, 2004.
5. S. Toledo. A Survey of Out-Of-Core Algorithms in Numerical Linear Algebra. In: *External Memory Algorithms*, James M. Abello and Jeffrey Scott Vitter, editors, *DIMACS Series in Discrete Mathematics and Theoretical Computer Science.* American Mathematical Society, 161–179, 1999.

A Hybrid Parallel Method for Large Sparse Eigenvalue Problems on a Grid Computing Environment Using Ninf-G/MPI

Tetsuya Sakurai[1,4], Yoshihisa Kodaki[2], Hiroaki Umeda[3,4], Yuichi Inadomi[3,4], Toshio Watanabe[3,4], and Umpei Nagashima[3,4]

[1] Department of Computer Science, University of Tsukuba, Tsukuba 305-8573, Japan
[2] Doctoral Program of Systems and Information Engineering, University of Tsukuba, Tsukuba 305-8573, Japan
[3] Grid Research Center, National Institute of Advanced Industrial Science and Technology
[4] Core Research for Evolutional Science and Technology, Japan Science and Technology Agency

Abstract. In the present paper, we propose a hybrid parallel method for large sparse eigenvalue problems in a grid computing environment. A moment-based method that finds several eigenvalues and their corresponding eigenvectors in a given domain is used. This method is suitable for master-worker type parallel programming models. In order to improve the parallel efficiency of the method, we propose a hybrid implementation using a GridRPC system Ninf-G and MPI. We examined the performance of the proposed method in an environment where several PC clusters are used.

1 Introduction

Eigenvalue problems are a very important class of linear algebra problems. The need for numerical solutions to these problems arises often in science and engineering. We consider a parallel method for the generalized eigenvalue problem

$$Ax = \lambda Bx,$$

where A and B are large sparse matrices. Several methods for such eigenvalue problems are building sequences of subspaces that contain the desired eigenvectors. Krylov subspace-based techniques are powerful tools for large-scale eigenvalue problems [1, 2, 7, 8]. The relationships among Krylov subspace methods, the moment-matching approach and Padé approximation are shown in [2].

In the present paper, we consider a parallel method for finding several eigenvalues and eigenvectors of a generalized eigenvalue problem in a grid environment. A moment-based method [10] to find all of the eigenvalues that lie inside a given domain is used. In this method, a small matrix pencil that has only the desired eigenvalues is derived by solving large sparse linear equations

I. Lirkov, S. Margenov, and J. Waśniewski (Eds.): LSSC 2005, LNCS 3743, pp. 438–445, 2006.

constructed from A and B. Because these equations can be solved independently, we solve them on remote servers in parallel. In [11], a parallel implementation of the method using a GridRPC system OmniRPC [12] is presented. This method is suitable for master-worker programming models. Moreover, the method has a good load balancing property.

In this method, matrix data are sent to each remote server, which causes a loss in efficiency when a large number of servers are employed. In order to improve the parallel efficiency of the method, we propose a hybrid implementation in which MPI is used to send data and to control procedures in a PC cluster with a high-speed network. We have implemented the method using Ninf-G [6] and MPI. Ninf-G provides a simple programming interface based on standard Grid protocols and the API for Grid computing and can support parallel execution of MPI.

The performance of the proposed method on PC clusters was evaluated. As a test problem, we used the matrices that arise in the calculation of molecular orbitals. The results demonstrate that the proposed method is efficient in a grid computing environment.

2 A Master-Worker Type Method

In this section, we describe a method for generalized eigenvalue problems presented in [10]. The proposed method finds several eigenvalues that are located inside a given circle.

Let $A, B \in \mathbb{C}^{n\times n}$, and let $\lambda_1, \ldots, \lambda_d$ $(d \le n)$ be finite eigenvalues of the matrix pencil $A - \lambda B$. The pencil $A - \lambda B$ is referred to as regular if $\det(A - \lambda B)$ is not identically zero for $\lambda \in \mathbb{C}$.

For nonzero vectors $\boldsymbol{u}, \boldsymbol{v} \in \mathbb{C}^n$, we define

$$f(z) := \boldsymbol{u}^H (zB - A)^{-1} \boldsymbol{v}.$$

Let K be the maximum size of Jordan blocks in the canonical form of B. Then, $f(z)$ is represented as

$$f(z) = \sum_{j=1}^{d} \frac{\nu_j}{z - \lambda_j} + g(z),$$

where $\nu_j \in \mathbb{C}$ and $g(z)$ is a polynomial of degree $K - 1$.

Let Γ be a circle with radius γ centered at ρ. Suppose that m distinct eigenvalues $\lambda_1, \ldots, \lambda_m$ are located inside Γ. Define the complex moments

$$\mu_k := \frac{1}{2\pi \mathrm{i}} \int_\Gamma (z - \gamma)^k f(z) dz, \quad k = 0, 1, \ldots. \tag{1}$$

By approximating the integral of (1) via the N-point trapezoidal rule, we obtain the following approximations for μ_k:

$$\mu_k \approx \hat{\mu}_k := \frac{1}{N} \sum_{j=0}^{N-1} (\omega_j - \gamma)^{k+1} f(\omega_j), \quad k = 0, 1, \ldots, \tag{2}$$

where

$$\omega_j := \gamma + \rho e^{\frac{2\pi \mathrm{i}}{N}(j+1/2)}, \quad j = 0, 1, \ldots, N-1.$$

Let the $m \times m$ Hankel matrices $\hat{H}_m$ and $\hat{H}_m^<$ be $\hat{H}_m := [\hat{\mu}_{i+j-2}]_{i,j=1}^m$ and $\hat{H}_m^< := [\hat{\mu}_{i+j-1}]_{i,j=1}^m$, and let $\hat{\zeta}_1, \ldots, \hat{\zeta}_m$ be the eigenvalues of the matrix pencil $\hat{H}_m^< - \lambda \hat{H}_m$. Then, the approximate eigenvalues are obtained by $\hat{\lambda}_j = \gamma + \hat{\zeta}_j, 1 \le j \le m$. The influence of the quadrature error is considered in [5, 9].

Let $\hat{W}_m$ be a matrix of which the column vectors are eigenvectors of $\hat{H}_m^< - \lambda \hat{H}_m$. The approximate vectors $\hat{\boldsymbol{q}}_1, \ldots, \hat{\boldsymbol{q}}_m$ for the eigenvectors $\boldsymbol{q}_1, \ldots, \boldsymbol{q}_m$ are obtained by

$$[\hat{\boldsymbol{q}}_1, \ldots, \hat{\boldsymbol{q}}_m] = [\hat{\boldsymbol{s}}_0, \ldots, \hat{\boldsymbol{s}}_{m-1}] \hat{W}_m, \tag{3}$$

where

$$\boldsymbol{y}_j = (\omega_j B - A)^{-1} \boldsymbol{v}, \quad j = 0, 1, \ldots, N-1, \tag{4}$$

and

$$\hat{\boldsymbol{s}}_k := \frac{1}{N} \sum_{j=0}^{N-1} (\omega_j - \gamma)^{k+1} \boldsymbol{y}_j, \quad k = 0, 1, \ldots. \tag{5}$$

The approximate residues $\hat{\nu}_1, \ldots, \hat{\nu}_m$ are evaluated by

$$[\hat{\nu}_1, \ldots, \hat{\nu}_m] = (\hat{\mu}_0, \ldots, \hat{\mu}_{m-1}) \hat{W}_m.$$

When the size of the Hankel matrices are larger than the exact number of eigenvalues in the circle, some of the approximate eigenvalues are located outside the circle, or the corresponding residue $\hat{\nu}_j$ is small. The criteria to find appropriate m were discussed in [3, 4] for the Hankel matrices. In the present paper, we select the approximate eigenvalues by checking their locations and residues. If the residue satisfies the condition $|\hat{\nu}_j| < \delta$ with small parameter δ, then we ignore $\hat{\lambda}_j$ as a ghost. The algorithm is shown in Figure 1.

Algorithm 1:

Input: $\boldsymbol{u}, \boldsymbol{v} \in \mathbb{C}^n$, N, m, γ, ρ

Output: m', $\hat{\lambda}_1, \ldots, \hat{\lambda}_{m'}$, $\hat{\boldsymbol{q}}_1, \ldots, \hat{\boldsymbol{q}}_{m'}$

1. Set $\omega_j \leftarrow \gamma + \rho \exp(2\pi \mathrm{i}(j+1/2)/N), j = 0, \ldots, N-1$
2. Solve $(\omega_j B - A)\boldsymbol{y}_j = \boldsymbol{v}$ for $\boldsymbol{y}_j$, $j = 0, \ldots, N-1$
3. Set $f(\omega_j) \leftarrow \boldsymbol{u}^H \boldsymbol{y}_j, j = 0, \ldots, N-1$
4. Compute $\hat{\mu}_k, k = 0, \ldots, 2m-1$ by (2)
5. Compute the eigenvalues $\hat{\zeta}_1, \ldots, \hat{\zeta}_m$ of the pencil $\hat{H}_m^< - \lambda \hat{H}_m$
6. Compute $\hat{\boldsymbol{q}}_1, \ldots, \hat{\boldsymbol{q}}_m$ by (3)
7. Set $\hat{\lambda}_j \leftarrow \gamma + \hat{\zeta}_j$, $j = 1, \ldots, m$
8. Select the approximate eigenvalues by checking their locations and residues, and set $\hat{\lambda}_1, \ldots, \hat{\lambda}_{m'}$, $\hat{\boldsymbol{q}}_1, \ldots, \hat{\boldsymbol{q}}_{m'}$.

Fig. 1. A moment-based method

In order to evaluate the value of $f(z)$ at $z = \omega_j$, $j = 0, \ldots, N-1$, we solve the following systems of linear equations;

$$(\omega_j B - A)\boldsymbol{y}_j = \boldsymbol{v}, \quad j = 0, 1, \ldots, N-1, \tag{6}$$

for $\boldsymbol{y}_j$, $j = 0, \ldots, N-1$. When matrices A and B are large and sparse, the computational costs to solve the linear systems are dominant in the algorithm. Since these linear systems are independent, they are solved on remote servers in parallel.

The procedure used to solve N systems in Step 2 of Algorithm 1 is performed on remote servers. Since A and B are common to each system of linear equations, we send these data to each server at the first time. In order to solve another equation on the same server, a scalar parameter ω_j is sent. In this approach, we do not need to exchange data between remote servers. Therefore, the presented method is suitable for master-worker programming models.

3 A Hybrid Implementation

In this section we describe a hybrid implementation of the algorithm that is applied to several circles. Let N_c be the number of circles, and let $\Gamma^{(1)}, \ldots, \Gamma^{(N_c)}$ be circles that include the desired eigenvalues. Let $\gamma^{(l)}$, $\rho^{(l)}$ be the center and the radius of the lth circle, respectively. The equidistributed points on $\Gamma^{(l)}$ are defined by

$$\omega_j^{(l)} = \gamma^{(l)} + \rho^{(l)} e^{\frac{2\pi \mathrm{i}}{N}(j+1/2)}, \quad j = 0, \ldots, N-1.$$

By applying Algorithm 1 for the lth circle, we obtain approximate eigenvalues and corresponding eigenvectors

$$\left(\hat{\lambda}_j^{(l)}, \hat{\boldsymbol{q}}_j^{(l)}\right), \quad j = 1, 2, \ldots, m_l,$$

where m_l is the number of approximate eigenvalues that are selected by checking their locations and residues.

Since the process to find eigenvalues in a circle is independent from those of other circles, we execute the method for these circles on remote servers by using Ninf-G. For the lth circle, the linear systems

$$(\omega_j^{(l)} B - A)\boldsymbol{y}_j^{(l)} = \boldsymbol{v}, \quad j = 0, \ldots, N-1,$$

are solved. We employ MPI processes for the solution of these linear systems. The hybrid algorithm using Ninf-G and MPI is shown in Figure 2.

In the hybrid algorithm, matrix data are sent from a client to a remote server when a Ninf-G process is started. Each Ninf-G process employs a specified number of MPI processes, and matrix data are delivered to all processes. By this approach, we can reduce the number of data transferred from a client to servers via a wide area network compared with the case in which all remote servers receive matrix data directly from a client.

for $l = 1, 2, \ldots, N_c$
  Ninf-G process
  for $j = 0, 1, \ldots, N-1$
    MPI process
    $\omega_j^{(l)} = \gamma^{(l)} + \rho^{(l)} e^{\frac{2\pi i}{N}(j+1/2)}$
    Solve $(\omega_j^{(l)} B - A)\boldsymbol{y}_j^{(l)} = \boldsymbol{v}$ for $\boldsymbol{y}_j^{(l)}$
    Set $f(\omega_j^{(l)}) \leftarrow \boldsymbol{u}^H \boldsymbol{y}_j$
  end
  Compute $\hat{\mu}_k^{(l)}, k = 0, \ldots, 2m-1$
  Compute the eigenvalues $\hat{\zeta}_1^{(l)}, \ldots, \hat{\zeta}_m^{(l)}$
  Compute $\hat{\boldsymbol{q}}_1^{(l)}, \ldots, \hat{\boldsymbol{q}}_{m^{(l)}}$
  Set $\hat{\lambda}_j^{(l)} \leftarrow \gamma^{(l)} + \hat{\zeta}_j^{(l)}, j = 1, \ldots, m$
  Select the approximate eigenvalues
end

Fig. 2. A hybrid implementation using Ninf-G and MPI

The maximum number of Ninf-G processes is limited to the number of circles N_c, and the maximum number of MPI processes is limited to N. Thus, the total number of processes is limited to $N_c \times N$.

When A and B are real matrices, the relation $f(\overline{z}) = \overline{f(z)}$ holds. Thus, we evaluated $N/2$ function values $f(\omega_0), \ldots, f(\omega_{N/2-1})$, and set the remaining function values using $f(\omega_{N-1-j}) = \overline{f(\omega_j)}$. In this case, we solve $N/2$ linear systems for each circle, and the maximum number of MPI processes is $N/2$.

4 Numerical Examples

In this section, we present a numerical example of the proposed method. In order to evaluate the performance of the method for the situation in which several PC clusters are employed via a wide area network, we regarded some parts of a large PC cluster system as distributed PC clusters.

Experiments were performed on AIST F32 Super Cluster of National Institute of Advanced Industrial Science and Technology. The node of the F32 cluster system is Xeon 3.06-GHz/512-MB cache 2 CPU-SMD with 2-GB RAM/nodes. The client machine was a Pentium 4 (3.0 GHz) with a memory of 2 GB, and it was connected to servers via a 100 Base-TX switching hub, as shown in Figure 3.

Since each process solves a large-scale linear system of equations, the algorithm has large granularity. In the hybrid algorithm, each MPI process communicates with other processes only in the same process group. Therefore, this test environment represents a model case in which several distributed PC clusters are used.

The test matrices were derived from computation of the molecular orbitals of lysozyme (129 amino-acid residues, 1961 atoms) with 11077 basis functions. Lysozyme is one of the most well-known proteins with respect to structure and

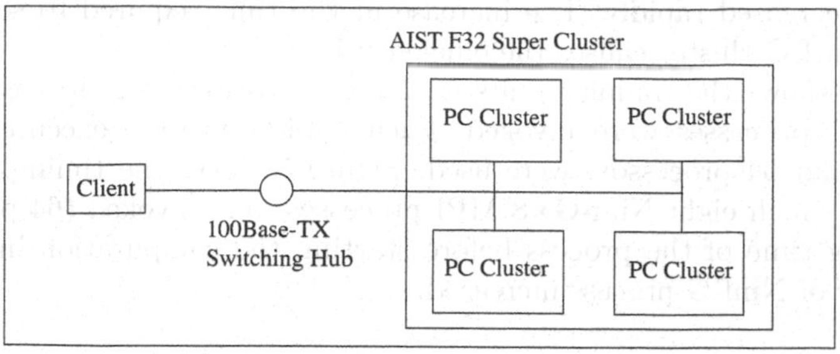

Fig. 3. Test environment

reaction mechanisms. The primary function of lysozyme is bactericidal action by removal of the cell wall. The structure of the lysozyme molecule has been determined experimentally, and we added counter-ions and water molecules around the lysozyme molecule in order to simulate in vivo conditions. The size of A and B was $n = 11077$, and the number of nonzero elements was 6169835. Both A and B were real symmetric.

Since the matrix $\omega_j B - A$ with complex ω_j is complex symmetric, the COCG method [13] with incomplete Cholesky factorization was used to solve the linear systems. The stopping criterion for the relative residual was 10^{-8}. Computation was performed in double-precision arithmetic. The elements of $\boldsymbol{u}$ and $\boldsymbol{v}$ were distributed randomly on the interval $[0, 1]$ by a random number generator.

The intervals $[-0.22, -0.16]$ and $[0.16, 0.22]$ were covered by 32 circles. These intervals include the energy levels of the highest occupied molecular orbitals (HOMO) and the lowest unoccupied molecular orbitals (LUMO) which are key factors in the amount of energy needed to add or remove electrons in a molecule. The parameters were chosen as $N = 32$ and $m = 12$. We obtained 22 eigenvalues and corresponding eigenvectors in these 32 circles.

We observed the wall-clock time in seconds with various combination of the number of Ninf-G processes and MPI processes. The time required to load the input matrices into the memory of the client and that required to start up the Ninf-G and MPI processes were not included. The results are shown in Table 1. The proposed method attains good performance with either four or eight Ninf-G processes. When the number of Ninf-G processes exceeds eight, the

Table 1. Wall-clock times in seconds

		# Ninf-G proc.					
		1	2	4	8	16	32
	1	2875	1445	737	397	246	218
# MPI	2	1666	836	438	241	175	169
proc.	4	847	436	233	143	113	126
	8	443	229	129	91	92	98
	16	237	128	78	64	76	

efficiency decreased rapidly. The increase in the time required to send matrix data to each PC cluster causes the efficiency loss.

Figure 4 shows the timing results for Ninf-G processes for the case in which four Ninf-G processes were invoked. Each Ninf-G process executed 16 MPI processes, and 64 processors were used. Figure 5 shows the timing results for the case in which eight Ninf-G×8 MPI processes were invoked (64 processors). The waiting time of the process before starting the computation increased as the number of Ninf-G process increased.

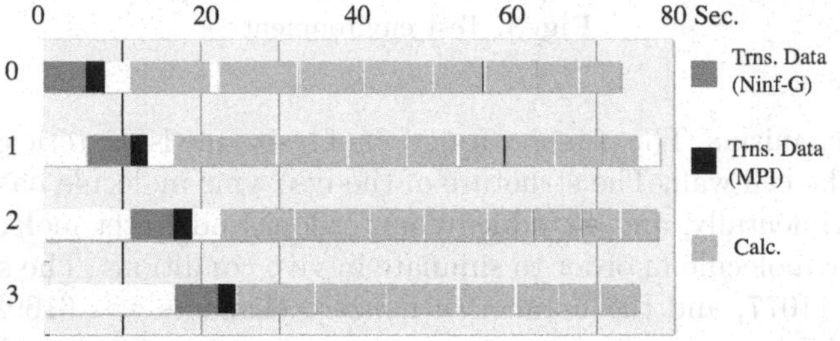

Fig. 4. Timing results for Ninf-G processes (Ninf-G:4, MPI:16, total:64)

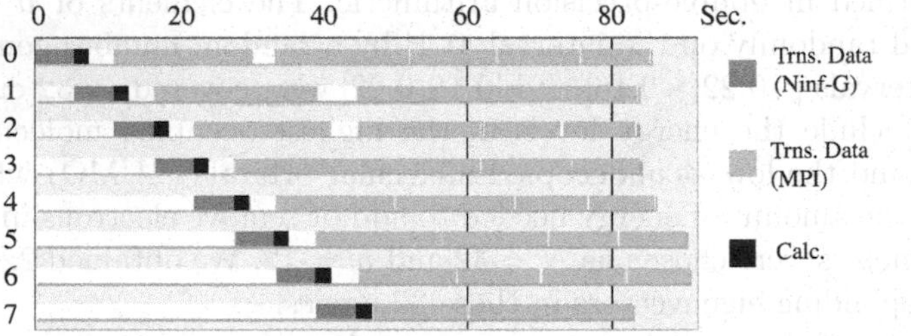

Fig. 5. Timing results for Ninf-G processes (Ninf-G:8, MPI:8, total:64)

5 Conclusions

In the present paper, we presented a hybrid algorithm to find eigenvalues and eigenvectors of generalized eigenvalue problems using a moment-based method. We implemented the method using a GridRPC system Ninf-G and MPI. The performance of the hybrid code was examined for a model case in which several PC clusters were employed from a client machine.

The obtained results demonstrate that the hybrid approach is efficient with respect to avoiding the efficiency loss caused by the communication time to send initial data. In the presented method, each linear system is solved on a single processor. The use of the parallel solver for a linear system will improve the scalability of the method.

Acknowledgements

This work was supported in part by CREST of the Japan Science and Technology Agency (JST), and by a Grant-in-Aid for Scientific Research (No. 15560049) from the Ministry of Education, Culture, Sports, Science, and Technology of Japan.

References

1. Arnoldi, W. E.: The principle of minimized iteration in the solution of the matrix eigenproblem, Quarterly of Appl. Math. **9** (1951) 17–29
2. Bai, Z.: Krylov subspace techniques for reduced-order modeling of large-scale dynamical systems, Appl. Numer. Math. **43** (2002) 9–44
3. Golub, G. H., Milanfar, P., Varah, J.: A stable numerical method for inverting shape from moments, SIAM J. Sci. Comp. **21**(4) (1999) 1222–1243
4. Kravanja, P., Sakurai, T., Van Barel, M.: On locating clusters of zeros of analytic functions, BIT **39** (1999) 646–682
5. Kravanja, P., Sakurai, T., Sugiura, H., Van Barel, M.: A perturbation result for generalized eigenvalue problems and its application to error estimation in a quadrature method for computing zeros of analytic functions, J. Comput. Appl. Math. **161** (2003) 339–347
6. Ninf-G: http://ninf.apgrid.org/
7. Ruhe, A., Rational Krylov algorithms for nonsymmetric eigenvalue problems II: matrix pairs, Linear Algevr. Appl. **197** (1984) 283–295
8. Saad, Y.: Iterative Methods for Large Eigenvalue Problems, Manchester University Press, Manchester, 1992
9. Sakurai, T., Kravanja, P., Sugiura, H., Van Barel, M.: An error analysis of two related quadrature methods for computing zeros of analytic functions, J. Comput. Appl. Math. **152** (2003) 467–480
10. Sakurai, T. and Sugiura, H.: A projection method for generalized eigenvalue problems, J. Comput. Appl. Math. **159** (2003) 119–128
11. Sakurai, T., Hayakawa, K., Sato, M., Takahashi, D.: A parallel method for large sparse generalized eigenvalue problems by OmniRPC in a grid environment, Proc. PARA'04, *Lecture Notes in Computer Science*, Springer-Verlag (to appear)
12. Sato, M., Boku, T., Takahashi, D.: OmniRPC: a Grid RPC System for parallel programming in cluster and grid environment, Proc. CCGrid 2003, (2003) 206–213
13. van der Vorst, H. A., Melissen, J. B. M.: A Petrov-Galerkin type method for solving $Ax = b$, where A is a symmetric complex matrix. IEEE Trans. on Magnetics, **26** (1990) 706–708

Eigenvalue Computation with NetSolve Global Computing System

S.-A. Shahzadeh-Fazeli[1], Nahid Emad[1], and Jack Dongarra[2]

[1] Laboratoire PR*i*SM, Université Versailles St-Quentin, 45, av. des États-Unis, 78035 Versailles Cedex, France
{sas, emad}@prism.uvsq.fr

[2] Computer Science Department, University of Tennessee, Knoxville, USA
dongarra@cs.utk.edu

Abstract. To compute a few eigenpairs of a large sparse matrix we use the hybrid Multiple Explicitly Restarted Arnoldi Method (MERAM). This method is a technique based upon a multiple projection of ERAM and accelerates its convergence. The MERAM updates the restarting vector of an ERAM by taking the interesting eigen-information obtained by the other ones into account. This method presents two main levels of parallelism which are intra-ERAM and inter-ERAM processes. The high level parallelism between ERAMs can be exploited by a network of heterogeneous machines. In MERAM the communications inter ERAM processes are totally asynchronous. The MERAM is fault tolerant and well adapted to GRID-type environments. In this paper, we propose an algorithm of MERAM for NetSolve global computing system. We point out that this kind of systems and their necessary centralism of the communicating information impose to adapt the concerned algorithms. The presented experiments show that a good acceleration of the convergence compared to ERAM can be obtained. We show that the MERAM-like hybrid methods are well suited for the GRID computing environments.

1 Introduction

The hybrid methods combine several different numerical methods or several differently parameterized copies of the same method to solve efficiently some linear algebra problems [4, 3, 7]. The hybrid Multiple Explicitly Restarted Arnoldi Method (MERAM) permits to compute a few eigenpairs of a large sparse non-Hermitian matrix [2]. It makes use of several differently parameterized ERAM for the benefit of the same application. In this paper we present an algorithm of MERAM for NetSolve global computing system. Some particularities of NetSolve and the systems like it imposing the adaptation of the concerned algorithms are presented. We show that this is due to the necessary centralism of the communicating information in this kind of systems. We present then an adaptation of MERAM for NetSolve and show that we can obtain a good acceleration of the convergence with respect to the explicitly restarted Arnoldi method.

Section 2 describes the context of the used numerical algorithms. Section 3 introduces the asynchronous Multiple ERAM. We point out some characteristics

I. Lirkov, S. Margenov, and J. Waśniewski (Eds.): LSSC 2005, LNCS 3743, pp. 446–453, 2006.

of NetSolve-type systems to implement the asynchronous algorithm of MERAM and an adaptation of this algorithm for NetSolve in section 4. An evaluation of this algorithm by a set of test matrices coming from various application problems is presented in section 5. The concluding remarks in section 6 contain our perspective on the problem.

2 General Purpose

Let A be a large non-Hermitian matrix of dimension $n \times n$. We consider the problem of finding a few eigenpairs (λ, u) of A :

$$Au = \lambda u \quad \text{with} \quad \lambda \in \mathbb{C} \quad \text{and} \quad u \in \mathbb{C}^n. \tag{1}$$

Let $w_1 = v/\|v\|_2$ be an initial guess, m be an integer with $m \ll n$. A Krylov subspace method allows to project the problem (1) onto a m-dimensional subspace $\mathbb{K} = span(w_1, Aw_1, \cdots A^{m-1}w_1)$. The well-known Arnoldi process is a projection method which generates an orthogonal basis $w_1, \cdots, w_m$ of the Krylov subspace $\mathbb{K}$ by using the Gram-Schmidt orthogonalization process. Let $\texttt{AR}(input : A, m, v; output : H_m, W_m)$ be such Arnoldi reduction. The $m \times m$ matrix $H_m = (h_{i,j})$ and the $n \times m$ matrix $W_m = [w_1, \cdots, w_m]$ issued from AR algorithm and the matrix A satisfy the equation:

$$AW_m = W_m H_m + f_m e_m^H \tag{2}$$

where $f_m = h_{m+1,m} w_{m+1}$ and e_m is the mth vector of the canonical basis of $\mathbb{C}^m$. The s desired Ritz values (with largest/smallest real part or largest/smallest magnitude) $\Lambda_m = (\lambda_1^{(m)}, \cdots, \lambda_s^{(m)})$ and their associate Ritz vectors $U_m = (u_1^{(m)}, \cdots, u_s^{(m)}) = (W_m y_1^{(m)}, \cdots, W_m y_s^{(m)})$ can be computed by a Basic Arnoldi Algorithm[1]. Let $\texttt{BAA}(input : A, s, m, v; output : r_s, \Lambda_m, U_m)$ be this Algorithm. If the accuracy of the computed Ritz elements by BAA is not satisfactory the projection can be restarted onto a new $\mathbb{K}$. This new subspace can be defined with the same subspace size and a new initial guess. The method is called the Explicitly Restarted Arnoldi (ERAM). Starting with an initial vector v, it computes BAA. If the convergence does not occur, then the starting vector is updated and a BAA process is restarted until the accuracy of the approximated solution is satisfactory (using appropriate methods on the computed Ritz vectors). This update is designed to force the vector in the desired invariant subspace. This goal can be reached by some polynomial restarting strategies proposed in [3] and discussed in section 3.1. Let $\texttt{ERAM}(input : A, s, m, v, tol; output : r_s, \Lambda_m, U_m)$ be an algorithm of Explicitly Restarted Arnoldi Method with tol a tolerance value and g a function which defines the stopping criterion of iterations. The function g can be defined by $g(r_s) = \|r_s\|_\infty$ or by $g(r_s) = \sum_{j=1}^{s} \alpha_j \rho_j$ where α_j are scalar values and $r_s = (\rho_1, \cdots, \rho_s)^t$ with $\rho_i = \|(A - \lambda_i^{(m)} I)u_i^{(m)}\|_2$.

[1] We suppose that the eigenvalues $\lambda_i^{(m)}$ and their corresponding eigenvectors $y_i^{(m)}$ of H_m are re-indexed so that the first s Ritz pairs are the desired ones.

3 Multiple Explicitly Restarted Arnoldi Method

The Multiple Explicitly Restarted Arnoldi Method is a technique based upon an ERAM with multiple projection processes to accelerate its convergence. In this method several differently parameterized ERAM co-operate to efficiently compute a solution of a given eigen-problem. The MERAM allows to update the restarting vector of an ERAM by taking the interesting eigen-information obtained by the other ones into account. The ERAMs begin with several subspaces spanned by a set of initial vectors and a set of subspace sizes. If the convergence does not occur for any of them, then the new subspaces will be defined with initial vectors updated by taking the solutions computed by all the ERAM processes into account. Each of these differently sized subspaces is defined with a new initial vector v. To overcome the storage dependent shortcoming of ERAM, a constraint on the subspace size of each ERAM is imposed. More precisely, it has to belong to the discrete interval $I_m = [m_{min}, m_{max}]$. The bounds m_{min} and m_{max} may be chosen in function of the available computation and storage resources and have to fulfill $m_{min} \leq m_{max} \ll n$. Let $M = (m_1, \cdots, m_\ell)$ be a set of ℓ subspace sizes with $m_i \in I_m$ $(1 \leq i \leq \ell)$, $m_1 \leq \cdots \leq m_\ell$ and $V^\ell = [v^1, \cdots, v^\ell]$ be the matrix of ℓ starting vectors. Let `Send_Eigen_Info` represents the task of sending results from an ERAM process to all other ERAM processes, `Receiv_Eigen_Info` be the task of receiving results from one or more ERAM processes by the current ERAM process and finally, `Rcv_Eigen_Info` be a boolean variable that is true if the current ERAM process has received results from the other ERAM processes. A parallel asynchronous version of MERAM to compute s $(s \leq m_1)$ desired Ritz elements of A is the following:

Asynchronous MERAM Algorithm.

---

1. `Start.` Choose a starting matrix V^ℓ and a set of subspace sizes $M = (m_1, \cdots, m_\ell)$.
2. `Iterate.` For $i = 1, \cdots, \ell$ do in parallel (`ERAM` process):
   - Computation process
     - `(a)` Compute a `BAA`$(input : A, s, m_i, v^i; output : r_s^i, \Lambda_{m_i}, U_{m_i})$ step.
     - `(b)` If $g(r_s^i) \leq tol$ stop all processes.
     - `(c)` Update the initial guess with
       if (`Rcv_Eigen_Info`) then hybrid restart strategy
       else simple restart strategy
   - Communication process
     - `(d)` `Send_Eigen_Info`
     - `(e)` `Receiv_Eigen_Info`

---

where r_s^i is the vector of the residual norms at the ith iteration. All ERAM processes defined in step 2 of the above algorithm are independent and can be run in parallel. Each of them is constituted by a computation part and a communication part. The computation and the communication can be overlapped inside of an ERAM process. The updating of the initial vector v^i can be done by taking the most recent results of the ERAM processes into account. We recall that, in the above MERAM algorithm, the ℓ last results are necessarily the results of the ℓ ERAM processes.

The above algorithm is fault tolerant. A loss of an ERAM process during MERAM execution does not interfere with its termination. It has a great potential for dynamic load balancing; the attribution of ERAM processes of MERAM to the available resources can be done as a function of their subspace size at run time. The heterogeneity of computing supports can be then an optimization factor for this method [2]. Because of all these properties, this algorithm is well suited to the GRID-type environments. In a such environment, the ℓ ERAM processes constituting a MERAM can be dedicated to ℓ different servers. Suppose that the ith ERAM process is dedicated to the server S_i. This server keeps the execution control of the ith ERAM process until the convergence which occurs, in general, by the fastest server. The computation and communication parts of the algorithm can be overlapped.

3.1 Restarting Strategies

To restart an iteration of ERAM with a vector preconditioning so that it has to be forced in the desired invariant subspace a polynomial restarting strategy can be used [3]. One appropriate possibility to define this polynomial is to compute the restarting vector with a linear combination of s desired Ritz vectors.

Let $u_k^{(m)}$ be the kth Ritz vector calculated at the current iteration. To compute the starting vector for the next iteration, we propose to make use of the following linear combination:

$$v = \sum_{k=1}^{s} l_k(\lambda) u_k{}^{(m)} \tag{3}$$

where s coefficients $l_k(\lambda)$ are defined by: $l_k(\lambda) = \prod_{\substack{j=1\\j\neq k}}^{s} (\frac{\lambda - \lambda_j{}^{(m)}}{\lambda_k{}^{(m)} - \lambda_j{}^{(m)}})$, with $\lambda = (\lambda_{min} + \bar{\lambda} - \frac{\lambda_{min}}{n})/2$, $\bar{\lambda} = \frac{\sum_{k=1}^{s} \lambda_k{}^{(m)}}{s}$ and λ_{min} is the eigenvalue with the smallest residual norm. In the experiments of the next section, we made use of this strategy (i.e., equation (3)) to update the initial vector of ERAM as well as the ones of the ERAM processes of MERAM. In the case of MERAM this equation becomes

$$v^i = \sum_{k=1}^{s} l_k^{(best)}(\lambda) u_k^{(best)} \tag{4}$$

where $u_k^{(best)}$ is "the best" k^{th} eigenvector computed by the ERAM processes of MERAM and $l_k^{(best)}$ is its associate coefficient. We suppose that $u_j^{(m_p)}$ is "better" than $u_j^{(m_q)}$ if $\rho_j^p \leq \rho_j^q$.

4 Asynchronous MERAM on NetSolve Global Computing System

NetSolve system is a Grid Middleware based on the concepts of *Remote Procedure Call (RPC)* that allows users to access both hardware and software computational resources distributed across a network. NetSolve provides an environment that monitors and manages computational resources and allocates the services they provide to NetSolve enabled client programs. NetSolve uses a load-balancing strategy to improve the use of the computational resources available. Three chief components of NetSolve are clients, agents and servers. The semantics of a NetSolve client request are: 1/client contacts the agent for a list of capable servers, 2/ client contacts server and sends input parameters, 3/ server runs appropriate service, 4/ server returns output parameters or error status to client.

There are many advantages to using a system like NetSolve which can provide access to otherwise unavailable software/hardware. In cases where the software is in hand, it can make the power of supercomputers accessible from low-end machines like laptop computers. Furthermore, NetSolve adds heuristics that attempt to and the most expeditious route to a problem's solution set. NetSolve currently supports the C, FORTRAN, MATLAB, and Mathematica as languages of implementation for client programs. To solve a problem using NetSolve, a problem description file (PDF) corresponding to the problem has to be defined [6].

4.1 Asynchronous MERAM on NetSolve System

The servers of NetSolve system can not communicate directly to each other. Consequently, contrarily to MERAM algorithm presented in the previous section a server can't keep the control of an ERAM process until the convergence. Indeed, when we run the asynchronous MERAM algorithm on a system which servers communicate directly, each server runs the steps 2.(a), 2.(b) and 2.(c) of an ERAM and communicates with the other servers by running the steps 2.(d) and 2.(e) of the algorithm. This can not be done on a system such as NetSolve where a server can not send/receive directly its information to/from the other servers. To adapt the asynchronous MERAM algorithm to NetSolve system a control process centralizing the information and corresponding to a client component of NetSolve has to be defined. This process has to request to the computation servers of the system to run the step 2.(a) of ERAM processes of MERAM in RPC mode. The running of the step 2.(a) of an ERAM occurs asynchronously with respect to the execution of the same step of the other ERAMs as well as with the execution of the rest of the client algorithm. Once the control process receives the results of an BAA step, it tests the convergence by running the step 2.(b) of the algorithm. If the convergence is not reached then it updates the initial guess with the available eigen-information on this control/client server. An adaptation of the asynchronous Multiple Explicitly Restarted Arnoldi Method for NetSolve is the following:

MERAM-NS(*input* : A, s, M, V^{ℓ}, tol; *output* : r_s, Λ_m, U_m)

---

1. **Start.** Choose a starting matrix V^{ℓ} and a set of subspace sizes $M = (m_1, \cdots, m_{\ell})$.
   Let $it_i = 0$ (for $i = 1, \ell$).
2. For $i = 1, \cdots, \ell$ do :
   (a) Compute a BAA(*input* : A, s, m_i, v^i; *output* : $r_s^i, \Lambda_{m_i}, U_{m_i}$) step in RPC mode.
3. **Iterate.** For $i = 1, \cdots, \ell$ do :
   – If (ready_results) then $it_i = it_i + 1$
     (e$'$) Receive results.
     (b) If $g(r_s^i) \leq tol$ stop all processes.
     (c) Update the initial guess in function of the available eigen-information.
     (a) Compute a BAA(*input* : A, s, m_i, v^i; *output* : $r_s^i, \Lambda_{m_i}, U_{m_i}$) step in RPC mode.
   – End if
4. **End.** $it = \max(it_1, \cdots, it_{\ell})$

Where ready_results is a boolean variable which is true if the outputs of the current BAA algorithm are ready. In other words, if the server computing the ith BAA in RPC mode is ready to send its outputs. We notice that in this implementation the step 2.(d) of the asynchronous MERAM algorithm is not necessary and the step 2.(e) is replaced by 3.(e$'$) which consists to receive *all* eigen-information on the control process. Instead, we notice that in each computation request in RPC mode, the client program has to send all inputs to the computation server which accepts this task. That means, in MERAM-NS algorithm, for each restart (i.e., iteration) of every ERAM process, the client program has to send the n-order matrix A, and an n-size initial vector to a computation server. This engenders an intense communication between the client and computation servers. But this communication is overlapped by the running of the rest of the algorithm. We can notice that when a computational server finishes the step 2.(a) or 3.(a), it has to return $s + 2$ n-size output vectors to the client process.

In asynchronous MERAM algorithm, at the end of an iteration each ERAM sends $s + 2$ n-size vectors to $\ell - 1$ other processes. That means, each ERAM has to communicate $(\ell - 1) \times (s + 2) \times n$ data to other processes. The reception of $s + 2$ n-size vectors by a process is not determinism and not quantifiable.

5 Numerical Experiments

The experiments presented in this section have been done on a NetSolve system whose computation servers have been located in France (at the university of Versailles and the Institute of Technology of Vélizy sites) and in U.S.A. and interconnected by Internet. We implemented ERAM and MERAM (i.e., MERAM-NS) algorithms using C and MATLAB for some real matrices on NetSolve system. The client applications are written in MATLAB while the programs having to run in RPC mode (i.e., ERAM processes) are written in C. The stopping criterion is $g(r_s^i) = \|r_s^i\|_\infty$ where $r_s^i = (\rho_1^i, \cdots, \rho_s^i)$ and ρ_j^i is normalized by $\rho_j^i = \rho_j^i / \|A\|_F$ for all $j \in 1, \cdots, s$ and $i \in 1, \cdots, \ell$. The initial vector is $v = z_n = (1, \cdots, 1)/\sqrt{n}$ and the initial matrix is $V^\ell = [v^1 = z_n, \cdots, v^\ell = z_n]$. We search a number $s = 2$ or $s = 5$ of the eigenvalues with the largest magnitude. The used matrices are $af23560$, $mhd4800b$, gre_1107, and $west2021$ which are taken from the matrix market [1] Their number of non zero elements are 484256, 16160, 5664, and 7353 respectively. In our experiments, we run MERAM-NS with $\ell = 3$ ERAM processes where the steps 2 and 3.(a) are computed in RPC nonblocking mode. The efficiency of our algorithms on NetSolve are measured in terms of the number *it* of the restarts. The number of iterations of MERAM is the number of iterations of the ERAM process which reaches convergence. It is generally the ERAM process with the largest subspace size.

5.1 MERAM-NS Versus ERAM

We denote by MERAM$(m_1, \cdots, m_l)$ a MERAM with subspaces sizes $m_1, \cdots, m_l$ and by ERAM(m) an ERAM with subspace size m. Table 1 presents the results obtained with the ERAM and MERAM algorithms on NetSolve. It also presents a comparison between the results obtained by these methods in term of the number of restarts. The results of our experiments indicate that in term of the number of the restarts, MERAM is considerably more efficient than ERAM.

Table 1. Comparison of ERAM(m) and ERAM$(m_1, \cdots, m_\ell)$

			ERAM			MERAM		
Matrix	s	ℓ	m	it	Res.Norms	$m_1, \ldots, m_\ell$	it	Res. Norms
af23560.mtx	2	3	10	240	No converge	5, 7, 10	74	9.329017e-10
mhd4800b.mtx	2	3	10	41	8.127003e-10	5, 7, 10	6	4.016027e-09
mhd4800b.mtx	5	3	20	19	4.089292e-15	10, 15, 20	4	2.999647e-15
gre_1107.mtx	2	3	30	46	3.389087e-09	5, 10, 30	32	6.753314e-09
west2021.mtx	2	3	10	18	1.742610e-09	5, 7, 10	14	6.267924e-09

6 Conclusion

In order to improve the overall performance of Arnoldi type algorithm, we proposed an adaptation of the multiple explicitly restarted Arnoldi method for NetSolve system. We have seen that the multiple explicitly restarted Arnoldi method accelerates the convergence of explicitly restarted Arnoldi method. The numerical experiments have demonstrated that this variant of MERAM is often much more efficient than ERAM. In addition, this concept may be used in some Krylov subspace type method for the solution of large sparse non symmetric eigenproblem.

We have shown that the MERAM-type asynchronous algorithms are very well adapted to the global computing systems such as NetSolve. Meanwhile, one of the major problems remains the transfer of the matrix from the client server towards the computation servers. Moreover, the classical evaluation of performances is no more valid in this kind of systems. For example, the execution response time can not be a good measure of performance for MERAM nor for a comparison between MERAM and ERAM. This is due to the execution time which is dependent to the Internet load, to the volatility and the transparency of the servers and to the implementation of the ERAM on NetSolve which introduces some artificial communications.

In order to have a rapid response time the use of a classical parallel supercomputer seems to be more interesting. But the supercomputers are not easily accessible. Moreover, the use of a global computing system allows to take advantage of the otherwise unavailable software and/or hardware resources.

References

1. Z. Bai, D. Day, J. Demmel, and J. Dongara, *A Test Matrix Collection for Non-Hermitian Problems*, http://math.nist.gov/MatrixMarket.
2. N. Emad, S. Petiton, and G. Edjlali, Multiple Explicitly Restarted Arnoldi Method for Solving Large Eigenproblems, *SIAM Journal on Scientific Computing*, **27** 1, 253–277, 2005.
3. Y. Saad, *Numerical Methods for Large Eigenvalue Problems*, Manchester University Press, 1993.
4. Y. Saad, Variations on Arnoldi's Method for Computing Eigenelements of Large Unsymmetric Matrices, *Linear Algebra Applications*, **34**, 269–295, 1980.
5. Arnold, D., Agrawal, S., Blackford, S., Dongarra, J., Miller, M., Seymour, K., Sagi, K., Shi, Z., and Vadhiyar, S., *Users' Guide to NetSolve v1.4.1*, ICL Technical Report, ICL-UT-02-05, June 25, 2002. http://icl.cs.utk.edu/netsolve
6. H. Casanova and J. Dongarra, NetSolve's Network Enabled Server: Examples and Applications, *IEEE Computational Science and Engineering*, **5** 3, 57–67, 1998.
7. G. L. G. Sleijpen and H. A. Van der Vorst, Jacobi-Davidson Methods (Section 4.7). In *Templates for the Solution of Algebraic Eigenvalue Problems: A Practical Guide*, Zhaojun Bai, James Demmel, Jack Dongarra, Axel Ruhe, and Henk van der Vorst, editors, http://www.cs.utk.edu/~dongarra/etemplates/book.html, SIAM, Philadelphia, 2000.

Computation of High-Precision Mathematical Constants in a Combined Cluster and Grid Environment

Daisuke Takahashi, Mitsuhisa Sato, and Taisuke Boku

Graduate School of Systems and Information Engineering, University of Tsukuba,
1-1-1 Tennodai, Tsukuba, Ibaraki 305-8573, Japan
{daisuke, msato, taisuke}@cs.tsukuba.ac.jp

Abstract. The computation of high-precision mathematical constants in a combined cluster and grid environment is presented. Mathematical constants (e.g., π and e) are computed from their series expansions. The binary splitting method recursively reduces the calculation of the sum of the series by effectively splitting the problem into two halves and performing the same calculation on each half. By using grid computing for part of the binary splitting process, the supercomputer computation time, which is very expensive, can be reduced. We implemented the independent binary splitting process in a grid environment using a grid RPC system called OmniRPC. We successfully achieved nearly linear speedup for larger digits on an 8-node PC cluster used over a wide-area network.

1 Introduction

Many computations of mathematical constants (e.g., π and e) have been performed with high precision [2, 9, 12, 1].

Brent presented algorithms for computing elementary functions [4]. In particular, Brent [4] and Salamin [13] independently discovered an approximation algorithm based on elliptic integrals. This algorithm yields quadratic convergence to π.

Borweins discovered a general technique for obtaining even higher-order convergent algorithms for π [3]. Borweins' quartically convergent algorithm is as follows: Let $a_0 = 6 - 4\sqrt{2}$ and $y_0 = \sqrt{2} - 1$. Iterate the following calculations:

$$y_{k+1} = \frac{1 - (1 - y_k^4)^{1/4}}{1 + (1 - y_k^4)^{1/4}},$$

$$a_{k+1} = a_k(1 + y_{k+1})^4 - 2^{2k+3} y_{k+1}(1 + y_{k+1} + y_{k+1}^2).$$

Then, a_k converges quartically to $1/\pi$. Each iteration must be performed to a level of numeric precision that is at least as high as that desired for the final result. This algorithm requires $O(\log N M(N))$ operations, where $M(N)$ is the number of operations for an N-digit multiplication.

A key operation in fast multiple-precision arithmetic is the multiplication, which consumes a significant percentage of the total computation time. The

I. Lirkov, S. Margenov, and J. Waśniewski (Eds.): LSSC 2005, LNCS 3743, pp. 454–461, 2006.

multiple-precision multiplication of N-digit numbers requires $O(N^2)$ operations by using an ordinary multiplication algorithm [11]. Karatsuba's algorithm [10, 11] is known to reduce the number of operations to $O(N^{\log_2 3})$.

It is known that the multiple-precision multiplication of N-digit numbers can be performed in $O(N \log N \log\log N)$ operations by using the Schönhage-Strassen algorithm [16], which is based on the fast Fourier transform (FFT) [7].

Mathematical constants are computed from their series expansion, such as:

$$\pi = 16 \arctan \frac{1}{5} - 4 \arctan \frac{1}{239}, \quad \arctan \frac{1}{q} = \sum_{k=0}^{\infty} \frac{(-1)^k}{(2k+1)q^{2k+1}}.$$

These series expansions have a computational complexity of $O(N^2)$ for computing N digits of the series using a classical approach.

On the other hand, the binary splitting method offers an efficient way to evaluate many series containing only rational numbers [5, 3, 9]. By using the binary splitting method, π can be computed in $O((\log N)^2 M(N))$ and e can be computed in $O(\log N M(N))$, where $M(N)$ is the number of operations for an N-digit multiplication.

The computation of π to more than 1.24 trillion decimal digits was performed by using both the binary splitting method and the arctangent series on a Hitachi SR8000/MPP parallel computer [1].

The BBP (Bailey-Borwein-Plouffe) formula computes the N-th hexadecimal digit of a variety of mathematical constants (e.g., π and $\log 2$) [2].

The BBP formula can be stated as follows:

$$\pi = \sum_{i=0}^{\infty} \frac{1}{16^i} \left(\frac{4}{8i+1} - \frac{2}{8i+4} - \frac{1}{8i+5} - \frac{1}{8i+6} \right)$$

The BBP formula enables computation of a specific bit in π without computing all the previous bits. PiHex was a distributed computing project which calculated the quadrillionth bit of π by using a BBP-like formula and a worldwide network of 1,734 PCs [12].

Although the BBP formula is suitable for distributed computing, it requires $O(N^2 \log N)$ operations to calculate N full-digits of π.

The Brent-Salamin algorithm and Borweins' quartically convergent algorithm require $O(\log N M(N))$ operations, which is the least number of asymptotic arithmetic operations possible for computing π. Since these algorithms require N full-digits computation at each iteration, they are not suitable for distributed computing.

Therefore, the binary splitting method is considerably better than both the Brent-Salamin algorithm and Borweins' quartically convergent algorithm, because no full-precision multiplication is involved.

In this paper, we present the computation of high-precision mathematical constants in a combination of parallel and distributed computing environments. By using grid computing for part of the binary splitting process, the supercomputer computation time, which is very expensive, can be reduced.

Section 2 describes the binary splitting method. Section 3 describes the parallelization of the binary splitting method. Section 4 presents the experimental results. We provide some concluding remarks in Section 5.

2 Binary Splitting Method

Haible and Papanikolaou proposed fast multiprecision evaluation of a series of rational numbers [9].

The general form of the linearly convergent series is given as follows:

$$S = \sum_{k=0}^{\infty} \frac{a_k}{b_k} \frac{\prod_{j=0}^{k} p_j}{\prod_{j=0}^{k} q_j}$$

where a_k, b_k, p_j and q_j are $O(\log k)$-bit integers.

If bounds $l \le k < n$ are given, the partial sum of S can be written as follows:

$$S_{l,n-1} = \sum_{k=l}^{n-1} \frac{a_k}{b_k} \frac{\prod_{j=0}^{k} p_j}{\prod_{j=0}^{k} q_j}.$$

Then, we have

$$P_{l,n-1} = \prod_{j=l}^{n-1} p_j,$$

$$Q_{l,n-1} = \prod_{j=l}^{n-1} q_j,$$

$$B_{l,n-1} = \prod_{j=l}^{n-1} b_j,$$

$$T_{l,n-1} = B_{l,n-1} Q_{l,n-1} S_{l,n-1}.$$

For $l \le m < n$, we have

$$\begin{aligned}
P_{l,n-1} &= P_{l,m-1} P_{m,n-1}, \\
Q_{l,n-1} &= Q_{l,m-1} Q_{m,n-1}, \\
B_{l,n-1} &= B_{l,m-1} B_{m,n-1}, \\
T_{l,n-1} &= B_{m,n-1} Q_{m,n-1} T_{l,m-1} + B_{l,m-1} P_{l,m-1} T_{m,n-1}.
\end{aligned}$$

Finally, we have

$$S_{0,n-1} = \frac{T_{0,n-1}}{B_{0,n-1}Q_{0,n-1}}.$$

For $\exp(x)$ (where $x = u/v$ is a rational number), $a_k = 1$, $b_k = 1$, $p_0 = q_0 = 1$, and $p_k = u$, $q_k = kv$ for $k > 0$. The bit complexity is $O(\log N M(N))$ [9].

For rational $|x| < 1$, $\arctan(x)$ can be computed to $a_k = 1$, $b_k = 2k+1$, $q_k = 1$, $p_0 = x$ and $p_k = -x^2$ for $k > 0$. The bit complexity is $O((\log N)^2 M(N))$ [9].

The binary splitting method for arctan series can be written as follows:

```
binsplit(l, n)
{
    if (l == n) {
      P(l,n) = 1;
      Q(l,n) = (2 * l + 1) * (q * q);
      R(l,n) = -(2 * l + 1);
    } else {
      m = floor((l + n) / 2);
      binsplit(l, m);
      binsplit(m, n);
      P(l,n) = P(l,m) * Q(m,n) + R(l,m) * P(m,n);
      Q(l,n) = Q(l,m) * Q(m,n);
      R(l,n) = R(l,m) * R(m,n);
    }
}
```

Then, we have the sum to n terms of $\arctan(1/q) = P_{0,n-1}/Q_{0,n-1}$ with `binsplit(0,n-1)`. The above operations can be performed recursively. For smaller digit multiplication (e.g., less than 1,000 decimal digits), an ordinary $O(N^2)$ multiplication algorithm or Karatsuba's $O(N^{\log_2 3})$ algorithm may be used. For larger digit multiplication (e.g., more than 1,000 decimal digits), FFT-based multiplication may be used.

The binary splitting method recursively reduces the calculation of the sum of the series by effectively splitting the problem into two halves and performing the same calculation on each half. It requires $O(\log N)$ stages for evaluating the sum of the series.

3 Parallelization of Binary Splitting Method

Haible and Papanikolaou pointed out that the binary splitting method is suitable for parallel computation [9].

If p processors are available, the partial sum of the series can be performed as follows:

$$\begin{array}{ll} \text{Processor } 0: & S_{0,\lceil \frac{n}{p} \rceil -1} \\ \text{Processor } 1: & S_{\lceil \frac{n}{p} \rceil, 2\lceil \frac{n}{p} \rceil -1} \\ \vdots & \\ \text{Processor } i: & S_{i\lceil \frac{n}{p} \rceil, (i+1)\lceil \frac{n}{p} \rceil -1} \\ \vdots & \\ \text{Processor } (p-1): & S_{(p-1)\lceil \frac{n}{p} \rceil, n-1} \end{array}$$

The independent binary splitting process can be performed in a grid computing environment. For implementation, a thread-safe remote procedure call (RPC) system, called OmniRPC [14], was used for the cluster and grid environments.

3.1 The OmniRPC System

OmniRPC [14] is a grid RPC system which allows seamless parallel programming from a PC cluster to a grid environment. OmniRPC inherits its API and basic architecture from Ninf [15]. A client and the remote computational hosts which execute the remote procedures are connected via a network.

The OmniRPC system supports a local environment with "`rsh`," a grid environment with Globus [8], and remote hosts with "`ssh`."

OmniRPC efficiently supports typical master/worker parallel grid applications such as parametric execution programs.

3.2 Parallel Implementation of Binary Splitting Method Using OmniRPC

The client program which assigns a partial sum to the calculation node of a PC cluster using RPC can be written as follows:

```
. . .
      call OmniRPC_Init
. . .
      do j=0,nproc-1
         . . .
         call OmniRPC_Call_Async(ireq(j),'binsplit*', . . .)
         . . .
      end do
      call OmniRPC_Wait_All(nproc,ireq)
. . .
      call OmniRPC_Finalize
```

The OmniRPC APIs used above are described as follows:

- `OmniRPC_Init` initializes the system and reads the registry files in the remote hosts to make the database that associates the entry names of the remote functions with remote executables.

– `OmniRPC_Call_Async` makes a non-blocking remote procedure and returns the handle for the request.
– `OmniRPC_Wait_All` blocks until the specified non-blocking request has been completed.
– `OmniRPC_Finalize` releases any resource being used by the OmniRPC system.

For the first $O(\log(N/p))$ stages, each partial sum of the series can be performed completely independently on a distributed computing environment which has p processors. Then, each partial sum of the series may be saved to files.

Finally, for the last $O(\log p)$ stages, parallel multiple-precision arithmetic [17] should be done on distributed-memory parallel computers (e.g., clusters of PCs).

Since all stages can be performed on distributed-memory parallel computers, $O(\log N)$ stages are needed. Thus, by using distributed computing for the sum of the series, the computation time of the expensive parallel computers can be reduced.

4 Computational Results

To evaluate the implemented parallel binary splitting method, we performed the partial sum of the series in a grid computing environment. The decimal digit N of π and the number of worker nodes p were varied. We averaged the elapsed times obtained from 10 executions. It should be noted that the elapsed times include the disk I/O time.

The specifications for the platforms used are shown in Table 1. The client node and the worker nodes were connected through a 1 Gbps wide-area network. For the authentication system of OmniRPC, "`ssh`" was used. The built-in Round-Robin scheduler of OmniRPC was used as the job scheduler.

All routines were written in Fortran. The Intel Fortran Compiler (version 8.1) was used. The optimization option was specified as "`-O3 -xW`."

We used the following Chudnovsky's formula [6]:

$$\frac{1}{\pi} = 12 \sum_{k=0}^{\infty} \frac{(-1)^k (6k)!(13591409 + 545140134k)}{(3k)!(k!)^3\, 640320^{3k+3/2}}$$

Table 1. Specification of machines

Platform	Client node	Worker nodes
Number of Nodes	1	8
CPU Type	Pentium4 3.2 GHz	dual Xeon 2.4 GHz
L1 Cache	I-Cache: 12 Kuops D-Cache: 16 KB	I-Cache: 12 Kuops D-Cache: 8 KB
L2 Cache	1 MB Unified	512 KB Unified
Main Memory	1 GB	1 GB
OS	Linux 2.4.20-8smp	Linux 2.4.20-20.7smp

Table 2. Performance of the parallel binary splitting method. The Time columns are in seconds.

N	1 worker		2 workers		4 workers		8 workers	
	Time	Speedup	Time	Speedup	Time	Speedup	Time	Speedup
2^{20}	6.34	1.00	3.59	1.77	2.05	3.09	1.45	4.37
2^{21}	15.80	1.00	8.24	1.92	4.46	3.54	2.69	5.87
2^{22}	38.29	1.00	19.37	1.98	10.16	3.77	5.55	6.90
2^{23}	90.73	1.00	45.51	1.99	23.31	3.89	12.23	7.42
2^{24}	208.15	1.00	104.24	2.00	53.96	3.86	27.84	7.48
2^{25}	476.28	1.00	238.45	2.00	121.83	3.91	63.04	7.56

Table 2 shows the averaged execution times of the parallel binary splitting method. The column with the N heading shows the decimal digits. The next eight columns contain the average elapsed time in seconds and the speedup ratio.

We can see that the speedup of the parallel binary splitting is nearly a linear speedup for larger digits.

We found a load imbalance in the processing time of each partial sum. This is because the number of digits of each partial sum is different.

5 Conclusion

In this paper, we proposed the computation of high-precision mathematical constants in a combined cluster and grid environment.

We implemented an independent binary splitting process in a grid environment using the grid RPC system called OmniRPC. We succeeded in obtaining a speedup of up to 7.56 times on an 8-node PC cluster.

Previous computations of high-precision π were performed on expensive (parallel) supercomputers.

But, by using grid computing for part of the binary splitting process, the supercomputer computation time, which is very expensive, can be reduced. A combined cluster and grid environment enables the computation of trillions of digits of mathematical constants.

References

1. Bailey, D.H.: Some background on kanada's recent pi calculation. `http://www.crd.lbl.gov/~dhbailey/dhbpapers/dhb-kanada.pdf` (2003)
2. Bailey, D., Borwein, P., Plouffe, S.: On the rapid computation of various polylogarithmic constants. Math. Comput. **66** (1997) 903–913
3. Borwein, J.M., Borwein, P.B.: Pi and the AGM — A Study in Analytic Number Theory and Computational Complexity. Wiley, New York, NY (1987)
4. Brent, R.P.: Fast multiple-precision evaluation of elementary functions. Journal of the ACM **23** (1976) 242–251
5. Brent, R.P. The Complexity of Multiple-Precision Arithmetic. University of Queensland Press, Brisbane (1976)

6. Chudnovsky, D.V., Chudnovsky, G.V. Approximations and complex multiplication according to Ramanujan. Academic Press Inc., Boston (1988) 375–396 and 468–472
7. Cooley, J.W., Tukey, J.W.: An algorithm for the machine calculation of complex Fourier series. Math. Comput. **19** (1965) 297–301
8. Foster, I., Kesselman, C.: Globus: A metacomputing infrastructure toolkit. International Journal of Supercomputer Applications and High Performance Computing **11** (1997) 115–128
9. Haible, B., Papanikolaou, T.: Fast multiprecision evaluation of series of rational numbers. In: Proc. 3rd International Symposium on Algorithmic Number Theory (ANTS-III). *Lecture Notes in Computer Science*, **1423**, Springer-Verlag (1998) 338–350
10. Karatsuba, A., Ofman, Y.: Multiplication of multidigit numbers on automata. Doklady Akad. Nauk SSSR **145** (1962) 293–294
11. Knuth, D.E.: The Art of Computer Programming, Vol. 2: Seminumerical Algorithms. 3rd edn. Addison-Wesley, Reading, MA (1997)
12. Percival, C.: PiHex — A distributed effort to calculate Pi. (`http://www.cecm.sfu.ca/projects/pihex/`)
13. Salamin, E.: Computation of π using arithmetic-geometric mean. Math. Comput. **30** (1976) 565–570
14. Sato, M., Boku, T., Takahashi, D.: OmniRPC: a grid RPC system for parallel programming in cluster and grid environment. In: Proc. 3rd IEEE/ACM International Symposium on Cluster Computing and the Grid (CCGrid 2003). (2003) 206–213
15. Sato, M., Nakada, S., Sekiguchi, S., Matsuoka, S., Nagashima, U., Takagi, H.: Ninf: A network based information library for global world-wide computing infrastructure. In: Proc. International Conference and Exhibition on High Performance Computing and Networking Europe (HPCN Europe 1997). *Lecture Notes in Computer Science*, **1225**, Springer-Verlag (1997) 491–502
16. Schönhage, A., Strassen, V.: Schnelle Multiplikation grosser Zahlen. Computing (Arch. Elektron. Rechnen) **7** (1971) 281–292
17. Takahashi, D.: Implementation of multiple-precision parallel division and square root on distributed-memory parallel computers. In: Proc. 2000 International Workshop on Parallel Processing (ICPP'00 Workshops). (2000) 229–235

Monte Carlo Valuation of Multidimensional American Options Through Grid Computing

Ioane Muni Toke[1] and Jean-Yves Girard[2,*]

[1] Ecole Centrale Paris, Laboratoire de Mathématiques Appliqués aux systèmes, Grande Voie des Vignes F-92295 Chatenay-Malabry, France
munitoke@mas.ecp.fr

[2] IBM Grid Design Center for eBusiness on demand, rue de la Vieille Poste F-34000 Montpellier, France
girardjy@fr.ibm.com

1 Introduction

We investigate several ways to implement a financial algorithm on a Grid architecture. The chosen algorithm is used to value an American stock option on the maximum of several assets. Such an evaluation has been a standard case in financial mathematics for the last years. These stock options become more and more common but cannot be easily valuated: the complexity of the usual algorithms grows exponentially with some parameters (number of assets involved, number of exercise date). Algorithms based on simulation (Broadie and Glasserman, [2, 3]) often need prohibitive computational efforts. Fu and al. [4] show that for a option on five assets, some methods do not terminate in less than ten hours of computational time (tests made on a Sun Ultra5 in 2000), whereas a trader in a financial institution doesn't have more than a few minutes to deal with the valuation of such an option. As a consequence, recent papers tend to explore parametrization methods for the space state or the exercise frontier. Longstaff and Schwartz 's algorithm [8], proposed in 2001, belongs to this trend. However, results seems to be very sensitive to the parameters and the choice of basis functions. Investigation on the loss of precision must be made.

The uninterrupted growth of the computational power potentially used by traders and in particular the birth of grid technologies show us a new way to implement such algorithms. Using parallelism and grid computing, we can aim at obtaining fast and liable results through a better use of idle machines and without sacrificing precision. To our knowledge, the stochastic mesh method (cf. [1]) is the only one that was investigated from a point of view of computing parallelism. Results presented there are promising but they were all obtained on massively parallel machines, which are rarely used in financial institutions that need these valuations.

So we have investigated these methods in the framework of Grid computing standards: we have achieved parallelism for a valuation method and implemented it with middleware such as the well-known Globus Toolkit ([9]), which allows

* This work was partially supported by an IBM Ph.D. Fellowship.

I. Lirkov, S. Margenov, and J. Waśniewski (Eds.): LSSC 2005, LNCS 3743, pp. 462–469, 2006.

a calculus to be well-distributed among a group of machines even if they differ from each other as for the localization, the hardware or software used in them.

For this study we have chosen an algorithm presented in 2002 by Ibanez et Zapatero [7]. In a first part, we briefly describe this method and justify our choice. Then follows the description of three implementations based on the Globus Toolkit middleware [9]. Our second part describes the implementations of three possible Grid methods and show their respective advantages and drawbacks. We finally make several performance tests and conclude our study.

2 Valuation of an American Option on the Maximum of Several Assets

2.1 Theoretical Introduction

We consider the case of a financial stock market fitting Black & Scholes framework. This market is composed of M assets, and their prices at time t are $S_t = (S_t^1, \ldots, S_t^M)$. The assets dynamics can be written as follows:

$$\frac{dS_t}{S_t} = r\,dt + \sigma\,dW_t \tag{1}$$

where W_t is a M-dimensional Brownian motion.

An American option with time of expiry T is a contract which can be activated at any time $t \in [0, T]$ and then pays the exercise value $I(t, S_t)$. In our example of an option on a maximum of M assets, we have:

$$I(t, S_t) = \max_{i=1,\ldots,M} (S_t^i - K)^+. \tag{2}$$

Formally, the value $Q(0, S_0)$ of the option at time 0 is:

$$Q(0, S_0) = \max_{\tau}[I(\tau, S_\tau)] \tag{3}$$

where τ is a stopping time. If we consider the case of a discrete time scale for exercise dates $\{0 = t_0 < t_1 < \ldots < t_N = T\}$, then the option at time t_n in the state x has the following value:

$$Q(t_n, x) = \max\left(I(t_n, x), \mathbb{E}[Q(t_{n+1}, S_{t_{n+1}} | S_{t_n} = x]\right) \tag{4}$$

For each date, states x in which both terms in the max function are equal are points of the exercise frontier. Such a description of options of American type leads immediately to a backward induction algorithm for the evaluation.

2.2 Evaluation Through the Computation of the Exercise Frontier

The algorithm chosen for our grid implementation has initially been presented in [7], where a detailed explanation can be found. Following the backward induction presented in section 2.1, pricing is done through the explicit determination of the exercise frontier of our American option: each asset is successively considered to determine its own frontier which is function of the other assets. Fixed point algorithm, Good Latice Points and Monte Carlo simulations are used.

Let us briefly describe its structure in the case of an option on M underlying assets:

Step 1. Compute J Good Lattice points in dimension $M-1$.

Step 2. For each asset, compute the exercise frontier for every J points knowing its full expression for dates $t_{n+1}, \ldots, T$. This requires a fixed point search through a Monte Carlo valuation at each iteration.

Step 3. For each asset, compute an approximation of the frontier thanks to a polynomial regression.

Step 4. Repeat steps 2 and 3 for time $t_n = T-1, \ldots, 0$.

Step 5. Knowing the frontier at each date, compute the option value through a straightforward Monte Carlo simulation.

This algorithm seems to be well-adapted for an implementation on a grid architecture. Firstly, given a time t, the computation of the frontier value at one point is independent from the $J-1$ other computations. Secondly, the final Monte Carlo valuation is naturally adapted for a parallel implementation.

As a consequence, our implementation exports two types of computations towards distant machines: one program calculating the frontier on a given point (`EvalPtFront`) and one program of Monte Carlo simulation calculating the price of the option with N_p simulations given the complete frontier (`EvalFin`). Remaining calculus (generation of Good Lattice Points et polynomial regression) remain conveniently on the local side.

3 Three Grid Implementations

Once the topology of the grid is known, any implementation of our algorithm must successively transfer the executable towards the computing nodes, transfer all data needed for the computation, run the computation and retrieve the results. Our three implementations differ in the methods used to achieve these goals. Table 1 summarizes the main points.

Table 1. Comparative table of the implementations. Though implementations 1 and 2 use the same version of the middleware, implementations 2 and 3 are closer and share the "service" approach as well as the 1..n relationship for instantiation-invocation.

Implementation	Provisioning	Data Transfer	Security	Instantiation-Invocation
GT2.4 Basic	Grid-FTP	File transfer	SSL	1..1
GT2.4 Service	Grid-FTP	Socket streams	SSL	1..n
IGTv3 Web Service	Grid Archive deployment	Web Services protocols (SOAP)	not tested	1..n

3.1 Scheduler Oriented Globus Toolkit 2.4 Implementation

This first implementation uses the Globus tools in the simplest way. One job is one execution of one of our exported programs (see 2.2). Each program `<prog>` needs data from an input file `<prog>_in.txt` and writes its output in `<prog>_out.txt`. The first step is called provisioning and consists in transferring the executables towards the computation nodes. This is done using the Globus command `globus-url-copy`, the client tool of the Grid FTP module. This command is a wrapper for `globus_gass_copy`, which means we use the "Global Access to Secondary Storage" (GASS) for the executable. Scheduling is handled with scripts submitted to the command `globus-job-run` of the API "Globus Resource Allocation Manager" (GRAM), and file transfer are handled with the option `-stagein` et `-stageout`.

This first implementation is a basic use of Globus Toolkit 2.4: no modification of the original code is necessary. Note that the instantiation-invocation relation is unary: each data set needs a new instantiation of the executable to be treated[1].

3.2 Service Oriented Globus Toolkit 2.4 Implementation with Globus_io API

We try here to have a service oriented approach within the same Globus Toolkit 2.4 framework. The provisioning step remains unchanged, while the data transfer is enhanced. Our implementation doesn't copy the input from the client machines to the nodes and output files in the reverse sense. It makes benefits from the use of sockets managed through the functions of the globus_io API. This interface is a simple Globus wrapper of the usual POSIX sockets API. Commands are identical and written `globus_io_tcp_XXX` (accept, bind, connect, listen...). As a consequence, our executables can be considered as daemons permanently waiting for data sets to treat. Finally note that the instantiation-invocation relationship is no more unary, since each executable is up with only one GRAM call and then listen to the socket for successive data sets. As a consequence, we expect the GRAM API overhead to be reduced.

3.3 Grid Service Oriented Globus Toolkit 3 — IBM GridToolBox Implementation

In this implementation, we follow a grid service approach. The provisioning step is the deployment of our services. Our original code is compiled as a library, Java interfaces are compressed in GAR files (Grid Archive) and deployed with Globus 3 tools. Communications and invocations are handled through Web Services SOAP protocols.

This Globus Toolkit 3 implementation needs deeper modifications of the original codes. Besides the library of the original code and the Java interface, we need a description of our service in WSDD (Web Services Deployment Descriptor),

[1] The executable being cached, "instantiation" is to be understood as an API GRAM call and not as the provisioning itself.

a description of the interface in WSDL (Web Services Description Language) and a Java Archive of all the Globus 3 stubs of our services. The Ant tool of the Apache Software Foundation generates these files automatically (see [5] for more details).

4 Performance Comparison

4.1 Known Results

We valuate an American option on a maximum of five assets. This is a standard choice to test our implementations, especially since the work of Broadie and Glasserman [3]. We use exactly parameters proposed by [7]. Monte Carlo simulations for the frontier computation use 5000 paths, final Monte Carlo uses 4.000.000 paths.

At the time of their publication, known results were obtained after several hours of computation (about ten hours in November 2002 according to [7]). Note that all our implementations return the expected result with the precision announced by the authors of the method. This being checked, we will focus on computation performances, keeping in mind that our reference is 911 seconds for our local sequential implementation on a 2.8 GHz CPU.

4.2 Our Results

- **Globus Toolkit 2.4 with file transfer**
  Our testbed consists in one pool of three hyperthreaded [2] bi-processors Xeon 2.8 GHz machines on site (IBM Montpellier, France) and another pool of five 500 MHz hyperthreaded CPUs in the United States (IBM Poughkeepsie, USA).

  Tests of this implementation show that gains are very small: when six hyperthreaded on site processors are used, speedup is ... 2.47 ! Furthermore, the use of the two complete pools simultaneously doesn't improve the computation time at all, on the contrary. This is perfectly understandable if we consider network latency: this waiting time might be 30 seconds while the first submitted jobs require a few tenth to a few tens of seconds. In these conditions, it is obvious that distant CPUs won't improve anything[3].

[2] Hyperthreading, also called simultaneous multithreading, is a technology for processors microarchitecture allowing one CPU to treat simultaneously two streams of instructions and as a consequence to be seen by the operating system as *two* distinct processors, called logical processors. As many elements on the chipset are shared by the different streams, performances are obviously not expected to be as good as those obtained with two real processors.

[3] In a very general case of distant pools, an optimal architecture seems to appear here: the client scheduler should not communicate with distant machines, but should send all data sets to a distant replica, closer to the computing nodes, and which would redistribute the jobs at a lower time expense.

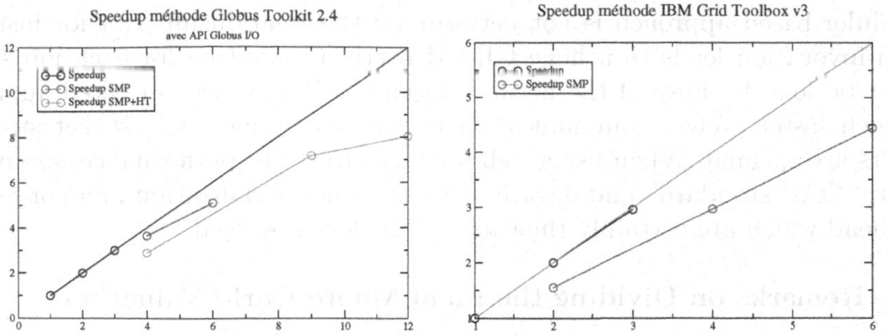

Fig. 1. Speedups for GT2.4 I/O and IGTv3 implementations on 6 processors

- **Globus Toolkit 2.4 with globus_io API**
  We have the same testbed. As shown above, it is obvious that distant machines will not be useful. This implementation allows numerous successive invocations of one computing instance. This is very efficient in bypassing the GRAM overhead: using 6 hyperthreaded CPUs, speedup is 8.06 (5.12 without hyperthreading) (cf. figure 1). Computation time has fallen to 120 seconds and speedup is close to the one obtained with MPI. When considering hyperthreading, it even seems to be better.
- **IBM Grid ToolBox — Globus Toolkit 3**
  Our testbed is in this case reduced to three 2 GHz bi-processors machines. Results are given on figure 1. We observe a curious loss when using simultaneously two processors on the same machine. This slowing effect on SMP is not understood for the moment.

4.3 Best Performances

Results are summarized on figure 2. As expected, the MPI implementation is our reference value which cannot be "beaten" (without hyperthreading). Our first

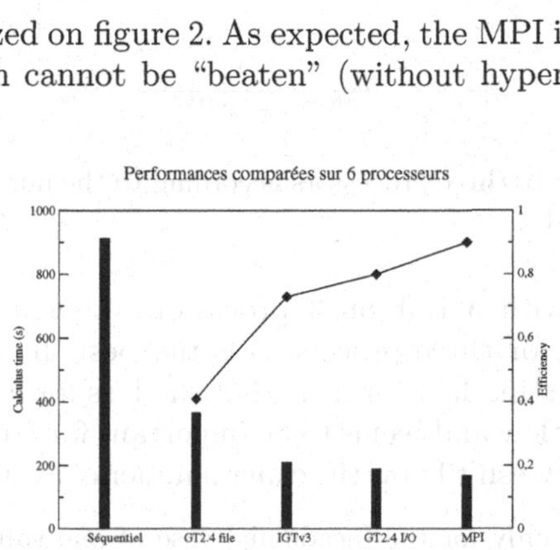

Fig. 2. Best results on 6 real processors. Remember that testbeds were different, so these results are corrected values after CPU comparison (one unitary test job is submitted and measured on each CPU). They are not measured values.

scheduler-based approach is not performant: the relationship $1..n$ for instantiation-invocation leads to a huge GRAM overhead (waiting for each job status might be long because of the network latency). The Service oriented approach is much faster. When communication is handled by globus_io socket streams, results are optimal. When using web services protocols, performances are not so good: SOAP standards and Java interfaces induce serialization and conversion overhead which are certainly the cause of the loss observed.

4.4 Remarks on Dividing the Final Monte Carlo Valuation

During the first parallelized step, the chosen parameters lead to the dispatching of 128 independent jobs and don't allow us to make another partition of the computational effort. So is this step perfectly parallelized if the grid has 128 processors or less. If more processors were available, we would choose to compute extra random points of our exercise frontier rather than leaving computational power unused.[4]

As for the second parallelized step, it is obvious that we have to divide our final Monte Carlo in as many jobs as we have available processors. Such a division optimize the computational time. If we have less jobs, some processors will remain unused. If we have more jobs, we will inefficiently increase the amount of communications, since CPUs will have to treat several jobs successively.

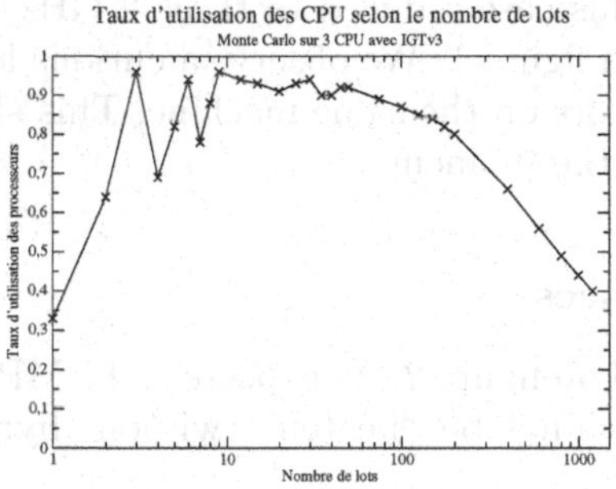

Fig. 3. Occupation rate of three processors according to the number of jobs for a Monte Carlo valuation. See 4.4.

We confirm this with a test on 3 processors on figure 3. Obvious results appeared: three jobs for three processors is the best division, four jobs is the worst one. The interesting fact for us is that the loss for a division in numerous jobs may remain very low and seems to be important for 50 jobs or more only. As a consequence, if one doesn't know the exact number of available CPU on the grid

[4] This remark applies only for the (academic) case of one valuation on a fully available grid of identical nodes. In the case of an implementation on a production site (a financial institution), there is a natural parallelization in quantity: N traders requesting P valuations would launch N*P*128 jobs EvalPtfront at the same time. No adaptation would be required.

(which will always be the case on a grid), it is better to divide the computation on a number of jobs larger than the expected number of CPUs. In our example, computation will be made on a dozen of processors and we decide to divide our final Monte Carlo of 4000000 simulations in 20 jobs of 20000 simulations each.

5 Conclusion

Finally, we showed that Monte Carlo algorithms for financial applications may be easily adapted to a Grid architecture using standard middleware. Flexibility for standard parallelization is shown to be efficient. Observed performances of "Service oriented" implementations are very promising, as the time loss compared to an optimized MPI program is limited. We also showed that "Web Services" standardisation may be too constraining in terms of high performance computing. Finally, note that such speedups allow us to test deeply implemented algorithm and parameters sensibility and obtain essential financial results.

References

1. Avramidis T., Zinchenko Y., Coleman T., Verma A., *Efficiency Improvements for Pricing American Options with a Stochastic Mesh: Parallel Implementation*, Financial Engineering News, no. 19, 2000
2. Broadie M. and P. Glasserman, *Monte Carlo Methods for Pricing High-Dimensional American Options: an Overview*, Columbia University, 1997
3. Broadie M. and P. Glasserman, *A Stochastic Mesh Method for Pricing High-Dimensional American Options*, Columbia University, 1997
4. Fu M.C. et al., *Pricing American Options: A Comparison of Monte Carlo Simulation Approaches*, University of Maryland, 2000
5. Girard J.Y., Muni Toke I., Lepesant J.L., Prost J.P., *Deploy a C application as a grid Service: How to easily implement a Grid Service from an existing application*, IBM developerWorks, 2004
6. Haber S., *Parameters for Integrating Periodic Functions of Several Variables*, Mathematics of Computation, 41, 115–129, 1983
7. Ibanez A. and Zapatero F., *Monte Carlo Valuation of American Options Through computation of the Optimal Exercise Frontier*, University of Southern California, 2002
8. Longstaff F.A. and Schwartz E.S., *Valuing American Options by Simulation: A Simple Least-Square Approach*, The Review of Financial Studies, 2001
9. *www-unix.globus.org/toolkit*

An Efficient and Easily Parallelizable Algorithm for Pricing Weather Derivatives

Yusaku Yamamoto

Department of Computational Science & Engineering,
Nagoya University, Nagoya, 464-8603, Japan
yamamoto@na.cse.nagoya-u.ac.jp
http://www.na.cse.nagoya-u.ac.jp/~yamamoto/

Abstract. We present a fast and highly parallel algorithm for pricing CDD weather derivatives, which are financial products for hedging weather risks due to higher-than average temperature in summer. To find the price, we need to compute the expected value of its payoff, namely, the CDD weather index. To this end, we derive a new recurrence formula to compute the probability density function of the CDD. The formula consists of multiple convolutions of functions with a Gaussian distribution and can be computed efficiently with the fast Gauss transform. In addition, our algorithm has a large degree of parallelism because each convolution can be computed independently. Numerical experiments show that our method is more than 10 times faster than the conventional Monte Carlo method when computing the prices of various CDD derivatives on one processor. Moreover, parallel execution on a PC cluster with 8 nodes attains up to six times speedup, allowing the pricing of most of the derivatives to be completed in about 10 seconds.

1 Introduction

Most business activities are greatly affected by weather conditions, especially temperature. For example, higher-than-average temperature in winter will result in significant energy cost savings for department stores or railway companies, but may decrease the revenue of electricity companies. Lower-than-average temperature in summer will reduce the cost of air-conditioning, but may adversely affect the sales of leisure industry.

To manage these risks associated with weather, new financial instruments called weather derivatives have been developed and are actively traded in the market. Among them, the most popular variants are the CDD (Cooling Degree Days) and HDD (Heating Degree Days) derivatives. The CDD derivative is a derivative security whose payoff depends on the CDD, a weather index that measures the extent to which the temperature in a specified period in summer is higher than a reference temperature. Thus a company exposed to weather risks due to higher-than-normal temperature in summer can buy a CDD derivative to compensate for the possible loss. The HDD derivative is defined similarly and enables the buyer to hedge against lower-than-average temperature in winter.

I. Lirkov, S. Margenov, and J. Waśniewski (Eds.): LSSC 2005, LNCS 3743, pp. 470–477, 2006.

To find the rational price of this derivative, one needs to compute the expectation value of the payoff under some stochastic time series model of the future temperature [4, 6, 12]. This has traditionally been done using the Monte Carlo method [5, 7], but, as it is well known, its convergence is quite slow. In fact, computing the price of a weather derivative to 4-digit accuracy often requires more than 10^8 sample paths and takes several minutes on a modern PC. So it is not suitable for real-time pricing or pricing a portfolio consisting of thousands of weather derivatives.

In this paper, we propose an alternative efficient algorithm for pricing CDD weather derivatives [11]. Our strategy is similar to the one successfully adapted to path-dependent options in [2, 3]; we consider the probability distribution function of the CDD explicitly and derive a recursion formula to compute it. The formula consists of multiple convolutions of functions with a Gaussian distribution and can be computed efficiently using the fast Gauss transform [8]. The resulting algorithm has proved much faster and more accurate than the Monte Carlo method. In addition, it is quite easy to parallelize the algorithm.

This paper is structured as follows: in Section 2, we formulate the pricing problem of CDD derivatives. Section 3 and 4 present our new algorithm based on the fast Gauss transform and a parallelization strategy for it, respectively. Numerical results that show the effectiveness of our approach are given in Section 5. Section 6 gives some conclusion.

2 Problem Formulation

We consider a weather derivative whose payoff depends on the daily temperatures at a city during a given period of N days. Let the temperature on the n-th day be denoted by T_n. Then the CDD weather index is defined as

$$CDD = \sum_{n=1}^{N} \max(0, T_n - \bar{T}), \tag{1}$$

where $\bar{T}$ is the prespecified reference temperature. The CDD weather derivative is a derivative security whose payoff is given by

$$P_{call} = k \cdot \max(CDD - K, 0) \quad \text{(CDD call)} \quad \text{or} \tag{2}$$

$$P_{put} = k \cdot \max(K - CDD, 0) \quad \text{(CDD put)}, \tag{3}$$

where k and K are constants called the tick value and the strike, respectively.

To find the price of this derivative, we need some stochastic model for predicting $\{T_n\}_{n=1}^{N}$. Many models have been proposed for this purpose so far [4, 6, 12]. In this paper, we use the Dischel D1 model [6], in which the daily temperature is assumed to evolve the equation:

$$T_n = (1-\beta)\Theta_n + \beta T_{n-1} + \epsilon_n, \tag{4}$$

where β is a constant, Θ_n is the temperature of the n-th day in an average year (also a constant) and ϵ_n's are a sequence of i.i.d. random variables that follow the normal distribution $N(\mu, \sigma^2)$. Though it is a simple model, it can reproduce the time series of daily temperatures fairly well and is widely used for pricing weather derivatives.

The price Q of the CDD derivative is computed by taking the expectation value of the payoff under the stochastic model (4) and adding to it some premium e determined by the issuer. That is, for a CDD call,

$$Q = E[P_{call}] + e. \tag{5}$$

Our main task is to construct an algorithm for computing this expectation value efficiently and accurately.

3 A New Pricing Algorithm Based on the Fast Gauss Transform

3.1 The Basic Idea

To compute the expectation value in eq. (5), we first define the partial CDD on the n-th day by

$$C_n = \sum_{i=1}^{n} \max(0, T_i - \bar{T}). \tag{6}$$

The CDD is equal to C_N by definition.

Let $P_n(T_n|T_{n-1})$ denote the conditional probability density function (pdf) of T_n given the temperature of the $(n-1)$-th day. Then from eq. (4) we have

$$P_n(T_n|T_{n-1}) = \frac{1}{\sqrt{2\pi}\sigma} \exp\left\{-\frac{(T_n - \mu_n)^2}{2\sigma^2}\right\}, \tag{7}$$

where

$$\mu_n = (1 - \beta)\Theta_n + \beta T_{n-1} + \mu. \tag{8}$$

We next consider the joint conditional pdf $p_n(T_n, C_n|T_{n-1}, C_{n-1})$. Eq. (6) implies that $C_n = C_{n-1} + (T_n - \bar{T})$ if $T_n \geq \bar{T}$ and $C_n = C_{n-1}$ otherwise. Hence,

(i) if $T_n < \bar{T}$,

$$\begin{aligned} p_n(T_n, C_n|T_{n-1}, C_{n-1}) &= P_n(T_n|T_{n-1})\delta(C_n - C_{n-1}) \\ &= \frac{1}{\sqrt{2\pi}\sigma} \exp\left\{-\frac{(T_n - \mu_n)^2}{2\sigma^2}\right\} \delta(C_n - C_{n-1}), \end{aligned} \tag{9}$$

(ii) else if $T_n \geq \bar{T}$,

$$\begin{aligned} p_n(T_n, C_n|T_{n-1}, C_{n-1}) &= P_n(T_n|T_{n-1})\delta(C_n - (C_{n-1} + (T_n - \bar{T}))) \\ &= \frac{1}{\sqrt{2\pi}\sigma} \exp\left\{-\frac{(T_n - \mu_n)^2}{2\sigma^2}\right\} \\ &\quad \times \delta(C_n - (C_{n-1} + (T_n - \bar{T}))). \end{aligned} \tag{10}$$

Using eqs. (9) and (10), we can compute the joint pdf of T_n and C_n as follows:

(i) if $T_n < \bar{T}$,

$$\begin{aligned} & p_n(T_n, C_n) \\ &= \int_{-\infty}^{+\infty} dT_{n-1} \int_0^{+\infty} dC_{n-1}\, p_n(T_n, C_n | T_{n-1}, C_{n-1})\, p_{n-1}(T_{n-1}, C_{n-1}) \\ &= \int_{-\infty}^{+\infty} dT_{n-1} \frac{1}{\sqrt{2\pi}\sigma} \exp\left\{ -\frac{(T_n - \mu_n)^2}{2\sigma^2} \right\} \times p_{n-1}(T_{n-1}, C_n), \end{aligned} \tag{11}$$

(ii) else if $T_n \geq \bar{T}$,

$$\begin{aligned} & p_n(T_n, C_n) \\ &= \int_{-\infty}^{+\infty} dT_{n-1} \int_0^{+\infty} dC_{n-1}\, p_n(T_n, C_n | T_{n-1}, C_{n-1})\, p_{n-1}(T_{n-1}, C_{n-1}) \\ &= \int_{-\infty}^{+\infty} dT_{n-1} \frac{1}{\sqrt{2\pi}\sigma} \exp\left\{ -\frac{(T_n - \mu_n)^2}{2\sigma^2} \right\} \times p_{n-1}(T_{n-1}, C_n - (T_n - \bar{T})). \end{aligned} \tag{12}$$

We use eqs. (11) and (12) as recurrence formulas to compute the joint pdf $p_N(T_N, C_N)$ from the initial joint pdf:

$$p_1(T, C) = \begin{cases} \delta(T - T_1)\, \delta(C) & (T_1 < \bar{T}), \\ \delta(T - T_1)\, \delta(C - (T_1 - \bar{T})) & (T_1 \geq \bar{T}). \end{cases} \tag{13}$$

Once $p_N(T_N, C_N)$ has been obtained, we can compute the expectation value of the payoff as

$$\begin{aligned} E[P_{call}] &= \int_{-\infty}^{+\infty} dT_N \int_0^{+\infty} dC_N\, P_{call}\, p_N(T_N, C_N) \\ &= \int_{-\infty}^{+\infty} dT_N \int_0^{+\infty} dC_N\, k \cdot \max(C_N - K, 0)\, p_N(T_N, C_N). \end{aligned} \tag{14}$$

Thus we have established an algorithm that computes the price of a CDD derivative based on the recurrence formulas of the joint pdf of T_N and C_N.

3.2 Acceleration by the Fast Gauss Transform

To evaluate the integrals (11) and (12), we discretize T and C with step size h. Let $T^i \equiv \bar{T} + ih$ and $C^j \equiv jh$ and denote the joint pdf at the (i, j)-th grid point on the n-th day by $p_n^{i,j} \equiv p_n(T^i, C^j)$. Then the discrete version of the recursion formula (11) and (12) can be written as

$$p_n^{i,j} = \begin{cases} \tilde{p}_n^{i,j} & (T_1 < \bar{T}), \\ \tilde{p}_n^{i,j-i} & (T_1 \geq \bar{T}), \end{cases} \tag{15}$$

$$(-M_T \leq i \leq j,\ 0 \leq j \leq M_C)$$

and

$$\tilde{p}_n^{i,j} = \sum_{i'=-M_T}^{M_T} \frac{w^{i'}}{\sqrt{2\pi}\sigma} \exp\left\{ -\frac{(T^i - \mu_n^{i'})^2}{2\sigma^2} \right\} p_{n-1}^{i',j}, \qquad (16)$$

$$(-M_T \leq i \leq M_T, \ 0 \leq j \leq M_C)$$

where $w^{i'}$ is the weight of the quadrature formula for the sample point $T^{i'}$ and

$$\mu_n^{i'} = (1-\beta)\Theta_n + \beta T^{i'} + \mu. \qquad (17)$$

Note also that we assumed that the joint pdf can be neglected outside the area of $\bar{T} - M_T h \leq T \leq \bar{T} + M_T h$ and $0 \leq C \leq M_C h$, where M_T and M_C are some positive integers.

For a fixed value of j, the direct computation of $\tilde{p}_n^{i,j}$ $(i = -M_T, \ldots, M_T)$ via eq. (16) requires $O(M_T^2)$ operations. However, noting that (16) has the form of discrete convolution of a sequence with a Gaussian distribution, we can apply the fast Gauss transform proposed by Greengard and Strain [8, 10, 1]. This will reduce the cost of each convolution to $O(M_T)$ and the total computational cost at each time step to $O(M_T M_C)$. This is the outline of our algorithm. Note that the use of the fast Gauss transform has proved successful in the pricing of various types of financial derivatives such as American options [2] and path dependent options [3].

In practice, the initial condition (13) has a δ-function like singularity, so we have to separate this part and integrate it analytically. The details are given in our former paper [11], which details the sequential version of our algorithm. With this modification, if $M_T \sim M_C \sim M$ and the Simpson's rule [9] is used as the quadrature rule, it can be shown that our algorithm has a pricing error of $O(1/M^2)$ [11]. Because the computational time is $O(M^2)$ in this case, it follows that the error E decreases with the computational time τ as $E = O(1/\tau)$. In contrast, for the MC method, the error decreases only as $E = O(1/\sqrt{\tau})$. Thus we can expect that our algorithm is asymptotically faster and more accurate than the MC method.

4 Parallelization

As it is clear from eq. (16), the computation of $\tilde{p}_n^{i,j}$'s for different values of j can be done completely independently. We use this parallelism to execute our algorithm on a distributed-memory parallel machine. More specifically, we distribute the 2-dimensional arrays $p_{n-1}^{i',j}$ and $\tilde{p}_n^{i,j}$ among the processors using block cyclic partitioning (with block size L) in the j-direction. For each time step n, in the computation phase, each processor computes the convolution in eq. (16) for the values of j allocated on it. In the communication phase, the processors exchange the results of the computing phase according to eq. (15) so that each processor can get the correct part of $p_n^{i,j}$ needed to perform the computing phase at the next time step.

When the number of processors is R, the computational work for each processor is $O(M_T M_C/R)$, while the amount of data sent or received by each processor is also $O(M_T M_C/R)$. However, since the constant behind O in the former is fairly larger than that in the latter, our algorithm is expected to achieve good speedup unless the speed of data transfer is much slower than the speed of computation.

5 Numerical Results

We implemented our algorithm using the C language and MPI on a Linux PC cluster and evaluated its performance. Each node of the cluster consists of a 533 MHz Alpha 21164A processor and 128 MB of memory and the nodes are connected via 100-Base T network. We used GNU C++ compiler and MPICH.

As the target problems, we used CDD call derivatives with monitoring period $N = 5$, 10 and 20 and $\bar{T}$=24(C). The other parameters are as follows:

$$k = 1(\$/\mathrm{C}),\ \ T_0 = 24(\mathrm{C}),\ \ \beta = 0.7763,\ \ \mu = 0.0896 \text{ and } \ \sigma = 2.3734. \qquad (18)$$

The values of β, μ and σ were obtained by fitting the model (4) to Tokyo's daily temperature by the least squares method [11]. The strike temperatures are $K = 10$, 20 and 40(C) for the case of $N = 5$, 10 and 20, respectively.

5.1 Comparison with the Monte Carlo Method

First, we compared the execution time and accuracy of our method with that of the Monte Carlo method. The results for $N = 5$, 10 and 20 are shown in

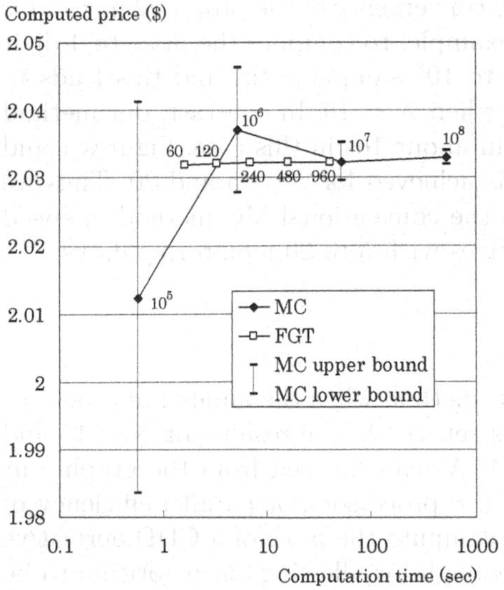

Fig. 1. Convergence of the MC method and the proposed method ($N = 5$)

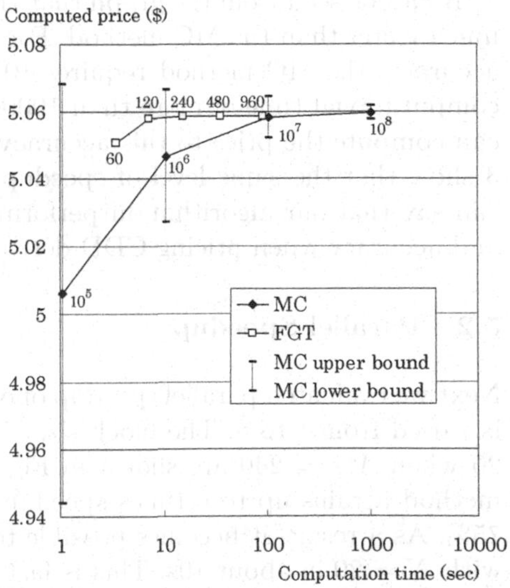

Fig. 2. Convergence of the MC method and the proposed method ($N = 10$)

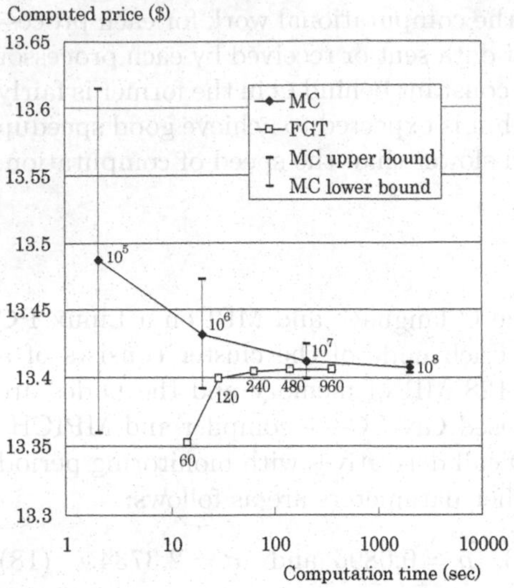

Fig. 3. Convergence of the MC method and the proposed method ($N = 20$)

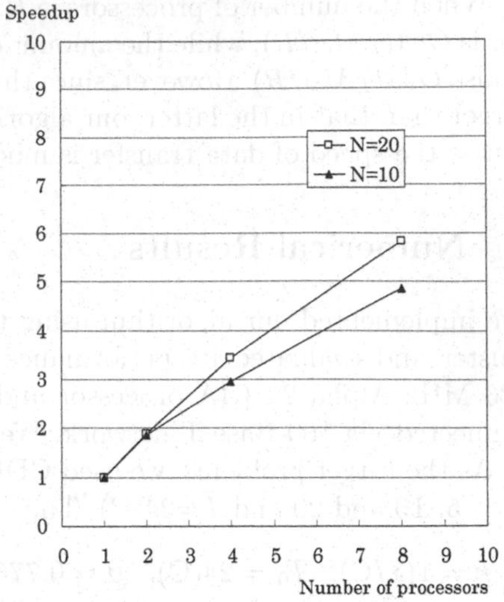

Fig. 4. Parallel speedup of the proposed method ($M_T = 240$)

Figures 1, 2 and 3, respectively. The numbers of sample paths for the MC method and the values of M_T for our method are also shown in the graph. M_C is set to $2M_T$ when $N = 5$ and 10 and to $4M_T$ when $N = 20$, while h is set to $60/M_T$. The vertical lines in the graph denote 95% confidence interval for the MC results.

It can be seen from the graph that the convergence of the proposed method is much faster than the MC method. For example, to compute the price to 4-digit accuracy, the MC method requires 10^7 to 10^8 sample paths and this leads to computational time of more than 1000s when $N = 10$. In contrast, our method can compute the price to this accuracy in about 16s in this case. Figures 1 and 3 show that the same level of speedup is achieved for $N = 5$ and 20. Thus we can say that our algorithm outperforms the conventional MC method in speed and accuracy when pricing CDD derivatives with 5 to 20 monitoring dates.

5.2 Parallel Speedup

Next we evaluated parallel speedup of our method when the number of processor is varied from 1 to 8. The block size L is set to 10. The results for $N = 10$ and 20 when $M_T = 240$ are shown in Fig. 4. As can be seen from the graph, our method attains up to 6 times speedup on 8 processors, or parallel efficiency of 75%. As a result, it becomes possible to compute the price of a CDD derivative with $N = 20$ in about 10s. This is fast enough for allowing our algorithm to be used for real time pricing. Furthermore, the parallel efficiency could be improved by overlapping the communication needed to distribute the 2-dimensional array $p_n^{i,j}$ with the computation of eq. (16). We are now investigating such possibility.

6 Conclusion

In this paper, we proposed a fast and highly parallel algorithm for pricing CDD weather derivatives. To compute the expectation value of the payoff, our algorithm uses a recursion formula consisting of multiple convolutions of functions with a Gaussian distribution. These convolutions can be computed efficiently using the fast Gauss transform. Moreover, parallelization is easy because each convolution can be computed independently.

Numerical experiments show that our method is considerably faster and more accurate than the conventional Monte Carlo method when pricing CDD weather derivatives with 5 to 20 monitoring dates on a single processor. On a PC cluster with 8 processors, our method attained up to 6 times speedup and computed the price of a CDD derivative with 20 monitoring dates in about 10 seconds. Thus our algorithm opens a way to applications that have been impractical due to long computational time, such as real time pricing or pricing of a portfolio consisting of thousands of weather derivatives.

Future work includes application to other types of weather derivatives and enhancement of the parallel efficiency by various optimization techniques.

References

1. Baxter, B., Roussos, G.: A New Error Estimate of the Fast Gauss Transform. SIAM Journal on Scientific Computing **24** (2002) 257–259
2. Broadie, M., Yamamoto, Y.: Application of the Fast Gauss Transform to Option Pricing. Management Science **49** (2003) 1071–1088
3. Broadie, M., Yamamoto, Y.: A Double-Exponential Fast Gauss Transform Algorithm for Pricing Discrete Path-Dependent Options. Operations Research **53** 5 (2005) 764–779
4. Cao, M., Wei, J.: Pricing Weather Derivative: An Equilibrium Approach. Department of Economics, Queen's University, Kingston, Ontario, Working Paper (1999)
5. http://www.climetrix.com
6. Dischel, B.: The D1 Stochastic Temperature Model for Valuing Weather Futures and Options. Applied Derivatives Trading (1999)
7. http://www.fea.com/web/home
8. Greengard, L., Strain, J.: The Fast Gauss Transform. SIAM Journal on Scientific and Statistical Computing **12** (1991) 79–94
9. Press, W., Flannery, B., Teukolsky, S., Vetterling, W.: Numerical Recipes in FORTRAN. Cambridge University Press (1992)
10. Strain, J.: The Fast Gauss Transform with Variable Scales. SIAM Journal on Scientific and Statistical Computing **12** (1991) 1131–1139
11. Yamamoto, Y., Egi, M.: Valuation of Weather Derivatives Using the Fast Gauss Transform (in Japanese). Journal of the Information Processing Society of Japan **45** (2004) 176–185
12. Zeng, L.: Pricing Weather Derivatives. The Journal of Risk Finance **1** (2000) 72–78

Part IX

Environmental Modelling

On the Sulphur Pollution over the Balkan Region*

Hristo Chervenkov, Dimiter Syrakov, and Maria Prodanova

National Institute of Meteorology and Hydrology, Bulgarian Academy of Sciences,
66 Tzarigradsko chaussee, 1784 Sofia, Bulgaria
hristo.tchervenkov@meteo.bg, dimiter.syrakov@meteo.bg

Abstract. The EMAP (Eulerian Model for Air Pollution) model is used to estimate the sulphur pollution over the Balkan region for 1995. A subdomain of the standard EMEP grid is chosen containing all 12 Balkan countries. The computational grid in this domain has a space step of 25 km, twice finer than the EMEP grid. The meteorological input to EMAP is the operational DWD "Europa-Model" product. This information is processed in a special way as to obtain input information to the PBL model YORDAN built in EMAP. The source input is the official sulphur emission data. This information is processed further as to obtain 25x25 km data for both Large-Point and Area sources. The calculations are made month by month having the last moment fields from the previous month as initial conditions for the next month. The boundary conditions are set to zero so the influence of the other European sources is not accounted for in this study. According to the EMEP methodology multiple runs are made setting every time the sources of various countries to zero. The impact of every Balkan country in sulphur pollution of all other countries for 1995 is estimated and commented. The results of calculation are compared with measurements.

Keywords: Air pollution, dispersion modelling, PBL- model, sulphur dioxideblame matrix.

1 Introduction

The EMAP model is used to estimate the sulphur deposition over Southeast Europe for 1995 due to sources from 12 countries: Albania (AL), Bosnia and Herzegovina (BH), Bulgaria (BG), Croatia (HR), Greece (GR), Hungary (HU), Moldavia (MO), Romania (RO), Slovenia (SL), Serbia and Montenegro (YU), the FYR Macedonia (MK) and the bigger part from Turkey (TR). As only sources from these countries are handled, the results can be considered as an estimate of their impact on the acid pollution of the region as well as an estimate of the reciprocal pollution.

* The present work is supported by the BULAIR project (EVK2-CT-2002-80024) under FP5 and ACCENT Network of Excellence (GOCE-CT-2002-500337) under FP6.

I. Lirkov, S. Margenov, and J. Waśniewski (Eds.): LSSC 2005, LNCS 3743, pp. 481–489, 2006.

2 Short Description of the EMAP Model

EMAP [2, 4] is a simulation model that allows describing the dispersion of multiple pollutants. The processes of horizontal and vertical advection, horizontal and vertical diffusion, dry deposition, wet removal, gravitational settling (aerosol version) and the simplest chemical transformation (sulphur version) [12] are accounted for in the model. Within EMAP, the semi-empirical diffusion-advection equation for scalar quantities is treated. The governing equations are solved in terrain-following coordinates. Non-equidistant grid spacing is settled in vertical directions. The numerical solution is based on discretization applied on staggered grids using the splitting approach. Conservative properties are fully preserved within the discrete model equations. Advective terms are treated with the TRAP scheme, which is a Bott-type one. Displaying the same simulation properties as the Bott scheme (explicit, conservative, positive definite, transportability, limited numerical dispersion), the TRAP scheme proves to be several times faster [5, 7]. The advective boundary conditions are zero at income flows and "open boundary" - at outcome ones. Turbulent diffusion equations are digitized by means of the simplest schemes — explicit in horizontal, and implicit in vertical direction. The bottom boundary condition for the vertical diffusion equation is the dry deposition flux, the top boundary condition is optionally "open boundary" and "hard lid" type. The lateral boundary conditions for diffusion are "open boundary" type. In the surface layer (SL), a parameterization is applied permitting to have the first computational level at the top of SL. It provides a good estimate for the roughness level concentration and accounts also for the action of continuous sources on the earth surface [9]. A similarity theory based PBL model [10] is built in the model producing 3D velocity and turbulence fields on the base of minimum meteorological information — the wind and temperature at geostrophic level and the surface temperature. The model is evaluated and validated during the ETEX-II intercalibration study - ranged 9th among 34 models [8]. It is validated on the database of 1996 EMEP/MSC-E intercalibration of heavy metal models [6].

3 Model Domain, Parameterisation and Input Data

The aim of this modeling is to estimate the sulphur pollution in the region of Southeast Europe, taking a territory of 38x28 EMEP 50x50 km^2 grid cells with Bulgaria in the center (see figure 1). Every cell is divided to four 25x25 km^2 cells. The chosen territory includes entirely all 12 countries of interest and partly other territories. In the created versions of EMAP, a 5-layer vertical structure is used. The first four layers have representative levels at 50, 200, 650 and 1450 m with layer boundaries 20-100, 100-375, 375-995, 995-1930 m. The 5th layer accounts parametrically for the free atmosphere. The volume of this layer is so big that the concentration tends to be zero there, although it can contain some mass. Two kinds of input data are necessary for EMAP performance: sources and meteorological data.

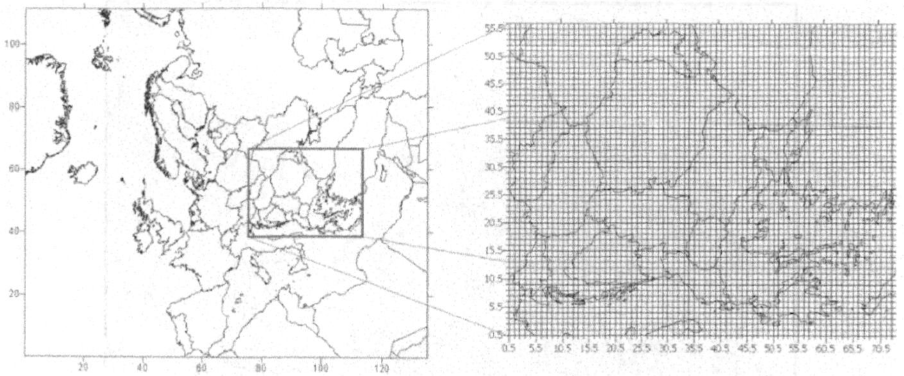

Fig. 1. Nesting the model domain in the EMEP-domain

3.1 Emissions

The sources are determined through an emission inventory based on the CORINAIR methodology [11]. They correspond to the official 50x50 km^2 data reported by the corresponding governmental authorities to EMEPs MSC-West and can be downloaded from its web site (http://www.emep.int/). Additional redistribution of these data is made over the finer grid of 25-km space resolution. The emissions are provided in mass units per second. All sources are divided in two classes: high sources (like high and very strength industrial stacks etc.) called Large Point Sources (LPS) and area sources (AS) — the sum of all low and diffusive sources in the given grid cell. As all LPS are supplied with high stacks, the emission of these sources is prescribed to be released in layer 2, i.e., between 100 and 375 m.. In previous works [1] data concerning emissions obtained through the EMEP inventarization scheme before 2000 have been used, where the sources are divided only into two groups: AS and LPS. On the territory of Turkey, for example, there were gaps both in the data for AS, and LPS. In this paper the much more sophisticated present scheme of EMEP has been applied. In it the sources are divided in 11 SNAP (Selected Nomenclature for reporting of Air Pollutants) sectors and the data gaps are filled. Each sector is treated as AS or LPS. The monthly emissions are obtained using the following dimensionless annual variation coefficients, recommended by MSC-E. The distribution of the AS and the LPS in the model domain is shown in Fig. 2 and Fig. 3.

3.2 Meteorological Data

An important advantage of the used model is that, due to the built in PBL model, it utilizes only numerical analysis and forecast data from the world weather centers, distributed via the Global Telecommunication System of the World Meteorological consists of the sequence of analyzed $U_{850}, V_{850}, T_{850}$ and T_{surf} fields and 6-hour forecast for precipitation from the standard 50x50 km^2 output of the former " Europa-Model " of the German Met. Service (DWD). On

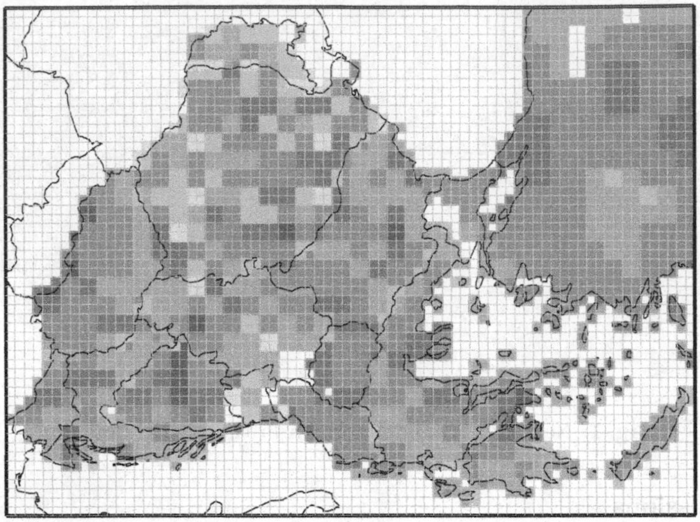

Fig. 2. Distribution and strength (unit: $\mu g(S)/s$) of the area sources in the model domain 10 % black: below 10^4, 20% black: $10^4 - -10^5$; 30% black: $10^5 - -10^6$; 40% black: $10^6 - -10^7$; 50% black $10^7 - -10^8$, black over 10^8)

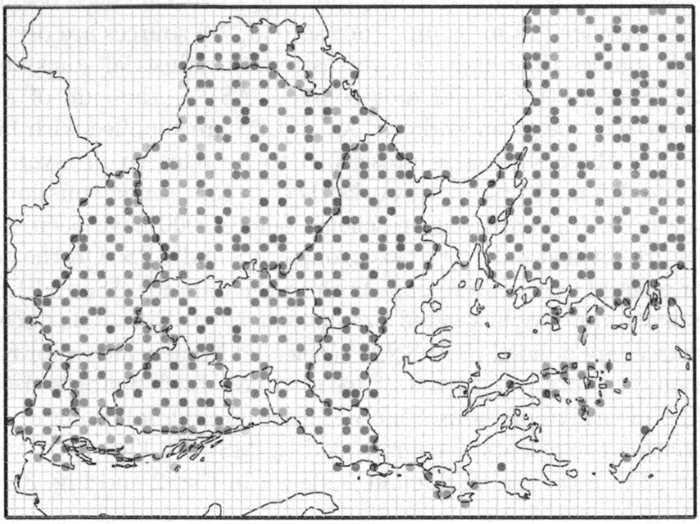

Fig. 3. Same as Fig. 2, but for the LPS

this base, the PBL model calculates U-, V-, W- and K_z- profiles at each grid point. The roughness and the Coriolis parameter fields are pre-set additional input to the PBL model. Orography height, surface type (sea-land mask) and roughness height are to be provided for each grid location. Initial concentration field is optionally introduced (spin-up fields).

4 Calculation Results

Monthly runs with the above mentioned sources are performed. The output consists of the following fields: monthly dry (DD), wet (WD) and total (TD) depositions, monthly concentration mean in air (CA) and additionally, monthly sum of the precipitation (CP). Then, it is calculated the annual total deposition (Figure 4), the mean annual surface concentration in air (Figure 5) and the mean concentration in the precipitation (Figure 6). All these pictures are quite similar: The most polluted area is in Southeast Bulgaria, the place around the most powerful source -the thermal power plant "Maritsa – Iztok". In its vacinity the total deposition reaches $18,9g/m^2$, the concentration in air $9\mu g(S)/m^3$), and the concentration in pracipitation 21,9 mg/l. In Table 1, the deposition budget matrix for oxidised sulphur is presented. It is obtained after multiplication of the total deposition field to array containing the distribution of the territory of each country in the model domain. It shows the impact of each country to the sulphur pollution of the other countries. The last three rows shows respectively the total deposited quantity due to the sources from the country in the column header, the

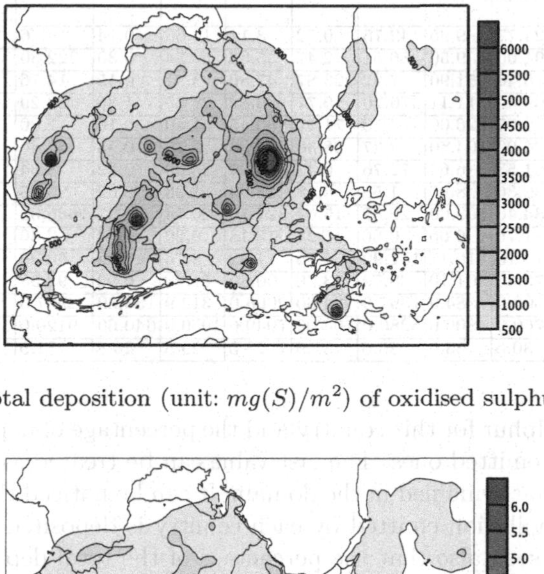

Fig. 4. Total deposition (unit: $mg(S)/m^2$) of oxidised sulphur in 1995

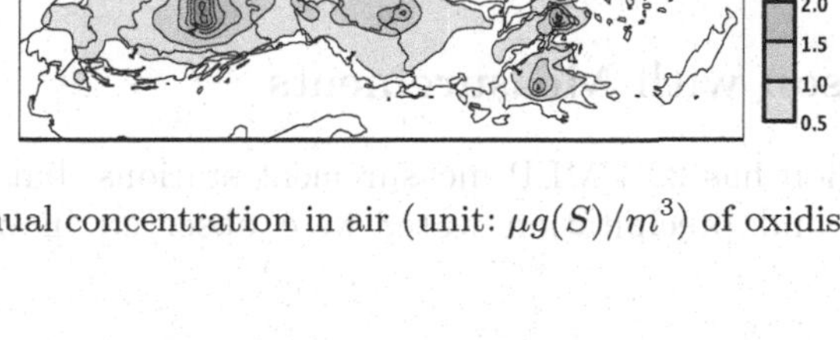

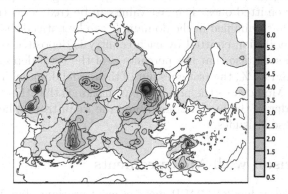

Fig. 5. Mean annual concentration in air (unit: $\mu g(S)/m^3$) of oxidised sulphur in 1995

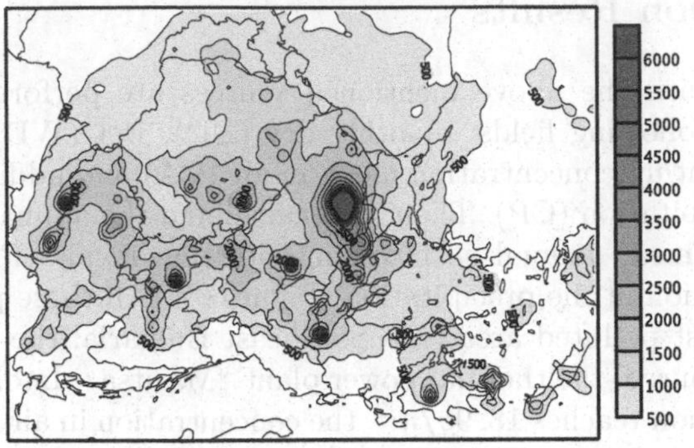

Fig. 6. Mean annual concentration in the precipitation (unit: $\mu g(S)/l$) of oxidised sulphur in 1995

Table 1. Deposition budget matrix (unit 100 t). The diagonal elements show the deposition quantity for each country due to its own sources.

emiter / receiver	AL	BG	BH	GR	HR	HU	MK	MO	RO	SL	TR	YU
AL	73.13	24.77	9.95	42.75	0.72	3.11	16.63	0.14	4.76	0.71	0.59	10.66
BG	9.67	2091.00	39.50	56.51	2.47	35.86	50.24	5.30	322.30	3.95	20.66	98.79
BH	4.69	15.46	551.90	7.22	23.87	47.80	3.31	0.15	14.76	14.98	0.20	81.92
GR	21.08	309.40	11.11	470.20	0.77	6.82	28.35	1.35	29.20	1.09	12.70	15.11
HR	1.58	9.97	126.00	3.73	71.27	50.45	1.40	0.10	7.80	40.56	0.14	36.03
HU	2.09	26.09	104.80	4.37	28.38	579.30	3.15	0.53	62.40	37.36	0.44	84.26
MK	21.27	81.53	6.64	72.76	0.41	4.07	85.09	0.22	8.34	0.49	0.97	13.79
MO	0.47	32.36	8.56	1.58	0.68	13.94	1.08	38.62	95.85	1.24	1.46	12.45
RO	8.88	434.40	208.60	22.35	15.64	324.10	20.54	32.27	1685.00	22.23	8.11	359.10
SL	0.21	1.76	9.60	0.86	15.25	13.18	0.20	0.04	2.30	76.55	0.02	5.75
TR	4.63	313.10	10.17	151.90	0.76	11.32	8.76	3.32	60.26	1.51	905.10	16.93
YU	31.17	176.30	301.10	38.73	13.06	100.80	51.49	0.92	93.58	11.44	1.53	568.30
tot. dep.	222.70	4503.0	1584.0	1397.0	224.60	2006.0	312.9	133.30	2907.0	323.6	1160.00	1487.0
tot. emitt.	720.00	14760.0	4800.0	5280.0	704.00	7049.6	1050.0	640.60	9120.0	1250.0	5461.66	4620.0
%	30.9	30.5	33.0	26.5	31.9	28.5	29.8	20.8	31.9	25.9	21.2	32.2

total emitted sulphur for this country and the percentage of deposited quantities from the yearly emitted ones. The last value can be treated as the relative part of the sulphur that remained in the domain. It can be noticed that the main part of the sulphur pollution emitted by each country is deposited over the country itself. It can be seen also that the percentage of the total deposited quantity is between 20 % and 35 %, the rest goes out of the model domain. The minimum of this value is for Moldova — the possible explanation is that this territory is close to the east border of the model domain and the main tropospheric transport is west east in these latitudes.

5 Comparison with Measurements

The Balkan Region has 22 EMEP measurement stations, but in 1995 sulphur products in air and precipitation water were observed in only 11 of them

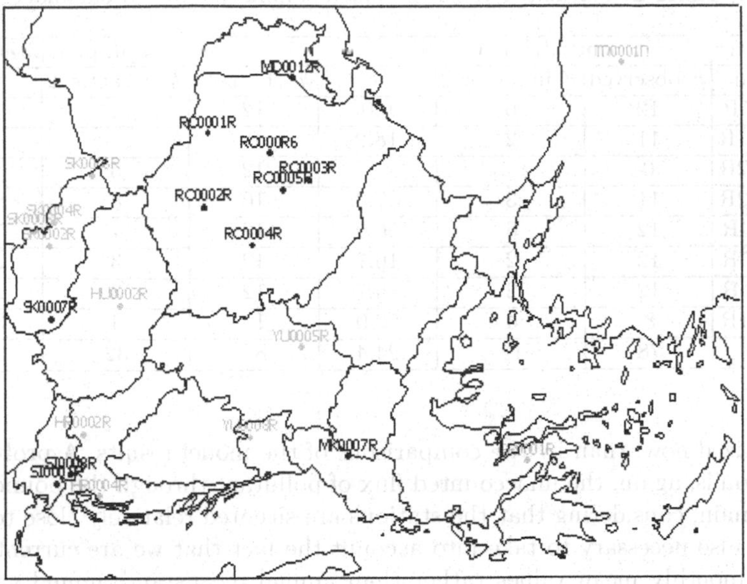

Fig. 7. Position of the EMEP-stations in the model domain. The stations with measurements of sulphur products are shown in grey.

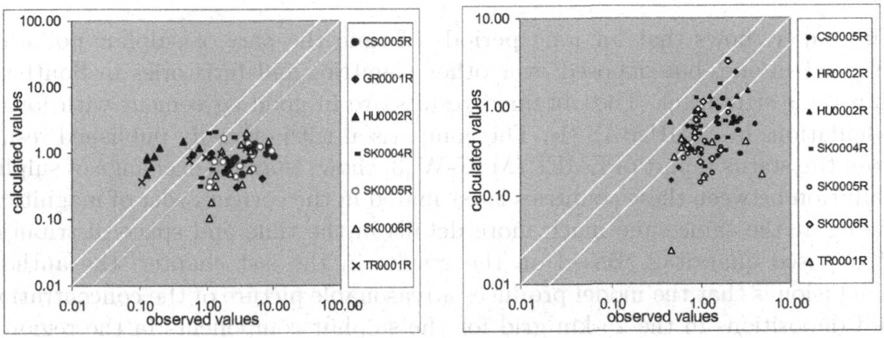

Fig. 8. Observed versus calculated concentration in air (left)(unit: $\mu g(S)/m^3$) and concentration in precipitation (right)(unit: mg (S)/l) of sulphur

(Fig. 7). The data obtained from measurement station at an altitude of over 1300 m are inadequate for such comparisons. Mean monthly values of sulphur concentration in air and precipitation are published (http://www.emep.int/). These two parameters are compared to the results obtained for the corresponding of each station gridcell through the above mentioned monthly runs. The results are presented graphically in Fig. 8 and sumarized in Table 2.

The comparisons show that, generally, the model *underestimates* the values and the sulphur concentration in air and precipitation. This fact is more

Table 2. Calculated versus observed sulphur parameter summary

parameter	concentration in air			concentration in precipitation		
station	# observed	# in factor 2	%	# observed	# in factor 2	%
CS0005R	12	0	0.0	12	2	16.7
GR0001R	11	2	18.2	0		
HR0002R	0			12	5	41.7
HU0002R	11	3	27.3	10	8	80.0
SK0004R	12	6	50.0	12	7	58.3
SK0005R	12	2	16.7	12	3	25.0
SK0006R	12	2	16.7	12	3	25.0
TR0001R	8	4	50.0	11	4	36.4
total	78	19	24.4	81	32	39.5

emphasized now, than by the comparisons of the model results. A probable explanation is, again, the unaccounted flux of pollutant through the boundaries of the domain, considering that the stations are situated relatively close to them.

It is also necessary to take into account the fact that we are currently comparing monthly mean values rather than annual means and annual values for which the random errors are higher than the absolute value.

6 Conclusion

The paper shows that for long periods of time the part of sulphur pollution, released in one, but deposed over other countries and territories in Southeast Europe is significant. The obtained results are in good agreement with former calculations in NIMH-BAS [1]. The comparison with officially published results from the status report of EMEP/MSC-W [3] shows that the exchange of sulphur pollution between these countries is estimated in the correct order of magnitude, giving at the same time much more details in the time and space distribution of deposed quantities. Based on the results in the last chapter, the author's conclusion is that the model produces a reasonable picture of the concentrations and depositions in the 25-km grid for the sulphur components in the region of Balkan region. The results of such calculations can be used in decision-making in negotiating and contamination strategies development.

References

1. B.EN.A annual reports 2004, On the Exchange of Sulphur Pollution over Southeast Europe, Chervenkov, H.
2. BC-EMEP 1994, 1995, 1996, and 1997: Bulgarian contribution to EMEP, Annual reports for 1994, 1995, 1996, 1997, NIMH, EMEP/MSC-E, Sofia-Moscow.
3. Barret, K., Berge, E., (Eds) Transboundary Air Pollution in Europe, EMEP/MSC-W status report 1997, Part One and Two: Numerical Adenddum, Norwegian Meteorological Institute, Oslo, Norway, B3–B46, 1996.

4. Syrakov, D., On a PC-oriented Eulerian Multi-Level Model for Long-Term Calculations of the Regional Sulphur Deposition, in Gryning S. E. and Schiermeier F. A. (eds), Air Pollution Modelling and its Application XI, NATO — Challenges of Modern Society, Vol. 21, Plenum Press, N.Y. and London, 645–646, 1995.
5. Syrakov, D., On the TRAP advection scheme — Description, tests and applications, in Geernaert G., Walloe-Hansen A. and Zlatev Z., Regional Modelling of Air Pollution in Europe. Proceedings of the first REMAPE Workshop, Copenhagen, Denmark, September 1996, National Environmental Research Institute, Denmark, 141–152, 1996.
6. Syrakov, D., Galperin, M., A model for airborne poli-dispersive particle transport and deposition, Proc. of 22nd NATO/CCMS International Technical Meeting on Air Pollution Modelling and its Application, 2-6 June 1997, Clermont-Ferrand, France, 111–118, 1997.
7. Syrakov, D., Galperin, M., On a new Bott-type advection scheme and its further improvement, in Proc. of the first GLOREAM Workshop, Aachen, Germany, Hass H. and Ackermann I. J. Eds., Ford Forschungszentrum Aachen, 103–109, 1997.
8. Syrakov, D., Prodanova, M., Bulgarian Emergency Response Models — Validation against ETEX First Release, Atmos. Environ., 32, No. 24, 4367–4375, 1998.
9. Syrakov, D., Yordanov, D., On the surface layer parameterization in an Eulerian multi-level model, Proceedings of the 4th Workshop on Harmonisation within Atmospheric Dispersion Modelling for Regulatory Purposes, 6-9 May 1996, Oostende, Belgium, v.1, 25–31, 1996.
10. Syrakov, D., Yordanov, D., Parameterization of SL Diffusion Processes Accounting for Surface Source Action, Proc. of 22nd NATO/CCMS International Technical Meeting on Air Pollution Modelling and its Application, 2-6 June 1997, Clermont-Ferrand, France, 118–125, 1997.
11. Proceedings of the eghth and ninth International Conference on Harmonisation within Athmosheric Dispersion Modelling for Regulatory Purpouses, 2002; 2004
12. Seinfeld, J. Athmospheric Chemistry and Physics of Air Pollution, 1986

Long-Term Atmospheric Aerosol Simulations — Computational and Theoretical Challenges

Adolf Ebel, Michael Memmesheimer, Elmar Friese, Hermann J. Jakobs, Hendrik Feldmann, Christoph Kessler, and Georg Piekorz

Rhenish Institute for Environmental Research (RIU) at the University of Cologne, Cologne, Germany
eb@eurad.uni-koeln.de

Abstract. Numerical simulations of aerosols in the atmospheric boundary can computationally be highly demanding since processes governing the formation and dynamics of atmospheric particles are extremely complex. It is necessary to apply reasonable simplifications to make the computations tractable. This paper gives an outline of the approach to the problem chosen for the EURAD model system. A special module MADOC (**M**odal **A**erosol **D**ynamics with **O**rganic compounds and **C**louds) has been developed for aerosol simulations employing EURAD. Annual runs for a region in Central Europe are shown and discussed with regard to model content, performance and reliability.

1 Introduction

Atmospheric aerosols have gained increasing importance for the assessment of air quality and climate forcing. There is increasing evidence that health effects of fine particles may be of greater relevance than believed a decade or so ago. As regards the impact of aerosols on climate change there is no doubt that they play a crucial role for global temperature modifications though quantification of the radiative forcing effect still seems to be partly controversial. This paper is focusing on particulate matter in the lowest layers of the troposphere where research of aerosols is mainly motivated through their impact on human health and possible ecological effects. The complexity of particle composition, chemistry and dynamics makes the theoretical and numerical treatment of particulate matter a demanding problem. Experimental work shows that there are still many questions to be answered before the understanding of atmospheric aerosols has come to a point that gaps of our knowledge and thus surprises for those who dare to design complex aerosol transport models. In addition, large uncertainties exist regarding sources of primary aerosol and secondary aerosol precursor emissions exist. Yet despite ongoing experimental research there are strong reasons to develop and apply such models in parallel with progressing laboratory and field work. On the one hand, one of the most important facts is the need of environmental planning and policy for more information on particulate matter,

I. Lirkov, S. Margenov, and J. Waśniewski (Eds.): LSSC 2005, LNCS 3743, pp. 490–497, 2006.

its distribution and changes of atmospheric load on which mitigation and adaptation strategies can be based. For instance, the EU directives for the control of particulate matter with sizes equal to or less than 10 μm (PM10) require the application aerosol and chemistry transport models for the analysis of the general state of air quality in larger domains, i. e. states or countries, industrial regions or cities. On the other hand, it is essential for comprehensive aerosol research to study particles in an integral way in the atmosphere. Here simple and complex models are serving as a convenient and indispensable tool.

Dealing with aerosol dynamics, chemistry and transport in full detail in numerical models would lead to extremely demanding algorithms which could not be treated on presently available computers. Evidently, simplifications and parameterizations are needed which make the problem numerically tractable. The way, computational demands are reduced to realistic size essentially depends on the problem to be treated. In this study, we describe a module which has been designed to simulate particulate matter below a given size. According to the relevance for the EU directives usually 10 μm (PM10) is chosen, but other aerosol sizes (e. g. PM2.5 or PM1.0) are applicable as well. The following section contains a description of the aerosol module and gives an impression of the simplifications needed for a computationally efficient algorithm allowing long-term simulation as needed by environmental agencies. Then some applications showing the potential as well as limitations of the chosen approach are presented and discussed. The final section is devoted to conclusions about the ongoing and future work on aerosol modelling in our group.

2 The Atmospheric Aerosol Module

The atmospheric particles may be grouped into three modes according to their size. The fraction of finest particles makes up the Aitken mode (below about 0.1 μm). The largest particles are found in the coarse mode with large and giant particles (above 1.0 μm). In between one has the accumulation mode (about 0.1-1.0 μm). Numerical representation of these modes which are variable in space and time require simplifying assumptions about the size distribution.

As regards standardised size distributions five principal approaches are found in aerosol models [26] (1) Mono-disperse approximation: fixed particle sizes represent the modes. (2) Modal approximation: suitable distribution functions of the individual modes are employed. (3) Sectional approximation: use of a larger number of intervals with constant aerosol parameters. (4) Splines. (5) Dense discrete values. The last method would be the most accurate but requires an unaffordable amount of computer time.

The approach chosen in this study is the approximation of the size spectrum by three log-normal distributions. This is realized in the Modal Aerosol Dynamics Model for Europe (MADE [1, 2] which was originally developed for the EURAD model system (EURAD: European Atmospheric Dispersion Model). Basic structure and content of EURAD have been described in various publications (e. g. [5, 10, 12, 15]) so that it is only briefly characterized here. It is a hemispheric to regional model system consisting of a meteorological model (MM5 [9]) and a chemical transport model (EURAD-CTM, based on RADM [3]) employing the chemical gas phase mechanisms RADM2 [25], RACM [24] or RACM-MIM [8, 14]. The latter mechanism has been applied to the simulations demonstrated in Section 3. EURAD supports nested calculations of the distribution of air pollutants [12] with horizontal resolution down to 1 km. Up to 23 layers are used in the vertical extending from the Earth's surface up to the lower stratosphere. Emissions are simulated by the EURAD Emission Model (EEM [18]).

The module MADE has been extended to include the estimation of secondary organic aerosols (module SORGAM [22]) and aerosol/cloud interactions [7]. We will refer to the extended module as MADOC (Modal **A**erosol **D**ynamics with **O**rganic compounds and **C**louds). It is connected to RACM-MIM [8] as chemical mechanism for the treatment of gas phase chemistry.

The temporal change of particle number n (representing an ensemble average in order to avoid the treatment of small, i. e. turbulent, fluctuations) is described by equation (1):

$$\begin{aligned}\frac{\partial n}{\partial t} = & \frac{1}{2}\int_0^v \beta(\bar{v}, v-\bar{v})n(\bar{v})n(v-\bar{v})d\bar{v} - \int_0^\infty \beta(\bar{v}, v)n(\bar{v})n(v)d\bar{v} \\ & - \frac{\partial v}{\partial I} - \nabla \cdot \vec{u}n + (\frac{\partial t}{\partial n})_{diff} - F_{dep}\end{aligned} \tag{1}$$

(e. g. [21, 23]). Particles of different size and composition are interacting with each other (coagulation, integral terms, where β is the rate of particle coagulation and v the particle volume). Nucleation and particle growth through uptake or loss of water vapour or other volatile substances (condensation/evaporation) is given by I (droplet current). Changes due to emission, chemistry and cloud effects including wash and rain out can be taken into account. Particles are transported by wind ($\vec{u}$ ensemble average of atmospheric motion) and turbulent diffusion (second last right term) and may be lost to the ground by dry deposition due to Brownian motion and sedimentation (last right term).

The log-normal distribution function has been chosen to represent the three aerosol modes. Each mode contains three independent variables so that a distribution with three modes is characterized by a set of only 9 parameters, and only 9 PDEs are needed for solving a system consisting of chemically identical aerosols. If the width of the modes is predefined and constant as in the case of usual MADOC applications, only six PDEs remain. This retains sufficient complexity of the size distribution, but considerably reduces computational requirements when compared to larger sectional approaches. The set of PDEs has to be solved for each chemical mode separately.

It is assumed that coagulation can occur within both fine modes and that nucleation only plays a role in the Aitken mode. Furthermore, these processes and condensation are neglected for the coarse mode whereas sedimentation is only taken into account for coarse particles. Like emission additional processes modifying the aerosol can be treated by proper formulation. The process of diffusion includes dry deposition as a lower boundary value. For the treatment of secondary aerosol chemistry thermodynamic equilibrium has been assumed

[1, 22]. Sulphate and organic particles may originate from the gas phase (gas-to-particle conversion). The main source of inorganic secondary aerosols is the formation of ammonium sulphate and ammonium nitrate particles:

$$H_2SO_4(g) \rightarrow H^+ + HSO_4^- \leftrightarrow 2H^+ + SO_4^{2-} \quad (2)$$

$$NH_3(g) + H^+ \leftrightarrow NH_4^+ \quad (3)$$

$$HNO_3(g) \leftrightarrow H^+ + NO_3^- \quad (4)$$

$$NH_3(g) + HNO_3(g) \leftrightarrow NH_4NO_3(s) \quad (5)$$

SORGAM calculates anthropogenic (ASOA) and biogenic (BSOA) components of the secondary organic aerosol. It parameterizes the transformation of reactive organic gases (ROG) into products which have sufficiently low vapour pressure, so that they can exist in the gas and liquid (aerosol) phase. The mechanism is described by

$$ROG + X^* \rightarrow \sum_i \alpha_{xi} C_{xi} \quad (6)$$

where α_{Xi} is a stoichiometric coefficient for the compound C_{Xi} which is generated by this reaction and exists in the gas phase ($C_{Xi,gas}$) and particle phase ($C_{Xi,aer}$) under equilibrium conditions. X* stands for the OH radical, the nitrate radical NO_3^- and ozone. The organic precursors ROG provided by CTM (with RACM-MIM) and EEM for the module MADOC are grouped in four classes containing aromatic compounds and higher alkanes and alkenes from anthropogenic sources and two classes containing α-pinene and limonene from biogenic sources. Four anthropogenic and four biogenic reaction products combined in the classes ASOA and BSOA, respectively, are calculated. Finally, the module MADOC provides size resolved concentrations of these products in addition to those of secondary inorganic compounds ($SO_4^{2-}, NO_3^-, NH_4^+$), aerosol liquid water and primary aerosol species (EC, OC, unidentified particulate matter).

As demonstrated in the following section nested simulations of hourly values of PM10 and total suspended particle mass (TSP) have been performed for several years with the aim to study the annual and seasonal variability as well as future trends in Europe and some sub-domains, in particular the state of Northrhine-Westphalia (NRW) in Germany [6, 16, 17]. Long-term nested aerosol calculations are also carried out in the framework of daily air quality forecasts with the EURAD model system [13] for Europe, Central Europe and NRW using a reduced version of MADOC to keep the computational time in a reasonable range required for operational use of EURAD.

3 Long-Term Aerosol Simulations: Selected Results and Discussion

Selected model applications are presented in Figures 1 and 2. An example of the distribution of individual aerosol components in the lowest model layer

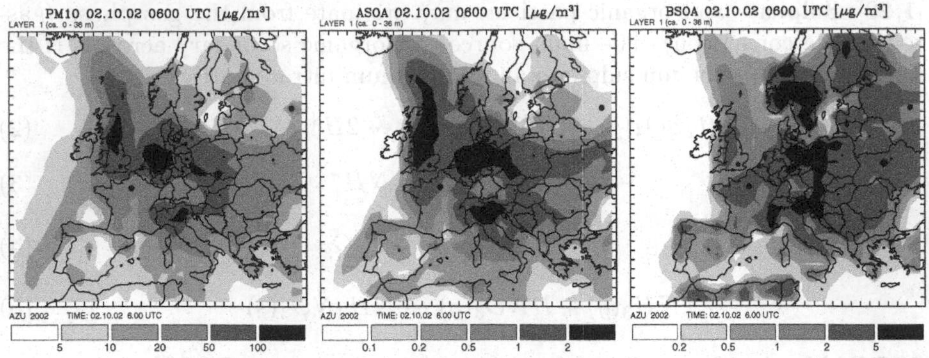

Fig. 1. Surface concentration ($\mu g/m^3$) of aerosol components, October 2, 2002, 06 UTC. From left to right: PM10, anthropogenic secondary organic aerosol, biogenic secondary organic aerosol [16, 17].

(40 m high) is shown in Fig. 1. A day in 2002 with anticyclonic weather conditions and high aerosol load over Central Europe has been chosen for demonstration. Interesting details of particle distribution can be identified in such figures. For instance, relatively large concentrations of secondary organic aerosol of biogenic origin (BSOA, right panel) are found in southern Scandinavia where strong emissions of biogenic precursors are found. In contrast, anthropogenic secondary aerosols (ASOA, middle panel) are highest in industrial regions and their vicinity. Yet the contribution of both components to total aerosol mass remains small and seldom reaches values near or above 10 %. The main contribution to total PM10 (left) comes from ammonium, nitrate and sulphate as evident from the respective panels in Fig. 1.

For model evaluation only PM10 and/or TSP observations are usually available. Comparison of the daily averages with calculated TSP or PM10 shows that the gross variations of these parameters are met quite reasonably. Yet it is rare that measurements of composition in such detail as given by the model are avail-

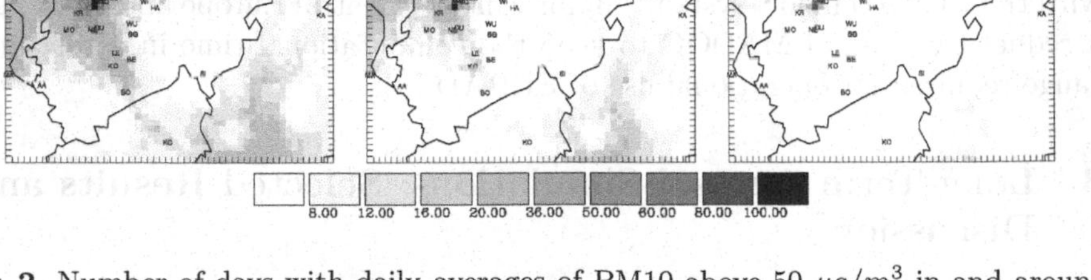

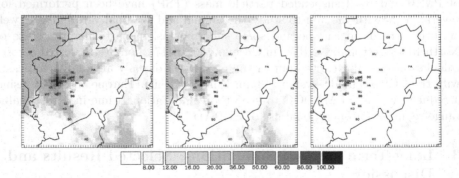

Fig. 2. Number of days with daily averages of PM10 above 50 $\mu g/m^3$ in and around NRW. Base case (left) is year 2002 with estimates of actual emissions of air pollutants. 2005 (middle) and 2010 (right) with predicted emission reduction [16].

able even for shorter episodes, so that comprehensive evaluation of composition simulations is not possible until now. This deficit constitutes an important challenge for future aerosol research since the study of health effects under realistic atmospheric conditions depends on reliable information about the composition, in particular, for lower size particles.

The 2002 run has been employed to estimate the possible reduction of aerosol mass load in Europe due to decreasing emissions predicted for the years 2005 and 2010. Fig. 2 exhibits the number of days with daily PM10 averages above 50 $\mu g/m^3$ in NRW. At present EU directives for air quality allow 35 days/year. The calculations point to considerable improvements till 2010, but also show that there will remain problem areas in industrialized regions. It should be noted that the calculations show results for grid boxes of 5x5 km^2. Street canyons in cities which may suffer from heavy traffic can not be resolved by EURAD. Evidently, the situation there may be much more problematic than can be demonstrated by such grid box results.

4 Final Remarks

On the one hand, the model provides considerably more information about the behaviour of the atmospheric aerosol than can be demonstrated in this short paper. For instance, the calculated vertical distribution of aerosols is rarely exploited. In the future this will possibly change when the role of aerosols for climate forcing will be studied with the help of detailed simulations. On the other hand, despite its complexity the model is limited in its present form with regard to the range of its application, e. g. in epidemiological and toxicological studies of the effects of fine and ultra-fine particles. There are strong indications that this fraction of atmospheric aerosol is particularly harmful to health [11, 19, 20]. Modifications of MADOC will be needed if the EURAD system will be applied to this problem of aerosol research as planned for the future.

Comparison of several aerosol chemistry and transport models by Hass et al. [10] has shown that the models provide reasonably similar results on larger regional scales but that evaluation of simulation may be hampered by deficiencies of available observations. Nevertheless, it is important to check the reliability and validity of an atmospheric aerosol model continuously through comparison with observations. The challenge is to get closer to the truth through skilful use of efficient evaluation procedures [4]. Results of the evaluation of EURAD/MADOC may be found in Memmesheimer et al. [16].

Acknowledgements

This work was supported by the Environmental Agency (LUA) of Northrhine-Westphalia (NRW), particularly within the projects ANABEL, AZUR, and ELAN, and by the BMBF within the AFO2000 programme. Fruitful discussions with S. Wurzeler (LUA) and the IDEC project group in AFO200 are gratefully acknowledged. We thank EMEP, TNO, UBA and LUA for providing emission data, and ECMWF and DWD for giving access to global meteorological data. LUA also provided monitoring data for NRW. Land use data were gratefully received from IfU/Research Center Karlsruhe. The numerical calculations have been supported by the ZAIK/RRZK, University of Cologne, and the Research Center Juelich (NIK/ZAM).

References

1. Ackermann, I. J., H. Hass, M. Memmesheimer, A. Ebel, F. S. Binkowski, and U. Shankar, Modal aerosol dynamics model for Europe: Development and first applications. *Atmos. Environment*, **32**, 2981–2999, 1998.
2. Binkowski, F. S., and U. Shankar, The regional particulate matter model, 1. Model description and preliminary results. *J. Geophys. Res.*, **100**, 26191–26209, 1995.
3. Chang, J. S., Brost, R. A., Isaksen, I. S. A., Madronich, S., Middleton, P., Stockwell, W. R., and Walcek, C. J., A three-dimensional acid deposition model: Physical concepts and formulation. *J. Geophys. Res.*, **92**, 14681–14700, 1987.
4. Ebel, A., Evaluation and reliability of meso-scale air pollution simulations. In: *Large-scale Scientific Computing*, S. Margenov, J. Wasniewski, P. Yalamov eds., *Lecture notes in computer science*, **2179**, Springer-Verlag, Berlin Heidelberg New York, 255–263, 2001.
5. Ebel, A., Changing atmospheric environment, changing views and an air quality model.s response on the regional scale. In: Proceedings of the 25th NATO/CCMS ITM on Air Pollution Modelling and its Application, C. Borrego and G. Schayes eds., Kluwer Academic/Plenum Publ., New York, 25–36, 2002.
6. Friese, E., H. J. Jakobs, M. Memmesheimer, H. Feldmann, C. Kessler, G. Piekorz, and A. Ebel, ANABEL - Ausbreitungsrechnungen für Nordrhein-Westfalen zur Anwendung im Rahmen der Beurteilung der Luftqualität nach EU-Richtlinien. Report, Landesumweltamt NRW and Rhenish Inst. Envir. Res., University of Cologne, 2002.
7. Friese, E., M. Memmesheimer, I. J. Ackermann, H. Hass, A. Ebel, and M. Kerschgens, A study of aerosol-cloud interactions with a comprehensive air quality model. *J. Aerosol Sci.*, **31**, S54–S55, 2000.
8. Geiger, H., I. Barnes, I. Bejan, T. Benter and M. Spittler, The tropospheric degradation of isoprene: an updated module for the regional atmospheric chemistry mechanism. *Atmos. Environm.*, **37**, 1503–1519, 2003.
9. Grell, G. A., J. Dhudia, and D. R. Stauffer, A description of the fifth-generation Penn State/NCAR mesoscale model (MM5). Technical Note TN-398, NCAR, Boulder, CO, 1994.
10. Hass, H., M. van Loon, C. Kessler, R. Stern, J. Matthijsen, F. Sauter, Z. Zlatev, J. Langner, V. Foltescu and M. Schaap, Aerosol modelling: Results and Intercomparison from European regional-scale modelling systems. Special Report, EUROTRAC-2, Intern. Sci. Secr., Munich, 2003. (http://www.trumf.fu-berlin.de/veranstaltungen/events/gloream)
11. Hoek, G., B. Brunekreef, S. Goldbohm, P. Fischer and P. A. van den Brandt, Association between mortality and indicators of traffic-related air pollution in the Netherlands: a cohort study; Lancet, **360** 9341, 1203–1209, 2002.
12. Jakobs, H. J., H. Feldmann, H. Hass, M. Memmesheimer: The use of nested models for air pollution studies: an application of the EURAD model to a SANA episode. *J. Appl. Met.*, **34**, 1301–1319, 1995.

13. Jakobs, H. J., E. Friese, M. Memmesheimer and A. Ebel, A real time forecast system for air pollution concentrations. In: Proceedings from the EUROTRAC-2 Symposium 2002, P. M. Midgley, and M. Reuter eds., Markgraf-Verlag, Weikersheim, 2002.
14. Karl, M., T. Brauers, H.-P. Dorn, F. Holland, M. Komenda, D. Poppe, F. Rohrer, L. Rupp, A. Schaub, A. Wahner, Kinetic study of the OH-isoprene and O3-isoprene reaction in the atmosphere simulation chamber SAPHIR. Geophys. Res. Lett., 31, LO5117, doi: 10.1029/ 2003GLO 19189, 2004.
15. Memmesheimer, M., E. Friese, A. Ebel, H. J. Jakobs, H. Feldmann, C. Kessler and G. Piekorz, Long-term simulations of particulate matter in Europe on different scales unsing sequential nesting of a regional model. *Int. J. Environm. Poll.*, **22**, 108–132, 2004.
16. Memmesheimer, M., E. Friese, H. J. Jakobs, C. Kessler, H. Feldmann, G. Piekorz and A. Ebel, AZUR, Final Project Report, Rhen. Inst. Evironmental Research (RIU), Cologne, Germany, 2004.
17. Memmesheimer, M., E. Friese, H. J. Jakobs, C. Kessler, H. Feldmann, G. Piekorz and A. Ebel, ELAN, Final Project Report, Rhen. Inst. Evironmental Research (RIU), Cologne, Germany, 2004.
18. Memmesheimer, M., J. Tippke, A. Ebel, H. Hass, H. J. Jakobs, and M. Laube, On the use of EMEP emission inventories for European scale air pollution modelling with the EURAD model. Proc. EMEP Workshop on Photo-Oxidant Modelling for Long-range Transport in Relation to Abatement Strategies, Berlin, 307–324, 1991.
19. Oberdoerster, G., Pulmonary effects of inhaled ultrafine particles. *Int. Arch. Occup. Environ. Health.* **74** 1, 1–8, 2001.
20. Peters A., S. von Klot, M. Heier, I. Trentinaglia, A. Hörmann, H. E. Wichmann, H. Löwel, Exposure to Traffic and the Onset of Myocardial Infarction. *N Engl J Med*, **351**, 721–1730, 2004.
21. Russell, A., and R. Dennis, NARSTO critical review of photochemical models and modelling. *Atmos. Environm.*, **24**, 2283–2324, 2000.
22. Schell, B., I. J. Ackermann, H. Hass, F. S. Binkowski, and A. Ebel, Modeling the formation of secondary aerosol within a comprehensive air quality modeling system. *J. Geophys. Res.*, **106**, 28275–28293, 2001.
23. Seinfeld, J. H., *Atmospheric Chenistry and Physics of Air Pollution*, John Wiley and Sons, New York, 1986.
24. Stockwell, W. R., F. Kirchner, M. Kuhn, and S. Seefeld, A new mechanism for regional atmospheric chemistry modelling. *J. Geophys. Res.*, **102**, 25847–25879, 1997.
25. Stockwell, W. R., P. Middleton, and J. S. Chang, The second generation regional acid deposition model chemical mechanism for regional air quality modeling. *J. Geophys. Res.*, **95**, 16343–16367, 1990.
26. Whitby, E. R., P. H. McMurray, U. Shankar and F. S. Binkowski, Modal aerosol dynamics modeling. Rep. 600/3-91/020, Atmos. Res. and Exposure Assess. Lab., US Envir. Prot. Agency, 1991.

The Use of MM5-CMAQ for an Incinerator Air Quality Impact Assessment for Metals, PAH, Dioxins and Furans: Spain Case Study

Roberto San José[1], Juan L. Pérez[1], and Rosa M. González[2]

[1] Environmental Software and Modelling Group,
Computer Science School,
Technical University of Madrid – UPM,
Campus de Montegancedo, Boadilla del Monte 28660 Madrid, Spain
roberto@fi.upm.es,
http://artico.lma.fi.upm.es

[2] Department of Meteorology and Geophysics, Faculty of Physics,
Complutense University of Madrid – UCM,
Ciudad Universitaria, 28040 Madrid, Spain
rgbarras@fis.ucm.es

Abstract. In this contribution we show the implementation of a modified version of MM5-CMAQ (Hutzell W.T., 2002) for carrying on an air quality impact analysis for installing an incinerator in the Basque Country Area (Spain). The modified CMAQ model (EPA USA, 2004) includes Poly-Chlorinated Dibenzo-p-Dioxins and Dibenzo-Furans (dioxins and furans). The model represents their cogeners as divided between gaseous and aerosol forms that exchange mass based on theoretical coefficients for gas to particle portioning. Modelled metals are included in CMAQ as part of the non-reactive aerosol component. Metals modelled are: As, Cd, Ni and Pb. In additional Benzo(a)pyrene (PAH) is also modelled. The model is implemented in a cluster platform in order to be used as a real-time air quality forecasting system by using the ON-OFF approach. The emissions of the projected incinerator in the ON run are incorporated by using the height of the chimney, the prescribed exit gas velocity, diameter of chimney and the limit emission values (worst scenario) prescribed in the Directive/2000/76/CE. The OFF run is done by using EMEP POP and PAH emission inventory. The system is mounted over one mother domain of 400 x 400 km with 9 km spatial resolution and two nesting levels: 100 km model domain with 3 km spatial resolution and 24 km with 1 km spatial resolution. All model domains have 23 vertical layers. The highest level is located at 100 mb. The architecture domain system is centered at the UTM coordinates assigned for the projected incinerator. EPER EU industrial emissions (May, 2004) of the surrounding large point industrial sources are used. Results are compared with the target values included in the proposal for a Directive of the European Parlament and of the Council (ENV 194 CODEC 439).

Keywords: Industrial air quality impact, MM5, CMAQ, air quality forecasts, software tools.

I. Lirkov, S. Margenov, and J. Waśniewski (Eds.): LSSC 2005, LNCS 3743, pp. 498–505, 2006.

1 Introduction

The advances in sensitivity of air quality models has allowed using much more sophisticated model for evaluating the air quality impact expected for future installations of industrial plants. In recent years, we have carried out several studies to know the expected impact of combined cycle power plants and incinerators. The use of more sophisticated models to determine the expected impact on air concentrations due to the emissions of industrial plants is becoming more common and recommended. Our group has conducted several of these studies in the last years by using last generation of mesoscale meteorological and air quality models. In this contribution we will show results concerned on a complementary studied carried out over the Basque Country Area (Txingudi, Spain) to analuze the impact of projected emission on the surrounding area for benzene, lead, nickel, cadmium, arsenic, benzo(a)pyrene and dioxins and furans (PCDD/F). The Cadmium, Arsenic, Nickel and Benzo(a)pyrene are expected to be limited in air concentrations by EU Directives in the next months. The Lead and Benzene are already limited by EU Directives 1999/30/CE and 2000/69/CE.

The air quality impact of industrial plants has been a key issue on air quality assessment and modelling since the 70's. Nowadays, the increased capacity on computer power and progress on air pollution science provide a powerful and reliable tool to measure the air quality impact on industrial emissions. In the last decade a considerable effort to incorporate the industrial production processes in an integrated environmental evaluation in on-line mode has been done. In a parallel way, a considerable increase of citizen concern has been detected particularly in those areas where important industrial point sources are present together with highly populated areas (urban areas). This case is particularly sensitive for refineries, waste city incinerators, etc.

In this contribution we will show the results of a preliminary modelling experiment to build a so-called TEAP tool [3,4] (A tool to evaluate the air quality impact of industrial emissions). This tool is designed to be used by the environmental impact department at the industrial site. The tool provides a response to air quality impact to industrial emissions in the form of surface patterns and lineal time series for specific geographical locations into the model domain. The model domain is designed in a way that the industrial source point is located approximately in the center of the model domain. The model domain can be as extense as wished but a specific nesting architecture should be designed for each case together with balanced computer architecture.

In this contribution we will use the MM5-CMAQ modeling system. MM5-CMAQ is a representative of the last generation of AQMS (third generation) developed by EPA (USA) in 2000 [1]. The model uses a full modular structure with the last advances on computer programming (FORTRAN-95). In essence many of the features of MM5-CMAQ are similar to OPANA but the programming and modularity is more advanced. MM5-CMAQ is not a limited area model and it can run over large domains (even at global level although a CMAQ global version is not existing yet). The model domains are obviously closely related to model forecast horizon so that the nesting capability (in a similar way that it was done in OPANA) plays an essential role to have reliable simulations over city and regional domains. Another representatives of third generation of AQMS are CAMx (Environ Co., USA) and EURAD (European Ford Research Group and university of Cologne (Germany)). MM5 is the well known Non-hydrostatic Mesoscale Meteorological Model developed in the Pennsylvania State University and NCAR starting on 1983. The MM5 is today one of the most robust and reliable meteorological models. In both cases and in all Eulerian models, the input datasets are key elements to work down to have reliable and accurate simulations. These datasets are: DEM (Digital elevation model), Land use data (usually satellite data, AVHRR/NOAA, Landsat, Spot, etc.), Initial and boundary meteorological conditions, initial and boundary air concentration profiles and finally, emission data sets. The emission datasets are usually the bottle neck on this type of applications since the uncertainty involved is important. In spite of this limitation, the TEAP tool extracts the most important benefits from the relative difference between a simulation with full emission inventory and a second simulation with an emission inventory without the industrial plant to be studied.

2 Methodology

We have implemented the MM5-CMAQ modeling system in a nesting architecture. The MM5 Mesoscale Meteorological Model (PSU/NCAR) and the Community Multiscale Air Quality Model (CMAQ) [1] from EPA (USA) (third generation of air quality modelling systems) are used as mainframe platform. The MM5 is built over a mother domain with 36 x 36 grid cells (81 km spatial resolution) and 23 vertical levels. This makes a domain of 2916 x 2916 km. The nesting MM5 level 1 model domain is built over a 69 x 66 grid cells (27 km spatial resolution) and 23 vertical levels, which makes a model domain of 1863 s 1782 km centered over the Iberian Peninsula. CMAQ model domains are 30 x 30 grid cells for mother domain and 63 x 60 over the nesting level 1 model domain. CMAQ mother domain lower left corner is located at (-1215000 m, -1215000 m) at the reference locations (-3.5W, 40N) and the first and second standard parallels (30N, 60N). The CMAQ nesting level 1 lower left corner is located at (-891000, -810000) with the same reference locations. The 9 km MM5 spatial resolution model domain has 54 x 54 grid cells, the 3 km MM5 spatial resolution model domain has 33 x 39 grid cells and finally the 1 km MM5 spatial resolution model domain has 30 x 30 grid cells. Figure 1 shows the domain architecture used and implemented in this specific case and Figure 2 shows the 3D view of the nesting level 3 (1 km spatial resolution). The corresponding CMAQ model domains are: 48 x 48 km, reference (-216000, -189000) in Lambert Conformal projection with 9 km spatial resolution; 27 x 33 grid cells, reference (-45000, -54000) with 3 km spatial resolution and finally, 24 x 24 grid cells, reference (-12000, 12000) with 1 km spatial resolution. All these Lambert coordinates are refered to the point (1,845W, 43,335N).

The projected incinerator is located in the center of this architecture design. We conducted a full air quality impact assessment study during 2003 by using periods of time corresponding to 120 hours per month along the 12 months of 2002. The periods were selected in a way that the air pollution monitoring data – from stations located in the model domain – values were compared with the simulated values. So that we avoided explicitly the very polluted days and the clear days since the objective was to evaluate the impact of the expected emissions on average days. In this complementary simulation we have simulated 6 periods of 120 hours in 2003.

The emission database is obtained from the EMIMO model [2] – a new emission model for large domains based on global emission ineventories such as GEIA, EMEP, EDGAR and the Digital Chart of the World. MM5-CMAQ is initiated by using global data sets from MRF (NOAA/NCEP, USA) and mother, nesting levels 1 and 2 provide the boundary conditions for running MM5-CMAQ for nesting level 3 over the Madrid Community Area with 27 x 33 grid cells (3 km) which makes 81 x 99 km. The OPANA model runs over a domain of 80 x 100 km with UTM coordinates and using only EMIMA data sets. EMIMA datasets are also used for MM5-CMAQ over nesting level 3. EMIMO data set is used for mother and nesting levels 1 and 2 for MM5-CMAQ. We have performed two simulations (MM5-CMAQ), one simulation with the industrial plant emissions and the second one *without* the industrial plant emissions. EMIMO model uses EMEP 2002 annual emission inventory data and EPER 2004 (point industrial emissions). The CBM-IV chemical scheme has been used to simulate all the additional pollutants except Benzene. Benzene has been simulated by using SAPRAC-99 chemical scheme. Dioxins and furans refer to compounds that have two benzene rings linked by two or one atoms of oxygen. The model has been designed by using the ON/OFF mode approach. This approach is based on the full simulation without the emissions expected from the incinerator (OFF) and substract the results from the ON scenario which is based on the simulations obtained when added the emissions at the OFF scenario. A main goal of the adapted model calculates not the concentration of one congener but the I-TEQ from all congeners. Deposition is the primary interest because it provides the main route into the food chain for dioxins and furans. I-TEQ equals a congener's International Toxicity Equivalency Factor (I-TEF_i) multiplied by its air concentration or deposition, M_i;

$$\text{I-TEQ} = \sum_{i=1}^{N} (\text{I-TEF}_i)^* \text{M}_i .$$

Nonzero values of I-TEF_i determine the number of congeners, N, that need to be modeled to calculate I-TEQ. I-TEF_i measures health effects from a congener relative to the most toxic congener, TCDD. The number, N, equals seventeen. A balance emission and removal processes determine air concentrations

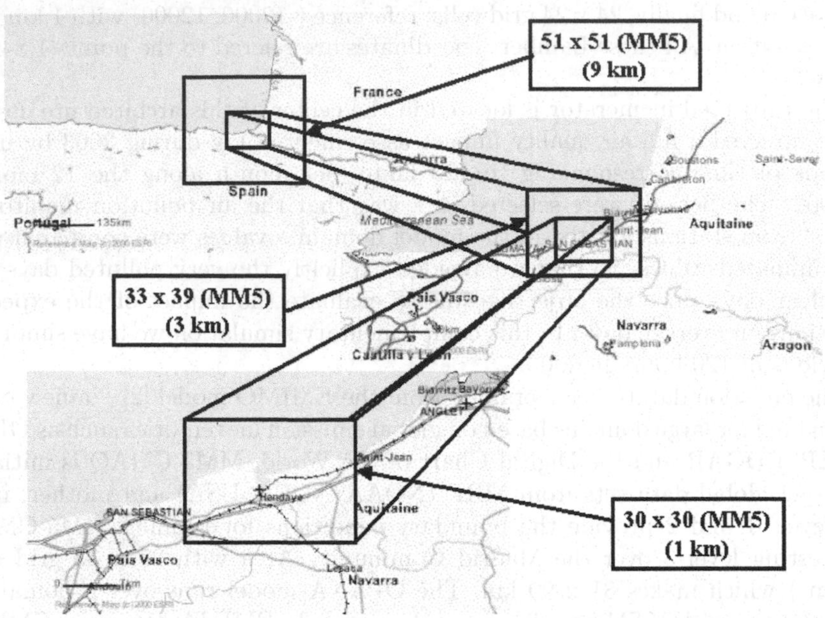

Fig. 1. Mother and nesting levels on this experiment with TEAP tool (MM5-CMAQ)

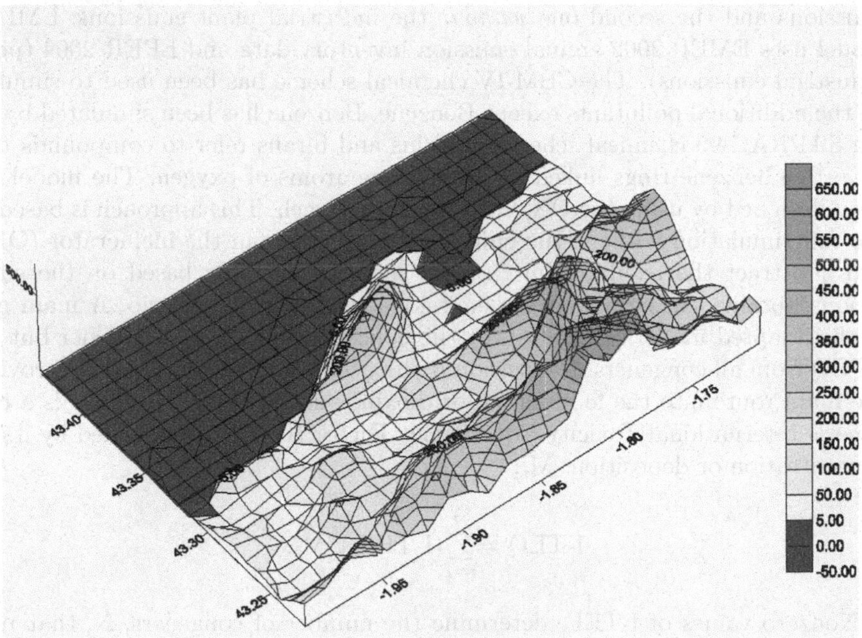

Fig. 2. Nesting level 3 (1 km spatial resolution)

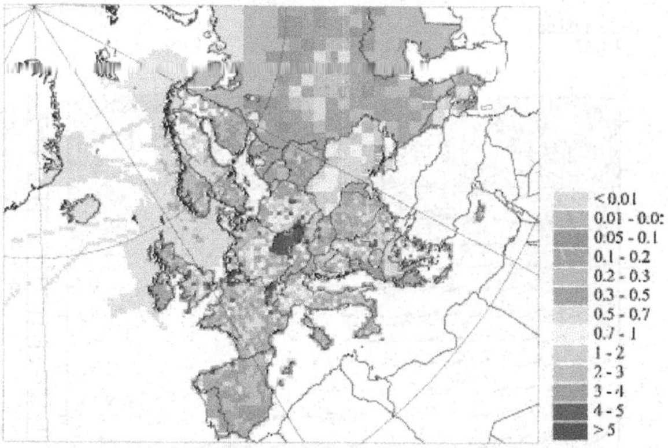

Fig. 3. Spatial distribution of emission flux of the 17 toxic congeners of PCDD/Fs in 2001

and deposition from dioxin and furan emissions. Anthropogenic combustion is believed the largest source. Waste incinerators, paper-pulp mill, chemical and metal sinter plants have the largest emission rates for point sources (Cleverly et. al, 1997). A total of 7 dioxins and 10 furans have been simulated. Metals (Cadmium, Arsenic, Lead and Nickel) are simulated as unspecified material into CMAQ as part of the PM2.5 part of the coarse mode. CMAQ uses three modes as suggested by Whitby (1978): 1) coarse particles; 2) Aitken particles (smaller group, I-mode) and 3) accumulation particles or mode (smaller group, j-mode). Primary emissions may also be distributed between I-mode and j-mode. The two modes interact with each other through coagulation, each mode may grow through condensation of gaseous precursors and each mode is subject to wet and dry deposition. The smaller model may grow into the larger mode and partially merge it. Figure 3 shows the emissions provided by EMEP for dioxins and furans, [5] and treated with EMIMO for high spatial and temporal resolution disagregation.

3 Results and Discussion

In Figure 4 we observe the YZ wind component and temperature in height (layers) (23 layers up to 100 mb not equally spaced) at 175500 m where O3 maximum is produced. In Figure 5 we observe the temperature and surface wind speed for the same domain and time showed in Figure 4. The complexity of the wind patters and strong gradient on wind direction in the area where the maximum is produced (175500,-40500) is shown. The tool allows a full analysis of the maximum concentrations for each analyzed pollutant together with meteorological conditions which were held on that time. The full analysis requires important computer power to generate several of thousand of figures to illustrate the different impacts (relative and absolute) for the different pollutants.

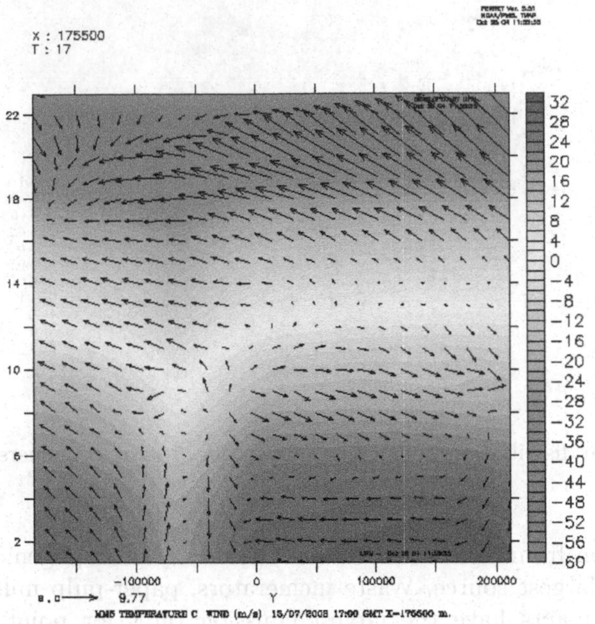

Fig. 4. Temperature (C) and YZ wind component (m/s)n at 17h00 GMT on July, 13, 2003 for the 9 km spatial resolution domain at X = 175500 m

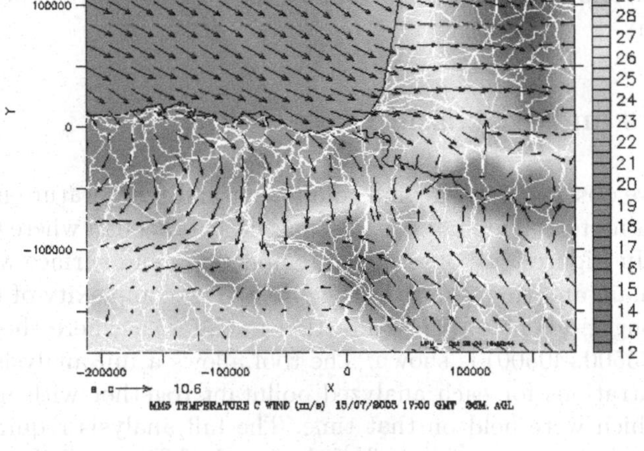

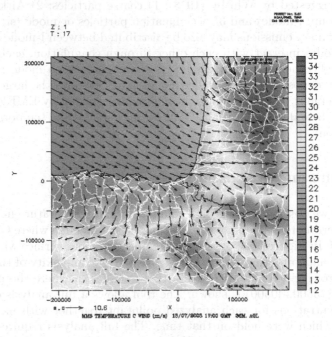

Fig. 5. Surface temperature (C) and wind speed (m/s) at 17h00 GMT, July, 13, 2004 over the 9 km spatial resolution model domain

4 Conclusions

The TEAP tool (based on the MM5-CMAQ modeling system) has been found to be a reliable tool to evaluate the air quality impact of industrial plants – in this case a waste incinerator.

The sensitivity analysis has shown that the system is capable to detect very small fractions of changes in concentrations due to the emissions of the analyzed industrial plant. The adapted CMAQ model for dioxins and furans (Hutzell W.T., 2002) and the adaptation for metals and B[a]P done for this work has been found to be an excellent tool to evaluate the impact of B[a]P, metals, dioxins and furans.

Acknowledgements

We would like to thank EUREKA network for approving the TEAP proposal and the Spanish Science and Technology Ministry for funding the Spanish section of the proposal. Also, we would like to thank the Spanish Centre for the Development of Industrial Technology for the full support received for this proposal. We also would like to thank EPA, PSU and NCAR for providing the codes for CMAQ and MM5 respectively. We also acknowledge the funding from Txingudi – Txingudiko Zerbitzuak S.A. to carry out this work.

References

1. San José R., Prieto J.F., Castellanos N., and Arranz J.M., Sensititivity study of dry deposition fluxes in ANA air quality model over Madrid mesoscale area, *Measurements and Modelling in Environmental Pollution*, Ed. San José and Brebbia, 119–130, 1997.
2. Grell G.A., Dudhia J., and Stauffer D.R., A description of the Fifth-Generation Penn State/NCAR Mesoscale Model (MM5). NCAR/TN- 398+ STR. NCAR Technical Note, 1994.
3. San José R., Cortés J., Moreno J., Prieto J.F., and González R.M., Ozone modelling over a large city by using a mesoscale Eulerian model: Madrid case study, *Development and Application of Computer Techniques to Environmental Studies, Computational Mechanics Publications*, Ed. Zannetti and Brebbia, 309–319, 1996.
4. Stockwell W., Kirchner F., Kuhn M., and Seefeld S., A new mechanism for regional atmospheric chemistry modeling. *J. Geophys. Res.*, **102**, 25847–25879, 1977.
5. Pacyna J.M., Breivik K., and Wania F., Final report for Project POPCYCLING-Baltic. EU DGXII, Environment and Climate Program ENV4-CT96-0214. Available on CD-rom including technical report, the emission and environmental databases as well as the POPCYCLING-Baltic model. NILU, P.O. Box 100, N-2027 Kjeller, Norway, 1999.

Part X

Large Scale Computation of Engineering Problems

Numerical Solution of the Heat Equation in Unbounded Domains Using Quasi-uniform Grids

Miglena N. Koleva

Center of Applied Mathematics and Informatics,
University of Rousse, 8 Studentska str., Rousse 7017, Bulgaria
mkoleva@ru.acad.bg

Abstract. Numerical solutions of the heat equation on the semi-infinite interval in one dimension and on a strip in two dimensions with nonlinear boundary conditions are investigated. At the space discretization with respect to the variable on the semi-infinite interval, we use quasi-uniform mesh with finite number of intervals. Convergence results are formulated. Numerical experiments demonstrate the efficiency of the approximations. The results are compared with those, obtained by the well known method of artificial boundary conditions.

1 Introduction

Differential problems on unbounded domains require specific techniques for their numerical treatment. When solving numerically a problem formulated on an unbounded domain, one typically truncates this domain, which necessitate setting the artificial boundary conditions (ABCs) at the newly formed external boundary. This method is often used in problems of acoustic, electro-dynamics, solid and fluid mechanics [1,10,15,16]. For almost any problem formulated on an unbounded domain, there are many different ways of classing its counterpart. In other words, the choice of the ABCs is never unique. But, very often the construction of ABCs is not easy [9,11,15,16]. Also, the minimal necessary requirement of ABCs is to ensure the solvability of the truncated problem, which leads to additional computational work. Therefore it is reasonable to be developed another methods.

In this paper we derive second order approximations (with respect to space variables), using quasi-uniform meshes (QUMs) for 1D and 2D problems with dynamical or Neumann nonlinear boundary conditions on the bounded part of domain boundary. The algorithm, we develop is effective also for blow-up solutions, because it uses decreasing time step, corresponding to the growth of the solution.

In the one-dimensional case, we consider the problem:

$$u_t = au_{xx} \quad \text{for} \quad x > 0, \quad t > 0, \tag{1}$$

$$c_0 u_t - u_x = f(u) \quad \text{for} \quad x = 0, \quad t > 0, \tag{2}$$

$$\lim_{x \to \infty} u(x,t) = u_\infty(t), \tag{3}$$

$$u(x,0) = u_0(x) \geq 0 \quad \text{for} \quad x > 0; \quad -u_0'(0) = f(u_0(0)). \tag{4}$$

I. Lirkov, S. Margenov, and J. Waśniewski (Eds.): LSSC 2005, LNCS 3743, pp. 509–517, 2006.

The function $f(u)$ often is positive and tends to infinity as $u \to \infty$. Thus, in heat flow interpretation, the condition (2) is an absorption law [14], which makes heat flow in the body (in the present paper the body is infinite), [2,14]. In [2,4,7,8], existence and nonexistence in large time of solutions of such problems are studied.

We also consider the two-dimensional problem

$$u_t = a\Delta u, \quad (x,y,t) \in \Omega = \{0 < x < l, y > 0, 0 < t < \infty\}, \tag{5}$$

$$u(0,y,t) = f_1(y,t), \quad u(l,y,t) = f_2(y,t), \quad y > 0, \quad 0 < t < \infty, \tag{6}$$

$$c_0 u_t - u_y = f(u), \; y = 0; \; \lim_{y\to\infty} u(x,y,t) = u_\infty(y,t), \; 0 < x < l, 0 < t < \infty, \tag{7}$$

$$u(x,y,0) = u_0(x,y), \quad y \geq 0, \quad 0 < x < l, \tag{8}$$

where l, a and c_0 are real numbers, f, f_1 and f_2 are given functions. This problem in comparison with the one dimensional one is less studied.

The remainder part of this paper is organized as follows. In Section 2 we define QUM and present some derivative approximations. Also, we derive second order approximation with respect to x for problem (1)-(4) and formulate two theorems for convergence. For blow-up solutions we use decreasing variable time step. The numerical experiments, given in this section illustrate the efficiency of the algorithm. The next section is devoted to the two-dimensional problem (5)-(8). Finally, we give some conclusions.

2 The 1D Heat Problem (1)-(4)

In this section we will provide an $O(\tau + N^{-2})$ approximation to the solution of the continuous problem (1)-(4).

2.1 Quasi-uniform Mesh and Space Discretization

Definition 1. [1] *Let $x(\xi)$, $\xi \in [0,1]$, $x \in [a,b]$ is strong monotone sufficiently smooth function. Then the mesh $w_N = \{x_i = x(\frac{i}{N}),\ 0 \leq i \leq N\}$ in $[a,b]$ we call quasi-uniform.*

We shall implement to our problems the meshes [1]

$$x(\xi) = -c\ln(1-\xi), \quad h_0 = x_1 - x_0 \simeq \frac{c}{N}, \quad x_{N-1} = c\ln N, \tag{9}$$

$$x(\xi) = c\xi/(1-\xi)^m, \quad h_0 = \simeq \frac{c}{N}, \quad x_{N-1} \simeq cN^m, \quad m > 0, \tag{10}$$

where $c > 0$ is controlling parameter. The choice of c is a result from the fact that the half of intervals are in domain with length $\sim c$. The last interval of (9) and (10), $[x_{N-1}, x_N]$, is infinite, but the point $x_{N-1/2}$ is finite, because the non integer nodes are given by $x_{i+\alpha} = x(\frac{i+\alpha}{N})$, $|\alpha| < 1$. Therefore, QUM transforms the infinite domain in to finite number of intervals and **states the original boundary condition directly on infinity**.

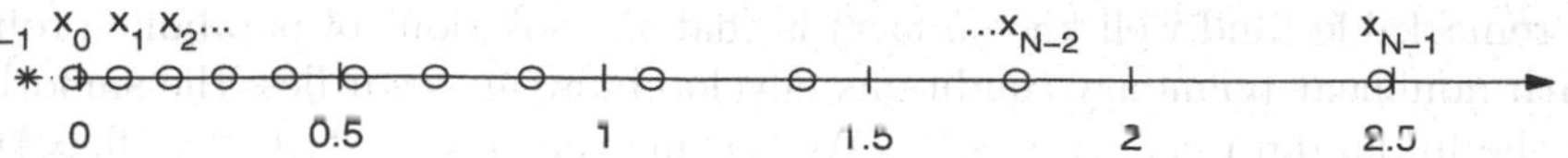

Fig. 1. Offset grid, QUM is (9), $N = 12$, $c = 1$, x_0 falls halfway between x_{-1} and x_1

We shall use the following derivative approximations

$$(\frac{\partial u}{\partial x})_{i+1/2} \approx \frac{u_{i+1} - u_i}{2(x_{i+3/4} - x_{i+1/4})},$$
$$(\frac{\partial^2 u}{\partial x^2})_i \approx \frac{1}{x_{i+1/2} - x_{i-1/2}}[(\frac{\partial u}{\partial x})_{i+1/2} - (\frac{\partial u}{\partial x})_{i-1/2}]. \quad (11)$$

The formulas contain $u_N = u_\infty(t)$, but not $x_N = \infty$. The truncation errors are of order $O(N^{-2})$. In order to obtain a second order approximation for the nonlinear boundary condition (2), we use an offset grid, see Figure 1.The point x_{-1}, outside of the domain, is fictitious. Using (11), and the second order approximation

$$(\frac{\partial u}{\partial x})_0 \approx \frac{1}{2}[(\frac{\partial u}{\partial x})_{1/2} + (\frac{\partial u}{\partial x})_{-1/2}] \quad (12)$$

(with residual term $\frac{1}{N^2}[\frac{x_{\xi\xi\xi}}{6x_\xi}u_x + \frac{x_{\xi\xi}}{2}u_{xx} + \frac{(x_\xi)^2}{6}u_{xxx}] + O(N^{-4})$), we obtain the difference equation in the node $x_0 = 0$. After an implementation of the above derivative approximations (11),(12) in (1)-(4), we obtain the next ordinary differential equations system:

$$\dot{z}_0 = \frac{2a}{H_0^1}[\frac{z_1 - z_0}{2H_0^2} - f(z_0)], \quad (13)$$
$$\dot{z}_i = \frac{a}{H_i^1}[\frac{z_{i+1} - z_i}{2H_i^2} - \frac{z_i - z_{i-1}}{2H_i^3}], \quad i = 1, ..., N-1 \quad (14)$$
$$\dot{z}_N = u_\infty(t), \quad (15)$$

where $z_i = z(x_i) \approx u(x_i, t)$, $H_0^1 = 2x_{1/2}$, $H_i^1 = x_{i+1/2} - x_{i-1/2}$, $i = 1, ..., N-1$, $H_i^2 = x_{i+3/4} - x_{i+1/4}$, $H_i^3 = x_{i-1/4} - x_{i-3/4}$, $i = 0, ..., N$.

In problem (1)-(4) a reaction term $f(u)$ at the boundary is considered and if for example $f(u) = u^p$, $p > 1$, then blow-up phenomena occurs in the sense that there exist a finite time T, such that $\lim_{t\to T} \|u(.,t)\|_\infty = +\infty$ for convenient initial data [2,4,6-8,12]. Then, the solution of the semidiscrete problem (13)-(15) also blows-up in finite time T_h. For numerical approximations of blow-up problems on **bounded domains** we refer to [5,6,12,13], the survey [3] and the references therein. In the frame of the present work we are not able to discuss for our problems such interesting questions as convergence of T to T_h, when $N \to \infty$, asymptotic behavior of the semidiscrete numerical approximations, etc.

A remarkable (and well known fact) is that the solutions of parabolic problems with nonlinear boundary conditions develop blow-up regardless the smoothness of the initial data u_0, [2,4,6-8,12]. We assume that $u \in C^{4,2}((0,\infty),(0,\infty))$ and we shall call such solution **regular**.

Theorem 1. *Let u be a regular solution of (1)-(4) and z is it's numerical approximation given by (13)-(15). Then for $\forall\ 0 < \tau < T$ there exists a constant C, independent of N, such that*

$$\max_i \max_{0\le t\le T-\tau} |u(x_i,t) - z_i(t)| \le CN^{-2}.$$

This theorem also covers the case of global existence of regular solution to problem (1)-(4).

Theorem 2. *Let T and T_h be the blow up times for u and z respectively. Then* $\lim\limits_{N\to\infty} T_h = T$.

2.2 Time Discretization

We introduce a nonuniform mesh grid in time

$$t_0 = 0, \quad t_n = t_{n-1} + \Delta t_{n-1} = \sum_{k=0}^{n-1} \Delta t_n \quad (\Delta t_k \ge 0, k = 0,1,...).$$

The choice of the time step Δt_n in the system (13)-(15) depends on the growth of the solution z, [3,13]. For the sake of simplicity, we take $f(u) = u^p$, $p > 0$. The time increment Δt_n is chosen to be variable

$$\Delta t_n = \tau \times \min\{1, \frac{H_0^1 + 2ac_0}{2a\,\|z\|_{\infty(or2)}^{p-1}}\}, \quad \tau = \max_{0\le k\le n-1} \Delta t_k.$$

The full discretization of the problem (1)-(4), $f(u) = u^p \approx (z^n)^p z^{n+1}$, $i = 1,...,N-1$ is as follows:

$$\begin{aligned}
&[1 + \frac{a\Delta t_n}{H_0^1 + 2ac_0}(\frac{1 - 2(z_0^n)^{p-1}H_0^2}{H_0^2})]z_0^{n+1} - \frac{a\Delta t_n}{H_0^2(H_0^1 + 2ac_0)} z_1^{n+1} = z_0^n, \\
&-\frac{a\Delta t_n}{2H_i^1 H_i^3} z_{i-1}^{n+1} + (1 + \frac{a\Delta t_n}{2H_i^1}(\frac{1}{H_i^2} + \frac{1}{H_i^3}))z_i^{n+1} - \frac{a\Delta t_n}{2H_i^1 H_i^2} z_{i+1}^{n+1} = z_i^n, \quad (16) \\
&z_N^{n+1} = u_\infty(t^{n+1}).
\end{aligned}$$

2.3 Computational Results

The aim of the numerical experiments is to show the convergence rate and to compare the accuracy of the algorithms, using QUM and ABCs (Example 1); also, to test the efficiency of both QUMs (9) and (10) (Example 2).

A method of ABCs, is presented in [16] for one-dimensional parabolic equation defined in semi-infinite interval with Dirichlet boundary condition at the left end. We developed this algorithm to the case of nonlinear (dynamical and Neumann) boundary conditions for 1D and 2D problems. The details are given in the submitted paper [11]. Here we draw the analogy between QUM method (16) (QUMM) and the exact artificial boundary method, [11], using QUM for computing the truncated problem (ABCQUMM). The approximation of ABC we derive in the same manner as for (2). The examples are chosen such that the construction of ABCs is possible in sense of [11,16], i.e. $u_\infty = 0$ and supp $u_0 < \infty$.

Example 1. The test problem is (1)-(4), where $f(u) = u^2 + \tilde{f}(t)$, $a = 1$, $u_0(x) \equiv 0$, $u_\infty = 0$, $\tilde{f}(t)$ is chosen such that the exact solution is $u(x,t) = \mathrm{erfc}(\frac{x+2}{2\sqrt{t}})$, $\mathrm{erfc}(x) = \frac{2}{\sqrt{\pi}} \int_x^\infty e^{-\lambda^2} d\lambda$, $\tilde{f}(t)$ is different for $c_0 = 0$ and $c_0 = 1$. Let denote by E_∞^N ($E_\infty^N = \|E_i\|_c$) and E_2^N the errors in corresponding discrete max and L_2 norms, obtained using mesh with $N+1$ nodes. Instead of the standard error E_2^N, for method of QUM we use the estimate $E_2^N = [\frac{1}{2} H_0^1(E_0)^2 + \sum_{i=1}^{N-1} H_i^1(E_i)^2]^{1/2}$. The ratio $\tau N^2 = 1$ is fixed. The computations are performed up to time $t = 1$ with QUM (9). The convergence rate (CR) is valued by the quantities $CR_{\infty(2)} = \log_2[E_{\infty(2)}^N / E_{\infty(2)}^{2N}]$.

Having exact solution, we compare the efficiency of the algorithms, imposing the exact boundary condition in x_{N-1}: $u(x_{N-1}, t) = \mathrm{erfc}(\frac{x_{N-1}+2}{2\sqrt{t}})$ on QUM (DQUM). In Table 1 we give the results for the case of dynamical ($c_0 = 1$) boundary conditions. There is no essential difference with the case $c_0 = 0$. The computations confirm the efficiency of QUMM and it's convergence order 2. The approximation of ABC (ABCQUMM) uses the discrete values of the artificial boundary from all previous time levels. That's way, the accumulation of round off error is possible. Nevertheless, ABCQUUMM is more stable in time than QUMM. For example, at t=2: $E_\infty^{24} = 6.926715e-4$ for QUMM, $E_\infty^{24} = 3.063709e-4$ for ABCQUMM and $E_\infty^{24} = 1.526854e-4$ for DQUM.

For $\tilde{f}(t) \equiv 0$ the solution blows-up in finite time, [8]. Numerical experiments for such solutions are discussed in [11].

Table 1. Global errors in different norms, $c_0 = 1$, QUM is (9)

N	QUMM		ABCQUMM		DQUM	
	E_∞^N	E_2^N	E_∞^N	E_2^N	E_∞^N	E_2^N
12	3.683837e-4	3.523132e-4	3.687771e-4	3.79614902e-4	3.681274e-4	3.369317e-4
24	9.323924e-5	8.858261e-5	9.324177e-5	9.206816e-5	9.323728e-5	8.703128e-5
48	2.332366e-5	2.206979e-5	2.332367e-5	2.273758e-5	2.332365e-5	2.202332e-5
96	5.830883e-6	5.517726e-6	5.830883e-6	5.587760e-6	5.830883e-6	5.507968e-6
192	1.473248e-6	1.394342e-6	1.473248e-6	1.397895e-6	1.457999e-6	1.377982e-6

Example 2. We use QUM (10) for solving the problem from Example 1. All input datum are the same, except the mesh. We give the results in Table 2.The efficiency of (9) and (10), $m = 1$ are close for this problem.

Table 2. Global errors in different norms and convergence rate, $c_0 = 1$, QUM is (10)

N	$m = 1$				$m = 2$			
	E_∞^N	CR_∞	E_2^N	CR_2	E_∞^N	CR_∞	E_2^N	CR_2
12	3.236310e-4		2.604571e-4		4.526466e-4		4.588175e-4	
24	8.221002e-5	1.9770	6.501900e-5	2.0021	1.066523e-4	2.0855	1.165572e-4	1.9769
48	2.056667e-5	1.9990	1.627574e-5	1.9981	2.645880e-5	2.0111	2.909814e-5	2.0020
96	5.141504e-6	2.0000	4.069750e-6	1.9997	6.604643e-6	2.0022	7.273963e-6	2.0001
192	1.285694e-6	1.9996	1.017630e-6	1.9997	1.654158e-6	1.9974	1.817768e-6	2.0006

3 The 2D Heat Problem

In this section we construct second order (with respect to space variable) approximation of the differential problem (5)-(8) by QUMM.

3.1 Discretization of (5)-(8)

The mesh in x direction is classical: uniform grid with $M + 1$ nodes and mesh step size k, while in y direction we use QUM (9) or (10) with $N + 1$ mesh points, see Figure 2. As in the one-dimensional case, we use fictitious nodes: $M - 1$ numbers, situated under the bottom boundary, h_0 away. The elimination is standard. Using (11) for $\frac{\partial^2 u}{\partial y^2}$, central approximation for $\frac{\partial^2 u}{\partial x^2}$ and (12) for $\frac{\partial u}{\partial y}$, we obtain the following semidiscretization of the problem (5)-(8):

$$z_{0j} = f_{10j}, \quad z_{Mj} = f_{2Mj}, \qquad j = 1, ..., N - 1, \tag{17}$$

$$z_{iN} = u_\infty(x_i, t), \qquad i = 1, ..., M, \tag{18}$$

$$\dot{z}_{ij} = \frac{a(z_{ij+1} - z_{ij})}{2H_j^2 H_j^1} - \frac{a(z_{ij} - z_{ij-1})}{2H_j^3 H_j^1} + \frac{a(z_{i+1j} - 2z_{ij} + z_{i-1j})}{k^2}, \tag{19}$$

$$\dot{z}_{i0} = \frac{aH_0^1 \left[z_{i+1,0} - 2z_{i0} + z_{i-1,0}\right]}{(H_0^1 + 2ac_0)k^2} + \frac{2a}{(H_0^1 + 2ac_0)} \left[\frac{z_{i1} - z_{i0}}{2H_0^2} + f(z_{i0})\right], \tag{20}$$

$$i = 1, ..., M - 1, \qquad j = 1, ..., N - 1.$$

Now, $z_{ij} = z(x_i, y_j) \approx u(x_i, y_j, t)$ and H_j^s, $s = 1, 2, 3$ are the same as in (13)-(15), but $x \leftrightarrow y$ $(i \leftrightarrow j)$.

Results of convergence, analogical to Theorem 1,2, hold for the 2D problem.

Remark 1. The QUMM can be applied without difficulties to the cases of more complicated boundary conditions. For example, if instead of $u(0, y, t) = 0$, we have $c_1 u_t - u_x = g(u)$, we will use $M + N$ fictitious nodes to obtain second order

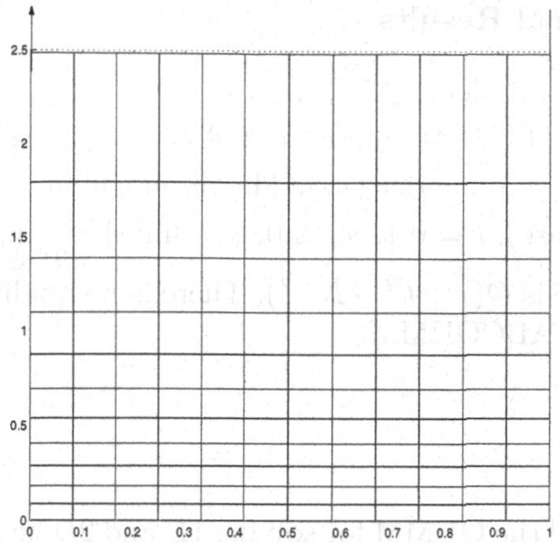

Fig. 2. QUM (9), N=M=12, $c = 1$

Table 3. Global errors in different norms and CR, $M = N$, $c_0 = 1$, QUM is (9)

N/CR	E_∞^N	E_2^N
12	4.058624e-3	2.070943e-3
24	1.084095e-3	5.283785e-4
CR	1.9045	1.9706
48	2.731855e-4	1.326197e-4
CR	1.9885	1.9943
96	6.864776e-5	3.318021e-5
CR	1.9926	1.9989
192	1.713742e-5	8.283924e-6
CR	2.0020	2.0019

approximation. For approximation of the solution in the corner node $x = 0$, $y = 0$ we use two fictitious nodes. Thus for $i = 0$ and $j = 0, ..., N-1$ we have

$$\dot{z}_{00} = \frac{2kaH_0^1}{kH_0^1 + 2a(c_1H_0^1 + c_0k)}\left[\frac{z_{10} - z_{00}}{k^2} + \frac{g(z_{00})}{k} + \frac{z_{01} - z_{00}}{2H_0^1H_0^2} + \frac{f(z_{00})}{H_0^1}\right],$$

$$\dot{z}_{0j} = \frac{2a}{k+2ac_1}\left[\frac{z_{1j}-z_{0j}}{k} + g(z_{0j})\right] + \frac{ak}{H_j^1(k+2ac_1)}\left[\frac{z_{0j+1}-z_{0j}}{2H_0^2} - \frac{z_{0j}-z_{0j-1}}{2H_0^3}\right].$$

Now, it's very easy to obtain the full discretization in the case $f(u) = u^p \approx (z^n)^{p-1}z^{n+1}$, $p > 1$ ($g(u) = u^q \approx (z^n)^{q-1}z^{n+1}$, $q > 1$) by substituting $\dot{z}$ with $\frac{z^{n+1}-z^n}{\Delta t_n}$ and write all unknown values z on the $n+1-th$ time level.

The time step in (17)-(19) is chosen as follows

$$\Delta t_n = \tau \times \min\{1, \frac{H_0^2(H_0^1 + 2ac_0)}{a\,\|z\|_{\infty(or2)}^{p-1}}\}, \quad \tau = \max_{0 \le k \le n-1} \Delta t_k.$$

3.2 Computational Results

Example 3. The test problem is (5)-(8), $f_1(y,t) = \mathrm{erfc}(\frac{y}{2\sqrt{t}})$, $f_2(y,t) = \mathrm{erfc}(\frac{l+y}{2\sqrt{t}})$, $u_\infty = 0$ and instead of (7), we impose $u(x,0,t) = \mathrm{erfc}(\frac{x}{2\sqrt{t}})$, $l = 1$, $a = 0.5$. The exact solution is $u(x,y,t) = \mathrm{erfc}(\frac{x+y}{2\sqrt{t}})$. The results are in Table 3 (L_2 error is on the analogy of 1D case), $t = 0.1$, $u_0 \equiv 0$, $\tau = [\min\{k, \min_{0 \le i \le N-1} h_i\}]^2$. Obviously, the convergence rate is $O(\tau + k^2 + N^{-2})$. There is no qualitative difference with results, obtained by ABCQUMM.

4 Conclusions

In this paper we used the QUMM for solving 1D and 2D heat problems with nonlinear dynamical ($c_0 = 1$ or $c_1 = 1$) or Neumann ($c_0 = 0$ or $c_1 = 0$) boundary conditions. We draw the analogy between two algorithms: QUMM and ABCQUMM. Both methods are with convergence order 2 in space and 1 in time and could be applied efficiently for solving problems, defined on unbounded domains. But it is easy to obtain higher order convergence rate, also with respect to time variable, using for example Crank-Nickolson (including half nodes), [16] or Rozenbrok-Vanner (two-step) schemes, see [1].

The efficiency of the QUMM is very close to those of the ABCQUMM. Indisputably, the most important advantage of the QUMM is that the method is easy applicable for a wide class of problems (including nonlinear problems).

Acknowledgments

This research is supported by Bulgarian National Fund of Science under Project HS-MI-106/2005.

References

1. Alshina, E., Kalitkin, N., Panchenko, S.: Numerical solution of boundary value problem in unlimited area. Math. Modelling 14 , **11** (2002) 10–22 (in Russian)
2. Andreuci, D., Gianni, R.: Global existence and blow up in a parabolic problem with nonlocal dynamical boundary conditions. Adv. Diff. Eqn. 1 **5** (1996) 729–752
3. Bandle, C., Brunner, H.: Blow-up in diffusion equations: A survey J. Comput. Appl. Math. **97** (1998) 3–22
4. Bonder, J., Rossi, J.: Life span for solution of the heat equation with a nonlinear boundary conditions. Tsukuba J. of Math. 25 **1** (2001) 215–220
5. Dimova, S., Kaschiev, M., Koleva, M., Vassileva, D.: Numerical analysis of the blow-up regimes of combustion of two-component nonlinear heat-conducting medium. J. Vych. Math. Math. Phys. **35** (1994) 380–399 (in Russian)
6. Ferreira, R., Groisman, P., Rossi, J.: An adaptive numerical scheme for a parabolic problem with blow-up. IMA J. Numer. Anal. **23** 3 (2003) 439–463

7. Fila, M. and Quittner, P.: Large time behavior of solutions of semilinear parabolic equations with nonlinear dynamical boundary conditions, in: Topics in Nonlinear Analysis. Progr. Nonl. Diff. Eq. Appl. **35** (1999) 251–272
8. Galaktionov, V., Levine, H.: On critical Fujita exponents for heat equations with nonlinear flux conditions of the boundary. Israel J. of Math. **94** (1996) 125–146
9. Givoli, D.: Numerical methods for problems in infinite domains. Elsevier, Amsterdam (1992)
10. Han, D., Huang, Z.: A class of artificial boundary conditions for heat equation in unbounded domains. Comp. Math. Appl. **43** (2002) 889–900
11. Koleva, M., Vulkov, L.: Numerical solution of the heat equation with nonlinear boundary conditions in unbounded domains. (submitted)
12. Koleva, M., Vulkov, L.: On the blow-up of finite difference solutions to the heat-diffusion equation with semilinear dynamical boundary conditions. Appl. Math. Comput. **161** (2005) 69-91
13. Nakagava, T.: Blowing up of a finite difference solution to $u_t = u_{xx} + u^2$. Appl. Math. Opt. **2** (1976) 337–350
14. Samarskii, A., Tihonov, A.: Eqnuation in Mathematical Physics. MGU Moskow (1999) (in Russian)
15. Tsynkov, S.: Numerical solution of problems on unbounded domains. A review. Appl. Numer. Math. **27** (1989) 465–532
16. Wu, X., Sun, Z.: Convergence of difference scheme for heat equation in unbounded domains using artificial boundary conditions Appl. Numer. Math. **50** (2004) 261–277

On Efficient Distribution of Data in Multicast Networks: QoS in Scalable Networks

Francis Lowu and Venansius Baryamureeba

Faculty of Computing & IT, Makerere University, P.O. Box 7062, Uganda
lowfrancs@yahoo.com, barya@cit.mak.ac.ug, barya@ii.uib.no
http://www.ii.uib.no/barya, http://www.cit.ac.ug

Abstract. Multimedia applications or real-time applications such as audio and video on demand, teleconferencing and whiteboard sharing require Quality of Service (QoS) guarantee assurance. QoS constraints, namely required bandwidth, end-to-end delay and delay jitter are the major parameters that need to be satisfied in order to have quality assurance in a dynamic multicast network.

In this paper we investigate the effect of the QoS constraints on multicast network applications and services during the processes of routing. Lastly, we propose linear routing tree algorithms for required bandwidth, end-to-end delay and delay jitter.

1 Introduction

1.1 Background

When Internet technology develops and grows, applications become more distributed and the need for group communication arises in the networks. This allows nodes in the network to share information. However there is a need for efficient and equally distributed QoS in the network during sessions, which can be provided within a multicast network that is responsible for multipoint information delivery. Internet Protocol(IP) and Asynchronous Transfer Mode (ATM) support multipoint information delivery in scaling networks, since most multicast networks use IP based services with support of User Datagram Protocol (UDP), the best-effort multicast service [11, 12]. It was found out [12] that best-effort multicast services do not support QoS guarantee such as message delivery, ordering, throughput and transmission delay. Transmission control protocol is reliable compared to UDP in ensuring that the delivered data stream is sequential. However it lacks the multicast applications requirement which support multipoint information delivery, but supports only peer-to-peer connections providing almost no support for latency control and uses only the go-back-n algorithm [11] for any error occurrence and recovery which brings about misuse of network resources. The receivers can also send information to the sender (source) giving an information about the packets that have not been received or that are lost due to error. After the information sent by the receiver has been verified and processed, a proof of whether the packets were lost is done and also whether there is need

I. Lirkov, S. Margenov, and J. Waśniewski (Eds.): LSSC 2005, LNCS 3743, pp. 518–525, 2006.

for a retransmission [11]. The previous research in multicast transport focused on providing reliable and ordered delivery sequence to distributed systems or groups [12]. Membership in these multicast groups is highly dynamic with group members leaving without permission or explicit knowledge of other hosts, while others join the groups. This changing of membership in the groups makes the integrity of the data transmitted along the multicast network lack efficiency. In case of many-to-many and one-to-many multicast and multimedia applications, TCP/IP and ATM network services become more inadequate [15]. Multicast protocols are important since they are used to bridge the gap between the demanding applications and the inadequate multicast networks. These protocols are needed to provide ordering, reliability, group management, error recovery and also to compensate bandwidth, latency receiver feedback rate and flow control requirements of multimedia applications [12]. Multimedia applications also must deal with the scalability and heterogeneity problems usually found in the distributed multimedia applications that use multicast technology in large-scale network infrastructure. If the scalability problem is dealt with, it will help in solving the QoS requirements for the dynamic multicast networks.

1.2 Overview

In Section 2 we formulate the problem statement; in Section 3 we discuss related work; in Section 4 we present the results and lastly in Section 5 we give concluding remarks.

2 Problem

In this paper we consider multicast networks running distributed applications and services that face a problem of QoS in scalable networks. This problem impacts greatly on the efficiency of distributed applications and services in large-scale networks. Multicast groups are dynamic and therefore the group members have different QoS requirements for any scaling network. Therefore there is need for an efficient routing tree algorithm that could improve data distribution in scaling multicast networks.

3 Related Work

Many multicast routing protocols have been proposed. However the QoS requirements in the network for efficient routing have not been fully solved. There is a need for more efficient routing algorithms for distributed multicast networks. Jose [11] focused on how to provide a full reliable data transmission to a large number of receivers, bandwidth saving and maintaining a suitable performance in heterogeneous environment. For reliable data transfer all receivers get an error free copy of the transmitted data. Here multicast flow control and congestion control issues were not well discussed. Jasleen *et al.* [9] worked on the effect of TCP on assigning higher priority to traffic requesting expected forwarding (EF) services in a differentiated services network. Here EF traffic occupied different fractions of link bandwidth and is bursty at different time-scale. Work on evaluating the effect of assigning higher priority to EF traffic on TCP throughput was left as an area that needed more research. Stefan *et al.* [14] proposed the budget based network admission control (NAC) and categorized it into four basically distinct approaches depending on the complexity and efficiency. Here the network size, connectivity and the internal structure of the network have a significant impact on the resource efficiency. Vijay *et al.* [15] worked on the problem of dealing with the great heterogeneity in QoS of wide area clients. It was found out that it is important because longer transmission times lead to tying up of kernel resources at the high-speed recipients. This limits the performance of the high bandwidth recipients. However the work didn't addressed how QoS and heterogeneity behave on the scaling multicast group. Alvarez *et al.* [1] describes the new minimum λ-tree multicast protocol. The periodic activation of the protocol allows maintaining a multicast tree that spans the group of users. In this protocol a specified QoS parameter λ along the route from the core to each member is optimized. An algorithm was proposed which partitioned a set of recipients on the basis of the available bandwidth. The algorithm separates transmission to each set of recipients sharing a common QoS. Gregory [7] worked on the management of QoS at the application level. The work considered all the system components involved in the application such as the communication network in order to manage the QoS for distributed multimedia applications. Here the scalability QoS management for the application involving a very large number of members or users and working on possibly many servers was not addressed. Bin Wang *et al.* [4] worked on the comprehensive overview of the existing multicast routing algorithm, protocols and their QoS extensions. Classification of multicast routing problems according to their optimization function and performance constraints was considered and also presented routing algorithms. Here a gap was left on the QoS-driven multicast routing in large-scale, real network to provide some insights into the trade off between the design complexity of QoS-driven protocol and the resulting performance. Shigang *et al.* [13] present a scalable QoS multicast Routing Protocol that shares the adaptive path-branching idea of QoS Multicast Routing Protocol (QMRP), but it eliminates the temporary use of per-group-per-join routing state. Jinquan *et al.* [10] propose a simpler and more effective mechanism of QoS guaranteed multicast routing in which a feasible multicast tree routing multiple QoS constraints can be constructed in a distributed fashion using local states at the routers. Dean *et al* . [6] investigated the problem of optimal resource allocation for end-to-end QoS requirements on unicast paths and multicast trees. Their work considered a framework in which resource allocation is based on local QoS requirements at each network link, and associated with each link is a cost function that increases with the severity of the QoS requirement. It addressed the problem of how to partition an end-to-end QoS requirement into local requirements. Hung-Ying *et al.* [8] address the problems of constructing both source-based and core-based many-to-many multicast trees for applications with delay and delay jitter constraints. Here the source-destination delay bound and the inter-destination delay jitter bound are used as the QoS requirement and formulate the delay and delay jitter constrained Many-to Many Multicast Tree problems. Dean *et al.* [5] solved the problems related to supporting unicast and multicast connections with QoS requirements. They investigated the problem of optimal routing and resource allocation in the context of performance dependent costs. Also considered was the "additive" QoS constraints like the delay and general link costs. Ariel Orda *et al.* [2] proposed the precomputation-based methods as an instrument to facilitate scalability, improve response time, and reduce computation load on network elements. The key idea was to effectively reduce the time needed to handle an event by performing a certain amount of computations in advance, that is prior to the event's arrival. However the work did not consider the delays that may be encountered during the event. Bin Wang *et al.* [3] study QoS routing for quorumcast, a generalized form of multicast communication. Here the need of quorumcast communication arises in a number of distributed applications such as distributed synchronization and resource discovery, but little work has been done on routing quorumcast messages with or without QoS constraints Stefan *et al.* [14] worked on the budget based network admission control (NAC). Their work was categorized into four basically distinct approaches depending on their complexity and efficiency. Comparison was made on the resource utilization in different network scenarios. It was found out that the network size; connectivity and the internal structure of the network have a significant impact on the resource efficiency. The performance of the NAC is limited by the topology while the other when offered large enough load they can achieve a very high utilization.

4 Results

The network is represented by an undirected graph $G(V, E)$ of V vertices and E edges, where $|V|$ represents the number of nodes while $|E|$ represents the number of edges. The communication links are denoted by set of E edges. We consider that for every n nodes there are $n-1$ links. The links are queued and therefore the delay measure on the link $l \in E$ is defined by d_l and the delay bound in the entire multicast tree is represented by Δ on any path in the tree. We consider QoS constraints registered at any link or path by the group leader at the router node. If the registered QoS are received by the source node from the entire multicast tree, a depth first search (DFS) is done for each node in the network. For every visited or searched node, the delay from the source to that specific node $v_i.delay$ and $u_i.delay$, for in this case called the current node, is computed for the i_{th} nodes v_i and u_i respectively. A node can be visited once or more than once depending on its position in the multicast network. Calculations of the delay bound Δ is made for every visited nodes to compare with the link delays.

Following below are the proposed algorithms for end-to-end delay, required bandwidth and delay jitter respectively.

4.1 Algorithm 1: End-to-End Delay

We consider that given an undirected graph $G(V, E)$ with a set of V vertices and E edges, we find a path with an end-to-end delay measure d_l that satisfies the condition

$$\sum_{\forall l \in E} d_l \leq \Delta$$

and for which the delay functions $u_i.delay$ and $v_i.delay$ are minimum for the ith nodes. Here Δ denotes the end-to-end delay bound.

Input: A graph $G(V, E)$, number of nodes, a set of QoS constraints, delay bound

Output: A path that satisfies the QoS requirement.

```
begin
for all vertices V do
    u_i.delay = ∞, v_i.delay = ∞
end for
for all (u_i, v_i) ∈ V where i = 1 to n − 1 do
    while l(u_i, v_i) ∈ E do
      DFS(u_i)
    label (u_i, v_i) visited
        if u_i = visited and v_i = visited
            record u_i.delay and v_i.delay
                elseif d_l + u_i.delay ≤ v_i.delay and
                    d_l + u_i.delay ≤ Δ or
                        v_i.delay > Δ then
                d_l + u_i.delay < v_i.delay
        else discard the message
            return a path with minimum delay
end for
```

In the above end-to-end delay algorithm each message sent along the link is assigned a minimum and a maximum delay bound (duration) at which it should arrive at the destination node in the multicast tree. If the message (packet) exceeds the maximum duration then it is discarded or labeled as a nullified message and therefore a retransmission is made along the same path.

4.2 Algorithm 2: Required Bandwidth

Here we also consider an undirected graph $G(V, E)$ of V vertices and E edges, we are interested in finding a path with maximum residue bandwidth, such that

$$c_l - b_l \leq r_b$$

where c_l is the capacity of the link, b_l is the link bandwidth and r_b is the required bandwidth.

Input: A graph $G(V, E)$, a set of QoS constraints, required link bandwidth (min, max), capacity of the link.
Output: The best residue path.

begin
for all vertices V and edges E do
  $b_l := 0$, and $c_l := \infty$
end for
for all $(u_i, v_i) \in V$ where $i = 1$ to $n - 1$ do
    while $l(u_i, v_i) \in E$ do
      visit u_i and v_i
  if $u_i := labelvisited(), v_i := labelvisited()$
    determine link $l(u_i, v_i) \in E$
      find the capacity of the link $l(u_i, v_i)$
        assign bandwidth to link $l(u_i, v_i)$
          elseif the assigned bandwidth is min
        and the capacity of the link is max
          $l(u_i, v_i)$:= residue path;
        residue path := $c_l - b_l$
        else stop
    return residue path
end for

The algorithm above proposes a way in which we can choose a path with the residue bandwidth which is labeled as the residue path. The algorithm stops if the assigned bandwidth is greater than the capacity of the link.

4.3 Algorithm 3: Delay Jitter

For any given least delay path in an undirected graph $G(V, E)$ of V vertices and E edges

$$delay - jitter = \sum_{\forall l \in E} (delay - jitter)$$

Input: A graph $G(V, E)$, number of nodes, source node v_s, a set of QoS constraints, delay bound, delay requirement.
Output: The link or path satisfies required QoS.

begin
for all $(u_i, v_i) \in V$ where $i = 1$ to $n - 1$ do
  $u_i.delay := \infty$, $v_i.delay := \infty$
end for
for all $l(u_i, v_i) \in E$ do
    if current node (u_i, v_i) is in the tree
        DFS(u_i)
      label (u_i, v_i) visited or unvisited
        compute $u_i.delay$ and $v_i.delay$
      compute $u_i.t$ and $v_i.t$

else if $d_l + u_i.delay \leq v_i.delay$ then
$v_i.t \leq v_i.delay$ and $v_i.delay \leq \Delta$
$jitter := v_i.t$
else
node not in the tree
return jitter;
end for

4.4 Discussion of Results

The end-to-end delay algorithm is proposed in this paper. The algorithm identifies a path from the source to the destination node, which is in this case taken as the current node. The delay measure of the link d_l is calculated in the process and also the delay of the packets from the source to the current i_{th} nodes u_i and v_i is calculated, that is $u_i.delay$ and $v_i.delay$. The delay bound Δ for each path from the source to the destination is known since the constraint is an additive parameter. The algorithm finds a path with a link delay d_l plus the delay to the current node, $v_i.delay$ and $u_i.delay$ that does not exceed the delay bound Δ for each link created in the tree and whose nodes are labeled as visited by the depth first search (DFS).

In the residual bandwidth algorithm we focus on the maximum residue path in the network where a given bandwidth is assigned to the link depending on the capacity of the link. The residual bandwidth is either maximum or minimum. We use the preorder tree traversal (equivalent to DFS) to search all the nodes, the searched nodes are labeled as visited. For all the visited i nodes a determination of the link $l(u_i, v_i)$ takes place and then the capacity of the link is determined after which assignment of bandwidth on that link is done. Comparission of whether the assigned bandwidth on the link is not greater than the capacity of the link is done and if its found that the link capacity is greater then the residual bandwidth is computed else the algorithm stops. This process is for all nodes in the multicast tree.

The delay jitter algorithm proposed finds a path with the least transmission time basing on the delay algorithm. We consider that after the nodes u_i and v_i have under gone a search to show that they are in the tree, calculations for the link delay d_l, the delay functions $u_i.delay$ and $v_i.delay$ are done. The functions $u_i.t$ and $v_i.t$ which represents the time taken for the packets to reach the current nodes u_i and v_i are compared each with the delay from the source to the destination. If its found that $u_i.t$ is less than the function $u_i.delay$; then delay jitter is computed else the algorithm stops which implies that the node is either not in the tree or has not yet fully joined the tree. We focus on the inner **for loop** which determines the frequency of each line in the algorithms proposed, during the process of execution. The complexity of all the proposed algorithms is linear and given by $\bigcirc(n)$ (a linear complexity) implying that the algorithms route data efficiently throughout a scaling multicast network.

5 Concluding Remarks

The QoS constraints, i.e. end-to-end delay, delay jitter and required bandwidth have been discussed in relation to scalable networks. Routing tree algorithms for end-to-end delay, delay jitter and required bandwidth have been proposed. All the algorithms have a time complexity of $\bigcirc(n)$ implying that they have a linear complexity for a linear search throughout the network.

References

1. Alvarez-Hamelin Ignacio, J., and Fraigniard P.: M λ T: A Multicast Protocol with QoS Support: Center for research in wireless mobility and Networking, Department of Computer Science and Engineering University of Texas at Arlington. (1999).
2. Areiel, O., and Alexander, S.: Precomputation schemes for QoS routing, IEEE/ACM transactions on networking. (2003).
3. Bin, W., and Jennifer Hou, C.: An Efficient QoS Routing Algorithm for Quorumcast Communication. (2003).
4. Bin, W., and Jennifer Hou, C.: Multicast Routing and Its QoS Extension: Problems, Algorithms and Protocol. *Department of Electrical Engineering. Ohio State University Columbus, OH 43210.*(2000).
5. Dean Lorenz, H., Ariel, O., and Danny, R.: Optimal partition on QoS requirements for many-to-many connections, *IEEE INFOCOM 2003. The conference on computer communications, March.* **22** (2003).
6. Dean Lorenz, H., Ariel, O.: Optimal Partition on QoS requirements on Unicast paths and Multicast tree, IEEE/ACM transactions on networking. (2002).
7. Gregory Bechmann: QoS Management at the Application level. *University of Ottawa Canada.* (1998).
8. Hung-Ying, T., Jennifer, C., Hou B. W.: Many to Many Multicast Routing With Temporal Quality of Service Guarantees *IEEE transactions on computers, July.* **52** (2003).
9. Jasleen, K., Harrick Vin, M.: Effect of Higher Priority EF Traffic on TCP Throughput and Fairness *Technical report Tr2001-1 Department of Computer Science University of Texas at Austin.* **1** (2002).
10. Jinquan, D., Hung Keng, P., Touchai, A.: A multicast Routing Protocol Supporting QoS Parameters. (2002).
11. Jose de Rezende, F.: Scalability Issue for Reliable Multicast Protocol, University of Pierre at Marie Curie: Place jussier-75252 Paris Cedex OS France. (1999).
12. Lucas Mathew, T.: Efficient data distribution in large-scale multicast networks *A PhD Thesis.* (1998) 104–149.
13. Shigang, C., Yoval, S.: A Scalable Distributed QoS Multicast Routing Protocol. *Technical Report, University of Florida 32611 USA.* (2004).
14. Stefan, M., Kopt, S., Jen, M.: Impact of Network Topology on the Performance of Budget Based Network Admission Control Methods. Technical Report. (2003).
15. Vijay, G., Campbell, R.: Reliable Sender-initiated Multicast for Improved QoS. (2002).

A Computational Approach on the Multitask Optimization of Inclined Slider Bearing Performance with Upper-Surface-Waviness

B. Turker Ozalp[1] and A. Alper Ozalp[2]

[1] Department of Industrial Engineering, Uludag University, Bursa, Turkey
tozalp@uludag.edu.tr
[2] Department of Mechanical Engineering, Uludag University, Bursa, Turkey
aozalp@uludag.edu.tr
http://www.geocities.com/aaozalp/

Abstract. The purpose of this paper is to propose an optimistic upper surface design, by applying a wavy pattern, for the slider bearing lubrication environment, without going beyond the geometric limits of the complete flow volume. Continuity, momentum and energy equations are handled simultaneously by interpreting the relation of lubricant motion and pressure distribution through a *Transfer Matrix*; the temperature dependent character of viscosity is considered in the computations with a convergence criterion of 0.01% for two consecutive temperature distributions within the implemented iterative approach. Numerical investigations are carried with wave amplitude and wave number ranges of 0-200 μm and 5-105 respectively and the pumping pressures are 1.01-3.01 times the exit value. The computational results point out an optimum upper surface design with a wave number range of 25-45, which not only increases the load capacity but also decreases the power requirement.

1 Introduction

During recent years, the interest in computer modeling of thermo-hydrodynamic-lubrication [9] has continuously increased, since the needs have expanded enormously, including the requirements like increased capacity, lowered power consumption and creative designs which will meet the necessities of the present technology. As the behaviors of non-linear designs often demonstrate unexpected output patterns, analysis on various sliding surface definitions is important in predicting the system responses. Application of a wavy pattern to the flow surfaces is a new approach and complicates the numerical-lubrication simulations, as the waviness is defined by 2 independent variables: the amplitude and the wavelength.

Effects of surface waviness on the lubrication process have been handled numerically in a few recent studies. Van Ostayen et al. [14] investigated the performance of a hydro-support and determined significant variations in lubricant flow rate by up to 60% due to the influence of random waviness. Honchi et al. [3] applied a micro-waviness model to an air slider bearing, where the contact force

I. Lirkov, S. Margenov, and J. Waśniewski (Eds.): LSSC 2005, LNCS 3743, pp. 526–534, 2006.

records were of fluctuating type. Kwan and Post [5] proposed that the load value of the aerostatic bearings increases with higher wave amplitude and decreases with increased number of waves. Harsha et al. [2] developed a numerical model that takes the sources of nonlinearities, such as surface waviness, into account for ball bearing applications. Influence of waviness on cylindrical sliding element was considered by Rasheed [11], who proposed a critical wave number range of 1-9 for improved operating conditions. Mehenny and Taylor [8] also studied journal bearings and found out that the maximum pressure increased with wave number; however the power loss appeared not to be influenced. On the other hand Ai et al. [1] showed that the lubricant film thickness decreased with waviness in journal bearings. The numerical model for journal bearing systems of Liu et al. [7] did not converge efficiently for waviness amplitudes above 6 μm. Roller bearings are studied by Sottomayor et al. [12] for various waviness amplitude values and augmented friction coefficients are recorded at higher amplitude cases.

Although the lubricant viscosity is highly dependent on flow temperature, the recent studies on waviness, presented above, considered the viscosity to be constant in the complete flow domain due to the complexity of the numerical structure when both the surface definition and the viscous character are numerically involved. However there have been many studies that focus on the effects of temperature and lubricant properties on the performance of bearings. A linearly narrowing slider bearing, with heat conduction to the stationary lower surface was investigated numerically by Kumar et al. [4]. Temperature dependency of lubricant viscosity was handled by imposing the temperature distribution of the previous solution set on the nodal viscosity values, until a convergence of 0.05% was achieved, for each node, between two successive solutions. A similar approach was used by Pandey and Ghosh [10] on both sliding and rolling contacts. Their convergence criterion for temperature distribution was less sensitive (0.1 %) and a unique viscosity value, which corresponds to the average lubricant temperature, was used for the complete flow volume. On the other hand, Yoo and Kim [15] took temperature dependent viscosity into consideration more precisely with a convergence criterion of 0.001 %. But in this study, to decrease the computation time, convergence was not applied to each temperature value in the flow direction, but for the sum of the complete temperature set.

Optimization of slider bearing lubrication is also included in the scope of numerical studies: As Lin [6] tried to get an optimum flow cavity for one-dimensional porous curved slider bearing, Stokes and Symmons [13] performed a multi-dimensional optimization on the plasto-hydrodynamic drawing of wires, where the deformation process occurs in a stepped cavity filled with a viscous fluid. The aim of the present optimization study is to generate a novel upper surface design, with the implementation of waviness, without varying the volume of the flow cavity, thus the physical limits of the complete lubrication structure. The proposed numerical approach first models the thermo-hydrodynamic flow-environment by a transfer matrix and then the dependence of viscosity on temperature and the so occurring variations in other flow properties are also determined by an iterative mechanism. To produce a complete overview, computations are performed with 2 pumping pressures, for 6 amplitudes and 11 wavelengths. Performance optimization outputs are discussed through streamwise pressure and wall shear stress variations, and with load capacity and power consumption data for various amplitude and wavelength cases.

2 Governing Equations and Computational Approach

Performance analysis of plane slider bearings covers the investigation of both momentum and energy transfer in the flow volume, thus velocity (u), pressure (P) and temperature (T) distributions are the primary concern of the fundamental theory. The outputs of the continuity, momentum and energy equations can be the focused items of the work, but generally the results of the former in the calculation order generates the input set for the following, which puts forth the simultaneous handling of the three equations. As the non-dimensional momentum equation in -x direction (Eq. (1a)) interprets the relation of viscous shear stress and the thermodynamic pressure, the Reynolds equation for 1-dimensional lubricant flows of slider bearings is given by Eq. (1b). Lubricant viscosity is effective in either of the flow and energy equations and the pressure and temperature terms are non-dimensionalized as $P^* = Ph_{ex}^2/\mu V_1 L$, $T^* = Th_{ex}^2 C_p \rho/\mu V_1 L$ ([4, 6]) respectively, thus the Newtonian viscosity-temperature relation is characterised by Vogel's rule of Eq. (1c).

$$\frac{d^2u^*}{dy^{*2}} = \frac{dP^*}{dx^*} \qquad \frac{d}{dx^*}(h^{*3}\frac{dP^*}{dx^*}) = 6(V_u^*+1)\frac{dh^*}{dx^*} \qquad \mu = \rho k e^{\frac{b}{T+\xi}} \qquad (1a-c)$$

The velocity profile (Eq. (2a)) can be obtained by imposing the boundary conditions of $y^* = 0 \rightarrow u^* = 1$ and $y^* = \frac{h}{h_{ex}} = h^* \rightarrow u^* = \frac{V_u}{V_1} = V_u^*$ to Eq. (1a) and integration of the velocity profile gives the volumetric flow rate (q_x^*) per unit width (Eq. (2b)).

$$u^* = \frac{dP^*}{dx^*}\frac{y^*}{2}(y^* - h^*) + (V_u^* - 1)h^* y^* + 1 \qquad q_x^* = \frac{-1}{12}\frac{dP^*}{dx^*}h^* + (V_u^* + 1)\frac{h^*}{2} \qquad (2a-b)$$

The streamwise pressure distribution (Eq. (3)) can be evaluated by integrating the Reynolds equation (Eq. (1b)) twice, where the integration involves both streamwise (-x) and pitchwise (-y) directions.

$$P^* = \int_0^{x^*} \frac{\int 6(V_u^* + 1)dh^*}{h^{*3}}dx^* - \frac{\int_0^1 \frac{\int 6(V_u^*+1)dh^*}{h^{*3}}dx^*}{\int_0^1 \frac{1}{h^{*3}}dx^*}\int_0^{x^*} \frac{1}{h^{*3}}dx^* + P_{in}^* \qquad (3)$$

The load-carrying capacity (W^*) is obtained by the streamwise integration of the film pressure (Eq. (4a)). The wall shear stress (τ^*), which is measure of surface wear, and the pumping power (Ω^*), necessary to supply the lubrication oil are defined by Eqs. (4b-c).

$$W^* = \int_0^1 P^* dx^* \qquad \tau^* = \mu^* \frac{du^*}{dy^*}|_{y^*_{wall}} \qquad \Omega^* = q_x^* P_{in}^* \qquad (4a-c)$$

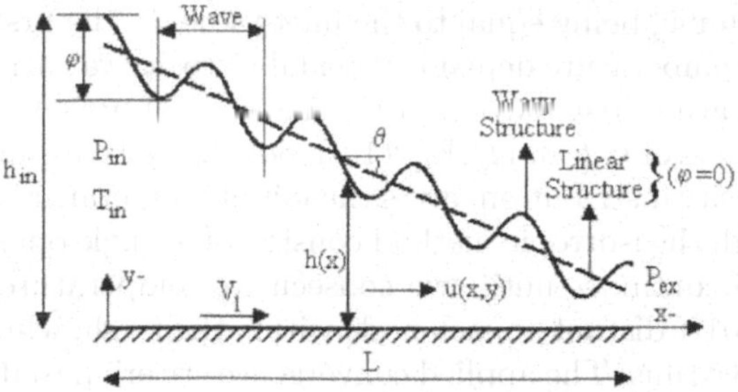

Fig. 1. Schematic layout of slider bearing with/without wavy structure at the upper surface

To solve the continuity, momentum and energy equations in harmony, the geometric domain of Fig. 1 is divided into 1000 sequential cells, where higher numbers, in the early stages of the code development, appeared to increase only the run times not the accuracy of the streamwise convergence. 2nd order finite difference marching procedure, in the streamwise direction with a constant cell width (Δx) of L/K, where for the sake of generality number of cells are denoted by K, is applied for the simulation of 1-dimensional, incompressible lubricant flow. Since the volumetric flow rate (q_x^*) is constant in flow direction, equating the derivative of Eq. (2*b*) to zero forms a system of $K-1$ linear equations, which completely represent the relation of geometric structure, static pressure and velocity distributions. The new implementation is a simple sigma notation, which consists of 2 coefficient matrices whose elements are mainly defined by the groove geometry and the upper and lower surface velocities of the bearing. As the nodes $i = 1$ and $i = K+1$ represent the inlet and exit planes, the explicit form of the $K-1$ equations constitute the "*Transfer Matrix*" of the system. Since left hand side is a banded matrix with a bandwidth of 3, Thomas algorithm is used in the evaluation procedure, where the outputs are the scope of continuity and Reynolds equations for nodes $i = 2$ to K.

In addition to the inlet conditions and surface velocities, results of the transfer matrix, especially the volumetric flow rate and the streamwise pressure gradient also participate as inputs when the temperature variation is under inspection. Superimposing the finite difference logic into the energy equation (Eq. (5*a*)) and rearranging the terms brings up a thermal relation within two consecutive nodes in the mesh, which in return displays the lubricant temperature distribution in the flow direction (Eq. (5*b*)).

$$q_x^* \frac{dT^*}{dx^*} = V_u^* \tau_u^* + \tau_1^* - q_x^* \frac{dP^*}{dx^*} \qquad T_{i+1}^* = \Delta x^* [\frac{V_u^* \tau_u^* + \tau_1^*}{q_x^*} - \frac{dP^*}{dx^*}]_i + T_i^* \qquad (5a\text{–}b)$$

Evaluation of the temperature dependent nature of viscosity covers both the traditional isotropic method and the present iterative transfer matrix approach, where the classical solution generates the initial set of guesses for the first iteration step. For the isotropic approach, lubricant viscosity is kept constant in the complete flow volume, being equal to the inlet value. In the first step of the iterative method the temperature dependent nodal viscosity variation is calculated by using the temperature distribution of the isotropic approach, together with the viscosity parameters of k, b and ξ (Eq. (1c)). Solving the transfer matrix, with the so obtained viscosity distribution, gives the initial temperature set of the iterative method. Although the isotropic method consists of a single operation loop, the iterative approach continues until two consecutive temperature distributions are not more than 0.01% distant at each node within the mesh, which may contribute up to 15 successive runs. The applied convergence criterion (0.01%) is more sensitive than that of Kumar et al. [4] and Pandey and Ghosh [10]. On the other hand, although Yoo and Kim's [13] criteria (0.001%) appears to be more precise, the present method differs from theirs by imposing the convergence on each nodal temperature, not on the sum of the complete set. This application makes the streamwise temperature determinations more reliable, since $K + 1$ times more control loops, for every iteration step, exist in the current approach.

3 Theoretical Model

The theoretical model involves the combined definitional necessities of both the physical and the thermo-fluid structural information of the complete lubrication environment. As the input data for the Reynolds and energy equations cover knowledge on upper-lower surface velocities of the bearing, the lubricant type and the inlet-exit boundary conditions. To generate a realistic overview, necessary compounds are chosen from the available recent numerical studies. The main sketch of the slider bearing is similar to that of Ostayen et al. [14] and Kwan and Post [5], that is narrowing in linear style (Fig. 1). As in most of the industrial applications, upper surface of the bearing is kept stationary ($V_u = 0$ m/s), which is also the case in the study of Honchi et al. [3]. On the other hand lower surface velocity data appear in a wide range: such as 2.55-10.21 m/s of Liu et al. [7] and 8.79 m/s of Ai et al. [1]; the current value is chosen as $V_l = 5$ m/s, which is the mean value of the most frequent data. Bearing length is selected as 10 mm, being between the choices of Honchi et al. [3] (1.25 mm) and Ai et al. [1] (14.5 mm). The most frequent slider bearing pad height and journal bearing clearance values appear in the range of 1-0.0175 mm, like those of Lin [6] and Liu et al. [7]. The inlet (h_{in}) and exit heights (h_{ex}) are selected as 1 mm and 0.125 mm, resulting in a mean pad inclination (Θ) of 5° (Fig. 1). The analysis is based on the fact that, unused lubricant is pumped in and emerges to atmosphere, therefore inlet (P_{in}) and exit (P_{ex}) oil pressure values are decided to be 101-301 kPa and 100 kPa respectively, which propose the $\Psi = P_{in}/P_{ex}$ cases of 1.01 & 3.01. Investigations are carried out with SAE 20 type lubricant that has comparable viscosity values of Mehenny and Taylor's [8] application, and with the inlet temperature (T_{in}) of 20°C, which is close to that of Sottomayor et al. [12] (24°C).

$$h(x) = h_{in} - x tan\Theta + \frac{\varphi}{2}[\cos(2\pi \frac{i}{K+1}\lambda)] \tag{6}$$

The upper surface of the bearing is defined, in a convenient way to fit the main aim of the work, by Eq. (6), where the wave amplitude and the number of waves are indicated by φ and λ respectively. The cosine curve is implemented to the streamwise-structure with the relative position of nodes (i) by comparing with the total meshing scale ($K+1$). To visualize the effects of φ and λ on the bearing performance, wave amplitude data are rated in the range of 0-200 μm, covering those of Harsha et al. [2], Rasheed [11] and Liu et al. [7]; the imposed number of waves is kept within the limits of 5-105, including the most recurrent values of literature (Honchi et al. [3], Kwan and Post [5] and Sottomayor et al. [12]).

4 Results and Discussion

Numerical experiments are performed with φ of 0-200 μm, λ values are in the range of 5-105, the pumping pressures are 1.01-3.01 times (Ψ) the exit value, and the results are displayed in non-dimensional form ($*$) for the sake of generality. Streamwise pressure variations, of slider bearings with λ of 5-45, are demonstrated for φ of 40 & 200 μm in Figs. 2(a)-(b) respectively. In the upstream sections of the cavity ($x^* < 0.4$) neither λ nor φ have influence on the nodal P^* values, however both λ and φ significantly influenced the lubricant pressure towards downstream ($x^* > 0.6$). On the other hand increasing either λ or φ results in higher maximum pressure values within the flow volume, moreover the effect of λ on P^* becomes more remarkable for higher φ cases (Fig. 2(b)); moreover the location of maximum pressure shifts downstream as the wave number is increased.

Shear stress (Eq. (4b)) is a measure of friction and possible wear on the flow surfaces and streamwise variations are demonstrated in Fig. 3. For the φ of 40 μm, (Fig. 3(a)) λ has no influence on shear stress and the τ^* values decreased towards downstream up to the point of P^*_{max} ($x^* \approx 0.84$); then followed by a sharp increase where the pressure values are determined to decrease (Fig. 2(a)). However as the amplitude is increased to 200 μm (Fig. 3(b)) the effect of λ

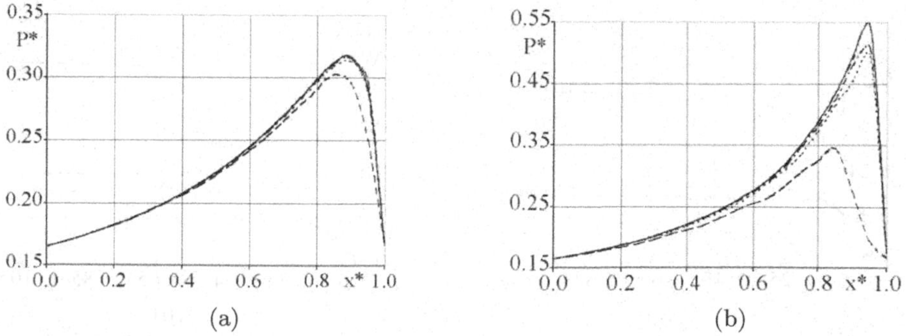

(λ: ---=5, ······=15, -·-·- =25, -··-=35, —— =45)

Fig. 2. Streamwise pressure distribution for φ of (a) 40 μm and (b) 200 μm

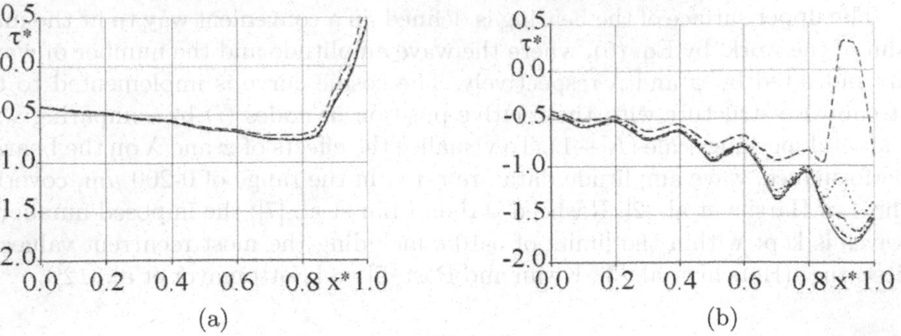

(λ: ---=5, ······=15, -·-·- =25, -··- =35, —=45)

Fig. 3. Streamwise shear stress distribution for φ of (a) 40 μm and (b) 200 μm

becomes more impressive in the flow direction, which is accompanied by fluctuations in τ^* values. As the strength of this structure increases towards the cavity exit, they are determined to be independent of λ, where the peak values of different λ cases are positioned at the identical locations. While the maximum pressure values moved downstream ($0.83 < x^* < 0.95$) with higher λ (Fig. 2(b)), this varying character also sensed in τ^* variations, where the most augmented gap, among the extreme τ^* values, extents to a value of ~2.1 which interprets the hydrodynamic instabilities in the wall neighborhood.

Showing parallelism with the P^* discussions, load (W^*) values (Fig. 4) are directly affected with φ but the number of waves creates an exceptional impact, where initially a sharp increase in W^* are recorded for the $5 < \lambda < 25$ range, thereafter the trend becomes asymptotic with negligible increase rates, among the $\lambda = 45$ and $\lambda = 105$ cases, of 1% and 0.7% for $\Psi = 1.01$ (Fig. 4(a)) and $\Psi = 3.01$ (Fig. 4(b)) respectively. Additionally, W^* of the $\Psi = 3.01$ case are above that of the $\Psi = 1.01$ investigation by 2.27 and 2.13 times for φ of 40 μm and 200 μm, denoting the augmented effects of pumping pressure on W^* at lower φ. Moreover

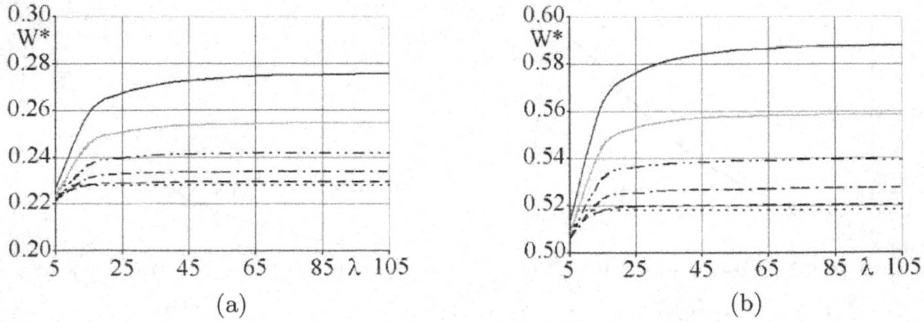

($\varphi(\mu m)$: ······ =0,--- =40,-·-· =80,-·-- =120,— =160,— =160)

Fig. 4. Variation of load carrying capacity with λ for Ψ of (a) 1.01 and (b) 3.01

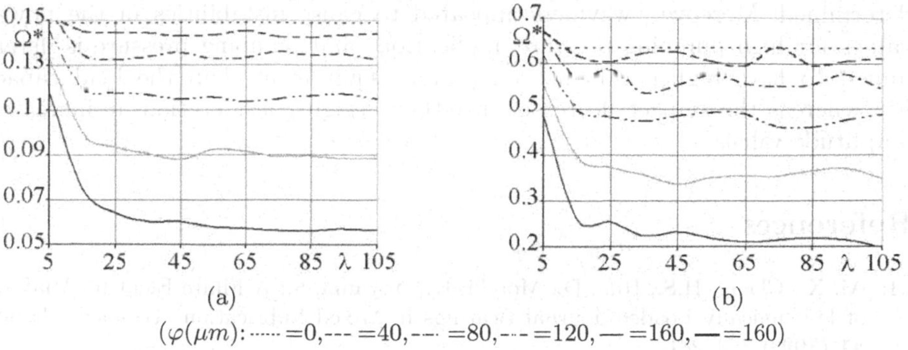

Fig. 5. Variation of input power requirement with λ for Ψ of (a) 1.01 and (b) 3.01

the influence of φ on W^* is higher in the lower pumping pressure case ($\Psi = 1.01$) with ratio of $\frac{W^*_{\varphi=200\mu m}}{W^*_{\varphi=0}} = 1.21$, whereas the same quantity is 1.13 for Ψ of 3.01.

Contrary to W^* results, as shown in Fig. 5, the power requirement (Ω^*) of the system decreases with both φ and λ, however fluctuations are determined with higher λ cases. These instabilities are limited to lower φ values for the lower pumping pressure case of $\Psi = 1.01$ (Fig. 5(a)), however the Ω^* outputs of even the highest amplitude of $\varphi = 200$ μm do not converge to a unique value at higher λ tasks in the project with Ψ of 3.01 (Fig. 5(b)). This record implies that the application of waviness can produce continuous instabilities if the demand in pumping pressure increases. Similar to W^* results, λ comes out to be most influential on Ω^* in the range of $5 - 25$ and the lowest stable Ω^* values are attained for λ of $25 - 45$, for the complete φ set. The power values of the $\Psi = 3.01$ project are 4.25 and 3.50 times above those of the $\Psi = 1.01$ case for φ of 40 μm and 200 μm respectively, indicating the augmented influence of Ψ at lower wave amplitudes.

5 Conclusion

The application of Upper-Surface-Waviness to inclined slider bearings is modeled numerically by taking the streamwise decay of lubricant viscosity also into account. The cumulative output implies that the overall system performance of inclined slider bearings can be improved, without going beyond the geometric limits, by imposing a wavy character on the upper surface, which in return not only increases the load carrying capacity but also decreases the power requirement. A close relationship is determined among the streamwise variations of shear stress and lubricant pressure, moreover the surface wave number is found out to become more influential on shear stress values for the designs with higher wave amplitudes. The computational results point out an optimum upper surface design with a wave number range of $\lambda = 25 - 45$ for the pumping pressure and wave amplitude intervals of $\Psi = 1.01 - 3.01$ and $\varphi = 40 - 200$ μm respectively, where λ can be fixed for each system on the basis that φ and Ψ are determined. Moreover, waviness appeared to cause instabilities in the power values for high pumping pressure applications and pumping pressure is determined to have higher influence on power requirement than the load capacity, whereas the impact decreases in either design consideration at increased amplitude values.

References

1. Ai, X., Cheng, H.S., Hua, D., Moteki, K., Aoyama, S.: A Finite Element Analysis of Dynamically Loaded Journal Bearings in Mixed Lubrication. Tribology Trans. **41** (1998) 273–281
2. Harsha, S.P., Sandeep, K., Prakash, R.: The Effect of Speed of Balanced Rotor on Nonlinear Vibrations Associated with Ball Bearings. Int. J. Mech. Sci. **45** (2003) 725–740
3. Honchi, M., Kohira, H., Matsumoto, M.: Numerical Simulation of Slider Dynamics During Slider-Disk Contact. Tribology Int. **36** (2003) 235–240
4. Kumar, B.V.R, Rao, P.S., Sinha, P.: A Numerical Study of Performance of a Slider Bearing with Heat Conduction to the Pad. Fin. Elem. Ana. Des. **37** (2001) 533–547
5. Kwan, Y.B.P., Post, J.B.: A Tolerancing Procedure for Inherently Compensated, Rectangular Aerostatic Thrust Bearings. Tribology Int. **33** (2000) 581–585
6. Lin, J.R.: Optimal Design of One-Dimensional Porous Slider Bearings Using the Brinkman Model. Tribology Int. **34** (2001) 57–64
7. Liu, W.K., Xiong, S., Guo, Y., Wang, Q.J., Wang, Y., Yang, Q., Vaidyanathan, K.: Finite Element Method for Mixed Elastohydrodynamic Lubrication of Journal-Bearing Systems. Int. J. Num. Meth. Eng. **60** (2004) 1759–1790
8. Mehenny, D.S., Taylor, C.M.: Influence of Cicumferential Waviness on Engine Bearing Performance. Proc Instn Mech. Engrs Part C **214** (2000) 51–61
9. Ozalp, A.A., Ozel, S.A.: An Interactive Software Package for the Investigation of Hydrodynamic-Slider Bearing-Lubrication. Comp. Appl. Eng. Educ. **11** (2003) 103–115
10. Pandey, R.K., Ghosh, M.K.: A Thermal Analysis of Traction in Elastohydrodynamic Rolling/Sliding Line Contacts. Wear **216** (1998) 106–114
11. Rasheed, H.E.: Effect of Surface Waviness on the Hydrodynamic Lubrication of a Plain Cylindrical Sliding Element Bearing. Wear **223** (1998) 1–6
12. Sottomayor, A., Campos, A., Seabra, J.: Traction Coefficient in a Roller-Inner Ring EHD Contact in a Jet Engine Roller Bearing. Wear **209** (1997) 274–283
13. Stokes, M.R., Symmons, G.R.: Numerical Optimisation of the Plasto-Hydrodynamic Drawing of Narrow Strips. J. Mat. Proc. Tech. **56** (1996) 733–742
14. van Ostayen, R.A.J., van Beek, A., Ros, M.: A Parametric Study of the Hydro-Support. Tribology Int. **37** (2004) 617–625
15. Yoo, J.G., Kim, K.W.: Numerical Analysis of Grease Thermal Elastohydrodynamic Lubrication Problems Using the Herschel-Bulkley Model. Tribology Int. **30** (1997) 401–408

A Genetic Algorithm for Scheduling of Jobs on Lines of Press Machines

S. Ayse Ozalp

Department of Industrial Engineering, Uludag University,
16059 Gorukle Bursa, Turkey
ayseozalp@uludag.edu.tr
http://www20.uludag.edu.tr/~ayseozalp/

Abstract. This paper introduces a Genetic Algorithm (GA) based solution technique for press machines scheduling problem of a car manufacturing factory. Firstly, the problem at hand, and the application of the GA in terms of coding, chromosome evaluation, crossover and mutation operators, are described in detail. After that, the GA is experimentally evaluated through some test problems. As the objective of the problem is the minimization of the completion time of the jobs, the GA based solution is compared with the Longest Processing Time (LPT) rule, and it is observed that the GA always produces better schedules than the LPT rule in a reasonably short amount of CPU time.

1 Introduction

Production planning and scheduling activities affect the productivity of manufacturing organizations. Effective usage of resources and on time delivery of orders can only be achieved by good production plans and schedules. In general, scheduling can be defined as allocating scarce resources (e.g., manpower, machine, energy, etc.) over time to perform a set of tasks being parts of some processes such as manufacturing, computation, etc. As scheduling plays an important role over the efficiency of manufacturing organizations, numerous researches have been performed to solve this problem.

One of the earliest studies about scheduling of manufacturing processes is [17] in which all kinds of scheduling problems were examined in detail, and a classification for these problems were made. In [1, 2, 9, 10, 16, 17] computational complexity of each class of scheduling problems were determined, and suitable solution techniques for each problem class were discussed. It was shown that most of the scheduling problems belong to NP-Complete problem class, and heuristics should be employed for solving "hard" scheduling problems. Dispatching rules, neighborhood search techniques, tabu search, simulated annealing, and genetic algorithms are among the most popular heuristic based techniques.

Genetic Algorithms, powerful and broadly applicable search and optimization techniques, were first introduced by Holland [8] in 1960s. The basic idea behind the GA is to find a (sub)optimal solution by sampling the solution space that has high probability of leading to a better solution. This technique imitates the

I. Lirkov, S. Margenov, and J. Waśniewski (Eds.): LSSC 2005, LNCS 3743, pp. 535–543, 2006.

natural selection process such that, in each generation, the fittest individuals have a better chance to produce off-springs whereas; the worst individuals are most likely to die. In general, a genetic algorithm consists of five basic steps [11]: (a) a genetic representation of solutions to the problem, (b) a way to create an initial population of solutions, (c) an evaluation function for rating solutions according to their fitness, (d) genetic operators that alter the genetic composition of solutions during reproduction, (e) a way to populate the next generation. The implementation of each of these steps may differ from one problem situation to other, which means that, genetic algorithms are customized according to the case under study. In [11, 7] it was shown that the performance of genetic algorithms in terms of the optimality of the solution found is highly dependent on the realization of each of the above steps, and the values of the GA parameters, that are number of solutions in the initial population (i.e., population size), number of generations (i.e., termination criterion), and probability values for genetic operators (i.e., crossover and mutation probabilities).

This study presents a GA based solution technique to find (sub)optimal schedules of jobs on a car manufacturer's press shop. Although scheduling of manufacturing processes is a hot research topic, scheduling of press lines has not taken much attention, and this study is the first one that applies GA to this problem. One of the only two studies in the literature on press lines scheduling problem belongs to Emel and Tasci [5] in which the press lines scheduling problem is solved by using a mixed integer programming model that is established in order to fill in the rectangular time-machine frame with small rectangular time-operation job elements without any overlap among elements. Emel and Tasci [5] solved the mixed integer programming model by using LINDO, and found optimal schedules for small sized press shops. As the size of the press shop (i.e., number of press lines, number of press machines in each line, and number of jobs to be scheduled) increases, the number of variables and equations to be solved also increases which makes mixed integer programming unsuitable for solution of large problems. In the other study [13], dispatching rules (e.g., SPT, LPT, EDD) based heuristic solution, which is applicable for all press shop sizes, is employed and it was shown that dispatching rules produce good schedules in terms of press machine utilization in a short amount of time. In this study, after the GA based solution technique for the press lines scheduling problem is described, solution found by the GA is compared with those generated by LPT dispatching rule in terms of optimality and CPU time required.

The organization of this paper is as follows: in the following section, the press lines scheduling problem is introduced, in the third section, the GA based solution for the problem is explained, experimental results are presented in the fourth section, and finally, section five concludes this study.

2 Problem Description

The press shop of a car manufacturing company produces major car body parts that are made of sheet steel by performing multiple and succeeding operations in

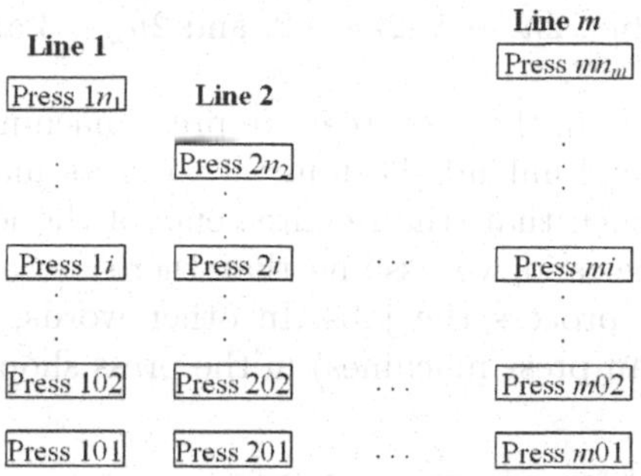

Fig. 1. Layout for press lines and press machines

sheet metal forming press lines. Press machines have material-handling equipment, and feeding/unloading devices to provide automated and synchronized operations to improve productivity. Each press machine in the press shop is capable of producing any part of the car body by just making quick die changes. Press machines are located to form a number of parallel lines as shown in Fig. 1. According to Fig. 1, a press shop consists of m press lines, where a line i contains n_i press machines. A typical press line consists of 3 to 8 press machines. These machines have identical capacity and speeds for synchronized operation and interchangeability.

A job to be scheduled in this shop consists of a minimum of 2 and a maximum of $\max_{1 \leq i \leq m}(n_i)$ operations. Each operation requires a separate die to form the sheet metal, and takes equal cycle time. Each job covers the production of a car body part, and parts are produced in lots and use one press machine for each operation, thus, press machines in a line are occupied simultaneously until the job is completed. Here it is assumed that for each job to be scheduled; the values of lot size, cycle time and setup time for machines are known in advance. Thus the processing time for a job is computed as *processing time* = (*lot size*) $*$ (*cycle time*) + (*total setup time*). According to the sample data provided in Table 1, processing time for job 1 is $(1400 * 0.15) + 30 = 240$ minutes. In the press shop, each job can only be loaded to press machines by starting from one of the both ends of a press line. For example job 1 in Table 1 can be loaded to line 2 in two ways; either starting from Press 201, or from Press $2n_2$. In the first case, the job occupies presses 201, 202, 203, and 204; in the second

Table 1. A set of jobs to be scheduled on press shop

Job No	Lot Size	Oper. No	Cycle Time	Setup Time
1	1400	4	0.15	30
2	2500	3	0.23	14
3	1000	6	0.79	100
4	400	5	0.15	26
5	2500	2	0.23	10

case press machines $2n_2$, $2n_2 - 1$, $2n_2 - 2$, and $2n_2 - 3$ are loaded until the job becomes completed.

In a typical press shop, the resources are press machines, dies (to form sheet metal), workers (to load-unload sheet metal to press machines), and safes (to store formed metal) such that, the requirements of the jobs for these resources are known in advance, and we assume that there are always enough amount of these resources to process the jobs. In other words, there are no resource constraints (other than press machines) in the press shop.

3 Genetic Algorithm-Based Solution

In this study, the press shop scheduling problem is modeled as "N jobs M non-identical parallel machines scheduling problem" where M takes values between m and $2m$; and m is the number of press lines in the shop, and the objective of the scheduling problem is the minimization of the total processing time of the jobs (i.e., makespan C_{max}). We modeled the problem as non-identical parallel machines problem because each press line may contain different number of press machines from other lines, and each job may have different number of operations from other jobs in the shop.

According to [18], identical parallel machine scheduling problem for minimizing the makespan is an NP-hard problem, and in [14], it is proved that the press lines scheduling problem, modeled as N jobs m to $2m$ non-identical parallel machines problem, is NP-Complete, which makes the use of heuristic methods inevitable. In [12, 3] GA-based solutions for scheduling identical parallel machines with single objective arc described, in [4] on the other hand, two-stage multi-population genetic algorithm is proposed to solve identical parallel machine scheduling problems with multiple objectives. Also, [15] employed GA to schedule jobs with fuzzy processing times on parallel machines.

The GA used to solve the press lines scheduling problem consists of the following steps among which steps 3.3 to 3.7 are repeated by *gen_size* (i.e., number of generations) times, to find a (sub)optimal solution.

3.1 Coding

In the press lines scheduling problem with N jobs and m lines, a chromosome consists of a randomly generated job sequence $J = (j_1, j_2, ..., j_N)$ where j_i (for $i = 1, 2, ..., N$) is a randomly generated integer number between 1 and N that represents the job number such that each j_i is different from all the remaining jobs in J.

3.2 Generation of Initial Population

The initial population consists of randomly generated *pop_size* (i.e., number of chromosomes in the population) chromosomes.

3.3 Evaluation of a Population

For each chromosome in the population, jobs are loaded to the press lines in the job order given in the chromosome, according to the algorithm presented in Fig. 2, and C_{max} value for the chromosome is determined by taking the maximum of the completion times of the jobs.

3.4 Reproduction

For the reproduction step, *fitness-proportionate* selection process [7] is employed in which a biased roulette wheel divided into *pop_size* slots, each with a size proportional to the individual's fitness, is spinned *pop_size* times, and at each time an individual is selected. In this selection process, a chromosome x is assigned a probability p_x as follows: Let f_x be the fitness value of chromosome x, such that $f_x = 1/C_{max}$ where C_{max} is determined according to the algorithm in Fig. 2, then p_x is computed by dividing the fitness value of the chromosome x to the sum of the fitness values of all the chromosomes in the population (Eq. (1a)). After that, for each chromosome in the population, cumulative probabilities (i.e., c_x) are obtained by adding up the fitnesses of the preceding population members (Eq. (1b)).

$$p_x = \frac{f_x}{\sum_{i=1}^{pop\_size} f_i} \qquad c_x = \sum_{i=1}^{x} p_i \qquad x = 1, 2, ..., pop\_size \qquad (1a - b)$$

```
Algorithm: Loading Jobs on Press Lines

Input: A sequence of jobs (i.e., a chromosome) to be scheduled J,
       Number of press lines, and number of press machines on each line.
Output: Schedule, Cmax.

for each unscheduled job ji in the list J do
    Get the number of operations of ji
    for each press line in the shop do
        i. Starting from the first line in the shop, find the
           available press machines that the job can be loaded.
        ii. if available lines exist,
         load the job to the first available line.
        iii. else if the number of empty press machines is less than
             the operation number of the job ji
                      Find the first unscheduled job j from J that can
                      be loaded to the line, and Load the job j.
if all of the jobs are scheduled then,
    output the schedule and the Cmax value.
```

Fig. 2. Algorithm for evaluating a chromosome by computing its C_{max} value

Finally, a random number r uniformly distributed in $[0,1]$ is drawn *pop_size* times and each time the ith chromosome is selected if $c_{i-1} < r \leq c_i$. When $0 \leq r \leq c_1$ the first chromosome is chosen.

3.5 Crossover

Crossover is a genetic operator that produces two children out of the parents. The two parental chromosomes are selected randomly among the chromosomes that are chosen at the reproduction step. A random number r uniformly distributed in $[0,1]$ is drawn and compared with the crossover probability (i.e., $p(C)$) that is determined beforehand, if $r \leq p(C)$, the parents are combined through the crossover; otherwise no crossover is performed. In this study, one-point crossover is used.

3.6 Mutation

Mutation operator performs local modification of the chromosomes. After the crossover operation, the resulting chromosomes are subjected to the mutation operation such that for each job (i.e., gene) in the chromosome, a random number r from the interval $[0,1]$ is generated and compared with the predetermined mutation probability (i.e., $p(M)$). If $r \leq p(M)$, the gene is mutated. In this study, adjacent swap of the genes is employed as the mutation operator.

3.7 Determination of the New Generation

After the crossover and mutation operations, the resulting chromosomes are evaluated according to step 3.3, and all the chromosomes that are in the new and in the previous generations are sorted according to their C_{max} value, and the best *pop_size* chromosomes are selected as the next generation. Therefore the best chromosomes found in each generation are kept without changing through the solution process (i.e., elitist strategy).

4 Experimental Results

The above genetic algorithm is used for minimizing the total processing time of N jobs over the press lines where N takes values from 5 to 30, and the press shop consists of 3 press lines such that the first and second lines contain 8, and the third line has 5 press machines. The total processing time of the jobs takes values between 56 to 896 minutes and the number of operations that the jobs require changes between 2 to 8. The data used in the experiments are real data that are obtained from the car manufacturer's press shop. The GA is implemented in Pascal language under Windows XP operating system, and the software is run on a PC having a Pentium III 450 Mhz processor and 256 MByte main memory.

Due to the fact that the values of genetic algorithm parameters have important role in the optimality of the results found, first of all, we tried to determine the best values for crossover and mutation probabilities by running the algorithm

Table 2. Minimum and average C_{max} values for different p(C) and p(M) values

P(M)	**0.1**		**0.2**		**0.3**	
P(C)	Min	Average	Min	Average	Min	Average
0.5	2713	2739	2711	2734	2720	2744
0.6	2711	2730	2718	2739	2714	2745
0.7	2713	2743	2720	2731	2713	2732
0.8	2712	2732	2720	2737	2710	2742
0.9	**2705**	**2724**	2709	2730	2708	2740

for 30 jobs with *pop_size* = 20, *gen_size* = 40, and crossover and mutation probabilities take values from 0.1 to 0.9 with 0.1 step interval. At the end of this experiment, we observed that as the crossover probability increases ($p(C) \geq 0.5$) and the mutation probability decreases ($p(M) \leq 0.3$), the fitness of the chromosomes found also increases. We run the algorithm 10 times for each parameter combinations that are given in Table 2, and reported the best and the average C_{max} values obtained for each combination. According to the Table 2, the best values for $p(C)$ and $p(M)$ are 0.9 and 0.1, respectively.

In the second part of the experiment, we run the algorithm for 30 jobs with fixed crossover and mutation probabilities (i.e., $p(C) = 0.9$ and $p(M) = 0.1$), and we changed the values of *pop_size* and *gen_size* parameters. The results for these experiments are presented in Table 3, in which variance is the deviation of the average C_{max} value from the best result found so far (i.e., $C_{max} = 2690$), improvement is the progress made by the new parameter settings according to the previous parameter settings in the table, and time is the CPU time (in seconds) required to run the algorithm with the given parameter values. According to Table 3, in the cases for fixed crossover and mutation probabilities, as population and generation sizes are grown from 20 to 100 and from 20 to 200 respectively, the quality of the solution is determined to improve, however the processing time of the algorithm is observed to rise from 0.73 sec to 30.9 sec. The processing time, for the computations performed on the problem where *pop_size* = 20 and *gen_size* = 40, is 1.37 sec, and the output deviates from the best solution found so far by just 1.26%. Also, when the population size is large (i.e., *pop_size* = 100), increasing the generation size (from 100 to 200) improves the quality of the

Table 3. C_{max} values, improvements of the results, and the CPU time (sec) for different population and generation sizes

pop_size	*gen_size*	**Min(C_{max})**	**Ave(C_{max})**	**Variance**	**Improvement**	**Time**
20	20	2738	2770	0.0297	-	0.73
20	40	2705	2724	0.0126	0.0171	1.37
40	40	2703	2724	0.0126	0.0000	2.66
40	100	2695	2713	0.0085	0.0041	6.43
100	100	2691	2698	0.0030	0.0055	15.73
100	200	2690	2697	0.0026	0.0004	30.89

Table 4. Performance of GA vs. LPT for varying number of jobs

	GA		LPT		Improvement
N	Min(C_{max})	Time	C_{max}	Time	of GA
5	502	0.66	502	< 0.1	–
10	968	0.76	1020	< 0.1	5.1%
15	1048	0.83	1068	< 0.1	1.9%
20	1433	0.88	1603	< 0.1	11%
25	2096	0.99	2237	< 0.1	6.3%
30	2705	1.37	2818	< 0.1	4%

solution by just 0.4%, however the processing time of the algorithm doubles (from 15.73 seconds to 30.89) seconds.

Finally, we run our genetic algorithm with fixed GA parameters (i.e., $p(C) = 0.9$, $p(M) = 0.1$, $pop\_size = 20$, and $gen\_size = 40$) but varying number of jobs, and compared the results in terms of optimality of the solutions found, and the CPU time required with the solutions of LPT dispatching rule which is used for minimizing the total processing time [16, 6]. As presented in Table 4, GA always produces better solutions for the problem at hand, and the CPU time required for the GA is not too much (less than 1.37 seconds) with respect to that for LPT.

The best solution found so far for scheduling of 30 jobs over the 3 press lines has C_{max} value which equals 2690 minutes, and this solution is obtained when the crossover and mutation probabilities are 0.9 and 0.1, respectively, and the population and generation sizes are set to 100 and 200, respectively. The utilizations of the press lines 1, 2 and 3 for this solution are 95%, 91% and 84%, respectively.

5 Conclusion

In this paper, press lines scheduling problem of a car manufacturing factory is studied, and a genetic algorithm based solution is developed and implemented. The genetic algorithm based solution technique generates good schedules in a few seconds when suitable GA parameters are selected. As it is expected, the GA always produces better schedules than LPT when the objective is the minimization of the makespan.

References

1. Blazewicz, J., Ecker, K.H., Pesch, E., Schmidt, G., Weglarz, J.: Scheduling Computer and Manufacturing Processes. Springer-Verlag, Berlin (1996)
2. Blazewicz, J., Cellary, W., Slowinski, R., Weglarz, J.: Scheduling Under Resource Constraints: Deterministic Models. Annals of Operations Research **7** (1986) 11–93
3. Cheng, R., Gen, M.: Parallel Machine Scheduling Problems Using Memetic Algorithms. Comp. Ind. Eng. **33** (1997) 761–764

4. Cochran, J.K., Horng, S. M., Fowler, J.W.: A Multi-Population Genetic Algorithm to Solve Multi-Objective Scheduling Problems for Parallel Machines. Computers & Operations Research **30** (2003) 1087–1102
5. Emel, E., Tasci, H.B.E.: An Optimization Approach for Scheduling of Jobs on Lines of Press Machines. The 6th Industrial Eng. Research Conference. Florida (1997)
6. Garey, M.R., Graham, R.L., Johnson, D.S.: Performance Guarantees for Scheduling Algorithms. Operations Research **26** (1978) 3–21
7. Goldberg, D. E.: Genetic Algorithms in Search, Optimization and Machine Learning. Addison Wesley, (1989)
8. Holland, J.: Adaptation in Natural and Artificial Systems. University of Michigan Press, Ann Arbor, MI, 1975; MIT Press, Cambridge, MA, (1992)
9. Lenstra, J.K.: Sequencing by Enumerative Methods. Mathematical Centre Tracts **69** (1985) 18–39
10. Lenstra, J.K., Rinnooy Kan, A.H.G.: Complexity of Scheduling Under Precedence Constraints. Operations Research **26** (1978) 22–35
11. Michalewicz, Z.: Genetic Algorithm+Data Structure=Evolution Programs. 3rd edn. Springer-Verlag, New York (1996)
12. Min, L., Cheng, W.: A Genetic Algorithm for Minimizing the Makespan in the Case of Scheduling Identical Parallel Machines. Artif. Intel. Eng. **13** (1999) 399–403
13. Ozalp, S. A.: Modeling and Scheduling of Jobs on Lines of Press Machines:A Heuristic Approach. 5th EUROSIM Congress on Modeling and Simulation. ESIEE Paris (2004)
14. Ozel, S.A.: Programming of Production Scheduling Algorithms. MSc Thesis, Uludag University, Industrial Engineering Department, Bursa (1999)
15. Peng, J., Liu, B.: Parallel Machine Scheduling Models with Fuzzy Processing Times. Information Sciences **166** (2004) 49–66
16. Pinedo, M.: Scheduling: Theory, Algorithms and Systems. Prentice Hall, New Jersey (1995)
17. Rinnooy Kan, A.H.G.: Machine Scheduling Problems: Classification, Complexity and Computations. Martinus Nijhoff, The Hague (1976)
18. Sethi, R.: On the Complexity of Mean Flow Time Scheduling. Mathematics of Operations Research **2** (1977) 320–330

A Fictitious Domain Method for Particle Sedimentation

C. Veeramani[1], P.D. Minev[2], and K. Nandakumar[1]

[1] Department of Chemical and Materials Engineering,
University of Alberta, Edmonton,
Alberta, Canada T6G 2G6
[2] Department of Mathematical and Statistical Sciences,
University of Alberta, Edmonton,
Alberta, Canada T6G 2G1
veeramani@ualberta.ca, minev@ualberta.ca,
kumar.nandakumar@ualberta.ca

Abstract. This paper presents an improved Fictitious Domain formulation for particle sedimentation using a global Lagrange multiplier. Although the rigid body motion is enforced weakly similarly to the DLM (Distributed Lagrange Multiplier) method of Glowinski et al. [2], there are some essential differences between the two formulations. The present formulation uses a Lagrange multiplier which is defined globally (as opposed to the locally defined multipliers in the DLM method) which makes it computationally more efficient. The present formulation was derived and discussed in detail in [1]. Here we discuss some improvements in the formulation and show the results of a validation problem.

1 Introduction

There are two main classes of methods for direct simulations of multiphase flows. The first type are methods which discretize the Navier-Stokes equations on a fixed, Eulerian grid and move the particles in this grid. In case of rigid particles Glowinski and his collaborators introduced a method for a weak imposition of the rigid body motion via distributed Lagrange multipliers (see Glowinski et al. [2, 3] and the references therein). The second type of methods uses moving grids and usually requires remeshing and re-interpolation. Examples of such techniques are the one developed by Hu [4] and Johnson and Tezduyar [5]. Each approach has its merits and drawbacks. We choose the first approach because it allows to solve the dynamic equations on fixed (usually structured) grids which is very convenient for the eventual parallelization. As usual, the price paid for this convenience is mostly in accuracy, although the second approach has its accuracy problems too.

In the remainder of the paper we briefly discuss the fictitious domain formulation with a global Lagrange multiplier and discuss some possibilities to improve it. The method is also thoroughly validated on some sedimentation problems.

I. Lirkov, S. Margenov, and J. Waśniewski (Eds.): LSSC 2005, LNCS 3743, pp. 544–551, 2006.

2 Fictitious Domain Formulation

Let us consider a bounded domain Ω_1 with an external boundary Γ filled with a Newtonian liquid of density ρ_1 and viscosity μ_1. Within this liquid we consider n rigid particles occupying a domain $\Omega_2 = \cup_{i=1}^n \Omega_{2,i}$ and having densities $\rho_{2,i}, i = 1,\dots,n$. Let us also denote the interface between Ω_1 and Ω_2 by Σ and the entire domain filled with the fluid and the particles by $\Omega = \Omega_1 \cup \Omega_2$. The equations of motion of the fluid in Ω_1 are the Navier-Stokes equations

$$\rho_1 \frac{D\hat{\mathbf{u}}_1}{Dt} = \nabla \cdot \hat{\sigma}_1, \quad \nabla \cdot \hat{\mathbf{u}}_1 = 0 \text{ in } \Omega_1 \tag{1}$$

where $\hat{\sigma}_1$ is the stress tensor defined by $\hat{\sigma}_1 = \hat{p}_1 \delta + 2\mu_1 D[\hat{\mathbf{u}}_1]$ with $\hat{p}_1$ being the pressure in the liquid phase, $D[\hat{\mathbf{u}}_1] = 0.5[\nabla\hat{\mathbf{u}}_1 + (\nabla\hat{\mathbf{u}}_1)^T]$ being the rate-of-strain tensor, and δ being the Kronecker tensor. A variety of boundary conditions on Γ can be considered but the different conditions do not change the fictitious domain formulation and therefore here we assume homogeneous Dirichlet conditions. On the internal boundary of Ω_1, Σ, we presume a no slip condition for the velocity. The equations of motion of a rigid particle are usually written in terms of the velocity of its centroid $\mathbf{U}_i$ and its angular velocity $\boldsymbol{\omega}_i$, $i = 1,\dots,n$

$$M_i \frac{d\mathbf{U}_i}{dt} = M_i \mathbf{g} + \mathbf{F}_i \tag{2}$$

$$\mathbf{I}_i \frac{d\boldsymbol{\omega}_i}{dt} + \boldsymbol{\omega}_i \times \mathbf{I}_i \boldsymbol{\omega}_i = \mathbf{T}_i \tag{3}$$

where M_i is the particle mass, $\mathbf{g}$ is the gravity acceleration, $\mathbf{F}_i$ is the total hydrodynamic force acting on it, $\mathbf{I}_i$ is its tensor of inertia, and $\mathbf{T}_i$ is the hydrodynamic torque about its centre of mass.

Diaz-Goano et al. [1] have shown that the momentum equation (1) can be properly extended to the entire domain Ω after the introduction of a global Lagrange multiplier λ, which is used to impose equations (2,3) inside Ω_2. The angular velocity of each particle can then be recovered from the no-slip boundary condition on the particles boundaries using the Stokes theorem. In case of spherical particles ($\mathbf{I}_i$ being an identity tensor), and if we denote the extension of $\hat{\mathbf{u}}_1$ and $\hat{p}_1$ to the entire domain Ω by $\mathbf{u}_1$ and p_1 correspondingly, the resulting modified system of equations for computation of $\mathbf{u}_1$, $\mathbf{U}_i$ and $\boldsymbol{\omega}_i$ reads

$$\rho_1 \frac{D\mathbf{u}_1}{Dt} = -\nabla p_1 + \mu_1 \nabla^2 \mathbf{u}_1 + \alpha\boldsymbol{\lambda} - \mu_1 \nabla^2 \boldsymbol{\lambda}, \quad \nabla \cdot \mathbf{u}_1 = 0 \text{ in } \Omega \tag{4}$$

$$\Delta M_i \frac{d\mathbf{U}_i}{dt} = \Delta M_i \mathbf{g} - \int_{\Omega_{2,i}} \alpha \boldsymbol{\lambda} d\Omega - \mu_1 \int_{\partial\Omega_{2,i}} \frac{\partial \boldsymbol{\lambda}}{\partial \mathbf{n}} ds, \quad i = 1,\dots,n \tag{5}$$

$$V_{\Omega_i} \boldsymbol{\omega}_i = 0.5 \int_{\Omega_{2,i}} \nabla \times (\mathbf{u}_1 - \mathbf{U}_i) d\Omega, \quad i = 1,\dots,n \tag{6}$$

$$\mathbf{U}_i(t) + \boldsymbol{\omega}_i(t) \times (\mathbf{x} - \mathbf{X}_i(t)) = \mathbf{u}_1, \text{ in } \Omega_{2,i}, \quad i = 1,\dots,n \tag{7}$$

where V_{Ω_i} is the volume of the i-th particle and $\Delta M_i = (\rho_{2,i} - \rho_1)V_{\Omega_i}$. The angular velocity equation (6) can be generalized to non-spherical particles but in this article we consider spherical particles only. This set of PDE's can be discretized by means of any discretization technique. In the next section we discuss its finite element discretization via an operator splitting technique.

3 Discretization

The PDE system above is first discretized in time by means of the following splitting algorithm.

Substep 1 (advection).
At this substep we approximate the solution of

$$\frac{\partial \tilde{\mathbf{u}}^k}{\partial t} = (\tilde{\mathbf{u}}^k \cdot \nabla)\tilde{\mathbf{u}}^k, \tag{8}$$

on the time interval (t^k, t^{n+1}) using the initial condition $\tilde{\mathbf{u}}^k(t^k) = \mathbf{u}^k$, k being equal to n or $n-1$. The physical meaning of this substep is that we advect the solutions from time levels $n, n-1$ to level $n+1$. This advection subproblem is resolved by the method of characteristics as described in [6]. This choice was determined by the good stability properties of the method in comparison to most other explicit methods.

The centre of mass of the i-th particle $\mathbf{X}_i^{n+1}$ is approximated with a predictor-corrector procedure, the predictor $\mathbf{X}_i^{p,n+1}$ being given by

$$\mathbf{X}_i^{p,n+1} = \mathbf{X}_i^{n-1} + 2\delta t \mathbf{U}_i^n \tag{9}$$

where δt is the time step. The corrector step is performed after the computation of the fluid and particles velocities at time level $n+1$ and is given by equation (15).

Substep 2 (diffusion).
If we set $\tau_0 = 3/(2\delta t), \tau_1 = -2/\delta t, \tau_2 = 1/(2\delta t)$, then we compute the solution on the next substep from the following Backward Euler discretization of the momentum equation in (4).

$$\begin{aligned} &\rho_1 \tau_0 \mathbf{u}_1^* - \mu_1 \nabla^2 \mathbf{u}_1^* = -\rho_1(\tau_1 \tilde{\mathbf{u}}_1^n - \tau_2 \tilde{\mathbf{u}}_1^{n-1}) - \nabla p^n, \text{ in } \Omega \\ &\mathbf{u}_1^* = 0 \text{ on } \Gamma \end{aligned} \tag{10}$$

Note that Substeps 1 and 2 constitute a second order splitting scheme for an advection-diffusion equation and therefore they give a consistent approximation to the solution of (4). The so computed velocity, however, is not divergence free since the pressure is only extrapolated from the previous time level. Therefore, the stability of the scheme can be guaranteed only if the predicted velocity field $\mathbf{u}_1^*$ at level $n+1$ is projected onto a solenoidal space on the next step of the algorithm.

Substep 3 (incompressibility).

$$\begin{aligned} \tau_0(\mathbf{u}_1^{**} - \mathbf{u}_1^{*}) &= -\nabla(p_1^{n+1} - p_1^n) \text{ in } \Omega \\ \nabla \cdot \mathbf{u}_1^{**} &= 0 \text{ in } \Omega \\ \mathbf{u}_1^{**} \cdot \mathbf{n} &= 0 \text{ on } \Gamma, \end{aligned} \tag{11}$$

n being the outward normal to Γ.

Note that after completion of Substeps 1-3 we end up with a solution of (4) with $\lambda = 0$ i.e. the so computed fluid velocity does not "feel" the presence of the rigid particles. In fact this can be considered as the zero-th iterate of an iteration for imposition of the fact that the velocity field in the particle domain Ω_2, $\mathbf{u}_2^{n+1}$, corresponds to a rigid body i.e. $\mathbf{u}_2^{n+1} = \mathbf{U}_i^{n+1} + \boldsymbol{\omega}^{n+1} \times (\mathbf{x} - \mathbf{X}_i^{p,n+1})$. This iteration is described in the next substep.

Substep 4 (rigid body constraint).

Set the 0-th iterates for $\boldsymbol{\lambda}^{n+1}$, $\mathbf{u}_1^{n+1}$, $\mathbf{U}_i^{n+1}$, $\boldsymbol{\omega}_i^{n+1}$ and $\mathbf{u}_2^{n+1}$ as

$$\begin{aligned} \boldsymbol{\lambda}^{0,n+1} &= 0 \\ \mathbf{u}_1^{0,n+1} &= \mathbf{u}_1^{**} \\ \tau_0 \mathbf{U}_i^{0,n+1} &= -\tau_1 \mathbf{U}_i^n - \tau_2 \mathbf{U}_i^{n-1} + \mathbf{g} \\ V_{\Omega_i} \boldsymbol{\omega}_i^{0,n+1} &= 0.5 \int_{\Omega_i} \nabla \times \mathbf{u}_1^{0,n+1} d\Omega \\ \mathbf{u}_2^{0,n+1} &= \mathbf{U}_i^{0,n+1} + \boldsymbol{\omega}^{0,n+1} \times (\mathbf{x} - \mathbf{X}_i^{p,n+1}). \end{aligned}$$

Let us denote the difference between two subsequent iterations for a quantity Q by δQ i.e. $\delta Q^{k+1} = Q^{k+1} - Q^k$. Then the subsequent iterates are computed for $k \geq 0$ by

$$\begin{cases} (1 + \frac{\rho_1}{\rho_{2,i} - \rho_1})(\alpha I - \mu_1 \nabla^2) \delta \boldsymbol{\lambda}^{k+1,n+1} & \\ = -(\rho_1 \tau_0 I - \mu_1 \nabla^2)(\mathbf{u}_1^{k,n+1} - \mathbf{u}_2^{k,n+1}) & \text{in } \Omega_{2,i}, \quad i = 1, \ldots, n \\ (1 + \frac{\rho_1}{\rho_{2,i} - \rho_1})(\alpha I - \mu_1 \nabla^2) \delta \boldsymbol{\lambda}^{k+1,n+1} = 0 & \text{in } \Omega_1 \\ \delta \boldsymbol{\lambda}^{k+1,n+1} = 0 & \text{on } \Gamma, \end{cases} \tag{12}$$

$$\begin{cases} (\rho_1 \tau_0 I - \mu_1 \nabla^2) \delta \mathbf{u}_1^{k+1,n+1} = (\alpha I - \mu_1 \nabla^2) \delta \boldsymbol{\lambda}^{k+1,n+1} & \text{in } \Omega \\ \delta \mathbf{u}_1^{k+1,n+1} = 0 & \text{on } \Gamma \end{cases} \tag{13}$$

$$\begin{cases} \Delta M_i \tau_0 \delta \mathbf{U}_i^{k+1,n+1} = -\frac{\rho_{2,i} - \rho_1}{\rho_{2,i}} \int_{\Omega_{2,i}} \alpha (\mathbf{u}_1^{k,n+1} - \mathbf{u}_2^{k,n+1}) d\Omega & \\ + \mu_1 \frac{\rho_{2,i} - \rho_1}{\rho_{2,i}} \int_{\partial \Omega_{2,i}} \frac{\partial (\mathbf{u}_1^{k,n+1} - \mathbf{u}_2^{k,n+1})}{\partial \mathbf{n}} ds & i = 1, \ldots, n \\ V_{\Omega_i} \boldsymbol{\omega}_i^{k+1,n+1} = 0.5 \int_{\Omega_i} \nabla \times \mathbf{u}_1^{k+1,n+1} d\Omega & i = 1, \ldots, n \\ \mathbf{u}_2^{k+1,n+1} = \mathbf{U}_i^{k+1,n+1} + \boldsymbol{\omega}_i^{k+1,n+1} \times (\mathbf{x} - \mathbf{X}_i^{p,n+1}) \text{ in } \Omega_{2,i} & i = 1, \ldots, n \\ \mathbf{u}_2^{k+1,n+1} = 0 \text{ in } \Omega_1 & \end{cases} \tag{14}$$

where I is the identity operator. The iteration (12-14) is a Richardson iteration for computation of λ and the imposition of the constraint (7) on the velocity

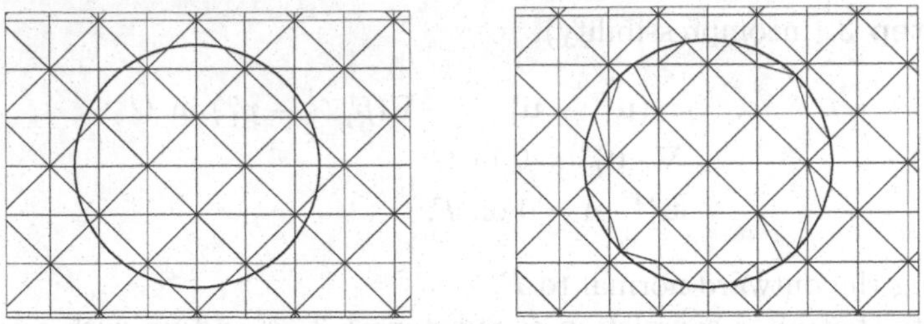

Fig. 1. A cross section of the basic tetrahedral grid (left) and the boundary fitted grid (right)

field inside the rigid particles (for more details see [1]). Note that this iteration is somewhat different than the iteration suggested in [1] since the update of $\delta \mathbf{U}_i^{k+1,n+1}$ is performed using the difference $(\mathbf{u}_1^{k,n+1} - \mathbf{u}_2^{k,n+1})$ directly rather than the Lagrange multiplier increment $\delta \boldsymbol{\lambda}^{k+1,n+1}$. The numerical experience showed that this iteration provides much more accurate results for the validation problems discussed below.

Upon convergence for some $k = N$ we set $\mathbf{u}_1^{n+1} = \mathbf{u}_1^{N+1,n+1}$, $\mathbf{U}_i^{n+1} = \mathbf{U}_i^{N+1,n+1}$, $\boldsymbol{\omega}^{n+1} = \boldsymbol{\omega}^{N+1,n+1}$, $\boldsymbol{\lambda}^{n+1} = \boldsymbol{\lambda}^{N+1,n+1}$. Finally, the position of the centre of mass of the particles is corrected according to

$$\mathbf{X}_i^{n+1} = \mathbf{X}_i^n + 0.5\delta t(\mathbf{U}_i^{n+1} + \mathbf{U}_i^n). \tag{15}$$

The equations (10-14) are further discretized in space by means of $P_2 - P_1$ tetrahedral elements. The computation of the integrals over $\Omega_{2,i}$ involved in the iteration above is not very straightforward since the finite element grid does not fit in general the surfaces of the particles. In [1] we used a Gauss quadrature with modified weights but the inaccuracy in the computation of the integrals leads to relatively large oscillations in the angular velocity and relatively inaccurate velocity of the centroids of the particles. Here we use an element subdivision procedure which subdivides every tetrahedron which is intersected by a particle boundary into tetrahedra so that the particle boundary is exactly fit by second order element faces. Note that the subdivided grid is used only for the computation of the integrals, and not for the discretization of the problem. So, we can still employ structured grids and thus allow for an efficient parallelization of the method. This advantage of the fictitious domain methods is probably the main reason for their existence and should not be compromised. A cross section of a 3D tetrahedral grid and the corresponding boundary-fit grid is shown in Figure 1. As shown in the next section, this integration procedure greatly improves the results for the angular velocity of the particle.

4 Numerical Examples

The first validation example considered here is for the sedimentation of a rigid sphere towards a horizontal wall. This problem has only one characteristic spatial

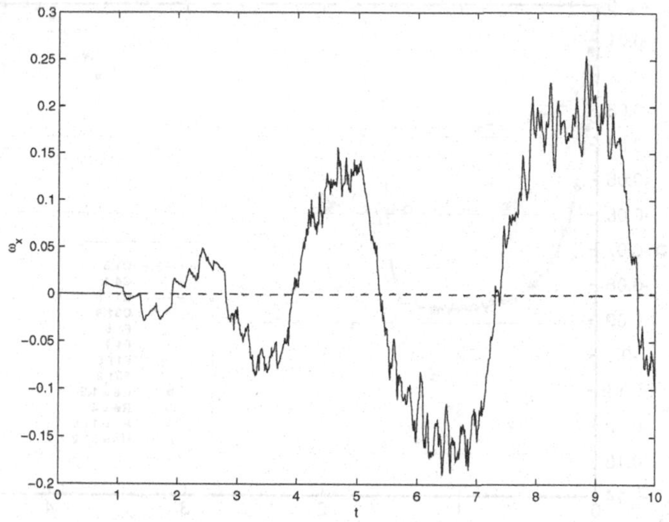

Fig. 2. The angular velocity of a settling particle with a body-fitted grid refinement for integration (dashed line) and integration on the base grid (continuous line)

scale until the gap between the particle and the wall becomes very small. Then a thin film between the particle and the wall is formed and unless the grid is sufficiently refined there, we cannot expect to obtain good results in the vicinity of the wall. Since such a refinement is unreasonably expensive we used a subgrid modelling approach adopting the subgrid modelling force suggested in [7]. It has the form

$$F_w = -6\pi r_p U_\perp \mu_1 \left(\frac{r_p}{\hat{h}} - \frac{r_p}{h} \right), \quad \text{if} \quad \hat{h} < h$$

where r_p is the radius of the particle, $U_\perp$ is its velocity component perpendicular to the wall, h is the grid size of the base grid, and $\hat{h}$ is the gap between the wall and the particle. In Figure 2 we show the results for the angular velocity of a settling particle at $Re = 1.5$ (based on the particle diameter), with and without surface fitting for the integral computations. Since at this Reynolds number the flow is perfectly axisymmetric, the angular velocity should equal zero. Clearly, the surface fitting greatly improves the results.

In the next figure we show a comparison with the experimental results of [7] for the settling velocity and the particle's centroid position for a variety of Reynolds numbers using two different grids: a coarser grid containing 134139 nodes (with a grid spacing around the sphere's path equal to 0.2) and a finer grid of 467261 nodes (with a grid spacing around the sphere's path of 0.1). The dimensions of the container in the experiments are: depth 100 mm, width 100 mm, height 160 mm, and the diameter of the particle is 15 mm. The numerical experiments were run in a dimensionless cavity $7 \times 7 \times 11$ (with respect to the particle diameter). Since in this simulation no lubrication force was used, the

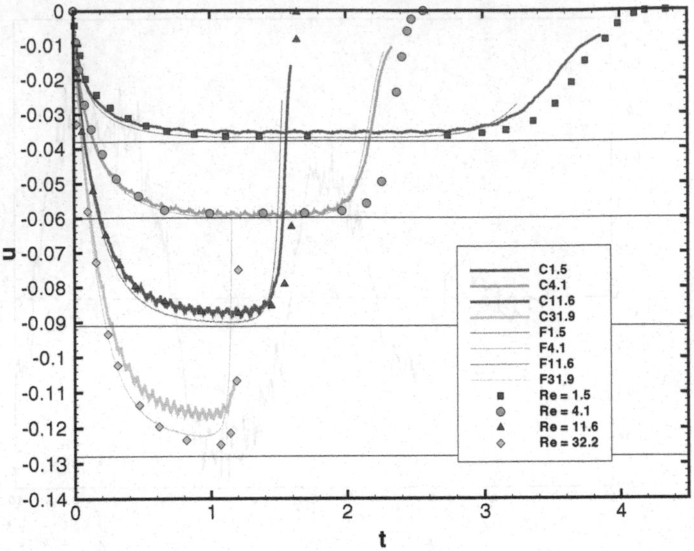

Fig. 3. Comparison of the vertical settling velocity of a particle with experimental data: the symbols represent the experimental results for various Reynolds numbers Re, the thick lines represent the corresponding numerical results on a course grid of 134139 nodes and the thin lines represent the corresponding numerical results on a finer grid of 467261 nodes

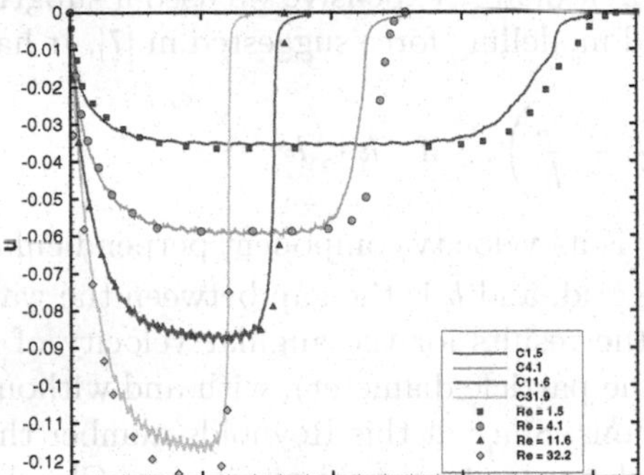

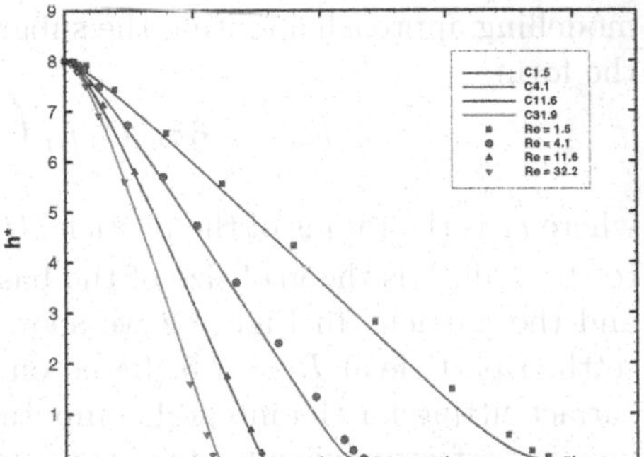

Fig. 4. Comparison with available experimental data of the velocity (left graph) and particle path (right graph) of a settling particle at various Reynolds numbers, Re. The symbols represent the experimental results, the lines represent the corresponding numerical results on a course grid of 134139 nodes. The subgrid force is included.

particle does not come to a complete stop once it hits the bottom wall. Clearly, the numerical results converge to the experimental data with the refinement of the grid.

In the next figure we show the same comparison but with the lubrication force included (using the coarser grid only). The lubrication force influences only the final stage of the particle trajectory. Both figures show a relatively good agreement of our numerical results and the experimental data.

5 Conclusions

An improved version of the global Lagrange multiplier (GLM) fictitious domain method of [1] is presented in this paper. The improvement comes from a slight change in the iteration imposing the rigid body motion and the grid refinement for computation of the integrals involved in the weak formulation of the problem. The method is validated on two problems and the results of the first problem are compared to available experimental data showing a good agreement.

Acknowledgments

The authors would like to acknowledge the support, under a Discovery Grant, of the National Science and Engineering Research Council of Canada (NSERC).

References

1. Diaz-Goano, C., Minev, P.D. and Nandakumar, K.: A fictitious domain/finite element method for particulate flows. J. Comp. Phys. **192** (2003) 105–123.
2. Glowinski, R., Pan T., Hesla T., Joseph D. and Periaux J.: A fictitious domain approach to the direct numerical simulation of incompressible viscous flow past moving rigid bodies: Application to particulate flow. J. Comput. Phys. **169** (2001) 363–426.
3. Glowinski, R., Pan T., Hesla T., and Joseph D.: A distributed Lagrange multiplier/fictitious domain method for particulate flows. Int. J. Muliphase Flow **25** (1999) 755–794.
4. Hu, H.H.: Direct simulation of flows of solid-liquid mixtures. Int. J. Multiphase Flow **22** (1996) 335–352.
5. Johnson, A.A. and Tezduyar: 3D simulation of fluid-particle interactions with the number of particles reaching 100. Comp. Meth. Appl. Mech. Eng. **145** (1997) 301–321.
6. Minev, P.D. and Ethier, C.R.: A Characteristic/Finite Element Algorithm for the 3-D Navier-Stokes Equations Using Unstructured Grids. Comp. Meth. Appl. Mech. Eng. **178** (1999) 39–50.
7. ten Cate, A., Nieuwstad, C.H., Derksen, J.J., and Van den Akker, H.E.A.: Particle imaging velocimetry experiments and lattice-Boltzmann simulations on a single sphere settling under gravity. Phys. Fluids **14** (2002) 4012–4025.

Part XI

Numerical Methods for the Schrödinger Equation and Application

Efficient Dynamical Simulation of Strongly Correlated One-Dimensional Quantum Systems

Stephen R. Clark and Dieter Jaksch

Clarendon Laboratory, University of Oxford, Oxford OX1 3PU, United Kingdom
s.clark@physics.ox.ac.uk
http://physics.ox.ac.uk/qubit

Abstract. Studying the unitary time evolution of strongly correlated quantum systems is one of the most challenging theoretical and experimental problems in physics. For an important class of one-dimensional (1D) systems dynamical simulations have become possible since the advent of the time-evolving block decimation (TEBD) algorithm. We study the computational properties of TEBD using the Bose-Hubbard model (BHM) as a test-bed. We demonstrate its efficiency and verify its accuracy through comparisons with an exactly solvable small system and via the convergence of one- and two-particle observables in a larger system.

1 Introduction

For many interesting and important quantum systems a description in terms of weak effective interactions is not possible. Instead these systems possess strong correlations which demand a description in terms of the full many-body quantum problem. Typically strongly correlated systems are described by simplified models, such as the Hubbard or Heisenberg Hamiltonians, which are believed to capture some of their essential physics. The simplicity of these models has allowed some insight to be gleaned by analytical approximations in certain limits. However the absence of a dominant exactly solvable contribution over the entire parameter range of these models has ultimately limited the applicability and controlled reliability of conventional perturbative methods. For this reason numerical techniques have become an essential tool in handling the inherent complexity of strongly correlated systems.

The standard numerical methods are well represented by three main approaches; namely, exact diagonalization (ED), quantum Monte Carlo (QMC), and density matrix renormalization group (DMRG) methods. In principle all physical quantities can be accurately and directly determined via ED [9]. However this approach fundamentally suffers from an *exponential* scaling with the number of quantum subsystems M composing the total system for both the storage and computational speed for the dynamical evolution of an arbitrary state. While the use of approximation schemes such as Lanczos iterations can extent the usefulness of ED this intractable scaling ultimately limits it to very small or restricted systems which are often not adequate to describe experimental situations. For this reason other techniques have been developed. Using QMC

I. Lirkov, S. Margenov, and J. Waśniewski (Eds.): LSSC 2005, LNCS 3743, pp. 555–563, 2006.

techniques [4] the ground state properties of a wide-ranging class of many-body Hamiltonians can be efficiently evaluated for moderately sized systems in any dimensions. One of the most attractive features of ground state QMC calculations is its efficient scaling, where the computational speed scales cubically in the number of particles. Extensions of QMC to dynamical time evolution have been proposed [11], however the numerical controllability and computational scaling are still to be determined in detail. For 1D systems in particular, DMRG has enabled the low-energy equilibrium properties to be computed with unprecedented precision [18, 12]. More recently DMRG has also been extended to yield accurate low energy spectra [8] and to calculate the time evolution of 1D systems for short times [1, 10].

In this paper we focus on the recently devised TEBD simulation algorithm. Despite arising from studies of weakly entangled quantum computations [16], it was quickly recognized that TEBD enables the dynamics of 1D systems with nearest-neighbor interactions to be computed efficiently and accurately for long times [17]. Its importance is broadened further by its close connections to extensions of DMRG, as highlighted by the demonstration that the scaling and typical accuracy of TEBD are the same as those of DMRG [3]. Indeed many of its novel features have now been incorporated with well established DMRG techniques, such as good quantum numbers, state targeting and White's 'State prediction' method, resulting in new time-dependent DMRG methods [3, 19]. Very recently TEBD/DMRG has been extended to describe mixed state dynamics of 1D systems opening up the possibility of simulating finite temperature effects, decoherence and dissipation [20, 15].

Here we investigate the computational properties of the TEBD algorithm using the 1D BHM as a test-bed. The importance of this model has increased after recent experiments with ultracold atomic gases confined in an optical lattice demonstrated its clean realization with tunable interaction parameters [7, 5, 13]. This has opened up the possibility of experimentally exploring coherent time-dependent phenomena, and so there is an increasing need to simulate the time evolution of the BHM in order to understand its novel properties [2], and its possible applications to quantum computing [6]. Through comparisons with an exact calculation of a small system we address the types of errors which arise in TEBD. This allows *suitable* simulation parameters for the BHM to be identified and these are verified by the convergence of one- and two-particle observables for larger systems. Finally we confirm the efficient computational scaling of TEBD by comparison to the actual performance of our implementation.

2 Time-Evolving Block Decimation

2.1 Decomposition of a State

Let us consider a 1D quantum system composed of a finite number M of subsystems (or sites) each described by a local n-dimensional Hilbert space and basis $|j_m\rangle$ with $j \in \{1, \ldots, n\}$ and $m \in \{1, \ldots, M\}$. An arbitrary state of this system can then be written as $|\psi\rangle = \sum_{j_1=1}^{n} \cdots \sum_{j_M=1}^{n} c_{j_1 \cdots j_M} |j_1, \ldots, j_M\rangle$ in this

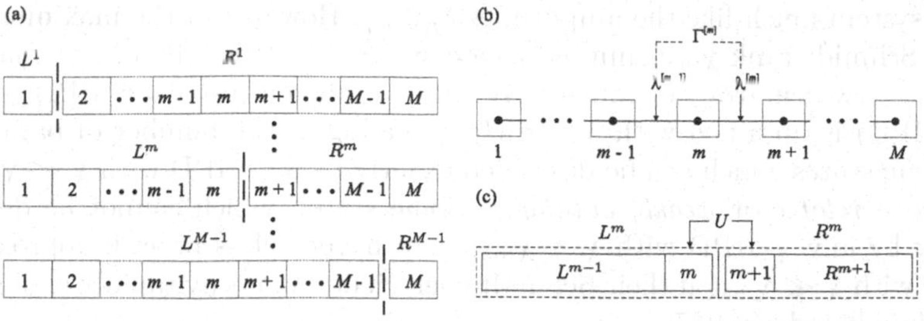

Fig. 1. (a) The sequence of contiguous partitions of the system in which the SD are computed. (b) A depiction of the Γ tensors associated to lattice sites and λ tensors associated to a partition between those sites. (c) The two-blocks-two-sites configuration used within TEBD for the application of a two-site unitary U to sites m and $m+1$.

product basis requiring n^M complex amplitudes $c_{j_1 \cdots j_M}$. The TEBD algorithm is based on approximating $|\psi\rangle$ via the matrix product ansatz in which the number of parameters describing a state scales only *linearly* with M.

Suppose we cut the system into two contiguous parts after site m as L^m composed of the first m sites and R^m composed of the last $M-m$ sites. For any state $|\psi\rangle$ a Schmidt decomposition (SD) can be performed with respects to this splitting which renders the state in the form $|\psi\rangle = \sum_{\alpha=1}^{\chi_m} \lambda_\alpha^{[m]} |L_\alpha^m\rangle |R_\alpha^m\rangle$, where χ_m is the Schmidt rank of the SD, $\lambda_\alpha^{[m]}$ are the Schmidt coefficients and $|L_\alpha^m\rangle$ and $|R_\alpha^m\rangle$ are the corresponding Schmidt states of the left and right subsystems respectively. The Schmidt rank χ_m is then a useful measure of the entanglement between the two subsystems L^m and R^m [16]. To begin the construction SDs are performed for the sequence of contiguous partitions of the system after site $m \in \{1, \cdots, M-1\}$, as depicted in Fig. 1(a). The information contained within $\lambda_\alpha^{[m]}$ and the states $|L_\alpha^m\rangle$ and $|R_\alpha^m\rangle$ from these SDs is then reformulated as set of Γ and λ tensors [16] which are a matrix product decomposition (MPD) of the state $|\psi\rangle$ in the fixed product basis [3], as

$$c_{j_1 \cdots j_M} = \sum_{\alpha_1, \ldots, \alpha_{M-1}} \Gamma_{\alpha_1}^{[1]j_1} \lambda_{\alpha_1}^{[1]} \Gamma_{\alpha_1 \alpha_2}^{[2]j_2} \lambda_{\alpha_2}^{[2]} \cdots \lambda_{\alpha_{M-1}}^{[M-1]} \Gamma_{\alpha_{M-1}}^{[M]j_M}, \quad (1)$$

where α_m is the Schmidt index of the m-th partition and sums from 1 to its respective Schmidt rank χ_m. The set of Γ and λ tensors, depicted in Fig. 1(b) together give a localized representation of all $(M-1)$ SD such that every Schmidt state can be readily reconstructed as

$$|R_{\alpha_m}^m\rangle = \sum_{j_{m+1}, \ldots, j_M} \sum_{\alpha_{m+1}, \ldots, \alpha_M} \Gamma_{\alpha_m \alpha_{m+1}}^{[m+1]j_{m+1}} \lambda_{\alpha_{m+1}}^{[m+1]} \cdots \lambda_{\alpha_{M-1}}^{[M-1]} \Gamma_{\alpha_{M-1}}^{[M]j_M} |j_{m+1}, \ldots, j_M\rangle,$$

$$|L_{\alpha_m}^m\rangle = \sum_{j_1, \ldots, j_m} \sum_{\alpha_1, \ldots, \alpha_{m-1}} \Gamma_{\alpha_1}^{[1]j_1} \lambda_{\alpha_1}^{[1]} \cdots \lambda_{\alpha_{m-1}}^{[m-1]} \Gamma_{\alpha_{m-1}\alpha_m}^{[m]j_m} |j_1, \ldots, j_m\rangle. \quad (2)$$

Under the circumstances described the MPD in Eq. (1) is exact and so the number of parameters stored within the MPD can grow exponentially with the size of the system much like the amplitudes $c_{j_1 \cdots j_M}$. However, if the maximum value of any Schmidt rank χ_m is limited to a constant χ, rather than its formal maximum $\chi_{\max}$ which grows exponentially with M, then a restricted MPD (denoted as χ-MPD) is obtained with a $O(nM\chi^2)$ scaling in the number of parameters. Quantum states which can be described exactly by a χ-MPD with $\chi \ll \chi_{\max}$ are *finitely correlated* or *weakly entangled*. Some states, which cannot be described exactly by any χ-MPD with $\chi < \chi_{\max}$, can nevertheless be well approximated by one with $\chi \ll \chi_{\max}$ if their Schmidt coefficients λ_α decay rapidly enough with the Schmidt index α [17].

2.2 Time Evolution

By dealing directly with the MPD of a state TEBD can implement two-site unitaries on neighboring sites exactly [16]. To do so, for two sites m and $m+1$, the MPD is rewritten in terms of a two-blocks-two-sites configuration involving $[1 \cdots m-1][m \quad m+1][m+2 \cdots M]$, as in Fig. 1(c), such that the two sites are expressed in the product basis as

$$|\psi\rangle = \sum_{l,r} \sum_{j_m, j_{m+1}} \lambda_l^{[m-1]} \left(\sum_{\alpha_m} \Gamma_{l\alpha_m}^{[m]j_m} \lambda_{\alpha_m}^{[m]} \Gamma_{\alpha_m r}^{[m+1]j_{m+1}} \right) \lambda_r^{[m+1]} |l\rangle |j_m\rangle |j_{m+1}\rangle |r\rangle , \tag{3}$$

where $|l\rangle = \left|L_l^{m-1}\right\rangle$ and $|r\rangle = \left|R_r^{m+1}\right\rangle$. It is at this point that the strong overlap of TEBD with DMRG emerges since this representation is identical to that used during a finite-system sweep in DMRG [19]. In this form any two-site operator U expressed in the product bases of the pair of sites can be applied to yield

$$U|\psi\rangle = \sum_{l,r} \sum_{j_m, j_{m+1}} \lambda_l^{[m-1]} \Theta_{j_m j_{m+1}}^{l,r} \lambda_r^{[m+1]} |l\rangle |j_m\rangle |j_{m+1}\rangle |r\rangle . \tag{4}$$

The application of U alters the Schmidt states of all subsystems in the $(M-1)$ SDs which contain these sites. These changes are entirely captured by updating the tensors $\Gamma_{\alpha_{m-1}\alpha_m}^{[m]j_m} \lambda_{\alpha_m}^{[m]} \Gamma_{\alpha_m \alpha_{m+1}}^{[m+1]j_{m+1}}$ local to these sites [16]. To compute these new tensors, and so return $U|\psi\rangle$ in Eq. (4) to a MPD, the density matrix of a subsystem composed of one site and one block, such as $R^m = [m+1][m+2 \cdots M]$, must be re-diagonalized to yield the new Schmidt coefficients $\lambda_{\alpha_m}^{[m]}$ and Schmidt states $\left|R_{\alpha_m}^m\right\rangle$. This is the most computational intensive part of the update and its scaling follows from standard diagonalization algorithms as $O(n^3\chi^3)$, given that the dimensions of the R^m density matrix is $(n\chi) \times (n\chi)$ at most. Of the resulting $n\chi$ Schmidt states $\left|R_{\alpha_m}^m\right\rangle$ only χ states with the largest Schmidt coefficients $\lambda_{\alpha_m}^{[m]}$ are kept. The resulting *Truncation error* is proportional to the sum of the discarded density matrix eigenvalues [17]. Since the Schmidt states $|r\rangle$ of R^{m+1} are left unaltered Eq. (2) can be used to extract the $\Gamma_{\alpha_m \alpha_{m+1}}^{[m+1]j_{m+1}}$ from $\left|R_{\alpha_m}^m\right\rangle$. The relation $\left\langle R_{\alpha_m}^m \middle| U\psi \right\rangle = \lambda_{\alpha_m}^{[m]} \left|L_{\alpha_m}^m\right\rangle$ then gives the corresponding left subsystem Schmidt states from which $\Gamma_{\alpha_{m-1}\alpha_m}^{[m]j_m}$ can be extracted in an identical way [16].

Since the χ-MPD of $U|\psi\rangle$ can be computed efficiently for any two-site unitary U our implementation of the TEBD algorithm computes the time evolution using a Suzuki-Trotter (ST) decomposition [14] of the propagator $\exp[-iH(t)\delta t]$. Consider an 1D nearest-neighbor Hamiltonian $H(t) = \sum_{m=1}^{M-1} H_{m,m+1}(t)$, where $H_{m,m+1}(t)$ are two-site terms. This type of Hamiltonian can be separated into $H(t) = F + G$ where F contains all terms with m odd, and G with m even. Given that no terms within F involve the same sites they all commute amongst themselves enabling $\exp(-iF\delta t)$ to be computed exactly as a product of two-site unitaries $\exp(-iF\delta t) = \prod_{n \text{ odd}} \exp[-iH_{n,n+1}(t)\delta t]$. These unitaries can be applied exactly with TEBD. The same is also true for G. The complications in computing the time-evolution arise from the fact that F and G do not in general commute. To overcome this we approximate $\exp[-i(F+G)\delta t]$ using a ST expansion. Ignoring their non-commutativity would constitute a 1-st order expansion in δt. If we define $s_2(F, G, y) = \exp(-iFy/2)\exp(-iGy)\exp(-iFy/2)$ then the 2-nd order expansion follows when $y = \delta t$. For the numerical simulations performed in this paper the 4-th order expansion [14] was used, which consists of a product of five 2-nd order type terms $\exp[-i(F+G)\delta t] = \prod_{l=1}^{5} s_2(F, G, q_l \delta t) + O(\delta t^5)$, where the parameters q_l are defined as $q_1 = q_2 = q_4 = q_5 \equiv q = (4 - 4^{1/3})^{-1}$ and $q_3 = 1 - 4q$. Thus to time evolve a state multiple applications of two-site unitaries are required, each of which locally truncates the resulting state to a χ-MPD. For sufficiently small δt this procedure will generate a good approximation to the χ-MPD of the complete ST time-evolved state with the decimated Hilbert space describing the state adapted to 'follow' the time evolution. For a p-th order ST decomposition the error made in one time-step δt is $O(\delta t^{p+1})$. Performing evolution to a given time τ requires $\tau/\delta t$ time-steps so the propagation of the *Trotter error* is, at worst, bounded as $O(\delta t^p \tau)$.

The usefulness of TEBD for studying strongly correlated systems is based on a crucial observation that for sufficiently regular, but otherwise arbitrary, 1D nearest-neighbor Hamiltonians the ground state and low-lying excitations have Schmidt coefficients $\lambda_\alpha^{[m]}$ that decay (roughly) exponentially with α [17]. Consequently the subspace containing the low-energy dynamics of these systems is accurately described by the class of weakly entangled states that TEBD can efficiently time-evolve.

3 Bose-Hubbard Model Dynamics

3.1 The Physics

The BHM is one of the simplest toy-models in physics. It consists of a kinetic energy term, with matrix element J, describing the hopping of particles from one site to the next and an interaction term, with matrix element U, which accounts for the repulsion or attraction of two particles occupying the same site. Competition between these two terms results in a transition at temperature $T = 0$ when $U/2J \approx 5.8$ in 1D from the superfluid (SF) to the Mott insulator (MI) phase. In terms of bosonic destruction (creation) b_m ($b_m^\dagger$) operators for a site m the BHM for an M site system has the form (with $\hbar = 1$)

$$H(t) = -J(t) \sum_{m=1}^{M-1} (b_m^\dagger b_{m+1} + \text{h.c.}) + \frac{U(t)}{2} \sum_{m=1}^{M} b_m^\dagger b_m^\dagger b_m b_m. \tag{5}$$

When U is sufficiently larger than J and small filling factors of the lattice are used this system can be described by M sites each with Fock states $|n_m\rangle$ where the bosonic occupancy n_m limited to n at most. For our considerations, which were restricted to commensurately filled systems with $N = M$ particles and box boundary conditions in all cases, $n = 4$ was sufficient.

3.2 Analysis for a Small System

The two main sources of error within TEBD are (a) *Trotter errors* caused by incorrect higher order terms in the ST expansion, and (b) *Truncation errors* which can potentially occur and accumulate after every two-site unitary. To investigate the accuracy of TEBD and how these two errors interplay we consider a small BHM system with $M = 7$ where an exact numerical solution can be readily found. Firstly, from the exact ground state in the SF regime, where $U/2J = 2$, we find that λ_α for the SD of the system partitioned as $[1 \ldots 3][4 \ldots 7]$ do indeed display, in Fig. 2(a), the (roughly) exponential decay with α required for it to be accurately approximated by some χ-MPD. The ground state was then computed with TEBD via imaginary time evolution for χ between 2 and 11. All these states where then time evolved with $U/2J$ being ramped from $U/2J = 2$ (SF regime) to $U/2J = 20$ (MI regime) and back again over a time $t = 143$ $[J^{-1}]$. To draw a close connection to experiments [5] we consider this transition for ^{87}Rb atoms trapped in a $\lambda = 826$ [nm] optical lattice [7] where $J^{-1} = 0.56$ [ms] for $U/2J = 2$ and $t = 80$ [ms], as shown in Fig. 2(b). In Fig. 2(c) the infidelity $\epsilon(t) = 1 - |\langle\tilde{\psi}(t)|\psi(t)\rangle|$, between the exact $|\tilde{\psi}(t)\rangle$ and TEBD $|\psi(t)\rangle$ many-body states, is shown over time. In the MI regime where the state of the system is well approximated by a Fock state described exactly by a χ-MPD with $\chi = 1$, we find in Fig. 2(c) that ϵ for all χ simulations roughly converge to the same accuracy determined by the Trotter error. The clear dependence of curves (i) $\chi = 3$ and (ii) $\chi = 5$ on the profile of the time evolution indicates that their errors are truncation dominated in SF regime. In contrast the curves (iii) $\chi = 9$ and (iv) $\chi = 11$ display no major dependence on the form of the dynamics. This indicates that for $\delta t = 0.01$ $[J^{-1}]$ and $\chi > 9$ the Trotter error dominates at all times causing the growth of ϵ observed. Using $\chi = 9$ we investigated the dependence of $\epsilon(\tau)$ on δt for the same evolution up to a time $\tau = 20$ [ms] which brings the system just above the transition indicated in Fig. 2(b). The results in Fig. 2(d) have a sharp and significant increase in ϵ around a small region centered on $\delta t \approx 0.45$ $[J^{-1}]$, but otherwise fit a power law δt^p with $p > 4$ expected for the Trotter error. The peak is an expected numerical resonance caused by δt entering a range of coincidence with a timescale associate to an excitation of the system with energy $2/U \sim 0.45$ $[J^{-1}]$ during the ramping.

The efficiency of our TEBD implementation was confirmed by computing the time evolution for $\tau = 20$ [ms], firstly for a variety of system sizes M from 7 to 21, all with $\chi = 5$, and secondly for a system with $M = 7$ for χ between

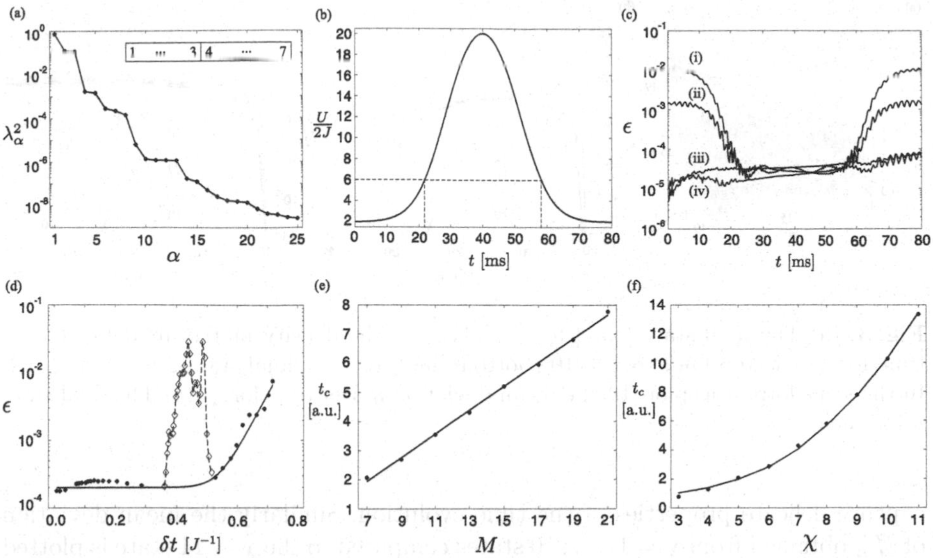

Fig. 2. (a) The decay of λ_α^2 with α for the splitting $[1\dots3][4\dots7]$. (b) The time profile of $U/2J$ used. (c) The infidelity ϵ of the dynamical simulation performed with (i) $\chi=3$, (ii) $\chi=5$, (iii) $\chi=9$, and (iv) $\chi=11$. (d) The scaling of ϵ with δt for $\chi=9$. (e) The linear scaling of the computational time t_c with the system size M. (f) The cubic scaling of t_c with χ.

3 and 11, as shown in Fig. 2(e) and Fig. 2(f) respectively. These confirm the expected linear scaling with M and cubic scaling with χ.

3.3 Analysis for a Large System

To consider systems closer to those studied in experiments [5,13] we applied the last half of $U/2J$ ramping in Fig. 2(b) on initial Fock states with $N=M=49$ particles and $\chi=1,\dots,11$. This state has the advantage of being represented exactly by all χ-MPD. Since an exact numerical solution is not available we instead study the convergence of one- and two-particle observables through the one-particle density matrix $\rho_{m,l}=\langle b_m^\dagger b_l\rangle$ and the average of the site occupancy squared $\Pi_m=\langle b_m^\dagger b_m b_m^\dagger b_m\rangle$. In Fig. 3(a) $|\rho_{m,l}|$ for the final state with $\chi=11$ is shown. To examine the accuracy of TEBD for larger systems we computed the infidelity $\zeta(\rho,\sigma)=1-\mathrm{tr}(\sqrt{\sqrt{\rho}\sigma\sqrt{\rho}})$ between the one-particle density matrices obtained for $\chi=1,\dots,10$ and the $\chi=11$ state over the first half of the ramping shown in Fig. 3(b). The inset Fig. 3(c) then shows ζ for the final states for each χ demonstrating its near-exponential improvement with χ. As the system is ramped to the SF regime phase coherence, indicated by off-diagonal elements in $|\rho_{m,l}|$, is established. Since the initial Fock state used here contains no phase coherence these must arise purely from quantum fluctuations during the dynamics, making this a very stringent test of TEBDs ability to

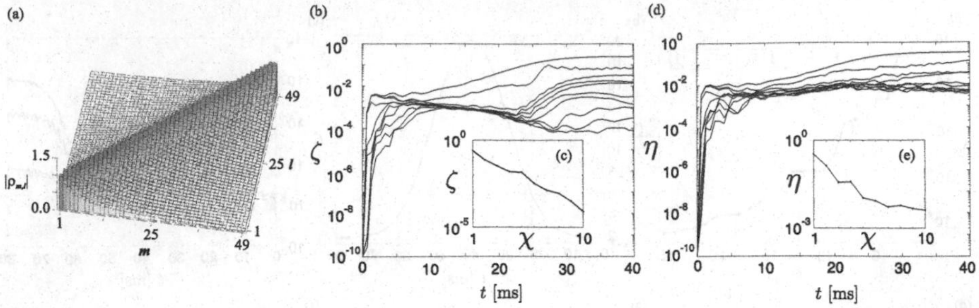

Fig. 3. (a) The final state $|\rho_{m,l}|$ for $\chi = 11$. (b) The density matrix infidelity ζ over time for $\chi = 1$ (top line) to $\chi = 10$ (bottom line). (c) The final state ζ for each χ. (d) In the same format as plot (b) the mean deviation η over time for χ. (e) The final state η for each χ.

capture delicate properties of the time evolution. Similarly the mean deviation of Π_m obtained from $\chi = 1, \ldots, 10$ states compared to the $\chi = 11$ state is plotted in Fig. 3(d) along with the inset Fig. 3(e) of the final deviations with χ. This demonstrates the rapid convergence of this two-particle observable with χ. In both cases $\chi = 9$ identified from the small system yields accurate results for one- and two-particle observables despite the possibility that the actual many-body state has not yet reached a similar degree of convergence in its infidelity.

4 Conclusions

We have reviewed the TEBD algorithm and emphasized its importance for studying strongly correlated 1D systems. Using a small BHM system we have illustrated the interplay between the errors within TEBD and shown our implementation to be efficient. Finally we studied the convergence of one- and two-particle observables for a larger system with χ and demonstrated that suitably accurate simulation parameters follow from those of the smaller system.

Acknowledgments

This work is supported by the EPSRC (UK) through the QIP IRC (www.qipirc.org) (GR/S82176/01) and the project EP/C51933/1, as well the OLAQUI project.

References

1. Cazalilla, M.A., Marston, J.B.: Time-dependent density-matrix renormalization group: A systematic method for the study of quantum many-body out-of-equilibrium systsms. Phys. Rev. Lett. **88** 256403 (2002)
2. Clark, S.R., Jaksch, D.: Dynamics of the superfluid to Mott-insulator transition in one dimension. Phys. Rev. A **70** 043612 (2004)

3. Daley, A.J., Kollath, C., Schollwöck, U., Vidal, G.: Time-dependent density-matrix renormalization-group using adaptive effective Hilbert spaces. J. Stat. Mech.: Theor. Exp. P04005 (2004)
4. Foulkes W. M. C., Mitas L., Needs R.J., Rajagopal G.: Quantum Monte Carlo simualtion of solids. Rev. Mod. Phys. **73** 1 (2001)
5. Greiner, M., Mandel, O., Esslinger, T., Hänsch, T.W., Bloch, I.: Quantum phase transition from a superfliud to a Mott-insulator in a gas of ultracold atoms. Nature (London) **415** 39 (2002)
6. Jaksch, D., Briegel, H.-J., Cirac, J.I., Gardiner, C.W., Zoller, P.: Entanglement of atoms via cold controlled collisions. Phys. Rev. Lett. **82** 1975 (1999)
7. Jaksch, D., Bruder, C., Cirac, J.I., Gardiner, C.W., Zoller, P.: Cold bosonic atoms in optical lattices. Phys. Rev. Lett. **81** 3108 (1998)
8. Jeckelmann, E.: Dynamical density-matrix renormalization group method. Phys. Rev. B **66** 045114 (2002)
9. Laflorencie N., Poilblanc D.: Simulations of pure and doped low-dimensional spin-1/2 gapped systems. Lect. Notes Phys. **645** 227 (2004)
10. Luo, H.G., Xiang, T., Wang, X.Q.: Comment on 'Time-Dependent Density-Matrix Renormalization Group: A Systematic Method for the Study of Quantum Many-Body Out-of-Equilibrium Systems'. Phys. Rev. Lett. **91** 049701 (2003)
11. Ostilli M., Presilla C.: Exact Monte Carlo time dynamics in many-body lattice quantum systems. J. Phys. A **38** 405 (2005)
12. Schollwöck, U.: The density-matrix renormalization group. Rev. Mod. Phys. **77** 259 (2005)
13. Stöferle, T., Moritz, H., Schori, C., Køhl, M., Esslinger, T.: Transition from a strongly interacting one-dimensional superfluid to a Mott-insulator. Phys. Rev. Lett. **92** 130403 (2004)
14. Suzuki, M.: Fractal Decomposition of Exponential Operators with Applications to Many-Body Theories and Monte Carlo Simulations. Phys. Lett. A **146**, 6 (1990); General theory of fractal path integrals with applications to many-body theories and statistical physics. J. Math. Phys. **32** 2 (1991)
15. Verstraete, F., Garía-Ripoll, J.J., Cirac, J.I.: Matrix product density operators: Simulations of finite-temperature and dissipative systems. Phys. Rev. Lett. **93** 207204 (2004)
16. Vidal, G.: Efficient Classical simulation of slightly entangled quantum computations. Phys. Rev. Lett. **91** 147902 (2003)
17. Vidal, G.: Efficient simulation of one-dimensional quantum many-body systems. Phys. Rev. Lett. **93** 040502 (2004)
18. White, S.R.: Density-matrix formulation for quantum renormalization groups. Phys. Rev. Lett. **69** 2863 (1992); Density-matrix algorithms for quantum renormalization groups. Phys. Rev. B **48** 10345 (1993)
19. White, S.R., Feiguin, A.E.: Real-time evolution using density matrix renormalization group. Phys. Rev. Lett. **93** 076401 (2004)
20. Zwolak, M., Vidal, G.: Mixed-state dynamics in one-dimensional quatum lattice systems: A time-dependent superoperator renormalization algorithm. Phys. Rev. Lett. **93** 207205 (2004)

Data Reduction Schemes in Davidson Subspace Diagonalization for MR-CI

Wilfried N. Gansterer[1], Wolfgang Kreuzer[1,*], and Hans Lischka[2,*]

[1] Institute of Distributed and Multimedia Systems, University of Vienna
[2] Institute for Theoretical Chemistry, University of Vienna

Abstract. In this paper, we investigate several data reduction schemes to improve the computational efficiency in the multi reference configuration interaction (MR-CI) method, one of the main quantum chemical approaches for solving the electronic Schrödinger equation. The basic idea is to take advantage of the often relatively low accuracy requirements on the solution of the resulting large eigenvalue problem, whose dimension may reach several hundred millions or even more. We will discuss some approaches to reduce the amount of data to be accessed and to be transferred within the Davidson subspace diagonalization method. We also show experimental results achieved with the COLUMBUS code.

1 Introduction

Quantum chemical methods provide very important procedures for the computation of molecular properties and for the computer simulations of chemical reactions which cannot be solved exactly. They are based on the electronic Schrödinger equation; a complicated many-particle differential equation, which cannot be solved analytically. Therefore, numerical approximations have to be used. These are usually very involved, extremely time consuming and require very large amounts of data flow.

In this paper, we focus on the *multireference configuration interaction* (*MR-CI*) approach [9]. It allows for accurate calculations of molecular systems based on the original many-particle Schrödinger equation. The MR-CI method is especially important in "difficult" cases, e. g., for the calculation of dissociation processes and electronically excited states.

A basis expansion of the many-particle wave function leads to an eigenvalue problem of enormous size. Although the Hamiltonian matrix H of this eigenproblem is very sparse, its dimension n can easily reach several hundred millions or even billions. H tends to be diagonally dominant and usually very few eigenpairs (or only a single one) need to be computed.

Given this setup, the Davidson method is a suitable approach for solving problems of this type numerically. However, in contrast to the matrix H, the subspace vectors and the Ritz vectors arising in this process are in general *dense*, and not sparse. Due to their enormous dimension, simply storing and handling

* Supported by the project P14817-N03 of the Austrian Science Fund FWF.

I. Lirkov, S. Margenov, and J. Waśniewski (Eds.): LSSC 2005, LNCS 3743, pp. 564–571, 2006.

those vectors becomes a major performance bottleneck. This is especially severe when solving the eigenproblem in parallel (which is required for large problems of interest), because intensive memory accesses and transfer of huge amounts of data leads to an often prohibitive communication overhead. The data reduction approach developed in this paper will help to overcome these problems.

Related Work. In earlier work, Dachsel and Lischka [2] have developed first ideas for a data compression approach in the Davidson subspace diagonalization. They propose bitwise data reduction in the subspace vectors based on an estimator for the resulting error. Their approach assumes diagonal dominance of the matrix H. Alternative approaches based on fixed-point truncation schemes have been developed by Harrison and Handy [4], Knowles [5], and Olson [8].

In this paper, we analyze and extend the approach of Dachsel and Lischka, we compare it with various newly developed error estimators, and also discuss aspects related to the implementation of such schemes. Moreover, based on numerical experiments with the COLUMBUS code [6, 7], we also give some insight into how much data reduction can be achieved in practical situations.

2 Multi Resolution Configuration Interaction

The stationary, nonrelativistic, clamped-nuclei, electronic Schrödinger equation is given as

$$\mathcal{H}\Psi = E\Psi$$

with the total Hamiltonian $\mathcal{H} = \sum_i h_i + \sum_{i<j} g_{ij} + V_{kk}$, where h_i is the one-electron operator for electron i containing the kinetic energy and Coulomb attraction, g_{ij} is the electron-electron repulsion term, V_{kk} is the nuclear repulsion and Ψ is the many-electron wave function.

Expanding Ψ into a many-electron basis (configurations state functions $\{\Phi_i\}$) and applying the Ritz variational principle leads to the matrix eigenproblem

$$Hc = Ec \tag{1}$$

with symmetric $H \in \mathbb{R}^{n\times n}$, where n tends to be extremely large (several hundred millions up to a billion). H is sparse, but it is prohibitively costly (in terms of computation as well as in terms of storage requirements) to construct it explicitly ($H_{st} = \langle\Phi_s|\mathcal{H}|\Phi_t\rangle$). Usually, the lowest eigenpair (or a few of the lowest eigenpairs) of H need to be computed.

2.1 Davidson Subspace Diagonalization

Given the properties of the eigenproblem (1), *Davidson's method* [3, 1] is a suitable approach for solving it. The basic structure of this method is shown in Algorithm 1. Starting with an initial vector v_0 or subspace V_0, a basis for a (small) subspace V is constructed, in which approximations $(\bar{E}, u)$ of the desired eigenpairs (E, c) can be computed cheaply (r denotes the associated residual).

Davidson [3] suggested to expand this basis in each iteration by the solution of the correction equation shown in Algorithm 1. This procedure is very successful for strongly diagonally dominant H.

Algorithm 1. Basic structure of Davidson subspace diagonalization

$V = V_0, W = HV_0, v = v_0$
repeat
  $w = Hv$, $V = [V, v]$, $W = [W, w]$
  $\bar{H} = V^\top HV = V^\top W$ ... projected matrix
  $[\bar{E}, \alpha] = \text{eig}(\bar{H}, V^\top V)$ ... Ritz value and Ritz vector
  $u = V\alpha$, $r = Hu - \bar{E}u$
  solve $\left[\text{diag}(H) - \bar{E}\right] v = r$ for v ... correction equation
  reduce V, W if necessary
until $|r| < tol$
return $\bar{E} \approx E, u \approx c$

2.2 Limitations

Davidson's method requires the storage and retrieval of the n-dimensional vectors spanning the subspaces V and W. Due to their size, these vectors are either stored on external disk devices or—in particular for parallel calculations—distributed over the memory of the individual nodes. Storage and retrieval of such large amounts of data constitutes a serious communication bandwidth bottleneck. Moreover, the subspace vectors and their memory requirements are the origin of serious limitations in parallel calculations. Although I/O requirements have been reduced significantly [10], the problem still remains. Therefore, reduction of the memory requirements for storage of the subspace vectors by means of data reduction schemes as described in this paper will lead to substantial alleviation of the bandwidth problem and, in parallel computations, will free memory needed for the local computation of the matrix-vector product Hv.

3 Data Reduction

The basic idea investigated in this paper is the following: Based on an accuracy tolerance τ determined by the user, reduce the amount of data to be stored for the v and w vectors in Davidson's method. The larger τ the more significant the data reduction. This reduction of data to be stored obviously also corresponds to a reduction of data accesses and transfers, which is especially important in parallel calculations.

In order to guarantee satisfactory accuracy of the computed spectral information of H, we develop error estimators and bounds for controlling the errors in E and c due to suggested perturbations in v and w. The main focus in this paper is on controlling the error in the energy E, analysis of the corresponding error in c is the subject of ongoing work.

3.1 Error Estimators and Bounds

One approach for deriving estimators of the effect of the data reduction on the computed energy is to examine the difference $\Delta\bar{E} := \bar{E} - E$. By definition,

$$\Delta\bar{E} = \frac{u^\top H u}{u^\top u} - \frac{c^\top H c}{c^\top c}.$$

Neglecting third and higher order terms in the Taylor series expansion with respect to u around c ($u = c + \Delta u$) leads to (first order terms are zero!)

$$\Delta\bar{E} \approx \Delta u^\top (H - EI)\Delta u \,. \tag{2}$$

From this relation, several estimators can be derived. In the following, we use the notation $e := (1, \ldots, 1)$ and for a matrix A we define $|A|$ by taking absolute values elementwise, that is, $|A|_{ij} := |A_{ij}|$.

Dachsel and Lischka [2] have proceeded by replacing the unknown eigenvalue E by the Ritz value $\bar{E}$ and the matrix H by its diagonal D in (2) which yields

$$\Delta\bar{E} \approx \mathrm{Est}_{DaLi} := \Delta u^\top (D - \bar{E}I)\Delta u = \sum_i \sum_j (D_i - \bar{E}I)\Delta u_i \Delta u_j$$

With an upper bound on the componentwise perturbation in the vector u, $|\Delta u_i| \leq \beta$, this leads to the bound

$$|\Delta\bar{E}| \approx |\mathrm{Est}_{DaLi}| \leq \beta^2 \sum_i |D_i - \bar{E}I|$$

Asking for $\Delta\bar{E} < \tau$ consequently corresponds to requiring

$$\beta < \sqrt{\frac{\tau}{e^\top \left|D - \bar{E}I\right| e}} \,. \tag{3}$$

Alternatives. The estimator constructed in [2] relies on two approximations: E is replaced by the Ritz value $\bar{E}$ in (2) (which may not always be appropriate because this is precisely the error to be estimated), and H is replaced by its diagonal D (which requires diagonally dominant H). Pursuing a more general approach, we can rewrite (2) as

$$\Delta\bar{E} \approx \Delta u^\top \left(H - \left(\bar{E} - \Delta\bar{E}\right) I\right) \Delta u,$$

leading to the estimator Est_H

$$\Delta\bar{E} \approx \mathrm{Est}_H := \frac{\Delta u^\top (H - \bar{E}I)\Delta u}{1 - \Delta u^\top \Delta u} \,. \tag{4}$$

In the following we assume that $1 - \Delta u^\top \Delta u > 0$ which implies $|\Delta u_i| < \sqrt{1/n}$. With this additional constraint we derive analogously to before an upper bound for the allowable componentwise perturbation in u:

$$\beta < \sqrt{\frac{\tau}{e^\top \left|H - \bar{E}I\right| e + n\tau}} \,. \tag{5}$$

If the matrix H happens to be diagonally dominant (as it tends to be the case in the MR-CI context), we can derive an estimator Est_D by replacing H with its diagonal D in (4), leading to the bound

$$\beta < \sqrt{\frac{\tau}{e^\top \left|D - \bar{E}I\right| e + n\tau}} . \tag{6}$$

Finally, we describe a fourth approach. Using the fact that due to the construction of the Davidson eigenvectors from a generalized eigenvalue problem $||u|| = 1$, we have

$$|\Delta\bar{E}| = |c^\top Hc - u^\top Hu| = |(u + \Delta u)^\top H(u + \Delta u) - u^\top Hu| . \tag{7}$$

Based on (7) we require

$$\left|\Delta\bar{E}\right| = |\mathrm{Est}_q| := \left|2\Delta u^\top Hu + \Delta u^\top H\Delta u\right| < \tau .$$

Analogously to before, this corresponds to requiring

$$\beta < -\frac{e^\top|Hu|}{e^\top|H|e} + \sqrt{\left(\frac{e^\top|Hu|}{e^\top|H|e}\right)^2 + \frac{\tau}{e^\top|H|e}}. \tag{8}$$

3.2 Implementation

Several aspects related to the efficient implementation of the concepts developed so far deserve some more detailed discussion.

Evaluation of Bounds. For performance reasons, the Hamiltonian matrix H is not constructed explicitly in the context of MR-CI calculations. For the same reasons, we introduce some additional simplifications in order to reduce the computational effort for evaluating the estimators and bounds derived before.

Our formulation of the bounds (3), (5), (6), and (8) is based on applying the triangle inequality, which leads to expressions with sums of absolute values to be evaluated in every iteration (because $\bar{E}$ changes). To further improve performance, we approximate the sums of absolute values by the absolute value of the sums in those bounds. Based on this simplification, we have to evaluate these sums only *once* during the entire process, and information from the huge matrix H only needs to be accessed once. This approximation tends to work well in our context since we need to approximate the *smallest* eigenvalue(s) E of H. Due to the fact that almost all diagonal entries D_i of H tend to have the same sign and due to the diagonal dominance of H, $\sum_i |D_i - \bar{E}| \approx |\sum_i D_i - \bar{E}|$.

Implementation of Data Reduction. The information about allowable perturbations in the u vectors contained in the bounds (3), (5), (6), and (8) can be translated into *bitwise* reductions in the mantissas of the entries in the v and w vectors as described in [2]. In this paper, we compare this approach with an *elementwise* data reduction strategy, where vector entries with absolute value below a certain threshold (determined by the bounds) are eliminated, whereas all entries above this threshold are stored with full precision.

The advantage of elementwise reduction is that it can be implemented more easily and more efficiently, the disadvantage is that the data reduction achieved is always lower than or at best equal to that of bitwise reduction. Experimental evidence presented in Section 4 shows that the memory requirements with elementwise data reduction can actually be significantly higher than with bitwise data reduction.

4 Experiments

In this section we present some experimental evidence that the amount of data reduction achieved in practice is significant.

Test Case. We evaluated the effect of data reduction with the different estimators and bounds (3), (5), (6), (8) for a "small" but representative test problem of dimension $n = 14558$ computing the MR-CI energy and wave function for the Ethylene molecule C_2H_4.

The calculations were performed with the COLUMBUS program system, a collection of programs for high-level ab initio molecular electronic structure calculations [6, 7]. Double precision accuracy, i. e., a mantissa length of 52 bits, was used, the threshold for the stopping criterium for the Davidson subspace method was set to tol $= 10^{-4}$, and the tolerance τ for the data reduction was set to $\tau = 10^{-8}$.

Convergence History. Fig. 1 illustrates that the artificial perturbations introduced in the vectors v and w of Davidson's method hardly influence its convergence behavior. In our test case, only one additional iteration step is required until convergence for all versions of data reduction. Moreover, it can be observed that for the test case considered the various estimators and bounds hardly differ.

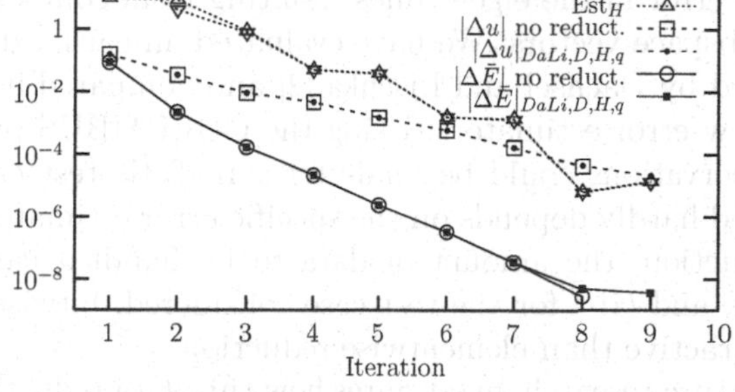

Fig. 1. Convergence history of Davidson's method with data reduction (based on various estimators and bounds) as well as without data reduction

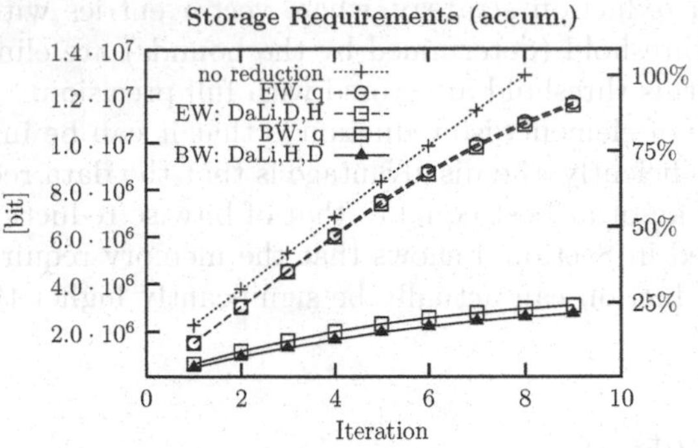

Fig. 2. Accumulative representation of the number of bits to be stored for different variants of data reduction (elementwise "EW", bitwise "BW", various estimators and bounds) as well as without data reduction

Storage Savings. In this example, we focus on achieving maximum storage efficiency, which means that per vector only one additional compression factor, representing the maximum difference between a vector entry and its value after data reduction, is stored. The actual length of the mantisse of each vector entry can then be calculated using this factor and the exponent of the entry. The starting vector for the Davidson procedure is created by diagonalizing a reference subspace. Most of the entries in this vector are zero. Therefore, this vector is submitted to the data reduction process at the beginning of the calculation as well.

Fig. 2 illustrates very high savings in storage requirements from bitwise data reduction. In contrast, for this test case the benefits of elementwise data reduction are much smaller.

5 Conclusions

A data reduction strategy for reducing the data transfer during the solution of the high dimensional eigenvalue problem arising in the MR-CI method for solving the electronic Schrödinger equation has been investigated. It is based on the estimation of the error in the eigenvalues resulting from componentwise perturbations in the subspace vectors. We have evaluated an earlier data compression scheme introduced by Dachsel and Lischka [2] and compared it to new alternatives based on new error estimators. Using the COLUMBUS program package, the following observations could be made for a realistic test case: (*i*) the data reduction achieved hardly depends on the specific error estimator used, (*ii*) with bitwise data reduction, the amount of data to be handled can be reduced to about one fourth, and (*iii*) for the test case considered, bitwise data reduction is much more attractive than elementwise reduction.

Current and future research investigates how this data reduction is best translated into performance gains in terms of actual runtime reductions. First experiences indicate that even for relatively fast communication, runtime reductions may be achieved if data reduction is beyond (roughly) 50% (even earlier for slower communication). We need to point out that the potential benefits from our technique presented here do not only come from reducing runtimes, but, sometimes even more important, from making the solution of problems feasible which could not be handled before due to storage restrictions. Nevertheless, we will also investigate alternative (and more efficient) implementation strategies beyond the elementwise and bitwise reduction discussed here.

References

1. Crouzeix, M., Philippe, B., Sadkane, M.: The Davidson method. SIAM J. Sci. Comput. **15** (1994) 62–76
2. Dachsel, H., Lischka, H.: An efficient data compression method for the Davidson subspace diagonalization scheme. Theor. Chim. Acta **92** (1995) 339–349
3. Davidson, E.R.: The iterative calculation of a few of the lowest eigenvalues and corresponding eigenvectors of large real symmetric matrices. J. Comput. Phys. **17** (1975) 87–94
4. Harrison, R.J., Handy, N.C.: Full CI calculations on BH, H_2O, NH_3 and HF. Chem. Phys. Lett. (1983) 386–391
5. Knowles, P.J.: Very large configuration interaction calculations. Chem. Phys. Lett. (1989) 513–517
6. Lischka, H., Shepard, R., Pitzer, R.M., Shavitt, I., Dallos, M., Müller, T., Szalay, P.G., Seth, M., Kedziora, G.S., Yabushita, S., Zhang, Z.: High-level multireference methods in the quantum-chemistry program system COLUMBUS: Analytic MR-CISD and MR-AQCC gradients and MR-AQCC-LRT for excited states, GUGA spin-orbit CI and parallel CI density. Phys. Chem. Chem. Phys. **3** (2001) 664–673
7. Lischka, H., Shepard, R., Shavitt, I., Pitzer, R.M., Dallos, M., Müller, T., Szalay, P.G., Brown, F.B., Ahlrichs, R., Böhm, H.J., Chang, A., Comeau, D.C., Gdanitz, R., Dachsel, H., Ehrhardt, C., Ernzerhof, M., Höchtl, P., Irle, S., Kedziora, G., Kovar, T., Parasuk, V., Pepper, M.J.M., Scharf, P., Schiffer, H., Schindler, M., Schüler, M., Seth, M., Stahlberg, E.A., Zhao, J.G., Yabushita, S., Zhang, Z.: COLUMBUS, an ab initio electronic structure program, release 5.9 (2004)
8. Olson, J., Jørgensen, P., Simons, J.: Passing the one-billion limit in full configuration-interaction (FCI) calculations. Chem. Phys. Lett. (1990) 463–472
9. Shavitt, I.: The method of configuration interaction. In: Methods of Electronic Structure Theory. H. F. Schaefer, III, ed. Plenum Press, New York (1977) 189–275
10. Shepard, R., Shavitt, I., Lischka, H.: Reducing I/O costs for the eigenvalue procedure in large-scale configuration interaction calculations. J. Comp. Chem. **23** (2002) 1121–1125

Efficient Calculation of Quasi-bound States for the Simulation of Direct Tunneling

Markus Karner, A. Gehring, S. Holzer, and H. Kosina

Institute for Microelectronics, TU Vienna,
Gußhausstraße 27–29, A-1040 Wien, Austria
Karner@iue.tuwien.ac.at

Abstract. We study the calculation of quasi-bound states (QBS) in MOS inversion layers, which represent the major source of tunneling electrons. The calculation of QBS is performed by the perfectly matched layer (PML) method. Introducing a complex coordinate stretching enables us to apply artifical absorbing layers at the boundaries. This allows us to determine the QBS as the eigenvalues of a linear non-Hermitian Hamiltonian where the QBS lifetimes are directly related to the imaginary part of the eigenvalues. The PML formalism has been compared to other established methods and it has proven to be as an elegant, numerically stable, and efficient method to calculate QBS lifetimes.

1 Introduction

The continuous progress in the development of MOS field-effect transistors within the last years goes hand in hand with down-scaling the device feature size and therefore, they feature gate oxide thicknesses below two nanometers. Thus, quantum mechanical tunneling has significant effects on the characteristics of state-of-the-art electrical devices. The major source of tunneling electrons in the inversion layers of MOS-structures is due to quasi-bound states (QBS) [4, 1]. Since QBS in the potential well have a finite lifetime, they give rise to a quantum mechanical tunneling current out of the well into the oxide. The contribution of the QBS to the tunneling current is given by

$$J_{\mathrm{2D}} = \frac{\mathrm{k_B} T \mathrm{q}}{\pi \hbar^2} \sum_{i,\nu} \frac{g_\nu m_\parallel}{\tau_\nu(\mathcal{E}_{\nu,i}(m_\mathrm{q}))} \ln\left(1 + \exp\left(\frac{\mathcal{E}_\mathrm{F} - \mathcal{E}_{\nu,i}}{\mathrm{k_B} T}\right)\right) \tag{1}$$

where $\mathcal{E}_\mathrm{F}$ denotes the Fermi level, g_ν the valley degeneracy, $m_\parallel$ the parallel mass, m_q the quantization masses, and $\tau_\nu(\mathcal{E}_{\nu,i})$ the lifetime of the quasi-bound state $\mathcal{E}_{\nu,i}$. For the calculation of direct tunneling in silicon MOS structures, assuming [100] orientation, (1) have to be evaluated two times using $g_\nu = 2$, $m_\parallel = m_\mathrm{t}$, $m_\mathrm{q} = m_\mathrm{l}$, and $g_\nu = 4$, $m_\parallel = \sqrt{m_\mathrm{l} m_\mathrm{t}}$, $m_\mathrm{q} = m_\mathrm{t}$.

Hence, the calculation of direct tunneling currents is based on the accurate determination of the QBS. Several approaches have been established for the calculation of the lifetime broadening of QBS but there are still no generally accepted and efficient algorithms. Within this work, a semi-classical approximation as well as a method based on scanning the derivative of the reflection

I. Lirkov, S. Margenov, and J. Waśniewski (Eds.): LSSC 2005, LNCS 3743, pp. 572–577, 2006.

coefficient are compared with a new method based on the perfectly matched layer formalism.

2 Calculation of Quasi-bound States

The potential well used for the further considerations was acquired from a self-consistent quantum-mechanical Schrödinger-Poisson solver. The calculation has been performed for an n-MOS structure with a poly gate doping of $N_{\mathrm{D}}=1\times 10^{18}\mathrm{cm}^{-3}$ and a substrate doping of $N_{\mathrm{A}}=1\times 10^{17}\mathrm{cm}^{-3}$ at a gate bias of 1.5 V resulting in a potential well as displayed in Fig. 1.

The calculation of tunneling currents is based on the accurate determination of the QBS which are obtained from the time-independent effective mass Schrödinger equation:

$$-\frac{\hbar^2}{2}\nabla\cdot\left(\tilde{m}^{-1}\nabla\Psi(\mathbf{x})\right)+V(\mathbf{x})\Psi(\mathbf{x})=\mathcal{E}\Psi(\mathbf{x})\,. \tag{2}$$

Here, $V(x)$ denotes the potential energy, and $\tilde{m}^{-1}$ denotes the inverse effective mass tensor. For the simulation of silicon within the effective mass approximation, two different quantization masses occur, and therefore (2) has to be solved twice. Note that three different valley sorts occur, but for one-dimensional simulations two of them have the same quantization mass. Then, the QBS are determined as the eigenstates of the Hamiltonian. The QBS lifetimes are directly related to the imaginary parts of the eigenvalues: $\tau_i=\hbar/2\mathcal{E}_i$. Assuming closed boundaries, we obtain a linear Hermitian eigenvalue problem. Some of the eigenstates are displayed in Fig. 1. Since the wavefunction of a closed quantum system does not carry current, open boundary conditions have to be applied for an accurate description of tunneling electrons.

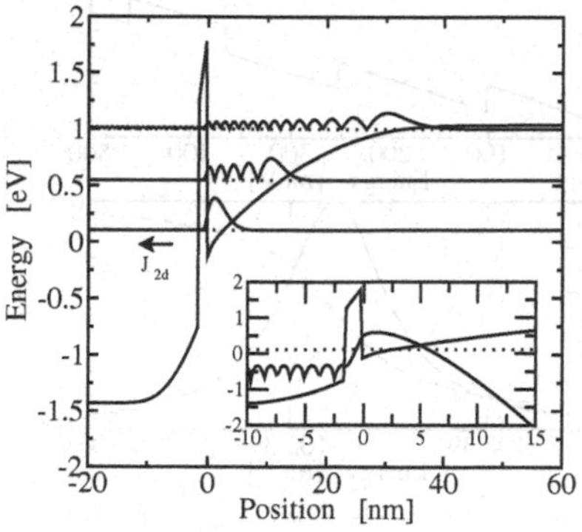

Fig. 1. The potential well that arises at an nMOS inversion layer and its eigenstates assuming closed boundary conditions. The inset displays the wavefunction of the first QBS at a logarithmic scale.

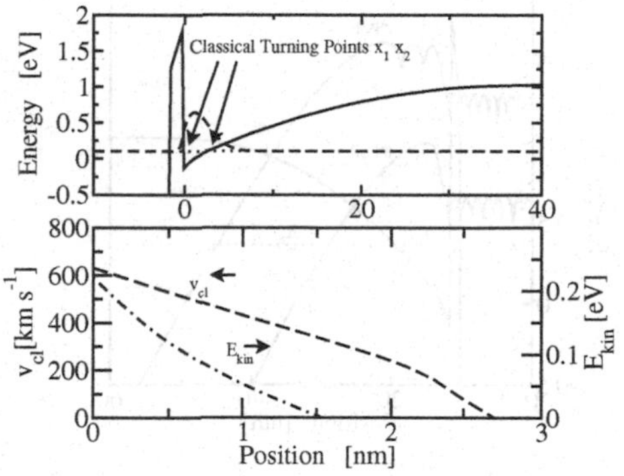

Fig. 2. The energy level and the classical turning points x_1, x_2 of the first eigenstate. In the lower figure, the kinetic energy and the velocity (assuming classical behavior of the particles) are displayed.

2.1 Semiclassical Approximation

However, it has been found that in practice the closed boundary eigenvalues are close to the eigenvalues of the open system [4]. This allows us to take the closed boundary eigenvalues, and determine a particle velocity [2] for each eigenstate under the assumption of classical behavior of the particle as displayed in Fig. 2. This implies that the electrons are bouncing between two classical turning points x_1 and x_2. At the oxide, the electrons are partially reflected according the transmission coefficient of the barrier. This gives rise to an exponential decay in time. For the ith QBS, the classical velocity $v_{\mathrm{cl},i}$ and the QBS lifetimes τ_i follow

$$v_{\mathrm{cl},i}(x) = \sqrt{\frac{\mathcal{E}_i - V(x)}{m(x)}} \quad (3) \qquad \tau_i = \frac{1}{TC(\mathcal{E}_i)} \int_{x_{1,i}}^{x_{2,i}} \frac{\mathrm{d}x}{v_{\mathrm{cl},i}(x)} \quad (4)$$

where the transmission coefficient $TC(\mathcal{E}_i)$ can be determined using the quantum transmitting boundary method (QTBM) or a WKB approximation.

2.2 The QTB Method

A more rigorous way to determine the energetic position and the lifetime broadening of the QBS is the quantum-transmitting boundary method (QTBM) [3] where a scanning of the transmission coefficient yields the lifetimes. Since the transmission coefficient is zero for half open system, it has been shown [5], that the phase of the reflection coefficient $\phi(\mathcal{E})$, defined as $\mathrm{RC}(\mathcal{E}) = \exp(\imath\phi(\mathcal{E}))$, follows $\phi(\mathcal{E}) \approx c + 2\arctan(\Gamma/2(\mathcal{E} - \mathcal{E}_i))$ near to the resonances and therefore its derivative has Lorentzian form

$$\frac{\partial \phi}{\partial \mathcal{E}} = \frac{\Gamma}{(\mathcal{E} - \mathcal{E}_i)^2 + \Gamma^2/4} \ . \quad (5)$$

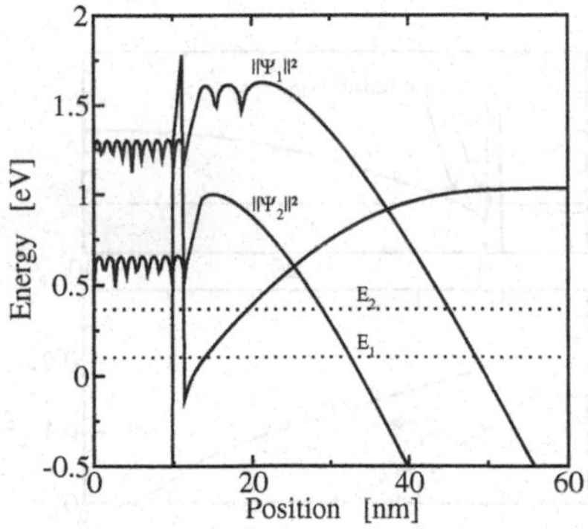

Fig. 3. The potential well and the energy levels of the first and third resonance which represent QBS, calculated with the QTB method. The wavefunctions at the resonances are displayed in a logarithmic scale.

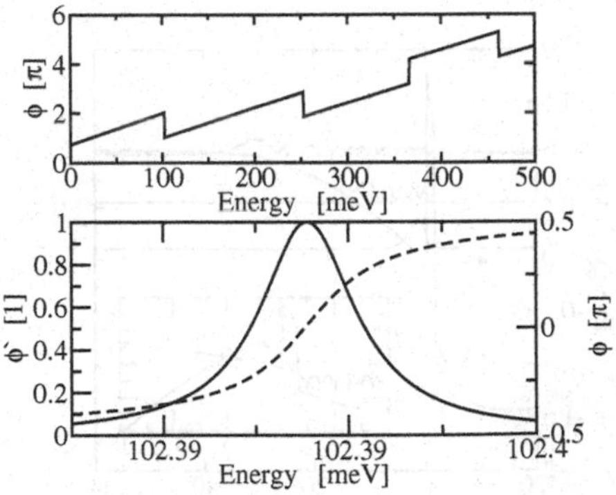

Fig. 4. The upper plot shows the phase of the reflection coefficient as a function of the electron energy. The second plot zooms in the first resonance peak and displays the phase and the derivative of the phase coefficient.

This is shown in Fig. 4. However, for the energy barriers of MOS capacitors, energy resolutions in the peV regime are necessary to accurately resolve the full-width half maximum (FWHM) value, which is required to calculate the QBS lifetime: $\tau_i = \hbar/\text{FWHM}_i$.

2.3 The PML Method

Within this work, a novel method to determine QBS, proposed in [6] is applied, using the perfectly matched layers formalism which is often used in electromagnetic theory. The idea is to add non-physical absorbing layers at the boundary of the simulation region (physical region). This procedure prevents reflection at the boundary of the physical region.

The newly introduced artifical absorbing layers allow application of Dirichlet boundary conditions, and the QBS are determined by the eigenvalues of the non-Hermitian, but linear Hamiltonian of the system. Some of the resulting eigenenergies and the corresponding wavefunctions are displayed in Fig. 5. The absorbing property of the PML region is achieved by introducing stretched coordinates in (2). The evaluation of the ∇ in one dimension yields:

$$\tilde{x} = \int_0^x s_x(\tau)\,\mathrm{d}\tau \qquad (6) \qquad\qquad \frac{\partial}{\partial \tilde{x}} = \frac{1}{s_x(x)}\frac{\partial}{\partial x} \qquad (7)$$

Within the PML region, the stretching function $s_x(x)$ is given as $s_x(x) = 1+(\alpha+\imath\beta)x^n$, with $\alpha = 1$, $\beta = 1.4$, and $n = 2$, while it is unity in the physical region as displayed in Fig. 6. Assuming a constant potential $V(z)$ within the PML regions, the wavefunctions can be written as a plane wave $\Psi(x) = \Psi_0 \exp(\imath\tilde{k}_x x)$ with the wave vector $\tilde{k}_x = k_x/s_x$. Considering two points in the PML region x_1, $x_2 = x_1 + \mathrm{d}x$ the wave vector at the point x_2 can be approximated as

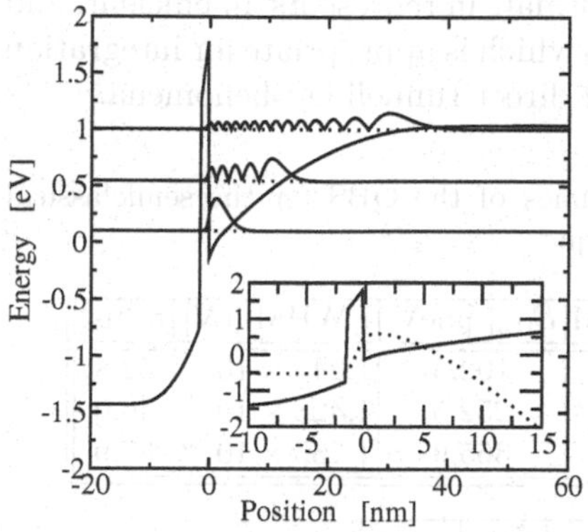

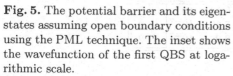

Fig. 5. The potential barrier and its eigenstates assuming open boundary conditions using the PML technique. The inset shows the wavefunction of the first QBS at logarithmic scale.

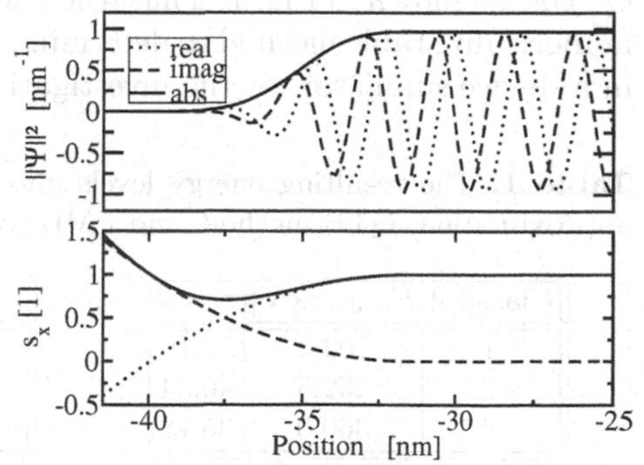

Fig. 6. The wavefunction of the first QBS and the complex stretching function are displayed in the perfectly matched layer region as well as its transition to the physical region

$$k_x(x_2) \approx \frac{s_x(x_2)}{s_x(x_1)} k_x(x_1) = (1 + (\alpha + \imath\beta)\,\mathrm{d}x) \quad . \tag{8}$$

The parameter α scales the phase velocity of the plane wave, while β acts as a damping parameter. Since this damping coefficient is greater than zero within the absorbing region the envelope of the wavefunctions is pinned to zero as shown in Fig. 6. These parameters, as well as the thickness of the absorbing layer can be varied over a wide range with virtual no influence on the results.

However, using QTBM or assuming closed boundary conditions yields a superposition of two moving plane waves in opposite directions. This can be seen as an envelope of the resulting wavefunctions displayed in the insets of Fig. 1 and Fig. 3. In contrast, using PML techniques yields no reflected wave into the well, and therefore the wavefunction is a plane traveling wave with constant envelope function as displayed in Fig. 5. The QBS, however, are reproduced correctly.

3 Results and Conclusions

Three methods for determining energetic positions and the lifetime broadening of QBS have been presented. The results are compared in Table 1. For the semiclassical method, it is still necessary to solve the closed system in order to get the eigenenergies of the well. Using closed eigenvalues together with the assumption of a classical velocity might result in inaccurate lifetimes. This is also true for the QTBM method since good initial values are needed in order to find the resonances. However, there is a perfect agreement with eigenvalues obtained from PML techniques. The PML formalism yields the desired eigenvalues in a natural way without any additional assumptions. Solving the eigenvalue problem directly yields the energetic positions and the lifetime of the QBS. Although the dimension of the system increases due to the additional points in the PML region, the computational effort of the PML has shown to be lower compared to QTBM as shown in Fig. 7. Thus, the PML formalism represents an efficient, and numerically stable method to determine QBS which is appropriate for integration in a device simulator for the investigation of direct tunneling phenomena.

Table 1. The resulting energy levels and lifetimes of the QBS for the semiclassical approximation, QTB method, and PML technique

Classical	E_{real} [meV]	τ_l [ns]
1	102.3	57.10
2	252.5	40.34
3	360.0	35.78

QTBM	E_{real} [meV]	FWHM [eV]	τ_l [ns]
1	102.3	1.245×10^{-6}	52.8
2	252.5	1.428×10^{-6}	46.0
3	365.9	1.735×10^{-6}	37.9

PML	E_{real} [meV]	E_{imag} [eV]	τ_l [ns]
1	102.3	-6.14×10^{-7}	53.5
2	252.5	-7.12×10^{-7}	46.2
3	365.9	-8.83×10^{-7}	37.2

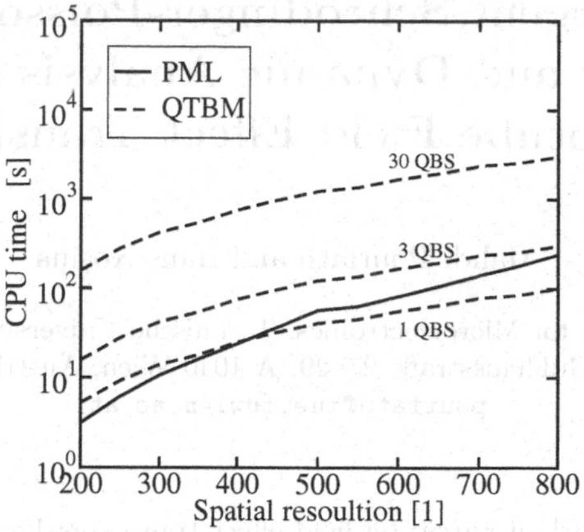

Fig. 7. Comparison of the CPU demand for the PML, and QTB methods with a given set of QBS

Acknowledgments

We thank S. Odermatt for bringing the PML method to our attention. Financial support from the special research project IR-ON (F25) and the Network of Excellence SINANO (506844) is acknowledged.

References

1. E. Cassan, P. Dollfus, S. Galdin, and P. Hesto, Semiclassical and Wave-Mechanical Modeling of Charge Control and Direct Tunneling Leakage in MOS and H-MOS Devices with Ultrathin Oxides, *IEEE Trans.Electron Devices*, **48** 4, 715–721, 2001.
2. A. Dalla Serra, A. Abramo, P. Palestri, L. Selmi, and F. Widdershoven, Closed- and Open-Boundary Models for Gate-Current Calculation in n-MOSFETs, *IEEE Trans.Electron Devices*, **48** 8, 1811–1815, 2001.
3. C. L. Fernando and W. R. Frensley, An Efficient Method for the Numerical Evaluation of Resonant States, *J.Appl.Phys.*, **76** 5, 2881–2886, 1994.
4. A. Gehring and S. Selberherr, On the Calculation of Quasi-Bound States and Their Impact on Direct Tunneling in CMOS Devices, in *Proc. Intl. Conf. on Simulation of Semiconductor Processes and Devices*, München, 2004, 25–28.
5. G. Gildenblatt, B. Gelmont, and S. Vatannia, Resonant Behavior, Symmetry, and Singularity of the Transfer Matrix in Asymmetric Tunneling Structures, *J.Appl.Phys.*, **77** 12, 6327–6331, 1995.
6. S. Odermatt, M. Luisier, and B. Witzigmann, Bandstructure Calculation Using the k·p Method for Arbitrary Potentials with Open Boundary Conditions, *J.Appl.Phys.*, **97** 4, 046104–1, 2005.

Fast Convergent Schrödinger-Poisson Solver for the Static and Dynamic Analysis of Carbon Nanotube Field Effect Transistors

Mahdi Pourfath and Hans Kosina

Institute for Microelectronics, Technische Universität Wien,
Gußhausstraße 27–29, A-1040 Wien, Austria
pourfath@iue.tuwien.ac.at

Abstract. Carbon nanotube field-effect transistors have been studied in recent years as a potential alternative to CMOS devices, because of the capability of ballistic transport. In order to account for the ballistic transport we solved the coupled Poisson and Schrödinger equations for the analysis of these devices. Conventionally the coupled Schrödinger-Poisson equation is solved iteratively with appropriate numerical damping. Often convergence problems occur. In this work we show that this problem is due to inappropriate energy discretization, and by using an adaptive integration method the simulation time is reduced and most of the simulations converge in a few iterations. Based on this approach we investigated the static and dynamic behavior of carbon nanotube field effect transistors.

1 Introduction

Exceptional electronic and mechanical properties together with nanoscale diameter make carbon nanotubes (CNTs) a candidate for nanoscale field effect transistors (FETs). While early devices have shown poor device characteristics, high performance devices were achieved recently [17, 11, 5, 7, 14]. In short devices (less than 100 nm) carrier transport through the device is nearly ballistic [5, 6]. As described in the next section the coupled Poisson and Schrödinger equation system was solved to study the static response of CNTFETs. We show that by using an adaptive integration method for calculating carrier concentration and current density, simulations converge very fast while the results are very accurate. Based on the Quasi Static Approximation (QSA) the dynamic response of these devices is also investigated.

The contact between metal and CNT can be of Ohmic [6] or Schottky type [1]. In this work we focus on Ohmic contact CNTFETs which theoretically [4] and experimentally [5] show better performance than Schottky contact devices. In a p-type device with ohmic contacts holes see no barrier while the barrier height for electrons is E_g. By changing the gate voltage the transmission coefficient of holes through the device is modulated and as a result the total current changes [6].

I. Lirkov, S. Margenov, and J. Waśniewski (Eds.): LSSC 2005, LNCS 3743, pp. 578–585, 2006.

2 Approach

In this section the models which were used to study the static and dynamic response of CNTFETs are explained. As it will be shown in the next section a good agreement between simulation and experimental results is achieved.

2.1 Static Response

In order to account for the ballistic transport we solved the coupled Poisson and Schrödinger equations for CNTFETs.

$$\frac{\partial^2 V}{\partial \rho^2} + \frac{1}{\rho}\frac{\partial V}{\partial \rho} + \frac{\partial^2 V}{\partial z^2} = -\frac{Q}{\epsilon} \tag{1}$$

$$-\frac{\hbar^2}{2\mathrm{m}^*}\frac{\partial^2 \Psi_{s,d}^{n,p}}{\partial z^2} + (U^{n,p} - E)\Psi_{s,d}^{n,p} = 0 \tag{2}$$

We considered an azimuthal symmetric structure, in which the gate surrounds the CNT, such that the Poisson equation (1) is restricted to two-dimensions. In (1) $V(\rho, z)$ is the electrostatic potential, and Q is the space charge density.

In the Schrödinger equation (2) the effective mass is assumed to be $\mathrm{m}^* = 0.05\mathrm{m}_0$ for both electrons and holes [19]. In (2) superscripts denote the type of the carriers. Subscripts denote the contacts, where s stands for the source contact and d for the drain contact. For example, Ψ_s^n is the wave function associated with electrons that have been injected from the source contact, and U^n is the potential energy that is seen by electrons. The Schrödinger equation is just solved on the surface of the tube, and is restricted to one-dimension because of azimuthal symmetry.

The space charge density in (1) is calculated as:

$$Q = \frac{\mathrm{q}(p - n)\delta(\rho - \rho_{cnt})}{2\pi\rho} \tag{3}$$

where q is the electron charge, and n and p are total electron and hole concentrations per unit length. In (3) δ/ρ is the Dirac delta function in cylindrical coordinates, implying that carriers were taken into account by means of a sheet charge distributed uniformly over the surface of the CNT [8].

Including the source and drain injection components, the total electron concentration in the CNT is calculated as:

$$n = \frac{4}{2\pi}\int f_s|\Psi_s^n|^2 dk_s + \frac{4}{2\pi}\int f_d|\Psi_d^n|^2 dk_d$$
$$= \int \frac{\sqrt{2\mathrm{m}^*}}{\pi\hbar\sqrt{E_s}} f_s|\Psi_s|^2 dE_s + \int \frac{\sqrt{2\mathrm{m}^*}}{\pi\hbar\sqrt{E_d}} f_d|\Psi_d|^2 dE_d \tag{4}$$

where $f_{s,d}$ are equilibrium Fermi functions at the source and drain contacts. All our calculations assume a CNT with 0.5 eV band gap [5]. The total hole concentration in the CNT is calculated analogously.

The Landauer-Büttiker formula [3] is used for calculating the current:

$$I^{n,p} = \frac{4\mathrm{q}}{\mathrm{h}} \int [f_s^{n,p}(E) - f_d^{n,p}(E)] T_c^{n,p}(E) dE \tag{5}$$

where $T_c^{n,p}(E)$ are the transmission coefficients of electrons and holes, respectively, through the device. The factor 4 in (4) and (5) stems from the twofold band and twofold spin degeneracy.

Conventionally the coupled Schrödinger and Poisson equations are solved iteratively [20], by using an appropriate numerical damping factor α. At the $(k+1)^{\mathrm{th}}$ iteration the Schrödinger equation is solved using the electrostatic potential V^k from the last iteration and the new space charge density Q^{k+1} is calculated. The Poisson equation is then solved by using Q^{k+1} and an intermediate new electrostatic potential is calculated V_{int}^{k+1}. Finally V^{k+1} is calculated as:

$$V^{k+1} = \alpha V_{\mathrm{int}}^{k+1} + (1 - \alpha) V^k \tag{6}$$

where $0 < \alpha < 1$. Successive iteration continues until a convergence criterion is satisfied. In this work an adaptive damping factor was used [10]. The damping factor is initially set to $\alpha = 1$. If the potential update $|V^{k+1} - V^k|$ increases from one iteration to the next iteration or remains constant the damping factor decreases by a constant factor. We used $\alpha = \alpha \times 0.8$ as suggested by [10]. If a high damping factor is initially selected the simulations may oscillate and will not converge. Using a low damping factor will result in long simulation time. We show that by appropriate evaluation of the carrier concentration this problem can be avoided. The integration in (4) and (5) are calculated in an energy interval $[E_{\mathrm{min}}, E_{\mathrm{max}}]$. In the simplest way the interval is divided into equidistant steps. By using this method narrow resonances at some energies may be missed or may not be evaluated correctly. In successive iterations as the potential profile changes the position of the resonances will also change, and it is possible that a resonance point locates very near to one of the energy steps. In this case the carrier concentration suddenly changes and as a result the simulation would oscillate and not converge. To avoid this problem the accuracy of the integration should be independent of the location of resonances. By using an adaptive integration method the integrations in (4) and (5) can be evaluated with a desirable accuracy. Assume f is an integrable function, and $[a, b]$ is the interval of integration. To compute

$$I = \int_a^b f(x) dx \tag{7}$$

adaptively I is calculated with two different integration methods, I_1 and I_2. If the relative difference of the two approximations is less than a predefined tolerance the integration is accepted, otherwise the interval [a,b] is divided into two equal parts $[a, c]$ and $[c, b]$, where $c = (a + b)/2$, and the two respective integrals are computed independently.

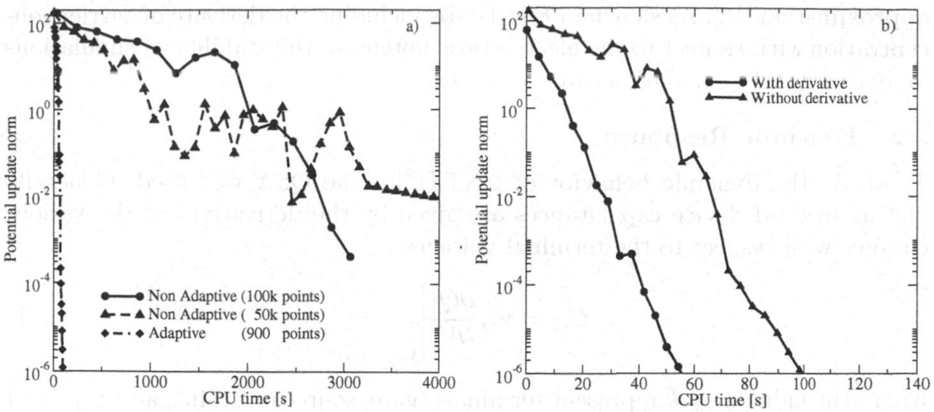

Fig. 1. The comparison of CPU time demand on an IBM-RS6000 for the same iterative simulation, but with different integration methods. The norm of the potential update is considered as a measure of convergence. a) Shows the results for adaptive and non adaptive integration. b) Show the result for adaptive integration with and without the derivative of carrier concentration with respect to the electrostatic potential.

$$I = \int_a^c f(x)dx + \int_c^b f(x)dx \tag{8}$$

The same procedure is performed for each of these integrals. The advantage of this methods is that the steps are non-equidistant, so there are many points around the resonances while in other regions there are few points. In this work an adaptive Simpson quadrature [15] is used. In this method the two successive Simpson approximates are calculated:

$$I_1 = \frac{h}{6}(f(a) + 4f(c) + f(b)) \tag{9}$$

$$I_2 = \frac{h}{12}(f(a) + 4f(d) + 2f(c) + 4f(e) + f(b)) \tag{10}$$

where $d = (a+c)/2$, and $e = (c+b)/2$. If $|I_1 - I_2| \leq \text{tol} \times |I_2|$ the integration is evaluated within one step of Romberg extrapolation: $I = I_2 + (I_2 - I_1)/15$. Fig. 1-a shows the CPU time demand on an IBM-RS6000 for the same iterative simulation using adaptive integration and non-adaptive integration with 5×10^4 and 10^5 points. Increasing the number of data points in the non adaptive method the simulation becomes more stable. Using adaptive integration method, only 9×10^2 points are required and most of the simulations can start with a high damping factor ($\alpha = 1$) and no or few oscillations occur. Therefore a high damping factor is used for all the iterations and as a result simulations converge very fast. It is also possible to make the simulations more stable by providing the derivate of carrier concentration with respect to the electrostatic potential for the Poisson solver [12, 2]. In general there is no exact form for this term, but $\partial n/\partial \phi \approx \text{q}\, \partial n/\partial E_F$ can be considered as a good approximation [12]. As shown in Fig. 1-b by including the derivate of carrier concentration with respect to the electrostatic potential, the stability of simulations increases and the simulation time decreases.

2.2 Dynamic Response

To study the dynamic behavior of CNTFETs, the QSA was used. Generally in this method device capacitances are given by the derivatives of the various charges with respect to the terminal voltages:

$$C_{ij} = \chi_{ij} \frac{\partial Q_i}{\partial V_j}\bigg|_{V_{k \neq j}=0} \tag{11}$$

where the indices i, j, k represent terminals (gate, source or drain), and $\chi_{ij} = -1$ for $i \neq j$ and $\chi_{ij} = +1$ for $i = j$. The differentiation of these expressions is performed numerically over steady state charges [18]. This method is widely used for the analysis of conventional semiconductor devices, where the charge in the semiconductor device is partitioned into two parts indicating the contribution of the source and drain contacts [18, 13]. For example, the gate-source capacitance is calculated by

$$C_{sg} = \frac{\partial Q_{se}}{\partial V_{gs}} + \frac{\partial Q_{st}}{\partial V_{gs}} = C_{se} + C_{sq} \tag{12}$$

where Q_{se} is total charge on the source contact and Q_{st} is the total charge on the tube injected from the source contact. As shown in (12) the total gate-source capacitance is split into two components, the first term indicates the electrostatic gate-source capacitance and the second term is usually referred to as quantum capacitance [9]. Therefore the capacitance matrix has a rank of 3, and due to quantum capacitances the matrix is not symmetric ($C_{ij} \neq C_{ji}$). In this work we assumed that only the gate voltage changes, whereas the voltages of the other terminals are kept constant. Therefore, the capacitance matrix simplifies to three components, and an equivalent circuit as shown in Fig. 2 is achieved [16]. In Fig. 2, g_m is the differential transconductance calculated by $g_m = \partial I_{ds}/\partial V_{gs}$ Based on the equivalent circuit in Fig. 2, the cutoff frequency of the device can be derived as

$$f_T = \frac{g_m}{2\pi C_{sg}\sqrt{1 + 2\frac{C_{dg}}{C_{sg}}}} \tag{13}$$

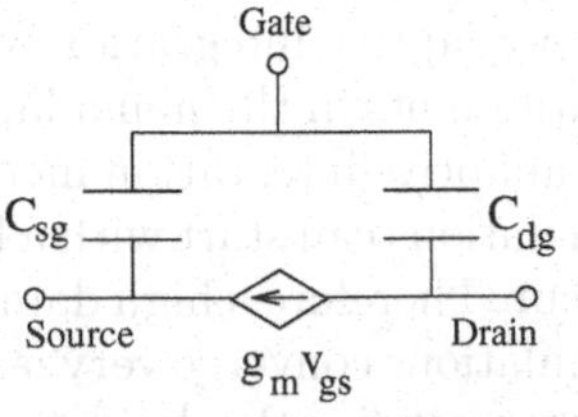

Fig. 2. Simplified equivalent circuit model for the dynamic response of CNTFETs. The model is based on the assumption that only the gate voltage changes.

3 Simulation Results

We consider a p-type ohmic device, where holes see no barrier while the barrier height for electrons is E_g. For a fair comparison with experimental results, we used the same material and geometrical parameters as reported in [5]. As shown in Fig. 3, there is a good agreement between simulation and experimental results despite the fact that the cylindrical structure is only an approximation of the real device structure.

The dynamic response of these devices has been also investigated. Fig. 4 shows the electrostatic and quantum capacitances associated with the source and drain contacts. It is clearly seen that the quantum capacitances unlike

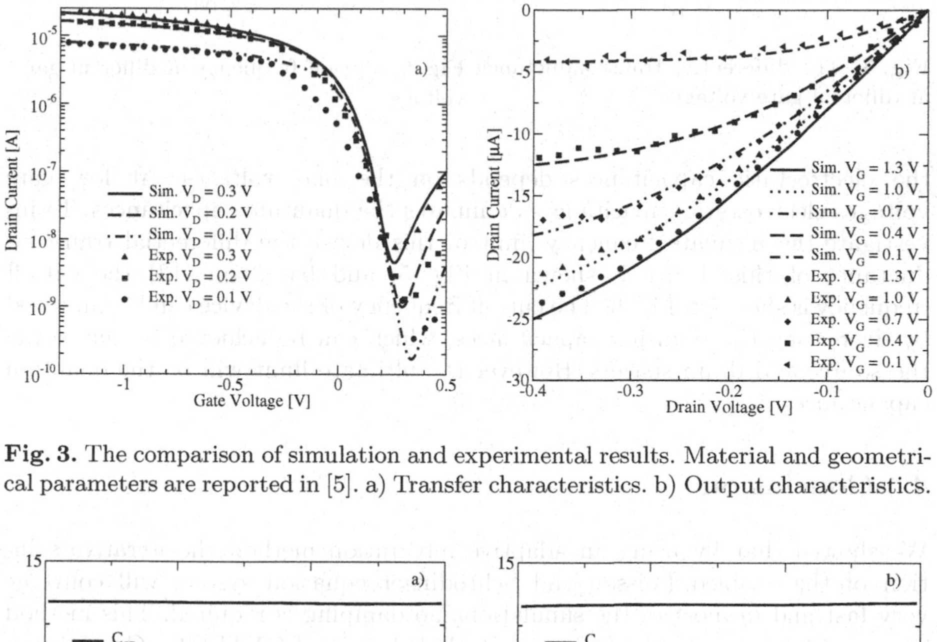

Fig. 3. The comparison of simulation and experimental results. Material and geometrical parameters are reported in [5]. a) Transfer characteristics. b) Output characteristics.

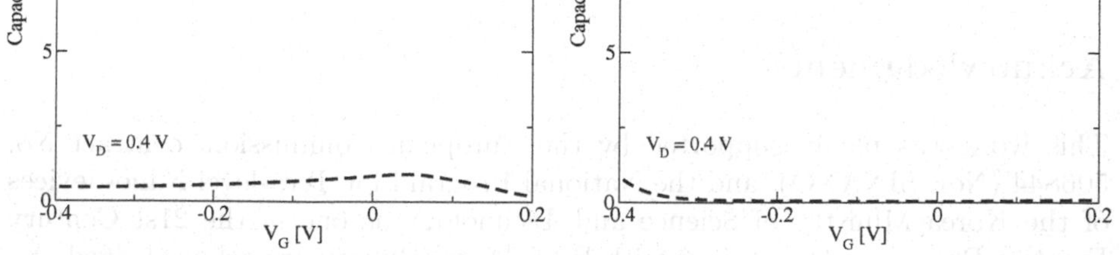

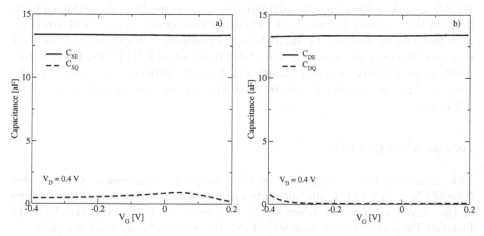

Fig. 4. Electrostatic and quantum capacitances associated with the a) Source contact, and b) Drain contact. Electrostatic capacitances dominate at low gate biases.

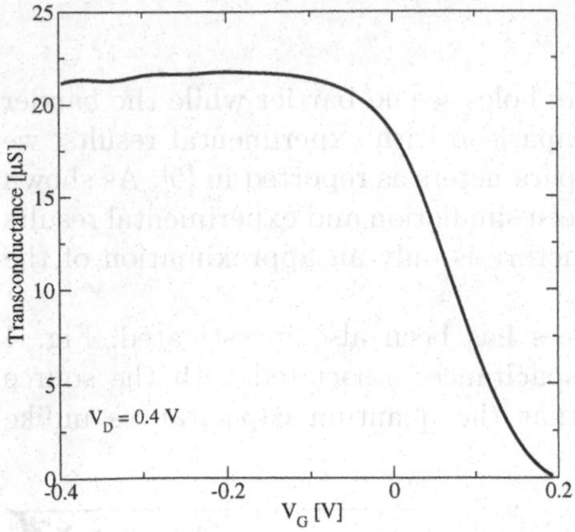

Fig. 5. The differential transconductance at different gate voltages

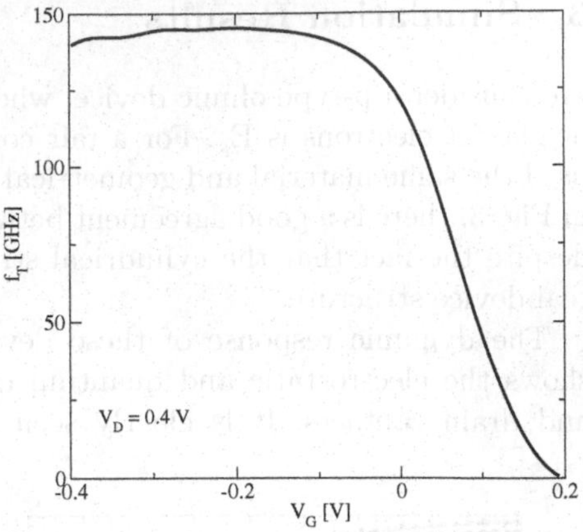

Fig. 6. Cut-off frequency at different gate voltages

the electrostatic capacitances depends on the bias voltages. At low gate voltages electrostatic capacitances dominates the quantum capacitances. To investigate the ultimate frequency limit of this device the differential transconductance of this device is shown in Fig. 5, and based on (13) the cut-off frequency is shown in Fig. 6. The cut-off frequency of the device can be improved by decreasing the parasitic capacitances, which can be achieved by increasing the source and drain spacers. However the ultimate limit will be the quantum capacitances.

4 Conclusion

We showed that by using an adaptive integration method the iterative solution of the coupled Poisson and Schrödinger equation system will converge very fast and in most of the simulations no damping is required. This method was used to study the dynamic and static behavior of CNTFETs. Good agreement between simulation and experimental results indicates the validity of the models. This methodology can be well applied for the optimization of the CNTFETs.

Acknowledgments

This work was partly supported by the European Commission, contract No. 506844 (NoE SINANO), and the National Program for Tera-level Nano-devices of the Korea Ministry of Science and Technology as one of the 21st Century Frontier Programs. Discussions with Prof. David Pulfrey are acknowledged.

References

1. Appenzeller, J., Radosavljevic, M., Knoch, J., Avouris, P.: Tunneling Versus Thermionic Emission in One-Dimensional Semiconductors. Phys.Rev.Lett. **92** (2004) 048301
2. Biegel, B.A.: Quantum Electronic Device Simulation. Dissertation, Stanford University (1997)
3. Datta, S.: Electronic Transport in Mesoscopic Systems. Cambridge University Press (1995)
4. Guo, J., Datta, S., Lundstrom, M.: A Numerical Study of Scaling Issues for Schottky Barrier Carbon Nanotube Transistors. IEEE Trans. Electron Devices **51** (2004) 172–177
5. Javey, A., Guo, J., Farmer, D.B., Wang, Q., Yenilmez, E., Gordon, R.G., Lundstrom, M., Dai, H.: Self-Aligned Ballistic Molecular Transistors and Electrically Parallel Nanotube Arrays. Nano Lett. **4** (2004) 1319–1322
6. Javey, A., Guo, J., Wang, Q., Lundstrom, M., Dai, H.: Ballistic Carbon Nanotube Field-Effect Transistors. Letters to Nature **424** (2003) 654–657
7. Javey, A., Tu, R., Farmer, D.B., Guo, J., Gordon, R.G., Dai, H.: High Performance n-Type Carbon Nanotube Field-Effect Transistors with Chemically Doped Contacts. Nano Lett. **5** (2005) 345–348
8. John, D., Castro, L., Pereira, P., Pulfrey, D.: A Schrödinger-Poisson Solver for Modeling Carbon Nanotube FETs. In: Proc. NSTI Nanotech. **3** (2004) 65–68
9. John, D.L., Castro, L.C., Pulfrey, D.L.: Quantum Capacitance in Nanoscale Device Modeling. J.Appl.Phys. **96** (2004) 5180–5184
10. Kerkhoven, T., Galick, A.T., Ravaioli, U., Arends, J.H., Saad, Y.: Efficient numerical simulation of electron states in quantum wires. J.Appl.Phys. **68** (1990) 3461–3469
11. Kim, B.M., Brintlinger, T., Cobas, E., Zheng, H., Fuhrer, M., Z.Yu, Droopad, R., Ramdani, J., Eisenbeiser, K.: High-Performance Carbon Nanotube Transistors on SrTiO3/Si Substrates. Appl.Phys.Lett. **84** (2004) 1946–1948
12. Lake, R., Klimeck, G., Bowen, R.C., Jovanovic, D., Blanks, D., Swaminathan, M.: Quantum Transport with Band-Structure and Schottky Contacts. Phys.stat.sol.(b) **204** (1997) 354–357
13. Laux, S.E.: Techniques for Small-Signal Analysis of Semiconductor Devices. IEEE Trans. Electron Devices **32** (1985) 2028–2037
14. Lin, Y.M., Appenzeller, J., Knoch, J., Avouris, P.: High-Performance Carbon Nanotube Field-Effect Transistor with Tunable Polarities. cond-mat/0501690 (2005)
15. Lyness, J.N.: Notes on the Adaptive Simpson Quadrature Routine. J.ACM **16** (1969) 483–495
16. Pulfrey, D.L., Castro, L., John, D., Pourfath, M., Gehring, A., Kosina, H.: Method for Predicting f_T for Carbon Nanotube Field-Effect Transistors. submitted to IEEE Tran. Nanotechnology (2005)
17. Radosavljevic, M., Appenzeller, J., Avouris, P., Knoch, J.: High Performance of Potassium n-Doped Carbon Nanotube Field-Effect Transistors. Appl.Phys.Lett. **84** (2004) 3693–3695
18. Rho, K.M., Lee, K., Shur, M., Fjeldly, T.A.: Unified Quasi-Static MOSFET Capacitance Model. IEEE Trans. Electron Devices **40** (1993) 131–136
19. Saito, R., Dresselhaus, G.D., Dresselhaus, M.S.: Physical Properties of Carbon Nanotubes. Imperial College Press (1998)
20. Stern, F.: Iteration Methods for Calculating Self-Consistent Fields in Semiconductor Inversion Layers. Phys.stat.sol.(b) **6** (1970) 56–67

A Modular Method for the Efficient Calculation of Ballistic Transport Through Quantum Billiards

S. Rotter, B. Weingartner, F. Libisch, F. Aigner, J. Feist, and J. Burgdörfer

Institute for Theoretical Physics, Vienna University of Technology,
A-1040 Vienna, Austria
stefan.rotter@tuwien.ac.at

Abstract. We present a numerical method which allows to efficiently calculate quantum transport through phase-coherent scattering structures, so-called "quantum billiards". Our approach consists of an extension of the commonly used *Recursive Green's Function Method (RGM)*, which proceeds by a discretization of the scattering geometry on a lattice with nearest-neighbour coupling. We show that the efficiency of the RGM can be enhanced considerably by choosing symmetry-adapted grids reflecting the shape of the billiard. Combining modules with different grid structure to assemble the entire scattering geometry allows to treat the quantum scattering problem of a large class of systems very efficiently. We will illustrate the computational challenges involved in the calculations and present results that have been obtained with our method.

1 Introduction

A major aim in ballistic transport theory is to simulate and stimulate experiments in the field of phase-coherent scattering through nano-scaled semiconductor devices [7, 4]. However, even for two-dimensional quantum dots ("quantum billiards") the numerical solution of the Schrödinger equation in an effective one-electron approximation has remained a computational challenge. This is partly due to the fact that many of the most interesting phenomena occur in parameter regimes which are difficult to handle from a computational point of view: (1) In the "semi-classical regime" of high Fermi energy E_F the de Broglie-wavelength of the electrons, $\lambda_D = 2\pi/\sqrt{2E_F}$, is much smaller than the linear dimensions of the scattering device, $\lambda_D \ll D$. To properly describe the continuum limit of the transport process, a large number of basis functions is necessary [14]. Eventually this requirement renders all available methods computationally unfeasible or numerically instable. (2) In the "quantum-Hall regime" of very high magnetic fields the magnetic length, $l_B = \sqrt{c/B}$ (in atomic units), is considerably smaller than the system dimensions, $l_B \ll D$. Methods based on the expansion in plane or spherical waves become invalid since diamagnetic contributions are generally neglected [14]. Methods employing a discretization on a grid do not allow the flux per unit cell to exceed a critical value and are therefore limited in the range of magnetic fields accessible [4, 2]. In this article we discuss a modification of the widely used *Recursive Green's Function Method* (RGM) [4] and illustrate

I. Lirkov, S. Margenov, and J. Waśniewski (Eds.): LSSC 2005, LNCS 3743, pp. 586–593, 2006.

how this modular extension of the RGM can bypass several of the limitations of conventional techniques.

2 Method

We consider a two-dimensional scattering geometry ("billiard") to which two semi-infinite waveguides ("leads") of width d are attached in different orientations. A constant flux of electrons is injected at the Fermi energy $E_F = \hbar^2 k_F^2/(2m_{\text{eff}})$ through one of the waveguides and can leave the cavity through either the entrance or the exit lead. We assume inelastic scattering processes to be absent, such that the electronic motion throughout the device region is ballistic and therefore determined by the shape of the billiard. The potential surface inside the boundary of the dot is allowed to have different shapes (flat, soft wall profile or disordered) and is infinitely high outside. Atomic units ($\hbar = |e| = m_{\text{eff}} = 1$) will be used, unless explicitly stated otherwise.

Our starting point is the *standard* recursive Green's function method (RGM). This approach is widely used in various fields of computational physics and consists in a discretization of the scattering geometry on a Cartesian grid. Setting up a tight-binding (tb) Hamiltonian on this grid,

$$\hat{H}^{\text{tb}} = \sum_i \varepsilon_i \, | \, i \rangle \langle \, i \, | + \sum_{i,j} V_{i,j} \, | \, i \rangle \langle \, j \, | \,, \tag{1}$$

the hopping potentials $V_{i,j}$ and the site energies ε_i are chosen such that the equation $\hat{H}^{\text{tb}}|\psi_m\rangle = E_m|\psi_m\rangle$ converges towards the continuum one-particle Schrödinger equation, $[-\Delta/2 + V(x,y)]|\psi_m\rangle = E_m|\psi_m\rangle$, in the limit of high grid density. The hopping potentials are non-zero only for nearest-neighbour coupling of grid-points (with spacing Δx , Δy) and result directly from a three-point difference approximation of the kinetic energy term in the free-particle Hamiltonian [4],

$$\varepsilon_i = \frac{1}{\Delta x^2} + \frac{1}{\Delta y^2}, \quad V^x_{i,i\pm 1} = \frac{-1}{2\Delta x^2}, \quad V^y_{j,j\pm 1} = \frac{-1}{2\Delta y^2} \,. \tag{2}$$

With the help of the eigenvectors $|\psi_m\rangle$ and the eigenvalues E_m of the Hamiltonian $\hat{H}^{\text{tb}}$ the Green's functions of one-dimensional tb strips are calculated,

$$G^{\pm}(\boldsymbol{x}, \boldsymbol{x}', E) = \lim_{\epsilon \to 0} \sum_m \frac{\langle \boldsymbol{x} | \psi_m \rangle \langle \psi_m | \boldsymbol{x}' \rangle}{E \pm i\epsilon - E_m} \,. \tag{3}$$

The different signs ($\pm$) denote the retarded and advanced Green's functions, respectively. The disconnected transverse tb-strips incorporate the boundary conditions at the top and the bottom (see Fig. 1a) as accurately as possible. The Green's functions of the strips are connected one at a time through recursive solutions of a matrix Dyson equation,

$$G = G^0 + G^0 \, V \, G \,, \tag{4}$$

where V is the hopping potential between the strips, G^0 and G denote the retarded Green's function matrices of the disconnected and the connected tb strips, respectively. The complete scattering structure can thus be assembled from the individual strips much like knotting a carpet. Note that in this procedure the number of transverse strips is equal to the number of recursions (each involving at least one matrix inversion). For very high electron energies E_F, the large number of strips required to simulate the continuum eventually renders transport calculations impractical. This is in part because of the very large size of the matrices which have to be inverted in the strip-by-strip recursion process.

The remedy which we have proposed [8, 9] to overcome such difficulties goes back to Sols *et al.* [11] and consists of an extension of the RGM. Starting point is the observation that the efficiency of the "conventional" discretization employed in the RGM can be increased considerably by taking the symmetry of the scattering problem into account. More specifically, if the two-dimensional nonseparable open quantum dot can be built up from simpler separable substructures ("modules"), one gains significantly in computational speed by calculating the Green's functions for each of these modules separately. We solve the tight-binding Schrödinger equation, $\hat{H}^{\mathrm{tb}}|\psi_m\rangle = E_m|\psi_m\rangle$, now for one module at a time. Employing symmetry-adapted tight-binding grids leads to the separability of the eigenfunctions $|\psi_m\rangle$ for the modules and allows to determine the Green's function for an entire module [according to Eq. (3)] fast and virtually exactly. For joining modules with each other we employ the technique of the RGM where the coupling between Green's functions is facilitated in terms of the corresponding hopping matrix elements of the tight-binding Hamiltonian. By solving one Dyson equation at each junction between the modules, the complete scattering structure can be assembled much like a jigsaw puzzle. In Fig. 1b we illustrate how the discretization of the circle billiard within the framework of this *modular recursive Green's function method (MRGM)* [8, 9] proceeds.

Note that, quite in contrast to the conventional RGM, the number of recursions (i.e., of matrix inversions) needed to obtain the Green's function of the total scattering problem is given by the number of separate modules required to build up the scattering structure. This number is independent of the de Broglie wavelength. The latter enters only in terms of the size of the matrices involved in the fixed number of recursions. Furthermore, for solving the transport problem, the module Green's functions have to be evaluated only on the subset of grid points which are coupled to grid points on neighbouring modules. Another advantage of the MRGM comes into play when solving the scattering problem at different Fermi energies E_F, since for all values of E_F the eigenvalue problem Eq. (1) has to be solved only once for each module. This is because the eigenvectors $|E_m\rangle$ and the eigenenergies E_m are independent of E_F. As a consequence of these advantages it is possible to incorporate a very high number of grid points in the calculations, which is the prerequisite to access the "semiclassical" as well as the "quantum Hall" regime. On the negative side, the most severe restriction of the MRGM is its restricted applicability to those scattering structures which can be assembled from separable modules. This includes random and soft-wall

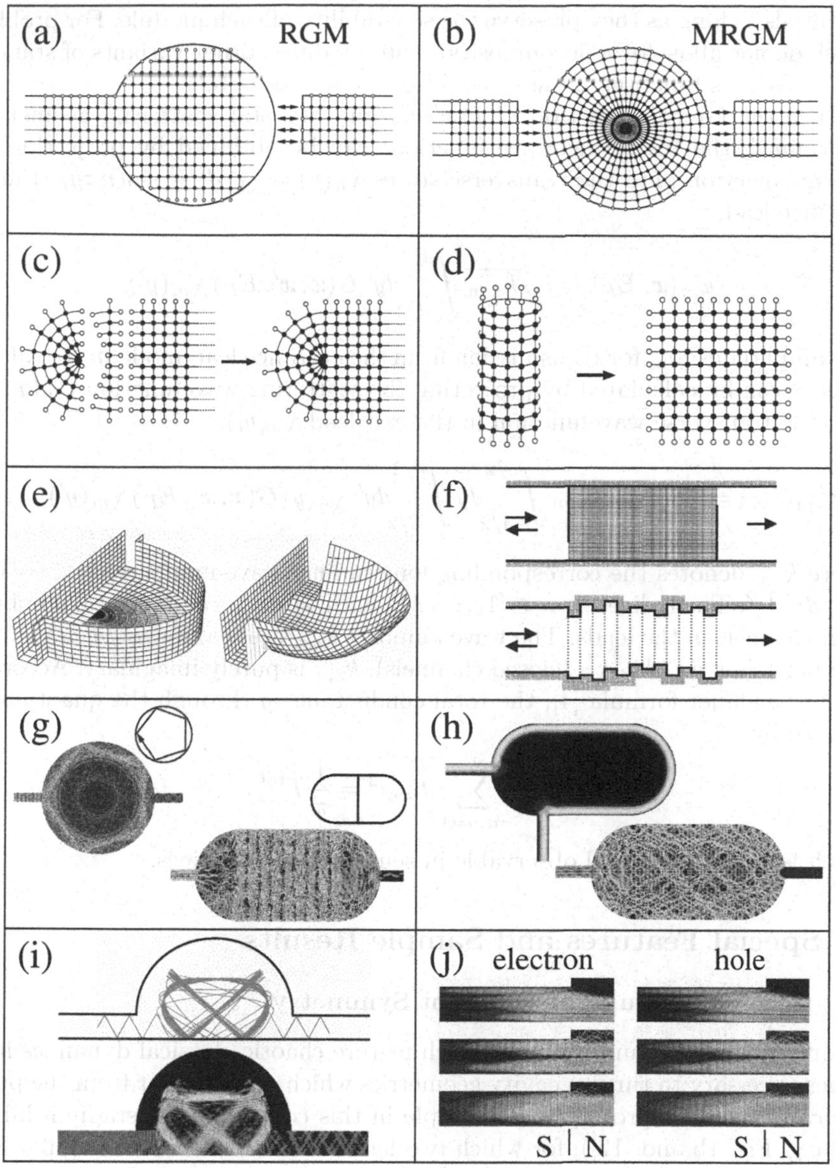

Fig. 1. (Color) Discretization of the circular billiard with leads employing (a) the conventional RGM and (b) the MRGM. (c)-(d) Features of the discretization with the MRGM (see text for details). (e) Hard wall vs. soft wall profile. (f) Bulk and surface disorder potentials. (g) Density of localized wavefunctions $|\psi(x,y)|^2$ in a regular vs. chaotic billiard. (h) Density of scattering wavefunctions in the high magnetic field vs. high energy limit. (i) "Trapped" trajectory in a soft wall billiard with a mixed classical phase space and the density of the corresponding quantum wavefunction. (j) Three bound electron-hole wavefunction densities in a square billiard with superconducting lead ("Andreev billiard").

potentials as long as they preserve the separability of each module. For problems which do not allow for a decomposition into modules, other variants of standard lattice Green's function methods could be employed [6, 15].

Once the Green's function $G(\boldsymbol{x}, \boldsymbol{x}', E)$ for the total scattering geometry is determined, the scattering wave functions can be obtained by projecting this Green's function onto the transverse states $\chi_n(y) = (2/d)^{1/2} \sin(n\pi y/d)$ in the entrance lead,

$$\psi_m(\boldsymbol{x}, E_F) = i\sqrt{k_{x,m}} \int_{-d/2}^{d/2} dy' \, G(\boldsymbol{x}, \boldsymbol{x}', E_F) \, \chi_m(y') \, . \tag{5}$$

The amplitudes t_{nm} for transmission from the entrance lead mode m to exit lead mode n can be calculated by projecting the scattering wavefunction $\psi_m(\boldsymbol{x}, E_F)$ onto the transverse wavefunction in the exit lead $\chi_n(y_i)$,

$$t_{nm}(E_F) = -i\sqrt{k_{x,n} k_{x,m}} \int_{-d/2}^{d/2} dy \int_{-d/2}^{d/2} dy' \, \chi_n^*(y) \, G(\boldsymbol{x}, \boldsymbol{x}', E_F) \, \chi_m(y') \quad , \tag{6}$$

where $k_{x,n}$ denotes the corresponding longitudinal wave numbers, $k_{x,n} = [k_F^2 - (n\pi/d)^2]^{1/2}$. The indices $n, m \in [1, \ldots, M]$, with M denoting the total number of open channels in the leads. The wave numbers $k_{x,n}$ are real for $n\pi/d < k_F$ (open channels). For $n\pi/d > k_F$ (closed channels), $k_{x,n}$ is purely imaginary. According to the Landauer formula [4], the total conductance g through the quantum dot is given by

$$g = \frac{1}{\pi} \sum_{m,n=1}^{M} |t_{nm}|^2 = \frac{1}{\pi} T^{\text{tot}} \, , \tag{7}$$

which is an experimental observable in semiconductor devices.

3 Special Features and Sample Results

3.1 Joining Modules of Different Symmetry

To investigate quantum billiards which feature chaotic classical dynamics it becomes necessary to turn to cavity geometries which are different from the purely separable cases. A prototypical example in this context is the stadium billiard (see e.g. Fig. 1h and [12]), for which two half-circular and one rectangular module are joined with each other. Using the MRGM we are confronted with the problem of how to properly connect the Cartesian grid of the rectangle with the polar grid structure of the half-circles without violating the Hermiticity of the tight-binding Hamiltonian at the junctions. To overcome this problem we insert additional link modules between the rectangle and the half-circles (see Fig. 1c). These link modules are essentially one-dimensional strips the site energies of which contain contributions from both adjacent grid structures [8].

In Fig. 1g we present two scattering wavefunctions for the circle and the stadium, respectively [9]. Note how the two wavefunctions in Fig. 1g are both localized along a typical classical trajectory in the respective cavity.

3.2 Magnetic Field

The MRGM allows to incorporate a magnetic field $\mathbf{B}$ oriented perpendicular to the 2D scattering geometry. The field enters the tb Hamiltonian of Eq. (1) by means of a Peierls phase factor [4, 2], with which the field-free hopping potential $V_{i,j}$ is multiplied: $V_{i,j} \rightarrow V_{i,j} \times \exp[i/c \int \mathbf{A}(\mathbf{x})\mathbf{dx}]$. The vector potential $\mathbf{A}(\mathbf{r})$ satisfies $\nabla \times \mathbf{A}(\mathbf{r}) = \mathbf{B}$. The Peierls phase will, of course, in most cases destroy the separability of the eigenfunctions of $\hat{H}^{\mathrm{tb}}$ in the modules. The resulting difficulties can be, in part, circumvented by exploiting the gauge freedom of the vector potential, i.e., $\mathbf{A} \rightarrow \mathbf{A}' = \mathbf{A} + \nabla\lambda$, where $\lambda(\mathbf{r})$ is a scalar function. By an appropriate choice of λ the wavefunction may remain separable on a given symmetry-adapted grid. Note, however, that even in the seemingly simple case of a rectangular module separability is destroyed by the magnetic field, no matter which gauge is chosen. The separability can however be restored by imposing periodic boundary conditions on two opposing sides of the rectangle. Topologically, this corresponds to folding the rectangle to the surface of a cylinder. The Green's function for the rectangle is finally obtained from the cylinder Green's function by a Dyson equation which is used here in "reversed" mode, i.e. for *disconnecting* tb grids (see Fig. 1d) [9].

The top part of Fig. 1h shows a stadium wavefunction in the very high magnetic field limit [9]. The magnetic field leads to the emergence of so-called "edge states" which creep along the boundary of the billiard [4]. The bottom part of Fig. 1h displays a field-free stadium wavefunction (in the high energy limit) which explores the entire cavity area.

3.3 Soft Potential Walls

Most theoretical investigations on quantum billiards focus on the two limiting cases of systems with either purely chaotic or purely regular classical dynamics [12]. However, neither of these cases is generic. For the semiconductor quantum dots that are realized in the experiment [4] a classical phase space structure with mixed regions of chaotic and regular motion is expected. This is due to the fact that the boundaries of such devices are typically not hard walls (as assumed in most theoretical investigations) but feature soft wall profiles [4] for which such a "mixed" phase space is characteristic. Soft walls can be incorporated in the MRGM, as long as their potential profiles do not violate the separability condition within a given module. In Fig. 1e we show an example of how this requirement can be fulfilled in the particular case of a half-stadium geometry (left part: hard wall case, right part: soft walls). Classical simulations reveal that the half-stadium with smooth boundaries indeed features a phase space within which regular and chaotic motion coexists. Characteristic for such "mixed systems" are very long trajectories that get "trapped" in the vicinity of regular islands of motion [12]. A typical example for such a trajectory is depicted in the top part of Fig. 1i. The quantum scattering wavefunction which corresponds to this orbit is shown right below [13].

3.4 Bulk and Surface Disorder

For many fundamental quantum transport phenomena the presence of disorder is essential and determines whether transport will be ballistic, diffusive or entirely suppressed by localization [4]. Whereas the (mostly unwanted) disorder is naturally present in the experiment, it is not straightforward to simulate disorder numerically. Bulk disorder can be viewed as random variations of the potential landscape through which the electron is transported. Within the framework of the MRGM the inclusion of such non-separable potential variations is not obvious, since separability is required in each of the modules. A way to overcome this difficulty is depicted in the top inset of Fig. 1f, where we decompose the cavity region into two square modules for each of which we choose a separable random potential $V(x, y) = V_1(x) + V_2(y)$. In order to destroy the unwanted separability, we combine two identical modules, however, rotated by 180° relative to each other [1]. The case of surface disorder can be treated by compiling a large number of rectangular modules of variable height (see bottom part of Fig. 1f). In this context the amount of spatial variation of the transverse module widths represents the strength of the disorder.

Since diffusion and localization in transport occur on very extended spatial scales, it is necessary to build up disorder regions of large length. For this purpose we employ an approach which allows to "exponentiate" the iteration process [10]: in the first step of the Dyson iteration we join individual rectangular modules with each other and employ the resulting combination of modules as the building block for the next step of the iteration. Repeating this procedure in each step of the iteration process, we manage to increase the size of the disordered region exponentially — with only a linear increase of computational time [3].

3.5 Andreev Billiards

The interface between a normal-conducting (N), ballistic quantum dot and a superconductor (S) gives rise to the coherent scattering of electrons into holes. This phenomenon is generally known as Andreev reflection. A N-S hybrid structure consisting of a superconducting lead attached to a normal cavity (see Fig. 1j) is commonly called an Andreev billiard. Such billiard systems attracted much attention recently, especially because of the unusual property that the classical dynamics in these systems features continuous families of periodic orbits, consisting of mutually retracing electron-hole trajectories. To learn more about the classical-to-quantum correspondence of Andreev billiards it is instructive to study the bound states in these billiards and the form of their wavefunctions. As shown in Fig. 1j, such wavefunctions indeed feature an electron and a hole part that (in most cases) closely resemble each other — in analogy to the classical picture of retracing electron-hole orbits. To obtain these quantum results numerically, we calculated the scattering states for the billiard with a normal conducting lead and entangled them in a linear superposition to construct the Andreev states [5]. Work on a more versatile approach is in progress which features coupled electron and hole tight-binding lattices explicitly.

4 Summary

We have given an overview of the modular recursive Green's function method (MRGM), which allows to calculate ballistic transport through quantum billiards efficiently. Key feature of the MRGM is the decomposition of billiard geometries into separable substructures ("modules") which are joined by recursive solutions of a Dyson equation. Several technical aspects of the method, as well as computational challenges and a few illustrative results have been presented.

Acknowledgments

Support by the Austrian Science Foundation (FWF Grant No. SFB-016, FWF-P17359, and FWF-P15025) is gratefully acknowledged.

References

1. Aigner, F., Rotter, S., and Burgdörfer, J.: Shot noise in the chaotic-to-regular crossover regime. Phys. Rev. Lett. **94** (2005) 216801
2. Baranger, H. U., DiVincenzo, D. P., Jalabert, R. A., Stone, A. D.: Classical and quantum ballistic-transport anomalies in microjunctions. Phys. Rev. B **44** (1991) 10637
3. Feist, J., Bäcker, A., Ketzmerick, R., Rotter, S., Huckestein, B., Burgdörfer, J. (to be published)
4. Ferry, D. K. and Goodnick, S. M.: Transport in Nanostructures. (Cambridge University Press, 1997) and references therein
5. Libisch, F., Rotter, S., Burgdörfer, J., Kormányos, A., and Cserti, J.: Bound states in Andreev billiards with soft walls. Phys. Rev. B **72** (2005) 075304
6. Mamaluy, D., Sabathil, M., and Vogl, P.: Efficient method for the calculation of ballistic quantum transport. J. Appl. Phys. **93** (2003) 4628
7. Marcus, C. M., Rimberg, A. J., Westervelt, R. M., Hopkins, P. F., and Gossard, A. C.: Conductance fluctuations and chaotic scattering in ballistic microstructures. Phys. Rev. Lett. **69** (1992) 506
8. Rotter, S., Tang, J.-Z., Wirtz, L., Trost, J., and Burgdörfer, J.: Modular recursive Green's function method for ballistic quantum transport. Phys. Rev. B **62** (2000) 1950
9. Rotter, S., Weingartner, B., Rohringer, N., and Burgdörfer, J.: Ballistic quantum transport at high energies and high magnetic fields. Phys. Rev. B **68** (2003) 165302
10. Skjånes, J., Hauge, E. H., and Schön, G.: Magnetotransport in a two-dimensional tight-binding model. Phys. Rev. B **50** (1994) 8636
11. Sols, F., Macucci, M., Ravaioli, U., and Hess, K.: Theory for a quantum modulated transistor. J. Appl. Phys. **66** (1989) 3892
12. Stöckmann, H.-J.: Quantum Chaos. (Cambridge University Press, 1999) and references therein
13. Weingartner, B., Rotter, S., and Burgdörfer, J.: Simulation of electron transport through a quantum dot with soft walls. Phys. Rev. B **72** (2005) 115342
14. Yang, X., Ishio, H., and Burgdörfer, J.: Statistics of magnetoconductance in ballistic cavities. Phys. Rev. B **52** (1995) 8219
15. Zozoulenko, I. V., Maaø, F. A., and Hauge, E. H.: Coherent magnetotransport in confined arrays of antidots. Phys. Rev. B **53** (1996) 7975

Quantum Correction to the Semiclassical Electron-Phonon Scattering Operator

V. Sverdlov[1], H. Kosina[1], C. Ringhofer[2], M. Nedjalkov[1], and S. Selberherr[1]

[1] Institute for Microelectronics, Technical University of Vienna, Gusshausstrasse 27-29, A 1040 Vienna, Austria
kosina@iue.tuwien.ac.at

[2] Department of Mathematics, Arizona State University, Tempe, AZ 85287-1804, USA

Abstract. A quantum kinetic equation approach is adopted in order to incorporate quantum effects such as collisional broadening due to finite lifetime of single particle states, and collisional retardation due to finite collision time. A quantum correction to the semiclassical electron distribution function is obtained using an asymptotic expansion for the quantum electron-phonon collision operator in its weak formulation. Based on this expansion, the evolution of a highly peaked, nonequilibrium distribution function in Si and Ge is analyzed. It is shown that in Ge and Si, where the electron-phonon interaction is weak, the quantum correction due to the finite collision time leads to an extra broadening of new replicas of the initial distribution function. As the observation time exceeds the collision duration, the quantum correction starts to diminish and the semiclassical solution for a particular replica is recovered.

1 Introduction

The semiclassical Boltzmann transport equation is successfully used for transport description and modeling in conventional semiconductor devices since the early development of semiconductor technology. A particular advantage of the Boltzmann equation is that it can be solved by a Markov-chain Monte Carlo algorithm which opens an immediate opportunity for direct transport process simulation. In the standard Monte-Carlo algorithm the carriers are moving on classical trajectories between the two consecutive collisions. Classical trajectories are characterized by the well defined values of coordinates $\mathbf{r}$ and momenta $\mathbf{p}$ which are related through the classical equations of motion. The scattering events are considered to be isolated from each other and instantaneous in both time and space. Locality of scattering events in time and space is one of the main assumptions underlying the semiclassical transport description based on the Boltzmann equation and should be re-evaluated in case of emerging quantum effects. Indeed, due to the quantum uncertainty principle a carrier may not have the well defined coordinate and momentum simultaneously. Therefore, the particle motion on the trajectory between collisions may not be described classically if the device dimension is comparable to the carrier de-Broglie wavelength. Similar, the scattering events may not be considered as local events in phase

I. Lirkov, S. Margenov, and J. Waśniewski (Eds.): LSSC 2005, LNCS 3743, pp. 594–601, 2006.

space. Locality of scattering in time may also be questioned. Due to the energy-time uncertainty, the energy conservation during scattering is justified only when the collision duration is large. This limit is usually referred to as the limit of a completed collision leading to the famous Fermi golden rule. When the duration of scattering is finite and the scattering may not be considered completed at an observation time moment t, the particle state at this moment t depends on the history of states the particle has assumed at all times $t' < t$, leading to the memory effect. This effect is in clear contradiction to the Markovian nature of the semiclassical Boltzmann equation and may not be described by the classical transport picture.

Going beyond the semiclassical approach in transport description becomes increasingly relevant. Indeed, with the 90 nm technology node being commercially implemented, the physical transistor gate length is already in the range of 45 nm. According to the International Technology Roadmap for Semiconductors, for the 32 nm technology node the physical gate length will be in the range of 10 nm, where quantum effects are expected to play a dominant role in determining the transport through the device.

Several advanced computational techniques for including the quantum effects were proposed recently. The method based on the Nonequilibrium Green's function formalism treats the quantum effects in the most complete and consistent way. However, due to its completeness, this method is rather complex and computationally costly [1]. Another approach is based on the solution of the Schrödinger equation using the modal analysis for an arbitrary 2D geometry (QDAME) [4]. Scattering can be included in this method by a Pauli master equation, and testing was successful for the resonant tunneling diode. Nevertheless, the QDAME applications to double-gate MOSFETs were so far limited to ballistic coherent regime [4].

An alternative method to address the quantum effects is the Wigner function approach [11]. Similar to the classical distribution function, the Wigner function depends on position and momentum simultaneously. Another attractive feature of the Wigner function approach is that it allows to include all scattering processes in the device via the Boltzmann scattering integral. It brings a unique opportunity to treat classical collisions on equal footing with the quantum scattering described by the quantum collision operator [3]. The question however rises as to whether the use of the classical Boltzmann scattering operator in the Wigner equation is justified. It is well known that the semiclassical transport theory based on the Boltzmann equation neglects several quantum mechanical effects such as collisional broadening due to the finite lifetime of single particle states, collisional retardation due to the finite collision duration, and intra-collisional field effects [10]. To answer this question, we shall begin from a complete quantum description of carrier scattering. The Levinson equation [5] which describes an interaction of a single electron with an equilibrium phonon bath represents a convenient starting point.

In this paper we analyze a quantum correction to the semiclassical scattering operator which is based on a recently obtained asymptotic expansion of

Levinson's scattering operator [9]. This method allows calculating a correction to the distribution function simultaneously with solving of the Boltzmann equation, which is the advantage as compared to the previously used techniques. An application of the algorithm to describe the transient processes in Si and Ge is investigated in details for the case of electron-phonon interaction. Taking a highly nonequilibrium initial distribution function which is sharply peaked around a certain energy as example, it is shown that for Si and especially for Ge the method adequately describes the quantum correction to the distribution function due to the finite collision time.

2 Basic Equations

A suitable quantum kinetic equation for the Wigner function describing the interaction of a single electron with an equilibrium phonon bath has been proposed by Levinson [5]. In case of vanishing electric field and spatially uniform semiconductor the Levinson equation has the following form:

$$\frac{\partial f}{\partial t} = \int_0^t dt' \int d\mathbf{p}' [S(p,p',t-t')f(p',t') - S(p',p,t-t')f(p,t')]. \quad (1)$$

Here, $\mathbf{p}$ denotes the momentum and $p = |\mathbf{p}|$ is its absolute value. The kernel $S(p,p',t)$ corresponding to an electron-optical phonon scattering is taken in the form

$$S(p,p',t) = \frac{2VF^2 n}{(2\pi\hbar)^3} \left\{ \cos\left[\frac{t}{\hbar}(E(p) - E(p') - \hbar\omega)\right] + \frac{n+1}{n} \cos\left[\frac{t}{\hbar}(E(p) - E(p') + \hbar\omega)\right] \right\}, \quad (2)$$

where $\hbar F$ denotes the electron-phonon interaction matrix element, $\hbar\omega$ the phonon energy, V the normalization volume, $n = (\exp(\beta\hbar\omega) - 1)^{-1}$ is the phonon occupation number corresponding to the temperature $k_B T = 1/\beta$, and $E(p)$ is the single particle energy. Due to an explicit time dependence of the kernel (2), the Levinson equation can fully describe the effects caused by the finiteness of the collision time. Numerical integration of the equation by Monte Carlo methods is, however, quite involved and can be performed for short evolution times only, due to fast growth of variances with time [2]. Our goal is to develop an approximate scheme which is computationally sound and less expensive. The path we would like to explore is based on the assumption of weak electron-phonon interaction. This assumption allows one to obtain an asymptotic expansion of the scattering operator (2) in powers of the dimensionless interaction constant. The principal term of this expansion reproduces the semiclassical Boltzmann scattering integral. The second term in the series describes the correction to the Boltzmann scattering integral due to the finite collision time. To derive it, we rewrite (1) in scaled variables, introduced in [9]. Measuring electron energies E in units of optical phonon energy $\epsilon = E/(\hbar\omega)$ and introducing the new variable $\widetilde{t} = t/t_0$, where $t_0 = (\lambda\omega)^{-1}$, the resulting equation takes the form [9]:

$$\frac{\partial f}{\partial \tilde{t}} = \int_0^{\tilde{t}/\lambda} d\tau \int d\varepsilon \rho(\epsilon')[s(\varepsilon,\varepsilon',\tau)f(\varepsilon',\tilde{t}-\lambda\tau) - s(\varepsilon,\varepsilon',\tau)f(\epsilon,\tilde{t}-\lambda\tau)], \quad (3)$$

$$s(\epsilon,\epsilon',\tau) = \sum_{\nu=\pm 1} a_\nu \cos(\tau(\epsilon-\epsilon'+\nu)), \quad a_{-1} = \frac{n}{n+1}, \quad a_1 = 1,$$

where $\rho(\epsilon) = 4\pi\sqrt{2\epsilon}$ is the density of states, corresponding to the parabolic dispersion $E = p^2/(2m)$. The dimensionless electron-phonon interaction constant λ is given by

$$\lambda = \frac{2VF^2 n}{(2\pi)^3}\sqrt{\frac{m^3}{\hbar^3\omega}}. \quad (4)$$

In the limit of small λ corresponding to weak interaction the time scale t_0 becomes much larger than the period of lattice vibrations. Considering an asymptotic behavior of the collision operator in the left-hand side of (3) for small λ when $\tilde{t}/\lambda \gg 1$, the Levinson equation becomes [9]:

$$\partial_{\tilde{t}} f = \Theta_0[f] + \lambda\Theta_1[\partial_{\tilde{t}} f] + o(\lambda), \quad (5)$$

where $\Theta_0[f]$ represents the Boltzmann scattering integral,

$$\Theta_0[f](p,t) = \sum_{\nu=\pm 1} \pi a_\nu \int d\varepsilon' \rho(\varepsilon')[\delta(\varepsilon-\varepsilon'+\nu)f(\varepsilon',t) - \delta(\varepsilon'-\varepsilon+\nu)f(\varepsilon,t)] \quad (6)$$

For the sake of brevity we omit the tilde and use the notation t for the scaled variable $\tilde{t}$ here and below, unless it is specified otherwise.

The correction scattering operator $\Theta_1[\partial_{\tilde{t}} f]$ formulated in a weak sense can be written as [9]

$$\int d\varepsilon \rho(\varepsilon)\Theta_1[\partial_t f](\phi(\varepsilon) = \sum_{\nu=\pm 1} a_\nu \int d\varepsilon \int d\varepsilon' \ln|\varepsilon'-\varepsilon+\nu| \times \frac{\partial}{\partial\varepsilon'}\frac{\partial}{\partial\varepsilon}\left\{\rho(\varepsilon')\rho(\varepsilon)\frac{\partial f(\varepsilon',t)}{\partial t}(\phi(\varepsilon)-\phi(\varepsilon'))\right\}, \quad (7)$$

where $\phi(\varepsilon)$ is a smooth test function. The weak formulation of the collision operator Θ_1 conserves mass locally since (7) is explicitely equal to zero for constant test functions.

In contrast to the the Boltzmann scattering integral being a functional of the distribution function, the correction scattering operator (7) depends on its time derivative $\partial_t f(\varepsilon', t)$. The time derivative is a consequence of memory effects present in the original Levinson equation. Expression (7) shows that the correction term $\Theta_1[\partial_t f]$ also depends on the values of the distribution function outside of the constant energy shell defined by the Fermi golden rule (6). These off-shell contributions are caused to the finiteness of the collision time and therefore provide the quantum mechanical correction to the Fermi golden rule.

3 Multiple Trajectories Monte Carlo Method

Equation (5) with scattering operators $\Theta_0[f]$ and $\Theta_1[\partial_t f]$, defined by (6) and (7), respectively, represents our starting point. We formally solve the equation (5) for $f(\varepsilon,t)$ using an iteration technique. We are looking for a solution of the form $f(\varepsilon,t)=f_0(\varepsilon,t)+\lambda f_1(\varepsilon,t)$, where $f_1(\varepsilon,t)$ is a correction term to the semiclassical solution $f_0(\varepsilon,t)$. The substitution into (5) allows to reduce the problem of finding an inverse of the operator $(I-\lambda\Theta_1)[\partial_t f]$, which is hardly treatable with Monte Carlo technique, to a simpler problem of computing the effect of the scattering operator Θ_1 acting on the first time derivative $\partial_t f_0(\varepsilon,t)$. Thus we arrive to the system of two coupled Boltzmann equations for f_0 and for the correction f_1:

$$\partial_t f_0 = Q_0[f_0], \qquad \partial_t f_1 = Q_0[f_1] + Q_1[\partial_t f_0],$$

with the initial conditions $f_0(\varepsilon,0)=\phi_0(\varepsilon)$ and $f_1(\varepsilon,0)=0$.

To obtain a forward Monte Carlo algorithm, we are following the standard procedure [7] writing the equations in integral form for an adjoint function $g_j(\varepsilon,t)$ defined in such a way that the average $(A,f_j)\equiv\int_0^\infty dt\int d\varepsilon\;\rho(\varepsilon)A(\varepsilon,t)f_j(\varepsilon,t)$ of an observable $A(\varepsilon,t)$ can be expressed as

$$(A,f_j)\equiv\int_0^\infty dt'\int d\varepsilon'\;\rho(\varepsilon')f_j^{(0)}(\varepsilon',t')g_j(\varepsilon',t'), \qquad j=0,1, \tag{8}$$

where the source-terms are defined as $f_1^{(0)}(\varepsilon,t)=\int_0^t dt' e^{-\kappa_0(\varepsilon)(t-t')}Q_1[\partial_t f_0](\varepsilon,t')$, $f_0^{(0)}(\varepsilon,t)=e^{-\kappa_0(\varepsilon)t}\phi_0(\varepsilon)$.

It can be shown [7] that the function $g_j(\varepsilon',t)$ must satisfy the adjoint equation

$$g_j(\varepsilon',t')=\int_0^\infty dt\int d\varepsilon\rho(\varepsilon)\;K(\varepsilon,t,\varepsilon',t')g_j(\varepsilon,t)+A(\varepsilon',t'), \qquad j=0,1, \tag{9}$$

with the free term given by the observable $A(\varepsilon',t)$. The kernel K in (9) is defined as $K(\varepsilon,t,\varepsilon',t')=H(t')H(t-t')e^{-\kappa_0(\varepsilon)(t-t')}S_0(\varepsilon,\varepsilon')$, where $\kappa_0(\varepsilon)$ is the total out-scattering rate: $\kappa_0(\varepsilon)=4\pi^2\sum_{\nu=\pm1}a_\nu\sqrt{2(\varepsilon-\nu)}H(\varepsilon-\nu)$.

It is important that neither the kernel nor the source term in (9) depend on j, so the index j can be omitted. Therefore, the sets of trajectories necessary to calculate f_0 and f_1 are similar. Different statistical averages are obtained by different terms $f_j^{(0)}$ in the inner product (8).

The solution of (9) is found through corresponding Neumann series. The forward Monte Carlo method is used to evaluate the series numerically. The algorithm is designed to compute f_0 together with its time derivative $\partial f_0/\partial t$. It allows to solve both equations for f_0 and f_1 simultaneously providing better computational scaling at large time compared to previously used techniques.

4 Results

We now apply the method developed to practically relevant semiconductors. In Si and Ge the constant of electron interaction with nonpolar optical phonons is expressed as [8]:

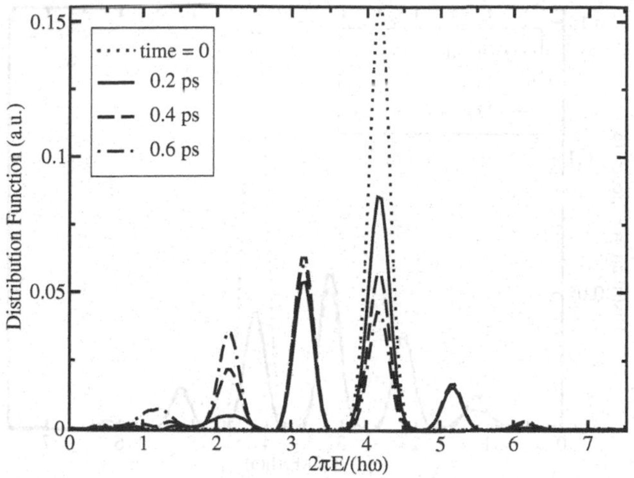

Fig. 1. Distribution $f_0(\varepsilon, t)$ with the quantum correction term $\lambda f_1(\varepsilon, t)$ added in Ge at the time instances $t_1 = 2 \times 10^{-13}$ s, $t_2 = 4 \times 10^{-13}$ s, $t_3 = 6 \times 10^{-13}$ s

$$F^2 = \frac{\hbar^2 (D_t K)^2}{2V \rho_0 \hbar\omega}, \tag{10}$$

leading to the dimensionless constant of electron-phonon interaction $\mu = \lambda\pi\rho(\varepsilon)$, which for $\varepsilon = 1$ is equal to $\mu \approx 0.055$. An even smaller value of the constant $\mu \approx 0.02$ is found in Ge. Because the values of the electron-phonon interaction constant in Si and Ge are comparable, results of simulations in Si and Ge are similar. Below we consider in details the case of Ge.

Simulation results of initial evolution of highly nonequilibrium distribution with the quantum correction taken into account are shown in Fig. 1. The appearance of additional replicas is clearly seen at energies which are multiples of the phonon energy, at first in the proximity of the initial energy distribution. With time passing the solutions propagate away from the initial energy ε_0 gradually creating more remote replicas. This is accompanied by an amplitude decrease of the initial peak. In contrast to zero-temperature results [6], our simulations were done at room temperature T=300 K. This leads to the possibility of phonon absorption which results in the creation of additional replicas with energies higher than the initial energy ε_0. However, the amplitude of these replicas is much smaller compared to those at energies lower than ε_0, reflecting the fact that phonon absorptions by nonequilibrium electrons is less probable than phonon emission.

In order to show the differences between the behavior with and without quantum corrections, a snapshot at time step $t = 6 \times 10^{-13}$ s is shown in Fig. 2. One can clearly see the additional broadening of more remote replicas when the quantum correction is taken into account in contrast to the classical solution. This is the result of scattering outside of an energy shell determined by the Fermi golden rule energy conserved delta-function, which is allowed due to the finiteness

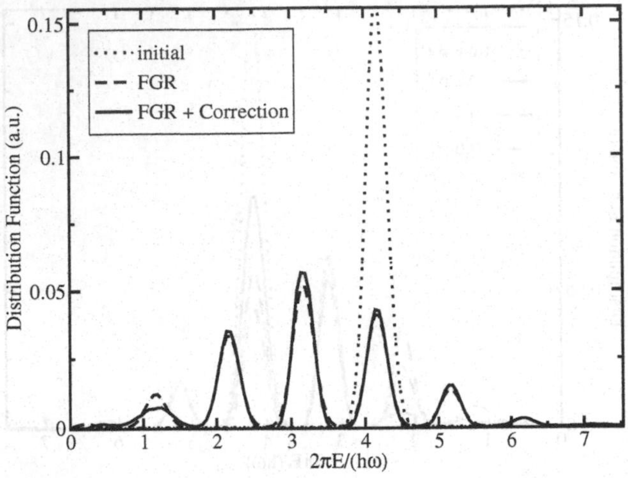

Fig. 2. Comparison of the semiclassical Boltzmann distribution $f_0(\varepsilon, t)$ and the distribution $f_0(\varepsilon, t)$ together with the quantum correction term $\lambda f_1(\varepsilon, t)$ added in Ge at the time $t = 6 \times 10^{-13}$ s. Additional broadening of remote replicas is due to the finiteness of the collision time.

of the collision time until the collision is complete. Similar collisional peak broadening was reported previously while solving the Levinson equation directly [6]. At the same time, neighboring replicas are becoming almost indistinguishable from those determined solely by the Fermi golden rule. This clearly shows that the quantum correction decreases at large time. In order to explain such a behavior, we note that the neighboring replicas correspond to emission (absorption) of only a single phonon. After some time, there will be almost no electrons left which did not emit (absorb) a phonon. For all electrons which emitted or absorbed a phonon, the collision process may be considered as completed, and the Fermi golden rule enforced energy conservation delta function is recovered. It is therefore expected that differences between the solutions with and without corrections will disappear, starting from replicas close to the initial peak.

5 Conclusion

Based on the asymptotic expansion of the Levinson equation, the quantum correction to the classical distribution function due to the finite collision time is analyzed. The Monte Carlo algorithm is developed in order to solve the Boltzmann equation simultaneously with the equation for the quantum correction. For the electron phonon interaction the method is also applied to calculate the quantum corrections in Si and Ge, where the electron phonon interaction is weaker than in GaAs. It is shown that for a highly nonequilibrium initial distribution peaked around a certain energy, the quantum correction leads to an extra broadening of replicas of the initial distribution peak appearing at frequencies shifted by a multiple of the phonon frequency. At the same time the quantum correction disappears for longer times when the limit of completed collision is recovered.

References

1. Bowen, R., Klimeck, G., Lake, R., Frensley, W., Moise, T.: Quantitative resonant tunneling diode simulation. J.Appl.Phys. **81** (1997) 3207–3213
2. Gurov, T., Whitlock, P.: An efficient backward Monte Carlo estimator for solving of a quantum-kinetic equation with memory kernel. Mathematics and Computers in Simulation **60** (2002) 85–105
3. Kosina, H., Nedjalkov, M., Selberherr, S.: Quantum Monte Carlo simulation of a resonant tunneling diode including phonon scattering, San Francisco, Computational Publications (2003) 190–193
4. Laux, S., Kumar, A., Fischetti, M.: Analysis of quantum ballistic electron transport in ultrasmall silicon devices including space-charge and geometric effects. J.Appl.Phys. **95** (2004) 5545–5582
5. Levinson, I.: Translational invariance in uniform fields and the equation for the density matrix in the Wigner representation. Soviet Phys.JETP **30** (1970) 362–367
6. Nedjalkov, M., Kosina, H., Kosik, R., Selberherr, S.: A space dependent Wigner equation including phonon interaction. J.Compt.Electronics **1** (2002) 27–31
7. Nedjalkov, M., Kosina, H., Selberherr, S., Ringhofer, C., Ferry, D.: Unified particle approach to Wigner-Boltzmann transport in small semiconductor devices. Physical Review B **70** (2004) 115319 1–16
8. Reggiani, L., ed.: Hot-Electron Transport in Semiconductors. Topics in Applied Physics, **58** Springer Verlag, Berlin, Heidelberg, New York, Tokyo (1985)
9. Ringhofer, C., Nedjalkov, M., Kosina, H., Selberherr, S.: Semi-classical approximation of electron-phonon scattering beyond Fermi's Golden Rule. SIAM J.Appl.Math. **64** (2004) 1933–1953
10. Schilp, J., Kuhn, T., Mahler, G.: Electron-phonon quantum kinetics in pulse-excited semiconductors: Memory and renormalization effects. Physical Review B **50** (1994) 5435–5447
11. Wigner, E.: On the quantum correction for thermodynamic equilibrium. Physical Review **40** (1932) 749–759

Efficient Solution of the Schrödinger-Poisson Equations in Semiconductor Device Simulations

Alex Trellakis, Till Andlauer, and Peter Vogl

Walter-Schottky Institute, TU München, D-85748 Garching, Germany

Abstract. This paper reviews the numerical issues arising in the simulation of electronic states in highly confined semiconductor structures like quantum dots. For these systems, the main challenge lies in the efficient and accurate solution of the self-consistent one-band and multiband Schrödinger-Poisson equations. After a brief introduction of the physical background, we first demonstrate that unphysical solutions of the Schrödinger equation due to the presence of material boundaries can be avoided by combining a suitable ordering of the differential operators with a robust discretization method like box discretization. Next, we discuss algorithms for the efficient solution of the resulting sparse matrix problems even on small computers. Finally, we introduce a predictor-corrector-type approach for the stabilizing the outer iteration loop that is needed to obtain a self-consistent solution of both Schrödinger's and Poisson's equation.

1 Physical Background

1.1 Introduction

Due to the rapid progress in semiconductor technology, the dimensions of electronic devices are approaching the nanometer scale. At this scale, electronic properties will be increasingly dominated by quantum effects, until at some point these effects become essential for device operation. For instance, the electronic states in highly confined semiconductor structures like quantum dots are entirely controlled by quantum mechanical size quantization, which can be exploited for entirely new device designs like quantum well or quantum dot lasers. At the same time, the physical equations that describe these quantum systems are much too complex to be solved analytically and efficient numerical solution methods need to be used instead. For this reason, we review in this paper the numerical issues arising in the simulation of electronic states in confined quantum systems, with focus on the efficient solution of the self-consistent Schrödinger-Poisson equations.

1.2 The k·p (Schrödinger) Equations

Since these quantum structures are about one to two orders of magnitude larger than a single lattice constant (which is about 0.5 nm), atomistic models like

I. Lirkov, S. Margenov, and J. Waśniewski (Eds.): LSSC 2005, LNCS 3743, pp. 602–609, 2006.

empirical tight binding theory or empirical pseudopotential methods are numerically not feasible and a continuum model with carefully tuned material parameters needs to be used. Such a continuum model is provided by the k·p-equations in the envelope function approximation [2, 5, 10].

In this model, we describe the electrons in the conduction band by an one-band k·p-equation which is called the effective-mass approximation. Here, the electron wave functions $\psi_n^{\mathrm{e}}(\mathbf{x})$ obey the Schrödinger equation

$$E_n^{\mathrm{e}}\, \psi_n^{\mathrm{e}}(\mathbf{x}) = \hat{H}_{\mathrm{eff}}\, \psi_n^{\mathrm{e}}(\mathbf{x}) \tag{1}$$

with the effective mass Hamiltonian

$$\hat{H}_{\mathrm{eff}} = \frac{1}{2}\hat{\mathbf{p}} \cdot \left(\frac{1}{m^*}\hat{\mathbf{p}}\right) + E_{\mathrm{C}}(\hat{\mathbf{x}}) + \hat{V}_{\mathrm{xc}}(\hat{\mathbf{x}}) \qquad \hat{\mathbf{p}} = \frac{\hbar}{\mathrm{i}}\boldsymbol{\nabla} \tag{2}$$

where E_n^{e} is the energy of wave function $\psi_n^{\mathrm{e}}(\mathbf{x})$, $E_{\mathrm{C}}(\mathbf{x})$ the conduction band edge, and $V_{\mathrm{xc}}(\mathbf{x})$ the exchange-correlation potential in the local density approximation. Finally, m^* is a 3×3-tensor describing the effective electron mass in the semiconductor.

For holes in the valence bands, the effective-mass approximation is usually not accurate enough anymore, and a multi-band approach treating the three main valence bands simultaneously is needed. Since each of the three valence bands has two spin components, we need here the 6-band k·p-equation

$$E_n^{\mathrm{h}}\, \psi_n^{\mathrm{h}}(\mathbf{x}) = \hat{H}_{6\times 6}\, \psi_n^{\mathrm{h}}(\mathbf{x}) \tag{3}$$

where $\psi_n^{\mathrm{h}}(\mathbf{x})$ is a 6-component hole wave function, and $\hat{H}_{6\times 6}$ a Hermitian 6×6-matrix operator

$$\hat{H}_{6\times 6} = \begin{pmatrix} \hat{H}_{3\times 3} & 0 \\ 0 & \hat{H}_{3\times 3} \end{pmatrix} + \hat{H}_{\mathrm{so}}^{6\times 6} \tag{4}$$

Here, $\hat{H}_{\mathrm{so}}^{6\times 6}$ is a momentum independent complex matrix that describes the spin-orbit interaction, and $\hat{H}_{3\times 3}$ a 3×3-matrix that for a Zincblende crystal structure oriented along the coordinate axes has the form [9]

$$\begin{aligned} \left(\hat{H}_{3\times 3}\right)_{ij} &= \left[E_{\mathrm{V}}(\hat{\mathbf{x}}) + \hat{V}_{\mathrm{xc}}(\hat{\mathbf{x}})\right]\delta_{ij} + \frac{1}{2m_0}\hat{\mathbf{p}}^2\delta_{ij} + \frac{N}{\hbar^2}\hat{p}_i\hat{p}_j \\ &+ \frac{L-N-M}{\hbar^2}\hat{p}_i^2\delta_{ij} + \frac{M}{\hbar^2}\hat{\mathbf{p}}^2\delta_{ij} \end{aligned} \tag{5}$$

where $E_{\mathrm{V}}(\mathbf{x})$ is the valence band edge, m_0 the vacuum mass of the electron, and L, M, and N the Dresselhaus parameters for the semiconductor [5].

Since the focus of our simulations are closed systems, we use homogeneous Dirichlet conditions ($\psi \equiv 0$) for all resulting Schrödinger equations. But note that the energies and wave functions of bound states in a confining potential are only weakly affected by their boundary conditions, as long as the simulation region in which Schrödinger's equation is solved is large enough.

1.3 Carrier Densities and the Poisson Equation

In thermal equilibrium, the electron and hole densities $n(\mathbf{x})$ and $p(\mathbf{x})$ are calculated from the quantum states $\{E_n, \psi_n\}$ as [15]

$$n(\mathbf{x}) = 2\sum_n |\psi_n^{\mathrm{e}}(\mathbf{x})|^2 F\left(\frac{E_n^{\mathrm{e}} - E_{\mathrm{F}}}{k_{\mathrm{B}}T}\right) \tag{6}$$

$$p(\mathbf{x}) = 2\sum_n \left|\psi_n^{\mathrm{h}}(\mathbf{x})\right|^2 F\left(\frac{E_{\mathrm{F}} - E_n^{\mathrm{h}}}{k_{\mathrm{B}}T}\right) \tag{7}$$

where k_{B} is the Boltzmann constant, T the temperature of the system, and $F(E)$ a suitable distribution function. This $F(E)$ is in three dimensions equal to the Fermi function

$$F(E) = \frac{1}{1 + \exp(E)} \tag{8}$$

while in one and two dimensions $F(E)$ is proportional to the complete Fermi-Dirac integrals $\mathcal{F}_0(-E)$ and $\mathcal{F}_{-1/2}(-E)$. Furthermore, in addition to these carrier densities there may also be space charges

$$N_{\mathrm{D}}^+(\mathbf{x}) = N_{\mathrm{D}}^+[E_{\mathrm{C}}(\hat{\mathbf{x}})] \qquad N_{\mathrm{A}}^-(\mathbf{x}) = N_{\mathrm{A}}^-[E_{\mathrm{V}}(\hat{\mathbf{x}})] \tag{9}$$

present due to doping with electron donors (N_{D}^+) and acceptors (N_{A}^-).

All these charge densities add up to a total charge

$$\rho(\mathbf{x}) = e\left[p(\mathbf{x}) - n(\mathbf{x}) + N_{\mathrm{D}}^+(\mathbf{x}) - N_{\mathrm{A}}^-(\mathbf{x})\right] \tag{10}$$

which then results through Poisson's equation

$$\boldsymbol{\nabla}\cdot[\epsilon(\mathbf{x})\boldsymbol{\nabla}\phi(\mathbf{x})] = -\rho(\mathbf{x}) \tag{11}$$

into an electric potential $\phi(\mathbf{x})$ that warps the band edges E_{C} and E_{V} as

$$E_{\mathrm{C}}(\mathbf{x}) \longrightarrow E_{\mathrm{C}}[\phi](\mathbf{x}) = E_{\mathrm{C}}(\mathbf{x}) - e\phi(\mathbf{x}) \tag{12}$$

$$E_{\mathrm{V}}(\mathbf{x}) \longrightarrow E_{\mathrm{V}}[\phi](\mathbf{x}) = E_{\mathrm{V}}(\mathbf{x}) - e\phi(\mathbf{x}) \tag{13}$$

As consequence, the charge density ρ becomes potential dependent and the Poisson equation (11) nonlinear in ϕ

$$\boldsymbol{\nabla}\cdot(\epsilon\boldsymbol{\nabla}\phi) = -e\left(p - n + N_{\mathrm{D}}^+[\phi] - N_{\mathrm{A}}^-[\phi]\right) \tag{14}$$

Note that the carrier densities n and p depend here indirectly on ϕ, since the band edges $E_{\mathrm{C}}[\phi]$ and $E_{\mathrm{V}}[\phi]$ enter the Hamiltonian, which of course makes also all wave functions and energies potential dependent. As a consequence, the nonlinear Poisson equation (14) and the Schrödinger equations become a coupled system of differential equations for which we need to find a self-consistent solution.

1.4 Operator Ordering and Symmetrization

An important issue we have ignored so far is that the material parameters in (2) and (5) are position dependent in heterostructures. This is a major problem since both Hamiltonians are strictly speaking only valid for a homogeneous semiconductor with slowly varying band edges $E_C(\mathbf{x})$ and $E_V(\mathbf{x})$. Some numerical experimentation now shows that even in the presence of material induced discontinuities in the band edges these Hamiltonians still deliver physically reasonable results. This empirical observation can also be partially justified using Burt's exact envelope function theory [3, 4].

Unfortunately, variations in m^* and L, M, N are much more problematic, since here we need also to define the operator ordering between these material parameters and the momentum operator $\hat{\mathbf{p}}$. While it is clear that any symmetrization procedure used must yield a Hermitian Hamiltonian

$$H(\hat{\mathbf{x}}, \hat{\mathbf{p}}) = H(\hat{\mathbf{x}}, \hat{\mathbf{p}})^\dagger \tag{15}$$

this requirement is not sufficient for uniquely defining $H(\hat{\mathbf{x}}, \hat{\mathbf{p}})$. Furthermore, a poorly chosen $H(\hat{\mathbf{x}}, \hat{\mathbf{p}})$ will result in wave functions that become unphysical at material interfaces and lead to highly incorrect solutions [7].

For the effective mass approximation (2), we can use in analogy to Poisson's equation (11) the Hamiltonian [11]

$$H_{\text{eff}}(\hat{\mathbf{x}}, \hat{\mathbf{p}}) = \frac{1}{2}\hat{\mathbf{p}} \cdot \left[\frac{1}{m^*(\hat{\mathbf{x}})}\hat{\mathbf{p}}\right] + E_C(\hat{\mathbf{x}}) + \hat{V}_{\text{xc}}(\hat{\mathbf{x}}) \tag{16}$$

where compared to (2) we have just replaced the effective mass m^* by the operator $m^*(\hat{\mathbf{x}})$. This Hamiltonian is flux conserving and leads to wave functions that are continuous at material interfaces and smooth everywhere else. Since most alternative Hamiltonians do not share these properties [11], Hamiltonian (16) is widely used in quantum simulations.

For the 6-band k·p-Hamiltonian (4-5) finding a suitable symmetrization is more difficult since $\hat{H}_{6\times6}$ as 6×6-matrix requires only

$$\left(\hat{H}_{6\times6}\right)^\dagger_{ij} = \left(\hat{H}_{6\times6}\right)_{ji} \tag{17}$$

in order to become Hermitian. Here, one can show that starting again from Burt's exact envelope function theory a suitable 6-band k·p-Hamiltonian $\hat{H}_{6\times6}(\hat{\mathbf{x}}, \hat{\mathbf{p}})$ can be constructed using the rather complicated operators [7, 8]

$$\begin{aligned}[H_{3\times3}(\hat{\mathbf{x}}, \hat{\mathbf{p}})]_{ij} = {} & \left[E_V(\hat{\mathbf{x}}) + \hat{V}_{\text{xc}}(\hat{\mathbf{x}})\right]\delta_{ij} + \frac{1}{2m_0}\hat{\mathbf{p}}^2\delta_{ij} + \hat{p}_i\frac{N(\hat{\mathbf{x}})}{\hbar^2}\hat{p}_j \\ & + \hat{p}_i\left[\frac{L(\hat{\mathbf{x}}) - N(\hat{\mathbf{x}}) - M(\hat{\mathbf{x}})}{\hbar^2}\right]\hat{p}_i\delta_{ij} + \hat{\mathbf{p}} \cdot \left[\frac{M(\hat{\mathbf{x}})}{\hbar^2}\hat{\mathbf{p}}\right]\delta_{ij} \\ & - \hat{p}_i\frac{M(\hat{\mathbf{x}})}{\hbar^2}\hat{p}_j + \hat{p}_j\frac{M(\hat{\mathbf{x}})}{\hbar^2}\hat{p}_i\end{aligned} \tag{18}$$

where the two last terms vanish in a bulk semiconductor. The resulting Hamiltonian $H_{6\times6}(\hat{\mathbf{x}}, \hat{\mathbf{p}})$ is again flux conserving and yields wave functions that are continuous at material interfaces and smooth everywhere else.

2 Numerical Issues

2.1 Discretization

Both the Poisson equation and the Schrödinger equations are discretized using box integration on a nonuniform tensor product grid. Such a grid can be seen as the tensor product of D nonuniform one-dimensional grids $\{x_i\}$ along each coordinate axis, where D is the dimension of the simulation domain. Such tensor grids have the advantage that no complicate meshing algorithm is required, and the discretization can be easily and efficiently implemented. In our implementation, the user provides the desired density of grid lines at selected locations along the coordinate axes. The resulting tensor product grid is then calculated by interpolating these line densities.

For the box integration finite difference method [14–page 184], we cover the simulation domain with N non-overlapping rectangular boxes, where each box is centered on a grid node. We then integrate our differential equations over each box volume Ω to obtain an integral equation which can then be discretized. For instance, the Poisson equation (14) becomes after integration over Ω

$$\int_{\partial\Omega} \mathbf{n}(\mathbf{x}) \cdot [\epsilon(\mathbf{x})\boldsymbol{\nabla}\phi(\mathbf{x})] = -\int_{\Omega} \rho(\mathbf{x}) \tag{19}$$

which after discretization then yields the familiar three, five, or seven-point stencils for ϕ.

Compared to standard finite differences, box integration has the advantage that discontinuities in a material parameter A are naturally taken into account for second order differential operators $\hat{\partial}_i A(\hat{\mathbf{x}})\,\hat{\partial}_j$, since this method is in first order flux conservative due to the Gauss theorem. This makes box integration a natural choice for discretizing the Poisson equation (14) and the Schrödinger equations resulting from (16) and (18).

After the discretization, all operators $\hat{\partial}_i A(\hat{\mathbf{x}})\,\hat{\partial}_j$ in these equations become banded sparse matrices. Such matrices can be easily stored and manipulated using banded storage schemes that store only the non-zero diagonals. In specific, the Schrödinger equation belonging to the Hamiltonians (16) or (18) becomes after discretization a generalized symmetric matrix eigenvalue problem of the form

$$H\psi_n = E_n D\psi_n \qquad D_{ij} = \delta_{ij}\mathrm{Vol}\,(\Omega_i) \tag{20}$$

with D being a diagonal matrix containing the volumes of the box regions Ω_i on the main diagonal. Similarly, the Poisson equation (14) as a boundary value problem yields a linear system, if we ignore the dependence of the charge density $\rho\,(\phi)$ on the potential ϕ for now.

2.2 Solving the Nonlinear Poisson Equation

At this point, solving each of the two differential equations offers its own set of challenges. For solving the nonlinear Poisson equation (14)

$$f\,(\phi) \equiv \nabla \cdot (\epsilon\nabla\phi) + \rho(\phi) = 0 \tag{21}$$

we can immediately use Newton-Raphson iteration with inexact line search as

$$\tilde{\phi}_k(\lambda) \equiv \phi_k - \lambda \, \left[(D_\phi f)(\phi_k) \right]^{-1} \cdot f(\phi_k) \tag{22}$$

$$\phi_{k+1} \equiv \tilde{\phi}_k(\lambda) \quad \text{with} \quad \left\| f\left(\tilde{\phi}_k(\lambda)\right) \right\| \equiv \min \qquad \lambda = 1, \frac{1}{2}, \frac{1}{4}, \frac{1}{8}, \ldots \tag{23}$$

in order to find a solution of the nonlinear problem (experience shows that local minima in $\|f(\phi)\|$ or multiple solutions do not occur for nanostructures).

We also note that, except for the extra diagonal term $D_\phi \rho$, the Jacobian matrix in (22)

$$(D_\phi f)(\phi) = \widehat{\nabla} \cdot (\epsilon \widehat{\nabla}) + (D_\phi \rho)(\phi) \tag{24}$$

is identical to the Poisson operator $\widehat{\nabla} \cdot (\epsilon \widehat{\nabla})$. Since $\widehat{\nabla} \cdot (\epsilon \widehat{\nabla})$ is positive definite and $D_\phi \rho$ is positive for all physical $\rho(\phi)$, the Jacobian $D_\phi \rho$ must be positive definite as well. Therefore, we can employ the preconditioned conjugate gradient method [12–chapter 6] for solving the linear system in (22). As preconditioner we use here the very fast fill-in free Dupont-Kendall-Rachford method [6] in order to achieve rapid convergence of the conjugate gradient iteration.

2.3 Solving the Schrödinger Equations

Solving the matrix eigenvalue systems resulting from the discretization of the Schrödinger equations is computationally much more demanding. Especially the 6-band k·p-Hamiltonian (18) results here into extremely large sparse matrices that prohibit the use of standard diagonalization routines. Luckily enough, all distribution functions $F(E)$ that are used in calculating the electron and hole densities (6) and (7) fall off exponentially as

$$F(E) \propto \exp(-E), \qquad E \gg 0. \tag{25}$$

For this reason, only relatively few quantum states at the lower end of the energy spectrum are needed to calculate the densities, which allows us to use highly efficient iterative eigenvalue methods like the Lanczos or Arnoldi iteration [12].

One software package that implements such iteration methods for both real and complex matrices is ARPACK [1]. In the authors' experience, ARPACK appears quite reliable and moderately fast for the matrices that occur in the simulation of quantum structures. However, ARPACK is still not fast enough for the huge system matrices that arise in three-dimensional quantum simulations and an additional preconditioner is needed.

Such a preconditioner can be easily obtained by utilizing the Chebyshev polynomials $T_n(x)$ for a spectral transformation. Here we use that if the Hermitian matrix $\hat{H}$ has the spectrum $\{E_n\}$, the matrix $T(\hat{H})$ must have the spectrum $\{T(E_n)\}$. Since the polynomials $T_n(x)$ increase very rapidly for $|x| > 1$, we can use these $T_n(x)$ to make the desired eigenvalues at the lower end of the spectrum very large, which accelerates the convergence of the Arnoldi iteration in ARPACK tremendously.

As a quick test for this preconditioned ARPACK code, we solve the eigenvalue problem for test matrices obtained by discretizing an one-band problem (16). For these matrices, we get for a 3 GHz PC and 15-th order Chebyshev polynomials as solution times

grid size	# states	CPU time
$50 \times 50 \times 50$	30	17 s
$50 \times 50 \times 50$	200	357 s
$100 \times 100 \times 100$	30	269 s

This test shows that memory exhaustion now surpasses excessive CPU time as the main concern for quantum simulations.

2.4 Solution of the Coupled System

As the last step, we need to tackle the self-consistent solution of the coupled nonlinear Poisson and Schrödinger equations. Here, a simple iteration does not converge due to the strong coupling between the equations. Similarly, under-relaxation stabilizes the outer iteration only very poorly, and we still observe strong charge oscillations from one iteration step to the next that interfere with convergence [13].

This situation can be much improved by partially decoupling both partial differential equations using a predictor-corrector-type approach. In order to do this, we replace the exact carrier densities $n(\mathbf{x})$ and $p(\mathbf{x})$ from (6) and (7) by the ϕ-dependent predictors [13]

$$\tilde{n}[\phi](\mathbf{x}) = 2\sum_n |\psi_n^{\mathrm{e}}(\mathbf{x})|^2 F\left(\frac{E_n^{\mathrm{e}} - E_{\mathrm{F}} + e[\phi(\mathbf{x}) - \phi_{\mathrm{prev}}(\mathbf{x})]}{k_{\mathrm{B}}T}\right) \tag{26}$$

$$\tilde{p}[\phi](\mathbf{x}) = 2\sum_n \left|\psi_n^{\mathrm{h}}(\mathbf{x})\right|^2 F\left(\frac{E_{\mathrm{F}} - E_n^{\mathrm{h}} - e[\phi(\mathbf{x}) - \phi_{\mathrm{prev}}(\mathbf{x})]}{k_{\mathrm{B}}T}\right) \tag{27}$$

where ϕ_{prev} is the electrostatic potential from the previous outer iteration step. These predictors for the quantum densities n and p are then used in the nonlinear Poisson equation

$$\boldsymbol{\nabla}\cdot(\epsilon\boldsymbol{\nabla}\phi) = -e\left(\tilde{p}[\phi] - \tilde{n}[\phi] + N_{\mathrm{D}}^{+}[\phi] - N_{\mathrm{A}}^{-}[\phi]\right) \tag{28}$$

in order to determine the new potential $\phi(\mathbf{x})$. Using this $\phi(\mathbf{x})$ we update the band edges $E_{\mathrm{C}}(\mathbf{x})$ and $E_{\mathrm{V}}(\mathbf{x})$ in the Hamiltonians (16) and (18), and solve the respective Schrödinger equations for a new set of energies and wave functions.

Obviously, once the iteration has converged we have $\phi = \phi_{\mathrm{prev}}$ and therefore with $n = \tilde{n}[\phi]$ and $p = \tilde{p}[\phi]$ also the correct densities. The numerical experiment then shows [13] that this approach leads to rapid convergence with the residuals in the quantum densities n and p decreasing by about one order of magnitude. No further steps are necessary to ensure convergence [13].

References

1. ARPACK libraries and related publications http://www.caam.rice.edu/software/ARPACK/.
2. Bastard, G.: Superlattice band structure in the envelope-function approximation. Phys. Rev. B **24** (1981) 5693–5697
3. Burt, M.G.: The justification for applying the effective-mass approximation to semiconductors. J. Phys.: Condens. Matter **4** (1992) 6651–6690
4. Burt, M.G.: Fundamentals of envelope function theory for electronic states and photonic modes in nanostructures. J. Phys.: Condens. Matter **11** (1999) R53–R83
5. Dresselhaus, G., Kip, A.F., Kittel, C.: Cyclotron Resonance of Electrons and Holes in Silicon and Germanium Crystals. Phys. Rev. **98** (1955) 368–384
6. Dupont, T., Kendall, R.P., Rachford, H.H.: An approximate factorization procedure for solving self-adjoint elliptic difference equations. SIAM J. Numerical Analysis **5** (1968) 559–573
7. Foreman, B.A.: Effective-mass Hamiltonian and boundary conditions for the valence bands of semiconductor microstructures. Phys. Rev. B **48** (1993) 4964–4967
8. Foreman, B.A.: Elimination of spurious solutions from eight-band k·p theory. Phys. Rev. B. **56** (1997) R12748–12751
9. Kane, E.O.: The k·p method. In: *Semiconductors and Semimetals: III–V compounds,* vol. 1, ed by R.K. Willardson, A.C. Beer (Academic Press, New York 1966) 75–100
10. Luttinger, J.M., Kohn, W.: Motion of Electrons and Holes in Perturbed Periodic Fields. Phys. Rev. **97** (1955) 869–883
11. Morrow, R.A., Brownstein, K.R.: Model effective-mass Hamiltonians for abrupt heterojunctions and the associated wave-function-matching conditions. Phys. Rev. B **30** (1984) 678–680
12. Trefethen, L.N. , Bau III, D.: *Numerical Linear Algebra* (SIAM Society for Industrial & Applied Mathematics, Philadelphia 1997)
13. Trellakis, A., Galick, A. T., Pacelli, A., Ravaioli, U.: Iteration scheme for the solution of the two-dimensional Schrödinger-Poisson equations in quantum structures. J. Applied Phys. **81** (1997) 7880-7884
14. Varga, R. S.: *Matrix Iterative Analysis* (Prentice-Hall, Englewood Cliffs 1962)
15. Yu, P.Y., Cardona, M.: *Fundamentals of Semiconductors* (Springer, Berlin Heidelberg New York 1996)

Part XII

Contributed Talks

Postprocessing and Improved Accuracy of the Lowest-Order Mixed Finite Element Approximation for Biharmonic Eigenvalues

Andrey Andreev[1], Raytcho Lazarov[2], and Milena Racheva[3]

[1] Department of Informatics, Technical University of Gabrovo, 5300 Gabrovo, Bulgaria
[2] Department of Mathematics, Texas A & M University, College Station, TX 77843, USA
[3] Department of Mathematics, Technical University of Gabrovo, 5300 Gabrovo, Bulgaria

Abstract. The mixed finite element method for the biharmonic eigenvalue problem using linear or bilinear finite elements is considered. The paper is based on approach described by the same authors in [1], where polynomials of degree n, $n \geq 2$, were used. The case of linear finite elements was studied by Ishihara in [5], where an error estimate of rate $\mathcal{O}(h^{\frac{1}{2}})$ for the eigenvalues and the eigenfunctions was established. Using postprocessing we derive an improved convergence rate for the approximate eigenvalues, namely $\mathcal{O}(h)$. This result is confirmed by model numerical experiments.

1 Introduction

We consider the biharmonic eigenvalue problem (related to eigenmodes of bending of homogeneous isotropic plates):

$$\Delta^2 u(x) \equiv \frac{\partial^4 u}{\partial x_1^4} + 2\frac{\partial^4 u}{\partial x_1^2 \partial x_2^2} + \frac{\partial^4 u}{\partial x_2^4} = \lambda u(x), \quad x \in \Omega, \tag{1}$$

with homogeneous Dirichlet boundary conditions

$$u(x) = 0 \quad \text{and} \quad \frac{\partial u}{\partial \nu}(x) = 0, \quad x \in \Gamma. \tag{2}$$

Here Ω is a bounded two-dimensional domain with boundary Γ, Δ is the Laplace operator, and $\frac{\partial}{\partial \nu}$ denotes differentiation in the normal direction to Γ.

In mixed form (1) is rewritten as a system of two equations of second order (see, e.g. [2, 3]) subject to the boundary conditions (2):

$$-\Delta u = \sigma, \quad -\Delta \sigma = \lambda u.. \tag{3}$$

The eigenvalue problem (3) has infinitely many solutions $(\lambda_i, (\sigma_i, u_i))$ with $\sigma_i = -\Delta u_i$, $i = 1, 2, \ldots$ so that if $(\lambda, (\sigma, u))$ is an eigenpair of (3), then (λ, u)

I. Lirkov, S. Margenov, and J. Waśniewski (Eds.): LSSC 2005, LNCS 3743, pp. 613–620, 2006.

is an eigenpair of (1), (2) and $\sigma = -\Delta u$ (see, e.g. [2]). The advantage of this formulation is that for its finite element approximation one can use C^0-elements.

In [1] a new post-processing procedure for improving the convergence rate of mixed finite element approximations of biharmonic eigenvalue problem (1) has been introduced and studied. The analysis was done for polynomials of degree $n \geq 2$ and thus leaving the case of linear finite elements (i. e. $n = 1$) open for investigation. Linear elements were used for mixed biharmonic problems in [5, 6, 7]. The aim of this paper is to extend the technique from [1] to the case of linear finite elements.

Let $H^k(\Omega)$, $k = 0, 1, \ldots$, be Sobolev space with a norm denoted by $\|\cdot\|_{k,\Omega}$ (see, e.g. [4]). As usual for $k = 0$ we have the standard $L_2(\Omega)$ space with a norm denoted by $\|\cdot\|_{0,\Omega}$. The space $H_0^k(\Omega)$ is the closure in the H^k–norm of the set of all infinitely smooth functions with compact support in Ω.

The weak form of (3) is obtained by multiplying the first equation of (3) by $\psi \in H^1(\Omega)$, the second equation by $v \in H_0^1(\Omega)$, and integrating by parts over Ω. Thus, we get the integral identities:

$$\int_\Omega \nabla\sigma \cdot \nabla v \, dx = \lambda \int_\Omega uv \, dx, \quad \int_\Omega \nabla u \cdot \nabla \psi \, dx = \int_\Omega \sigma\psi \, dx, \tag{4}$$

which are well defined for $u \in H_0^1(\Omega)$ and $\sigma \in H^1(\Omega)$. The finite element discretization studied here is based on the weak mixed form (4). Following Babuška and Osborn [2] we recast it in an abstract form. This form involves the real Hilbert spaces V, Σ and H with inner products and norms $(\cdot,\cdot)_V$, $\|\cdot\|_V$, $(\cdot,\cdot)_\Sigma$, $\|\cdot\|_\Sigma$, $(\cdot,\cdot)_H$ and $\|\cdot\|_H$, respectively, and the bilinear forms $a(\cdot,\cdot)$ and $b(\cdot,\cdot)$, defined on $H \times H$ and $\Sigma \times V$, respectively. We assume that:

1. $V \subset H$ and $\Sigma \subset H$;
2. the forms $a(\cdot,\cdot)$ and $b(\cdot,\cdot)$ are continuous, i.e.
$$b(\psi, v) \leq C_1 \|\psi\|_\Sigma \|v\|_V, \quad \forall \psi \in \Sigma, \ \forall v \in V,$$
$$a(\sigma, \psi) \leq C_2 \|\sigma\|_H \|\psi\|_H, \quad \forall \psi, \sigma \in H;$$
3. $a(\cdot,\cdot)$ is symmetric, $a(\sigma,\sigma) > 0, \forall 0 \neq \sigma \in H$, and
$$\sup_{\psi\in\Sigma} |b(\psi, u)| > 0, \ \forall\, 0 \neq u \in V.$$

We identify $H = L_2(\Omega)$, $\Sigma = H^1(\Omega)$, $V = H_0^1(\Omega)$,

$$b(\sigma, v) = \int_\Omega \nabla\sigma \cdot \nabla v \, dx \quad \text{and} \quad a(\sigma, \psi) = \int_\Omega \sigma\psi \, dx,$$

and check easily that the above assumptions are satisfied. Then the weak form of (4) reads as follows (see [2], p. 752): find $(\sigma, u) \in \Sigma \times V$, $(\sigma, u) \neq (0, 0)$ and $\lambda \in \mathbf{R}$ such that

$$-a(\sigma, \psi) + b(\psi, u) + b(\sigma, v) = \lambda a(u, v), \quad \forall (\psi, v) \in \Sigma \times V. \tag{5}$$

If $(\lambda; (\sigma, u))$ is an eigenpair of (5) then $(\lambda; u)$ is an eigenpair of (1), (2) and $\sigma = -\Delta u$. The problem (5) has infinitely many real eigenvalues (see [2])

$$0 < \lambda_1 \leq \lambda_2 \leq \ldots \nearrow \infty$$

with corresponding eigenfunctions (σ_i, u_i) and $\sigma_i = -\Delta u_i$, $i = 1, 2, \ldots$.

Now we consider the finite element approximation of problem (5). For simplicity we assume that Ω is a polygonal domain. Let τ_h be a partition of Ω into triangular and/or rectangular finite elements depending on a small positive parameter h, the largest size (diameter) of the finite elements. We also assume that τ_h is regular and quasi-uniform (see, Ciarlet [4], Miyoshi [8]). Associated with the triangulation τ_h we introduce the finite element spaces $V_h \subset V$ and $\Sigma_h \subset \Sigma$ of continuous piecewise linear/bilinear polynomials.

The approximate eigenpairs $(\lambda_h; (\sigma_h, u_h))$ of (5) are solutions of the mixed finite element problem: find $(\sigma_h, u_h) \in \Sigma_h \times V_h$ and $\lambda_h \in \mathbf{R}$ such that

$$-a(\sigma_h, \psi) + b(\psi, u_h) + b(\sigma_h, v) = \lambda_h a(u_h, v), \quad \forall (\psi, v) \in \Sigma_h \times V_h. \tag{6}$$

For the latter problem, Ishihara [5] proved the following convergence rate for the eigenvalues and eigenfunctions:

$$|\lambda - \lambda_h| \leq C\lambda^2 h^{\frac{1}{2}}, \tag{7}$$

$$\|u - u_h\|_{0,\Omega} \leq C h^{\frac{1}{2}}. \tag{8}$$

Remark 1. Buckling of a plate under pure compression (see, e.g. [6])

$$\Delta^2 u = -\lambda \Delta u, \quad \text{in } \Omega, \quad u = \frac{\partial u}{\partial \nu} = 0, \quad \text{on } \Gamma$$

could be studied by the mixed finite element method in the same manner.

The paper is organized as follows. In Section 2 we introduce the main ingredients of the proposed postprocessing method, Theorem 1 that establishes the relevance of the source problem and Theorem 2 that gives a practical way to use the source problem for more accurate recovery of the eigenvalues. Further, in Section 3 we formulate our postprocessing algorithm and in Section 4 we present some numerical experiments.

2 Main Result

Our aim is to present an inexpensive postprocessing procedure which will lead to a better accuracy of the mixed finite element eigenvalues. An important ingredient of this procedure is the corresponding source problem: for a given $f \in L_2(\Omega)$ find $(\tau, w) \in \Sigma \times V$ such that

$$-a(\tau, \psi) + b(\psi, w) + b(\tau, v) = a(f, v) \quad \forall (\psi, v) \in \Sigma \times V. \tag{9}$$

The solution (τ, w) of this problem defines two-component solution operators:

$$S : L_2(\Omega) \to \Sigma, \quad Sf = \tau, \quad T : L_2(\Omega) \to V, \quad Tf = w.$$

Then the solution $(\lambda; (\sigma, u))$ of the eigenvalue problem (5) satisfies the following relations: $\sigma = \lambda Su$ and $u = \lambda Tu$.

The corresponding finite element approximation of (9) is: find $(\tau_h, w_h) \in \Sigma_h \times V_h$ such that

$$-a(\tau_h, \psi) + b(\psi, w_h) + b(\tau_h, v) = a(f, v) \quad \forall(\psi, v) \in \Sigma_h \times V_h. \tag{10}$$

Similarly, the finite element solution $(\tau_h, w_h) \in \Sigma_h \times V_h$ defines the discrete component solution operators

$$S_h : L_2(\Omega) \to \Sigma_h, \quad S_h f = \tau_h, \quad T_h : L_2(\Omega) \to V_h, \quad T_h f = w_h.$$

Miyoshi [7] has proved that the mixed finite element approximation (τ_h, w_h) of (τ, w) (defined by (9)) satisfies the following error estimate:

$$\|w - w_h\|_{1,\Omega} + \|\tau - \tau_h\|_{0,\Omega} \le Ch^{\frac{1}{2}} \|f\|_{0,\Omega}. \tag{11}$$

Regarding (9) and (10) the following results have been established in [1, 2]:

1. If $f \in L_2(\Omega)$ then
$$-a(Sf, \psi) + b(\psi, Tf) + b(Sf, v) = a(f, v), \qquad \forall(\psi, v) \in \Sigma \times V,$$
$$-a(S_h f, \psi) + b(\psi, T_h f) + b(S_h f, v) = a(f, v), \qquad \forall(\psi, v) \in \Sigma_h \times V_h,$$
i.e., $(S_h f, T_h f)$ is the "Ritz projection" of (Sf, Tf) on the finite element space $\Sigma_h \times V_h$.
2. The solution operators T and T_h are compact and $\|T - T_h\| \to 0$ as $h \to 0$.
3. T and T_h are symmetric in the inner product defined by the bilinear form $a(\cdot, \cdot)$, i.e. $a(u, Tv) = a(Tu, v)$ and $a(u, T_h v) = a(T_h u, v)$.
4. λ is an eigenvalue of (5) if and only if λ^{-1} is an eigenvalue of T. Similarly, λ_h is an eigenvalue of (6) if and only if λ_h^{-1} is an eigenvalue of T_h.

Now we shall present the basic ideas and ingredients of our postprocessing technique. Assume that a finite element solution $(\lambda_h; (\sigma_h, u_h))$ of (6) is already found. Then, we consider the elliptic problem (10) with a right-hand side u_h and solution $(\widetilde{\tau}, \widetilde{w})$:

$$-a(\widetilde{\tau}, \psi) + b(\psi, \widetilde{w}) + b(\widetilde{\tau}, v) = a(u_h, v) \quad \forall(\psi, v) \in \Sigma \times V. \tag{12}$$

Obviously the solution of this problem could be written as $(\widetilde{\tau}, \widetilde{w}) = (Su_h, Tu_h)$. Denote by

$$\widetilde{\lambda} = a(u_h, Tu_h)^{-1} = a(u_h, \widetilde{w})^{-1}. \tag{13}$$

Theorem 1. *Let the finite element subspace $\Sigma_h \times V_h \subset \Sigma \times V$ contain piecewise linear polynomials, $(\lambda; (\sigma, u))$ be a solution of (5), and $(\lambda_h; (\sigma_h, u_h))$ be its finite element approximation obtained from (6). If the eigenfunctions u and u_h are normalized by $\|u\|_{0,\Omega} = \|u_h\|_{0,\Omega} = 1$, then*

$$|\lambda - \widetilde{\lambda}| \leq Ch. \tag{14}$$

Proof. Using the symmetry of the operator T as well as the normalization $a(u,u) = a(u_h, u_h) = 1$ we easily get

$$\begin{aligned}
\frac{1}{\lambda} - \frac{1}{\widetilde{\lambda}} &= a(u, Tu) - a(u_h, Tu_h) \\
&= a(u, Tu) - a(u_h, Tu_h) + a(u - u_h, T(u - u_h)) - a(u - u_h, T(u - u_h)) \\
&= 2a(u, Tu) - 2a(u_h, Tu) - a(u - u_h, T(u - u_h)) \\
&= \frac{1}{\lambda} a(u - u_h, u - u_h) - a(u - u_h, T(u - u_h)).
\end{aligned}$$

Since $a(u - u_h, T(u - u_h)) \leq \|T\| \ \|u - u_h\|^2_{0,\Omega}$, we obtain

$$|\lambda - \widetilde{\lambda}| \leq \lambda\widetilde{\lambda}\left(\frac{1}{\lambda} + \|T\|\right) \|u - u_h\|^2_{0,\Omega} \leq C\|u - u_h\|^2_{0,\Omega}$$

and estimate (8) proves the theorem.

Next we consider the question how to find an appropriate approximation to $\widetilde{\lambda}$. We introduce additional finite element spaces $\widetilde{\Sigma}_h \times \widetilde{V}_h$ such that $\Sigma_h \times V_h \subset \widetilde{\Sigma}_h \times \widetilde{V}_h \subset \Sigma \times V$, which contain continuous piecewise quadratic polynomials. Next we solve for $(\widetilde{\tau}_h, \widetilde{w}_h) \in \widetilde{\Sigma}_h \times \widetilde{V}_h$ such that

$$-a(\widetilde{\tau}_h, \psi) + b(\psi, \widetilde{w}_h) + b(\widetilde{\tau}_h, v) = a(u_h, v), \quad \forall(\psi, v) \in \widetilde{\Sigma}_h \times \widetilde{V}_h, \tag{15}$$

and define

$$\widetilde{\lambda}_h^{-1} = a(u_h, \widetilde{w}_h). \tag{16}$$

Obviously, the following orthogonality condition holds

$$-a(\widetilde{\tau}_h - \widetilde{\tau}, \psi) + b(\psi, \widetilde{w}_h - \widetilde{w}) + b(\widetilde{\tau}_h - \widetilde{\tau}, v) = 0, \quad \forall(\psi, v) \in \widetilde{\Sigma}_h \times \widetilde{V}_h. \tag{17}$$

In the theorem below we show that $\widetilde{\lambda}_h$ is a better approximation to λ.

Theorem 2. *Let the conditions of Theorem 1 hold and $\widetilde{\lambda}_h$ be determined by (16), where $(\widetilde{\tau}_h, \widetilde{w}_h) \in \widetilde{\Sigma}_h \times \widetilde{V}_h$ is the solution of (15). Then the following estimate is valid:*

$$|\widetilde{\lambda} - \widetilde{\lambda}_h| \leq Ch. \tag{18}$$

Proof. By the definition of $\widetilde{\lambda}$ and $\widetilde{\lambda}_h$ and using (17) we get

$$\begin{aligned}\frac{1}{\widetilde{\lambda}} - \frac{1}{\widetilde{\lambda}_h} &= a(u_h, \widetilde{w}) - a(u_h, \widetilde{w}_h) \\ &= a(u_h, \widetilde{w}) + a(u_h, \widetilde{w}_h) - 2a(u_h, \widetilde{w}_h) \\ &= 2b(\widetilde{\tau} - \widetilde{\tau}_h, \widetilde{w} - \widetilde{w}_h) - a(\widetilde{\tau} - \widetilde{\tau}_h, \widetilde{\tau} - \widetilde{\tau}_h).\end{aligned}$$

Then

$$\begin{aligned}|\frac{1}{\widetilde{\lambda}} - \frac{1}{\widetilde{\lambda}_h}| &\le 2|b(\widetilde{\tau} - \widetilde{\tau}_h, \widetilde{w} - \widetilde{w}_h)| + |a(\widetilde{\tau} - \widetilde{\tau}_h, \widetilde{\tau} - \widetilde{\tau}_h)| \\ &\le 2\|\widetilde{\tau} - \widetilde{\tau}_h\|_{1,\Omega}\|\widetilde{w} - \widetilde{w}_h\|_{1,\Omega} + \|\widetilde{\tau} - \widetilde{\tau}_h\|^2_{0,\Omega}.\end{aligned}$$

Since the right-hand side u_h of (12) belongs to $H^1(\Omega)$ then the regularity of the solution $\widetilde{w}$ is at least $H^3(\Omega)$. Note that according to the estimate (10) piecewise linear finite elements are not applicable for the equation (15). The estimate (18) follows from the standard estimate for $\|\widetilde{\tau}-\widetilde{\tau}_h\|_{1,\Omega}$, $\|\widetilde{w}-\widetilde{w}_h\|_{1,\Omega}$, and $\|\widetilde{\tau}-\widetilde{\tau}_h\|^2_{0,\Omega}$ while using quadratic finite elements.

Finally, an immediate consequence of (14) and (18) is

$$|\lambda - \widetilde{\lambda}_h| \le |\lambda - \widetilde{\lambda}| + |\widetilde{\lambda} - \widetilde{\lambda}_h| \le Ch.$$

Remark 2. Another possible choice for $\widetilde{\Sigma}_h \times \widetilde{V}_h$ is discussed and studied in [1]. This choice is based on the idea of Xu and Zhou [10] of two-grid approximations of the eigenvalues of elliptic problems. Namely, $\widetilde{\Sigma}_h$ and $\widetilde{V}_h$ are spaces of piecewise linear functions defined on a mesh $\widetilde{\tau}_h$ that has characteristic size h^2.

3 Algorithm

Following the results from the presented above two theorems we propose an inexpensive algorithm for computing the eigenvalues of biharmonic boundary value problems with an improved accuracy:

(i) Find an approximate eigenpair $(\lambda_h; (\sigma_h, u_h))$ from the mixed problem (6) with $a(u_h, u_h) = 1$ using linear/bilinear finite elements;

(ii) Find the solution $(\widetilde{\tau}_h, \widetilde{w}_h) \in \widetilde{\Sigma}_h \times \widetilde{V}_h$ of the problem (15);

(iii) Define $\widetilde{\lambda}_h = a(u_h, \widetilde{w}_h)^{-1}$; this is the postprocessed eigenvalue that according to Theorem 2 is a better approximation of λ.

4 Numerical Examples

4.1 Example 1

Consider a model problem of a long thin bar of length l with unit flexural rigidity and density, which is simply supported at its endpoints. The natural frequencies of the bar are determined by the eigenvalues of the problem:

$$u^{IV} = \lambda u, \quad \text{in } (0, l), \quad u(0) = u^{II}(0) = u(l) = u^{II}(l) = 0.$$

When $l = \pi$, the exact eigenvalues are $\lambda_j = j^4$, $j = 1, 2, \ldots$.

Table 1. The eigenvalues computed by the mixed finite element method

N	$\lambda_{1,h}$	$\lambda_{2,h}$	$\lambda_{3,h}$	$\lambda_{4,h}$
32	1.001607542	16.10310555	82.17866457	262.6585415
64	1.000401668	16.02572067	81.29324023	257.6671733
128	1.000100403	16.00642668	81.07322345	256.4473224
256	1.000025100	16.00160645	81.01829971	256.1044307

Table 2. The eigenvalues obtained after applying the postprocessing algorithm

N	$\widetilde{\lambda}_{1,h}$	$\widetilde{\lambda}_{2,h}$	$\widetilde{\lambda}_{3,h}$	$\widetilde{\lambda}_{4,h}$
32	1.000000387	16.00009942	81.00255901	256.025916
64	1.000000024	16.00000620	81.00015906	256.003286
128	1.000000002	16.00000039	81.00001002	256.004112
256	1.000000000	16.00000002	81.00000062	256.000130

The numerical results for the first four eigenvalues computed by the standard finite element method with N linear elements are given in Table 1. This example shows higher than predicted by the theory of [5, 6, 7] convergence rate. Further, in Table 2 we present the results of the same four eigenvalues computed by our postprocessing method. Note that the accuracy of our computations is substantially higher than those of standard finite element method.

4.2 Example 2

To illustrate our theoretical results we report numerical results for two-dimensional biharmonic eigenvalue problem. Let Ω be a square domain:

$$\Omega: \quad -\frac{\pi}{2} < x_i < \frac{\pi}{2}, \quad i = 1, 2.$$

Table 3. Comparison between the eigenvalues computed by the mixed finite element method (FEM) and those obtained after applying the postprocessing algorithm (PP)

h		λ_1	λ_2	λ_3	λ_4
$\pi/5$	FEM	17.1341	76.5380	95.4773	219.9627
	PP	13.9849	69.5796	70.5793	155.4328
$\pi/6$	FEM	16.0015	71.0237	82.9916	194.2231
	PP	13.7453	65.6851	66.2851	144.1984
$\pi/7$	FEM	15.0288	66.9950	76.0354	177.7987
	PP	13.6111	63.2219	63.5193	136.1713
$\pi/8$	FEM	14.7871	63.9766	66.8246	161.4620
	PP	13.4731	59.8513	59.8587	128.8196
bounds:					
lower		13.2820	55.2400	55.2400	120.0070
upper		13.3842	56.5610	56.5610	124.0740

Consider the following model problem:

$$\Delta^2 u = \lambda u \quad \text{in } \Omega,$$

$$u = \frac{\partial u}{\partial \nu} = 0 \quad \text{on } \partial\Omega.$$

The exact eigenvalues for this problem are not known. So, we use their lower and upper bounds obtained by Weinstein and Stenges [9] (see also Ishihara [3]). In Table 3 the results from our numerical experiments for the first four eigenvalues are given.

5 Comments and Conclusion

Comparing the results proved in the previous sections as well as the numerical results for the eigenvalues of the mixed variant of the biharmonic eigenvalue problem with so-called "optimal" case (see [5]), we may conclude that:

- The global postprocessing method presented and studied here gives an effective and accurate algorithm for calculating the biharmonic eigenvalues using the lowest order (linear or bilinear) finite elements;
- The presented above postprocessing analysis can be directly extended to more general fourth-order eigenvalue problems, namely to buckling problems and other boundary conditions.

References

1. Andreev, A.B., Lazarov, R.D., Racheva, M.R.: Postprocessing and higher order convergence of mixed finite element approximations of biharmonic eigenvalue problems. J. Comput. Appl. Math., **182 (2)** (2005) 333–349.
2. Babuška, I., Osborn, J.: Eigenvalue Problems. In Handbook of Numerical Analysis Vol. II (Eds. P. G. Lions and Ciarlet P.G.) Finite Element Methods (Part 1) North-Holland, Amsterdam (1991) 641–787
3. Canuto, C.: Eigenvalue approximation by mixed methods. RAIRO Anal. Numer. R3, **12** (1978) 27–50
4. Ciarlet, P.G.: The Finite Element Method for Elliptic Problems. North-Holland Amsterdam New York Oxford (1978)
5. Ishihara, K.: A mixed finite element method for the biharmonic eigenvalue problem of plate bending. Publ. Res. Institute of Mathematical Sciences, Kyoto University, **14** (1978) 399–414
6. Ishihara, K.: On the mixed finite element approximation for the buckling of plates. Numer. Math. **33** (1979) 195–210
7. Miyoshi, T.: A finite element method for the solution of fourth-order partial differential equations. Kumamoto J.Sci. (Mathematics) **9** (1973) 87–116
8. Miyoshi, T.: A mixed finite element method for the solution of the von Karman equations. Numer. Math. **26** (1976) 255–269
9. Weinstein, A., W. Stenger: Methods of intermediate problems for eigenvalues, theory and applications. Academic Press, 1972
10. Xu, J., A. Zhou, A two-grid discretization scheme for eigenvalue problems, Math. Comp. **70** (2001) 17–25

Computation of Some Unsteady Flows over Porous Semi-infinite Flat Surface

Nataliya Atanasova and Iliya Brayanov

Department of Applied Mathematics and Informatics,
University of Rousse "A. Kanchev", Studentska str. 8,
7017 Rousse, Bulgaria
natanasova@ru.acad.bg, brayanov@ru.acad.bg

Abstract. Finite difference solutions of an one-dimensional unsteady convection-diffusion problem on semi-infinite interval is considered. An artificial boundary is introduced to make the computational domain finite. On the artificial boundary an exact boundary condition is applied to reduce the original problem to an initial-boundary value problem. A finite difference scheme is derived by the method of reduction of order. It is proved that the finite difference scheme is convergent in energy norm of order 2 in space and order 3/2 in time. Numerical experiments confirm the theoretical results and show the efficiency of the constructed scheme.

1 Introduction

Let us consider the boundary-layer flow of an electrically conducting incompressible fluid over a continuously moving flat surface. The boundary-layer equation for the flow field has the following form, see [8]

$$\frac{\partial u}{\partial t} - D\frac{\partial^2 u}{\partial x^2} - \nu\frac{\partial u}{\partial x} = f(x,t), \quad (x,t) \in \Omega = (-1,\infty) \times (0,\infty), \tag{1}$$

$$u(x,0) = \phi(x), \quad -1 \le x < \infty, \tag{2}$$

$$u(-1,t) = \psi(t), \quad 0 \le t < \infty, \tag{3}$$

$$u \to 0, \text{ when } x \to \infty, \tag{4}$$

where $D = R_0^{-1}$, R_0 is the suction Reynolds number and ν is the suction velocity. The functions $f(x,t), \phi(x)$ and $\psi(t)$ are sufficiently smooth and satisfy $\psi(0) = \phi(-1)$. The support of $f(x,t)$ is in the domain $\bar{\Omega}_0 = [-1,0] \times [0,\infty)$ and the support of $\psi(x)$ is in the interval $[-1,0]$.

Many application problems can be modeled by parabolic equations in unbounded domains, such as the heat transfer problems, fluid dynamics problems, see [7] for example, and recently various option pricing problems in the financial mathematics. For standard finite difference or finite element methods, that solve problems on unbounded domain, an artificial boundary must be introduced to make the computational domain finite. Then on the artificial boundary appropriate boundary condition is needed for accurate numerical computation. For parabolic problems, in [1, 2] is proposed an exact boundary condition to reduce

I. Lirkov, S. Margenov, and J. Waśniewski (Eds.): LSSC 2005, LNCS 3743, pp. 621–628, 2006.

the original problem to an initial-boundary value problem on a finite computational domain. Theoretical analysis for the numerical method of parabolic-reaction diffusion problem is given in [9]. In this paper we analyze this method for parabolic convection-diffusion problem.

For computing the numerical solution of problem (1)-(4) we introduce an artificial boundary $\Gamma_0 = \{x = 0\} \times [0, \infty)$. Then the domain $\bar{\Omega}$ is divided into a bounded part $\bar{\Omega}_0$ and an unbounded part $\bar{\Omega}_e = [0, \infty) \times [0, \infty)$. We first consider the restriction of the solution on the domain Ω_e. Then using techniques from [9] we obtain

$$\frac{\partial u}{\partial x}(0,t) - \tilde{\mu} u(0,t) = -\frac{e^{\tilde{\lambda} t}}{\sqrt{D\pi}} \int_0^t \frac{\frac{\partial}{\partial \lambda}\left(u(0,\lambda) e^{-\tilde{\lambda}\lambda}\right)}{\sqrt{t-\lambda}} d\lambda, \tag{5}$$

where $\tilde{\mu} = -\nu/2D$ and $\tilde{\lambda} = -\nu^2/4D$.

Thus we reduce the problem (1)-(4) to the problem (P_0): (1)-(3), (5) on the bounded domain $\bar{\Omega}_0$.

2 Numerical Approximation

In the finite domain $\bar{\Omega}_0$ we shall construct a mesh $\omega = \omega_\tau \times \omega_h$, that is uniform with respect to the time variable and possibly nonuniform with respect to the space variable.

$$\omega_\tau = \{t_j, t_j = t_{j-1} + \tau, j = 1, 2, \dots, t_0 = 0\},$$
$$\bar{\omega}_h = \{x_i, x_i = x_{i-1} + h_i, i = 1, 2, \dots, N, x_0 = -1, x_N = 0\}.$$

Let $w_i^j = w(x_i, t_j)$ be a grid function of the discrete argument $(x_i, t_j) \in \bar{\omega}$. We shall use further the following notations

$$\hat{h}_i = \frac{h_i + h_{i+1}}{2}, \; w^j_{i-1/2} = \frac{w_i^j + w_{i-1}^j}{2}, \; \delta_x w^j_{i-1/2} = \frac{w_i^j - w_{i-1}^j}{h_i},$$

$$\delta_{\overset{\circ}{x}} w_i^j = \frac{w_{i+1}^j - w_{i-1}^j}{2\hat{h}_i}, \; \delta_{\hat{x}} w_i^j = \frac{w_{i+1}^j - w_i^j}{\hat{h}_i}, \; \delta_x^2 w_i^j = \frac{\delta_x w^j_{i+1/2} - \delta_x w^j_{i-1/2}}{\hat{h}_i},$$

$$w_i^{j-1/2} = \frac{w_i^j + w_i^{j-1}}{2}, \; \delta_t w_i^{j-1/2} = \frac{w_i^j - w_i^{j-1}}{\tau}, \; \|w^j\|_A = \sqrt{\sum_{i=1}^N h_i (w^j_{i-1/2})^2},$$

$$\|\delta_x w^j\| = \sqrt{\sum_{i=1}^N h_i (\delta_x w^j_{i-1/2})^2}, \; \|w^j\|_\infty = \max_{0 \le i \le N} |w_i^j|.$$

The following lemma will be used in the derivation of the difference scheme, see [9].

Lemma 1. *Suppose $f(t) \in C^2[0, t_j]$. Then*

$$\left| \int_0^{t_j} \frac{f'(t)dt}{\sqrt{t_j - t}} - \sum_{k=1}^{j} \frac{f(t_k) - f(t_{k-1})}{\tau} \int_{t_{k-1}}^{t_k} \frac{dt}{\sqrt{t_j - t}} \right| \le \frac{10\sqrt{2} - 11}{6} \max_{0 \le t \le t_j} |f''(t)| \tau^{3/2}.$$

2.1 Derivation of the Finite Difference Scheme

The difference scheme will be derived by the method of reduction of order. This method is introduced first by Sun [5, 6] for solution of some high-order partial differential equations. It has been proved there, that it is unconditionally stable, has a second order approximation both in space and time, that makes it preferable to implicit Euler and Crank-Nicolson based methods. The similar idea is used here to construct our difference scheme. First, we introduce a new variable: $v = \partial u/\partial x$. Then the problem (P_0) is equivalent to

$$\frac{\partial u}{\partial t} - D\frac{\partial v}{\partial x} - \nu v = f(x,t), \quad (x,t) \in \Omega_0, \tag{6}$$

$$v - \frac{\partial u}{\partial x} = 0, \quad (x,t) \in \Omega_0, \tag{7}$$

$$u(x,0) = \phi(x), \quad -1 \le x \le 0, \tag{8}$$

$$u(-1,t) = \psi(t), \quad 0 \le t < \infty, \tag{9}$$

$$v(0,t) - \tilde{\mu}u(0,t) = -\frac{e^{\tilde{\lambda}t}}{\sqrt{D\pi}}\int_0^t \frac{\frac{\partial}{\partial\lambda}\left(u(0,\lambda)e^{-\tilde{\lambda}\lambda}\right)}{\sqrt{t-\lambda}}d\lambda, \quad 0 \le t < \infty. \tag{10}$$

Define the grid functions:

$$U_i^j = u(x_i,t_j),\ V_i^j = v(x_i,t_j),\ 0 \le i \le N,\ j \ge 0.$$

Using Lemma 1 and noticing $U_N^0 = 0$, it follows from (10) that for all $j = 1,2,3,\dots$, we have

$$\begin{aligned} V_N^j - \tilde{\mu}U_N^j &= -\frac{e^{\tilde{\lambda}t_j}}{\sqrt{D\pi}}\sum_{k=1}^{j}\int_{t_{k-1}}^{t_k}\frac{\frac{\partial}{\partial\lambda}\left(u(0,\lambda)e^{-\tilde{\lambda}\lambda}\right)}{\sqrt{t_k-\lambda}}d\lambda \\ &= -\frac{2}{\sqrt{D\pi}}\left[a_0U_N^j - \sum_{k=1}^{j-1}\left(a_{j-k-1}-a_{j-k}\right)U_N^k e^{\tilde{\lambda}\tau(j-k)}\right] + O(\tau^{3/2}), \end{aligned}$$

where

$$a_j = \frac{1}{\sqrt{t_{j+1}}+\sqrt{t_j}} = \frac{1}{\sqrt{\tau}(\sqrt{j+1}+\sqrt{j})}, \quad j = 0,1,2,\dots. \tag{11}$$

Thus we obtain the following approximation of problem (6)-(10)

$$\delta_t U_{i-1/2}^{j-1/2} - D\delta_x V_{i-1/2}^{j-1/2} - \nu V_{i-1/2}^{j-1/2} = f_{i-1/2}^{j-1/2}, \quad 1 \le i \le N,\ j \ge 1, \tag{12}$$

$$V_{i-1/2}^{j-1/2} - \delta_x U_{i-1/2}^{j-1/2} = 0, \quad 1 \le i \le N,\ j \ge 1, \tag{13}$$

$$U_i^0 = \phi(x_i), \quad 0 \le i \le N, \tag{14}$$

$$U_0^j = \psi(t_j), \quad j \ge 1, \tag{15}$$

$$\begin{aligned} V_N^{j-1/2} - \tilde{\mu}U_N^{j-1/2} = -\frac{2}{\sqrt{D\pi}}\Big[a_0U_N^{j-1/2} - \\ - \sum_{k=1}^{j-1}(a_{j-k-1}-a_{j-k})U_N^{k-1/2}e^{\tilde{\lambda}\tau(j-k)}\Big], \quad j \ge 1. \end{aligned} \tag{16}$$

Using similar arguments as in [9] we can show that (12)-(16) is equivalent to the following scheme (P_1^h):

$$\alpha_i\delta_t U_{i+1/2}^{j-1/2} + \beta_i\delta_t U_{i-1}^{j-1/2} - D\delta_x^2 U_i^{j-1/2} - \nu\left(\alpha_i\delta_x U_{i+1/2}^{j-1/2} + \beta_i\delta_x U_{i-1/2}^{j-1/2}\right)$$
$$= \alpha_i f_{i+1/2}^{j-1/2} + \beta_i f_{i-1/2}^{j-1/2}, \quad 1 \le i \le N, \quad j \ge 1, \tag{17}$$
$$U_i^0 = \phi(x_i), \quad 0 \le i \le N, \tag{18}$$
$$U_0^j = \psi(t_j), \quad j \ge 1, \tag{19}$$
$$\delta_t U_{N-1/2}^{j-1/2} + \left(\frac{2D}{h_N} - \nu\right)\delta_x U_{N-1/2}^{j-1/2} + \left(\frac{4\sqrt{D}}{h_N\sqrt{\pi}} - \frac{2D\tilde{\mu}}{h_N}\right)U_N^{j-1/2} -$$
$$-\frac{4\sqrt{D}}{h_N\sqrt{\pi}}\sum_{k=1}^{j-1}(a_{j-k-1} - a_{j-k})U_N^{k-1/2}e^{\tilde{\lambda}\tau(j-k)} = f_{N-1/2}^{j-1/2}, \; j \ge 1, \tag{20}$$

where $\alpha_i = \frac{h_{i+1}}{2\hat{h}_i}$, $\beta_i = \frac{h_i}{2\hat{h}_i}$.

2.2 Analysis of the Difference Scheme

In order to analyze the convergence of difference scheme (P_1^h) we need the following lemma.

Lemma 2. *For any $F = \{F_0, F_1, F_2, \dots\}$, we have*

$$\sum_{l=1}^{j}\left[a_0 F_l e^{-\tilde{\lambda}t_l} - \sum_{k=1}^{l-1}(a_{l-k-1} - a_{l-k})F_k e^{-\tilde{\lambda}t_k}\right]F_l e^{\tilde{\lambda}t_l}$$
$$\ge \frac{e^{\tilde{\lambda}t_j}}{2\sqrt{2t_j}}\sum_{l=1}^{j}F_l^2, \quad j = 1, 2, 3, \dots,$$

where a_m is defined in (11).

The following theorem proves the convergence in energy norm of the derived difference scheme.

Theorem 1. *Let U_i^j and u_i^j, $\{0 \le i \le N, 0 \le j \le J = T/\tau\}$ be respectively the solution of the difference scheme (P_1^h) and the continuous problem (P_0) then for all $j = 0, 1, 2, \dots J$ hold the estimates*

$$\|u^j - U^j\|_A^2 + \frac{\tau D}{2}\sum_{l=1}^{j}\|\delta_x(u^{l-1/2} - U^{l-1/2})\|^2 \le C\left(N^{-2} + \tau^{3/2}\right)^2, \tag{21}$$

$$\|u^j - U^j\|_A^2 + \frac{\tau D}{2}\sum_{l=1}^{j}\|u^{l-1/2} - U^{l-1/2}\|_\infty^2 \le C\left(N^{-2} + \tau^{3/2}\right)^2. \tag{22}$$

Proof. Due to equivalency of problems (17)-(20) and (12)-(16), it is sufficient to consider the second one. Let U_i^j, V_i^j and u_i^j, v_i^j, $\{0 \le i \le N, 0 \le j \le J\}$ be respectively the solution of difference scheme (12)-(16) and the continuous problem (6)-(10). Let $\tilde{u}_i^j = U_i^j - u_i^j, \tilde{v}_i^j = V_i^j - v_i^j$. Using the Taylor expansion we have

$$\delta_t \tilde{u}_{i-1/2}^{j-1/2} - D\delta_x \tilde{v}_{i-1/2}^{j-1/2} - \nu \tilde{v}_{i-1/2}^{j-1/2} = P_{i-1/2}^{j-1/2}, \ 1 \le i \le N, \ j \ge 1, \tag{23}$$

$$\tilde{v}_{i-1/2}^{j-1/2} - \delta_x \tilde{u}_{i-1/2}^{j-1/2} = Q_{i-1/2}^{j-1/2}, \quad 1 \le i \le N, \ 0 \le j \le J, \tag{24}$$

$$\tilde{u}_i^0 = 0, \quad 0 \le i \le N, \tag{25}$$

$$\tilde{u}_0^j = 0, \quad 0 \le j \le J, \tag{26}$$

$$\tilde{v}_N^{j-1/2} - \tilde{\mu}\tilde{u}_N^{j-1/2} + \frac{2}{\sqrt{D\pi}}\Bigg[a_0\tilde{u}_N^{j-1/2} - \sum_{k=1}^{j-1}(a_{j-k-1} - a_{j-k})\tilde{u}_N^{k-1/2} e^{\tilde{\lambda}\tau(j-k)}\Bigg] = S^{j-1/2}, \ 0 \le j \le J, \tag{27}$$

where

$$P_{i-1/2}^{j-1/2} = O(\tau^2 + h_i^2), \ Q_{i-1/2}^{j-1/2} = O(\tau^2 + h_i^2), \ S^{j-1/2} = O\left(\tau^{3/2}\right), \quad 1 \le i \le N, \quad j \ge 1. \tag{28}$$

Multiplying (23) by $2h_i \tilde{u}_{i-1/2}^{j-1/2}$, (24) by $2Dh_i \tilde{v}_{i-1/2}^{j-1/2}$, summing over $i = 1, \ldots, N$, and taking into account (25)-(27), and Lemma 2, similarly to [9], we get the following a priory estimates

$$\frac{1}{\tau}\left(\|\tilde{u}^j\|_A^2 - \|\tilde{u}^{j-1}\|_A^2\right) + \frac{D}{2}\|\delta_x \tilde{u}^{j-1}\|_A^2 \le \frac{4}{D}\|P^{j-1/2}\|_A^2 + \frac{7D^2 + 8\nu}{2D}\|Q^{j-1/2}\|_A^2$$
$$+ 2D\tilde{u}_N^{j-1/2} S^{j-1/2} - \frac{4}{\sqrt{D\pi}}\left[a_0\tilde{u}_N^{j-1/2} - \sum_{k=1}^{j-1}(a_{j-k-1} - a_{j-k})\tilde{u}_N^{k-1/2} e^{\tilde{\lambda}\tau(j-k)}\right].$$

Multiplying the above inequality by τ, summing up for j, then noticing (25) and using Lemma 2, we have

$$\|\tilde{u}^j\|_A^2 + \frac{\tau D}{2}\sum_{l=1}^{j}\|\delta_x \tilde{u}^{j-1/2}\|^2 \le$$
$$\tau \sum_{l=1}^{j}\left(\frac{4}{D}\|P^{l-1/2}\|^2 + \frac{7D^2 + 8\nu}{2D}\|Q^{l-1/2}\|^2 + \frac{\sqrt{D\pi}}{\sqrt{2}e^{\tilde{\lambda}t_j}}\left(S^{l-1/2}\right)^2\right). \tag{29}$$

Now using (28),(29) we obtain (21). The estimate (22) follows from (21) and the discrete Sobolev inequality $\|u^{j-1/2}\|_\infty^2 \le \|\delta_x u^{j-1/2}\|^2$. □

2.3 Second Order Discretization in Space Combined with Crank-Nicolson Scheme

We can use the Crank-Nicolson discretization to obtain scheme that preserves $O(\tau^{3/2})$ order of convergence in time. To get second order approximation in space we shall use the monotone Samarskii scheme, see [4]

$$\delta_t U_i^{j-1/2} - \delta_{\hat{x}}\left(D\kappa_i \delta_x U_{i-1/2}^{j-1/2}\right) - \nu\delta_{\hat{x}} U_{i-1}^{j-1/2} = f_i^{j-1/2},\ 1 \le i \le N-1,\ j \ge 1, \quad (30)$$

where

$$\kappa_i = (1 + R_i)^{-1},\ R_i = h_i\nu/(2D),$$

and central difference scheme

$$\delta_t U_i^{j-1/2} - D\delta_x^2 U_i^{j-1/2} - \nu\delta_{\overset{\circ}{x}} U_i^{j-1/2} = f_i^{j-1/2},\ 1 \le i \le N-1,\ j \ge 1. \quad (31)$$

Using Lemma 1 at $x = 0$ we obtain the discretization

$$\delta_t U_N^{j-1/2} + \left(\frac{2D}{h_N} - \nu\right)\delta_x U_{N-1/2}^{j-1/2} + \left(\frac{4\sqrt{D}}{h_N\sqrt{\pi}} - \frac{2D\tilde{\mu}}{h_N}\right) U_N^{j-1/2}$$
$$-\frac{4\sqrt{D}}{h_N\sqrt{\pi}} \sum_{k=1}^{j-1} (a_{j-k-1} - a_{j-k}) U_N^{k-1/2} e^{\tilde{\lambda}\tau(j-k)} = f_N^{j-1/2},\ j \ge 1. \quad (32)$$

We denote by (P_2^h) the approximation: (18),(19),(30),(32), and by (P_3^h) the approximation: (18),(19),(31),(32). The scheme (P_2^h) is monotone when $\nu \le 2D/h_N$. It can be shown that both schemes preserve $O\left(\tau^{3/2} + N^{-2}\right)$ order of convergence. For problems with large diffusion coefficient D the schemes (P_1^h) and (P_3^h) generally give higher accuracy compare to the monotone Samarskii scheme (P_2^h). For convection-dominated problems, the schemes (P_1^h) and (P_3^h) produce an oscillations on uniform meshes and the monotone scheme (P_2^h) is preferable, but if special meshes are used with condensed near to boundary layers, then the schemes (P_1^h) and (P_3^h) could also be applied (see [3] for analysis of central difference scheme on layer adapted meshes).

3 Numerical Results

In order to demonstrate the effectiveness of the finite difference method using the artificial boundary conditions we consider the following problem

$$\frac{\partial u}{\partial t} - \frac{\partial^2 u}{\partial x^2} - \frac{\partial u}{\partial x} = 0,\ (x,t) \in \Omega, \quad u(x,0) = 0,\ -1 \le x < \infty,$$
$$u(-1,t) = \exp\left(-\frac{t}{4}\right) erfc\left(\frac{1}{2\sqrt{t}}\right),\ 0 \le t < \infty;\ u \to 0 \text{ when } x \to \infty,$$

where $erfc(x) = \frac{2}{\sqrt{\pi}}\int_x^\infty e^{-\lambda^2} d\lambda$. We know that the exact solution of the above problem is $u(x,t) = \exp\left(-(2(1+x)+t)/4\right) erfc\left((x+2)/(2\sqrt{t})\right)$.

Table 1. Error of the solution in maximum norm, $h = \tau$

Scheme\\N	$N = 8$	$N = 16$	$N = 32$	$N = 64$	$N = 128$	$N = 256$	$N = 512$	$N = 1024$
P_1^h	3.14e-3	7.54e-4	2.51e-4	5.42e-5	1.35e-5	5.08e-6	1.87e-6	6.83e-7
ρ_N	2.06	1.59	2.21	2.01	1.41	1.44	1.46	-
P_2^h	3.75e-3	7.83e-4	2.64e-4	5.79e-5	1.62e-5	5.76e-6	2.05e-6	7.27e-7
ρ_N	2.26	1.57	2.19	1.84	1.49	1.49	1.49	-
P_3^h	3.72e-3	7.82e-4	2.64e-4	5.78e-5	1.59e-5	5.70e-6	2.03e-6	7.23e-7
ρ_N	2.25	1.57	2.19	1.86	1.48	1.49	1.49	-

Table 2. Error of the solution in energy norm, $h = \tau$

Scheme\\N	$N = 8$	$N = 16$	$N = 32$	$N = 64$	$N = 128$	$N = 256$	$N = 512$	$N = 1024$
P_1^h	1.78e-3	3.20e-4	8.58e-5	2.15e-5	5.39e-6	2.21e-6	8.64e-7	3.27e-7
ρ_N	2.47	1.90	1.99	2.00	1.28	1.36	1.40	-
P_2^h	2.12e-3	3.55e-4	9.69e-5	2.40e-5	7.84e-6	2.84e-6	1.02e-6	3.66e-7
ρ_N	2.58	1.87	2.01	1.61	1.47	1.47	1.48	-
P_3^h	2.10e-3	3.54e-4	9.66e-5	2.40e-5	7.07e-5	2.65e-6	9.75e-7	3.55e-7
ρ_N	2.57	1.87	2.01	1.76	1.42	1.44	1.46	-

Tables 1 and 2 presents the global error correspondingly in maximum and energy norm when $h = \tau$, i.e. $M = N$. For large N we observe $O(h^{3/2})$ order of convergence in each norm. The convergence rate is taken to be

$$\rho_N = \log_2 (\|E_N\| / \|E_{2N}\|),$$

where $\|E_N\|$ is the error norm for the corresponding value of N. The results for the scheme (P_1^h) are slightly better than the ones for the schemes (P_2^h) and (P_3^h). Table 3 presents the global energy norm and maximum norms for fixed small $\tau = 0.00005$ and final time $T = 0.1$. The results in the table confirmed the second order of convergence with respect to the space variable. Thus the numerical results confirm the theoretical ones and show the effectiveness of the constructed scheme.

Table 3. Error of the solution in energy norm, $\tau = 0.00005$

Scheme\\N	$N = 8$	$N = 16$	$N = 32$	$N = 64$	$N = 128$	$N = 256$	$N = 512$	$N = 1024$
P_1^h	2.52e-4	6.39e-5	1.60e-5	4.01e-6	1.00e-6	2.51e-7	6.25e-8	1.55e-8
ρ_N	1.98	1.99	2.00	2.00	2.00	2.00	2.01	-
P_2^h	9.61e-5	2.55e-5	6.48e-6	1.63e-6	4.07e-7	1.02e-7	2.57e-8	6.58e-9
ρ_N	1.91	1.98	1.99	2.00	2.00	1.99	1.96	-
P_3^h	8.75e-5	2.31e-5	5.86e-6	1.47e-6	3.68e-7	9.22e-8	2.32e-8	5.98e-9
ρ_N	1.92	1.98	1.99	2.00	2.00	1.99	1.96	-

4 Conclusions

An exact artificial boundary condition is applied to reduce the unbounded domain problem into a finite domain problem. The method of reduction of order is used to derive a difference scheme for the reduced problem. It is proved that the difference scheme is unconditionally stable in energy norm and convergent with second order of convergence in space and 3/2 order of convergence in time.

Acknowledgments

This research is supported by Bulgarian National Fund of Science under Project HS-MI-106/2005.

References

1. Givoli, D.: Numerical methods for problems in infinite domains. Elsevier, Amsterdam (1992)
2. Han, H.D., Huang, Z.Y.: A class of artificial boundary conditions for heat equation in unbounded domains. Comput. Math. Appl. **43** (2002) 889–900.
3. Kopteva, N,V.: On the convergence, uniform with respect to the small parameter, of a scheme with central difference on refined grids, Comput. Math. Math. Phys. **39** No. 10 (1999) 1594–1610.
4. Samarskii, A.A.: Theory of difference schemes. Nauka, Moscow, (1983) (in Russian)
5. Sun, Z.-Z.: A new class of difference schemes for linear parabolic equations in 1-D. Chinese J. Numer. Math. Appl. **16** (3), 1–20 (1995)
6. Sun, Z.-Z.: A second order linearized difference scheme for the two dimensional Cahn-Hilliard equation. Math. Comp. **64**, 1463–1471 (1995)
7. Tichonov, A.N., Samarskii, A.A.: Equation of the mathematical physics, Nauka, Moscow 1972, (in Russian)
8. Vajzavelu, K., Rollins, D.: On solutions of some unsteday flows over continuous moving porous flat surface. J. Math. Anal. Appl. **153**, 52–53 (1990)
9. Wu, X., Sun, Z.-Z.: Convergence of difference scheme for heat equation in unbounded domains using artificial boundary conditions, Appl. Numer. Math. 50, 261–277 (2004).

On the Convergence of an Inexact Primal-Dual Interior Point Method for Linear Programming

Venansius Baryamureeba and Trond Steihaug

Department of Informatics, University of Bergen,
5020 Bergen, Norway
barya@ii.uib.no, trond@ii.uib.no
http://www.ii.uib.no/~barya
http://www.ii.uib.no/~trond

Abstract. The inexact primal-dual interior point method which is discussed in this paper chooses a new iterate along an approximation to the Newton direction. The method is the Kojima, Megiddo, and Mizuno globally convergent infeasible interior point algorithm. The inexact variation is shown to have the same convergence properties accepting a residue in both the primal and dual Newton step equation also for feasible iterates.

1 Introduction

For the standard primal-dual linear programming problem the optimality conditions are the Karush-Kuhn-Tucker (KKT) conditions:

$$F(x,y,z) \equiv \begin{pmatrix} Ax-b \\ A^Ty+z-c \\ XZe \end{pmatrix} = 0, \quad x \geq 0,\ z \geq 0, \tag{1}$$

where A is an m-by-n matrix of full rank m, b an m-vector, c an n-vector,z an n-vector, $X = \mathrm{diag}(x)$, $Z = \mathrm{diag}(z)$ and e is the vector of all ones in $\Re^n$.

Bellavia [1] proved global convergence of an inexact interior point method. Ito, Kelly, and Sachs [7] and Ito [6] discuss an inexact primal-dual interior point iteration for linear programs in functional spaces. Mizuno and Jarre [9] proved global and polynomial-time convergence of an infeasible interior point algorithm using inexact computation. Portugal *et al.* [11] presented a truncated primal-infeasible dual-feasible interior point algorithm for linear programming. Portugal *et al.* [10] presented a truncated primal-infeasible dual-feasible interior point algorithm for solving monotone linear complementarity problems. The methods suggested in [7, 10, 11] have a major drawback of remaining primal-feasible once they become primal-feasible.

Throughout this paper we use the following notation: For any vector x, x^k denotes x at the k-th (interior point) iteration. Similary for any matrix X or real number η, X^k and η^k denotes X and η at the k-th iteration. We have adopted the notation $(u,v,w) = (u^T, v^T, w^T)^T$, so for any vectors $x \in \Re^n, y \in \Re^m, (x,y) \in \Re^{m+n}$.

I. Lirkov, S. Margenov, and J. Waśniewski (Eds.): LSSC 2005, LNCS 3743, pp. 629–637, 2006.

2 Inexact Computation

The perturbed KKT conditions for (1) with a positive μ is the nonlinear system

$$F_\mu(x,y,z) \equiv \begin{pmatrix} Ax-b \\ A^Ty+z-c \\ XZe-\mu e \end{pmatrix} = 0, \quad x \geq 0,\ z \geq 0. \tag{2}$$

The parameter μ is referred to as the μ-complementarity parameter. Newton's method defines the equation of directional change (the Newton step equation)

$$F'_{\mu^k}(x^k,y^k,z^k) \begin{pmatrix} \Delta x^k \\ \Delta y^k \\ \Delta z^k \end{pmatrix} = -F_{\mu^k}(x^k,y^k,z^k). \tag{3}$$

If the linear system of equation is solved approximately, then equation (3) has a residual error r^k given by

$$F'_{\mu^k}(x^k,y^k,z^k) \begin{pmatrix} \Delta x^k \\ \Delta y^k \\ \Delta z^k \end{pmatrix} = -F_{\mu^k}(x^k,y^k,z^k) + r^k. \tag{4}$$

The residual will be partioned into the primal infeasibility $\bar{r}^k$, dual infeasibility $\hat{r}^k$, and deviation in complementarity $\tilde{r}^k$, $r^k = (\bar{r}^k, \hat{r}^k, \tilde{r}^k)$. Since $F' = F'_\mu$ and $F_\mu = F - \mu^k(0,0,e)$ an inexact Newton step of (4) is an approximate solution of the Newton step equation derived from (1)

$$F'(x^k,y^k,z^k) \begin{pmatrix} \Delta x^k \\ \Delta y^k \\ \Delta z^k \end{pmatrix} = -F(x^k,y^k,z^k) + r_g^k, \text{ where } r_g^k = \mu^k \begin{pmatrix} 0 \\ 0 \\ e \end{pmatrix} + r^k. \tag{5}$$

For the centering parameter β_1, typical choice of the μ-complementarity parameter at iteration k is $\mu_k = \beta_1 \frac{(x^k)^T z^k}{n}$, where $\beta_1 \in [0,1]$. Bellavia [1] observed that $\|F(x^k,y^k,z^k)\|_2 \geq \frac{(x^k)^T z^k}{\sqrt{n}}$. If $\|r^k\|_2 \leq \eta^k (x^k)^T z^k$ then

$$\|r_g^k\|_2 \leq \mu^k \sqrt{n} + \|r^k\|_2 \leq \frac{(x^k)^T z^k}{\sqrt{n}} (\beta_1 + \eta^k \sqrt{n}) \leq (\beta_1 + \eta^k \sqrt{n})\, \|F(x^k,y^k,z^k)\|_2.$$

Hence the sequence $\{\beta_1 + \eta^k \sqrt{n}\}$ can be regarded as a forcing sequence of inexact Newton methods [3] applied to the nonlinear system (1). Let

$$\|r^k\|_2 \leq \eta^k (x^k)^T z^k \ \text{ for } \ \eta^k < (1-\beta_1)/\sqrt{n}. \tag{6}$$

Since (6) is overly restrictive, we will later show that solving the linear system (4) with an accuracy $\|r^k\|_1 \leq \eta^k (x^k)^T z^k$ for $0 \leq \eta^k < 1 - \beta_1$ will be sufficient to achieve convergence. Bellavia [1] establishes global convergence results for an inexact interior point method by interpreting it as an inexact Newton method using the global inexact Newton method of Eisenstat and Walker [4]. The convergence theory and implementation [1, 2] is based on solving the inexact Newton equation (5) with a forcing sequence where $\eta^k < (1-\beta_1)/\sqrt{n}$. In this paper, we prove global convergence results based on the reduced linear systems (the augmented systems or the normal equations), and we bound the residual $r^k = (\bar{r}^k, \hat{r}^k, 0)$ by

$$\|r^k\| \le \eta^k (x^k)^T z^k \quad \text{for } 0 \le \eta^k < 1 \tag{7}$$

for suitable choices of the input parameters of the algorithm. The norm $\|\cdot\|$ can be any l_p norm or any norm in appropriate spaces that satisfies

$$\max\{\|\bar{r}^k\|, \|\hat{r}^k\|\} \le \|(\bar{r}^k, \hat{r}^k)\|. \tag{8}$$

It is evident that termination criterion (6) may be more demanding to satisfy than termination criterion (7). To simplify the notation, we follow [7] and introduce the residuals $\xi^k = b - Ax^k$ and $\zeta^k = c - A^T y^k - z^k$. Let G^k be defined by $G^k = (Z^k)^{-1} X^k$. Eliminating Δz^k in (3) leads to an indefinite augmented system

$$\begin{bmatrix} O & A \\ A^T & -(G^k)^{-1} \end{bmatrix} \begin{pmatrix} \Delta y^k \\ \Delta x^k \end{pmatrix} = \begin{pmatrix} \xi^k \\ z^k + \zeta^k - (X^k)^{-1} \mu^k e \end{pmatrix}, \text{ and} \tag{9}$$

$$\Delta z^k = (X^k)^{-1} (\mu^k e - Z^k \Delta x^k) - z^k. \tag{10}$$

An inexact solution of (9) satisfies

$$\begin{bmatrix} O & A \\ A^T & -(G^k)^{-1} \end{bmatrix} \begin{pmatrix} \Delta y^k \\ \Delta x^k \end{pmatrix} = \begin{pmatrix} \xi^k \\ z^k + \zeta^k - (X^k)^{-1} \mu^k e \end{pmatrix} + \begin{pmatrix} \bar{r}^k \\ \hat{r}^k \end{pmatrix}. \tag{11}$$

After computing $(\Delta y^k, \Delta x^k)$ Δz^k can be found from (10). The approximate step $(\Delta y^k, \Delta x^k, \Delta z^k)$ is thus an inexact Newton step of (4) with $r^k = (\bar{r}^k, \hat{r}^k, 0)$. Eliminating Δx^k from (9) gives the normal equations

$$AG^k A^T \Delta y^k = AG^k (z^k - (X^k)^{-1} \mu^k e + \zeta^k) + \xi^k. \tag{12}$$

An approximate solution of (12) satisfies

$$AG^k A^T \Delta y^k = AG^k (z^k - (X^k)^{-1} \mu^k e + \zeta^k) + \xi^k + \bar{r}^k. \tag{13}$$

After computing Δy^k then Δx^k and Δz^k are computed from

$$\Delta x^k = -G^k (c - A^T (y^k + \Delta y^k) - \mu^k (X^k)^{-1} e) \tag{14}$$

$$\Delta z^k = (X^k)^{-1} (\mu^k e - Z^k \Delta x^k) - z^k. \tag{15}$$

Satisfying the termination criterion used in [6, 7, 10, 11] for the iterative linear system solver is computationally expensive once the iterates become almost primal feasible or primal-dual feasible. Freund, Jarre and Mizuno [5] suggest to control the errors

$$\|\bar{r}^k\| \le \epsilon_p^k \text{ and } \|\hat{r}^k\| \le \epsilon_d^k \tag{16}$$

which may avoid the problem with exact solution when the iterates are feasible. The variables ϵ_p^k and ϵ_d^k are updated at every iteration k, but are not known a priori.These variables go to zero faster than $\|\xi^k\|$ and $\|\zeta^k\|$ respectively and thus requiring high accuracy close to a solution. Mizuno and Jarre [9] control the error $\|(\bar{r}^k, \hat{r}^k)\|$ using a semi norm. This norm is not computable, but accuracy requirement is shown to be slightly weaker than the relative error with respect to the right hand side in a modified linear system. This system requires a **QR** factorization of the matrix A and the accuracy requirement is thus not suitable for computation.

3 The Kojima-Megiddo-Mizuno Algorithm

The algorithm we discuss in this paper is a variant of the infeasible primal-dual interior point algorithm by Kojima, Megiddo, and Mizuno [8]. For any given accuracy $\epsilon > 0$ required for the total complementarity, any tolerance $\epsilon_p > 0$ for the primal feasibility, any tolerance $\epsilon_d > 0$ for the dual feasibility define

$$
\begin{aligned}
\mathcal{N} = \{(x,y,z) \in \mathcal{Q} :& \\
& x_j z_j \geq \gamma x^T z/n \text{ for } j = 1, 2, \ldots, n, && (17)\\
& x^T z \geq \gamma_p \|Ax - b\| \text{ or } \|Ax - b\| \leq \epsilon_p, && (18)\\
& x^T z \geq \gamma_d \|A^T y + z - c\| \text{ or } \|A^T y + z - c\| \leq \epsilon_d\} && (19)
\end{aligned}
$$

where $\mathcal{Q} = \{(x,y,z) \in \Re^{n+m+n} : x > 0, z > 0\}$. The constants $0 < \gamma < 1$, $\gamma_p > 0$, $\gamma_d > 0$ will in a weak sense depend on the starting point, but will be chosen so that the neighborhood $\mathcal{N}$ is as large as possible. Further, let ω^* be any large number and $\epsilon^* = \min\{\epsilon, \gamma_p\epsilon_p, \gamma_d\epsilon_d\}$. Then the neighborhood [8]

$$\mathcal{N}^* = \{(x,y,z) \in \mathcal{N} : \epsilon^* \leq x^T z^k \leq \omega^*\} \tag{20}$$

is a compact set. We will show that either $(x^k, y^k, z^k) \in \mathcal{N}^*$ or satisfies the *termination criteria*

$$\left(x^k\right)^T z^k \leq \epsilon, \|Ax^k - b\| \leq \epsilon_p \text{ and } \|A^T y^k + z^k - c\| \leq \epsilon_d \tag{21}$$

or

$$(x^k)^T z^k > \omega^* \tag{22}$$

after a finite number of steps. We choose any initial point $(x^1, y^1, z^1) \in \mathcal{Q}$ and parameters $\gamma, \gamma_p, \gamma_d$ and ω^* so that $(x^1, y^1, z^1) \in \mathcal{N}$ and $(x^1)^T z^1 \leq \omega^*$. Let $0 < \beta_1 < \beta_2 < \beta_3 < 1$. We state the algorithm below:

Algorithm 1. *Inexact Infeasible Primal-Dual Algorithm*
Step 1. *Set* $k = 1$. *Assume* $(x^1, y^1, z^1) \in \mathcal{N}$ *such that* $(x^1)^T z^1 \leq \omega^*$.
Step 2. *If (21) or (22) is satisfied then terminate.*
Step 3. *Let* $\mu^k = \beta_1 (x^k)^T z^k / n$.

Step 4. *Compute the inexact solution* $(\Delta x^k, \Delta y^k, \Delta z^k)$ *of (3).*
Step 5. *Let* $0 < \bar{\alpha}^k < 1$ *such that*

$$(x^k, y^k, z^k) + \alpha(\Delta x^k, \Delta y^k, \Delta z^k) \in \mathcal{N} \tag{23}$$

$$(x^k + \alpha \Delta x^k)^T (z^k + \alpha \Delta z^k) \le (1 - \alpha(1 - \beta_2))(x^k)^T z^k, \tag{24}$$

hold for every $\alpha \in (0, \bar{\alpha}^k]$.
Step 6. *Choose a primal step length* $\alpha_p^k \in [\bar{\alpha}^k, 1]$, *a dual step length* $\alpha_d^k \in [\bar{\alpha}^k, 1]$ *and a new iterate*

$$(x^{k+1}, y^{k+1}, z^{k+1}) = (x^k + \alpha_p^k \Delta x^k, y^k + \alpha_d^k \Delta y^k, z^k + \alpha_d^k \Delta z^k) \tag{25}$$

such that

$$(x^{k+1}, y^{k+1}, z^{k+1}) \in \mathcal{N}, \tag{26}$$

$$(x^{k+1})^T z^{k+1} \le (1 - \bar{\alpha}^k(1 - \beta_3))(x^k)^T z^k. \tag{27}$$

Step 7. *Increase k by 1. Go to Step 2.*

In this algorithm we take relatively short steps when the search directions are computed to a relatively low accuracy. In Section 4 we will show the existence of an $\bar{\alpha}^k > 0$ for all k as long as (21) is not satisfied.

4 Convergence

For $(x^k, y^k, z^k) \in \mathcal{N}$ the following inequalities will be used in the discussion of the algorithm.

$$(x^k)^T z^k \ge \gamma_p \|Ax^k - b\| \text{ or } \|Ax^k - b\| \le \epsilon_p \tag{28}$$

$$(x^k)^T z^k \ge \gamma_d \|A^T y^k + z^k - c\| \text{ or } \|A^T y^k + z^k - c\| \le \epsilon_d \tag{29}$$

$$(x^k)^T \Delta z^k + (\Delta x^k)^T z^k = -(1 - \beta_1)(x^k)^T z^k \tag{30}$$

$$x_i^k \Delta z_i^k + \Delta x_i^k z_i^k = \beta_1 (x^k)^T z^k / n - x_i^k z_i^k \tag{31}$$

$$\text{Let } 0 \le \eta_{\max} < \min\left\{\frac{\beta_1}{\max\{\gamma_p, \gamma_d\}}, 1\right\}. \tag{32}$$

The neighborhood $\mathcal{N}$ is made large by making γ_p and γ_d small which implies that in most cases $\eta_{\max} < 1$. The inexact Newton direction $(\Delta x^k, \Delta y^k, \Delta z^k)$ computed in (11) and (10) satisfies

$$\begin{bmatrix} A & 0 & 0 \\ 0 & A^T & I \\ Z^k & 0 & X^k \end{bmatrix} \begin{pmatrix} \Delta x^k \\ \Delta y^k \\ \Delta z^k \end{pmatrix} = \begin{pmatrix} b - Ax^k \\ c - A^T y^k - z^k \\ \mu^k e - X^k Z^k e \end{pmatrix} + \begin{pmatrix} \bar{r}^k \\ \hat{r}^k \\ 0 \end{pmatrix} \tag{33}$$

where the residual satsifies

$$\|(\bar{r}^k, \hat{r}^k)\| \leq \eta^k \left(x^k\right)^T z^k \ \text{ for } \ 0 \leq \eta^k \leq \eta_{\max}. \tag{34}$$

Note that $\eta^k (x^k)^T z^k = \frac{\eta^k n}{\beta_1} \mu^k$. The coefficient matrix F'_μ on the left hand side of (33) is nonsingular and continuous for $(x, y, z) \in \mathcal{N}^*$ defined in (20). Since $\|(\bar{r}^k, \hat{r}^k)\|$ in (34) is bounded for $(x^k, y^k, z^k) \in \mathcal{N}^*$ and $\eta^k \leq \eta_{\max}$ the inexact Newton direction $(\Delta x^k, \Delta y^k, \Delta z^k)$ determined in (33) of equations is well defined (in the sence that for a given residual solution of (33) is unique) and is bounded over the compact set $\mathcal{N}^*$. Hence there exists a positive constant τ such that the inexact Newton direction $(\Delta x^k, \Delta y^k, \Delta z^k)$ from (33) computed at Step 4 of every interior point iteration satisfies the inequalities

$$|\Delta x_i^k \Delta z_i^k - \gamma (\Delta x^k)^T \Delta z^k / n| \leq \tau \text{ and } |(\Delta x^k)^T \Delta z^k| \leq \tau. \tag{35}$$

We will now show that for an inexact Newton direction from Step 4 in Algorithm 1 we can find $0 < \bar{\alpha}^k < 1$ so that the conditions (23) and (24) in Step 5 are satisfied. It will be shown that for a given $\mathcal{N}^*$ there exists a $0 < \alpha^*$ so that $\alpha^* \leq \bar{\alpha}^k$. Let $\xi(\alpha)$ and $\zeta(\alpha)$ be defined by

$$\xi(\alpha) \equiv b - A(x^k + \alpha \Delta x^k) = \xi^k - \alpha A \Delta x^k \tag{36}$$

$$\zeta(\alpha) \equiv c - A^T(y^k + \alpha \Delta y^k) - z^k = \zeta^k - \alpha (A^T \Delta y^k + \Delta z^k). \tag{37}$$

For Δy^k and Δx^k given by (11), and Δz^k by (10) it follows from (33) that the expressions for $\xi(\alpha)$ and $\zeta(\alpha)$ simplify to

$$\xi(\alpha) = (1 - \alpha)\xi^k - \alpha \bar{r}^k \tag{38}$$

$$\zeta(\alpha) = (1 - \alpha)\zeta^k - \alpha \hat{r}^k. \tag{39}$$

Define the real-valued functions $f_i, i = 1, 2, \ldots, n$, g_p, g_d, and h as follows:

$$\begin{aligned}
f_i(\alpha) &= (x_i^k + \alpha \Delta x_i^k)(z_i^k + \alpha \Delta z_i^k) - \gamma (x^k + \alpha \Delta x^k)^T (z^k + \alpha \Delta z^k)/n \\
g_p(\alpha) &= (x^k + \alpha \Delta x^k)^T (z^k + \alpha \Delta z^k) - \gamma_p \|\xi(\alpha)\| \\
g_d(\alpha) &= (x^k + \alpha \Delta x^k)^T (z^k + \alpha \Delta z^k) - \gamma_d \|\zeta(\alpha)\| \\
h(\alpha) &= (1 - \alpha(1 - \beta_2))\left(x^k\right)^T z^k - (x^k + \alpha \Delta x^k)^T (z^k + \alpha \Delta z^k).
\end{aligned}$$

Consider Step 5 in the Inexact Infeasible Primal-Dual Algorithm. From the definition of the neighborhood $\mathcal{N}$ condition (23) is equivalent to: $f_i(\alpha) \geq 0$ $(i = 1, 2, \ldots, n)$; $g_p(\alpha) \geq 0$ or $\|\xi(\alpha)\| \leq \epsilon_p$; $g_d(\alpha) \geq 0$ or $\|\zeta(\alpha)\| \leq \epsilon_d$ for $0 < \alpha \leq \bar{\alpha}^k$. Similarly, (24) is equivalent to $h(\alpha) \geq 0$ for $0 < \alpha \leq \bar{\alpha}^k$.

Consider the function g_p and the simplified expression for ξ in (38). Then

$$g_p(\alpha) \geq \alpha \beta_1 (x^k)^T z^k + \alpha^2 (\Delta x^k)^T \Delta z^k - \gamma_p \alpha \|\bar{r}^k\| \tag{40}$$

using (18), and (30). By (34) and the property of the norm (8)

$$\|\bar{r}^k\| \leq \|(\bar{r}^k, \hat{r}^k)\| \leq \eta^k \, (x^k)^T z^k, \tag{41}$$

$$g_p(\alpha) \geq \alpha\,(\beta_1 - \eta_{\max}\,\gamma_p)\,(x^k)^T z^k + \alpha^2(\Delta x^k)^T \Delta z^k. \text{ for } \eta^k \leq \eta_{\max}$$

For $(x^k, y^k, z^k) \in \mathcal{N}^*$ and (35) $(x^k)^T z^k > \epsilon^*$ and $(\Delta x^k)^T \Delta z^k \geq \tau$

$$g_p(\alpha) \geq \alpha\,[(\beta_1 - \eta_{\max}\,\gamma_p)\,\epsilon^* - \alpha\tau]$$

Hence if $0 \leq \alpha \leq \frac{(\beta_1 - \eta_{\max}\,\gamma_p)\epsilon^*}{\tau}$ and $g_p(0) \geq$ then $g_p(\alpha) \geq 0$. Consider $g_p(0) < 0$. By definition of g_p this is equivalent to $\gamma_p\|Ax^k - b\| > (x^k)^T z^k$ and from (38)

$$\|A(x^k + \alpha\Delta x^k) - b\| \leq (1 - \alpha(1 - \beta_1))\|Ax^k - b\| \leq \|Ax^k - b\|$$

for the choice of η^k in (32). From the definition of $\mathcal{N}$ and Step 5 of Algorithm 1 we know the iterate (x^k, y^k, z^k) generated satisfies $(x^k, y^k, z^k) \in \mathcal{N}$ for all k. Thus $g_p(0) < 0$ implies that $\|Ax^k - b\| \leq \epsilon_p$. From (42) it follows that $\|A(x^k + \alpha\Delta x^k) - b\| \leq \epsilon_p$ if $g_p(0) < 0$ for $\alpha \leq 1$. Similarly, $\|\bar{r}^k\| \leq \|(\bar{r}^k, \hat{r}^k)\| \leq \eta^k (x^k)^k z^k$ and

$$g_d(\alpha) \geq \alpha(\beta_1 - \eta_{max}\gamma_d)(x^k)^T z^k + \alpha^2(\Delta x^k)^T \Delta z^k \geq \alpha\,[(\beta_1 - \eta_{max}\gamma_d)\epsilon^* - \alpha\tau]\,.$$

Hence, if $0 \leq \alpha \leq \frac{(\beta_1 - \eta_{max}\gamma_d)\epsilon^*}{\tau}$, and $g_d(0) \geq 0$ then $g_d(\alpha) \geq 0$. Next consider

$$f_i(\alpha) \geq \beta_1(1-\gamma)(\epsilon^*/n)\alpha - \tau\alpha^2 = \alpha[\beta_1(1-\gamma)(\epsilon^*/n) - \tau\alpha]. \quad (42)$$

Inequality (42) follows from application of (30), (31) and (35) on the expression for $f_i(\alpha)$ above. Finally, consider

$$h(\alpha) = \alpha(\beta_2 - \beta_1)(x^k)^T z^k + \alpha^2(\Delta x^k)^T \Delta z^k \geq \alpha[(\beta_2 - \beta_1)\epsilon^* - \alpha\tau]. \quad (43)$$

The inequality for $h(\alpha)$ (43) follows from application of (30) and (35) on $h(\alpha)$ above.

Let $\eta_{\max}$ be given by (32), $\bar{\gamma} = \max\{\gamma_p, \gamma_d\}$ and define

$$\alpha^* = \min\left\{1, \frac{(\beta_1 - \eta_{\max}\,\bar{\gamma})\epsilon^*}{\tau}, \frac{\beta_1(1-\gamma)\epsilon^*}{n\tau}, \frac{(\beta_2 - \beta_1)\epsilon^*}{\tau}\right\}. \quad (44)$$

For α^* defined in (44) we observe that

$$\begin{aligned} &f_i(\alpha) \geq 0 \; (i = 1, 2, \ldots, n) \\ &g_p(\alpha) \geq 0 \text{ if } g_p(0) = (x^k)^T z^k - \gamma_p\|Ax^k - b\| \geq 0 \\ &\|A(x^k + \alpha\Delta x^k) - b\| \leq \epsilon_p \text{ if } g_p(0) < 0 \end{aligned} \quad (45)$$

$$\begin{aligned} &g_d(\alpha) \geq 0 \text{ if } g_d(0) = (x^k)^T z^k - \gamma_d\|A^T y^k + z^k - c\| \geq 0 \\ &\|A^T(y^k + \alpha\Delta y^k) + (z^k + \alpha\Delta z^k) - c\| \leq \epsilon_d \text{ if } g_d(0) < 0 \\ &h(\alpha) \geq 0 \end{aligned} \quad (46)$$

hold for every $\alpha \in [0, \alpha^*]$. Next that $\eta^k\,\gamma_d < \beta_1$ for $\eta^k \leq \eta_{\max}$ and $\eta_{\max}$ defined by (32). By a similar argument as for (45), observation (46) follows.

Thus, we have shown that there exists an $\alpha^* > 0$ such that $\bar{\alpha}^k \geq \alpha^*$ in Algorithm 1 for pt all k. By the construction of the real-valued functions $f_i(i = 1, \ldots, n), g_p, g_d$, and h, this is equivalent to saying that (23) and (24) hold for every $\alpha \in (0, \alpha^*]$. Consider Step 6 in the Algorithm 1. Condition (26) and (27) are satisfied for a common primal and dual step length $\alpha_p^k = \alpha_p^k = \bar{\alpha}^k$ and since $\beta_2 < \beta_3$ (27) is satisfied with strict inequality. Further, note that the right hand side of (27) is constant with respect to α_p^k and α_d^k.

Theorem 1. *Define* $\psi^k = \max\{\|Ax^k - b\|, \|A^T y^k + z^k - c\|, (x^k)^T z^k\}$. *Let the norm of the residual* $(\bar{r}^k, \hat{r}^k)$ *in (11) satisfy* $\|(\bar{r}^k, \hat{r}^k)\| \leq \eta^k (x^k)^T z^k$, *for* $0 \leq \eta^k \leq \eta_{\max}$ *where* $\eta_{\max}$ *satisfy 32. Then there exist* $\tilde{\alpha} \in (0, 1)$ *so that*

$$\psi^{k+1} \leq (1 - \tilde{\alpha}\psi^k) \tag{47}$$

where $\tilde{\alpha} = \alpha^*(1 - \max\{\beta_3, \eta_{\max}\})$ *and* α^* *is given by (44).* ■

Theorem 1 implies that $\|\xi^k\| \to 0$, $\|\zeta^k\| \to 0$ and $(x^{k+1})^T z^{k+1} \to 0$ as $k \to \infty$. This contradicts assumption (20), i.e. $(x^k)^T z^k \geq \epsilon^*$ for all k. Thus as $k \to \infty$ there exists a k after which the algorithm generates a point $(x^{k+1}, y^{k+1}, z^{k+1})$ which is either an approximate optimal solution $(x^{k+1}, y^{k+1}, z^{k+1})$ satisfying (21) or satisfies (22). Therefore Algorithm 1 is globally convergent.

It follows from Theorem 1 that the sequences $\{\psi^k\}, \{(x^k)^T z^k\}$ and $\{\max\{\|Ax^k - b\|, \|A^T y^k + z^k - c\|\}\}$ are q−linearly convergent. For $u \in \Re^n$, $v \in \Re^n$ and $w \in \Re^n$ then $|||(u, v, w)||| = max\{\|u\|, \|v\|, \|w\|_1\}$ is a norm in $u \in \Re^{m+n+n}$. Theorem 1 shows that for the function F in (1) the KKT condition inequality (47) is equivalent to $|||F(x^{k+1}, y^{k+1}, z^{k+1})||| \leq (1 - \bar{\alpha})|||F(x^k, y^k, z^k)|||$. Classical stopping criterion [5] includes $\frac{\|Ax^k - b\|}{max\{1, \|b\|\}} + \frac{\|A^T y^k + z^k - c\|}{max\{1, \|c\|\}} + \frac{(x^k)^T z^k}{max\{1, \|b\| \|c\|\}} \leq \epsilon$.

References

1. Bellavia,S.: Inexact interior point method, Journal of Optimization Theory and Applications. **96** (1998) 109–121.
2. Bellavia,S.: Numerical performance of an inexact interior point method. R. De Leone et al. (eds), High Performance Algorithms and Software in Nonlinear Optimization. ©1998 Kluwer Academic Publishers, Boston. (1998) 43–51.
3. Dembo,R.S., Eisenstat,S.C., and Steihaug,T.: Inexact Newton Methods. SIAM J. Numer. Anal. **19** (1982) 400–408.
4. Eisenstat,S.c., and Walker,H.F.: Globally convergent inexact Newton methods, SIAM J. Optimization. **4** (1994) 393–422.
5. Freund, R.W., Jarre,F., Mizuno,S.: Convergence of a class of inexact interior point algorithms for linear programs. Math of Oper Research. **24** (1999) 50–71.
6. Ito,S.: Inexact implementation of interior point algorithms for optimal control problems. Lecture Notes in Num. Appl. Anal. **14** (1995) 245–248.
7. Ito,S., Kelley,C.T., and Sachs,E.W.: Inexact primal-dual interior-point iteration for linear programs in functional spaces. Computational Optimization and Applications. **4** (1995) 189–201.
8. Kojima,M., Megiddo,N., and Mizuno,S.: A primal-dual infeasible-interior point algorithm for linear programming, Math Programming. **61** (1993) 263–280.

9. Mizuno,S., and Jarre,F.: Global and polynomial-time convergence of an infeasible interior point algorithm using inexact computation. Mathematical Programming. **84** (1999) 105–122.
10. Portugal, L., Fernandes,L., and Judice,J.: A truncated Newton interior point algorithm for the solution of a multicommodity spatial equilibrium model. Complementarity and Variational Problems, edited by J.S. Pang and M. Ferris, SIAM. (1997) 315–344.
11. Portugal,L.F., Resende,M.G.C., Veiga, G., and Judice, J.: A truncated primal-infeasible dual-feasible interior point network flow method. Networks. **35** (2000) 91–108.

A Novel, Parallel PDE Solver for Unstructured Grids

Dulcenéia Becker and Chris Thompson

Cranfield University, Cranfield MK43 0AL, United Kingdom
{d.becker.2002, chris.thompson}@cranfield.ac.uk

Abstract. We propose a new parallel domain decomposition algorithm to solve symmetric linear systems of equations derived from the discretization of PDEs on general unstructured grids of triangles or tetrahedra. The algorithm is based on a single-level Schwarz alternating procedure and a modified conjugate gradient solver. A single layer of overlap has been adopted in order to simplify the data-structure and minimize the overhead. This approach makes the global convergence rate to vary slightly with the number of domains and the algorithm becomes highly scalable. The algorithm has been implemented in FORTRAN 90 using MPI and hence portable to different architectures. Numerical experiments have been carried out on a SUNFIRE 15K parallel computer and there have been shown superlinear performance in some cases.

1 Introduction

Domain decomposition methods refer to divide-and-conquer techniques to solve partial differential equations by iteratively solving subproblems defined on smaller subdomains. The earliest known iterative domain decomposition technique was proposed by H. A. Schwarz in 1870 to prove the existence of harmonic functions on irregular regions which are the union of overlapping subregions. Variants of Schwarz's method were later studied by several authors (see Chan *et al.* [2] for a survey). Domain decomposition methods have the capability of providing numerical solvers which are portable, efficient and highly parallelisable. An overview of the methods, their implementation, analysis and relation to multigrid methods may be found in the book by Smith *et al.* [12].

The convergence rate and, therefore, the overall parallel efficiency of single-level Additive Schwarz (AS) methods are often dependent on subdomain granularity. The number of iterations required for convergence tends to increase as the number of subdomains increases. The reason lies in the fact that the only means for communicating information between subdomains is through the overlap region.

A two-level method which employs a global coarse mesh solver to provide global communication is often used. For structured meshes, a grid hierarchy is naturally available. However, for an unstructured mesh the coarse grid may not be easy to find [3]. Thus, a procedure is needed that generates this grid, as well as the associated interpolation and restriction operators. Moreover, the amount

I. Lirkov, S. Margenov, and J. Waśniewski (Eds.): LSSC 2005, LNCS 3743, pp. 638–645, 2006.

of memory required to store a second, even coarser grid, is a disadvantage. Unlike recursive multilevel methods, a two-level Schwarz method may require too high resolution on the coarse grid, which makes it less scalable overall. Parallelizing the coarse grid solver is ultimately necessary for high performance [9].

The algorithm proposed in this paper is based on the additive form of a single-level Schwarz alternating procedure and a modified conjugate gradient (CG) solver. Preconditioned conjugate gradient methods are usually very efficient, albeit not always highly parallelisable. Their parallelism is mainly limited by the sequential component of the preconditioner. Schwarz methods are highly parallelisable and depend on a rapid local convergence to be effective. The main difficulty to use CG as the local solver on a Schwarz method is the perturbation on the right-hand-side of each subdomain system at each cycle, i.e. a new problem has to be solved at each iteration, slowing the local and, therefore, the overall convergence.

This paper is concerned mainly about a modified CG that computes each search direction from the previous one, except on the first cycle, even when the right-hand-side has been perturbed. We illustrate our algorithm with the Poisson equation since it is of wide application interest and is particularly demanding of parallel algorithms because the equations are especially simple.

2 Conjugate Gradient Method

The method of conjugate gradient, first introduced by Hestenes and Stiefel [7], is often the method of choice to solve systems of linear equations

$$Ax = b \tag{1}$$

where A is an $n \times n$ symmetric positive definite (SPD) sparse matrix. CG is remarkably fast when compared to other methods like steepest descent, does not require the specification of any parameters and, in the absence of round-off errors, will provide the solution in at most n steps.

The method may be described as Algorithm 1. For details about the derivation and convergence analysis see for instance [7, 6, 11] and [1]. The formula for β in Algorithm 1 is the well-known Fletcher-Reeves one [5]; for alternates β see for instance [10] and [4].

Algorithm 1. Conjugate Gradient

$r_0 = b - Ax_0$
$p_0 = r_0$
for i=0,1,... **do**
  $\alpha_i = (r_i, r_i)/(Ap_i, p_i)$
  $x_{i+1} = x_i + \alpha_i p_i$
  $r_{i+1} = r_i - \alpha_i Ap_i$
  $\beta_i = (r_{i+1}, r_{i+1})/(r_i, r_i)$
  $p_{i+1} = r_{i+1} + \beta_i p_i$
end for

3 Additive Schwarz Method

To obtain high parallelism, we use a Schwarz method which allocates one subdomain to each processor. We consider a single-level overlapping additive Schwarz method on matching grids. The method consists of partitioning the domain Ω into p overlapping subdomains Ω_i and approximate the solution by solving a sequence of boundary value problems in each subdomain.

3.1 Mesh Partitioning

The mesh has to be decomposed into p non-overlapping subdomains Ω_i so that the number of nodes assigned to each subdomain is almost the same and the number of adjacent elements assigned to different processor is minimized. The goals of the first and second conditions are, respectively, to balance the computation among processors and minimize communication. An overlap is introduced algebraically by enlarging these subsets to contain the vertices within one edge from the original subset. Note that even the initial domain partitioning is vertex-based; the final partitioning is element-based and overlapped.

Let us define $\partial\Omega$ as the boundary of the domain and Γ_i as the artificial boundary of each subdomain, i.e. the part of the boundary of Ω_i that is interior to Ω. If a node of Ω_i is on Γ_i it is called a *halo-node*, otherwise, it is called a *local-node*. Edges may have two local-nodes, one local- and one halo-node, or two halo-nodes. They are called *local-*, *halo-* and *remote-edges*, respectively.

3.2 Data Partitioning

The matrix A of (1) is row-wise partitioned, i.e. only element edges whose first node is local are kept. Assuming that n is the number of nodes of the mesh, p is the number of subdomains and n_i is the number of local nodes per subdomain, each subdomain is represented by an $n_i \times n$ matrix

$$A_i = [H_{i,1} \ \cdots \ H_{i,i-1} \ L_i \ H_{i,i+1} \ \cdots \ H_{i,p}],$$

where L_i is the matrix of local-edges and $H_{i,j}$, for $j = 1, \ldots, p$ and $j \neq i$, are matrices of halo-edges from neighbour j. If j is not a neighbour of i then $H_{i,j} = 0$. Note that each subdomain has all columns of A and there are no replicated data. Vectors contain only the elements corresponding to local-nodes, i.e. they are of size n_i.

3.3 Algorithm Description

The overall system is written as

$$A = \begin{bmatrix} \mathrm{A}_1 \\ \mathrm{A}_2 \\ \vdots \\ \mathrm{A}_p \end{bmatrix} = \begin{bmatrix} \mathrm{L}_1 & \mathrm{H}_{1,2} & \cdots & \mathrm{H}_{1,p} \\ \mathrm{H}_{2,1} & \mathrm{L}_2 & \cdots & \mathrm{H}_{2,p} \\ \vdots & \vdots & \ddots & \vdots \\ \mathrm{H}_{p,1} & \mathrm{H}_{p,2} & \cdots & \mathrm{L}_p \end{bmatrix} \qquad x = \begin{bmatrix} \mathrm{x}_1 \\ \mathrm{x}_2 \\ \vdots \\ \mathrm{x}_p \end{bmatrix} \qquad b = \begin{bmatrix} \mathrm{b}_1 \\ \mathrm{b}_2 \\ \vdots \\ \mathrm{b}_p \end{bmatrix}$$

Letting

$$\hat{H}_i = \left[H_{i,1} \cdots H_{i,i-1} \; 0 \; H_{i,i+1} \cdots H_{i,p} \right], \quad \hat{x}_i = \bigcup_{\substack{j=1 \\ j \neq i}}^{p} x_j$$

and defining halo-nodes as Dirichlet-type boundary nodes, the problem to be solved at each subdomain may be written as

$$\left[\begin{array}{c|c} \mathrm{L}_i & \hat{H}_i \\ \hline 0 & \mathrm{I} \end{array} \right] \left[\begin{array}{c} \mathrm{x}_i \\ \hline \hat{x}_i \end{array} \right] = \left[\begin{array}{c} \mathrm{b}_i \\ \hline \hat{x}_i \end{array} \right] \implies L_i x_i + \hat{H}_i \hat{x}_i = b_i$$

Assuming that $\hat{x}_i$ is known we can write:

$$L_i x_i = b_i - \hat{H}_i \hat{x}_i \tag{2}$$

$$= b_i - \left[H_{i,1} \cdots H_{i,i-1} \; 0 \; H_{i,i+1} \cdots H_{i,p} \right] \left[x_1 \, x_2 \cdots x_p \right]^T$$

$$= b_i - \sum_{\substack{j=1 \\ j \neq i}}^{p} H_{i,j} \, x_j \tag{3}$$

The solution is obtained by iteratively solving

$$L_i x_i^{(k)} = b_i - \hat{H}_i \hat{x}_i^{(k-1)} \tag{4}$$

where L_i, $\hat{H}_i$ and b_i are constant, and $\hat{x}_i^{(k-1)}$ is updated at each cycle by data exchange with the neighbours.

4 Distributive Conjugate Gradient Additive-Schwarz

Consider the solution of Equation (4) in one subdomain. For simplicity, the subdomain subscripts were dropped and Equation (4) is written in a general form as

$$Lx = b - \hat{H}\hat{x} \tag{5}$$

Initially, the artificial boundary nodes are set to zero, i.e. $Lx = b$ is approximated. After $\hat{x}$ has been updated the first time, a problem with a perturbed right-hand-side has to be solved. Note that solving each subdomain problem accurately does not necessarily mean that the overall solution is accurate.

The algorithm proposed in this paper, updates $\hat{x}$ at each iteration, i. e. the right-hand-side changes at each step. To solve (5) by using one iteration of a conjugate gradient method per step, a technique that uses previous information has to be adopted, otherwise, only steepest descent iterations would be performed, which degrade the convergence rate.

In the so-called additive Schwarz with distributive conjugate gradient (DCG-AS) procedure, at each step k, the initial approximate solution x, residual

r and search direction p are set as the previous ones. Either r and p are updated accordingly to the previous $\hat{x}$ before performing the iteration. The residual is updated such that

$$r^{(k+1)} = b - \hat{H}\hat{x}^{(k+1)} - Lx^{(k+1)}$$

Given $\tilde{p}^{(k)}$ the search direction computed before the perturbation, the updated search direction p is taken as an orthogonal vector of $\tilde{p}$ projected onto r:

$$p^{(k)} = \tilde{p}^{(k)} - \frac{\left(r^{(k+1)}, \tilde{p}^{(k)}\right)}{\left(r^{(k+1)}, r^{(k+1)}\right)} r^{(k+1)}$$

This leads to Algorithm 2.

This approach makes $p^{(k)}$ conjugate to $p^{(k+1)}$ but does not guarantee any other orthogonality. However, it improves the convergence rate quite significantly if compared to the steepest descent method. For a given accuracy, the additive Schwarz procedure with DCG as a subdomain solver achieves an overall convergence rate comparable to the CG method, as reported on Section 5.

Algorithm 2. Distributive Conjugate Gradient Additive-Schwarz

$r_0 = b - Lx_0$
$p_0 = r_0$
$\hat{x}_0 = 0$
for i=0,1,... **do**
  $\alpha_i = (r_i, r_i)/(Lp_i, p_i)$
  $x_{i+1} = x_i + \alpha_i p_i$
  $r_{i+1} = r_i - \alpha_i L p_i$
  Update $\hat{x}_{i+1}$
  $r_{i+1} = r_{i+1} - \hat{H}(\hat{x}_{i+1} - \hat{x}_i)$
  $p_i = p_i - (r_{i+1}, p_i)/(r_{i+1}, r_{i+1}) r_{i+1}$
  $\beta_i = -(r_{i+1}, Lp_i)/(p_i, Lp_i)$
  $p_{i+1} = r_{i+1} + \beta_i p_i$
end for

5 Numerical Results

In this section we present experimental results obtained with our parallel implementation of DCG-AS, using as test problems the linear systems derived from a finite element discretization (Galerkin scheme) of the Poisson equation on unstructured grids of triangles or tetrahedra. The test problems are stated as:

Problem 2D: $\nabla^2 \phi = -2\pi^2 \sin(\pi x)\sin(\pi y)$ in Ω
$\phi = 0$ on $\partial\Omega$
$\Omega = \{ (x, y) \in \mathbf{R}^2 : -1 < x, y < 1 \}$

Problem 3D: $\nabla^2 \phi = 12c(x^2 + y^2 + z^2)^{2c-1} + 4c(2c-1)(x^2 + y^2 + z^2)^{2c}$ in Ω
$\phi = (x^2 + y^2 + z^2)^{2c}$ on $\partial\Omega$
$\Omega = \{ (x, y, z) \in \mathbf{R}^3 : 0 < x, z < 20 \text{ , } 0 < y < 120 \}$

where c is a scalar.

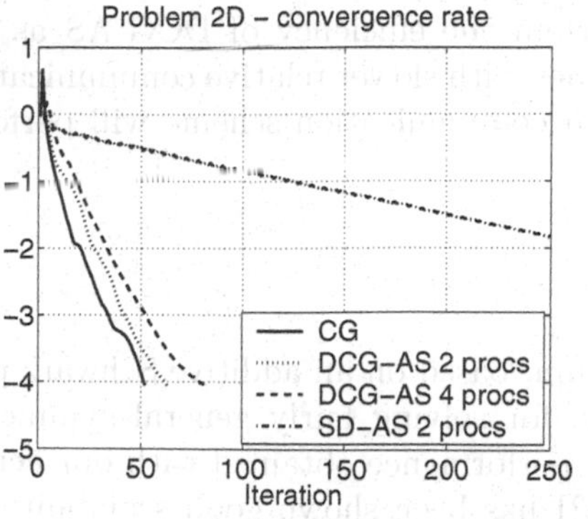

Fig. 1. Problem 2D, convergence rate

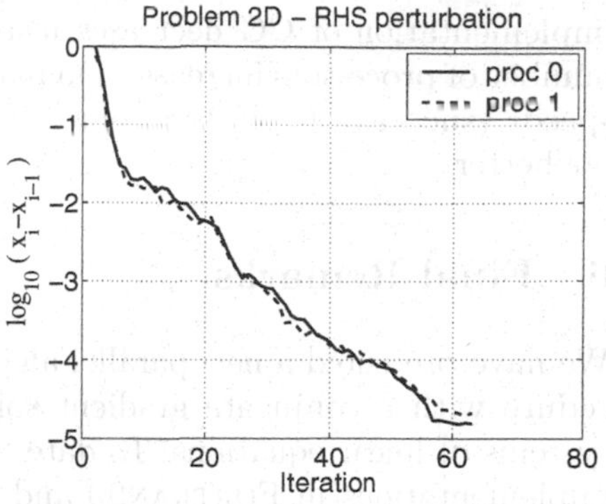

Fig. 2. Problem 2D, RHS perturbation

The algorithm has been implemented in FORTRAN 90 using MPI, making it portable to different architectures. The numerical experiments have been carried out on a SUNFIRE 15K parallel computer and the meshes were partitioned using Metis [8].

Figure 1 shows the convergence rate of Problem 2D solved by CG (1 processor), DCG-AS (2 and 4 processors) and SD-AS (2 processors), which is basically DCG-AS, except that p is always equals to r, i.e., a steepest descent iteration is performed at each step. It shows that DCG-AS converges much faster then SD-AS. Figure 2 shows the 2-norm of the perturbation on the right-hand-side at each step for the same problem in 2 subdomains. We note that it is almost monotonic and similar on both subdomains. Similar behaviour has been observed on more subdomains.

Table 1 shows the number of iterations, the run time (in seconds), the speed-up and the efficiency of Problem 3D obtained by our sequential and parallel implementation. The results presented show that the parallelization of the method led to a good overall performance. One can also notice that the speed-up is even superlinear in some cases. This is mainly due to the minimized amount of communication and data transfer scheme adopted.

A parallel implementation of CG can be as efficient as, or more efficient than, DCG-AS for a small number of processors. However, the efficiency of a parallel

Table 1. Efficiency of Problem 3D for n=294,518 and n=581,027

Procs	n=294,518				n=581,027			
	Iters	Run-time	Speed-up	Efficiency	Iters	Run-time	Speed-up	Efficiency
1	380	146.0	-	-	534	501.4	-	-
2	561	118.4	1.23	0.62	734	357.1	1.40	0.70
4	575	59.6	2.45	0.61	881	241.4	2.08	0.52
8	514	19.2	7.60	0.95	709	116.5	4.30	0.54
16	454	6.0	24.33	1.52	691	48.2	10.40	0.65
32	752	3.9	37.44	1.17	659	9.4	53.34	1.67

implementation of CG decreases faster than the efficiency of DCG-AS as the number of processors increases. On machines with slower relative communication speeds (such as PC clusters) our minimal communication scheme will perform far better.

6 Final Remarks

We have presented a new parallel algorithm, based on an additive Schwarz procedure with a conjugate gradient solver, for solving fairly general symmetric systems of linear equations. To date, the performance obtained with our actual implementations in FORTRAN90 and MPI has been shown good scalability for most of the problems.

Techniques to minimize the variation in the rate of convergence as the number of subdomains increase are under investigation and are expected to improve the overall performance. Preconditioning methods for the local solver and the use of DCG-AS as a preconditioner are also under development. Furthermore, we intend to adapt the procedure for the Bi-CGSTAB method that will allow other PDE problems to be considered.

Acknowledgments

The authors would like to thank CAPES - Coordenação de Aperfeiçoamento de Pessoal de Nível Superior - for partially funding this project and the Cambridge-Cranfield High Performance Computing Facility for allowing our use of its computing facilities. The authors are grateful to Professor Rudnei Dias da Cunha for valuable discussions.

References

1. O. Axelsson. Iteration number for the conjugate gradient method. *Mathematics and Computers in Simulation*, 61:421–435, 2003.
2. T. F. Chan and T. P. Mathew. Domain decomposition algorithms. In: *Acta Numerica 1994*, Cambridge University Press, 61–143, 1994.
3. T. F. Chan and B. F. Smith. Domain Decomposition and Multigrid Algorithms for Elliptic Problems on Unstructured Meshes. *Eletronic Transactions on Numerical Analysis*, 2:171–182, 1994.
4. X. Chen and J. Sun. Global convergence of a two-parameter family of conjugate gradient methods without line seach. *Journal of Computational and Applied Mathematics*, 146:37–45, 2002.
5. R. Fletcher and C. M. Reeves. Function minimization by conjugate gradients. *The Computer Journal*, 7:149–154, 1964.
6. G. H. Golub and C. F. Van Loan. *Matrix Computations.* Johns Hopkins University Press, Baltimore, MD, third edition, 1996.
7. M. R. Hestenes and E. Stiefel. Methods of Conjugate Gradient for Solving Linear Systems. *Journal of Research of the National Bureau of Standards*, 49:409–436, 1952.

8. G. Karypis. METIS home page. http://www-users.cs.umn.edu/~karypis/metis/, Last time accessed 16/03/2004.
9. D. E. Keyes. How Scalable is Domain Decomposition in Practice? In C.-H. Lai, P. E. Bjørstad, M. Cross, and O. B. Widlund, editors, *Eleventh International Conference on Domain Decomposition Methods*, 286–297, Bergen, 1999. Domain Decomposition Press.
10. B. S. N. Murty and A. Husain. Orthogonality correction in the conjugate-gradient method. *Journal of Computational and Applied Mathematics*, 9:299–304, 1983.
11. Y. Saad. *Iterative methods for sparse linear systems.* PWS Publishing Company, Boston, 1995.
12. B. F. Smith, P. E. Bjørstad, and W. D. Gropp. *Domain Decomposition: Parallel Multilevel Methods for Elliptic Partial Differential Equations.* Cambridge University Press, Cambridge, United Kingdom, 1996.

Parallel PCG Solver for Nonconforming FE Problems: Overlapping of Communications and Computations

Gergana Bencheva[1], Svetozar Margenov[1], and Jiří Starý[2]

[1] Institute for Parallel Processing, Bulgarian Academy of Sciences, Acad. G. Bontchev, Bl. 25A, 1113 Sofia, Bulgaria
gery@parallel.bas.bg, margenov@parallel.bas.bg

[2] Institute of Geonics, Academy of Sciences of the Czech Republic, Studentská 1768, 70800 Ostrava, Czech Republic
stary@ugn.cas.cz

Abstract. New results concerning a recently introduced parallel preconditioner for the solution of large nonconforming FE linear systems are presented. The studied algorithm is based on the modified incomplete Cholesky factorization MIC(0) applied to a locally constructed approximation of the original stiffness matrix. The overlapping of communications and computations is possible due to a suitable reordering of the computations applied in the MPI code. The obtained improvement of the real performance is illustrated by numerical tests on a Beowulf-type Linux cluster and on a Sun Fire symmetric multiprocessor.

1 Introduction

The nonconforming finite elements and the parallel algorithms are two advanced computational mathematics topics which have provoked a lot of publications during the last decades. This work is focused on the case where a second order elliptic boundary value problem is discretized by rotated bilinear nonconforming finite elements on quadrilaterals. The considered elements are firstly proposed in [5] as a cheapest stable Finite Element (FE) approximation of the Stokes problem.

Parallel preconditioned conjugate gradient (PCG) solver for the related elliptic FE system was recently introduced in [1]. Our aim is to improve the performance of its Message Passing Interface (MPI) implementation. The included preconditioner is based on the modified incomplete Cholesky factorization MIC(0) applied to a locally constructed approximation of the original stiffness matrix. The properties of the whole method have been analyzed already in [2], where also numerical results obtained on two Beowulf-type Linux clusters are compared with theoretically derived estimates of the execution times.

The structure of the paper is as follows. The needed background of the bilinear nonconforming FE discretization, the essence of MIC(0) and the preconditioning strategy, are given in Section 2. The parallel implementation of the resulting

I. Lirkov, S. Margenov, and J. Waśniewski (Eds.): LSSC 2005, LNCS 3743, pp. 646–654, 2006.

PCG algorithm is described in Section 3, where a suitable reordering of the computations is applied in the MPI code. This approach allows the usage of nonblocking send and receive operations, i.e. overlapping of communications and computations. The achieved improvement of the real performance is illustrated in Section 4 by numerical tests on a Beowulf-type Linux cluster and on a Sun Fire symmetric multiprocessor.

2 Preliminaries

We start the exposition in this section with formulation of the problem and its discretization. Further, we give the essence of MIC(0). It is shown in the end, how the well parallelizable structure of the resulting preconditioner is obtained.

Nonconforming FE discretization. We consider the isotropic two-dimensional elliptic problem associated with the bilinear form

$$a_h(u_h, v_h) = \sum_{e\in\omega_h} \int_e a(e) \sum_{i=1}^{2} \frac{\partial u_h}{\partial x_i} \frac{\partial v_h}{\partial x_i} dx \,. \tag{1}$$

Here, ω_h is a decomposition of the computational domain Ω into convex quadrilaterals denoted by e. The coefficient $a(e)$ is a constant on each $e \in \omega_h$ defined as an averaged value of the varying coefficient of the original boundary value problem.

The finite element space V_h corresponding to ω_h is obtained using the rotated bilinear nonconforming finite elements, as proposed in [5]. The reference element E is the unit square with sides Γ_j, $j = 1, \ldots, 4$, parallel to coordinate axes, and nodes $j = 1, \ldots, 4$ which are the mid-points of the sides. Mid-point and integral mid-value interpolation operators are implemented in the definition of the nodal basis functions $\varphi_i \in S_p$, $S_p = \text{span}\{1, x_1, x_2, x_1^2 - x_2^2\}$. This leads to two alternative constructions of V_h, referred as *Algorithm MP* and *Algorithm MV* respectively. The nodal interpolation conditions $\varphi_i(j) = \delta_{ij},\ i, j = 1, \ldots, 4$ are used in MP, where δ_{ij} is the Kronecker symbol. The conditions for MV are $\frac{1}{|\Gamma_j|}\int_{\Gamma_j} \varphi_i \, dx = \delta_{ij},\ i, j = 1, \ldots, 4.$

The standard finite element procedure continues with computation of the element stiffness matrices $A_e = \{\alpha_{ij}\}_{i,j}^4$, $e \in \omega_h$. It follows the isoparametric technique using the reference element E basis functions. The assembling of A_e, $e \in \omega_h$ leads to the linear system of equations

$$\mathbf{A}\mathbf{x} = \mathbf{b} \,. \tag{2}$$

The stiffness matrix $A = \{a_{ij}\}_{i,j=1}^N$ is sparse, symmetric and positive definite, with at most seven nonzero elements per row. Its structure is one and the same for MP and MV and hence the preconditioning strategy does not depend on the type of the nodal basis functions.

The essence of the **Modified Incomplete Cholesky** (MIC(0)) factorization of sparse matrices is given below. Some more details may be found in [3, 4].

Let us rewrite the real symmetric $N \times N$ matrix A as a sum of its diagonal (D), strictly lower $(-\widetilde{L})$ and upper $(-\widetilde{L})^t$ triangular parts

$$A = D - \widetilde{L} - \widetilde{L}^t . \tag{3}$$

Then we consider the approximate factorization of A with the following form

$$\mathcal{C}_{\mathrm{MIC}(0)}(A) = (X - \widetilde{L})X^{-1}(X - \widetilde{L})^t , \tag{4}$$

where $X = diag(x_1, \cdots, x_N)$ is a diagonal matrix determined by the condition of equal rowsums $\mathcal{C}_{MIC(0)}\mathbf{e} = A\mathbf{e}$, $\mathbf{e} = (1, \cdots, 1)^t \in \mathbf{R}^N$.

We say that the factorization (4) of A is a *stable* MIC(0) factorization if $X > 0$ and thus $\mathcal{C}_{\mathrm{MIC}(0)}(A)$ is positive definite. Concerning the stability of MIC(0) factorization, the following theorem holds.

Theorem 1. *Let $A = \{a_{ij}\}$ be a symmetric real $N \times N$ matrix and $A = D - \widetilde{L} - \widetilde{L}^t$ be the splitting (3) of A. Let us also assume that*

$$\widetilde{L} \geq 0 , \quad A\mathbf{e} \geq 0 , \quad A\mathbf{e} + \widetilde{L}^t\mathbf{e} > 0 , \quad \mathbf{e} = (1, \ldots, 1)^t \in \mathbf{R}^N ,$$

i.e., A is a weakly diagonally dominant matrix with nonpositive off-diagonal entries and that $A + \widetilde{L}^t = D - \widetilde{L}$ is strictly diagonally dominant. Then the relation

$$x_i = a_{ii} - \sum_{k=1}^{i-1} \frac{a_{ik}}{x_k} \sum_{j=k+1}^{N} a_{kj}$$

gives only positive values and the diagonal matrix $X = diag(x_1, \cdots, x_N)$ defines a stable MIC(0) *factorization of A.*

Preconditioning strategy. Here, we briefly present the approach proposed in [1]. We use the PCG method with a preconditioner based on the MIC(0) factorization of sparse matrices. For construction of the preconditioner, two steps are performed.

At first, a local modification with a diagonal compensation is applied to the stiffness matrix A. Its geometrical sense is the following. Each node P of the mesh is connected with the nodes from two neighbouring quadrilateral elements, as shown in Fig. 1 (a). Two of the links are "cut" (see Fig. 1 (b)) to introduce the locally modified element stiffness matrices. The resulting global matrix B allows for a stable MIC(0) factorization.

At the second step, MIC(0) factorization of B is performed, i.e. the preconditioner for A is $\mathcal{C} = \mathcal{C}_{\mathrm{MIC}(0)}(B)$. It has a specific well parallelizable block structure preserving the robustness of the pointwise incomplete factorization.

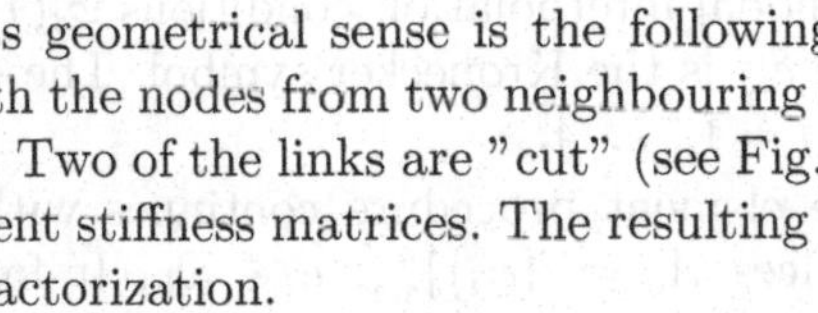

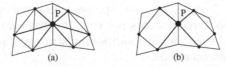

Fig. 1. The connectivity pattern

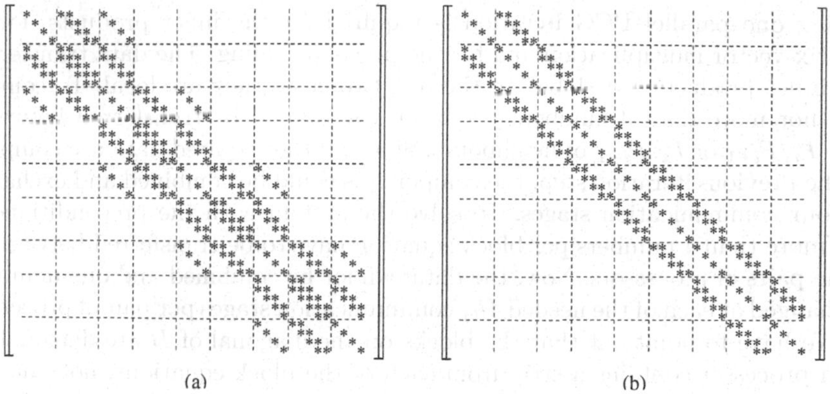

Fig. 2. The structure of the matrix A (a) and the locally modified matrix B (b)

The steps needed to solve a system with a preconditioner $\mathcal{C}$ are based on recursive computations and therefore the resulting PCG algorithm is inherently sequential. The locally constructed approximation B of the original stiffness matrix A is introduced to allow a possibility of parallel implementation.

The structures of the matrices A and B are shown in Fig. 2. They correspond to a problem in a square domain, discretized by a rectangular $n_1 \times n_2$ mesh with vertical node numbering. Then $N = n_1(2n_2 + 1) + n_2$ denotes the number of degrees of freedom.

A condition number model analysis of $B^{-1}A$ can be found in [1]. It is shown there that: a) the matrices A and B are spectrally equivalent; b) the conditions for a stable MIC(0) factorization hold for B; and c) the PCG convergence rates of the preconditioners $\mathcal{C}_{\mathrm{MIC(0)}}(B)$ and $\mathcal{C}_{\mathrm{MIC(0)}}(A)$ are very similar preserving the robustness of the pointwise incomplete factorization. The diagonal blocks of the matrix B allow a parallel implementation of the resulting PCG algorithm.

3 Parallel PCG Solver

At the beginning of this section, we review briefly the algorithm and its parallel properties. The exposition at this stage is based on the theoretical and experimental results in [1, 2]. Further, we focus our attention on the question, How to improve the performance of the considered parallel algorithm?". Therefore we introduce a proper reordering of the computations and communications.

Data distribution, computations and communications. Let us have N_p processors denoted by $P_0, P_1, \ldots, P_{N_p-1}$. The domain is partitioned into N_p horizontal strips with an approximately equal number of elements. This leads to a division of each of the block rows of the matrices A and B into strips with an almost equal number of equations. All the vectors are distributed in the same way. Each P_i deals with the data from its strip and the computations are equally distributed among the processors. The communication of concurrent processes during one parallel PCG iteration is required for the inner products, for the matrix-vector multiplication and for the preconditioning. The data transfers for the inner products are global, in the rest two cases they are local. To compute $A\mathbf{v}$, not more than 3 numbers per block row should be transferred between a pair P_i, P_{i+1} or P_i, P_{i-1} of neighbours. Since all the required data are computed at the previous iteration step, the components could be combined and exchanged at four communication stages. To solve the system with the preconditioner $\mathcal{C}$, not more than 4 numbers per block equation have to be transferred between the same pairs of processors. Now, the data cannot be combined and one number is exchanged at each of the needed $4n_1$ communication stages per pair of processors.

We have to point out that the blocks on the diagonal of B are diagonal and each processor contains a strip from each of the block equations, not one strip from the whole system. There is still recursion but it is for blocks and each block is handled in parallel.

The following theoretical estimate of the execution time for one iteration is derived in [2]:

$$T^{it}_{N_p} = T^{it}_a + T^{it}_{com} \approx 34 \frac{n_1(2n_2+1)+n_2}{N_p} .t_a + 8n_1.t_s + 14n_1.t_w \,.$$

Here, t_a is the average unit time to perform one arithmetic operation by one processor, t_s is the start-up time for communication, t_w is the incremental time necessary to send a word between two neighbouring processors. The crucial part of the needed communications is local and the performance of the algorithm weakly depends on the number of processors and the parallel architecture. The coefficients in front of t_s and t_w depend on the problem size. This is the reason why the performance depends substantially on the ratio between t_s, t_w and t_a.

A comparative analysis of the achieved performance and the real distribution of computation and communication times is made in [2] using two Beowulf-type Linux clusters – Thea and Parmac. It is experimentally shown there that: 1) the speed-up and efficiency coefficients are approximately one and the same for the variants MP and MV of the nodal basis functions (see Section 2); 2) the computation time decreases almost two times when the number of processors is increased twice; 3) the communication time does not depend on the number of processors; 4) the communications for the preconditioning is the most time consuming part.

Overlapping of computations and communications. How to reduce the influence of the communication time for distributed memory machines with large ratios $\frac{t_s}{t_a}$ and $\frac{t_s}{t_w}$? We reorder the computations in the forward and backward recurrence to solve the system with preconditioner as described below. This allows us to use nonblocking send and receive operations of MPI and to overlap computations with communications.

To apply the preconditioner $\mathcal{C}_{\mathrm{MIC}(0)}(B)\mathbf{w} \equiv (X - \widetilde{L})X^{-1}(X - \widetilde{L})^t\mathbf{w} = \mathbf{v}$ (see (3), (4)), one has to perform the following three steps: 1) find $\mathbf{y}$ from $L\mathbf{y} = \mathbf{v}$, where $L = X - \widetilde{L}$; 2) compute $\mathbf{z} := X\mathbf{y}$; and 3) find $\mathbf{w}$ from $L^t\mathbf{w} = \mathbf{z}$. Data transfers between neighbouring processors (see below and [2] for details) are

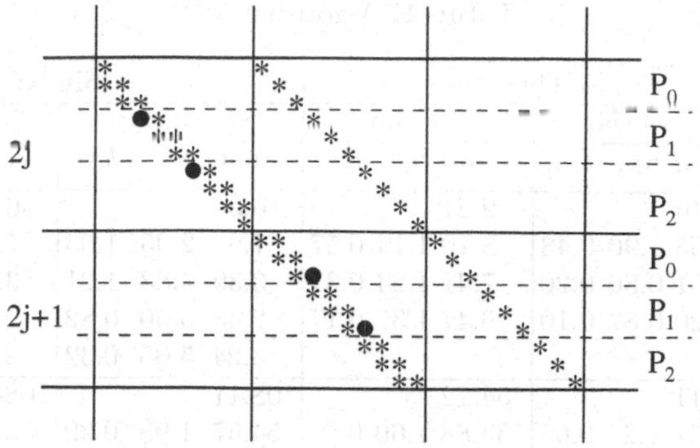

Fig. 3. The scheme of communications and computations for the solution of $L\mathbf{y} = \mathbf{v}$. Note that $N_p = 3,\ n_1 = n_2 = 9$.

needed at steps 1) and 3), while no communication is required at step 2). The important advantage of the matrix B is that all of its diagonal blocks are diagonal and hence the same holds for L.

The structure of two successive block rows ($2j$ and $2j+1$) of L and their distribution among the processors is presented in Fig. 3. The system $L\mathbf{y} = \mathbf{v}$ is solved by a standard forward recurrence. The block L_{11} is diagonal, divided into N_p strips – one per each processor. So $L_{11}\mathbf{y}_1 = \mathbf{v}_1$ is solved in parallel without any data transfers. Then we have to determine $\mathbf{y}_2$ from $L_{21}\mathbf{y}_1 + L_{22}\mathbf{y}_2 = \mathbf{v}_2$. The component $\mathbf{y}_1$ is computed at the previous step but each processor has a strip of $\mathbf{y}_1$. The block L_{21} is two-diagonal and hence the processor P_i has to send one component of $\mathbf{y}_1$ to P_{i+1} and to receive another one from P_{i-1}. These data are multiplied by entries denoted by • in Fig. 3, i.e. they participate only in the computation of the first element of $\widetilde{\mathbf{v}}_2 = \mathbf{v}_2 - L_{21}\mathbf{y}_1$ in each P_i. So each P_i calculates the first entry of $\widetilde{\mathbf{v}}_2$ at the end and the communication is overlapped with the computation of the rest components. Then $L_{22}\mathbf{y}_2 = \mathbf{v}_2 - L_{21}\mathbf{y}_1$ is handled concurrently without any other communications. After that, $L_{32}\mathbf{y}_2 + L_{33}\mathbf{y}_3 = \mathbf{v}_3$ has to be solved. Again, one component of $\mathbf{y}_2$ is transferred, but now to processor P_{i-1}. It participates in the last element of $\widetilde{\mathbf{v}}_3$ in each P_i and it is determined after the communication, overlapped with computation of the remaining entries, is completed. The procedure continues till the last block of $\mathbf{y}$ is computed. The third step is handled in similar way as the first step, applying the standard backward recurrence.

4 Numerical Tests

We compare in this section the performance of the parallel algorithm and the related C/MPI codes with and without overlapping of communications and computations on two parallel architectures. Namely, new results on a Beowulf-type Linux cluster referred as *Thea* and on a Sun Fire symmetric multiprocessor referred as *Simba* are presented.

Table 1. Algorithm MP

		Thea						Simba					
		no overlap			overlap			no overlap			overlap		
N_p	$\frac{n}{iter}$	cpu	S_{N_p}	E_{N_p}	cpu	S_{N_p}	E_{N_p}	cpu	S_{N_p}	E_{N_p}	cpu	S_{N_p}	E_{N_p}
1		9.16			9.21			16.36			16.36		
2	256	9.55	0.96	0.48	8.10	1.14	0.57	7.97	2.05	1.03	7.71	2.12	1.06
4	71	11.51	0.80	0.20	7.41	1.24	0.31	3.39	4.83	1.21	3.20	5.11	1.28
8		11.20	0.82	0.10	6.47	1.37	0.17	2.48	6.60	0.82	2.43	6.73	0.84
16								3.24	5.05	0.32	2.88	5.68	0.36
1		54.11			54.22			108.11			108.07		
2	512	41.91	1.29	0.65	33.88	1.60	0.80	54.57	1.98	0.99	53.46	2.02	1.01
4	104	41.35	1.31	0.33	24.97	2.17	0.54	29.06	3.72	0.93	29.13	3.71	0.93
8		36.47	1.48	0.19	20.65	2.63	0.33	15.38	7.03	0.88	14.86	7.27	0.91
16								11.10	9.77	0.61	9.84	10.98	0.69
1		286.91			287.51			646.95			647.40		
2	1024	212.41	1.35	0.68	192.32	1.49	0.75	325.05	1.99	1.00	323.91	2.00	1.00
4	148	155.52	1.84	0.46	107.49	2.67	0.67	170.77	3.79	0.95	167.87	3.86	0.96
8		125.01	2.30	0.29	71.09	4.04	0.51	88.85	7.28	0.91	86.49	7.49	0.94
16								52.37	12.35	0.77	51.95	12.46	0.78

Thea is located at the Institute of Geonics, Academy of Sciences of the Czech Republic, and consists of eight computing nodes plus, a file server and one interactive node. Each of the nodes is equipped by a single AMD Athlon processor at 1.4 GHz frequency and 1.5 GB of memory. The nodes are interconnected via two standard FastEthernet networks.

Simba is located at the Department of Information Technology, Uppsala University, Sweden. It has 36 UltraSPARC III+ processors at 900 MHz frequency and 36 GB of shared memory. The system is configured as a separate domain of a Sun Fire 15000 server based on the Sun Fireplane interconnect architecture with very fast data transfers up to 9.6 GB/s of the peak performance.

We consider a model Poisson equation in a unit square with homogeneous Dirichlet boundary conditions. The partitioning of the domain is uniform where $n_1 = n_2 = n$. The size of the discrete problem is $N = n_1(2n_2+1)+n_2 = 2n(n+1)$. The relative stopping criterion $(C^{-1}r^{n_{it}}, r^{n_{it}})/(C^{-1}r^0, r^0) < \varepsilon$ is used in the PCG algorithm, where r^i stands for the residual at the i-th iteration step, $(\cdot, \cdot)$ is the Euclidean inner product, and $\varepsilon = 10^{-6}$.

The obtained values of speed-up and efficiency for the parallel algorithm related to the variant MP (see Section 2) of the nodal basis functions are collected in Table 1. The number N_p of the processors is given in the first column. There are two numbers in each box of the second field – the number of elements n in each direction and the related number of iterations for that size of the discrete problem. The remaining columns are in two groups with similar structure – one per each computer introduced above. Each group has again two subgroups for the cases of nonoverlapping (no overlap) and overlapping (overlap) of computations and communications. The measured execution time in seconds is given in the first field of each subgroup. The rest two fields present the speed-up $S_{N_p} = T_1/T_{N_p}$ and the efficiency $E_{N_p} = S_{N_p}/N_p$. The execution time T_{N_p} is the best one obtained from 10 runs of the code on a given machine.

Let us see the results obtained on Thea cluster. We observe that: 1) for a given number of processors, the speed-up and respectively the efficiency grow up for larger size of the problem; 2) the speed-up and the efficiency without overlapping are far away from the upper bounds $S_{N_p} \leq N_p$, $E_{N_p} \leq 1$; 3) there is a significant improvement of S_{N_p} and E_{N_p} in the case with overlapping.

In contrast, the efficiency coefficients with and without overlapping on Simba are close to each other. They are larger in general than the highest ones obtained on Thea. The reason for this behaviour is hidden in the system parameters t_a, t_s, t_w. Comparing T_1 on both machines, we observe that the computation times on Simba are larger than those on Thea. At the same time, communications are faster on Simba since, e.g. for $n = 512$, T_4 and T_8 without overlapping on Simba are smaller than the corresponding values on Thea. The performance is also influenced by the nonuniform memory access. The superlinear speed-ups on Simba for relatively small problems are due to the size of the cache memory.

As expected, the results confirm that the performance of our code strongly depends on the system's parameters.

5 Concluding Remarks

The recently introduced scalable parallel MIC(0) preconditioner is studied in the paper. Its real performance for distributed memory machines is improved by overlapping of communications and computations.

Our future plans include a development and analysis of theoretical estimates of the parallel execution times for overlapping of each local communication with computations. Modifications of the code for shared memory machines with the aid of OpenMP as well as the generalization of the considered algorithm into 3D case and its implementation for linear elasticity problems are also among our interests.

Acknowledgements

This work has been supported in part by the Uppsala Multidisciplinary Center for Advanced Computational Science (UPPMAX) under the project p2004009 "Parallel Computing in Geosciences", by the bilateral IG AS – IPP BAS interacademy exchange grant "Reliable Modelling and Large Scale Computing in Geosciences", by the Bulgarian IST Center of Competence in 21st century — BIS-21++ funded by European Commission in FP6 INCO via grant 016639/2005, and by the Bulgarian NSF grant I1402/04. The scientific visit of G. Bencheva in Uppsala was possible due to the sponsorship of the Royal Swedish Academy of Engineering Sciences, IVA.

References

1. G. Bencheva and S. Margenov, Parallel incomplete factorization preconditioning of rotated linear FEM systems, *J. Comp. Appl. Mech.*, **4** 2 (2003), 105–117.
2. G. Bencheva and S. Margenov, Performance analysis of a parallel MIC(0) preconditioning of rotated bilinear nonconforming FEM systems, *Mathematica Balkanica*, **17** (2003), 319–335.
3. R. Blaheta, Displacement decomposition – incomplete factorization preconditioning techniques for linear elasticity problems, *Numer. Linear Algebra Appl.* **1** (1994) 107–126.
4. I. Gustafsson, Modified incomplete Cholesky (MIC) factorization, in: D.J. Evans, ed., *Preconditioning Methods; Theory and Applications,* (Gordon and Breach, 1984) 265–293.
5. R. Rannacher and S. Turek, Simple nonconforming quadrilateral Stokes Element, *Numer. Methods for PDEs* **8** 2 (1992) 97–112.

Systolic Architecture for Adaptive Censoring CFAR PI Detector*

Ivan Garvanov[1], Christo Kabakchiev[1], and Plamen Daskalov[2]

[1] Institute of Information Technologies,
Bulgarian Academy of Sciences,
Acad G. Bonchev, Bl. 2, 1113 Sofia, Bulgaria
igarvanov@yahoo.com, ckabakchiev@yahoo.com
[2] Multiprocessor Systems Ltd., 63 Shipchensky prohod Blvd.,
1574 Sofia, Bulgaria
Daskalov@mps.bg

Abstract. A new parallel algorithm for signal processing and a parallel systolic architecture of a robust Constant False Alarm Rate (CFAR) processor with post-detection integration and adaptive censoring (RACPI) is presented in the paper. This detector is effective in conditions of flow from strong impulse interference. The ACPI CFAR processor uses sorting and censoring algorithms. We offer the sorting algorithm to be realized on the basis of the odd-even transposition sort method. We propose the censoring algorithm to be used for obtaining of the noise level estimation and for estimation of the impulse interference parameters. These parameters are needed for automatically choosing the scale factor, which keeps the false alarm rate constant. The real-time implementation of this detection algorithm requires large computational resources because of the great volume and high speed of the incoming data. The time consumption of the sorting and censoring procedures is also very high and therefore the practical realization is difficult. For all these reasons, we choose systolic architectures in the considered case for being more effective than conventional multiprocessor architectures. The computational losses of the systolic architecture are estimated in terms of the number of the processor elements, the computational time and the speed-up needed for real-time implementation.

1 Introduction

The problem of detecting known signals in impulse interference with unknown parameters is common in sensor systems. Modern search radars, for example, often employ an adaptive detection threshold to maintain a constant false alarm rate (CFAR). There are a lot of methods for increasing the efficiency of CFAR processors in case of non-stationary and non-homogeneous interference. One of these methods, suggested by Rohling in [14], includes the use of ordered statistics

* This work is supported by IIT - 010059/2004, MPS Ltd. Grant "Digital Radar"/2005 and Bulgarian NF "SR" Grant No TH - 1305/2003.

I. Lirkov, S. Margenov, and J. Waśniewski (Eds.): LSSC 2005, LNCS 3743, pp. 655–662, 2006.

for estimating the interference level in the reference window. Another approach for estimating the interference level, proposed by Himonas in [7], is used in this paper. Later, Himonas suggests this method to be used for censoring of randomly arriving interference impulses from the reference window, when the test cell contains no impulse interference [6]. Behar offers in [2] an adaptive censoring PI CFAR detector in the presence of Poisson distribution pulse jamming. She considers more complex and more general situation than the one studied in [7, 6], in which the two-dimensional reference window and the test cells are corrupted by randomly arriving impulse interference and the model of the appearance of pulse jamming is more general. Censoring of the randomly arriving interference impulses both, in the test resolution cells and the reference cells, is used. Garvanov proposes in [5] a new modification of the parallel algorithm presented in [2]. This algorithm works successfully in the presence of total interference (thermal noise plus pulse jamming) that is binomially distributed according to the compound exponential law [1], with unknown average repetition frequency and magnitude.

When the probability for the appearance of impulse interference is high, the censoring algorithm, proposed in [6], is unsatisfactory [5, 4, 3]. In such a case the ACPI CFAR processor does not keep constant the false alarm probability, as it is in [6]. Similarly to Garvanov and Lazarov [4, 11, 12], we suggest the scale factor to be corrected automatically if there is any change in the estimation of impulse interference parameters. The scale factor is chosen automatically from a matrix of preliminary calculated values for different impulse interference parameters.

The systolic architectures of some CFAR algorithms based on ordered statistics filters are studied by Hwang in [9, 8, 13, 10]. These architectures are constructed by means of the sample-oriented approach, which maintains the window in sorted order in an array of processors and updates the ranks based on the sample values of arriving and departing data. The sample-oriented approach is a computationally more efficient technique, which takes advantage of the evaluated ranks in the current window for the evaluation of the ranks in the next window.

In contrast to Hwang, who used the known criteria for choosing of the noise level estimate from ordered statistics, we studied an adaptive censoring robust algorithm. In our case, the sample-oriented approach is not suitable.

In our work, the RACPI CFAR algorithm is proposed as parallel algorithm in which the sorting is realized on the basis of the odd-even transposition method.

The best matching between the statistical algorithm and the suggested new systolic architecture requires the using of all reference window cells in the censoring of the impulse interference. Therefore, we suggest a two-dimensional sorting algorithm in the reference array to be used. There are several sorting algorithms that can be implemented for such purposes, e.g. $LS3$ sort, 4 way mergesort, rotatesort, etc. However, we use a 2D odd-even transposition sort algorithm, because it is simple and easy for implementation. Unfortunately, this sorting method is quite slow. Nevertheless, it sorts the reference array in a snake-like order and this sorting direction is very suitable for further processing. The using of all reference window cells allows the censoring algorithm to separate the impulse interference from the "clean" cells and therefore the best possible noise level estimations and estimations of the impulse interference parameters are obtained. This is the reason for achieving of the most effective useful signal detection.

The suggested statistical algorithm for a RACPI CFAR detector is in fact a further development of the results achieved in [2, 5]. The proposed new systolic architecture allows its calculating structure to follow closely the statistical algorithm, which leads to increased quality of detection and keeping constant false alarm probability. This new algorithm is compared to the similar algorithms, presented in [2, 5] as they all solve the same problem, for the number of the calculating steps and the number of PE elements used. The new algorithm is not compared to algorithms proposed by other authors, because they are different and do not refer to the same task. There are systolic structures of sorting algorithms for different types of CFAR detectors. However, the most suitable for comparison in our case are the algorithms presented in [2, 5].

2 Robust Adaptive Censoring Post-detection Integration CFAR Processor

The RACPI CFAR processor is effective in conditions of flow from strong impulse interference. This algorithm consists of the following steps:

The elements of the reference window $\boldsymbol{x} = (x_1, x_2, ...x_{NL})$ and the test resolution cells $\boldsymbol{x_{0l}} = (x_{01}, x_{02}, ...x_{0L})$ are rank-ordered according to increasing magnitude. Each of the such ranked elements is compared to the adaptive threshold, according to the following rule:

$$x_{i+1}^{(1)} \geq s_i^x T_i^x,\ i = 1, ..., NL-1; \quad x_{0j+1}^{(1)} \geq s_j^{x_0} T_j^{x_0},\ j = 1, ..., L-1 \qquad (1)$$

where $s_i^x = \sum_{p=1}^{i} x_p^{(1)}$ and $s_j^{x_0} = \sum_{p=1}^{j} x_{0p}^{(1)}$. The scale factors T_i^x and $T_j^{x_0}$ are determined in accordance with the given level of probability of false censoring as in [6]:

$$P_{fa}^{cen} = \binom{NL}{i} \frac{1}{(1 + T_i^x(NL - i))^i}; \quad P_{fa}^{cen} = \binom{L}{j} \frac{1}{(1 + T_j^{x_0}(L - j))^j} \qquad (2)$$

The recursive procedure is stopped when the condition (1) becomes true. In this way, the reference and the test resolution cells are divided into two parts. The first part contains the "clean" elements, i.e. without interference. The noise level estimate V and the summed signal q_0 are the statistical medians of the "clean" elements from reference and test resolution cells [15]. We suggest the parameters of the interference to be estimated by using the second part of the reference window. The estimate of the average interference-to-noise ratio r_j and the probability for the appearance of impulse interference e, can be calculated as follows.

By using the estimates of the impulse interference parameters, the RACPI CFAR processor chooses automatically the scale factor T_{RACPI} from a matrix of preliminary calculated values. The target is detected according to the following rule.

$$\begin{cases} H_1 : \text{if } q_0 \geq H_D = T_{RACPI}V \\ H_0 : \text{if } q_0 < H_D = T_{RACPI}V \end{cases} \tag{3}$$

where H_1 is the hypothesis that the test resolution cells contain the echoes from the target and H_0 is the hypothesis that the test resolution cells contain the receiver noise only.

3 Parallel Architecture of a RACPI CFAR Processor

The new systolic parallel architecture of a robust CFAR processor with adaptive censoring and non-coherent integration is presented on Fig.1. The computational blocks for sorting of vectors are denoted as S_1 and S_2. The sorting algorithm for the test cells is realized on the basis of the well-known odd-even transposition sort method. The elements of the reference window are sorted with a 2D odd-even transposition sort method. In this algorithm, the elements are compared not only with their right and left neighbours, but also with their upper and lower neighbours. The sorting direction in the array is snake-like.

The systolic architectures of sorting computational blocks S_1 and S_2 are shown on Fig.2 and Fig.3.

The systolic architecture of the censoring algorithm is realized as a parallel algorithm by means of blocks B_1, B_2, B_3, processor elements PE_6 (for reference cells) and B_{01}, B_{02}, B_{03} (for test cells). The blocks B_1 and B_{01} consist of the processor elements PE_3. These blocks form the adaptive censoring thresholds. The blocks B_2 and B_{02} compare the censoring threshold with the sorted cells from the memories M_1 and M_{01}. These blocks consist of the processor elements PE_4. The blocks B_3 and B_{03} consist of the processor elements PE_5, and they

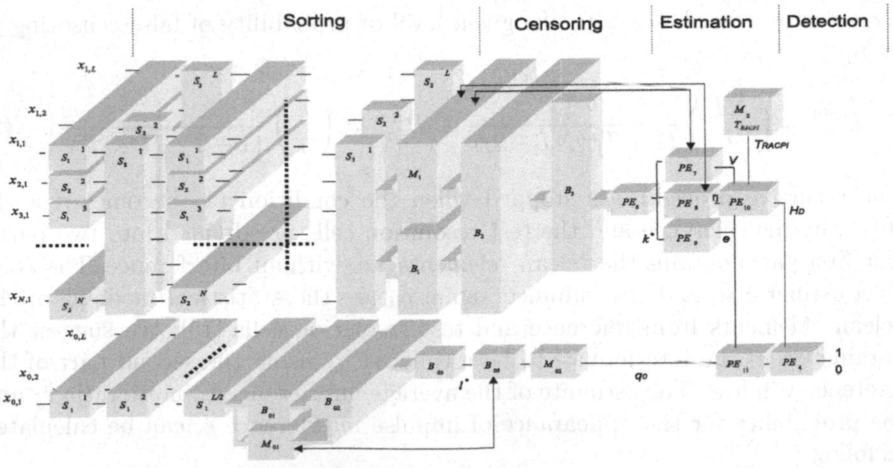

Fig. 1. Systolic architecture of a RACPI CFAR processor

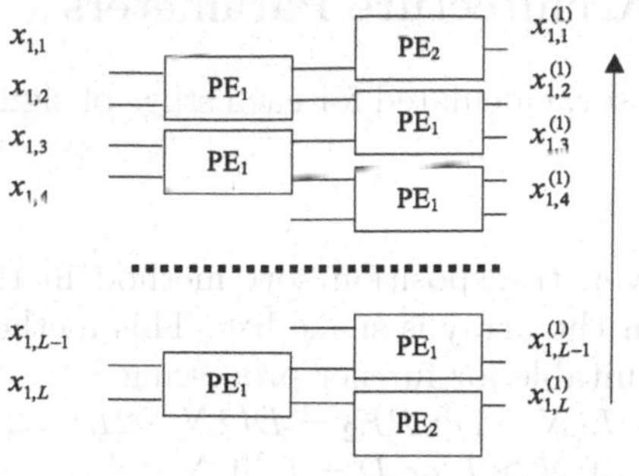

Fig. 2. Systolic architecture of the sorting algorithm S_1

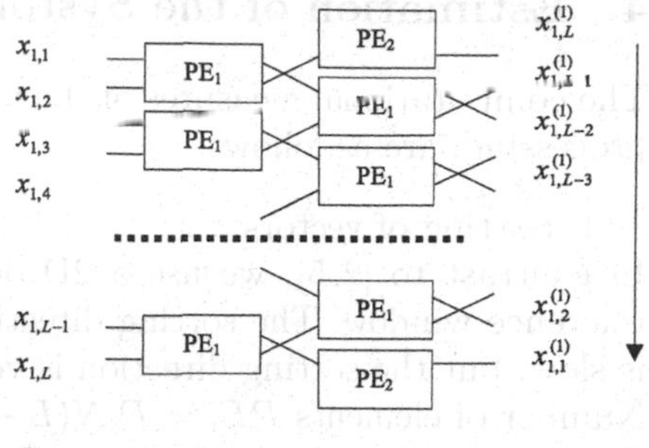

Fig. 3. Systolic architecture of the sorting algorithm S_2

are used for the obtaining of k^* and l^*. By using the processor elements PE_7, PE_8, PE_9 and PE_{11}, the RACPI CFAR processor obtains the estimates V, q_0 and the parameters of impulse interference (e and r_j). Through the medium of the processor element PE_{10} and the estimates of the impulse interference parameters, the RACPI CFAR processor chooses automatically the scale factor T_{RACPI} from memory of preliminary calculated values.

The analysis of the architecture shows that eleven types of processor elements are needed for the realization of a RACPI CFAR processor. The logical operations of the processor elements are shown in the following Table.

Processor elements	The logical operation of the processor elements
PE_1	$x^{out}_{1PE_1} = min(x^{in}_1, x^{in}_2)$ $x^{out}_{2PE_1} = max(x^{in}_1, x^{in}_2)$
PE_2	$x^{out}_{PE_2} = x^{out}_{PE_1}$
PE_3	$x^{out}_{PE_3} = T^x_i \sum_{j=1}^{i} x^{in}_j$
PE_4	$x^{out}_{PE_4} = \begin{matrix} 0, \text{if } x^{in}_1 < x^{in}_2 \\ 1, \text{if } x^{in}_1 > x^{in}_2 \end{matrix}$
PE_5	$x^{out}_{PE_5} = L - \sum_{j=1}^{L} x^{in}_j$
PE_6	$x^{out}_{PE_6} = \sum_{j=1}^{L} x^{in}_j$
PE_7	$x^{out}_{PE_7} = \begin{cases} x^{(1)}_{(x^{in}+1)/2}, & \text{if } x^{in} \text{is odd} \\ x^{(1)}_{(x^{in})/2} + x^{(1)}_{1+x^{in}/2}, & \text{if } x^{in} \text{is even} \end{cases}$
PE_8	$x^{out}_{PE_8} = \begin{cases} x^{(1)}_{(NL+x^{in}+1)/2}, & \text{if } (NL - x^{in}) \text{is odd} \\ x^{(1)}_{(NL+x^{in})/2} + x^{(1)}_{(NL+x^{in}+2)/2}, & \text{if } (NL - x^{in}) \text{is even} \end{cases}$
PE_9	$x^{out}_{PE_9} = (NL - x^{in})/NL$
PE_{10}	$x^{out}_{PE_{10}} = x^{in}_V x^{in}_T$
PE_{11}	$x^{out}_{PE_{10}} = x^{in}_{q_0}, \text{if} x^{in}_1 \neq 0$

4 Estimation of the Systolic Architecture Parameters

The computational measures of the processor, calculated for each stage of signal processing, are as follows:

1. Sorting of vectors:
In contrast to [2, 5], we use a 2D odd-even transposition sort method in the reference window. The sorting direction in the array is snake-like. This method is slow, but the sorting direction is very suitable for further processing.
Number of elements $PE_1 = D[N(L-1)+L(N-1)]$; $PE_2 = D(2N+2L)+2L$;
Number of steps: $T_1 = 4D$, where $D = N$, if $N \geq L$ or $D = L$, if $N < L$.

2. Censoring of vectors:
The censoring algorithm is realized as a parallel algorithm and in comparison with [2, 5]:
Number of elements $PE_3 = LN + L$; $PE_4 = LN + L$; $PE_5 = N + 1$; $PE_6 = 1$
Number of steps $T_2 = 4$.

3. Parameter estimation and scale factor T_{RACPI} formation:
In contrast to [2, 5], the noise level estimate is obtained after the censoring procedure. We estimate the impulse interference parameters as well in this paper. We suggest the scale factor to be chosen automatically from a matrix with preliminary calculated values.
Number of elements $PE_7 = 1$; $PE_8 = 1$; $PE_9 = 2$; $PE_{10} = 1$
Number of steps $T_3 = 2$.

4. Comparison:
Number of elements $PE_4 = 1$; $PE_{11} = 1$. Number of steps $T_4 = 1$.

Consequently, the computational measures of the systolic architecture are as follows: Total number of processor elements.

$$N_{PE} = \begin{cases} (3L+N)(N+1) + L(2N^2+L) + 9, & \text{if } N \geq L \\ L(N+1)(2L+3) + N + 9, & \text{if } N < L \end{cases} \tag{4}$$

Total number of computational steps.

$$T_0 = \begin{cases} 4N+7, & \text{if } N \geq L \\ 4L+7, & \text{if } N < L \end{cases} \tag{5}$$

The number of processor elements and computational steps of the API CFAR processor proposed in [2] are:

$$N_{PE}^{(1)} = LN(5N-1)/2 + L(3L+1) + 1; \quad T_0^{(1)} = \begin{cases} 2N+L+1, & \text{if } N \geq L \\ 3L+1, & \text{if } N < L \end{cases} \tag{6}$$

A similar systolic architecture of the API CFAR processor is offered in [5], but the detection probability of the new algorithm is better. The number of processor elements and computational steps of the API CFAR processor proposed in [5] are:

$$N_{PE}^{2} = (3N^2L + 6L^2 + L + 4)/2; \quad T_0^{(2)} = \begin{cases} 2(N+L)+1, & \text{if } N \geq L \\ 4L+1, & \text{if } N < L \end{cases} \tag{7}$$

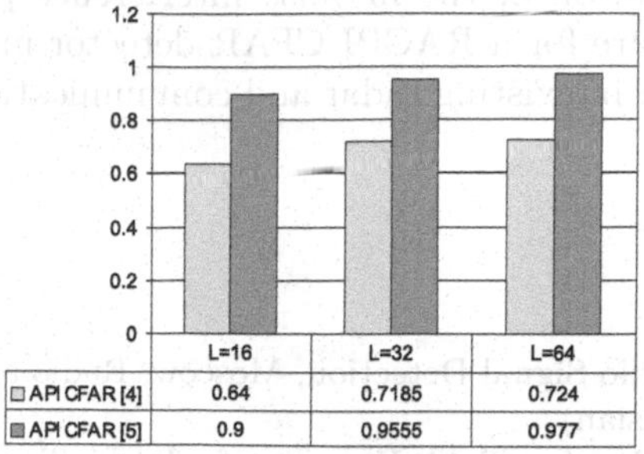

Fig. 4. Speed up of the computational process of RACPI CFAR algorithm toward algorithms in [2, 5]

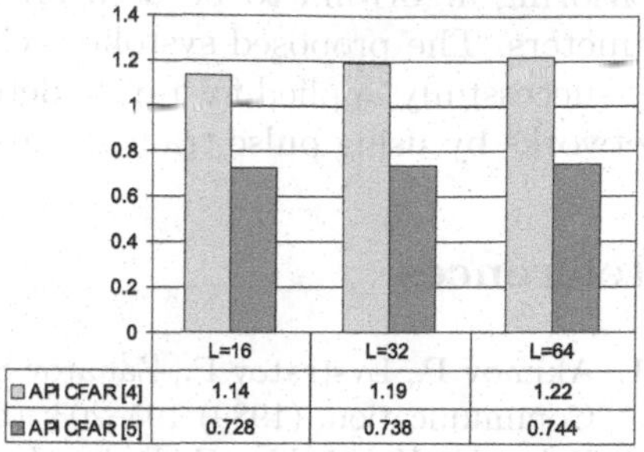

Fig. 5. The ratio between the number of processor elements in the RACPI CFAR algorithm and the algorithms in [2, 5]

The speed-up of the computational process of the systolic structure for a RACPI CFAR processor is obtained as:

$$K_{up} = T_0^{(*)}/T_0 \tag{8}$$

The speed-up of the computational process and the ratio between the number of processor elements in the RACPI CFAR algorithm and the algorithms in [2, 5], are shown on Fig.4 and Fig.5. The proposed systolic architecture for a RACPI CFAR processor contains more processor elements compared to the architectures presented in [2, 5], and the number of the computational steps is commensurable with the ones in [2, 5]. However, the new algorithm is continuously sorting all reference cells in the reference window and therefore better estimations of the noise level and the impulse interference parameters are achieved. The censoring algorithm is more effective. Due to that the detection probability of the proposed new RACPI CFAR algorithm is better.

5 Conclusions

A new parallel algorithm of a RACPI CFAR processor for target detection is proposed in this paper. The approach is similar to the parallel algorithms of the CFAR processors from [2, 5]. In contrast to the algorithms presented in [2, 5], the new algorithm is robust towards the changes of the impulse interference parameters.

By using a 2D odd-even transposition sort method in the reference cells, the sorting direction in the array is snake-like. In such cases the censoring algorithm uses all reference cells in the reference window. The described approach leads to the better performance of the RACPI CFAR processor compared to the ones proposed in [4, 5].

We propose the adaptive censoring algorithm to be realized as a parallel systolic architecture that reduces the computational time. We also suggest a censoring algorithm to be used for estimation of the impulse interference parameters. The proposed systolic architecture for a RACPI CFAR detector may be successfully applied for target detection in existing radar and communication networks by using pulse train signals.

References

1. Akimov P., Evstratov F., Zaharov S.: Radio Signal Detection, Moscow, Radio and Communication, (1989) 195–203, (in Russian).
2. Behar V., Kabakchiev C., Dukovska L.: Adaptive CFAR Processor for Radar Target Detection in Pulse Jamming, VLSI, **26** 11/12, (2000) 383–396.
3. Garvanov I., V. Behar, Chr. Kabakchiev: CFAR processors in pulse jamming, In: *Numerical methods and applications*, I. Dimov, I. Lirkov, S. Margenov, Z. Zlatev eds., *Lecture notes in computer sciences*, **2542**, Springer Verlag, (2003) 291–298.
4. Garvanov I., Chr. Kabakchiev: Sensitivity of API CFAR Detectors Towards Change of Input Parameters of Pulse Jamming, Proc. of IRS-2004, Warszawa, Poland, (2004) 233–238.
5. Garvanov I., Chr. Kabakchiev, P. Daskalov: Systolic architecture of adaptive post detection integration CFAR processor in binomial distribution pulse jamming, In: *Large scale scientific computing*, I. Lirkov, S. Margenov, J. Waśniewski, P. Yalamov eds., *Lecture notes in computer sciences*, **2907**, Springer-Verlag, (2004) 448–455.
6. Himonas S.: CFAR integration processors in randomly arriving impulse interference, IEEE Trans., vol. AES-30, 4, (1994) 809–817.
7. Himonas S., M. Barkat: Automatic censored CFAR detection for non-homogeneous environments, IEEE Trans., vol. AES-28, 1, (1992) 286–304.
8. Hwang J., J. Jong: Systolic architectures for 2-D ramk order filtering, Proc. of ICASAP, Sept. (1990) 90–99.
9. Hwang J., J. Ritcey: Systolic architectures for radar CFAR detectors, Proc. of ICASSP, Apr. (1990) 1025–1028.
10. Hwang J., J. Ritcey: Systolic architectures for radar CFAR detectors, IEEE Trans. vol., SP-39, 10, Oct., (1991) 2286–2295.
11. Lazarov A., Ch. Minchev: Antistealth ISAR Technology for Target Detection and Identification by Linear Frequency Modulated Signal, Proc of 2004 NASA ICNS Conference and Workshop, CD-R, (2004).
12. Lazarov A., Ch. Minchev: Algorithm for automatic determination of ISAR parameters, Proc. of 22 DASC Indianapolis, USA, Oct. 2003, CD-R, (2004).
13. Ritcey J., J. Hwang: Detection Performance and Systolic Architectures for OS-CFAR Detector, Proc. of Radar (1990) 112–116.
14. Rohling H.: Radar CFAR Thresholding in Clutter and Multiple Target Situations, IEEE Trans., vol. AES-19, No 4, (1983) 608–621.
15. http://mathworld.wolfram.com/OrderStatistic.html

On the Non Hierarchical Matrix Representation of the Negative, Non Integer Order Sobolev Norms*

Béla Kiss

Széchenyi István University, Győr, Egyetem tér 1, Hungary
bkiss@sze.hu

Abstract. In this paper a new cyclic matrix representation of the Sobolev norms H^a, $a \in (-1, 0)$ are presented. The matrix-vector multiplication by these matrices requires only $O(N \cdot \log(N))$ arithmetic operations, where N is the number of unknowns. The application of the new $H^{-1/2}$ norm representation as Schur complement preconditioning matrix requires only matrix-vector multiplication. The efficiency of the construction to elliptic problems has been verified by numerical tests.

1 Introduction

The Schur complement preconditioners play a determining role in the preconditioned conjugate gradient (PCG) algorithm based on fictitious domain and domain decomposition methods for elliptic problems [2, 4, 13, 15, 16]. Recently we have developed several simple and efficient cyclic Schur complement preconditioning matrix constructions [7, 8, 9]. The matrix constructions are based on the cyclic matrix representations of the Sobolev norm $H^{1/2}$. They can be applied to the preconditioning of frictional contact problems and can be combined with the multilevel techniques [10, 11, 12]. Their application as Schur complement preconditioning matrix, however, requires an approximate inverse matrix computation. This is a disadvantage of the constructions.

In this paper new cyclic matrix representations G_a of the Sobolev norms H^a, $a \in (-1, 0)$ are presented. The matrices have simple explicit form and can be used in the non cyclic case as well. The matrix–vector multiplication by these matrices can be computed by a very simple algorithm. The cost of this algorithm is $O(N \cdot \log(N))$ arithmetic operations for general, where N is the number of unknowns.

The application of the new matrix construction $G_{-1/2}$ as Schur complement preconditioning matrix requires only matrix-vector multiplication. The matrix $G_{-1/2}$ has the same computational complexity as the widely used BPX [13, 16] and H-matrix [1, 5, 6] type hierarchical preconditioning matrices, but its structure is much simpler. The preconditioning effect of $G_{-1/2}$ to elliptic problems has been verified by numerical tests.

* This work was partly supported by the Grant T043258 of the Hungarian National Research Found and by the PHARE CBC Award No. 2002 / 000-317-02-20.

I. Lirkov, S. Margenov, and J. Waśniewski (Eds.): LSSC 2005, LNCS 3743, pp. 663–670, 2006.

2 The New H^a, $a \in (-1,0)$ Norm Representations

Introducing the notation

$$C = C(c_0, c_1, \ldots, c_{n-1}, c_n, c_{n-1}, \ldots, c_1) = \begin{pmatrix} c_0 & c_1 & c_2 & \ldots & c_1 \\ c_1 & c_0 & c_1 & \ldots & c_2 \\ c_2 & c_1 & c_0 & \ldots & c_3 \\ \ldots & \ldots & \ldots & \ldots & \ldots \\ c_1 & c_2 & c_3 & \ldots & c_0 \end{pmatrix} \in \mathbb{R}^{N \times N}, \tag{1}$$

the piecewise linear finite element approximation of the H^1 Sobolev semi norm on the boundary $\partial\Omega$ of the unit circle Ω can be represented by the matrix

$$K = N \cdot C(k_0, k_1, \ldots, k_{n-1}, k_n, k_{n-1}, \ldots, k_1), \quad k_i = \begin{cases} 2 & \text{if} \quad i = 0 \\ -1 & \text{if} \quad i = 1 \\ 0 & \text{otherwise,} \end{cases} \tag{2}$$

where N is the number of unknowns and $n = [N/2]$.

The matrix representations of the H^a, $a \in (-1,1)$ Sobolev semi norms are defined by the formula

$$N^{-1+a} \cdot K^a, \tag{3}$$

and for the eigenvalues of these matrices the estimations

$$\lambda_{N^{-1+a} \cdot K^a, 0} = 0, \quad \frac{1}{10} \cdot \frac{i^{2a}}{N} < \lambda_{N^{-1+a} \cdot K^a, i} < 10 \cdot \frac{i^{2a}}{N}, \quad (i = 1, \ldots, n) \tag{4}$$

hold [2].

Our matrix representations of the H^a, $a \in (-1,0)$ Sobolev norms are

$$G_a = N^{-1+2a} \cdot C(g_0, g_1, \ldots, g_{n-1}, g_n, g_{n-1}, \ldots, g_1), \tag{5}$$

where

$$g_i = \begin{cases} \sum_{l=1}^{m} 2^{(-2a-1)l} & \text{if} \quad i = 0, \\ \sum_{l=j}^{m} 2^{(-2a-1)l} - 2^{(-2a-1)j} \cdot \left(\frac{i+1}{2^j} - 1\right) & \text{if} \quad 2^j - 1 \le i < 2^{j+1} - 1 \\ & \text{and} \quad 1 \le j < m, \\ 0 & \text{otherwise,} \end{cases}$$

with $2^{m-1} < N/2 \le 2^m$.

Since the symmetric cyclic matrices have a common eigenvector system [14], the spectral equivalence of the matrices G_a with the matrices $N^{-1+a} \cdot K^a$, $a \in (-1,0)$, in the $N = 2n = 2^{m+1}$ special case, follows from the next theorem.

Theorem 1. *If $N = 2n = 2^{m+1}$ and $m \ge 3$ then*

$$\frac{1}{32} \cdot (2^{a+1} - 1) \cdot \frac{i^{2a}}{N} < \lambda_{G_a, i} < 14 \cdot \frac{1}{1 - 2^a} \cdot \frac{i^{2a}}{N}, \quad (i = 0, 1, \ldots, n), \tag{6}$$

where the eigenvalues of G_a are of the form

$$\lambda_{G_a, i} = N^{-1+2a} \cdot \left(g_0 + 2 \cdot \sum_{k=1}^{2^m - 1} cos\left(\frac{ik\pi}{2m}\right) g_k + (-1)^i \, g_{2^m} \right).$$

This theorem can be extended for general N in a straightforward way, but it involves a large amount of calculations.

Let us introduce the notations

$$I_0 = g_0 + 2 \cdot \sum_{k=1}^{2^m-1} g_k + g_{2^m}, \quad I_1 = g_0 + 2 \cdot \sum_{k=1}^{2^m-1} g_k \cdot cos\left(\frac{k\pi}{2^m}\right) - g_{2^m}, \tag{7}$$

$$I_{2^i} = g_0 + 2 \cdot \sum_{k=1}^{2^{m+1-i}-1} g_k \cdot \cos\left(\frac{2^i k\pi}{2^m}\right) + g_{2^{m+1-i}}, \quad (1 \le i \le m-2), \tag{8}$$

$$I_{2^{m-1}} = g_0 + 2 \cdot \sum_{k=1}^{2^{m-1}-1} g_{2k} \cdot (-1)^k + g_{2^m}, \tag{9}$$

$$I_{2^m} = g_0 + 2 \cdot \sum_{k=1}^{2^m-1} g_k \cdot (-1)^k + g_{2^m}, \tag{10}$$

$$J_{2^i} = g_{2^{m+1-i}} + g_{2^m} + 2 \cdot \sum_{k=2^{m+1-i}+1}^{2^m-1} g_k \cdot \cos\left(\frac{k\pi}{2^{m-i}}\right), \quad (2 \le i \le m-2), \tag{11}$$

and

$$K_{2^i} = g_{2^m - 2^{m+1-i}} - g_{2^m} + 2 \cdot \sum_{k=1}^{2^{m+1-i}-1} g_{2^m - 2^{m+1-i}+k} \cdot \cos\left(\frac{k\pi}{2^{m+1-i}}\right), \tag{12}$$

$$(1 \le i \le m-1).$$

The proof of Theorem 1 is based on the following estimations of these quantities.

Lemma 1. *If $N = 2n = 2^{m+1}$ and $m \ge 3$ then the following estimations hold.*

$$\begin{aligned} &a)\ \tfrac{1}{4} \cdot 2^m < I_0 < \tfrac{2}{1-2^a} \cdot 2^m, \\ &\quad \tfrac{1}{16} \cdot \left(2^{a+1} - 1\right) \cdot 2^{(-2a)(m-i)} < I_{2^i} < \tfrac{16}{3} \cdot \tfrac{1}{1-2^a} \cdot 2^{(-2a)(m-i)}, \\ &\quad (i = 0, \dots, m), \\ &b)\ 0 < J_{2^i} < 2^{(-2a)(m-i)}, \quad (i = 2, \dots m-2), \\ &c)\ 0 < K_{2^i} < 2 \cdot 2^{(-2a)(m-i)}, \quad (i = 1, \dots, m-1). \end{aligned} \tag{13}$$

Proof. a) We prove the most complex case only, i.e. when $1 \le i \le m-2$. The form

$$\begin{aligned} I_{2^i} = {} & g_0 - 2 \cdot g_{2^{m-i}} + g_{2^{m+1-i}} \\ & + 2 \cdot \textstyle\sum_{k=1}^{2^{m-1-i}-1} \left(g_k - g_{2^{m-i}-k} - g_{2^{m-i}+k} + g_{2^{m+1-i}-k}\right) \cdot \cos\left(\frac{k\pi}{2^{m-i}}\right) \end{aligned}$$

of I_{2^i} is applied. Then the following estimations hold

$$\begin{aligned} I_{2^i} &> g_0 - 2 \cdot g_{2^{m-i}} + g_{2^{m+1-i}} \\ &\quad +2 \cdot \sum_{k=1}^{2^{m-2-i}-1} \left(g_k - g_{2^{m-i}-k} - g_{2^{m-i}+k} + g_{2^{m+1-i}-k}\right) \cdot \cos\left(\tfrac{\pi}{4}\right) \\ &> \sum_{k=0}^{2^{m-2-i}-1} \left(g_k - g_{2^{m-i}-k} - g_{2^{m-i}+k} + g_{2^{m+1-i}-k}\right) \\ &= \sum_{k=0}^{2^{m-2-i}-1} \left(g_k - g_{2^{m-i-1}+2^{m-i-2}+k}\right) \\ &\quad - \sum_{k=0}^{2^{m-2-i}-1} \left(g_{2^{m-i}+k} - g_{2^{m-i}+2^{m-1-i}+2^{m-2-i}+k}\right) \\ &> \sum_{k=0}^{2^{m-2-i}-1} \left(g_{2^{m-2-i}-1} - g_{2^{m-i}-1}\right) \\ &\quad - \sum_{k=0}^{2^{m-2-i}-1} \left(g_{2^{m-i}+k} - g_{2^{m-i}+2^{m-1-i}+2^{m-2-i}+k}\right) \\ &= \tfrac{1}{4}\left(\tfrac{1}{2^{(-2a-1)}} + \tfrac{1}{2^{(-2a-1)2}} - \tfrac{3}{4}\right) 2^{(-2a)(m-i)} > \tfrac{1}{16}\left(2^{a+1}-1\right) 2^{(-2a)(m-i)} \end{aligned}$$

and

$$\begin{aligned} I_{2^i} &< g_0 - g_{2^{m-i}} + 2 \cdot \sum_{k=1}^{2^{m-1-i}-1} \left(g_k - g_{2^{m-i}-k}\right) \\ &< 2 \cdot \sum_{k=0}^{2^{m-1-i}-1} \left(g_k - g_{2^{m-i}-1}\right) < 2 \cdot \sum_{k=0}^{m-2-i} 2^k \left(g_{2^k-1} - g_{2^{m-i}-1}\right) \\ &= 2 \cdot \sum_{k=0}^{m-2-i} 2^k \cdot \left(\sum_{l=k}^{m-i-1} 2^{(-2a-1)l}\right) \\ &= 2 \cdot \sum_{k=0}^{m-1-i} 2^k \cdot \left(\sum_{l=k}^{m-i-1} 2^{(-2a-1)l}\right) \\ &< 2 \cdot \sum_{k=0}^{m-1-i} 2^{(-2a-1)k} \left(\sum_{l=0}^{k} 2^l\right) \\ &= 4 \cdot \sum_{k=0}^{m-1-i} 2^{-2a} < \tfrac{16}{3} \cdot \tfrac{1}{1-2^a} \cdot 2^{(-2a)(m-i)}. \end{aligned}$$

b) Since for arbitrary linear function $a \cdot x + b$ the identity

$$\sum_{l=0}^{2j} \left(a \cdot \left(\frac{l\pi}{j}\right) + b\right) \cdot \cos\left(\frac{l\pi}{j}\right) = 0$$

holds, we obtain by the piecewise linearity of the matrix entries g_i that

$$\begin{aligned} J_{2^i} &= \sum_{k=m+1-i}^{m-1} \left(g_{2^{k+1}} - \tilde{g}_{2^{k+1}}\right) = \sum_{k=m+1-i}^{m-1} \left(-2^{(-2a-2)(k+1)} + 2^{(-2a-2)k}\right) \\ &= 2^{(-2a-2)(m+1-i)} - 2^{(-2a-2)m}, \end{aligned}$$

where

$$\tilde{g}_{2^{k+1}} = \sum_{l=k}^{m} 2^{(-2a-1)l} - 2^{(-2a-2)k}.$$

Consequently $J_{2^i} > 0$ and $J_{2^i} < 2^{(-2a)(m-i)}$.
c) The form

$$K_{2^i} = g_{2^m - 2^{m+1-i}} - g_{2^m} + 2 \cdot \sum_{k=1}^{2^{m-i}-1} \left(g_{2^m - 2^{m+1-i}+k} - g_{2^m-k}\right) \cdot \cos\left(\frac{k\pi}{2^{m+1-i}}\right),$$

of K_{2^i} is applied. Then the lower estimation is $K_{2^i} > 0$. The piecewise linearity of the construction implies

$$\begin{aligned} K_{2^i} &< 2 \cdot \sum_{k=0}^{2^{m-i}-1} \left(g_{2^m - 2^{m+1-i}+k} - g_{2^m-k}\right) \\ &= 2 \cdot \sum_{k=0}^{2^{m-i}-1} \left(g_{2^m - 2^{m+1-i}} - g_{2^m - 2^{m-i}}\right) \\ &= 2^{2m+1-2i} \cdot 2^{(-2a-2)(m-1)} < 2 \cdot 2^{(-2a)(m-i)}. \end{aligned}$$

Proof (Theorem 1.). When the index of an eigenvalue is zero or a power of two, the estimations follow from the identities

$$\lambda_{G_a,i} = N^{-1+2a} \cdot I_i, \quad (i = 0, 1, 2, 2^{m-1}, 2^m),$$

$$\lambda_{G_a,2^i} = N^{-1+2a} \cdot (I_{2^i} + J_{2^i}), \quad (i = 2, \ldots, m-2)$$

and from the estimations of Lemma 1. If the index of an eigenvalue is between two powers of two then it can be estimated as

$$\lambda_{G_a,2^{i+1}} < \lambda_{G_a,j} < \lambda_{G_a,2^i} + N^{-1+2a} \cdot K_{2^i},$$
$$(j = 2^i + 1, \ldots, 2^{i+1} - 1), \quad (i = 1, \ldots, m-1),$$

where the member K_{2^i} is used for estimating the last half cosine period in the odd j cases.

Remark 1. The H^a, $a \in (-1, 0)$ norm representations can be used in the non cyclic case as well. For example the calculation of an H^a semi norm of a vector $\underline{x}_\Gamma$, which is defined on $\Gamma \subset \partial\Omega$, can be performed by computing the H^a norm of the vector $\underline{x}_{\partial\Omega} = [\underline{\tilde{x}}_\Gamma, \underline{0}_{\partial\Omega\setminus\Gamma}]$ applying the matrix G_a, where $\underline{\tilde{x}}_\Gamma = \underline{x}_\Gamma - \frac{\sum_{i=1}^k x_{\Gamma,i}}{k} \cdot [1, 1, \ldots, 1]$.

3 Multiplication by the Matrices G_a, $a \in (-1, 0)$

When a matrix G_a is multiplied by a vector $\underline{x}$, a linear segment

$$(g_{2^j-1}, \ldots, g_k, \ldots, g_{2^{j+1}-2}) = \tag{14}$$

$$\left(a, \ldots, a - b \cdot \left(\frac{k+1}{2^j} - 1\right), \ldots, a - b \cdot \left(\frac{2^{j+1}-1}{2^j} - 1\right)\right)$$

G_a is multiplied by the segments

$$(x_{2^j-1+k-1}, \ldots, x_{2^{j+1}-2+k-1}), \quad (k = 1, \ldots, N) \tag{15}$$

of $\underline{x}$, where k is the row index and $x_j = x_{j-N}$ if $j > N$.

The products $p_k := \sum_{i=2^j-1}^{2^{j+1}-2} g_i \cdot x_{i+k}$ can be calculated by the following algorithm.

Let

$$p_1 := \textstyle\sum_{i=2^j-1}^{2^{j+1}-2} g_i \cdot x_i, \quad s_1 := \sum_{i=2^j}^{2^{j+1}-2} x_i,$$

and from $k := 2$ *to* N *let* (16)

$$p_k := p_{k-1} - \left(a - b \cdot \left(1 - \tfrac{1}{2^j}\right)\right) \cdot x_{2^{j+1}+k-2} + a \cdot x_{2^j+k-2} - \tfrac{b}{2^j} \cdot s_{k-1},$$

$$s_k := s_{k-1} + x_{2^{j+1}+k-2} - x_{2^j+k-1}.$$

The computational cost of this algorithm is $O(N)$ arithmetical operations. Since there are $O(\log(N))$ linear matrix segments, the total computational cost of a complete matrix-vector product computation is $O(N \cdot \log(N))$ arithmetic operations for general N.

4 Numerical Experiments

We have chosen the weak form of the following simple Neumann problem as a model problem: For $f \in L_2(\Omega)$ find $u \in H^1(\Omega)$ such that

$$\int_\Omega \left(\sum_{i=1}^{2} \partial_i u \partial_i v + uv \right) = \int_\Omega fv, \quad \forall v \in H^1(\Omega). \tag{17}$$

We solve this problem by the finite element method using equidistant triangulation with a grid size h and piecewise linear approximation. Let $V_h(\Omega)$ be the space of continuous piecewise linear functions defined on the given triangulation of Ω. $V_h(\partial\Omega)$ will denote the restrictions of $V_h(\Omega)$ to the boundary $\partial\Omega$ of Ω.

The approximation problem in $V_h(\Omega)$ is the following. Find $u_h \in V_h(\Omega)$ such that

$$\int_\Omega \left(\sum_{i=1}^{2} \partial_i u_h \partial_i v_h + u_h v_h \right) = \int_\Omega f v_h, \quad \forall v_h \in V_h(\Omega). \tag{18}$$

This approximation problem immediately leads to a system of linear algebraic equations

$$A u_h = f_h, \tag{19}$$

where

$$(Au_h, v_h)_{R^M} = \int_\Omega \left(\sum_{i=1}^{2} \partial_i u_h \partial_i v_h + u_h v_h \right), \quad (f_h, v_h)_{R^M} = \int_\Omega f v_h,$$

for all $v_h \in V_h(\Omega)$. M denotes the number of grid points of the triangulation of Ω. The matrix of this linear system is symmetric, positive definite and so it has a unique solution [3].

In the usual nodal basis $\Phi = \{\phi_1, \ldots, \phi_{M_\Omega}, \phi_{M_\Omega+1}, \ldots, \phi_{M_\Omega+M_{\partial\Omega}}\}$ the linear system can be rewritten in the block form

$$\begin{pmatrix} A_{\Omega,\Omega} & A_{\Omega,\partial\Omega} \\ A_{\partial\Omega,\Omega} & A_{\partial\Omega,\partial\Omega} \end{pmatrix} \begin{pmatrix} u_\Omega \\ u_{\partial\Omega} \end{pmatrix} = \begin{pmatrix} f_\Omega \\ f_{\partial\Omega} \end{pmatrix}, \tag{20}$$

where the indices "Ω" and "$\partial\Omega$" denote the nodes belonging to the interior of Ω and to the boundary $\partial\Omega$ of Ω, respectively.

The Schur complement part of A is

$$S_C = A_{\partial\Omega,\partial\Omega} - A_{\partial\Omega,\Omega} A_{\Omega,\Omega}^{-1} A_{\Omega,\partial\Omega}. \tag{21}$$

In the remainig part of this section we present some numerical results which show the effect of the optimal Schur complement preconditioner matrix $G_{-1/2}$.

The shape of the investigated test domains can be seen in Figure 1 and 2. The corresponding numerical results are reported in Table 1 and 2. The calculation of the matrix condition numbers was made by using MATLAB.

From our test results we can conclude that the new matrix $G_{-1/2}$ is an efficient preconditioning matrix. The advantage of this matrix construction is that it does not require a hierarchical level of discretization and the matrix–vector multiplication by this matrix can be computed by a very simple algorithm.

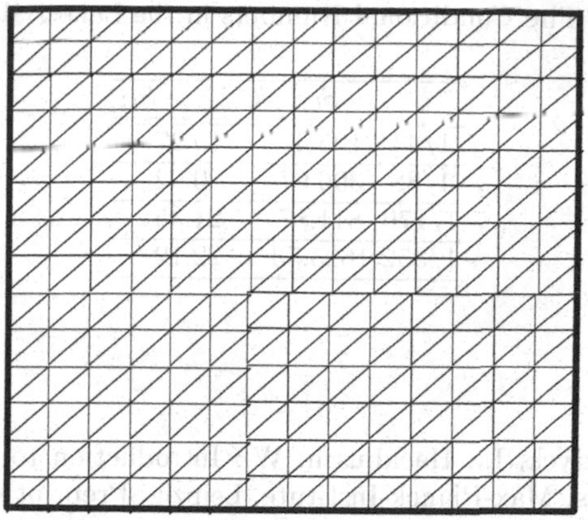

Fig. 1. A Uniform Discretization of the Unit Square

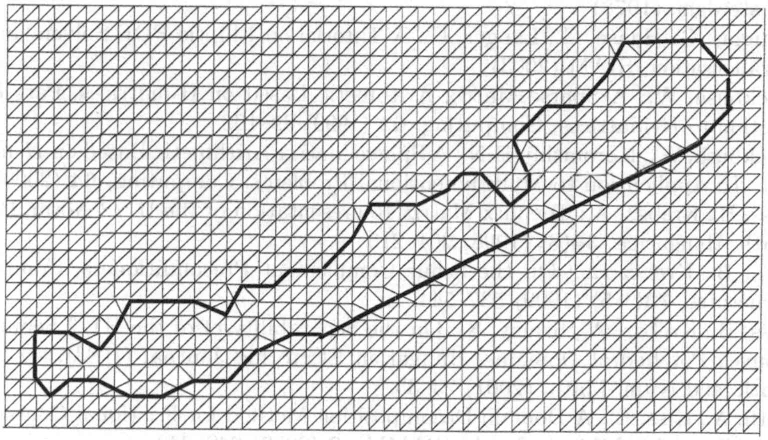

Fig. 2. A Uniform Discretization of the Lake Balaton

Table 1. The Conditional Numbers in the Unit Square Case

h	S_C	$G_{-1/2} \cdot S_C$
1/4	27.43	3.27
1/8	77.43	3.92
1/16	179.87	4.35
1/32	417.67	4.72
1/64	732.12	4.90
1/128	1448.56	5.08

Table 2. The Conditional Numbers in the Lake Balaton Case

h	S_C	$G_{-1/2} \cdot S_C$
1/44	195.43	19.27
1/88	396.67	20.92
1/176	811.87	21.35
1/352	1625.53	22.12

References

1. Börm, S., Grasedyck, L., Hackbusch, W.: Introduction to Hierarchical Matrices with Applications, Max-Planck-Institute, Leipzig, Preprint no. 18 (2002)
2. Bramble, J.H., Pasciak,J.E., Schatz.A.H.: The Construction of Preconditioners for Elliptic Problems by Substructuring I-IV. Math. of Comp. **47** (175) (1986) 103–134, **49** (184) (1987) 1–16, **51** (184) (1988) 415–430, **53** (187) (1989) 1–24
3. Ciarlet, P.G.: The Finite Element Method for Elliptic Problems, North-Holland, Amsterdam, (1978)
4. Douglas, C., Haase, G., Langer, U.: A Tutorial on Elliptic PDE Solvers and Their Parallelisation, SIAM (2003)
5. Hackbush, W.: A Sparse Matrix Arithmetic Based on H-matrices, Part I: Introduction to H-matrices, Computing, **62** (1999) 89–108
6. Hackbush, W., Koromskij, B.N.: A Sparse Matrix Arithmetic Based on H-matrices, Part II: Application to Multidimensional Problems, Computing, **64** (2000) 21–47
7. Kiss, B., Krebsz, A.: Toeplitz Matrix Representation of the H1/2 Norm, Periodica Math. Hungarica, **34**(3) (1998) 201–210
8. Kiss, B., Krebsz, A.: On the Schur Complement Preconditioners, Computers & Structures, **73**(1-5) (1999) 537–544
9. Kiss, B., Krebsz, A., Szalay, K.: On the Separability of of the H1/2 Seminorm on Convex Polyhedral Domains, Hungarian Electornic Journal (HEJ), ser. Applied Math. and Num. Math, (2001) 23–35.
10. Lotfi, A., Kiss, B.: A Domain Decomposition Method for Frictional Contact Problems, Proc. Appl. Math. Mech. (PAMM), **2** (2003) 242–243
11. Lotfi, A.: A Preconditioned Domain Decomposition Algorith for Frictional Contact Problem, Proc. of Int. Conf. on Numerical Analysis and Applied Math. (ICNAAM), (2004) 235–238
12. Miletics, P.E, Molnárka, G.: Taylor Series Methods with Numerical Derivatives, Hungarian Electornic Journal (HEJ), ser. Applied Math. and Num. Math. (2002) 1–15
13. Oswald, P.: Multilevel Finite Element Approximation, Theory and Application, B.G. Teubner, Stuttgart, (1994)
14. Pickering, P.: An Introduction to FFT Method for PDEs, with Applications, Research Studies Press, Letchworth, Herts, (1986)
15. Steinbach, O, Wendland, W.L.: The Construction of Some Efficient Preconditioners in Boundary Element Methods, Adv. Comput. Math. **9** (1998) 191–216
16. Tong, C.H., Chan, T.F, Kuo, C.J.: A Domain Decomposition Preconditioner Based on a Change to a Multilevel Nodal Basis, SIAM J. Sci. Stat. Comput. **12** (1991) 1486–1495

Parallel Computing of Thermoelasticity Problems

Roman Kohut, Jiří Starý, Radim Blaheta, and Karel Krečmer

Institute of Geonics, Academy of Sciences of the Czech Republic,
Studentská 1768, 70800 Ostrava-Poruba, Czech Republic
kohut@ugn.cas.cz, stary@ugn.cas.cz,
blaheta@ugn.cas.cz, krecmer@ugn.cas.cz

Abstract. The paper deals with a finite element solution of transient thermoelasticity problems. For each time step the system of linear algebraic equations is solved using a parallel solver based on the overlapping domain decomposition method. The time steps are chosen adaptively. The results of numerical tests on a large benchmark problem are presented.

1 Introduction

In this paper, we consider the thermoelasticity problem which is not fully coupled. The deformations are slow and do not influence temperature fields and we can compute them only in predefined time points as a post-processing to the solution of the heat equations. Thus the problem can be divided in two parts. Firtsly, the temperature distribution is determined by the solution of the nonstacionary heat equation, secondly, at given time points the linear elasticity problem is solved. The numerical solution of both problems leads to the repeated solution of large systems of linear equations and our aim is to find efficient and parallelizable iterative solution methods. For elasticity problems, we got an experience with the Schwarz methods. In this case, both theory and experiments showed good efficiency of two-level Schwarz methods (see [1]). For the evolution heat equations, the theory indicates (see [4]) that even one-level Schwarz methods could be efficient. We shall investigate this fact as well as some other aspects of the numerical solution by numerical tests on a large geotechnical problem arising from the assessment of nuclear waste repositories, see [2].

2 Thermoelasticity

The thermoelasticity problem is concerned with finding the temperature $\tau = \tau(x,t)$ and the displacement $u = u(x,t)$,

$$\tau : \Omega \times (0,T) \rightarrow R, \; u : \Omega \times (0,T) \rightarrow R^3$$

I. Lirkov, S. Margenov, and J. Waśniewski (Eds.): LSSC 2005, LNCS 3743, pp. 671–678, 2006.

that fulfill the following equations

$$\kappa\rho\frac{\partial\tau}{\partial t} = k\sum_i \frac{\partial^2\tau}{\partial {x_i}^2} + Q(t) \qquad in \quad \Omega\times(0,T)\,, \tag{1}$$

$$-\sum_j \frac{\partial\sigma_{ij}}{\partial x_j} = f_i \quad (i=1,\ldots,d) \quad in \quad \Omega\times(0,T)\,, \tag{2}$$

$$\sigma_{ij} = \sum_{kl} c_{ijkl}\left[\varepsilon_{kl}(u) - \alpha_{kl}(\tau-\tau_0)\right] \quad in \quad \Omega\times(0,T)\,, \tag{3}$$

$$\varepsilon_{kl}(u) = \frac{1}{2}\left(\frac{\partial u_k}{\partial x_l} + \frac{\partial u_l}{\partial x_k}\right) \quad in \quad \Omega\times(0,T), \tag{4}$$

together with the corresponding boundary and initial conditions.

The finite element method is based on a weak formulation of the heat conduction problem, which can be written after application of Green's theorem as follows:

find $\tau = \tau(x,t)$, $(x,t) \in \Omega\times(0,T)$, $\tau(.,t) \in V_D$ such that the following equations hold:

$$(\kappa\rho\,\dot\tau, v)_0 + a(t,\tau,v) = b(t,v) \quad \forall v \in V_0,\ t\in(0,T), \tag{5}$$

$$(\tau(x,0), v)_0 = (\tau_0, v)_0 \quad \forall v \in V_0. \tag{6}$$

In these equations

$$V_0 = \left\{v \in H^1(\Omega) : v = 0 \quad \text{on} \quad \Gamma_0\right\},$$

$$V_D = \left\{v \in H^1(\Omega) : v = \hat\tau \quad \text{on} \quad \Gamma_0\right\},$$

where $(\,,\,)_0$ is the scalar product in the space of square-integrable functions $L_2(\Omega)$, $H^1(\Omega) \subset L_2(\Omega)$ is the Sobolev space of functions having first weak derivatives in the space $L_2(\Omega)$. For a, b there holds

$$a(t,\tau,v) = \int_\Omega \sum_{ij} k_{ij}\frac{\partial\tau}{\partial x_j}\frac{\partial v}{\partial x_i}dx + \int_{\Gamma_2} H\tau v ds\,, \tag{7}$$

$$b(t,v) = \int_\Omega Q(t)v dx - \int_{\Gamma_1} q v ds + \int_{\Gamma_2} H\hat\tau_{out} v ds\,. \tag{8}$$

3 Time Discretization

We find $\tau = \tau(x,t), (x,t) \in \Omega\times(0,T), \tau(\cdot,t) \in V_D$ such, that (5), (6) hold. We transfer the problem to the FEM formulation. Consider the space $V_{0,h} = $ span $\{\varphi\} \subset V_0$. We find

$$\tau_h(x,t) = \hat\tau(x,t) + \sum_i \tau_i(t)\varphi_i(x), \tag{9}$$

where $\hat{\tau}(x,t)$ acquires the prescribed values on Γ_0. In our case we use the function $\hat{\tau}(x,t) = \tau_0(x)$, which represents the initial condition for the non-stationary problem. Substituting (9) into the equation (5), for the determination of the coefficient vector $\underline{\tau} = [\tau_i]$ we receive the system of linear differential equations

$$M_h\dot{\underline{\tau}}(t) + A_h(t)\underline{\tau}(t) = \underline{b}_h(t) \quad \forall t \in (0,T),$$
$$\frac{1}{\kappa\rho}M_h\underline{\tau}(0) = 0, \tag{10}$$

where

$$M_h = [(\kappa\rho\varphi_i, \varphi_j)_0],$$
$$A_h(t) = [a(t, \varphi_i, \varphi_j)], \tag{11}$$
$$\underline{b}_h(t) = [b(t,\varphi_i) - a(t,\hat{\tau},\varphi_i)].$$

We divide the interval $< 0, T >, 0 = t_0 < t_1 < ... < t_p = T, \Delta_i = t_i - t_{i-1}$. We find values $\underline{\tau}_i^j = \underline{\tau}_i(t_j)$. Using the time discretization we obtain

$$M_h\frac{1}{\Delta_j}(\underline{\tau}^j - \underline{\tau}^{j-1}) + \theta A_h(t_j)\underline{\tau}^j + (1-\theta)A_h(t_{j-1})\underline{\tau}^{j-1} = \varphi^j,$$
$$\underline{\tau}^0 = 0, \tag{12}$$

where $\theta \in< 0, 1 >, \varphi^j = \theta\underline{b}_h(t_j) + (1-\theta)\underline{b}_h(t_{j-1})$. In each time step we have to solve the system

$$[M_h + \Delta_j\theta A_h(t_j)]\underline{\tau}^j = [M_h - (1-\theta)\Delta_j A_h(t_{j-1})]\underline{\tau}^{j-1} + \Delta_j\varphi_j. \tag{13}$$

For $\theta = 0$ we obtain the so-called explicit Euler scheme, for $\theta = 1$ we obtain the backward Euler (BE) scheme, $\theta = 0.5$ gives the Crank-Nicholson(CN) scheme. In our case we will use the BE scheme. Because the matrix A_h is not time dependent, we can write $A_h(t_j) = A_h$. Then the system (13) is replaced by

$$[M_h + \Delta_j A_h]\underline{\tau}^j = M_h\underline{\tau}^{j-1} + \Delta_j\varphi_j, \tag{14}$$

where $\Delta_j\varphi_j = \Delta_j\underline{b}_h(t_j)$. If we substitute

$$\underline{\tau}^j = \underline{\tau}^{j-1} + \Delta\underline{\tau}^j$$

into (14), we obtain the system of equations for the increment of temperature

$$[M_h + \Delta_j A_h]\Delta\underline{\tau}^j = \Delta_j(\varphi_j - A_h\underline{\tau}^{j-1}). \tag{15}$$

To optimize the computation we use the adaptive choice of time-step. The BE method allows to use the adaptive time stepping scheme based on a local comparison of the BE and CN steps (see [2]). Let $\underline{\tau}^j$ is the solution of the system (13) for $\theta = 1$. If this solution is considered as the initial approximation for the solution of system (13) for $\theta = 0.5$ (CN scheme) by the simple Richardson's method, $\underline{\tau}_1^j = \underline{\tau}_0^j + r$, where $\underline{\tau}_0^j = \underline{\tau}^j$,

$$r^j = (M_h + 0.5\Delta_j A_h)\underline{\tau}^j - (M_h - 0.5\Delta_j A_h)\underline{\tau}^{j-1} - 0.5\underline{b}_h(t_j) - 0.5\underline{b}_h(t_{j-1}), \tag{16}$$

then the first iteration $\underline{T}_1^j$ of the Richardson's method presents an approximation of the solution $\underline{T}_{CN}^j$ of the system (13) for $\theta = 0.5$. Thus the time steps can be controlled with the aid of the ratio $\eta = \frac{\|\underline{T}^j - \underline{T}_{CN}^j\|}{\|\underline{T}^j\|} = \frac{\|r^j\|}{\|\underline{T}^j\|}$ according to the following algorithm ($k = 1, 2, \ldots$ denotes the adaptive changes, $\underline{T}^{j,k}$ corresponding solutions of system (13)):

for $k = 1, 2, \ldots$ **until** *stop* **do**
  solve system (13) $\rightarrow \underline{T}^{j,k}$, **compute** $r^{j,k}$ a η_k
  if $\eta_k < \varepsilon_{min}$ **then** $2\Delta_j \rightarrow \Delta_j$
  if $\eta_k > \varepsilon_{max}$ **then** $\Delta_j/2 \rightarrow \Delta_j$
  if $\eta_k \in < \varepsilon_{min}, \varepsilon_{max} >$ **or** $\eta_k < \varepsilon_{min}$ & $\eta_{k-1} > \varepsilon_{max}$ **then**
    $\Delta_{j+1} = \Delta_j, \underline{T}^j = \underline{T}^{j,k}$, *stop*
  if $\eta_k > \varepsilon_{max}$ & $\eta_{k-1} < \varepsilon_{min}$ **then**
    $\Delta_{j+1} = \Delta_j/2, \quad \underline{T}^j = \underline{T}^{j,k-1}$, *stop*
end

4 The Solution of Linear Equations

For the solution of the linear system $B_h \Delta \underline{T}^j = (M_h + \Delta_j A_h)\Delta \underline{T}^j = f_j$ (15) we shall use the preconditioned CG method. In the sequential code, the preconditioning is given by the incomplete factorization, in the parallel codes the preconditioning is given by the additive Schwarz method. In the case of Schwarz method the domain is divided into m subdomains Ω_k (in our case the domain is divided only in vertical direction), nonoverlaping subdomains Ω_k are then extended to domains Ω_k' in such a way that overlaping between the subdomains are given by two or more layers of elements. If B_{kk}' are the FE matrices corresponding to problems on Ω_k', I_k' and $R_k' = (I_k')^T$ are the interpolation and restriction matrices, respectively, then introduced matrices $B_{kk}' = R_k' B I_k'$ allow to define one-level additive Schwarz preconditioner G,

$$g = Gr = \sum_{k=1}^{m} I_k' {B_{kk}'}^{-1} R_k' r.$$

If the Schwarz method is used for elliptic problems, the efficiency of the preconditioner decreases with the increasing number of subproblems and it is necessary to use in the preconditioner the coarse mesh solution. For the parabolic problems it is proved in [4] that under the assumption that Δ_j/H^2 is reasonably bounded, the algorithms remain numerically scalable even if the coarse mesh space is eliminated. Here Δ_j is in order of the time stepsize and H is the diameter of the largest subdomain.

The system matrix B_h is composed from two matrices, the matrix M_h which is not M-matrix (all matrix elements are positive for linear FE basis functions) and the matrix $\Delta_j A_h$ which is M-matrix in many practical situations, e.g., if the heat flow is isotropic and the inner angles of tetrahedra do not exceed $\pi/2$. For small values of Δ_j the matrix $M_h + \Delta_j A_h$ is not M-matrix and the incomplete factorization fails for preconditioning. But it is possible to apply the incomplete factorization to the matrix $M_h^L + \Delta_j A_h$, where M_h^L is the lumped matrix to the matrix M_h, which means that its diagonal has diagonal elements equal to the sum of the elements on the corresponding row.

5 Numerical Tests

The numerical tests were realized on the KBS-3 benchmark problem which represents a model of prototype nuclear waste repository located at Äspö in Sweden [3]. A 3D model is shown in Figure 1. The computational domain has dimensions $200 \times 200 \times 100$ m. The model is discretized by linear tetrahedral FE with 15 088 320 tetrahedra, 2 586 465 DOF for heat equations and 7 759 395 DOF for elasticity computations. The time interval is selected to be 50 years, the adaptive time stepping begins with the step of 0.0001 year. Fore more details see [2].

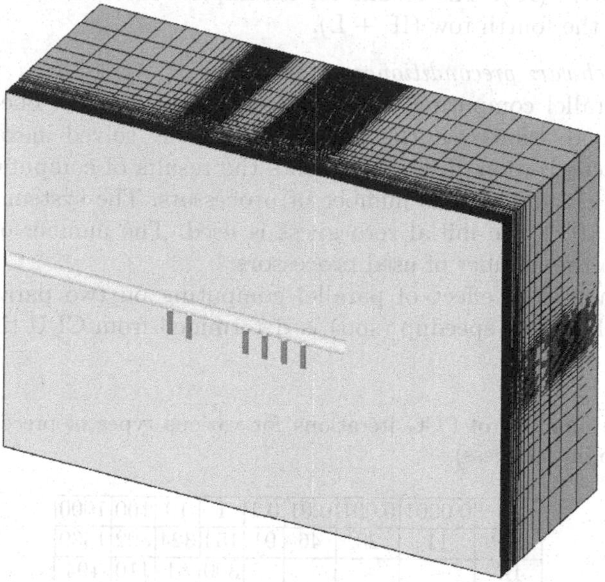

Fig. 1. The model problem KBS-3

The parallel computations were performed on the IBM xSeries 455 computer (symmetric multiprocessor (SMP), 8 processors) with Intel Itanium2 1.3 GHz 64bit Processor and the PC cluster THEA with 8 AMD Athlon 1.4 GHz, 1.28 GB RAM computer nodes, the sequential computations were performed on the IBM xSeries 455 computer. The parallel programming uses OpenMP and MPI paradigms on SMP computer and MPI paradigm on the PC cluster.

a) *Preconditioning by incomplete factorization*
We shall start with numerical investigation of the solution of linear system arising in each time step of the discretization of heat transfer problem. In Table 1, we present test results concerning the solution of system (15) with the preconditioning given by the MIC(0) incomplete factorization of both the original matrix and the lumped matrix, respectively, in dependence of the time step size. The results for diagonal preconditioning are also presented. The system is solved with the accuracy 10^{-6} for one time step, the initial zero guess is used. The tests used the sequential code. We can see that the condition number decreases with the decrease of Δt. The numbers of PCG iterations for both the original matrix and lumped matrix, respectively, are the same for $\Delta t > 1$. For small Δt the incomplete factorization of the original matrix cannot be used as a preconditioner.

The heading row of Table 1 contains the given values of tested time steps. The numbers of PCG iterations for diagonal preconditioning (DP) are presented in the second row. The third row corresponds to the incomplete factorization of the origin matrix (IF). The results for incomplete factorization of the lumped matrix are in the fourth row (IF + L).

b) *Additive Schwarz preconditioner*
To enable parallel computing, we shall test now the use of one-level additive Schwarz method. Moreover, the local problems are solved inexactly by using incomplete factorization. Table 2 presents the results of computations for various values of Δt and various number of processors. The system is solved with the accuracy 10^{-6}, the initial zero guess is used. The number of subproblems coresponds to the number of used processors.

Table 3 shows the effect of parallel computing on two parallel computers described earlier. The speedup (spd) is determined from CPU time for 1 PCG iteration.

Table 1. The numbers of PCG iterations for various types of preconditioning (one time step, zero initial guess)

Δt	0.0001	0.001	0.01	0.1	1	10	100	1000
DP	11	20	46	91	154	324	802	1339
IF	-	-	-	-	300	61	110	194
IF+L	12	12	17	27	40	61	110	194

Table 2. The dependence of the numbers of PCG iterations on the time step size and the numbers of subdomains (processors)

#P \ Δt	0.0001	0.001	0.01	0.1	1.0	10.0	100.0	1000.0
1	12	12	17	27	40	61	110	194
2	14	14	17	27	40	65	134	223
4	13	13	17	27	40	65	140	240
8	15	17	21	27	40	71	162	282

Table 3. The solution (one time step $\Delta t = 1$) using various parallel platforms

	1 processor			2 processors			4 processors			8 processors		
	it	CPU	spd	it	CPU	spd	it	CPU	spd	it	CPU	spd
SMP, OpenMP	40	98s	-	40	52s	1.88	40	28s	3.50			
SMP, MPI	39	115s	-	40	60s	1.95	40	31s	3.76			
cluster, MPI	39	133s	-	40	71s	1.91	40	37s	3.66	40	22s	5.94

Table 4. The accuracy of the solution vector for various values of parameters $\varepsilon_{min}, \varepsilon_{max}, n_t$

$< \varepsilon_{min}, \varepsilon_{max} >, n_t$	CPU t.	PCG it	time st.	$\Vert\tau - \tau_{ex}\Vert_{l_2}/\Vert\tau\Vert_{l_2}$	$\Vert\tau - \tau_{ex}\Vert_0$
$< 0.001, 0.01 >, 5$	542s	664	47	1.16 %	0.658
$< 0.001, 0, 01 >, 10$	626s	759	70	0.95 %	0.539
$< 0.001, 0.01 >, 20$	718s	836	112	0.91 %	0.515
$< 0.0001, 0.001 >, 5$	806s	967	115	0.48%	0.269
$< 0.0001, 0.001 >, 10$	867s	998	135	0.46 %	0.262
$< 0.0001, 0.001 >, 20$	947s	1104	176	0.29 %	0.166
$< 0.0001, 0.001 >, 50$	1188s	1299	296	0.13 %	0.076

c) *Solution of the full nonstacionary problem*
In the previous paragraph, we investigated solution in one time step. Now, we shall consider the whole sequence of time steps. In this, the system (15) is solved with accuracy 0.01 and with the initial guess taken from the previous time step. The adaptive choice of the time step size Δt is done . The time steps are controlled according to the algorithm shown in section 3. The decision about changing of Δt depends on input parameters $\varepsilon_{min}, \varepsilon_{max}$. The test for adaptivity is done after each n_t time steps (n_t is input parameter). In Table 4 we present the results of solution for various values of parameters $\varepsilon_{min}, \varepsilon_{max}, n_t$. The solution vector is compared with "exact" solution vector τ_{ex} (the solution for $\varepsilon_{min} = 0.00001, \varepsilon_{max} = 0.0001, n_t = 100$). The system (15) is solved with accuracy 0.001), the adaptive time step is starting with $\Delta t = 0.0001$. We can see that the presented relative errors (the fifth column) depend only slightly on parameter n_t. If we accept that for practical problems the relative errors close to one percent are suitable we can use input parameters $\varepsilon_{min} = 0.001, \varepsilon_{max} = 0.01, n_t = 5$. The test were performed on the IBM xSeries 455 computer (OpenMP) using four processors.

d) *The solution of the elasticity problem*
The displacements and stresses are determined at given time points and represent a post-processing for the heat conduction problem. The parallel computation of the system with 7759395 unknowns was done on the cluster Thea. The presented results (see Table 5) show that the usage of coarse grid makes the preconditioner more effective.

The determination of the displacements and stresses was done for four time points: 1, 10, 20, 50 years. The whole CPU time including the solution of nonstacionary heat problem was about 2400 seconds.

Table 5. The results of the parallel solution of the elasticity problem

	without coarse grid			with coarse grid		
n.proc.	it	CPU	CPU/it	it	CPU	CPU/it
2	560	4747s	8.46s	208	1836s	8.77s
4	606	2733s	4.5s	201	939s	4.64s
7	651	1798s	2.76s	206	620s	2.99s

6 Conclusion

In the paper, the solution of the thermoelasticity problem is described. The parallel computing with the adaptive time step size shows to be very efficient for the solution of large practical problems.

Acknowledgments

The work is supported by the Grant Agency of the Czech Republic under the project GAČR 105/04/P036 and by the Academy of Sciences of the Czech Republic under the project S3086102.

References

1. Blaheta, R.: Space decomposition preconditioners and parallel solvers. In: *Numerical Mathematics and Advanced Applications*, Feistauer, M. et al. (eds), Springer-Verlag, Berlin (2004) 20–38.
2. Blaheta, R., Byczanski, P., Kohut, R., Starý, J. : Algorithm for parallel FEM modelling of thermo–mechanical phenomena arising from the disposal of the spent nuclear fuel. In: Konečný, P. at al. (eds), to be published by A. A. Balkema, Amsterdam, 2005.
3. Börgesson, L., Hernelind, J.: Coupled thermo-hydromechanical calculations of the water saturation phase of a KBS-3 deposition hole, TR99-41 (1999), SKB Stockholm
4. Cai, X.-C.: Multiplicative Schwarz methods for parabolic problems, *SIAM Jounal on Scientific Computing*, **15** (1994), 587–603.

On Symmetric Part PCG for Mixed Elliptic Problems

Tamás Kurics

Department of Applied Analysis, ELTE University, H-1117 Budapest, Hungary
fantom@cs.elte.hu

Abstract. The CGM is studied for nonsymmetric elliptic problems with mixed boundary conditions. The mesh independence of the convergence is in focus when symmetric part preconditioning is applied to the FEM discretizations of the BVP. Computations in 2 dimensions are presented to illustrate and complete the theoretical results.

Keywords: Conjugate gradient method, preconditioning, symmetric part, mesh independence, mixed boundary conditions, experiments.

1 Introduction

The conjugate gradient method is a widespread way of solving large nonsymmetric linear algebraic systems arising from discretized elliptic problems, see the book [3] where a comprehensive summary is given on the convergence of the CGM. When for elliptic problems the discretization parameter tends to 0, the convergence estimates deteriorate under refinement, i.e., the number of iterations for prescribed accuracy tends to ∞. The remedy is suitable preconditioning [3], which sometimes relies on Hilbert space theory [15, 24]. Moreover, it has been shown in [15] that the preconditioned CGM can be competitive with multigrid methods.

This paper studies symmetric part preconditioning, which means using the symmetric part

$$\mathbf{S} = (\mathbf{A} + \mathbf{A}^T)/2$$

as preconditioner for the system matrix $\mathbf{A}$. It has been proved an efficient tool in this respect. It has been introduced and analysed in [13, 24], see also [2, 25], and efficiently applied to nonsymmetric elliptic problems (convection-diffusion equations). Linear convergence results for such PCG methods are included in the rigorously described framework of equivalent operators in Hilbert space [15, 21], which provides mesh independence of the estimates for the discretized problems. The CGM in Hilbert space has been studied in [7, 8]: superlinear convergence has been proved in Hilbert space and, based on this, mesh independence of the superlinear estimate has been derived for the discretized problems. The mentioned papers consider PDEs with Dirichlet boundary conditions.

I. Lirkov, S. Margenov, and J. Waśniewski (Eds.): LSSC 2005, LNCS 3743, pp. 679–686, 2006.

In this paper we consider an elliptic convection-diffusion equation with mixed boundary conditions:

$$\begin{cases} Lu \equiv -\Delta u + \mathbf{b} \cdot \nabla u + \alpha u = g \\ u_{|\Gamma_D} = 0, \qquad \frac{\partial u}{\partial \nu} + \beta u_{|\Gamma_N} = 0. \end{cases} \tag{1}$$

satisfying the following conditions (Assumptions BVP):

(i) $\Omega \subset \mathbf{R}^d$ is a bounded piecewise C^1 domain; Γ_D, Γ_N are disjoint open measurable subparts of $\partial\Omega$ such that $\partial\Omega = \overline{\Gamma}_D \cup \overline{\Gamma}_N$;
(ii) $\mathbf{b} \in C^1(\overline{\Omega})^d$, $\alpha \in L^\infty(\Omega)$, $\beta \in L^\infty(\Gamma_N)$ and $\alpha, \beta \geq 0$;
(iii) we have the coercivity properties $\hat{\alpha} := \alpha - \frac{1}{2}\operatorname{div}\mathbf{b} \geq 0$ in Ω and $\hat{\beta} := \beta + \frac{1}{2}(\mathbf{b} \cdot \nu) \geq 0$ on Γ_N;
(iv) $g \in L^2(\Omega)$;
(v) either $\Gamma_D \neq \emptyset$, or $\hat{\alpha}$ or $\hat{\beta}$ is not constant zero.

When one tries to consider mixed boundary value problems in the operator setting, care must be taken with the boundary conditions in the preconditioning operator, as turns out from the analysis in [21]. The presence of mixed boundary conditions raises certain difficulties as well when the symmetric part preconditioning is studied, hence the latter requires a more general weak approach, see in the recent paper [19].

The goal of this paper is to confirm and complete the theoretical results with numerical experiments. We first provide a brief summary of the theoretical results, mainly relying on [19]. Then several tables are given about the numerical experiments: the feasibility of the algorithm in [19] is illustrated, then similar numerical results are obtained for analogous cases of preconditioning that the theory does not yet cover.

2 Symmetric Part Preconditioning

2.1 Preconditioning and Boundary Conditions

Let us consider the complex Hilbert space $H = L^2(\Omega)$ with the usual inner product and define the operator L:

$$Lu \equiv -\Delta u + \mathbf{b} \cdot \nabla u + \alpha u$$

with the domain

$$D(L) := D \equiv \{u \in H^2(\Omega) : u_{|\Gamma_D} = 0, \ \tfrac{\partial u}{\partial \nu} + \beta u_{|\Gamma_N} = 0\}, \tag{2}$$

which is dense in $L^2(\Omega)$.

$$\langle u, v \rangle_L := \int_\Omega (\nabla u \cdot \nabla \overline{v} + (\mathbf{b} \cdot \nabla u)\overline{v} + \alpha u \overline{v}) \, \mathrm{d}x + \int_{\Gamma_N} \beta u \overline{v} \, \mathrm{d}\sigma \quad (u, v \in D(L)) \tag{3}$$

One can easily construct the weak symmetric part of L, see [19], which is the following bilinear form:

$$\langle u,v\rangle_S := \frac{1}{2}\big(\langle Lu,v\rangle_{L^2} + \langle u,Lv\rangle_{L^2}\big) =$$

$$= \int_\Omega \left(\nabla u\cdot\overline{\nabla v} + \left(\alpha - \frac{1}{2}\operatorname{div}\mathbf{b}\right)u\overline{v}\right)\,\mathrm{d}x + \int_{\Gamma_N}\left(\beta + \frac{1}{2}(\mathbf{b}\cdot\nu)\right)u\overline{v}\,\mathrm{d}\sigma =$$

$$= \int_\Omega (\nabla u\cdot\overline{\nabla v} + \hat\alpha u\overline{v})\,\mathrm{d}x + \int_{\Gamma_N}\hat\beta u\overline{v}\,\mathrm{d}\sigma \qquad (u,v\in D(L)), \tag{4}$$

and the energy space H_S - which has been defined as the completion of D under the inner product $\langle\cdot,\cdot\rangle_S$ - is

$$H_S = H^1_D(\Omega) := \{u\in H^1(\Omega) : u_{|\Gamma_D} = 0\}\,. \tag{5}$$

The operator $Q_S : H_S \to H_S$ has the form

$$\langle Q_S u,v\rangle_S = \frac{1}{2}\left(\int_\Omega (\mathbf{b}\cdot\nabla u)\overline{v}\,\mathrm{d}x - \int_\Omega u(\mathbf{b}\cdot\nabla\overline{v})\,\mathrm{d}x\right). \tag{6}$$

2.2 Finite Element Discretization

Now we consider finite element discretizations of problem (1). Let H_S be defined as in (5) and let $V_h = \operatorname{span}\{\varphi_1,\dots,\varphi_n\}\subset H_S$ be a given FEM subspace. The FEM solution $u_h\in V_h$ of equation (1) in V_h is $u_h = \sum_{j=1}^n c_j\varphi_j$, where $\mathbf{c} = (c_1,\dots,c_n)\in\mathbf{C}^n$ is the solution of the $n\times n$ system

$$\mathbf{L}_h\mathbf{c} = \mathbf{g} \tag{7}$$

where

$$(\mathbf{L}_h)_{i,j} = \int_\Omega (\nabla\varphi_i\cdot\overline{\nabla\varphi_j} + (\mathbf{b}\cdot\nabla\varphi_i)\overline{\varphi}_j + \alpha\varphi_i\overline{\varphi}_j)\,\mathrm{d}x + \int_{\Gamma_N}\beta\varphi_i\overline{\varphi}_j\,\mathrm{d}\sigma$$

and $\mathbf{g}_j = \int_\Omega g\overline{\varphi}_j$.

Let $\mathbf{S}_h$ and $\mathbf{Q}_h$ be the symmetric and antisymmetric parts of $\mathbf{L}_h$, i.e., $\mathbf{S}_h = \frac{1}{2}(\mathbf{L}_h + \mathbf{L}_h^*)$, $\mathbf{Q}_h = \mathbf{L}_h - \mathbf{S}_h$. Then using (3) and (4),

$$(\mathbf{S}_h)_{i,j} = \int_\Omega (\nabla\varphi_i\cdot\overline{\nabla\varphi_j} + \hat\alpha\varphi_i\overline{\varphi}_j)\,dx + \int_{\Gamma_N}\hat\beta\varphi_i\overline{\varphi}_j\,d\sigma = \langle\varphi_i,\varphi_j\rangle_S$$

Using the symmetric part $\mathbf{S}_h$ as preconditioner, system (7) is replaced by

$$\mathbf{S}_h^{-1}\mathbf{L}_h\mathbf{c} = (\mathbf{I}_h + \mathbf{S}_h^{-1}\mathbf{Q}_h)\mathbf{c} = \mathbf{S}_h^{-1}\mathbf{g}. \tag{8}$$

Theorem 1. *Let problem (1) satisfy Assumptions BVP. Then*

$$\sum_{m=1}^k |\lambda_m(\mathbf{S}_h^{-1}\mathbf{Q}_h)| \le \sum_{m=1}^k |\lambda_m(Q_S)| \qquad (k=1,\dots,n), \tag{9}$$

where $\lambda_m(Q_S)$ $(m=1,\dots,\infty)$ are the ordered eigenvalues of the operator Q_S.

The proof for Theorem 1 can be found in [19], for Theorem 2 see the book [3].

2.3 The Preconditioned Conjugate Gradient Method

The generalized conjugate gradient, least square (GCG-LS) method is defined in [2]. Two versions are discussed: the *full version* which uses all previous search directions, whereas the *truncated version* uses only $s+1$ previous search directions (denoted by GCG-LS(s)), where s is a nonnegative integer. Similar versions have been constructed for the GMRES method , see [23]. The conjugate gradient method for symmetric positive definite systems has been formulated in a Hilbert space H [14, 17]. Similarly, the generalized CG methods can be formulated in H. The PCG algorithm for our problem reads as follows:

$$
\left\{
\begin{array}{ll}
(a) & \text{Let } u_0 \in V_h \text{ be arbitrary, and let} \\
& r_0 \in V_h \text{ be the solution of problem} \\
& \qquad \langle r_0, v\rangle_S = \langle u_0, v\rangle_L - \langle g, v\rangle_{L^2} \qquad (\forall v \in V_h); \\
& d_0 = -r_0; \\
& \quad \text{for any } k \in \mathbf{N}: \text{ when } u_k, d_k, r_k \text{ are obtained, let} \\
(b1) & p_k \in V_h \text{ be the solution of problem} \\
& \qquad \langle p_k, v\rangle_S = \langle d_k, v\rangle_L \qquad (\forall v \in V_h); \\
& \gamma_k = \|p_k\|_S^2, \quad \alpha_k = -\frac{1}{\gamma_k}\langle p_k, r_k\rangle_S; \\
(b2) & u_{k+1} = u_k + \alpha_k d_k; \\
(b3) & r_{k+1} = r_k + \alpha_k p_k; \\
(b4) & \beta_k = \frac{1}{\gamma_k}\langle r_{k+1}, p_k\rangle_L; \\
(b5) & d_{k+1} = -r_{k+1} + \beta_k d_k.
\end{array}
\right. \tag{10}
$$

Theorem 2. *Let problem (1) satisfy Assumptions BVP. Then the preconditioned CG algorithm (10) yields*

$$\left(\frac{\|e_k\|_{\mathbf{S}_h}}{\|e_0\|_{\mathbf{S}_h}}\right)^{1/k} \leq \frac{2}{k}\sum_{i=1}^{k}|\lambda_i(Q_S)| \qquad (k \in \mathbf{N}) \tag{11}$$

where $e_k = u_k - u_h$ is the error vector, $\lambda_m(Q_S)$ $(m = 1, \ldots, \infty)$ are the ordered eigenvalues of the operator Q_S, and hence the sequence on the right-hand side is independent of the subspace V_h.

Remark 1. The sequence on the right-hand side tends to 0 and taking the numerical errors into consideration remains bounded.

3 Numerical Illustration

The main goal of this paper is to confirm and complete the previously cited theoretical results with numerical experiments. The following numerical experiments are the first steps in the way that leads to numerical results and proofs for superlinear convergence considering inequality (11). The aim of this paper is the numerical verification of a reduction of Theorem 2 as it is indicated in Remark 1.

On the one hand, the feasibility of the algorithm in [19] is illustrated. Moreover, similar numerical results are obtained when not the symmetric part of the operator L but another symmetric elliptic operator is used as preconditioner. The theory does not cover this case but the numerical results show much similar behaviour.

3.1 The Test Problem

Our test problem is the following elliptic convection-diffusion equation with two possible boundary conditions (a) and (b):

$$\begin{cases} Lu \equiv -\Delta u + \frac{\partial u}{\partial x} + \alpha u = g \\ (a) \qquad u_{|\Gamma_D} = 0. \\ (b) \qquad u_{|\Gamma_D} = 0, \qquad \frac{\partial u}{\partial \nu}_{|\Gamma_N} = 0. \end{cases} \tag{12}$$

This special modell problem satisfies the suitable conditions:

(i) $\Omega \subset \mathbf{R}^2$ is the unit square, $\Omega = [0,1] \times [0,1]$; Γ_D, Γ_N are disjoint open measurable subparts of $\partial\Omega$ such that $\partial\Omega = \overline{\Gamma}_D \cup \overline{\Gamma}_N$; in this case we have (a) $\Gamma_D = \partial\Omega$ and (b) $\Gamma_D = \{(x,y)\colon (x,y) \in \partial\Omega,\ x = 0 \text{ or } x = 1\}$.
(ii) $(1,0) = \mathbf{b} \in C^1(\overline{\Omega})^2$, $\alpha \in L^\infty(\Omega)$, $0 = \beta \in L^\infty(\Gamma_N)$, $\alpha, \beta \geq 0$;
(iii) we have the coercivity properties $\hat{\alpha} := \alpha - \frac{1}{2}\operatorname{div}\mathbf{b} = \alpha \geq 0$ in Ω and $\hat{\beta} := \beta + \frac{1}{2}(\mathbf{b}\cdot\nu) = \beta = 0 \geq 0$ on Γ_N;
(iv) $g \in L^2(\Omega)$;
(v) $\Gamma_D \neq \emptyset$.

3.2 Experiments

Numerous experiments have been performed in connection with the test problem. First of all, the mesh independence of the convergence has been investigated in two cases. In the first set of experiments the equation (12) has been considered with boundary conditions (a), and the second part of Table 1 shows the results for the mixed problem (b). Our goal is to study the $(\|e_k\|_{\mathbf{S}_h})/(\|e_0\|_{\mathbf{S}_h})$ quotient, their values appear in the last four columns of the tables.

Experiment 1. Let $\alpha = 1$ in the operator L and in the preconditioning operator S as well. Exact integrations have been used in the algorithm (10) at the computation of the right-hand side of the first finite element subroutine.

Analysing the results, the mesh independence is recognisable and the type of the boundary condition has no effect to the rate of convergence just slows down a bit the scale of decrease between two steps; four iterations are enough to reach the 10^{-4} accuracy in both cases. In a real-life problem to determine the

Table 1. Boundary conditions (a) and (b), exact integration

1/h	1st iteration	2nd iteration	3rd iteration	4th iteration
32	0.0790	0.0049	0.0003	0.0001
64	0.0792	0.0049	0.0003	0.0000
128	0.0792	0.0050	0.0003	0.0000
256	0.0792	0.0050	0.0003	0.0000
32	0.1029	0.0089	0.0008	0.0001
64	0.1031	0.0089	0.0007	0.0001
128	0.1031	0.0089	0.0007	0.0000
256	0.1031	0.0089	0.0007	0.0000

Table 2. Dirichlet boundary conditions (a), trapezoid rule

1/h	1st iteration	2nd iteration	3rd iteration	4th iteration
32	0.0790	0.0049	0.0009	0.0008
64	0.0792	0.0049	0.0004	0.0002
128	0.0792	0.0049	0.0003	0.0001
256	0.0792	0.0050	0.0003	0.0000

exact integrals is often impossible, therefore we tried to apply a simple numerical method (2 dimensonal trapezoid rule), and the result is nearly the same, the rate of the reduction is similar to the previous ones, but the order of the numerical integration has small but noticeable effect.

Experiment 2. Let $\alpha = 1$ in L and $\alpha = 0$ in S, so S is not the symmetric part of the operator L. The numerical result shows that the algorithm works with this modification (Table 3). This change can be useful in practice, the computation of the stiffness matrix and α_k, γ_k become more simple. Moreover, the same result is shown (Table 4), when the value of α has been transposed, i. e. $\alpha = 0$ in L and $\alpha = 1$ is S.

Experiment 3. As it is mentioned before, care must be taken with the boundary conditions of the preconditioning operator (see [21]). The last task is to prove the importance of the proper boundary conditions of S with respect to the given operator L. Let us consider equation (12) with boundary conditions (b). Let S be the symmetric part of L, but with the *different* boundary conditions (a).

Table 3. Dirichlet boundary conditions (a), L: $\alpha = 1$, S: $\alpha = 0$

1/h	1st iteration	2nd iteration	3rd iteration	4th iteration
32	0.0780	0.0049	0.0003	0.0002
64	0.0782	0.0050	0.0003	0.0001
128	0.0782	0.0050	0.0003	0.0000
256	0.0782	0.0050	0.0003	0.0000

Table 4. Dirichlet boundary conditions (a), L: $\alpha = 0$, S: $\alpha = 1$

1/h	1st iteration	2nd iteration	3rd iteration	4th iteration
32	0.0831	0.0053	0.0003	0.0002
64	0.0833	0.0054	0.0003	0.0000
128	0.0833	0.0054	0.0003	0.0000
256	0.0833	0.0054	0.0003	0.0000

Table 5. Mixed boundary conditions in L, Dirichlet b. c. in S

1/h	1st iteration	2nd iteration	3rd iteration	4th iteration
32	0.1356	0.0771	0.0763	0.0762
64	0.1359	0.0773	0.0764	0.0764
128	0.1360	0.0773	0.0765	0.0765
256	0.1360	0.0774	0.0765	0.0765

The result in table 5 shows that the algorithm does not work in this case, as theoretical results predicted.

References

1. Adams, R.A.: Sobolev Spaces. Academic Press, 1975.
2. Axelsson, O.: A generalized conjugate gradient least square method. Numer. Math. **51** (1987), 209–227.
3. Axelsson, O.: Iterative Solution Methods. Cambridge University Press, 1994.
4. Axelsson, O., Barker, V.A.: Finite Element Solution of Boundary Value Problems. Academic Press, 1984.
5. Axelsson, O., Kaporin, I.: On the sublinear and superlinear rate of convergence of conjugate gradient methods. Mathematical journey through analysis, matrix theory and scientific computation. (Kent, OH, 1999) Numer. Algorithms **25** 1–4, (2000), 1–22.
6. Axelsson, O., Karátson J.: On the rate of convergence of the conjugate gradient method for linear operators in Hilbert space. Numer. Funct. Anal. **23** 3–4, (2002) 285–302.
7. Axelsson, O., Karátson J.: Symmetric part preconditioning for the conjugate gradient method in Hilbert space. Numer. Funct. Anal. **24** 5–6, (2003) 455–474.
8. Axelsson, O., Karátson J.: Superlinearly convergent CG methods via equivalent precondition ing for nonsymmetric elliptic operators. Numer. Math. 2004, SpringerLink DOI: 10.1007/s00211-004-0557-2 (electronic)
9. Bank, R.E., Rose, D.J.: Marching algorithms for elliptic boundary value problems. I. The constant coefficient case. SIAM J. Numer. Anal. **14** 5, (1977), 792–829.
10. Börgers, C., Widlund, O. B.: On finite element domain imbedding methods. SIAM J. Numer. Anal. **27** 4, (1990), 963–978.
11. Ciarlet, P. G.: The finite element method for elliptic problems. North-Holland, Amsterdam, 1978
12. Ciarlet, Ph. G., Lions J.-L. (eds.): Handbook of numerical analysis. Vol. II. Finite element methods, Part 1, North-Holland, Amsterdam, 1991.

13. Concus, P., Golub, G.H.: A generalized conjugate method for non-symmetric systems of linear equations. (eds. Glowinski, R., Lions, J.-L.) *Lect. Notes Math. Syst.* **134** Springer, 1976, 56–65.
14. Daniel, J.W.: The conjugate gradient method for linear and nonlinear operator equations. SIAM J. Numer. Anal. **4** 1, (1967), 10–26.
15. Faber, V., Manteuffel, T., Parter, S.V.: On the theory of equivalent operators and application to the numerical solution of uniformly elliptic partial differential equations. Adv. in Appl. Math., **11** (1990), 109–163.
16. Hackbusch, W.: Elliptic differential equations. Theory and numerical treatment. *Springer Series in Computational Mathematics* **18**, Springer, Berlin, 1992.
17. Hayes, R.M.: Iterative methods of solving linear problems in Hilbert space. Nat. Bur. Standards Appl. Math. Ser. **39** (1954), 71–104.
18. Karátson J.: Mesh independence of the superlinear convergence of the conjugate gradient method. Appl. Math. (Prague) **50** 3 (2005), 277–290.
19. Karátson J.: Superlinear PCG algorithms: symmetric part preconditioning and boundary conditions. submitted
20. Manteuffel, T., Otto, J.: Optimal equivalent preconditioners. SIAM J. Numer. Anal. **30** (1993), 790–812.
21. Manteuffel, T., Parter, S. V.: Preconditioning and boundary conditions. SIAM J. Numer. Anal. **27** 3, (1990), 656–694.
22. Riesz F., Sz.-Nagy B. Vorlesungen über Funktionalanalysis. Verlag H. Deutsch, 1982.
23. Saad, Y., Schultz, M.H.: Conjugate gradient-like algorithms for solving nonsymmetric linear systems. Math. Comput. **44** (1985), 417–424.
24. Widlund, O.: A Lanczos method for a class of non-symmetric systems of linear equations. SIAM J. Numer. Anal., **15** (1978), 801–812.
25. van der Vorst, H. A.: Iterative solution methods for certain sparse linear systems with a nonsymmetric matrix arising from PDE-problems. J. Comput. Phys. **44** 1, (1981), 1–19.

Taking into Account the Third Kind Conditions in Weight Estimates for Difference Schemes

Volodymyr L. Makarov and Lyubomyr I. Demkiv

Institute of Mathematics, NAS of Ukraine
makarov@imath.kiev.ua, demkivl@ukrpost.net

Abstract. The main goal of this paper is to generalize the results, obtained in [1] to the case of quasi-linear ordinary differential equations of the second order with the third kind boundary conditions. It is a continuation of the paper series in which we have obtained weight a priori estimates of accuracy for difference schemes for linear parabolic type equations in one-dimensional [2] and two-dimensional [3] cases, quasi-linear parabolic type equations [4] and quasi-linear elliptic equations with conditions of the first kind [5]. In this paper it is shown that on approaching to the left or right boundary of the domain the rate of convergence of solution or it's first derivative, correspondingly, increases. The second accuracy order difference scheme of the special form has been used for this purpose.

The paper is completed by numerical experiment, which results confirm theoretical statements.

1 A Priori Weight Estimates of the Difference Schemes Taking into Account Boundary Effect

Let us consider the following boundary value problem with the third kind condition in the left boundary point of the interval $[0, 1]$ and Dirichlet condition in the right boundary point

$$\begin{aligned} s''(x) &= -g(x, s(x)), \quad x \in (0,1) \\ s'(0) - \alpha s(0) &= \mu_1, \quad s(1) = \mu_2 \\ & \qquad\qquad \alpha > 0. \end{aligned}$$

Let us make some transformations. One can make the following change of function $s(x) = \varphi(x) u(x)$, then obtain the following boundary conditions

$$\varphi(0) u'(0) + [\varphi'(0) - \alpha\varphi(0)] u(0) = \mu_1$$

Let us look for function $\varphi(x)$ such that equality

$$\varphi'(0) - \alpha\varphi(0) = 0 \tag{1}$$

is satisfied.

I. Lirkov, S. Margenov, and J. Waśniewski (Eds.): LSSC 2005, LNCS 3743, pp. 687–694, 2006.

It is obvious, that it can be taken in the following form:

$$\varphi(x) = \alpha x + 1$$

As a result we obtain the following problem:

$$\begin{gathered} u''(x) = -f(x, u(x)), \; x \in (0,1) \\ u'(0) = \mu_1, \quad u(1) = \tilde{\mu}_2, \end{gathered} \tag{2}$$

where $f(x, u(x)) = g(x, (\alpha x + 1) u(x)), \tilde{\mu}_2 = \frac{\mu_2}{1+\alpha}$.

So, one can see that there exists such change of variables that it is possible to pass from the third kind condition to Neumann condition. So, further we will consider only the problem (2), where function $f(x, u(x))$ satisfies Lipschitz condition

$$|f(x,u) - f(x,v)| \leq L|u - v|, \quad u, v \in \mathbb{R}, \quad x \in [0,1]. \tag{3}$$

Let us rewrite the problem (2)-(3) in the form of the boundary value problem for the system of the differential equations of the first order

$$\begin{cases} u_1'(x) = u_2(x), \\ u_2'(x) = -f(x, u_1(x)), \\ u_2(0) = \mu_1, \quad u_1(1) = \tilde{\mu}_2. \end{cases} \tag{4}$$

We introduce the following grid for the approximation of this problem by a difference scheme:

$$\omega_h = \{x = ih, \quad i = \overline{1, N-1}, \quad h = 1/N\}$$

We assign the problem (4) to the difference scheme on the grid ω_h

$$\begin{cases} y_{1,x}(x) - y_2(x) = -\frac{h}{2} f(x, y_1(x)), \\ y_{2,x}(x) = -f\left(x + \frac{h}{2}, y_1^+(x)\right), \quad x \in \omega_h^- \\ y_2(0) = \mu_1, \quad y_1(1) = \tilde{\mu}_2 \end{cases}$$

where $\omega_h^- = \omega_h + \{0\}$, $y_i^+ = \frac{1}{2}(y_i(x) + y_i(x+h)), i = \overline{1,2}$, functions $y_i(x)$ are a grid approximation of the corresponding functions $u_i(x), i = \overline{1,2}$.

We introduce error functions for determination of the weight a priori accuracy estimates of the difference scheme

$$z_1(x) = y_1(x) - u_1(x), \quad z_2(x) = y_2(x) - u_2(x)$$

For these functions we obtain the following difference scheme:

$$\begin{cases} z_{1,x}(x) - z_2(x) = -u_{1,x}(x) + u_2(x) - \frac{h}{2} f(x, y_1(x)), \\ z_{2,x}(x) = -u_{2,x}(x) - f\left(x + \frac{h}{2}, y_1^+(x)\right), \\ z_2(0) = 0, \quad z_1(1) = 0. \end{cases} \tag{5}$$

We separate the local truncation errors for each equation in (5). At first let us consider the first equation. One can obtain the following equality:

$$z_{1,x}(x) - z_2(x) = -u_{1,x}(x) + u_2(x) - \frac{h}{2} f(x, y_1(x)) =$$
$$= \psi_1(x) + \frac{h}{2}\left[f(x, u_1(x)) - f(x, y_1(x))\right],$$

where $\psi_1(x)$ is a local truncation error of the first equation of the system:

$$\psi_1(x) = u_1'(x) - \frac{u_1(x+h) - u_1(x)}{h} - \frac{h}{2} f(x, u_1(x)) =$$
$$= u_1'(x) - \frac{1}{h}\left\{\left[u_1(x) + h u_1'(x) + \frac{h^2}{2} u_1''(x) + \frac{h^3}{6} u_1'''(\tilde{x})\right] - u_1(x)\right\}$$
$$+ \frac{h}{2} u_1''(x) = -\frac{h^2}{6} u_1'''(\tilde{x}), \quad \tilde{x} \in (x, x+h).$$

Now we pass to the second equation of the system

$$z_{2,x}(x) = -u_{2,x}(x) - f\left(x + \frac{h}{2}, y_1^+(x)\right) =$$
$$= \psi_2(x) + \left[f\left(x + \frac{h}{2}, u_1^+(x)\right) - f\left(x + \frac{h}{2}, y_1^+(x)\right)\right],$$

where ψ_2 is a local truncation error of the second equation in the system

$$\psi_2(x) = u_2'\left(x + \frac{h}{2}\right) - \frac{u_2(x+h) - u_2(x)}{h} +$$
$$+ f\left(x + \frac{h}{2}, u_1\left(x + \frac{h}{2}\right)\right) - f\left(x + \frac{h}{2}, u_1^+(x)\right).$$

The following estimate

$$|\psi_2(x)| \le \left| u_2'\left(x + \frac{h}{2}\right) - \frac{1}{h}\left\{\left[u_2\left(x + \frac{h}{2}\right) + \frac{h}{2} u_2'\left(x + \frac{h}{2}\right) + \right.\right.\right.$$
$$\left.\left.\left. + \frac{h^2}{8} u_2''\left(x + \frac{h}{2}\right) + + \frac{h^3}{24} u_2'''(\tilde{x})\right] - u_2(x)\right\}\right| +$$
$$+ L\left| u_1\left(x + \frac{h}{2}\right) - \frac{u_1(x+h) + u_1(x)}{2}\right| = \left| u_2'\left(x + \frac{h}{2}\right) - \right.$$
$$- \frac{1}{h}\left\{\left[u_2\left(x + \frac{h}{2}\right) + \frac{h}{2} u_2'\left(x + \frac{h}{2}\right) + \frac{h^2}{8} u_2''\left(x + \frac{h}{2}\right) + \frac{h^3}{24} u_2'''(\tilde{x})\right] - \right.$$
$$\left.\left. - \left[u_2\left(x + \frac{h}{2}\right) - \frac{h}{2} u_2'\left(x + \frac{h}{2}\right) + \frac{h^2}{8} u_2''\left(x + \frac{h}{2}\right) - \frac{h^3}{24} u_2'''(\tilde{x})\right]\right\}\right| +$$
$$+ L\left| u_1\left(x + \frac{h}{2}\right) - \frac{1}{2}\left\{\left[u_1\left(x + \frac{h}{2}\right) + \frac{h}{2} u_1'\left(x + \frac{h}{2}\right) + \frac{h^2}{8} u_1''(\tilde{x})\right] + \right.\right.$$
$$\left.\left. + \left[u_1\left(x + \frac{h}{2}\right) - \frac{h}{2} u_1'(\tilde{x}) + \frac{h^2}{8} u_1''(\tilde{x})\right]\right\}\right| = \left|\frac{h^2}{12} u_1'''(\tilde{x})\right| + L\left|\frac{h^2}{8} u_1''(\tilde{x})\right|$$

holds true, where $\tilde{x} \in [x, x+h]$.

Thus it is possible to see that both local truncation errors have the second order of approximation when solution of the problem (4) is such that $u_i(x) \in C^3[0,1], i = 1,2$.

Therefore, difference scheme (5) can be written down in the following form:

$$\begin{cases} z_{1,x}(x) - z_2(x) = \psi_1(x) + \frac{h}{2}\left[f(x, u_1(x)) - f(x, y_1(x))\right], \\ z_{2,x}(x) = \psi_2(x) + f\left(x + \frac{h}{2}, u_1^+(x)\right) - f\left(x + \frac{h}{2}, y_1^+(x)\right) \\ z_2(0) = 0, \quad z_1(1) = 0. \end{cases} \tag{6}$$

Form the second equation of the system (6) we obtain

$$\sum_{\xi=0}^{x-h} \frac{z_2(\xi+h) - z_2(\xi)}{h} h = \sum_{\xi=0}^{x-h} h\psi_2(\xi) +$$
$$+ \sum_{\xi=0}^{x-h} h\left[f\left(\xi + \frac{h}{2}, u_1^+(\xi)\right) - f\left(\xi + \frac{h}{2}, y_1^+(\xi)\right)\right],$$

whence we have

$$|z_2(x)| \le x\,\|\psi_2\|_{0,\infty,\omega_h} + L\sum_{\xi=0}^{x-h} h\left|z_1^+(\xi)\right|. \tag{7}$$

Let us consider the first equality from (6)

$$\begin{aligned} \sum_{\xi=x}^{1-h} h z_{1,\xi}(\xi) = \sum_{\xi=x}^{1-h} h z_2(\xi) + \\ + \sum_{\xi=x}^{1-h} h\psi_1(x) + \frac{h}{2}\sum_{\xi=x}^{1-h} h\left[f(x, u_1(x)) - f(x, y_1(x))\right]; \end{aligned} \tag{8}$$

Let condition

$$\left(1 + \frac{h}{2}\right)L < 1 \tag{9}$$

be satisfied.

Then, taking into account (7), we have inequality

$$\begin{aligned} |z_1(x)| \le \|\psi_2\|_{0,\infty,\omega_h} \sum_{\xi=x}^{1-h} h\xi + \sum_{\xi=x}^{1-h} hL\sum_{\eta=0}^{\xi-h} h\left|z_1^+(\eta)\right| + (1-x)\,\|\psi_1\|_{0,\infty,\omega_h} + \\ + \frac{h}{2}L\sum_{\xi=x}^{1-h} h\left|z_1(x)\right| \le (1-x)\left(\|\psi_1\|_{0,\infty,\omega_h} + \|\psi_2\|_{0,\infty,\omega_h}\right) + \\ + \|z_1\|_{0,\infty,\omega_h}\left(L\sum_{\xi=x}^{1-h} h\sum_{\eta=0}^{\xi-h} h + \frac{h}{2}L\sum_{\xi=x}^{1-h} h\right) \le \\ \le (1-x)\left(\|\psi_1\|_{0,\infty,\omega_h} + \|\psi_2\|_{0,\infty,\omega_h} + L\left(1 + \frac{h}{2}\right)\|z_1\|_{0,\infty,\omega_h}\right). \end{aligned}$$

Whence it is possible to obtain the following inequality:

$$\left\| \frac{z_1(x)}{1-x} \right\|_{0,\infty,\omega_h} \leq \|\psi_1\|_{0,\infty,\omega_h} + \|\psi_2\|_{0,\infty,\omega_h} + L\left(1+\frac{h}{2}\right) \left\| \frac{z_1(x)}{1-x} \right\|_{0,\infty,\omega_h},$$

which yields the estimate

$$\left\| \frac{z_1(x)}{1-x} \right\|_{0,\infty,\omega_h} \leq \frac{1}{1-L\left(1+\frac{h}{2}\right)} \left(\|\psi_1\|_{0,\infty,\omega_h} + \|\psi_2\|_{0,\infty,\omega_h} \right),$$

where $\|y\|_{0,\infty,\omega_h} = \max\limits_{x\in\omega_h} |y(x)|$.

Let us now return to the estimate of the function $z_2(x)$. We have

$$\left\| \frac{z_2(x)}{x} \right\|_{0,\infty,\omega_h} \leq \|\psi_2\|_{0,\infty,\omega_h} + L \left\| \frac{z_1(x)}{1-x} \right\|_{0,\infty,\omega_h} \leq$$

$$\leq \|\psi_2\|_{0,\infty,\omega_h} + \frac{L}{1-L\left(1+\frac{h}{2}\right)} \left(\|\psi_1\|_{0,\infty,\omega_h} + \|\psi_2\|_{0,\infty,\omega_h} \right).$$

The following statement holds true:

Theorem 1. *Let the solution of the problem (2), (3) belongs to the class $C^{(4)}[0,1]$. Then, when condition (9) is fulfilled, the following weight estimates*

$$\left\| \frac{u(x)-y_1(x)}{1-x} \right\|_{0,\infty,\omega_h} \leq Ch^2,$$

$$\left\| \frac{u'(x)-y_2(x)}{x} \right\|_{0,\infty,\omega_h} \leq Ch^2,$$

hold true, where C is constant, independent of h.

Thus, one can see that on approaching to the boundary from the right, rate of convergence of the solution's first derivative increases and approaching to the domain boundary from the left rate of convergence of the solution itself increases. These rates of convergence are of the $O(h^2)$ order.

2 Numerical Experiment

Example 1. We are looking for the solution of the problem

$$u''(x) = -\frac{1}{2(1+u^2(x))} + 12x^2 + \frac{1}{2(1+x^8)} \qquad (10)$$

$$u'(0) = 0, \quad u(1) = 1$$

The exact solution of this problem is $u(x) = x^4$. Computations where held by means of Maple 9. Results of computations are shown in the table. Some graphics of the investigated values for the case $N = 8$ are presented below.

Table 1. Computation results

N	err_1	p_1	err_2	p_2	err_3	p_3	err_4	p_4
4	0.127		0.237		0.441e-1		0.545e-1	
8	0.351e-1	1.8544	0.673e-1	1.8204	0.118e-1	1.8943	0.150e-1	1.8611
16	0.919e-2	1.9364	0.177e-1	1.9212	0.303e-2	1.9694	0.386e-2	1.9574
32	0.234e-2	1.9708	0.455e-2	1.9635	0.762e-3	1.9927	0.973e-3	1.9884
64	0.592e-3	1.9861	0.115e-2	1.9825	0.190e-3	1.9986	0.243e-3	1.9969
128	0.148e-3	1.9932	0.290e-3	1.9913	0.476e-4	1.9999	0.610e-4	1.9992

For the practical estimation of the rate of convergence we consider the following quantities

$$err_1 = \|z_1(x)\|_{0,\infty,\omega_h}, \; err_2 = \left\| \frac{z_1(x)}{(1-x)} \right\|_{0,\infty,\omega_h},$$

$$err_3 = \|z_2(x)\|_{0,\infty,\omega_h}, \; err_4 = \left\| \frac{z_2(x)}{x} \right\|_{0,\infty,\omega_h},$$

$$p_1 = \log_2 \frac{\|z_1(x)\|_{0,\infty,\omega_h}}{\|z_1(x)\|_{0,\infty,\omega_h/2}}, \; p_2 = \log_2 \frac{\left\| \frac{z_1(x)}{(1-x)} \right\|_{0,\infty,\omega_h}}{\left\| \frac{z_1(x)}{(1-x)} \right\|_{0,\infty,\omega_h/2}},$$

$$p_3 = \log_2 \frac{\|z_2(x)\|_{0,\infty,\omega_h}}{\|z_2(x)\|_{0,\infty,\omega_h/2}}, \; p_4 = \log_2 \frac{\left\| \frac{z_2(x)}{x} \right\|_{0,\infty,\omega_h}}{\left\| \frac{z_2(x)}{x} \right\|_{0,\infty,\omega_h/2}}.$$

Thus, it is possible to see that numerical computations confirm theoretical results.

References

1. Makarov V. On a priori estimates of difference schemes giving an account of the boundary effect. *C.R. Acad. Bulgare Sci.*, **42** 5, 1989, 41–44.
2. Makarov V.L., Demkiv L.I. Accuracy estimates of the difference schemes for parabolic type equations taking into account initial-boundary effect. Dopov. Nats. Akad. Nauk Ukr., **2**, 2003, 26–32. (in Ukrainian)
3. Makarov V.L., Demkiv L.I. Improved accuracy estimates of the difference schemes for parabolic equations. Praci Ukr. Mat. Congresu — 2001. Kyiv: Inst. matematyky NAN Ukr., 2001, 36–47.
4. Makarov V.L., Demkiv L.I. Accuracy estimates of the difference schemes for quasi-linear parabolic equations taking into account the initial-boundary effect. *Computational methods in applied mathematics*, **3** 4, 2003, 579–595.
5. Makarov V.L., Demkiv L.I., Accuracy Estimates of Difference Schemes for Quasi-Linear Elliptic Equations with Variable Coefficients Taking into Account Boundary Effect. *Lecture notes in computer science*, **3401**, 2005, 80–90.

Appendix

Function error

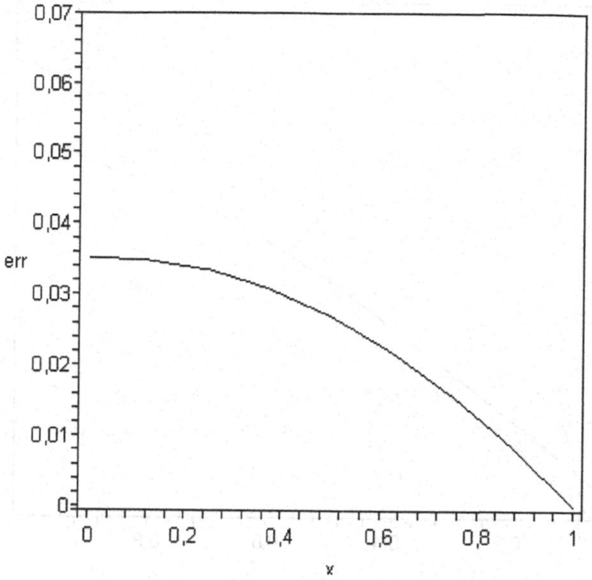

Weighted function error

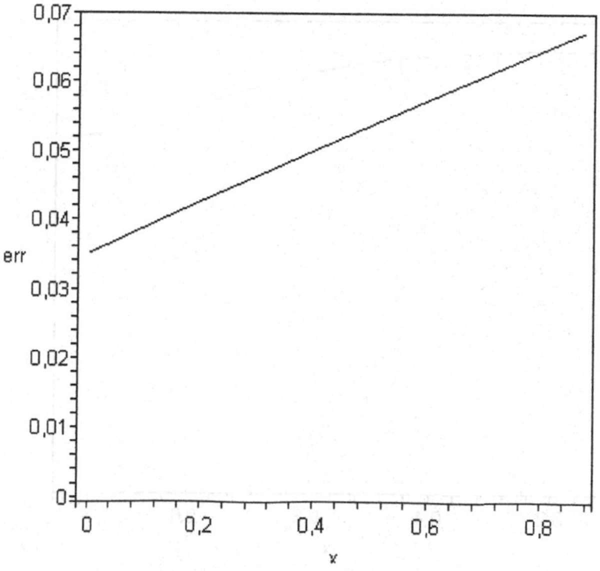

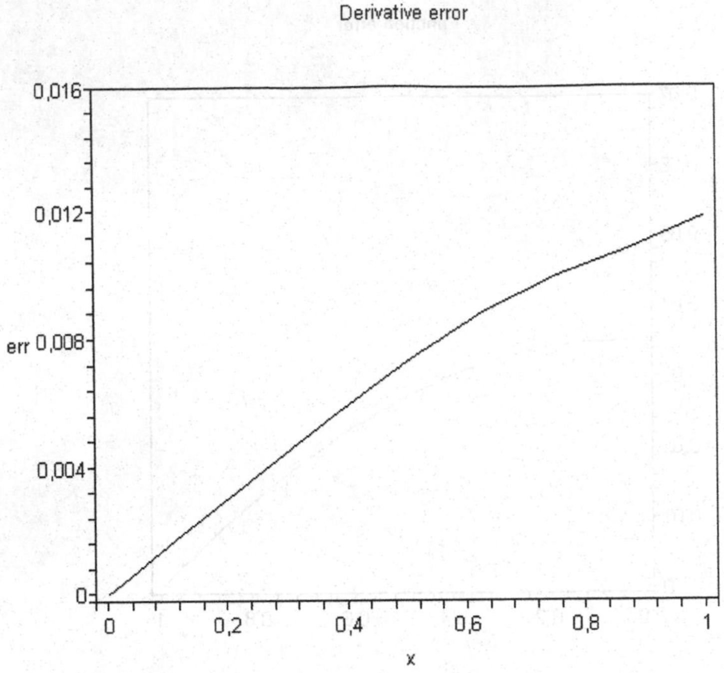
Derivative error
0,016
0,012
err 0,008
0,004
0
0
0,2
0,4
0,6
0,8
1
x

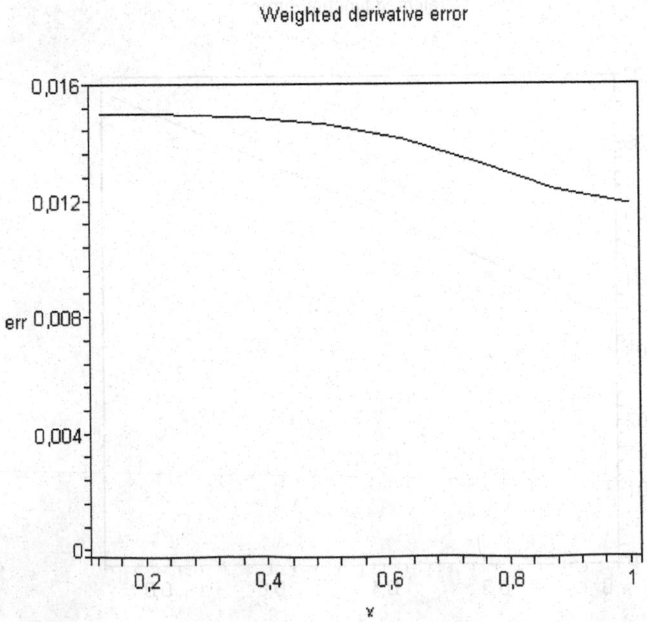
Weighted derivative error
0,016
0,012
err 0,008
0,004
0
0,2
0,4
0,6
0,8
1
x

On Weighted Structured Total Least Squares

Ivan Markovsky and Sabine Van Huffel

K.U. Leuven, ESAT-SCD, Kasteelpark Arenberg 10, B-3001 Leuven, Belgium
{ivan.markovsky, sabine.vanhuffel}@esat.kuleuven.ac.be
www.esat.kuleuven.ac.be/~imarkovs
www.esat.kuleuven.ac.be/sista/members/vanhuffel.html

Abstract. In this contribution we extend our previous results on the structured total least squares problem to the case of weighted cost functions. It is shown that the computational complexity of the proposed algorithm is preserved linear in the sample size when the weight matrix is banded with bandwidth that is independent of the sample size.

1 Introduction

The *total least squares (TLS)* method (Golub and Van Loan, [1], Van Huffel and Vandewalle, [2])

$$\min_{\Delta A, \Delta B, X} \left\| \begin{bmatrix} \Delta A \ \Delta B \end{bmatrix} \right\|_{\mathrm{F}}^2 \quad \text{subject to} \quad (A - \Delta A)X = B - \Delta B \ , \tag{1}$$

is a solution technique for an overdetermined system of equations $AX \approx B$, $A \in \mathbb{R}^{m \times n}$, $B \in \mathbb{R}^{m \times d}$. It is a natural generalization of the least squares approximation method when the data in both A and B is perturbed. The method has been generalized in two directions:

– *weighted total least squares*

$$\min_{\Delta A, \Delta B, X} \left\| \begin{bmatrix} \Delta A \ \Delta B \end{bmatrix} \right\|_{W}^2 \quad \text{subject to} \quad (A - \Delta A)X = B - \Delta B \ , \tag{2}$$

where $\|\Delta C\|_W^2 := \operatorname{vec}^\top(\Delta C^\top) W \operatorname{vec}(\Delta C^\top)$, $W > 0$, and

– *structured total least squares (STLS)*

$$\min_{\Delta A, \Delta B, X} \left\| \begin{bmatrix} \Delta A \ \Delta B \end{bmatrix} \right\|_{\mathrm{F}}^2 \quad \text{subject to} \quad (A - \Delta A)X = B - \Delta B \quad \text{and} \\ \begin{bmatrix} \Delta A \ \Delta B \end{bmatrix} \ \text{has the same structure as} \ \begin{bmatrix} A \ B \end{bmatrix} \ . \tag{3}$$

While the basic TLS problem allows for an analytic solution in terms of the singular value decomposition of the data matrix $C := \begin{bmatrix} A \ B \end{bmatrix}$, the weighted and structured TLS problems are solved numerically via local optimization methods.

In (Markovsky, Van Huffel, Pintelon[4]) we show that under a general assumption (see Assumption 1) about the structure, the cost function and first derivative of the STLS problem can be evaluated in $O(m)$ floating point operations (flops).

I. Lirkov, S. Margenov, and J. Waśniewski (Eds.): LSSC 2005, LNCS 3743, pp. 695–702, 2006.

This allows for efficient computational algorithms based on standard methods for local optimization. Via a similar approach, see (Markovsky et al., [5]), the weighted TLS problem can be solved efficiently when the weight matrix W is block-diagonal with blocks of size $n+d$.

In this paper, we extend our earlier results on the STLS problem by accounting for weighted cost function. Thus the weighted TLS problem becomes a special case of the considered weighted STLS problem when the data matrix is unstructured. In Sect. 2 we review the results of (Markovsky, Van Huffel, Pintelon [4]). Section 3 presents the necessary modifications for the weighted STLS problem and Sect. 4 discusses the implementation of the algorithm.

2 Review of Results for the STLS Problem

Let $\mathcal{S} : \mathbb{R}^{n_p} \to \mathbb{R}^{m\times(n+d)}$ be an injective function. A matrix $C \in \mathbb{R}^{m\times(n+d)}$ is said to be $\mathcal{S}$-structured if $C \in \text{image}(\mathcal{S})$. The vector p for which $C = \mathcal{S}(p)$ is called the parameter vector of the structured matrix C. Respectively, $\mathbb{R}^{n_p}$ is called the parameter space of the structure $\mathcal{S}$. The aim of the STLS problem is to perturb as little as possible a given parameter vector p by a vector Δp, so that the perturbed structured matrix $\mathcal{S}(p+\Delta p)$ becomes rank deficient with rank at most n.

Problem 1 (STLS). Given a data vector $p \in \mathbb{R}^{n_p}$, a structure specification $\mathcal{S} : \mathbb{R}^{n_p} \to \mathbb{R}^{m\times(n+d)}$, and a rank specification n, solve the optimization problem

$$\hat{X} = \arg\min_{X,\Delta p} \|\Delta p\|_2^2 \quad \text{subject to} \quad \mathcal{S}(p-\Delta p)\begin{bmatrix} X \\ -I_d \end{bmatrix} = 0 \ . \tag{4}$$

Let $\begin{bmatrix} A & B \end{bmatrix} := \mathcal{S}(p)$. Problem 1 makes precise the STLS problem formulation (3) from the introduction. In what follows, we often use the notation

$$X_{\text{ext}} := \begin{bmatrix} X \\ -I \end{bmatrix} .$$

The STLS problem is said to be *affine structured* if the function $\mathcal{S}$ is affine, i.e.,

$$\mathcal{S}(p) = S_0 + \sum_{i=1}^{n_p} S_i p_i, \quad \text{for all } p \in \mathbb{R}^{n_p} \text{ and for some } S_i,\ i = 1,\dots,n_p \ . \tag{5}$$

In an affine STLS problem, the constraint $\mathcal{S}(p-\Delta p)X_{\text{ext}} = 0$ is bilinear in the decision variables X and Δp.

Lemma 1. *Let $\mathcal{S} : \mathbb{R}^{n_p} \to \mathbb{R}^{m\times(n+d)}$ be an affine function. Then*

$$\mathcal{S}(p-\Delta p)X_{\text{ext}} = 0 \quad \Longleftrightarrow \quad G(X)\Delta p = r(X) \ ,$$

where

$$G(X) := \begin{bmatrix} \text{vec}\big((S_1 X_{\text{ext}})^\top\big) & \cdots & \text{vec}\big((S_{n_p} X_{\text{ext}})^\top\big) \end{bmatrix} \in \mathbb{R}^{md\times n_p} \ , \tag{6}$$

and

$$r(X) := \text{vec}\Big(\big(\mathcal{S}(p)X_{\text{ext}}\big)^\top\Big) \in \mathbb{R}^{md} \ .$$

Using Lemma 1, we rewrite the affine STLS problem as follows

$$\min_{X} \left(\min_{\Delta p} \|\Delta p\|_2^2 \quad \text{subject to} \quad G(X)\Delta p = r(X) \right) . \tag{7}$$

The inner minimization problem has an analytic solution, which allows to derive an equivalent optimization problem.

Theorem 1 (Equivalent optimization problem for affine STLS). *Assuming that $n_p \geq md$, the affine STLS problem (7) is equivalent to*

$$\min_{X} f(X) \quad \textit{where} \quad f(X) := r^{\top}(X)\Gamma^{\dagger}(X)r(X) \quad \textit{and} \quad \Gamma(X) := G(X)G^{\top}(X) .$$

The significance of Theorem 1 is that the constraint and the decision variable Δp in problem (7) are eliminated. Typically the number of elements nd in X is much smaller than the number of elements n_p in the correction Δp. Thus the reduction in the complexity is significant.

The equivalent optimization problem (1) is a nonlinear least squares problem, so that classical optimization methods can be used for its solution. The optimization methods require a cost function and first derivative evaluation. In order to evaluate the cost function f for a given value of the argument X, we need to form the weight matrix $\Gamma(X)$ and to solve the system of equations $\Gamma(X)y(X) = r(X)$. This straightforward implementation requires $O(m^3)$ flops. For large m (the applications that we aim at) this computational complexity becomes prohibitive.

It turns out, however, that for a special case of affine structures $\mathcal{S}$, the weight matrix $\Gamma(X)$ has a block-Toeplitz and block-banded structure, which can be exploited for efficient cost function and first derivative evaluations.

Assumption 1 (Flexible structure specification). *The structure specification $\mathcal{S} : \mathbb{R}^{n_p} \to \mathbb{R}^{m\times(n+d)}$ is such that for all $p \in \mathbb{R}^{n_p}$, the data matrix $\mathcal{S}(p) =: C =: \begin{bmatrix} A & B \end{bmatrix}$ is of the type*

$$\mathcal{S}(p) = \begin{bmatrix} C^1 & \cdots & C^q \end{bmatrix}, \textit{ where } C^l, \textit{ for } l = 1,\ldots,q, \textit{ is block-Toeplitz, block-Hankel, unstructured, or exact and all block-Toeplitz/Hankel structured blocks } C^l \textit{ have equal row dimension } K \textit{ of the blocks.}$$

Assumption 1 says that $\mathcal{S}(p)$ is composed of blocks, each one of which is block-Toeplitz, block-Hankel, unstructured, or exact. A block C^l that is exact is not modified in the solution $\hat{C} := \mathcal{S}(p - \Delta p)$, i.e., $\hat{C}^l = C^l$. Assumption 1 is the essential structural assumption that we impose on the STLS problem. It is fairly general and covers many applications.

We use the notation n_l for the number of *block* columns of the block C^l. For unstructured and exact blocks $\mathsf{n}_l := 1$.

Theorem 2 (Structure of the weight matrix Γ). *Consider the equivalent optimization problem (1) from Theorem 1. If in addition to the assumptions of*

Theorem 1, the structure $\mathcal{S}$ is such that Assumption 1 holds, then the weight matrix $\Gamma(X)$ has the block-Toeplitz and block-banded structure,

$$\Gamma(X) = \begin{bmatrix} \Gamma_0 & \Gamma_1^\top & \cdots & \Gamma_s^\top & & \mathbf{0} \\ \Gamma_1 & \ddots & \ddots & & \ddots & \\ \vdots & \ddots & \ddots & \ddots & & \Gamma_s^\top \\ \Gamma_s & \ddots & \ddots & \ddots & \ddots & \vdots \\ & \ddots & \ddots & \ddots & \ddots & \Gamma_1^\top \\ \mathbf{0} & & \Gamma_s & \cdots & \Gamma_1 & \Gamma_0 \end{bmatrix} \in \mathbb{R}^{md \times md} , \tag{8}$$

where $\Gamma_k \in \mathbb{R}^{dK \times dK}$, for $k = 0, 1, \ldots, s$, and $s = \max_{l=1,\ldots,q}(\mathbf{n}_l - 1)$.

3 Modifications for the Weighted STLS Problem

Next we consider the generalization of the STLS problem where the cost function is weighted.

Problem 2 (Weighted STLS). Given a data vector $p \in \mathbb{R}^{n_p}$, a positive definite weight matrix $W \in \mathbb{R}^{n_p \times n_p}$, a structure specification $\mathcal{S} : \mathbb{R}^{n_p} \to \mathbb{R}^{m \times (n+d)}$, and a rank specification n, solve the optimization problem

$$\hat{X}_{\mathrm{w}} = \arg\min_{X, \Delta p} \Delta p^\top W \Delta p \quad \text{subject to} \quad \mathcal{S}(p - \Delta p) \begin{bmatrix} X \\ -I_d \end{bmatrix} = 0 \ . \tag{9}$$

The counterpart of Theorem 1 for the case at hand is the following one.

Theorem 3 (Equivalent optimization problem for weighted STLS). *Assuming that $n_p \geq md$, the affine weighted STLS problem (9) is equivalent to*

$$\min_X f_{\mathrm{w}}(X) \quad \textit{where} \quad f_{\mathrm{w}}(X) := r^\top(X) \Gamma_{\mathrm{w}}^\dagger(X) r(X)$$
$$\textit{and} \quad \Gamma_{\mathrm{w}}(X) := G(X) W^{-1} G^\top(X) \ . \tag{10}$$

Proof. The equivalent optimization problem in Theorem 1 is obtained by solving a least squares problem. In the weighted STLS case, we solve the weighted least squares problem

$$\min_{\Delta p} \Delta p^\top W \Delta p \quad \text{subject to} \quad G(X)\Delta p = r(X) \ .$$

The optimal parameter correction as a function of X is

$$\Delta p_{\mathrm{w}}(X) = W^{-1} G^\top(X) \big(G(X) W^{-1} G^\top(X) \big)^\dagger r(X) \ ,$$

so that

$$f_{\mathrm{w}}(X) = \Delta p_{\mathrm{w}}^\top(X) W \Delta p_{\mathrm{w}}(X) = r^\top(X) \big(\underbrace{G(X) W^{-1} G^\top(X)}_{\Gamma_{\mathrm{w}}} \big)^\dagger r(X) \ . \qquad \square$$

In general neither the block-Toeplitz nor the block-banded properties of $\Gamma = GG^\top$ are present in $\Gamma_\mathrm{w} = GW^{-1}G^\top$. In the rest of this section, we show that in certain special cases these properties are preserved.

Assumption 2 (Block-diagonal weight matrix). *Consider the flexible structure specification of Assumption 1, let the blocks C^l, $l = 1,\dots,q$ be parameterized by parameter vectors $p_l \in \mathbb{R}^{n_{p,l}}$, and assume without loss of generality that $p = \mathrm{col}(p_1,\dots,p_q)$. The weight matrix W is assumed to be block-diagonal*

$$W = \mathrm{blk\,diag}(W^1,\dots,W^q), \qquad \textit{where} \quad W^l \in \mathbb{R}^{n_{p,l}\times n_{p,l}}.$$

Assumption 2 forbids cross-weighting among the parameters of the blocks $C^1,\dots,C^q$. Under Assumption 2 the effect of C^l on Γ_w is independent from those of the other blocks. Thus the problem of determining the structure of Γ_w, resulting from the flexible structure specification of C decouples into three independent problems: what is the structure of Γ_w, resulting from respectively an unstructured matrix C, a block-Hankel matrix C, and a block-Toeplitz matrix C.

In what follows "i-block-Toeplitz matrix" stands for block-Toeplitz matrix with $i \times i$ block size and "s-block-banded matrix" stands for a block-symmetric and block-banded matrix with upper/lower block-bandwidth i. Let $V := W^{-1}$ and $V_i := W_i^{-1}$.

Proposition 1. *Let G be defined as in (6) and let Assumptions 1 and 2 hold. If all blocks V^l corresponding to unstructured blocks C^l are $(n+d)K$-block-Toeplitz and all blocks V^l corresponding to block-Toeplitz/Hankel blocks C^l are dK-block-Toeplitz, $\Gamma_\mathrm{w} = GVG^\top$ is dK-block-Toeplitz.*

Proof. See the Appendix. □

For a particular type of weight matrices, the block-Toeplitz structure of Γ is preserved. More important, however, is the implication of the following proposition.

Proposition 2. *Let G be defined as in (6) and let Assumptions 1 and 2 hold. If W is p-block-banded, then $\Gamma_\mathrm{w} = GVG^\top$ is $(s+p)$-block-banded, where s is given in Theorem 2.*

Proof. See the Appendix. □

For block-banded weight matrix W, the block-banded structure of Γ is preserved, however, the block-bandwidth is increased by the block-bandwidth of W. In the following section, the block-banded structure of Γ (and Γ_w) is utilized for $O(m)$ cost function and first derivative evaluation.

Summary: We have established the following special cases:

V block-Toeplitz $\Longrightarrow \Gamma_\mathrm{w}$ block-Toeplitz (generally not block-banded),
V^l p-block-banded $\Longrightarrow \Gamma_\mathrm{w}$ $(s+p)$-block-banded (generally not block-Toeplitz),
W block-diagonal $\Longrightarrow \Gamma_\mathrm{w}$ s-block-banded (generally not block-Toeplitz).

The case W block-diagonal, i.e., $W^l = \mathrm{blk\,diag}(W^l_1,\dots,W^l_m)$, for $l = 1,\dots,q$, covers most applications of interest and will be considered in the next section.

4 Algorithm for Solving Weighted STLS Problem

In [3] we have proposed an algorithm for solving the STLS problem (4) with the flexible structure specification of Assumption 1. The structure of $\mathcal{S}(\cdot)$ is specified by the integer K, the number of rows in a block of a block-Toeplitz/Hankel structured block C^l, and the array $\mathtt{S} \in (\{\mathtt{T}, \mathtt{H}, \mathtt{U}, \mathtt{E}\} \times \mathbb{N} \times \mathbb{N})^q$ that describes the structure of the blocks $\{C^l\}_{l=1}^q$. The lth element $\mathtt{S}_l$ of the array $\mathtt{S}$ specifies the block C^l by giving its type $\mathtt{S}_l(1)$, the number of columns $n_l = \mathtt{S}_l(2)$, and (if C^l is block-Hankel or block-Toeplitz) the column dimension $t_l = \mathtt{S}_l(3)$ of a block in C^l. Therefore, the input data for the STLS problem is the data matrix $\mathcal{S}(p)$ (alternatively the parameter vector p) and the structure specification K and $\mathtt{S}$.

It is shown that the blocks Γ_k of Γ are quadratic functions of X

$$\Gamma_k(X) = (I_K \otimes X_{\text{ext}}^\top) S_k (I_K \otimes X_{\text{ext}}^\top)^\top , \quad k = 0, 1, \dots, s , \tag{11}$$

where the matrices $S_k \in \mathbb{R}^{K(n+d)\times K(n+d)}$ depend on the structure $\mathcal{S}$. The first step of the algorithm is to translate the structure specification $\mathtt{S}$ to the set of matrices S_k, $k = 0, 1, \dots, s$. Then for a given X, the Γ_k matrices can be formed, which specifies the Γ matrix.

For cost function evaluation, the structured system of equations $\Gamma(X)y(X) = r(X)$ is solved and the product $f(X) = r^\top(X)y(X)$ is computed. Efficiency is achieved by exploiting the structure of Γ in solving the system of equations. Moreover, as shown in [3], the first derivative $f'(X)$ can also be evaluated from $y(X)$ with $O(m)$ extra computations. The resulting solution method is outlined in Algorithm 1.

Algorithm 1. Algorithm for solving the STLS problem

Input: structure specification K, $\mathtt{S}$ and matrices A and B, such that $[A \; B] = \mathcal{S}(p)$.

1: Form the matrices $\{S_k\}$.

2: Compute the TLS solution X_{ini} of $AX \approx B$.

3: Execute a standard optimization algorithm, e.g., the BFGS quasi-Newton method, for the minimization of f_0 over X with initial approximation X_{ini} and with efficient cost function and first derivative evaluation.

Output: $\hat{X}$ the approximation found by the optimization algorithm upon convergence.

The changes for the case of weighted STLS problem are only in (11). Now the matrix Γ is replaced by Γ_{w}, which is no longer block-Toeplitz but is s-block-banded with block-elements

$$\Gamma_{ij}(X) = (I_K \otimes X_{\text{ext}}^\top) S_{ij} (I_K \otimes X_{\text{ext}}^\top)^\top , \tag{12}$$

where

$$S_{ij} := \begin{cases} \text{blk diag}(V_i^1, \dots, V_i^q) S_{i-j} & \text{if } 0 \le i-j \le s \\ S_{ji}^\top & \text{if } -s \le i-j < 0 \,. \\ 0 & \text{otherwise} \end{cases}$$

For an exact block C^l, with some abuse of notation, we define $W_i^l = 0$ for all i. (Our previous definition of W^l is an empty matrix since $p_l = 0$ in this case.)

5 Conclusions

We have extended the theory of Markovsky, Van Huffel, Pintelon ([4]) for the case of weighted STLS problems. The main question of interest is what properties of the Γ matrix in the equivalent optimization problem are preserved when the cost function is weighted. Block-Toeplitz inverse weight matrix V, results in corresponding Γ_{w} matrix that is also block-Toeplitz. More important for fast computational methods, however, is the fact that block-banded weight matrix W with block-bandwidth p leads to increase of the block-bandwidth of Γ with p. In particular W block-diagonal, results in Γ_{w} block-banded with the same bandwidth as Γ. This observation was used for efficient solution of weighted STLS problems.

Acknowledgments

Dr. Ivan Markovsky is a postdoctoral researcher and Dr. Sabine Van Huffel is a full professor at the Katholieke Universiteit Leuven, Belgium.

Our research is supported by Research Council KUL: GOA-AMBioRICS, GOA-Mefisto 666, several PhD/postdoc & fellow grants; Flemish Government: FWO: PhD/postdoc grants, projects, G.0078.01 (structured matrices), G.0407.02 (support vector machines), G.0269.02 (magnetic resonance spectroscopic imaging), G.0270.02 (nonlinear Lp approximation), G.0360.05 (EEG signal processing), research communities (ICCoS, ANMMM); IWT: PhD Grants; Belgian Federal Science Policy Office IUAP P5/22 ('Dynamical Systems and Control: Computation, Identification and Modelling'); EU: PDT-COIL, BIOPATTERN, ETUMOUR; HPC-EUROPA (RII3-CT-2003-506079), with the support of the European Community — Research Infrastructure Action under the FP6 "Structuring the European Research Area" Program.

References

1. Golub, G., Van Loan, C.: An analysis of the total least squares problem. SIAM J. Numer. Anal. **17** (1980) 883–893
2. Van Huffel, S., Vandewalle, J.: The total least squares problem: Computational aspects and analysis. SIAM, Philadelphia (1991)
3. Markovsky, I., Van Huffel, S.: High-performance numerical algorithms and software for structured total least squares. J. of Comput. and Appl. Math. **180** 2 (2005) 311–331
4. Markovsky, I., Van Huffel, S., Pintelon, R.: Block-Toeplitz/Hankel structured total least squares. SIAM J. Matrix Anal. Appl. **26** 4 (2005) 1083–1099
5. Markovsky, I., Rastello, M.L., Premoli, A., Kukush, A., Van Huffel, S.: The element-wise weighted total least squares problem. Comp. Stat. & Data Anal. **51** 1 (2005) 181–209

A Proof of Propositions 1 and 2

The G matrix, see (6), has the following structure $G = \begin{bmatrix} G^1 & \cdots & G^q \end{bmatrix}$, where $G^l \in \mathbb{R}^{md \times p_l}$ depends only on the structure of C^l (see Lemma 3.2 of [4]). For an unstructured block C^l,

$$G^l = I_m \otimes X_{\text{ext},l}^\top \,, \tag{13}$$

where $X_{\text{ext}} =: \text{col}(X_{\text{ext},1}, \ldots, X_{\text{ext},q})$, $X_{\text{ext},l} \in \mathbb{R}^{n_l \times d}$ and $\otimes$ is the Kronecker product. For a block-Toeplitz block C^l,

$$G^l = \begin{bmatrix} \mathbf{X}_1 & \mathbf{X}_2 & \cdots & \mathbf{X}_{\mathbf{n}_l} & 0 & \cdots & 0 \\ 0 & \mathbf{X}_1 & \mathbf{X}_2 & \cdots & \mathbf{X}_{\mathbf{n}_l} & \ddots & \vdots \\ \vdots & \ddots & \ddots & \ddots & & \ddots & 0 \\ 0 & \cdots & 0 & \mathbf{X}_1 & \mathbf{X}_2 & \cdots & \mathbf{X}_{\mathbf{n}_l} \end{bmatrix} \,, \tag{14}$$

where $\mathbf{X}_k := I_K \otimes X_{\text{ext},k}$, and for a block-Hankel block C^l,

$$G^l = \begin{bmatrix} \mathbf{X}_{\mathbf{n}_l} & \mathbf{X}_{\mathbf{n}_l-1} & \cdots & \mathbf{X}_1 & 0 & \cdots & 0 \\ 0 & \mathbf{X}_{\mathbf{n}_l} & \mathbf{X}_{\mathbf{n}_l-1} & \cdots & \mathbf{X}_1 & \ddots & \vdots \\ \vdots & \ddots & \ddots & \ddots & & \ddots & 0 \\ 0 & \cdots & 0 & \mathbf{X}_{\mathbf{n}_l} & \mathbf{X}_{\mathbf{n}_l-1} & \cdots & \mathbf{X}_1 \end{bmatrix} . \tag{15}$$

Due to Assumption 2, we have

$$\Gamma_{\text{w}} = GVG^\top = \sum_{l=1}^{q} \underbrace{G^l V^l (G^l)^\top}_{\Gamma_{\text{w}}^l} \,, \tag{16}$$

so that we need to consider the three independent problems: structure of Γ_{w}^l for respectively unstructured, block-Toeplitz, and block-Hankel block C^l. The statements of Propositions 1 and 2 are now easy to see by substituting respectively (13), (14), and (15) in (16) and doing the matrix products.

Author Index

Lecture Notes in Computer Science

For information about Vols. 1–3782

please contact your bookseller or Springer

Vol. 3884: B. Durand, W. Thomas (Eds.), STACS 2006. XIV, 714 pages. 2006.

Vol. 3879: T. Erlebach, G. Persinao (Eds.), Approximation and Online Algorithms. X, 349 pages. 2006.

Vol. 3878: A. Gelbukh (Ed.), Computational Linguistics and Intelligent Text Processing. XVII, 589 pages. 2006.

Vol. 3877: M. Detyniecki, J.M. Jose, A. Nürnberger, C. J. ‘. van Rijsbergen (Eds.), Adaptive Multimedia Retrieval: User, Context, and Feedback. XI, 279 pages. 2006.

Vol. 3874: R. Missaoui, J. Schmidt (Eds.), Formal Concept Analysis. X, 309 pages. 2006. (Sublibrary LNAI).

Vol. 3873: L. Maicher, J. Park (Eds.), Charting the Topic Maps Research and Applications Landscape. VIII, 281 pages. 2006. (Sublibrary LNAI).

Vol. 3872: H. Bunke, A. L. Spitz (Eds.), Document Analysis Systems VII. XIII, 630 pages. 2006.

Vol. 3870: S. Spaccapietra, P. Atzeni, W.W. Chu, T. Catarci, K.P. Sycara (Eds.), Journal on Data Semantics V. XIII, 237 pages. 2006.

Vol. 3869: S. Renals, S. Bengio (Eds.), Machine Learning for Multimodal Interaction. XIII, 490 pages. 2006.

Vol. 3868: K. Römer, H. Karl, F. Mattern (Eds.), Wireless Sensor Networks. XI, 342 pages. 2006.

Vol. 3863: M. Kohlhase (Ed.), Mathematical Knowledge Management. XI, 405 pages. 2006. (Sublibrary LNAI).

Vol. 3861: J. Dix, S.J. Hegner (Eds.), Foundations of Information and Knowledge Systems. X, 331 pages. 2006.

Vol. 3860: D. Pointcheval (Ed.), Topics in Cryptology – CT-RSA 2006. XI, 365 pages. 2006.

Vol. 3858: A. Valdes, D. Zamboni (Eds.), Recent Advances in Intrusion Detection. X, 351 pages. 2006.

Vol. 3857: M.P.C. Fossorier, H. Imai, S. Lin, A. Poli (Eds.), Applied Algebra, Algebraic Algorithms and Error-Correcting Codes. XI, 350 pages. 2006.

Vol. 3855: E. A. Emerson, K.S. Namjoshi (Eds.), Verification, Model Checking, and Abstract Interpretation. XI, 443 pages. 2005.

Vol. 3853: A.J. Ijspeert, T. Masuzawa, S. Kusumoto (Eds.), Biologically Inspired Approaches to Advanced Information Technology. XIV, 388 pages. 2006.

Vol. 3852: P.J. Narayanan, S.K. Nayar, H.-Y. Shum (Eds.), Computer Vision - ACCV 2006, Part II. XXXI, 977 pages. 2005.

Vol. 3851: P.J. Narayanan, S.K. Nayar, H.-Y. Shum (Eds.), Computer Vision - ACCV 2006, Part I. XXXI, 973 pages. 2006.

Vol. 3850: R. Freund, G. Păun, G. Rozenberg, A. Salomaa (Eds.), Membrane Computing. IX, 371 pages. 2006.

Vol. 3848: J.-F. Boulicaut, L. De Raedt, H. Mannila (Eds.), Constraint-Based Mining and Inductive Databases. X, 401 pages. 2006. (Sublibrary LNAI).

Vol. 3847: K.P. Jantke, A. Lunzer, N. Spyratos, Y. Tanaka (Eds.), Federation over the Web. X, 215 pages. 2006. (Sublibrary LNAI).

Vol. 3846: H. J. van den Herik, Y. Björnsson, N.S. Netanyahu (Eds.), Computers and Games. XIV, 333 pages. 2006.

Vol. 3845: J. Farré, I. Litovsky, S. Schmitz (Eds.), Implementation and Application of Automata. XIII, 360 pages. 2006.

Vol. 3844: J.-M. Bruel (Ed.), Satellite Events at the MoDELS 2005 Conference. XIII, 360 pages. 2006.

Vol. 3843: P. Healy, N.S. Nikolov (Eds.), Graph Drawing. XVII, 536 pages. 2006.

Vol. 3842: H.T. Shen, J. Li, M. Li, J. Ni, W. Wang (Eds.), Advanced Web and Network Technologies, and Applications. XXVII, 1057 pages. 2006.

Vol. 3841: X. Zhou, J. Li, H.T. Shen, M. Kitsuregawa, Y. Zhang (Eds.), Frontiers of WWW Research and Development - APWeb 2006. XXIV, 1223 pages. 2006.

Vol. 3840: M. Li, B. Boehm, L.J. Osterweil (Eds.), Unifying the Software Process Spectrum. XVI, 522 pages. 2006.

Vol. 3839: J.-C. Filliâtre, C. Paulin-Mohring, B. Werner (Eds.), Types for Proofs and Programs. VIII, 275 pages. 2006.

Vol. 3838: A. Middeldorp, V. van Oostrom, F. van Raamsdonk, R. de Vrijer (Eds.), Processes, Terms and Cycles: Steps on the Road to Infinity. XVIII, 639 pages. 2005.

Vol. 3837: K. Cho, P. Jacquet (Eds.), Technologies for Advanced Heterogeneous Networks. IX, 307 pages. 2005.

Vol. 3836: J.-M. Pierson (Ed.), Data Management in Grids. X, 143 pages. 2006.

Vol. 3835: G. Sutcliffe, A. Voronkov (Eds.), Logic for Programming, Artificial Intelligence, and Reasoning. XIV, 744 pages. 2005. (Sublibrary LNAI).

Vol. 3834: D.G. Feitelson, E. Frachtenberg, L. Rudolph, U. Schwiegelshohn (Eds.), Job Scheduling Strategies for Parallel Processing. VIII, 283 pages. 2005.

Vol. 3833: K.-J. Li, C. Vangenot (Eds.), Web and Wireless Geographical Information Systems. XI, 309 pages. 2005.

Vol. 3832: D. Zhang, A.K. Jain (Eds.), Advances in Biometrics. XX, 796 pages. 2005.

Vol. 3831: J. Wiedermann, G. Tel, J. Pokorný, M. Bieliková, J. Štuller (Eds.), SOFSEM 2006: Theory and Practice of Computer Science. XV, 576 pages. 2006.

Vol. 3829: P. Pettersson, W. Yi (Eds.), Formal Modeling and Analysis of Timed Systems. IX, 305 pages. 2005.

Vol. 3828: X. Deng, Y. Ye (Eds.), Internet and Network Economics. XVII, 1106 pages. 2005.

Vol. 3827: X. Deng, D.-Z. Du (Eds.), Algorithms and Computation. XX, 1190 pages. 2005.

Vol. 3826: B. Benatallah, F. Casati, P. Traverso (Eds.), Service-Oriented Computing - ICSOC 2005. XVIII, 597 pages. 2005.

Vol. 3824: L.T. Yang, M. Amamiya, Z. Liu, M. Guo, F.J. Rammig (Eds.), Embedded and Ubiquitous Computing – EUC 2005. XXIII, 1204 pages. 2005.

Vol. 3823: T. Enokido, L. Yan, B. Xiao, D. Kim, Y. Dai, L.T. Yang (Eds.), Embedded and Ubiquitous Computing – EUC 2005 Workshops. XXXII, 1317 pages. 2005.

Vol. 3822: D. Feng, D. Lin, M. Yung (Eds.), Information Security and Cryptology. XII, 420 pages. 2005.

Vol. 3821: R. Ramanujam, S. Sen (Eds.), FSTTCS 2005: Foundations of Software Technology and Theoretical Computer Science. XIV, 566 pages. 2005.

Vol. 3820: L.T. Yang, X.-s. Zhou, W. Zhao, Z. Wu, Y. Zhu, M. Lin (Eds.), Embedded Software and Systems. XXVIII, 779 pages. 2005.

Vol. 3819: P. Van Hentenryck (Ed.), Practical Aspects of Declarative Languages. X, 231 pages. 2005.

Vol. 3818: S. Grumbach, L. Sui, V. Vianu (Eds.), Advances in Computer Science – ASIAN 2005. XIII, 294 pages. 2005.

Vol. 3817: M. Faundez-Zanuy, L. Janer, A. Esposito, A. Satue-Villar, J. Roure, V. Espinosa-Duro (Eds.), Nonlinear Analyses and Algorithms for Speech Processing. XII, 380 pages. 2006. (Sublibrary LNAI).

Vol. 3816: G. Chakraborty (Ed.), Distributed Computing and Internet Technology. XXI, 606 pages. 2005.

Vol. 3815: E.A. Fox, E.J. Neuhold, P. Premsmit, V. Wuwongse (Eds.), Digital Libraries: Implementing Strategies and Sharing Experiences. XVII, 529 pages. 2005.

Vol. 3814: M. Maybury, O. Stock, W. Wahlster (Eds.), Intelligent Technologies for Interactive Entertainment. XV, 342 pages. 2005. (Sublibrary LNAI).

Vol. 3813: R. Molva, G. Tsudik, D. Westhoff (Eds.), Security and Privacy in Ad-hoc and Sensor Networks. VIII, 219 pages. 2005.

Vol. 3811: C. Bussler, M.-C. Shan (Eds.), Technologies for E-Services. VIII, 127 pages. 2006.

Vol. 3810: Y.G. Desmedt, H. Wang, Y. Mu, Y. Li (Eds.), Cryptology and Network Security. XI, 349 pages. 2005.

Vol. 3809: S. Zhang, R. Jarvis (Eds.), AI 2005: Advances in Artificial Intelligence. XXVII, 1344 pages. 2005. (Sublibrary LNAI).

Vol. 3808: C. Bento, A. Cardoso, G. Dias (Eds.), Progress in Artificial Intelligence. XVIII, 704 pages. 2005. (Sublibrary LNAI).

Vol. 3807: M. Dean, Y. Guo, W. Jun, R. Kaschek, S. Krishnaswamy, Z. Pan, Q.Z. Sheng (Eds.), Web Information Systems Engineering – WISE 2005 Workshops. XV, 275 pages. 2005.

Vol. 3806: A.H. H. Ngu, M. Kitsuregawa, E.J. Neuhold, J.-Y. Chung, Q.Z. Sheng (Eds.), Web Information Systems Engineering – WISE 2005. XXI, 771 pages. 2005.

Vol. 3805: G. Subsol (Ed.), Virtual Storytelling. XII, 289 pages. 2005.

Vol. 3804: G. Bebis, R. Boyle, D. Koracin, B. Parvin (Eds.), Advances in Visual Computing. XX, 755 pages. 2005.

Vol. 3803: S. Jajodia, C. Mazumdar (Eds.), Information Systems Security. XI, 342 pages. 2005.

Vol. 3802: Y. Hao, J. Liu, Y.-P. Wang, Y.-m. Cheung, H. Yin, L. Jiao, J. Ma, Y.-C. Jiao (Eds.), Computational Intelligence and Security, Part II. XLII, 1166 pages. 2005. (Sublibrary LNAI).

Vol. 3801: Y. Hao, J. Liu, Y.-P. Wang, Y.-m. Cheung, H. Yin, L. Jiao, J. Ma, Y.-C. Jiao (Eds.), Computational Intelligence and Security, Part I. XLI, 1122 pages. 2005. (Sublibrary LNAI).

Vol. 3799: M. A. Rodríguez, I.F. Cruz, S. Levashkin, M.J. Egenhofer (Eds.), GeoSpatial Semantics. X, 259 pages. 2005.

Vol. 3798: A. Dearle, S. Eisenbach (Eds.), Component Deployment. X, 197 pages. 2005.

Vol. 3797: S. Maitra, C. E. V. Madhavan, R. Venkatesan (Eds.), Progress in Cryptology - INDOCRYPT 2005. XIV, 417 pages. 2005.

Vol. 3796: N.P. Smart (Ed.), Cryptography and Coding. XI, 461 pages. 2005.

Vol. 3795: H. Zhuge, G.C. Fox (Eds.), Grid and Cooperative Computing - GCC 2005. XXI, 1203 pages. 2005.

Vol. 3794: X. Jia, J. Wu, Y. He (Eds.), Mobile Ad-hoc and Sensor Networks. XX, 1136 pages. 2005.

Vol. 3793: T. Conte, N. Navarro, W.-m.W. Hwu, M. Valero, T. Ungerer (Eds.), High Performance Embedded Architectures and Compilers. XIII, 317 pages. 2005.

Vol. 3792: I. Richardson, P. Abrahamsson, R. Messnarz (Eds.), Software Process Improvement. VIII, 215 pages. 2005.

Vol. 3791: A. Adi, S. Stoutenburg, S. Tabet (Eds.), Rules and Rule Markup Languages for the Semantic Web. X, 225 pages. 2005.

Vol. 3790: G. Alonso (Ed.), Middleware 2005. XIII, 443 pages. 2005.

Vol. 3789: A. Gelbukh, Á. de Albornoz, H. Terashima-Marín (Eds.), MICAI 2005: Advances in Artificial Intelligence. XXVI, 1198 pages. 2005. (Sublibrary LNAI).

Vol. 3788: B. Roy (Ed.), Advances in Cryptology - ASIACRYPT 2005. XIV, 703 pages. 2005.

Vol. 3787: D. Kratsch (Ed.), Graph-Theoretic Concepts in Computer Science. XIV, 470 pages. 2005.

Vol. 3786: J. Song, T. Kwon, M. Yung (Eds.), Information Security Applications. XI, 378 pages. 2006.

Vol. 3785: K.-K. Lau, R. Banach (Eds.), Formal Methods and Software Engineering. XIV, 496 pages. 2005.

Vol. 3784: J. Tao, T. Tan, R.W. Picard (Eds.), Affective Computing and Intelligent Interaction. XIX, 1008 pages. 2005.

Vol. 3783: S. Qing, W. Mao, J. Lopez, G. Wang (Eds.), Information and Communications Security. XIV, 492 pages. 2005.